Shallow Impurities in Semiconductors 1988

Proceedings of the Third International Conference held in
Linköping, Sweden, 10–12 August 1988

Edited by B Monemar

Institute of Physics Conference Series Number 95
Institute of Physics, Bristol and Philadelphia

This compilation © 1989 by IOP Publishing Ltd. All rights reserved. Multiple copying of the contents or parts thereof without permission is in breach of copyright but permission is hereby given to copy titles and abstracts of papers and names of authors. Permission is usually given upon written application to IOP Publishing Ltd to copy illustrations and short extracts from the text of individual contributions, provided that the source (and, where appropriate, the copyright) is acknowledged. Multiple copying is only permitted under the terms of the agreement between the Committee of Vice-Chancellors and Principals and the Copyright Licensing Agency. Authorisation to photocopy items for internal use, or the internal and personal use of specific clients in the USA, is granted by IOP Publishing Ltd for libraries and other users registered with the Copyright Clearance Center (CCC) Transactional Reporting Service, provided that the base fee of $2.50 per copy per article is paid direct to CCC, 27 Congress Street, Salem, MA 01970, USA.
0305-2346/89 $2.50+.00

CODEN IPHSAC 95 1–580 (1989)

British Library Cataloguing in Publication Data

International conference on Shallow Impurities in
    Semiconductors. (3rd. 1988. Linkoping, Sweden)
    Shallow impurities in semiconductors 1988.
    1. Impurities. Semiconductors
    I. Monemar, B. (Bo)
    537.6'22

ISBN 0-85498-189-6

Library of Congress Cataloging-in-Publication Data are available

Honorary Editor
    B Monemar

Published under The Institute of Physics imprint by IOP Publishing Ltd
Techno House, Redcliffe Way, Bristol BS1 6NX, England
242 Cherry Street, Philadelphia, PA 19106, USA

Printed in Great Britain by J W Arrowsmith Ltd, Bristol

Contents

xiii   Preface

Chapter 1: Shallow impurities in quantum structures

1–10   The δ-doping layer. Electronic properties of a Si dopant layer in the high density limit
A Zrenner and F Koch

11–20   Transport properties and FIR spectroscopy of the metal-insulator transition in GaAs/AlGaAs quantum well structures
H Sigg, K v Klitzing, M Hauser and K Ploog

21–26   A far infrared study of confinement effects on acceptors in GaAs/AlGaAs quantum wells
A A Reeder, B D McCombe, F A Chambers and G P Devane

27–32   Optical study of the excited states of the Be-acceptor confined in a 70 Å wide GaAs/AlGaAs quantum well
P O Holtz, M Sundaram, J L Merz and A C Gossard

33–38   Photoluminescence study of donors in selectively doped GaAs/AlGaAs quantum wells
X Liu, A Petrou, B D McCombe, J Ralston and G Wicks

39–44   Circular polarization of impurity luminescence in GaAs/AlGaAs quantum wells
P S Kop'ev, V P Kochereshko, I N Uraltsev and D R Yakovlev

45–50   The investigation of impurity distributions around an oval defect in molecular beam epitaxy AlGaAs/GaAs single quantum wells by transmission electron microscopy and cathodoluminescence
J Wang, J W Steeds and C W Tu

51–56   Shallow donor impurities and far-infrared absorption in GaAs-Ga$_{1-x}$Al$_x$As quantum well structures
F Bassani, S Fraizzoli and R Buczko

57–62   Heterointerface states of shallow impurities
P S Kop'ev, I N Uraltsev, V M Ustinov and A M Vasil'ev

63–68   Secondary ion mass spectrometry of Si in δ-doped GaAs
C Grattepain, A M Huber, E Barbier and G Gillmann

Chapter 2: Shallow impurities in silicon and germanium

69–78   Metal-insulator transition in Si:P
M A Paalanen and R N Bhatt

Contents

79–88   Excited states of deep defects in silicon
M Kleverman, J Olajos, G Grossmann and H G Grimmeiss

89–94   Zeeman splitting of double-donor spin-triplet levels in silicon
R E Peale, R M Hart, A J Sievers and F S Ham

95–100   Piezospectroscopy of boron impurity in silicon
R A Lewis, P Fisher and N A McLean

101–106   Shallow excited states of the deep silver donor in silicon
J Olajos, M Kleverman, B Bech Nielsen and H G Grimmeiss

107–112   Shallow acceptor bound and resonant states in Si
R Buczko and F Bassani

113–118   Stability of bound exciton state in highly stressed Si:B
H Nakata, A Uddin and E Otsuka

119–124   Phonoionization of A$^+$ states in Si under uniaxial stress: influence of the valence-band structure
R Haug and E Sigmund

125–130   Vibronic coupling in shallow excited states of optical centres in silicon
G Davies and M C do Carmo

131–136   Linear stark coupling to the ground state of effective mass acceptors
A Köpf, A Ambrosy and K Laßmann

137–142   Position of the chemical potential, near the melting temperature, in heavily shallow-impurity doped silicon
B E Sernelius

143–148   Uniaxial stress by substrate bending: application in a study of boron acceptors in Si MBE layers
J A Chroboczek and A Briggs

149–154   The stability of In-As pairs in silicon observed by the PAC-method
S Deubler, N Achtziger, P Dohlus, D Forkel, H Wolf and W Witthuhn

155–160   On the nature of defects formed during chemomechanical polishing of silicon
M Deicher, G Grübel, R Keller, E Recknagel, N Schulz, H Skudlik, Th Wichert, H Prigge and A Schnegg

161–166   New results in the theory of electron structure and spectra of shallow non-hydrogenlike impurities in semiconductors. I. Donors in germanium and silicon
I L Beinikhes and Sh M Kogan

167–172   Impurity impact ionization in indium doped germanium
U Rau, J Parisi, K M Mayer, W Clauß, J Peinke, B Röhricht and R P Huebener

Contents

173–178	One-phonon ionization of neutral donors in germanium *M Gienger, P Groß and K Laßmann*
179–184	Nonlinear effects in impurity pumping and in impurity ionization *T Theiler, F Keilmann and E E Haller*
185–190	Simulation of far-IR absorption and PTI-spectra of shallow impurities in compensated germanium *E Rotsaert, P Clauws, J Vennik and L Van Goethem*

Chapter 3: Thermal donors and related defects in silicon

191–200	The structure of the NL10 thermal donor in silicon *C A J Ammerlaan, T Gregorkiewicz and H H P Th Bekman*
201–210	ENDOR investigations on thermal donors in silicon *J Michel, N Meilwes, J R Niklas and J M Spaeth*
211–220	The kinetics of thermal donor formation in silicon *R C Newman and M Claybourn*
221–226	Molecular cluster and cyclic cluster calculations on models for the core of the 450 °C oxygen thermal donor in silicon *L C Snyder, P Deak, R Wu and J W Corbett*
227–232	Oxygen related photoluminescence lines in 450 °C annealed silicon *A Henry, H Weman and B Monemar*

Chapter 4: Shallow impurities in III–V compounds

233–242	Optically detected magnetic resonance study of Si donors in hetero-epitaxial layers of $Al_xGa_{1-x}As$ on GaAs *E Glaser, T A Kennedy and B Molnar*
243–248	Shallow donor time-resolved magnetospectroscopy using a nanosecond pulse-slice laser: measurements and theory for high purity n-InP *G L J A Rikken, P Wyder, J M Chamberlain and G A Toombs*
249–254	Optically detected magnetic resonance investigations of shallow donors in GaP *J J Lappe, B K Meyer and J-M Spaeth*
255–257	Search for the zero-field ODMR transition in the $J=2$ state of the bound exciton in GaP:N *M C J M Donckers and J Schmidt*
259–264	Electronic structure and optical property of a D^- centre in a strong magnetic field *A Natori and H Kamimura*

viii   *Contents*

265–270 Spectroscopic study of D⁻ state transitions in GaAs and InSb
S P Najda, A Natori, H Kamimura, J C Maan and R A Stradling

271–276 Line shape dependence on temperature for photothermal ionization of donors in GaAs
S D Baranovskii, B L Gel'mont, V G Golubev, V I Ivanov-Omskii and A V Osutin

277–282 Ionized impurities and the FIR-photoconductivity spectrum of n-GaAs
A van Klarenbosch, J Burghoorn, T O Klaassen, W Th Wenckebach and C T Foxon

283–288 Shallow positron traps in gallium arsenide
S Dannefaer, P Mascher and D Kerr

289–294 Quantitative determination of the carbon content in MOMBE p-GaAs by low temperature photoluminescence
S Ambros, M Kamp, K Wolter, M Weyers, H Heinecke, H Kurz and P Balk

295–300 Residual impurities in autodoped n-GaAs grown by MBE
M B Stanaway, R T Grimes, D P Halliday, J M Chamberlain, M Henini, O H Hughes, M Davies and G Hill

301–306 Electron beam doping of Zn into GaAs (impurity depth profiles in GaAs and Zn)
T Wada, A Takeda, M Ichimura, M Takeda and H Morikawa

Chapter 5: The DX centre in III–V compounds and related problems

307–314 The DX center in GaAs and $Al_xGa_{1-x}As$
T N Theis

315–324 Studies of the DX centre in heavily doped n^+GaAs
L Eaves, T J Foster, D K Maude, J C Portal, R Murray, R C Newman, L Dmowski, R B Beall, J J Harris, M I Nathan and M Heiblum

325–334 Metastable effective mass states of centers exhibiting large lattice relaxation
J E Dmochowski, J M Langer and W Jantsch

335–340 Crossover between shallow and deep levels in pressurized gallium arsenide
A Oshiyama

Chapter 6: Shallow impurities in II–VI compounds

341–350 Magnetospectroscopy in HgCdTe: shallow donors and localization
V J Goldman, M Shayegan, H D Drew and J B Choi

Contents

351–359 The intrinsic and extrinsic doping of mercury cadmium telluride grown by molecular beam epitaxy
M Boukerche and J P Faurie

361–370 Shallow impurities in diluted magnetic semiconductors
A K Ramdas

371–376 The shallow resonant acceptor in semimagnetic, zerogap $Hg_{1-x}Mn_xTe$
R G Mani and J R Anderson

377–382 Excited states and doublet structure of the shallow-acceptor Na bound excitons in ZnSe bulk and MOCVD film
Y Yamada, T Taguchi and A Hiraki

383–388 Excited states of acceptors in cubic semiconductors: central-cell effect in ZnTe
M A Kanehisa and M Said

389–393 Energy levels of phosphorus and arsenic acceptors in CdTe: interaction with deuterium and lattice defects
L Svob and Y Marfaing

Chapter 7: Hydrogen passivation of shallow impurities. Hydrogen-related defects

395–404 Mechanism of hydrogen passivation in silicon
T Sasaki and H Katayama-Yoshida

405–414 Hydrogen diffusion and passivation of shallow impurities in crystalline silicon
C G Van de Walle, P J H Denteneer, Y Bar-Yam and S T Pantelides

415–424 Hydrogen migration and complex formation in silicon
N M Johnson and C Herring

425–436 Hydrogen-related effects in crystalline semiconductors
E E Haller

437–446 Hydrogen passivation of shallow donors and acceptors in GaAs
B Pajot

447–452 The symmetry and properties of donor-H and acceptor-H complexes in Si from uniaxial stress studies
M Stavola, K Bergman, S J Pearton, J Lopata and T Hayes

453–458 Shallow acceptor action of a C-H pair in silicon and germanium
L V C Assali, V M S Gomes and J R Leite

459–464 *Ab-initio* calculations of the passivation of shallow impurities in GaAs
P Briddon and R Jones

465–470 Hydrogen passivation of deep and shallow levels in carbon-doped silicon
A E Jaworowski, J H Robison and S R Hayden

471–476 Hydrogen passivation of indium acceptors in silicon
A Baurichter, S Deubler, D Forkel, M Uhrmacher, H Wolf and W Witthuhn

477–481 Hydrogen passivation of shallow acceptor impurities and radiation defects in p-type silicon
Kh A Abdullin, B N Mukashev, M F Tamendarov, T B Tashenov, S Z Tokmoldin and E V Chikhrai

Chapter 8: Diffusion of shallow impurities

483–492 Diffusion of shallow impurities in silicon
P Fahey

493–498 Diffusion of shallow impurities in silicon
C S Nichols, C G Van de Walle and S T Pantelides

499–504 The effect of heavy doping on complex formation and diffusivity of Sb in Si
A Nylandsted Larsen, P Tidemand-Petersson, P E Andersen and G Weyer

Chapter 9: Shallow impurities: general properties and theory

505–514 Excited state spectroscopy of excitonic systems in semiconductors
M L W Thewalt, M Nissen, D J S Beckett and S Charbonneau

515–520 Interband Auger recombination in silicon
D B Laks, G F Neumark, A Hangleiter and S T Pantelides

521–526 One- and two-photon acceptor spectra in semiconductors
N Binggeli, A Baldereschi and A Quattropani

527–532 The ground state of the acceptor molecule A_2 in a cubic semiconductor at large interimpurity distance
Sh M Kogan and A F Polupanov

533–538 Influence of electron-phonon coupling on the properties of D^- centers
J Adamowski and S Bednarek

539–544 Valence charge distribution around shallow donors in semiconducting compounds
S Bednarek and J Adamowski

545–550 Monte Carlo simulation as a novel technique in the study of shallow centers
L Reggiani, P Lugli and V Mitin

Contents

551–554 Electron correlation and disorder effects on the spin susceptibility of doped semiconductors at finite temperature
A Ferreira da Silva

555–560 Dynamical resistivity from shallow-impurity scattering in polar semiconductor films, used as optical coatings
B E Sernelius and M Morling

561–564 Luminescence enhancement by impurity-impurity interaction in heavily iodine doped AgBr
A Testa, W Czaja, A Quattropani and P Schwendimann

565–570 Nitrogen passivation of shallow levels in amorphous silicon
L Martín-Moreno, J A Vergés, E San-Fabián and E Louis

571–576 Photoluminescence and tail states in amorphous semiconductors
T M Searle and W A Jackson

577–579 Author Index

Preface

The Third International Conference on Shallow Impurities in Semiconductors was held in Linköping, Sweden, 10–12 August 1988. The Conference attracted 135 attendants, of which 90% came from outside Sweden. A total of 99 research papers (out of about 140 submitted) were accepted for presentation at the conference, 24 as oral and 75 as posters.

The emphasis for the programme of this conference was on basic research related to shallow impurities in semiconductors. In spite of the fact that this area of research has been active for about three decades, many basic physical problems remain unsolved or incompletely understood. The number of researchers in the field appears to be growing, partly due to the fact that this area is still the backbone of semiconductor applications in electronics. Many active areas in the field are reflected in the programme, such as shallow impurities in quantum structures, electronic structure of shallow impurities in different semiconductors, lattice relaxation and metastability, hydrogen passivation, doping and epitaxy, and diffusion of shallow impurities. More specific topics were also given special sessions, such as the DX centres in III–V compounds, and the thermal donor problem in silicon.

The papers in this volume have been selected from the contributions presented at the Conference. A few papers have been left out to preserve originality, since the material in those cases was going to be published elsewhere. In addition a few papers were rejected in the referee procedure. The volume is organized by subjects, and the invited oral contributions (10 pages) are mixed with the poster papers (6 pages). Some overlap between subjects will occur, so the reader is encouraged to explore the list of contents carefully.

The Conference was organized by an International Organizing Committee with 16 members, chaired by B Monemar. This Organizing Committee also served as Programme Committee. We gratefully acknowledge the financial support for the conference from the following national sources: Swedish Natural Science Research Council; Swedish Board for Technical Development; Telefonaktiebolaget L M Ericsson, Stockholm, Sweden; ABB-HAFO AB, Järfälla, Sweden; and ABB Corporate Research, Västerås, Sweden.

Finally it is my pleasure to acknowledge the kind assistance of a number of people in the Local Committee for this Conference. In particular I would like to mention Anne Henry, Peder Bergman, Göran Rune, Johan Svensson and Helge Weman for hard work with practical problems before and during the Conference. Mrs Ingrid Nyman and Mrs Gunnel Åhsberg handled most of the practical planning and the correspondence, including registration and hotel booking, in a very efficient manner. The preparation of these Proceedings would not have been possible without the continuous cooperation with Anne Henry and Peder Bergman.

Bo Monemar
Conference Chairman

The δ-doping layer. Electronic properties of a Si dopant layer in the high density limit

A. Zrenner and F. Koch

Physik-Department, Technische Universität München, D-8046 Garching, FRG

ABSTRACT: We discuss the saturation of the free electron concentration with increasing doping density in δ-doped GaAs(Si). The individual subband density is determined by Shubnikov-de Haas measurements. The saturation mechanisms are explored by both variation of N_D and hydrostatic pressure. Saturation occurs if the depth of the Hartree potential is large enough (200meV) to make the energy of the DX-center line up with the Fermi energy of the 2D system. For samples grown at $T_s = 500°C$ the maximum free electron concentration is limited to $6-7 \times 10^{12} cm^{-2}$. At higer growth temperatures ($T_s = 600°C$) the depth of the Hartree potential can not reach 200meV because of severe dopant ion spreading. In this case the maximum free electron concentration is limited to $1.2 \times 10^{13} cm^{-2}$ by autocompensation or dopant ion clustering.

1. INTRODUCTION

The controlled incorporation of electrically activ impurities is a fundamental step in the production of electronic devices. With the highly elaborated epitaxial growth techniques available today, such as MBE, it is possible to introduce the dopants in controlled numbers already during the crystal growth and achieve theoretically an incorporation within a single atomic layer. Following the basic approach of Wood et al. 1980 such δ-doping layers have been prepared in GaAs by MBE in a stop and go procedure. For n-type doping Si-atoms are introduced on an As-terminated GaAs surface (in order to avoid autocompensation) while the Ga-flux and thus the crystal growth is interrupted. During the subsequent GaAs growth the Si-dopants are expected to be buried on Ga-sites within the monolayer of incorporation, forming thus a δ(z)-function like doping profile.

Unfortunately it turned out recently that the stop and go procedure is no guarantee for a δ(z)-function like doping profile. Instead Zrenner et al. 1987b, 1988a, 1988b and Beal et al. 1988 found that the dopants are distributed over a finite distance in the growth direction which is dependent on the growth parameters. A second drawback of the stop and go procedure turned out to be a severe saturation of the free electron concentration with increasing design donor density which again depends on the growth parameters (Zrenner et al. 1987a, 1988a, 1988b).

© 1989 IOP Publishing Ltd

In this work we attempt to provide insight into the basic physics of the δ(z)-doping layer, which is the interaction between a spacially confined arrangement of donor atoms and the electrons they are supplying.

2. EXPERIMENTAL FACTS AND THEIR INTERPRETATION

Either an insulating or a metallic system can be prepared by choosing the donor density N_D smaller or larger than some critical density N_c. For $N_D < N_c$ the δ-doping layer is a system of isolated impurities arranged randomly in a 2D plane. At T=0K the electrons occupy a 1s hydrogenic groundstate. For $N_D < 1 \times 10^{11} \text{cm}^{-2}$ the conductivity at finite temperatures is activated by roughly the donor binding energy. For $N_D > 1 \times 10^{11} \text{cm}^{-2}$ the δ-doping layer exhibits metallic conductivity at low temperatures, indicating that the critical density N_c for the metal insulator transition is in the $1 \times 10^{11} \text{cm}^{-2}$ regime for this system.

In presenting experimental results we concentrate in this work on δ-doping layers in the metallic regime. The confinement of the electrons to the donor sheet in the x-y plane is so strong that the electronic level spectrum splits into individual 2D subbands. The shape of the corresponding Hartree potential is given by both the spatial distribution of the positively charged donor ions and the subband electrons. The magnetotransport properties of such a system clearly reflect its 2D character despite of the high degree of randomnes in the x-y plane. The quantum Hall effect in δ-doping layers as observed by Gillmann et al. 1988 and Zrenner et al. 1988 is certainly the most prominent feature in fig. 1a.

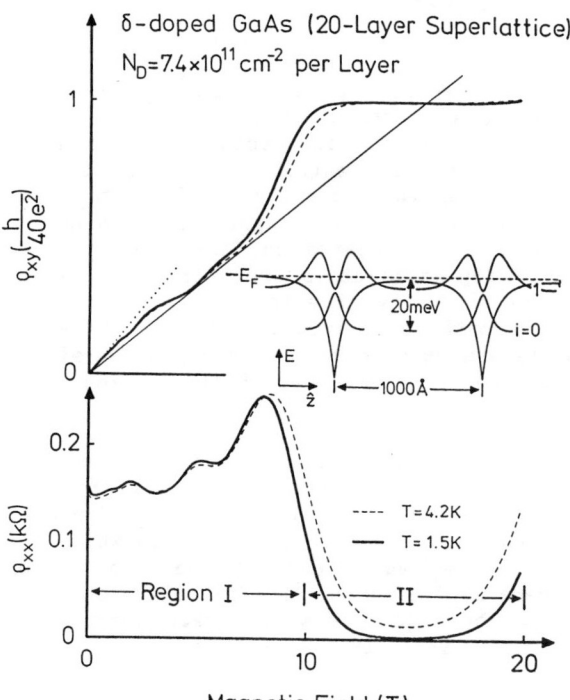
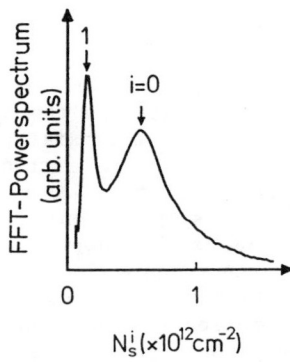

Fig. 1b: FFT of the 1/B periodic SdH-oscillations in region I of fig. 1a.

Fig. 1a: Quantum Hall effect in a superlattice of impurity sheets. Two subbands are occupied. The calculated potential and the charge distribution is shown in the insert.

We want to concentrate now however on the question how the correct number of subband electrons has to be determined in such a multisubband system.

There are two different ways in getting a carrier concentration from the data shown in fig. 1a. One is the analysis of the Shubnikov-de Haas (SdH) oscillations contained in ρ_{xx}. Analysing the periodicity in $1/B$ by means of a FFT algorithm (Zrenner et al. 1988a) we find two different periods as shown in fig. 1b, which correspond to subband densities of $5.7 \times 10^{11} cm^{-2}$ (i=0) and $1.7 \times 10^{11} cm^{-2}$ (i=1) per layer. Those add up to a total N_s of $7.4 \times 10^{11} cm^{-2}$ and compare nicely with the design donor density of $7.0 \times 10^{11} cm^{-2}$ per layer. The classical Hall line which corresponds to the experimental N_s (straight full line in fig. 1a) intersects the Hall plateau at its center as expected for a system with many coulomb attractive scatterers (Ando 1984).

An other way frequently used to obtain an electron concentration is the analysis of the low field Hall effect. The low field slope of ρ_{xy} (dotted straight line in fig. 1a) as measured on a mesa etched Hall bar corresponds nominally to $4.9 \times 10^{11} cm^{-2}$ electrons per layer only. This discrepancy between the results from the SdH effect and the low field Hall effect is due to the fact that different subbands have different mobilities (Zrenner et al. 1985). The low temperature mobility in a δ-doping layer is nearly exclusively controlled by ionized impurity scattering. Hence the spatial overlap of the individual wavefunctions with the ionized donors (see inset of fig. 1a) is reflected in the individual subband mobility. For the example shown in fig. 1a the mobilities are required to be $2000 cm^2/V \cdot sec$ (i=0) and $5000 cm^2/V \cdot sec$ (i=1) in order to explain the low field slope of the Hall effect within the framework of a two carrier model as developed in textbooks. At higher carrier concentration and negligible donor ion spread the mobility of the highly confined i=0 subband is so low compared to the one of the remaining subbands that the discrepancy between SdH and low field Hall measurement can be as high as a factor of two (see Zrenner et al. 1988b). An analysis of carrier concentration and mobility based on low field Hall measurements as presented for example by Schubert et al. 1987 does not reveal physically relevant quantities in the 2D world of δ-doping.

The donor ion spread is another quantity which has lead to some controversy in the past. Figure 2 shows the calculated energy levels and Hartree potentials of three doping layers for different amount of uniform donor spreading dz at a fixed density of $N_D = 8 \times 10^{12} cm^{-2}$. The free subband charge N_s is taken identical with N_D. We note that level energies E_i and occupations N_s^i sensitively depend on the value of dz only when the latter exceeds the classical turning points of the lowest subband. The level spacings for the case dz=150Å are nearly equidistant, the result that applies for a parabolic potential. Note that the subband energies for dz=5Å and dz=50Å are almost indistinguishable.

Figure 3 shows explicitely how the subband occupations N_s^i evolve with increasing dz. There is no measurable change in N_s^i for the first few 10Å. Just as in the previous figure, the distinct level spacings (and probability distribution) of the truely atomic-layer confined case are preserved over a finite range of dz. Trying to get dz by comparison between measurement and theory one has to make sure that the nonpara-

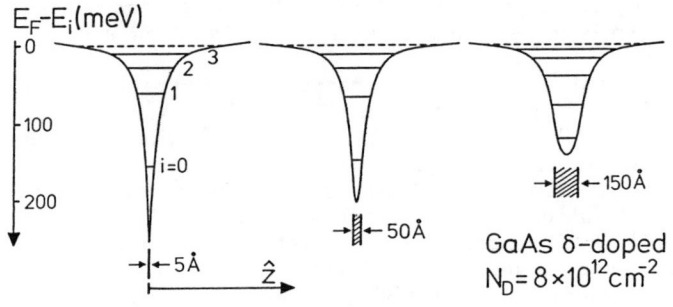

Fig. 2: Potential profiles and sub-band energy levels for different doping ion spread dz. Nonparabolicity effects have been included.

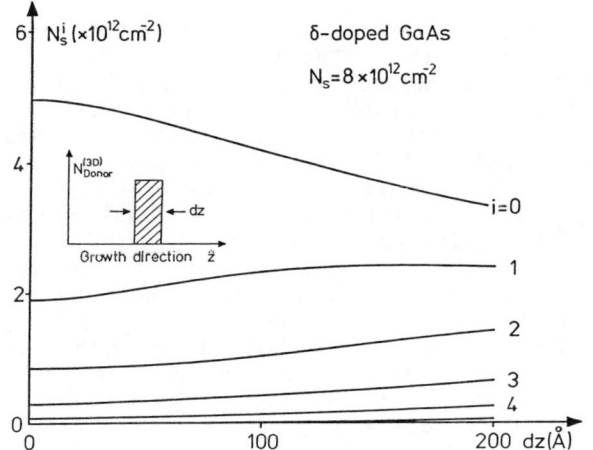

Fig. 3: Subband occupation N_s^i for $N_s = 8 \times 10^{12} \text{cm}^{-2}$ as a function of donor ion spread dz. A square shaped donor distribution has been assumed as shown in the inset.

bolicity of the GaAs conduction band is included. The nonparabolicity has been described and proofen with high accuracy (see for example Rössler 1984 and Suemoto et al. 1988). In δ-doping layers nonparabolicity adds typically a 6% correction to the occupancy of the i=0 subband for a total $N_s = 8 \times 10^{12} \text{cm}^{-2}$ (Zrenner et al. 1988a). This translates to a huge difference in dz for dz→0 as fig. 3 suggests. If nonparabolicity is included as in fig. 3 a comparison between theory and SdH-experiment can give dz in the best case with an error of 20-30Å for dz→0. For higher dz the accuracy is considerably better. The same statements are true for the analysis of CV-profiles in contradiction to the conclusions of Schubert et al. 1986 and 1988.

The systematic SIMS analysis of Beal et al. 1988 is a more straight forward approach in finding dz as a function of the growth parameters. But also in this case the obtainable accuracy in dz is only 30-40Å. Nevertheless finite donor ion spread can be detected for growth temperatures $T_s \geq 550°C$.

In fig. 4 the free electron concentration N_s and the corresponding donor ion spread dz as determined by an analysis of the SdH-effect is shown for various samples grown at $T_s = 600°C$. For low design donor

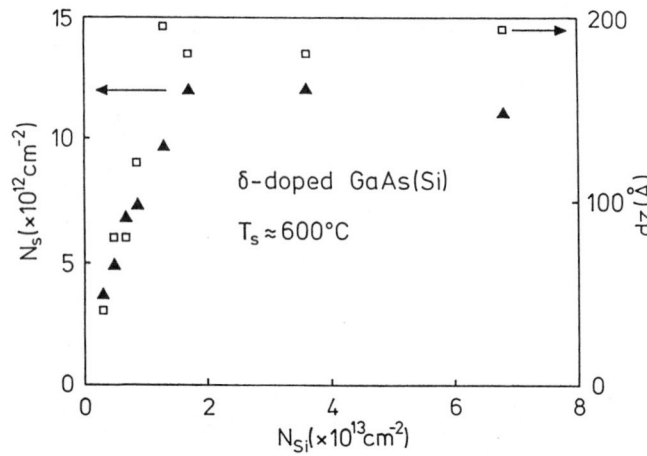

Fig. 4: Free electron concentration N_s and donor ion spread dz as a function of the design doping density N_{Si}. The data is obtained by an analysis of SdH results. The samples were grown in a stop and go procedure at $T_s \approx 600°C$.

densities ($N_{Si} < 10^{13} cm^{-2}$) both N_s and dz start increasing proportional. In the regime $N_{Si} > 10^{13} cm^{-2}$ N_s appears to be totally saturated and does not exceed $1.2 \times 10^{13} cm^{-2}$. It appears from our observations that the Si atoms deposited on the surface during the growth interruption are incorporated in the newly grown material at a 3D density which corresponds closely to the known solubility limit for the given growth temperature. From the experiments we obtain $N_D \approx 7 \times 10^{18} cm^{-3}$, with individual samples varying up to $\pm 2 \times 10^{18} cm^{-3}$ about this value. When N_{Si} exceeds $1.2 \times 10^{13} cm^{-2}$, the excess is electrically inactive and is supposed to remain behind on the original doping plane. The active portion of the dopants migrates at the growth front. The continuous incorporation of those dopants during the subsequent crystal growth (nominally undoped material) produces a uniformly doped n⁺ layer until the source of dopants on the surface is used up. The doping density is given by the solubility limit for the selected growth parameters (see Neave et al. 1983).

Samples grown at $T_s = 500°C$ exhibit a smaller donor ion spread as the previous ones shown in fig. 4. For $N_s = 6 \times 10^{12} cm^{-2}$ we detect a 30Å spread by SdH-analysis with an uncertainty as mentioned earlyer. The most striking effect of lowering T_s however is, that the saturation of the free electron concentration occurs now already at $7 \times 10^{12} cm^{-2}$.

3. SATURATION MECHANISMS IN 2D DOPING SHEETS

In search for explanations for the observed saturation we first quote, that SdH-experiments performed on δ-doped GaAs detect only electrons at the Γ-conduction band (CB) minimum. In magnetic fields $B_\perp$ which are experimentally available the condition $\mu_1 B_\perp > 1$ can be fullfilled for that particular group of electrons only. A saturation of N_s at the Γ-point for high doping densities might be caused by two different mechanisms.

The first type of mechanism, we call it structural saturation, can be regarded as autocompensation in the widest sense. It occurs when group IV Si atoms are incorporated on As sites, if Si-Si pairs are introduced on neighbouring Ga-As sites or if even larger Si clusters are formed (see Maguire et al. 1987). Critical doping densities for enhanced autocompen-

sation or clustering are expected to depend on the growth conditions which control the surface kinetics during the doping procedure.

The second type of saturation mechanism, the electronic saturation, is a bare bandstructure effect in GaAs. As fig. 5 is intended to show, it will happen when the Fermi energy in the δ-doping layer is high enough to reach new electronic states above the Γ-point. Those states could be sub-bandstates at the L- and X-point (Zrenner et al. 1987a) as well as deep donor states like the DX-center (Lang et al. 1977 and Nelson 1977). A high density of states at or near the L- or X-point would prevent the Fermi energy in the 2D system from rising further. A population of those states would therefore cause a saturation of the free electron concentration in the Γ-subbandsystem with increasing donor concentration, as observed in the SdH-experiment.

Trying to find the operative saturation effects we start to analyse subband states which are bound at the L-point. As pointed out in earlier work by Zrenner et al. 1988b the i=0 subband at the L-point starts to get occupied at a donor density of $1.6 \times 10^{13} cm^{-2}$, if donor ion confinement within a single monolayer is assumed. This result is obtained by a self-consistent subband calculation which takes into account the nonparabolic dispersion at the Γ-point and the appropriate effective mass and density of states at the L-point. The qualitative features of the saturation observed in experiment are well described by the calculation, the experimental threshold densities of $1.2 \times 10^{13} cm^{-2}$ (T_s=600°C) and $7 \times 10^{12} cm^{-2}$ (T_s=500°C) however are far off. This becomes even worse if the finite donor ion spread is included. The theoretical threshold density is shifted beyond $2 \times 10^{13} cm^{-2}$ if dz is assumed to be 200Å. A static population of subbandstates at the L-point (or X-point) as a saturation mechanism can therefore be excluded.

As indicated in fig. 5 the δ-doping layer is a 2D quantumstructure which is made by spatial impurity confinement. Because of the strong spatial overlapp between Si-donors and subband electrons additional inter-

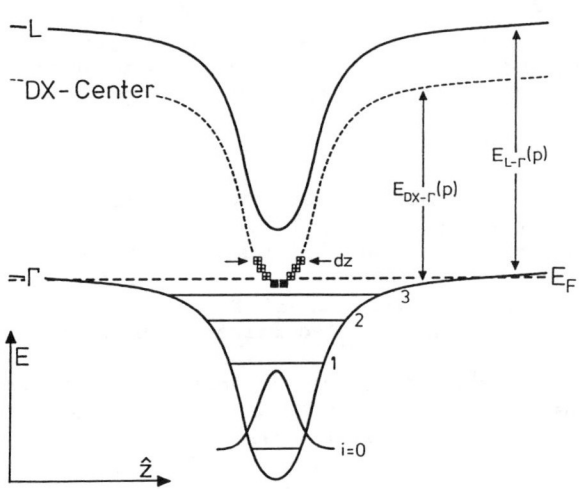

Fig. 5: Relevant electronic levels in a Si doping sheet in GaAs. Subband states are occupied at the Γ-point but have to be considered also at the L-point. The DX-center energy is defined locally in the Hartree potential and tracks the CB-edge over the width dz of the donor distribution in real space. DX-centers can be neutralized if their energy is lower than E_F.

action compared to conventional quantumstructures (confined by hetero barriers) is expected. Looking for an electronic saturation mechanism of the free electron concentration in 2D we concentrate on the DX-center, a metastable deep donor state above the Γ-CB edge. Since the DX-center is a non effective mass like state of a Si atom on a Ga site, the number of available deep donor states is equal to the number of subbandelectrons in our system. In 3D the DX-center was shown to limit the maximum free electron concentration in analytic (Morgan 1987) and experimental work (Maude et al. 1987 and Theis et al. 1988). As a deep level with small extent the energetic position of the DX-center in the δ-doping layer will track the local electrostatic potential provided by the ions and the screening subband charge (see fig. 5). As soon as the DX-center energy lines up with the Fermi energy of the 2D system an electron transfer to the DX-center in terms of $DX^+ + e^- = DX^0$ has to be considered.

4. SdH-EFFECT UNDER HYDROSTATIC PRESSURE

In order to determine which of the remaining saturation mechanisms is indeed relevant, we modify the bandstructure of GaAs by applying hydrostatic pressure. The energetic distance between the DX-center and the Γ-CB-minimum decreases with increasing pressure. At p≈20kbar the DX-center becomes the electronic groundstate in GaAs (Mizuta et al. 1985). If DX-centers are populated N_s at the Γ-point is expected to decrease with increasing pressure. If structural saturation is operative N_s should not change with pressure in the regime of interest. Three different samples with electron densities close to saturation have been choosen to explore the active saturation mechanisms. The samples were characterized to have Hartree potentials as shown in fig. 6 by SdH measurements at zero pressure in the condition $N_s = N_D^+$ (after brief illumination). The samples are labeled A,B and C from left to right. Sample A was grown at $T_s=500°C$, sample B and C at $T_s=600°C$. Both N_s and E_F in the i=0 subband increase from A to C, whereas the depth of the Hartree potential increases from C to A. The latter behaviour is due to the fact that the width dz of the donor distribution increases from A to C. The samples were mounted in a liquid clamp cell (Wasilewski et al. 1986). Once mounted the samples could not be exposed to light any more. A constant of $3-4 \times 10^{11} cm^{-2}$ electrons in each sample remain therefore transfered to surface and interface states (see Zrenner et al. 1988b for details).

As a main result we observe up to 12kbar a strong decrease of N_s in sample A, a smaller decrease in B and no decrease in C. The degree of de-

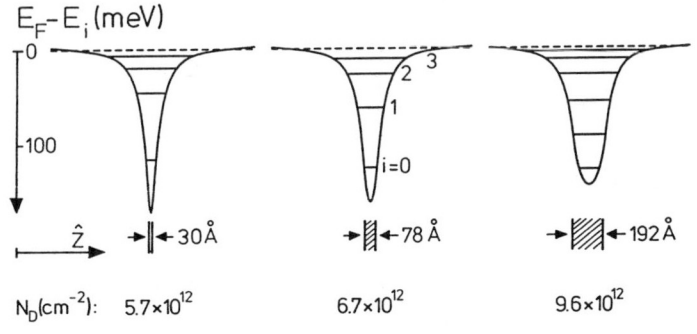

Fig. 6: Hartree potentials of the samples under investigation. The position of the energy levels and the width of the potentials is drawn to scale.

population seems to depend on the depth of the Hartree potential (measured with respect to the Fermi energy). In fig. 7 the Fourier transform of the 1/B-periodic SdH-oscillations is shown for sample A and B. For sample A we observe a gradual decrease of the individual subband density already for small pressures. Using the experimental subband densities we calculate the depth of the Hartree potential to be 195,187,174,159 and 120meV for p=0,3.1,8.0,11.8 and 17kbar (see Zrenner et al 1988b). At 17kbar 47% of all subband electrons have vanished. For sample B there is essentially no effect up to 8kbar. Between 8 and 12kbar however we record a pronounced decrease ($\approx$8%) of N_s. At 12kBar the depth of the Hartree potential is calculated to be 155meV.

The observed features are totally consistent with our model shown in fig. 6 and support the idea that the DX-center is the dominant saturation mechanism for sample A and B at finite pressure. At 12kBar the energetic distance between the DX-center and the Γ-CB-edge (the depth of the Hartree potential) is in both samples 155-160meV. The nonzero pressure coefficient at zero pressure in sample A suggests that the depth of the Hartree potential is limited to 195meV due to Fermi level pinning at the DX-center energy. Indeed none of the samples grown at T_s=500°C has a Hartree potential deeper than 200meV, even those with a design donor density twice as large as in sample A. Since none of the samples grown at T_s=600°C has a Hartree potential deeper than 180meV at zero pressure we can argue further that structural saturation is active under those growth conditions. In this case the DX-center enters only at finite hydrostatic pressure.

With increasing pressure the DX-centers at the bottom of the Hartree potential become neutralized first at the onset of depopulation (see

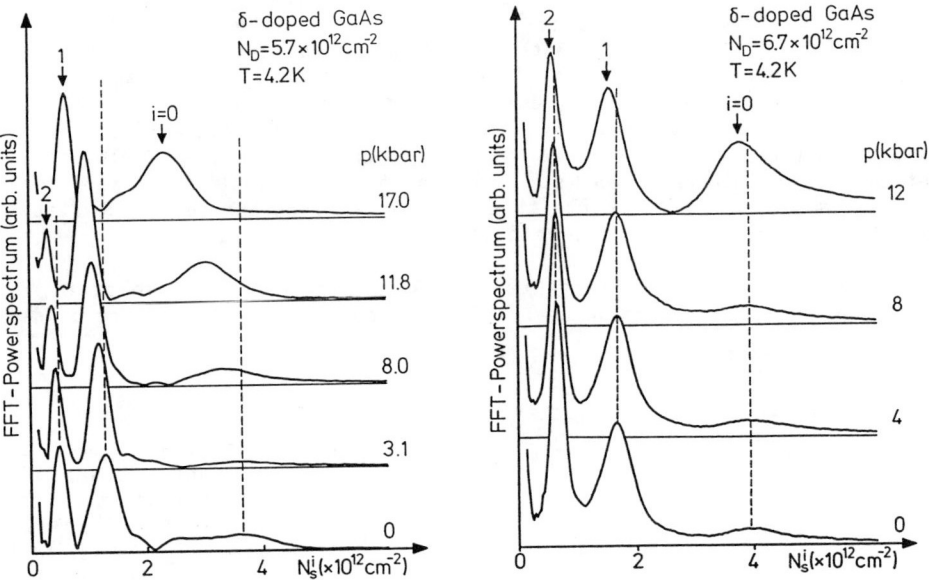

Fig. 7: Fourier spectrum of the 1/B-peridic SdH-effect for sample A (left) and sample B (right) for various amounts of hydrostatic pressure.

fig. 5). Ionic scatterers are removed selectively from the center of the potential where the probability distribution of the i=0 subband is largest, but where the wavefunctions have a node for the i=1,3,5.. subbands (or otherwise small amplitude). Therefore the mobility of the i=0 subband is expected to rise as soon as depopulation occurs. This feature is indeed observed in the experimental data. The strength of the i=0 FFT peak gets selectively enhanced when depopulation occurs, due to the fact, that the onset of that particular quantum oscillation moves to lower magnetic field.

5. THE DX-CENTER AS A LOCAL PROBE IN THE HARTREE POTENTIAL

We have shown in the previous chapter that the DX-center really can probe the depth of the Hartree potential on a quasi atomic length scale. This allows us to determine the donor ion spread on a length scale which is smaller than the subband confinement length. As shown in fig. 2 the depth of the potential is still sensitive on dz in this regime. On the other hand we can calculate a limit for the maximum free electron concentration possible in truely δ-doped GaAs(Si). As a result of our selfconsistent calculation (nonparabolicity included) we get a 200meV deep Hartree potential already at $N_s = N_D = 5.5 \times 10^{12} cm^{-2}$ for dz->0. From that point on additionl donors appear only as neutral DX-centers.

It has to be quoted however that the DX-center energy, as measured in our experiment, can be a single sharp energy level only if the donors are ordered in a sublattice on the x-y plane. Only in this case the energetic difference between the Fermi energy of the 2D system and each individual DX-level would be the same. In a real system the observable DX-level energy is smeared by the randomnes of the potential in x-y direction (in addition to temperature effects). The qualitative difference between the inset of depopulation as a function of pressure in sample B as compared to A may be explained by an enhancement of randomnes in x-y plane when the growth temperature is lowered from 600°C to 500°C. The potential fluctuations in x-y direction are reduced if pressure induced neutralization of DX-centers takes place, since the locally deepest centers are neutralized at first. This pressure induced ordering was shown already to work for the δ-doping layer in z-direction in the previous chapter. But also in a monoatomic sharp δ-doping layer as well as in a 3D system (Maude et al. 1987) pressure induced ordering should happen and is expected to make a contribution to the experimentally observed mobility enhancement.

We acknowledge support for this work by the Deutsche Forschungsgemeinschaft. The pressure experiments have been performed at the Imperial College for Science and Technology, London together with R. L. Williams and R. A. Stradling. Samples have been grown by K. Ploog and G. Weimann. The high field measurements were performed at the MPI Hochfeld Magnetlabor, Grenoble. A. Zrenner thanks the Siemens AG for financial support via the Ernst von Siemens Stipendium.

References

Ando T., 1982, J. Phys. Soc. Jpn. 53, 3126.
Beal R. B., J. B. Clegg and J. J. Harris, 1988, Semicond. Sci. Technol. 3, 612.
Gillmann G., P. Bois, E. Barbier, B. Vinter, D. Lavielle, M. Stohr, S. Najda, A. Briggs and J. C. Portal, 1988, Semicond. Sci. Technol. 3, 620.
Hjalmarson H. P., P. Vogl, D. J. Wolford and J. D. Dow, 1980, Phys. Rev. Lett. 44, 810.
Lang D. V. and R. A. Logan, 1977, Phys. Rev. Lett. 39, 635.
Maguire J., R. Murray and R. C. Newman, 1987, Appl. Phys. Lett. 50, 516.
Maude D. K., J. C. Portal, L. Dmowski, T. Foster, L. Eaves, M. Nathan, M. Heiblum, J. J. Harris and R. B. Beal, 1987, Phys. Rev. Lett. 59, 815.
Mizuta M., M. Tachikawa, H. Kukimoto and S. Minomura, 1985, Jpn. J. Appl. Phys. 24, L143.
Morgan T. N., 1987, Proc. of the 18th Int. Conf. on the Physics of Semiconductors, Stockholm 1986, ed. by O. Engström (World Scientific, Singapore 1987), p. 923.
Neave J. H., P. J. Dobson, J. J. Harris, P. Dawson and B. A. Joyce, 1983, Appl. Phys. A 32, 195.
Nelson R. J., 1977, Appl. Phys. Lett. 31, 351.
Rössler U., 1984, Solid State Commun. 49, 943.
Schubert E. F. and K. Ploog, 1986, Jpn. J. Appl. Phys. 25, 966.
Schubert E. F., J. E. Cunningham and W. T. Tsang, 1987, Solid State Commun. 63, 591.
Schubert E. F., J. B. Stark, B. Ullrich and J. E. Cunningham, 1988, Appl. Phys. Lett. 52, 1508.
Suemoto T., G. Fasol and K. Ploog, 1988, Phys. Rev. B37, 6397.
Theis T. N., P. M. Mooney and S. L. Wright, 1988, Phys. Ref. Lett. 60, 361.
Wasilewski Z., S. Porowski and R. A. Stradling, 1986a, J. Phys. E: Sci. Instrum. 19, 480.
Wood C. E., G. Metze, J. Berry and L. F. Eastman, 1980, J. Appl. Phys. 51, 383.
Zrenner A., H. Reisinger, F. Koch, and K. Ploog, 1985, Proc. of the 17th Int. Conf. on the Physics of Semiconductors, San Francisco, 1984, ed. by J. P. Chadi, and W. A. Harrison (Springer-Verlag, New York, 1985), p. 325.
Zrenner A. and F. Koch, 1987a, Proc. of the 18th Int. Conf. on the Physics of Semiconductors, Stockholm 1986, ed. by O. Engström (World Scientific, Singapore 1987), p. 1523.
Zrenner A., F. Koch and K. Ploog, 1987b, Inst. Phys. Conf. Ser. 91, 171.
Zrenner A. and F. Koch, 1988a, Surface Sci. 196, 671.
Zrenner A., F. Koch, R. L. Williams, R. A. Stradling, K. Ploog and G. Weimann, 1988b, Semicond. Sci. Technol. (to be published).
Zrenner A., F. Koch, J. Leotin, M. Goiran and K. Ploog, 1988c, Semicond. Sci. Technol. (to be published).

Transport properties and FIR spectroscopy of the metal-insulator transition in GaAs/AlGaAs quantum well structures

H. Sigg, K. v. Klitzing, M. Hauser and K. Ploog

Max-Planck-Institut für Festkörperforschung, Heisenbergstr. 1
D-7000 Stuttgart 80, Federal Republic of Germany

Abstract: Double quantum well structures with an undoped 5nm AlGaAs barrier were grown consisting of a 15nm GaAs well center doped with Si of concentration $10^{17} cm^{-3}$ and a 2nm well Si doped at various concentrations. On increasing this doping concentration the obtained 2D electron system could be tuned from a shallow impurity dominated insulator state to a high electron density metallic state. New results of the dc-magneto-transport, the FIR transmission and photoconductivity are obtained from samples in a state which is between the two limits, an insulator showing strong localization and metal with transport behaviour dominated by ionized impurity scattering.

1 Introduction

One of the interesting things one can do with artificially grown 2 dimensional (2D) semi-conductor systems - as e.g. the GaAs/AlGaAs system - is to modify the bandstructure and dope the material such that the charge carriers are either separated from their host donor ions or are very close to each other. Thereby, the binding energy and the overlap of the electron wavefunction with the ion can be moderated (Bastard 1981). By choosing the separation and doping concentration properly, the transport properties of such 2D electron systems (2DES) can be that of an insulator with activated conductivity or that of a degenerate metallic 2DES. Therefore, on systematically varying the growth parameters one should also obtain 2DES exhibiting the intermediate case, the so called metal-insulator (MI) transition.

The most convenient 2D systems with a large potential for application are modulation doped AlGaAs/GaAs heterostructures. This doping technique allows the realization of a high mobility 2DES metallic at already low carrier densities. Unfortunately, a reduction of the carrier concentration will drastically change the interface potential which complicates any systematic investigation of the thereby induced MI-transition (Raymond et al 1987). 2DES of much inferior mobilities are obtained in δ-doped GaAs (Zrenner et al 1988a). In samples with a rather low doping concentration a magnetic field induced MI-transition was observed (Zrenner 1988 b). However, the major drawback for studying this transition in such δ-doped samples is again the doping concentration dependence of the 2D confinement; moreover, in the limit of a vanishing doping concentration the 2D character of the system also vanishes.

© 1989 IOP Publishing Ltd

An alternative system, where the dimensionality of the quasi free and impurity-bound carriers does not depend on the doping concentration, is a quantum well (QW). Numerous theoretical work was already performed on these systems regarding the energy spectrum of a single isolated shallow impurity (Bastard 1981, Greene and Bajaj 1983). Experimentally, the impurity shifted CR, often referred to as $1s-2p^+$ transition, is found to be a very sensitive measure of the confinement dependent binding energy (Jarosik et al 1983). In agreement with the theory, binding energies in center doped QW were found to be increased with respect to the bulk GaAs value of about 5.8 meV; reduced values, which were only in qualitative agreement with theory, were observed in edge and barrier doped systems (Glaser et al 1987, Huant et al 1988). However, little is known about the dc-transport properties in these systems and, in particular, no systematic investigation of the MI-transition in these QW systems has yet been performed.

For this study a series of double QW structures were prepared consisting of a center doped well with low donor concentration and, close to this main well, a second one of narrow size. The doping concentration in this second well has been varied while for a given series of samples all other growing parameters remained unchanged. In the experiment we made use of the fact that due to the strong 2D confinement in the 20Å well the donors get ionized and the additional amount of carriers are transfered into the main well. On increasing the set back doping level from zero to its maximum level, 2DES were realized which in one limit consist of isolated shallow impurities and, in the other limit form a degenerate 2DES with the center doped impurities acting as ionized scatterers. Consequently, using a particular doping level a 2D system is obtained that exhibits the MI-transition. We carefully studied the FIR response and the dc-transport properties in all these systems, and could clearly observe the characteristics of the transport behaviour in 2DES when going this well defined way from the insulator state, through an intermediate to the metallic state.

More details on the sample preparation and the experimental set up are given in chapter two. Chapter three gives the experimental results: in section 3.1 the case of the isolated impurity is discussed - in sections 3.2 and 3.3 we discuss the case of the degenerate metallic and the intermediate system, respectively. Chapter four summarizes and concludes this experimental study of the transition from metal to insulator in 2DES.

2 Sample Preparation and Experimental Set Up

Figure 1 shows the dependence of the bandstructure on the growth direction and the location of the doping for the investigated $GaAs/Al_xGa_{1-x}As$ (x=0.33) quantum well structures. The samples were grown by molecular beam epitaxy at substrate

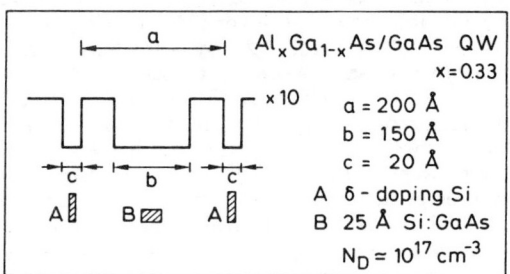

Figure 1: Schematic illustration of bandstructure and doping locations for the investigated 10 period quantum well (QW) structures

temperatures of 550°C. Each of the two double QW consists of a 150Å wide GaAs

well which, at the center for $\frac{1}{6}$ th of its width was doped with Si. The doping level was $N_D \simeq 10^{17} cm^{-3}$ giving a mean distance in the plane between the donors of more than six times the Bohr radius of $a_B \simeq 100$Å. Symmetrical with respect to the center of each well an undoped AlGaAs barrier of width 50Å and a 20Å small QW were grown. This small well was used for introducing the desired amount of additional doping which has been the parameter of our experiment and ranged from nominally undoped to a maximum level of $10^{12} cm^{-2}$. The electrical active layers were contacted by annealing of In at 400°C for 2 min. Carrier concentrations n_s were obtained from Hall measurements performed at room temperature and at various temperatures between 2.2 K and 77 K. The far infrared (FIR) response was measured in photoconductivity and/or transmission with the direction of incident radiation parallel to the magnetic field at temperatures up to 70 K. FIR sources used were a CO_2 laser pumped molecular gas laser system and a fast scan Fourier transform spectrometer. In order to prove the 2D character of the observed effects, measurements were performed where the sample has been tilted by angle Θ with respect to the direction parallel to the applied magnetic field.

3 Experimental Result and Discussion

3.1 Isolated 2D impurity system

Samples with transport properties dominated by isolated shallow impurities were obtained by leaving the 20Å well undoped. Characteristic for these only center doped samples is an activated transport behaviour as is illustrated in Figure 2. From the temperature dependent freeze out of the carriers an activation energy of $E_A \simeq 5 meV$ is deduced and we obtain from the carrier density at high temperatures $n_s \simeq 1.3 \cdot 10^{10} cm^{-2}$ per layer a lower bound for $N_D \geq 5 \cdot 10^{16} cm^{-3}$. From the analogy with the result of

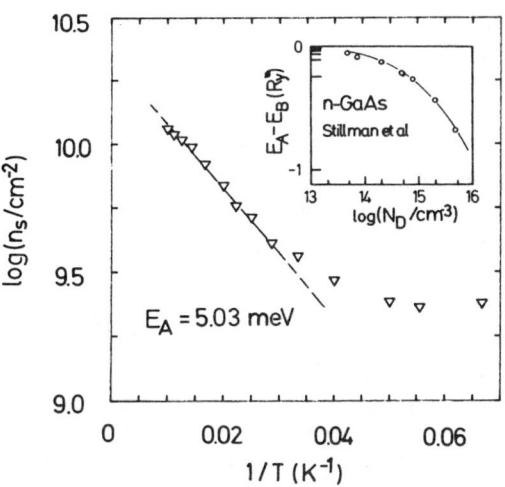

Figure 2: Measurement of the activation energy E_A for a 150Å quantum well. For comparison, the inset gives the E_A dependence on doping concentration N_D for a bulk n-GaAs taken from Stillman et al 1976

Stillman et al (1976) obtained for n-GaAs (see inset of Figure 2) we also expect E_A to increase for decreasing N_D for center doped QW systems. In a partially compensated material and for very low doping densities E_A should approach the value of the full binding energy E_B of an individual 2D confined donor. The reduced value for E_A as in the case of n-GaAs indicates the formation of an impurity band by the excited

donor states. Our measurements therefore show that the activation gap E_A between the donor groundstate and the impurity band is definitely larger in center doped QW systems than in correspondingly doped bulk systems.

An alternative to the investigation of the E_A dependence on N_D for the determination of the binding energy is offered by FIR spectroscopy. Here, we will concentrate on the study of the field dependent 1s - $2p^+$ transition energy. For the determination of the resonance position the technique of photoconductivity was found to be of great advantage. The energy versus magnetic field dependence of this transition for a 150Å wide QW measured at $\Theta = O°$ (∇) and 20° (Δ) is shown in Figure 3. For comparison, transition energies of the $1s - 2p^+$ and the CR obtained in a volume GaAs sample are included. The angular dependence of the QW impurity resonance shows its 2D character: only the component of the magnetic field perpendicular to the layer is found to be relevant. The impurity resonance in both the 2D system at $\Theta = 0$ and the 3D system is found to be shifted by an approximately constant energy with respect to the CR of unbound electrons. The strong shift of $\nabla \simeq 65 cm^{-1}$ obtained for the QW sample indicates a large binding energy. A quantitative evaluation of the binding energy can be obtained from a comparison with theory. While for a bulk shallow impurity

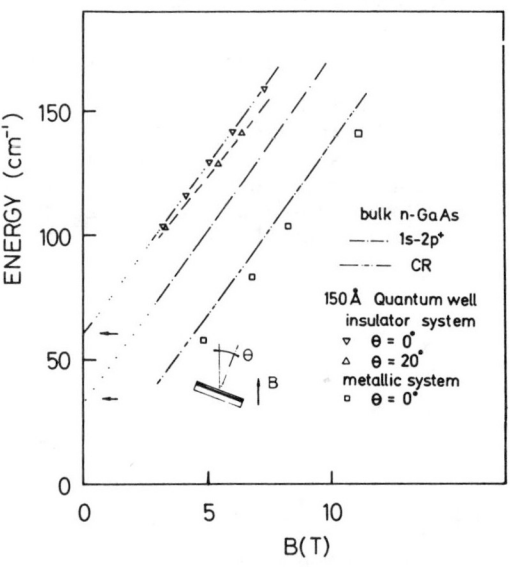

Figure 3: Resonance energies as a function of magnetic field for a 150Å QW and bulk n-GaAs systems. For the insulator type of QW one observes the $1s - 2p^+$ impurity resonances (∇, Δ), for the metallic system the free electron CR with $m_c \simeq 0.075 m_e$ was found ($\square$)

a shift of $\Delta \simeq 32 cm^{-1}$ corresponds to $E_B \simeq 5.9 meV$ (Larsen 1968), the comparison with the theory yields $E_B \simeq 11.5 meV$ for 150Å center doped QW systems (Greene and Bajaj 1983). This is shown in Figure 4, where the shift Δ obtained in several QW systems of different well width has been compared with Greene and Bajaj's values for the $1s - 2p^-$ transition at $B \simeq 6T$, which is known to be the almost field independent difference between the $1s - 2p^+$ and the CR transition energy. The agreement with our experiment is excellent for QW of width close to 150Å, for broader wells the theory predicts somewhat larger values than the observed ones. However, both techniques, the dc-transport and FIR spectroscopy, give a consistent picture of an insulator type of transport behaviour for these center doped QW systems. This means, that similar to what characterizes an insulator also in the bulk case we found a) activated conductivity with carrier freeze out following $E_A \simeq 5 meV$ and b) values for the binding energy E_B are found which - and that becomes important in the following - agrees with the theory of a 2D confined but isolated impurity.

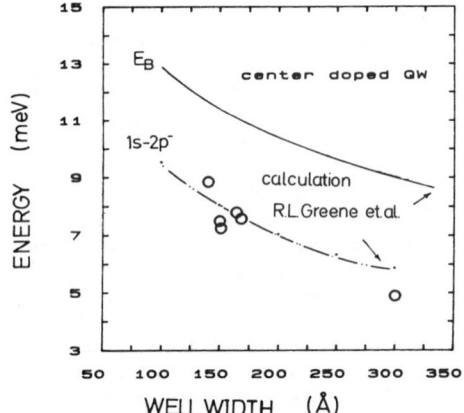

Figure 4: Comparison of the experimentally deduced shift Δ of the impurity resonance with the related $1s - 2p^-$ transition energy taken from the calculation by Greene and Bajaj (1985)

3.2 Metallic 2D System

2DES of high carrier density n_s were obtained by strongly doping the 20Å GaAs well. From the Hall measurements we deduced values for n_s up to $1.1 \cdot 10^{12} cm^{-2}$ per layer showing the efficient transfer of charge into the 150Å well. At these high n_s the 2DES is highly degenerated as follows from the observed weak temperature dependence of n_s and the strong Shubnikov-de Haas (SdH) oscillations shown by the trace at the bottom of Figure 5. These oscillations in the longitudinal resistance R_{xx} are a consequence of the magnetic field dependence of the density of states at the Fermi level E_F that determines the scattering rate. Consequently, the SdH-oscillations are very strong because of the discrete Landau level structure in 2D systems and an apparently strong scattering in these QW systems. Scattering dependent effects are also observed in the CR transmission. The upper traces in Figure 5 give the magnetic field dependent transmission at FIR wavelength between $63\mu m \leq \lambda \leq 170\mu m$. The corresponding resonance positions at the transmission minimum are depicted

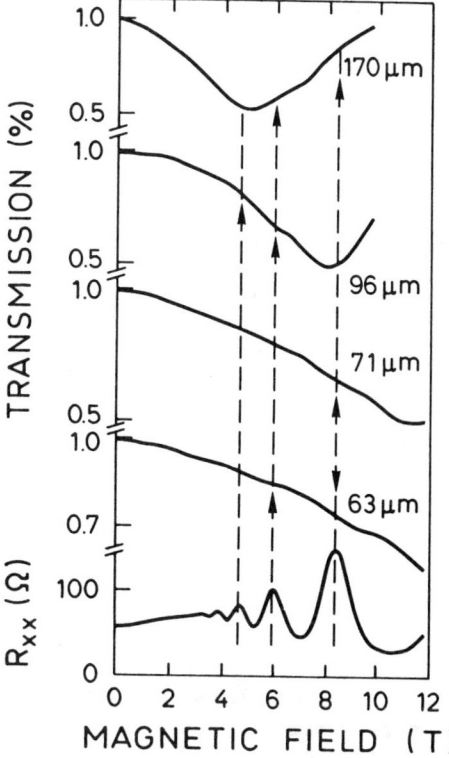

Figure 5: Magnetic field dependence of the longitudinal magneto resistance R_{xx} and CR transmission signal at different wavelength λ for a metallic QW sample

in Figure 2 (□) and from a linear fit to these data a CR mass of $m_c \simeq 0.075\, m_e$ is obtained. The resonances in these strongly doped QW systems are therefore not impurity shifted. The resonances even appear at higher magnetic fields than the free electron CR in bulk GaAs where $m_c \simeq 0.066 m_e$. The observed increase of m_c is in agreement with the known conduction band nonparabolicity in GaAs (Sigg et al 1987) if one considers the 2D confinement and Fermi energy of $E_F \simeq 40$ meV. This free electron CR does not show any specific features at the position of the impurity resonance. However, some additional structure in the lineshape related to the SdH-oscillations in R_{xx} was observed. In Figure 5 the magnetic field positions of the maxima R_{xx} are indicated; the arrows mark the small dips and peaks appearing in the lineprofile at the positions of the R_{xx} maxima. The dips appear on the high magnetic shoulder of the CR while the peaks appear on its low field shoulder. Similar CR line shapes with peaks and dips at respectively the low and high magnetic field side of the resonance were reported first for Si inversion layer systems (Abstreiter et al 1976) and were explained by the specific Landau level broadening induced by short range scatterers (Ando 1975). In GaAs/AlGaAs heterostructures SdH- oscillations of this type in the CR have been recently observed by Richter et al (1988) on systems which were, similar to our case, weakly Si doped at the interface. From the agreement with theoretical and experimental facts observed on these standard metallic systems it is evident that in our high density 2D system the conductivity is also metallic and the transport properties are dominated by the unbound electrons. However, the impurities in the well influence the scattering rate and cause the very pronounced SdH-oscillations in R_{xx} and the dips and peaks in the CR transmission line profile. For the following we would like to note that the sensitivity of the photoconductivity is very poor in such metallic systems.

4 2DES at the metal-insulator transition

In order to obtain QW samples with an intermediate type of transport behaviour, samples were grown with an increasing level of dopant in the 20 Å well. An example of dc-magneto-transport measurements for a sample with an additional doping level of about $1 \cdot 10^{11} cm^{-2}$ is shown in Figure 6. Up to $3T$, the Hall resistance R_{xy} depends

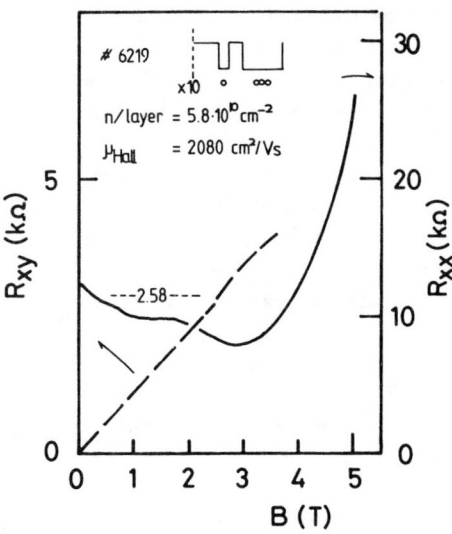

Figure 6: dc-magneto transport properties R_{xy} (scale to the left hand side) and R_{xx} (right hand side) showing a magnetic field induced MI-transition at $B \simeq 3T$

linearly on the field with slope corresponding to $n_s \simeq 5.8 \cdot 10^{10} cm^{-2}$/layer. This value

Shallow Impurities in Quantum Structures

of n_s is found to be almost temperature independent as in the case of a metallic sample. Similarly, the slow decrease of R_{xx} with increasing field is typical of a metallic behaviour. Above $3T$, however, R_{xx} strongly increases for increasing fields and the sample becomes an insulator. The temperature dependence of R_{xx} at magnetic fields above this apparently magnetic field induced MI-transition is shown in an Arrhenius plot in Figure 7. At temperatures below 10 K an approximately exponential dependence of R_{xx} on 1/T is observed from which we deduced activation energies. The inset of Figure

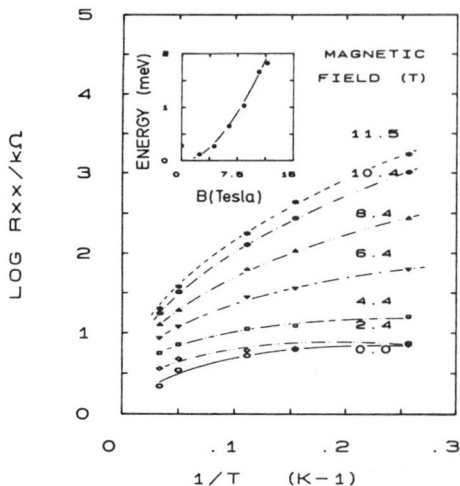

Figure 7: Arrhenius plot of R_{xx}. From the slope a magnetic field dependent "activation" energy is deduced (inset)

7 shows the thereby obtained values of energy as a function of the magnetic field. These energy values are not very practical for any immediate physical interpretation because they were obtained from R_{xx}. In a simple magnetic freeze out picture, for example, one would expect that not R_{xx}, but R_{xy} and, therefore, the carrier density would show an activated behaviour. In our case, however, n_s seems to be magnetic field independent and unaffected by the MI-transition. This could be established for sufficient low magnetic fields and high temperatures where the sample resistance was not too high for an accurate measurement of R_{xx}. Experiments are underway on samples with a better defined geometry to clarify this point. Figure 8 shows the magnetic field dependence of the photoconductivity signal obtained at different wavelength of the incident radiation. At low magnetic fields, where the sample behaves metallic like, the sensitivity of the PC signal is very low for all wavelengths. The signal for $\lambda = 103\mu m$ is peaked at about 7 T which corresponds to the position of the CR (marked by an arrow). Decreasing the wavelength moves the peak to higher magnetic fields but interestingly, the peak position thereby shifts away from the respective CR position with increasing strength. Therefore, at high magnetic fields both the FIR response and the dc-transport show a behaviour manifesting a surprising analogy to the isolated impurity system: i.e. the dc-conductivity shows some type of activated behaviour while the CR is shifted. Because the transport properties in these QW systems depend strongly on the magnetic field one prefers to analyse the FIR response for a given magnetic field as a function of the frequency. Therefore, we have studied similar samples with a Michelson interferometer and in this way obtained a transmission frequency spectrum.

Figure 9 (a) shows the transmission spectrum at $14T$ normalized by T_o, the spectrum obtained at B=0, for a sample with a MI-transition occuring at $4T$. Two distinct resonances are observed on top of a smooth background. One of these resonances almost coincides in position with the free electron CR (marked by an arrow) while the second resonance is clearly upshifted in energy. This second high energy resonance is

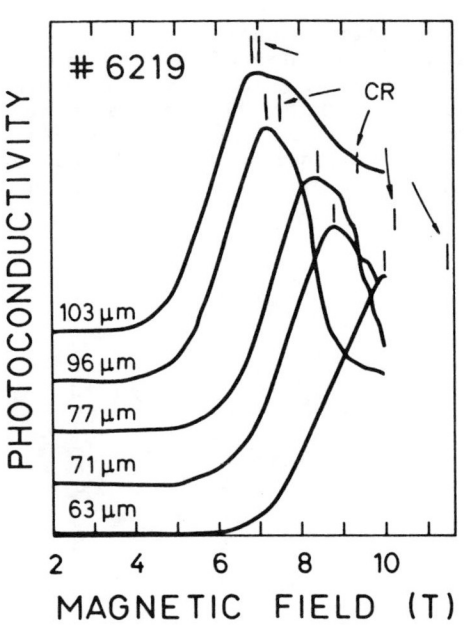

Figure 8: Magnetic field dependence of the photoconductivity signal. The arrows mark the position of the CR at the respective wavelength λ

also clearly observed in the PC measurements. For the low energy resonance the PC signal of this sample is weaker than expected from the strong transmission peak. It is worthwhile mentioning that we do not observe a third resonance which was reported before by Glaser et al (1987) and Huant et al (1988) on QW systems with doped barriers. In fact, we do not observe a resonance at the expected position of the $1s - 2p^+$ impurity transition. However, the appearance of two resonances reminds us of the two types of shallow impurities existing in our QW samples in the 20Å and in the 150Å well, respectively (c.f. Figure 1). Due to the large separation of ion and electron for the impurity in the 20Å well the corresponding $1s - 2p^+$ resonance might indeed be shifted from the CR by only a small amount. However, from the extensive discussion given above on the impurity resonance of the 2D confined center impurity, we know that the shift Δ is about $65 cm^{-1}$ and not $40 cm^{-1}$ as shown in Figure 9(a). Furthermore, the two resonances show an increasing linewidth for decreasing fields (Figure 9(b)) and the line positions, obtained from a fit procedure assuming Lorentzian line profiles, approach each other while shifting away from the position of the CR and center impurity resonance respectively (Figure 9(c)). Although the two resonances seem to merge into a single broad one, neither the fit assuming 2 resonances nor a single resonance could describe the transmission spectrum at low magnetic fields satisfactorily. Therefore we will not use the fitted results for fields below $6T$ and will rather give an analysis for the zero field data. Assuming that the transmission spectra at high magnetic fields consist solely of the two distinct high frequency peaks while they are flat at lower frequencies the zero field spectrum T_o would correspond to the difference between the spectrum generated by the Lorentzfit and the original specta T/T_o.

Applying this procedure for all normalized spectra measured at high magnetic fields ($B \geq 12T$) almost identical "zero field" spectra were generated. Further confidence in this procedure is gained from the reasonably looking spectrum at $B=6.6T$(c.f. inset Figure 9(b)) that is obtained by subtracting the averaged "zero field" spectrum from the original one of Figure 9(b). This "zero field" spectrum T_o is depicted in Figure 9(d) and shows a remarkable oscillator strength towards the low frequency end of the spectrum. In addition, we observe some absorption around $70 cm^{-1}$ which may be

Shallow Impurities in Quantum Structures

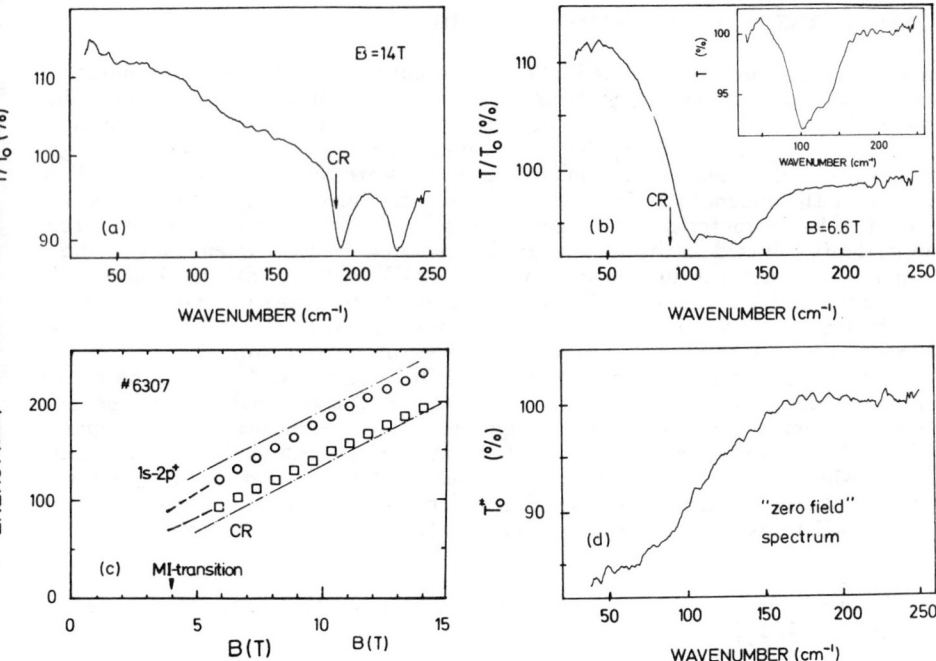

Figure 9: Transmission spectra measured at $B=14T$ (a) and $B=6.6T$ (b) normalized by T_o the spectrum taken at $B=0$. Fig. 9 (c) gives the peak positions obtained from a fit procedure assuming Lorentzian CR line profiles. The generated "zero field" spectrum T_o^* in Fig.9(d) is obtained from the high field spectra substracting the two distinct resonances (see text)

related to the two resonances resolved at high magnetic fields only. Despite the crude assumptions upon which the "zero field" spectrum is based we may conclude that dependent on the magnetic field, the FIR response shows both a metallic and insulator type of behaviour. Metallic like is the absorption at low frequency in accordance with the distinct dc-conductivity at low magnetic field and insulator like is, the shifted resonance at high magnetic field. However peculiarities are still evident e.g. the fact that the shift Δ does not agree with that of an isolated center impurity resonance. Therefore an accurate determination of the 2D impurity binding energy from the shift of the impurity resonance can only be obtained provided the 2DES is in the insulator state.

We like to note that FIR experiments performed on n-GaAs with dc-transport properties exhibiting a similar magnetic field induced MI-transition also show a magneto absorption which is reminiscent of an impurity resonance (Ming-Way Lee et al, 1988). This fact led the authors to suggest that the MI-transition in n-GaAs is connected to the impurity band. We are more hesitating about making such a statement for the QW systems investigated because,more than for the n-GaAs system, the transport properties observed here are a true mixture of both a metallic and insulator-like behaviour. A more complex analysis based on a memory function approach (Götze 1978) as has been shown by Gold and Götze (1986) predicts, at least qualitatively, such a well-balanced transport behaviour for 2DES at the MI-transition. However, no decisive quantitative comparison with our experiment was yet possible.

5 Summary and Conclusions

We have shown that for a well-defined QW system center doped with shallow impurities the transport properties are tunable from insulator to metallic like. The specific 2DES were defined by the single parameter of the dopant concentration outside the main well. For the only center doped QW system enhanced activation and binding energies due to the 2D confinement of the isolated impurity were found. Shubnikov-de Haas oscillations in the magneto resistance and in the free electron cyclotron resonance were found to be characteristic for degenerate and metallic 2DES obtained by strong additional doping. Based on the knowledge obtained in these two systems, a systematic description of the metal to insulator transition in QW systems with weak additional doping was given. A magnetic field induced MI-transition was observed exhibiting some particular transport features for both the metallic low field and insulator high field phase. At low magnetic fields we also observed some absorption at about $70 cm^{-1}$ in additon to the metallic like absorption for low FIR frequencies. Although the system becomes an insulator with strong increase of resistance at higher fields we got no clear evidence for either a temperature or a magnetic carrier freeze out and the impurity type of CR shift was shallower than that of the known 2D confined isolated impurity.

In conclusion, well defined 2D electron systems exhibiting the metal-insulator transition were obtained enlarging the experimental means for investigating this not yet well-understood transition and representing a critical test for its theoretical description.

Acknowledgement
Part of this work has been supported by the "Bundesministerium für Forschung und Technologie" (BMFT).

6 References

Abstreiter G Kotthaus J P Koch J F and Dorda G 1976 *Phys Rev B* **14** 2480
Ando T 1975 *J Phys Soc Japan* **38** 989
Bastard G 1981 *Phys Rev B* **24** 4714
Glaser E Shanabrook B V Hawkins R L Beard W Mercy J M Mc Combe B A and Musser D 1987 *Phys Rev B* **36** 8185
Gold A and Götze W 1986 *Phys Rev B* **33** 2495
Greene R L and Bajaj K K 1983 *Solid State Commun* **45** 825
Götze W 1978 *Solid State Commun* **27** 1393
Huant S Stepniekwski R Martinez G Thierry-Mieg V and Etienne B 1988 *Proc Int Conf Superlattices Trieste* to be published
Jarosik N C McCombe B D Shanabrook B V Comas I Ralston I and Wicks G 1985 *Phys Rev Lett* **54** 1283
Larsen D M 1968 *J Phys Chem Solids* **29** 271
Raymond A Piotrzkowski R Robert JJ Azema S Kubisa M Zawadzki W 1987 *Journal de Physique C5-243 Int Conf on Modulated Semicond Structures* ed Raymond and Voisin
Sigg H Perenboom J A A J Pfeffer P and Zawadzki W 1987 *Solid State Commun.* **61** 685
Stillmann G E and Wolfe C M 1976 *Thin Solid Films* **31** 69
Zrenner A Koch F and Ploog K 1988a *Surface Science* **196** 671
Zrenner A 1988b *thesis* unpublished

A far infrared study of confinement effects on acceptors in GaAs/AlGaAs quantum wells

A.A. Reeder and B.D. McCombe, Department of Physics,
Fronczak Hall, University of Buffalo, Buffalo, NY 14260

F.A. Chambers and G.P. Devane, Amoco Corporation,
Warrenville Road & Mill Street, Naperville, IL 60566

ABSTRACT
Beryllium acceptors doped in the centers of GaAs/AlGaAs quantum wells with widths between 300Å and 100Å, as well as in a "bulk" epitaxial layer of GaAs for comparison have been studied by far infrared magnetospectroscopy. Results clearly show the effects of confinement on the acceptor, and the observed increase in transition energy is in qualitative agreement with recent calculations. For the bulk acceptor the observed splittings in a magnetic field cannot be explained with calculated g-values.

INTRODUCTION

Shallow donors in GaAs/AlGaAs quantum wells have been studied extensively both theoretically Bastard (1985) and experimentally Jarosik (1985) over the past few years. The behavior of the donors is understood sufficiently well that dopant redistribution on a length scale of 10-20Å can be investigated[3]. The valence band of bulk GaAs is four-fold degenerate and has a cubic symmetry which must be taken into account to describe the acceptor states. Initial calculations of acceptor states in a GaAs/AlGaAs quantum well have been made by Masselink and Raman studies (Gammon, 1986) of transitions between s-like states are in reasonable agreement with calculations. At present, however, there are no reported calculations of the p-like acceptor states in quantum wells. In far infrared transmission experiments, the dominant features are due to transitions between the s-like ground states and higher p-like states.

In this paper we report a far infrared transmission study of Be acceptors in GaAs/Al$_{1-x}$Ga$_x$As (x=0.3) quantum wells with widths between 300Å and 100Å. A thick (~4μm) Be-doped GaAs epitaxial layer was also studied as a limiting case, and to provide a basis for understanding the quantum well results.

Due to the more complicated nature of the valence band of GaAs the acceptor states are complex, and it is helpful to consider these states in the bulk prior to introducing the confining quantum well potential. The top most valence band states of GaAs near the Γ point (Brillouin Zone Center) consist of 3 p-like (ℓ=1) states that are triply degenerate (6-fold with spin) in the absence of spin-orbit interaction. Including spin these states can be characterized by their total angular momentum quantum

number J and its projection along an axis, M_J, as J=3/2 (M_J=3/2,1/2,-1/2,-3/2) and J=1/2 (M_J=1/2,-1/2). With spin-orbit coupling included the J=3/2 and J=1/2 bands are split by Δ=340meV, with the 4-fold degenerate (at k=0) J=3/2 bands lying highest in energy. The J=3/2 bands are split into two Kramers degenerate bands, $M_J=\pm 3/2$ and $M_J=\pm 1/2$, usually referred to as the "heavy hole" and "light hole" bands, respectively.

An acceptor in bulk GaAs in lowest order can be thought of as a spin 3/2 particle in a Coulomb potential, essentially a one particle spin 3/2 atom. The total angular momentum, F, of the "atom" is obtained by adding the total spin of the hole (3/2) and the orbital angular momentum of the hydrogenic envelope, according to the usual quantum mechanical rules for addition of angular momenta; L, thus F takes the integer spaced values between L+3/2 and |L-3/2|. The valence band, however, has cubic symmetry which leads to splittings of the acceptor states of a spherically symmetric potential. Acceptor states must therefore be given an irreducible representation of the valence band point group to identify them uniquely. The lowest acceptor states are identified as $1s_{3/2}\Gamma_8$, $2p_{3/2}\Gamma_8$, $2s_{3/2}\Gamma_8$, $2p_{5/2}\Gamma_8$, $2P_{5/2}\Gamma_7$ and $2p_{1/2}\Gamma_6$. Notice that the F=5/2 states are split as a result of the lower cubic symmetry of the valence band.

Using the Luttinger Hamiltonian, which describes the dispersion of the valence band, the energy levels of an acceptor have been determined. In the simplest form Baldereschi (1973) only spin orbit coupling was considered, while in the later treatment Baldereschi (1974) the cubic term was added as a pertubation. Kirkman (1974) used a "zeroth order" wave function approach to determine the allowed optical transitions for acceptors in GaAs. For the transitions relevant to this discussion ($1s_{3/2}\Gamma_8$ to $2p_{5/2}\Gamma_8$ and $2p_{5/2}\Gamma_7$) the strongest absorption lines in the Faraday geometry arise from $\Delta m=\pm 1$ and weaker transitions from $\Delta m=\pm 3$. These authors also reported calculations of the g-values for the ground state and the p-like excited states which agree within 20% with their far infrared studies.

The only reported calculation of acceptor states in GaAs/AlGaAs quantum wells are those of Masselink. These workers used the Luttinger Hamiltonian with the addition of a confining potential

$$V(z) = \begin{cases} V=0, & |z|<\frac{W}{2} \\ V=V_o, & |z|>\frac{W}{2} \end{cases}$$

where V_o represents the valence band discontinuity, and W is the well width.

The presence of the quantum well potential reduces the symmetry of the valence band to D_{2d}. In this reduced symmetry the J=3/2 - like valence band states are split into "heavy" ($M_J=\pm 3/2$) and "light" ($M_J=\pm 1/2$) hole bands. "Heavy" and "light" refer to the curvature effective mass in the confinement direction. Additionally, the $1s_{3/2}\Gamma_8$ acceptor ground state and the $2p_{5/2}\Gamma_8$ excited state must be split accordingly into $1s_{3/2}(\Gamma_6+\Gamma_7)$ and $2p_{5/2}(\Gamma_6+\Gamma_7)$, respectively.

Fig. 1 Left, a schematic energy level diagram of an acceptor in bulk GaAs, T_d. Right, the expected splittings of an acceptor under D_{2d} symmetry of a quantum well.

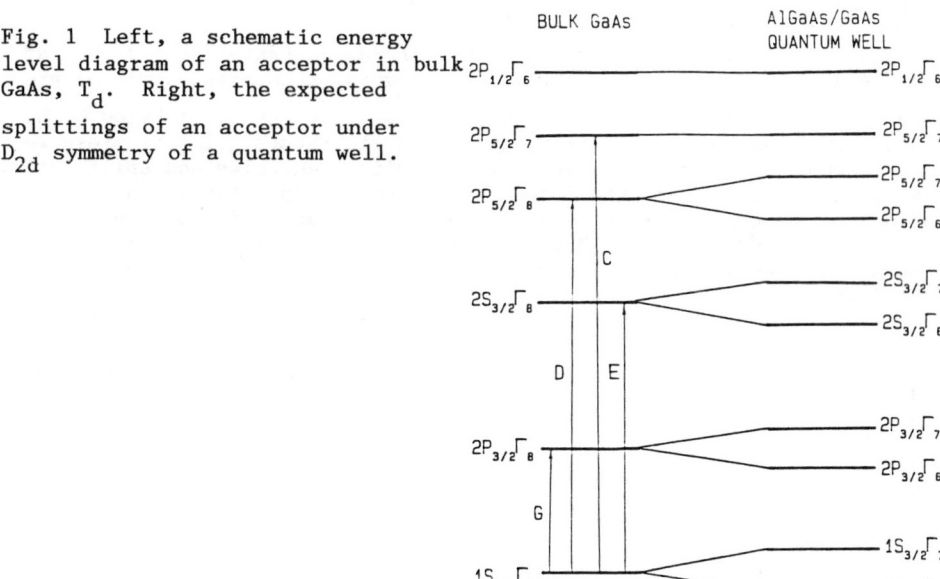

In Fig. 1 the expected splittings of the states for an acceptor at the center of a GaAs/AlGaAs quantum well are shown schematically. Masselink (1985) solved the Luttinger (1956) Hamiltonian variationally in the presence of the confining potential $V(\hat{z})$, with and without the Coulomb potential. With the Coulomb term omitted the resulting energy is the appropriate confinement subband energy. Their calculations show that as the quantum well width decreases from infinity (bulk) to 100Å the acceptor ground state splits by about 2.5meV, and the binding energy increases by about 4meV. As with a donor in a quantum well an increase in the binding energy of the acceptor states with decreasing well width is found. Since the acceptor ground state wave function in the bulk is much less extended than the donor ground state wave function, the percentage effect over this range of well-width is much smaller for acceptors. Confinement effects on the excited states are much greater due to their more extended wave functions.

EXPERIMENTAL

Far infrared 80-250 cm^{-1} absorption measurements were made with a Specac 40.000 or a Bomem DA3.02 Fourier Transform spectrometer. Conventional light-pipe optics were used to couple the interferometer to a 9T superconducting magnet and a Ge:Ga photoconductive detector operated at 4.2K. Samples were mounted in He exchange gas in the Faraday geometry in a two-position rotatable sample holder. An acceptor-doped sample and a section of semi-insulating GaAs, which was used to provide reference spectra, were mounted in this holder. Transmittance spectra were obtained by ratioing sample spectra to reference spectra taken under the same conditions during the same run. All samples were MBE-grown on semi-insulating GaAs substrates. Table 1 shows the characteristics of the samples studied. The GaAs quantum wells were doped over the center 1/3 of each well; the $Al_{0.3}Ga_{0.7}As$ barriers were 150Å thick and were nominally undoped.

RESULTS & DISCUSSION

The transmission spectrum of 3μm of Be-doped (3×10^{16} cm^{-3}) GaAs at zero field and 4.2K is shown in Fig. 2a. Several transmission minima are evident at 135, 167, 184 and 220 cm^{-1}. These features are labeled G, D, C and A, respectively, consistent with the literature. In the presence of a magnetic field the D line moved to higher energies and split into several components, and the C feature also moved up slightly in energy. Heating the sample to 20K caused new lines to appear at frequencies below the D feature (See Figs. 2b, 2c). At lower magnetic fields a similar temperature dependence was observed but with proportionally smaller splittings.

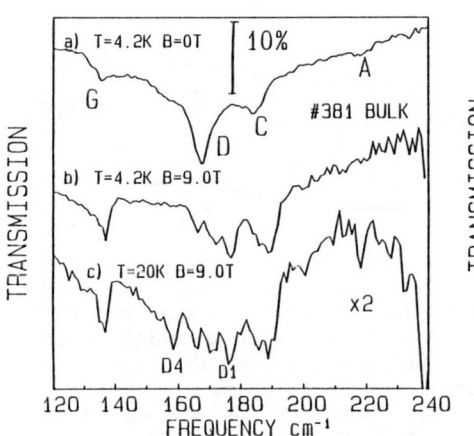

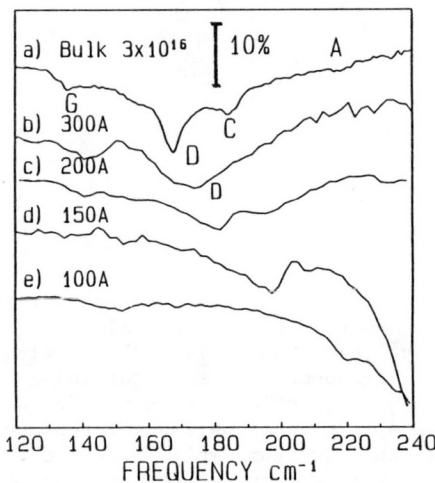

Fig. 2 Absorption spectra of 3μm of Be (3×10^{16} cm^{-3}) doped GaAs at a) 0T and 4.2K b) 9T 4.2K and c) 9T 20K. Note, at high magnetic field heating the sample causes a feature D4 to appear.

Fig. 3 Zero field transmission spectra taken at 4.2K of a) Be doped GaAs 3μm thick and GaAs/AlGaAs Quantum Wells Be doped are center 1/3 at 3×10^{17} cm^{-3} with widths b) 300Å c) 200Å d) 150Å and e) 100Å.

In Figs. 3a, b, c, d and e, the zero field transmission spectra of the bulk, 300Å, 200Å, 150Å and 100Å samples are shown. A series of absorption lines which move to higher energies as the well width decreases is evident. The dominant feature displays the same general magnetic field dependence as the bulk D line.

The relative separation of the features G, D, C and A is consistent with the photoconductive peaks observed by Kirkman. These lines arise from transitions between the $1s_{3/2}\Gamma_8$ ground state and $2p_{3/2}\Gamma_8$, $2p_{5/2}\Gamma_8$, $2p_{5/2}\Gamma_7$ and $2p_{3/2}\Gamma_6$ states, respectively. Since the energy of the acceptor ground state has a much larger chemical shift than the p-like states, the absolute position of these features depends on the chemical identity of the impurity while their relative separation is independent of impurity species.

The appearance of lower energy features associated with the D line at 20K at 9.0T is due to splitting of the ground state $1s_{3/2}\Gamma_8$ in the magnetic field. When k_BT is comparable to the field splitting, the higher energy states of the ground states multiplet are populated, and the absorption intensity for transitions originating in these states becomes appreciable. With the g-values given Kirkman (1978) or Bimberg (1978), we cannot explain the temperature dependence of the absorption found in the present work. Although the magnitudes of the splittings are consistent with the magnitude of the calculated g-values, the sign of the g-values appear to be in error; ie. the ordering of the levels of either the ground state $(1s_{3/2}\Gamma_8)$ or excited state $(2P_{5/2}\Gamma_8)$ multiplet is incorrect, similar discrepancies have been noted by Liu (1988).

The dominant absorption feature labeled D' in Fig. 3b displays a magnetic field dependence, approximately $1\ cm^{-1}/T$, similar to the highest component of the D line of the bulk acceptor. The relatively high dopant density $3 \times 10^{17} cm^{-3}$ used in the quantum wells would lead to very broad C and D features in bulk GaAs. Although the C and D features are much sharper in the quantum well than would be expected for a bulk sample with the same dopant density, they are still quite broad; and it is difficult to identify any fine structure as was the case with the present bulk sample.

As the quantum well width decreases from 300Å to 100Å the C and D features move to higher energies (Fig. 4).

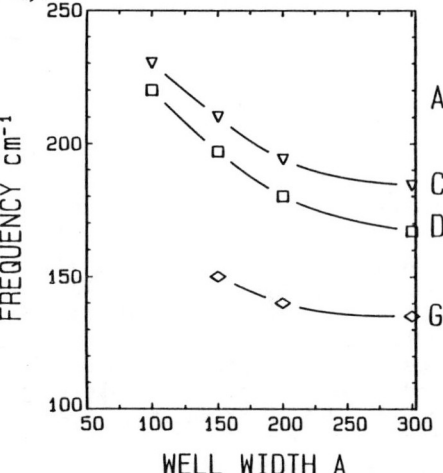

Fig. 4 Position of the zero field C,D,G features as a function of well width. As the well width decreases the transitions move to higher energies.

This is due to the quantum well "pushing" the hole closer to the negative core thereby increasing its binding energy. Thus the energy of all the states will increase at a rate that depends on the Bohr radius of the state and the orientation of the lobes of its wave function relative to the confinement axis. Clearly the transition energy will move to higher energy as the well width decreases. This behavior is similar to the donor system for which extensive calculations are in good agreement with experiment[2]. The calculations of Masselink (1985) are for the s-like ground and excited states from which only the E transition can be calculated. This transition is parity forbidden for the infrared experiments. However, confinement effects are expected to be of similar

magnitude for the D-like transition due to the similar Bohr radii of the $2s_{3/2}\Gamma_8$ and $2p_{5/2}\Gamma_8$ states Baldereschi (1973).

SUMMARY

The absorption spectra of Be acceptors in GaAs/AlGaAs quantum wells have been measured and clear effects of the confinement have been observed in the form of an increase in the transition energy with decreasing well width. This behavior is qualitatively similar to that observed for shallow donors. The increase of the transition energy of the dominant absorption line (D line) is of the same magnitude as predicted for the E transition. For more quantitative comparison with the present data, calculations of the p-like excited states are required. The splittings of the D feature in a magnetic field cannot be explained using the g-values of Kirkman (1978) or Bimberg (1978).

This work was supported by the Office of Naval Research and the Center for Electronic and Electro-Optic Materials at SUNY at Buffalo.

TABLE ONE

WELL WIDTH	NUMBER OF WELLS	DOPANT DENSITY cm-3	DOPANT POSITION
Bulk	---	3×10^{16}	Uniform
300Å	33	3×10^{17}	Center 1/3
200Å	50	3×10^{17}	Center 1/3
150Å	50	3×10^{17}	Center 1/3
100Å	100	3×10^{17}	Center 1/3

REFERENCES

Baldereschi, A. and Lipari, N.O., Phys. Rev. B8, 2697 (1973).
Baldereschi, A. and Lipari, N.O., Phys. Rev. B9, 1525 (1974).
Bastard, G., Phys. Rev., B24, 4714 (1985).
Bimberg, D., Phys. Rev. B18, 1794 (1978).
Gammon, D., Merlin, R., Masselink, W.T. and Morkoc, H., Phys. Rev. B33, 2919 (1986).
Jarosik, N.C., McCombe, B.D., Shanabrook, B.V., Comas, J., Ralston, J. and Wicks, G., Phys. Rev. Lett. 54, 1283 (1985).
Kirkman, R.F., Stradling, R.A. and Lin-Chung, P.J., J. Phys. C11, 419 (1978).
Liu, X., and Petrou, A.P., proceedings of the Superlattice, Microstructures and Microdevices Conference, Chicago, August 1987, published in Superlattices and Microstructures, 4(2), p141 (1988).
Luttinger, J.M., Phys. Rev. 102, 1030 (1956).
Masselink, W.T., Chang, Yia-Chung and Morkoc, H., Phys. Rev. B32, 5190 (1985).
Mercy, J-M., Jarosik, N.C., McCombe, B.D., Ralston, J. and Wicks, G., J. Vac. Sci. Technol., B4, 1011 (1986).

Optical study of the excited states of the Be-acceptor confined in a 70 Å wide GaAs/AlGaAs quantum well

P.O. Holtz*, M. Sundaram, J.L. Merz, and A.C. Gossard

Department of Electrical and Computer Engineering and Materials Department,
University of California at Santa Barbara, Santa Barbara, CA 93106, USA

ABSTRACT. The Be-acceptor confined in 70 Å wide GaAs/AlGaAs QW has been investigated by two spectroscopic methods: Resonant Raman scattering and two hole transitions of the bound exciton, both observed upon selective dye-laser excitation. The two transitions are interpreted as the transitions from the $1s_{3/2}[\Gamma_6]$ ground state to the $2s_{3/2}[\Gamma_6]$ and $2s_{3/2}[\Gamma_7]$ excited states of the confined Be-acceptor. The achieved experimental results are finally compared with theoretical predictions.

1. INTRODUCTION

The binding energy of the ground state and also the excited states of impurities confined in quantum wells (QWs), have been predicted in several theoretical papers. The original calculation by Bastard (1981) assumed hydrogenic impurity potentials and also infinite barrier heights, which resulted in a continuously increasing binding energy with decreasing well width. In later reports (Greene et al 1983, Maihilot et al 1982), where the calculations were extended to include e.g. more realistic finite heights for the QW barriers, it was shown that the binding energy goes through a maximum at a non-zero well width. Further extension of the calculations include coupling of adjacent QWs (Chaudhuri 1983) and the effect of external perturbations such as uniaxial stress (Masselink et al 1985), electric field (Masselink et al 1985, Brum et al 1985) and magnetic field (Masselink et al 1985, Greene et al 1985). Thus, there has been recent progress, what concerns the theoretical approach on this subject, while the experimental demonstrations on the properties of the confined impurities are quite limited. One of the reasons for the limited number of experimental reports is the difficulty to directly transfer the techniques, which have been used for corresponding studies on bulk material, to QWs. Some experimental progress has, however, been achieved with Resonant Raman Scattering (RRS), which has been performed on both confined donors (Shanabrok et al 1984) and acceptors (Gammon et al 1986). Also, far infrared magnetospectroscopy measurements on confined impurities have recently been reported (Reeder et al 1987).
In the present paper, we report on observed transitions from the ground state to the excited states of the confined Be-acceptor via two hole transitions (THT) of the acceptor bound exciton (ABE), observed with selective photoluminescence (SPL). This technique has frequently been used for studies of the excited states of acceptors in bulk material, but has not earlier been reported for QWs to the knowledge of the authors'. Also RRS has been used for the same QW structure, which has made it possible to observe the same transitions in the same series of measurements with two different techniques. Finally, the achieved results are compared with theoretical predictions by Masselink et al, 1985.

© 1989 IOP Publishing Ltd

2. SAMPLES AND SPECTROSCOPY

The samples used in this work were grown by molecular beam epitaxy (MBE) on semi-insulating (100) GaAs substrates. The width of the GaAs QWs were for all samples intentionally 70 Å, and it was shown by photoluminescence (PL) measured at different spots, that the homogeneity was usually of the order ±1 monolayer over the wafer. The composition of the $Al_xGa_{1-x}As$ barrier was nominally x=0.3. Single QWs (SQWs) as well as multiple QWs (MQWs) with up to 100 QWs were investigated. All samples were doped in the central 1/4 of the QWs with concentrations varying from $2x10^{16}$ to $2x10^{17}$ cm^{-3}.

An Ar ion laser was used as the excitation source for the (PL) measurements. The same laser pumping a dye laser with Styryl 8 dye was used for the SPL and PL excitation (PLE) experiments. The samples were cooled in a He bath cryostat, in which the temperature could be varied continuously from 1.4 K to room temperature. The emitted light was detected via a 1m spectrometer and a GaAs PM detector.

3. EXPERIMENTAL RESULTS

A typical PL spectrum with above-bandgap excitation of a 70 Å wide SQW, Be-doped in the central part at a concentration of $1x10^{17}$ cm^{-3}, is shown in Fig.1. In addition to the free exciton (FE), which usually dominates the PL spectrum in a QW, one can observe the exciton bound at the Be-acceptor (denoted BE in Fig.1), which is approximately of the same intensity and width as the FE. No additional PL related to the 70 Å QW can be observed with this excitation. The PLE spectra of this sample will be almost identical, when either the FE or the BE is detected. In both cases are the hh- and lh-states of the FE observed.

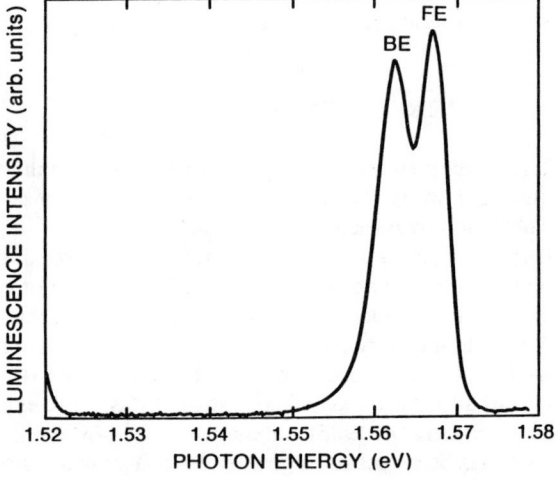

Fig. 1 PL spectrum at 1.5 K of a Be-doped SQW

When any of these states is excited resonantly, additional structure will appear in the SPL spectrum towards lower energy (Fig. 2). As can be observed in this spectrum, the exciton peak has broadened, making it impossible to resolve the FE and BE peaks. Further, a strong and fairly broad peak will show up at ≈1.55 eV, which has been interpreted as the free-to-bound (FB) transition (R.C. Miller, 1984). On the low energy side of this peak a threshold, denoted P1 in Fig. 2, can be seen and towards lower energy, another weak feature, denoted P2, is observed. At excitation, slightly off resonance with the FE states (Fig. 3), the strength of the FB peak is reduced relatively the P1 and P2 peaks, which thus are better resolved. The joint exciton peak is still too broad to resolve the individual FE and BE peaks, but an important observation is the significant shift of the joint exciton peak at the different excitations. Thus, the FE gains strength at resonant excitation, while the BE will dominate or is at least comparable with the FE for the non-resonant case. Obviously, there is an intensity

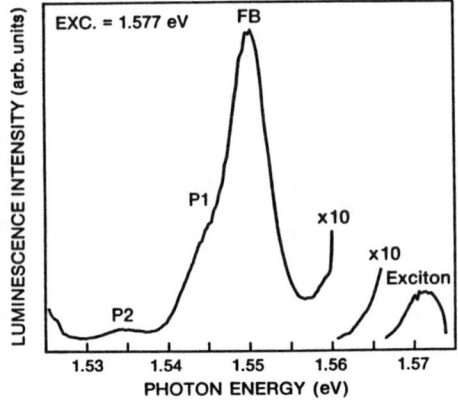

Fig. 2 SPL spectrum of the same sample as shown in Fig. 1 with excitation resonant with the hh-state of the FE

Fig. 3 SPL spectrum similar to the spectrum in Fig. 2, but with non-resonant excitation

correlation between the BE on one hand and the P1 and P2 peaks on the other hand, when the excitation energy is varied. This fact gives strong support to the interpretation of the P1 and P2 peaks as the THTs related to the exciton bound at the confined Be-acceptor.

In contrast to the case of bulk material, where the intensity of the THT peaks is enhanced upon excitation resonant with the BE, the P1 and P2 peaks are as strongest when the excitation is resonant with the FE states. This is, however, consistent with the PLE measurements, where just the FE states are observed even when the BE is detected. Also, the PL spectrum for a QW is usually dominated by the FE even at relatively high doping levels. The experimentally observed energies for the transitions from the ground state to the excited states of the confined Be-acceptor corresponding to the energy separation between the BE and the THT peaks (P1 and P2) are 29.5±1 and 37.5±1 meV, respectively.

When the corresponding SPL measurements are performed in a MQW (with 100 periods of QWs), doped in the same way as the SQW, additional peaks will appear at excitation resonant with or close to the BE state (Fig. 4). Two well-defined lines (denoted R1 and R2 in Fig. 4) with a halfwidth of less than 2 meV can be observed in approximately the same range as the P1 and P2 peaks. The R1 and R2 lines are, however, Raman-active and shift with the excitation energy within a limited

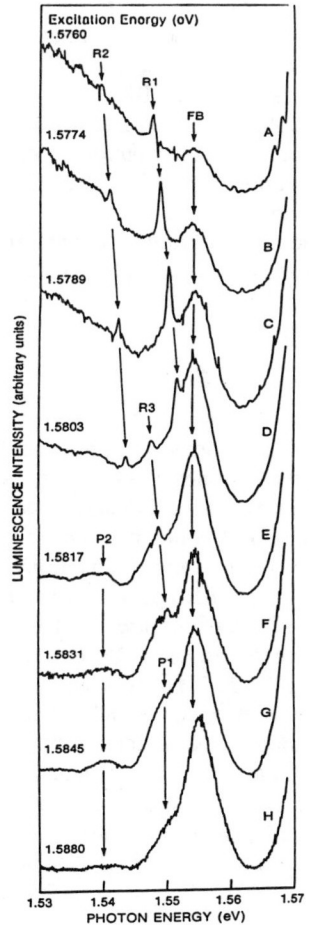

Fig. 4 A synopsis of SPL spectra for a Be-doped MQW, where the excitation energy is varied from 1.576 to 1.588 eV

range.
The dependence of the intensity of the R1 and R2 peaks on the excitation energy is plotted in Fig. 5. From this plot it can be seen that the cross-section for the R1 and R2 lines are almost identical and has a maximum at ≈1.579 eV, which corresponds to the position of the BE in PL for this sample. Also the halfwidth of this cross-section is similar to what is observed for the BE peak in PL. The energy separation between the excitation energy and the Raman-active lines, R1 and R2, are 29.0±0.5 and 36.5±0.5 meV, respectively. When the excitation energy is shifted slightly towards higher energy, a third Raman-active line, R3, can be observed (Fig. 4). The cross-section of the R3 line has a maximum at about 4 meV above the corresponding maximum for the R1 and R2 lines. The position for the maximum of the R3 line coincides with the FE peak in PL. The R3 line thus originates from RRS at excitation resonant with the FE (Holtz et al, 1988) in the same way as the R1 and R2 lines are due to scattering via the BE.

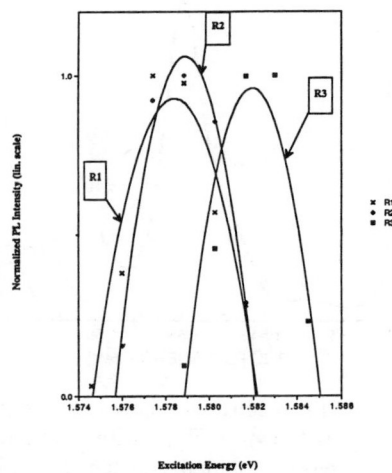

Fig. 5 The SPL intensity of the Raman-active lines versus the excitation energy

The P1 and P2 peaks show no detectable shift upon a variation of the excitation intensity over three orders of magnitude. Such a shift had been expected for a donor-acceptor pair (DAP) transition, which had been the alternative explanation for at least the P1 peak. Further, the thermal behaviours of the P1, P2, R1 and R2 peaks are very similar, with reduction of the intensity starting at about 10 K and completely quenched at ≈30 K. This temperature dependence is also almost identical with what is observed for the BE peak. Thus, the thermal behaviour of the P1, P2, R1 and R2 transitions seems to reflect the temperature dependence of the BE.

4. DISCUSSION

Two transitions, interpreted as the transitions from the ground state to different excited states of the confined Be-acceptor, have been observed in the same sample with two independent spectroscopic techniques. Both transitions are observed in SPL upon excitation in the exciton range and could thus be observed in the same series of measurements. However, the behaviour of these transitions differ in two important aspects: As described in the section above, the RRS peaks have an intensity maximum at excitation resonant with the ABE, while the THT peaks are as strongest, when any of the FE states are selectively excited. Furthermore, the RRS lines are shifting with the excitation energy and are also more narrow than the THT peaks. These properties are expected for Raman-active transitions. The shift is due to the nature of RRS: The photon energy of the Raman line is reduced with the energy needed to excite the acceptor due to inelastic scattering and is thus at a constant energy separation from the excitation energy. The width of the RRS line is in this case determined by the distribution of the acceptor binding energies, which in the case of a 70 Å wide QW, doped just in the central 1/4 is expected to be of the order 2 meV (Bastard, 1981 and Masselink et al, 1985). The halfwidth for the cross-section of the RRS line is expected to be

of the same order as the ABE peak. Both these predictions are in good agreement with what is observed. Transitions between states of the same parity are strongly favoured by the selection rules in RRS as well as THT in bulk material. Thus, transitions from the $1s_{3/2}$ ground state to the $2s_{3/2}$ excited state will dominate the spectrum. The same transition is observed to be the strongest in RRS spectra also for impurities confined in QWs (B.V. Shanabrok et al, 1984 and D. Gammon et al, 1986). Accordingly, the strongest lines, P1 in THT and R1 in RRS (Fig. 4), are interpreted as the $1s_{3/2}[\Gamma_6] - 2s_{3/2}[\Gamma_6]$ transitions of the confined Be-acceptor (Fig. 6).

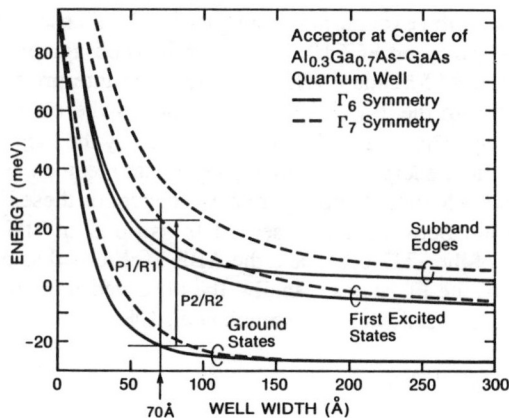

Fig. 6 Calculated binding energies for acceptors doped in the central part of a QW versus the well width

The observed transition energies, 29.5±1 meV from THT and 29.0±0.5 meV from RRS, are also in agreement with the theoretical calculations (Masselink et al, 1985), which predicts a transition energy of about 30 meV for an effective mass-like acceptor confined in a 70 Å wide QW.

To get a relevant comparison with the experimentally determined transition energy, a central-cell correction has to be added to the calculated effective mass-like value. If the same correction is used for the confined acceptor as for the Be-acceptor in bulk material (2 meV), the predicted $1s_{3/2}[\Gamma_6] - 2s_{3/2}[\Gamma_6]$ transition energy is 32 meV. The calculated ionization energy of the Be-acceptor in Γ_6 symmetry provided in the same way is 36 meV. This binding energy can alternatively be estimated from the energy separation between the bandgap and the FB peak in the same sample. If the binding energy of the FE is assumed to be 11 meV (P. Dawson et al, 1986), the position of the FB peak is about 38 meV below the bandgap.

For the case of single, substitutional acceptors, the symmetry is reduced from T_d in bulk to D_{2d} upon confinement in a QW. In the case of a confined impurity, also transitions between states of different symmetry are expected if higher-order terms are taken into account, which is a reasonable assumption, since RRS as well as THT are of resonant character (D. Gammon et al, 1986). The most reasonable interpretation for the second transition (P2 and R2) is then the transition from the $1s_{3/2}[\Gamma_6]$ ground state to the first s-like excited state in the Γ_7 symmetry, $2s_{3/2}[\Gamma_7]$ (Fig. 6). According to the theoretical work by Masselink et al, 1985, the transition energy to this state is about 41 meV including the central cell correction. Alternative interpretations like a transition to the higher excited state, $3s_{3/2}[\Gamma_6]$ is less plausible due to energy considerations (Masselink et al, 1985). For the case of a transition to the Γ_6 subband edge, a broader peak than P2 and R2 is expected.

5. SUMMARY

Two excited states of the Be-acceptor confined in a narrow GaAs/AlGaAs QW have been observed with two different spectroscopic techniques: THT and RRS. Selective dye-laser

excitation in the exciton range was used in both cases. The observed transition energies in RRS, 29.0±0.5 and 36.5±0.5 meV, are in good agreement with what is observed in THT, 29.5±1 and 37.5±1 meV. The transitions are interpreted as transitions of the confined Be-acceptor from the $1s_{3/2}[\Gamma_6]$ ground state to the excited states, $2s_{3/2}[\Gamma_6]$ and $2s_{3/2}[\Gamma_7]$, respectively. This interpretation is supported by the fact, that the two transitions are observed at very similar energies with two independent methods, where the same selection rules apply. The almost identical temperature dependence for these lines and the ABE also confirms the relationship between the observed transitions and the ABE. A correlation between the intensity of the THT peaks and the ABE is observed as the excitation energy is varied in the vicinity of the FE states. Finally, the observed transition energies are compared with the theoretical predictions by Masselink et al, 1985.

6. REFERENCES

*Permanent address: Department of Physics and Measurement Technology, Linköping University, S581 83 Linköping, SWEDEN

G. Bastard, *Phys. Rev. B24*, 4714 (1981)
A. Brum, C. Priester, and G. Allan, *Phys. Rev. B32*, 2378 (1985)
S. Chaudhuri, *Phys. Rev. B28*, 4480 (1983)
P. Dawson, K.J. Moore, G. Duggan, H.I. Ralph, and C.T.B. Foxon, *Phys.Rev. B34*, 6007 (1986)
D. Gammon and R. Merlin, *Phys. Rev. B33*, 2919 (1986)
R.L. Greene and K.K. Bajaj, *Solid State Commun. 45*, 852 (1983)
R.L. Greene and K.K. Bajaj, *Phys. Rev. B31*, 913 (1985) and *Phys. Rev. B31*, 4006 (1985)
P.O. Holtz, M. Sundaram, J.L.Merz, and A.C. Gossard, to be published
C. Mailhiot, Y.C. Chang, and T.C. McGill, *J. Vac. Sci. Technol. 21*, 519 (1982) and *Phys. Rev. B26*, 4449 (1982)
W.T. Masselink, Yia-Chung Chang, and H. Morkoç, *Phys. Rev. B32*, 5190 (1985)
W.T. Masselink, Yia-Chung Chang, H. Morkoç, D.C. Reynolds, C.W. Litton, K.K. Bajaj, and P.W. Yu, *Solid State Electr. 29*, 205 (1986)
R.C. Miller, *J. Appl. Phys. 56*, 1136 (1984)
A.A. Reeder, B.D. McCombe, F.A. Chambers, and G.P. Devane, *Third Int. Conf. on Superlattices, Microstructures and Microdevices*, Chicago, 1987
B.V. Shanabrok, J. Comas, T.A. Perry, and R. Merlin, *Phys. Rev. B29*, 7096 (1984)

Inst. Phys. Conf. Ser. No 95: Chapter 1
Paper presented at Int. Conf. Shallow Impurities in Semiconductors, Linköping, Sweden, 1988

Photoluminescence study of donors in selectively doped GaAs/AlGaAs quantum wells

X. Liu[*], A. Petrou[*], B. D. McCombe[*], J. Ralston[+] and G. Wicks[+]

[*]SUNY at Buffalo, Buffalo NY 14260, [+]Cornell University, Ithaca, NY 14853

ABSTRACT

The photoluminescence associated with silicon donors in GaAs/AlGaAs multiple quantum wells was studied at very low laser excitation intensity. Under these conditions two new impurity related features are resolved, at energies below and above the previously reported impurity photoluminescence feature. We provide a new interpretation for the observed spectra. With this assignment, binding energies of donors at the centers of the wells agree with values deduced from far-infrared magnetoabsorption experiments as well as with calculated values.

1. INTRODUCTION

Donors in GaAs/AlGaAs quantum wells have attracted considerable attention during the last few years. Photoluminescence (Shanabrook 1984a, Shanabrook 1986, Reynolds 1984, Nomura 1985), Raman (Shanabrook 1984b), and far-infrared (Jarosik 1985, Mercy 1988) spectroscopies have been used to study Si donors in these heterostructures. The binding energies of donors have been calculated as function of well width and position inside the well (Greene 1985, Mailhiot 1982). A donor-associated feature in the luminescence spectra from doped GaAs/AlGaAs quantum wells was first reported by Shanabrook and Comas (1984a). This impurity related feature, below the heavy hole exciton was attributed to transitions between electrons on Si donors at the centers of the wells and the top of the highest heavy hole confinement subband (Si(c)→VB). The binding energies of donors determined from this study were consistently lower than those deduced from far-infrared (FIR) magneto-absorption experiments (Jarosik 1985, Mercy 1988). The FIR values on the other hand are in good agreement with the calculations of Greene and Bajaj (1985) and Mailhiot et al (1982). We note here that the Rydberg used by Mailhiot et al (1982) was 5.3 meV, which is 9% lower than that used by Greene and Bajaj (1985). The latter agrees with the spectroscopic Rydberg from donors in bulk GaAs. When the appropriate modification is made in the Rydberg of Mailhiot et al (1982), the two calculations agree.

In this paper, we present a photoluminescence study of high quality Si-doped GaAs/AlGaAs quantum well samples carried out under experimental conditions that reveal two new features at energies slightly below and slightly above the previously reported impurity-related feature. The lowest energy feature is attributed to the Si(c)→VB transitions; the feature reported by Shanabrook and Comas (1984a) is ascribed to the ionized donor bound exciton (Si^+/X); finally, the new high energy impurity associated feature is identified as the neutral donor bound exciton (Si/X). Comparison of the photoluminescence spectra from quantum wells doped at the center with edge-doped structures having the

same dimensions support this interpretation. The binding energies of donors from this study are in agreement with FIR measurements on the same samples (Jarosik 1985, Mercy 1988), as well as with calculated values (Greene 1985, Mailhiot 1982). The observation of the (Si/X) feature also enables us to measure the binding energy of excitons on Si donors. The experimental values are found to be in good agreement with calculated values (Kleinmann 1983).

EXPERIMENTAL

Several MBE-grown GaAs/AℓGaAs multiple quantum well structures with well widths varying from 375 to 80Å were used in this study. The samples were doped with Si donors (1×10^{16} cm^{-3}) either in the central one third of the GaAs layers, or in the bottom or top one third of the wells. "Bottom" and "top" are defined as regions of the GaAs wells grown immediately after, and immediately before the AℓGaAs barrier, respectively. All samples had barrier widths of 125Å. Some sample characteristics are shown in table I. The samples were positioned in a variable temperature optical cryostat. The photoluminescence spectra were excited with the 6328Å line of a Helium-Neon laser, the 4880Å line of an Argon-ion laser, and the 3250Å line of a Helium-Cadmium laser. The emitted luminescence was analyzed by a double monochromator equipped with a photomultiplier tube and photon counting electronics. Laser powers smaller than 0.5 mW were used in these experiments in order to resolve all the impurity related features. The various luminescence components were better resolved at shorter laser wavelengths. A shorter wavelength photon samples fewer wells, resulting in smaller inhomogeneous broadening of the luminescence feature due to any fluctuation of well width vs depth.

2. RESULTS AND DISCUSSION

In Fig. 1, the photoluminescence spectra from sample 2, recorded using different laser power levels are shown. As the exciting laser power is reduced, considerable structure emerges from the broad impurity feature of Fig 1(a). In Fig 1(c) the various spectral components are labeled for convenience in the discussion. Feature (i) at 1525.0 meV is the heavy hole exciton (hhX). Feature (iii) at 1523.7 meV is the donor related feature observed for laser powers above 5 mW; this feature was reported earlier (Shanabrook 1984a, Shanabrook 1986) and interpreted as being the Si(c)→VB transition. Features (iv) at 1522.3 meV and

Table I

Sample	Well Width Å	Doping	E_{BD} (meV)	$E_{Si/X}$ (meV)
1	375	Bottom 1/3		1.1
2	210	Center 1/3	10.3	1.3
3	210	Center 1/3	10.3	1.3
4	210	Bottom 1/3	5.6	
5	210	Top 1/3		
6	140	Center 1/3	12.1	1.5
7	80	Center 1/3	3.0	1.9

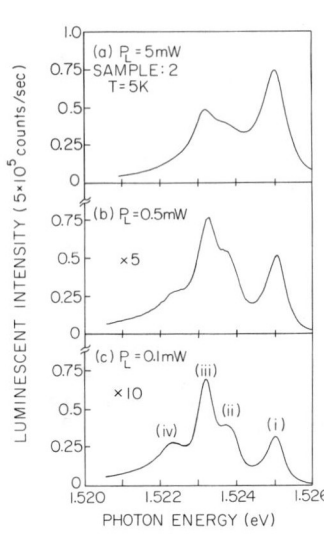

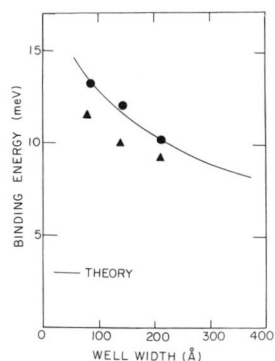

Fig. 2 Binding energies of donors at the centers of the GaAs wells vs well width. Circles: experimental values with feature (iv) in Fig. 1(c) assigned to the Si(c)→VB transition. Triangles: Values obtained with feature (iii) in Fig. 1(a) assigned to the Si(c)→VB transition. Curve: Greene, 1985.

Fig. 1 Photoluminescence Spectra from sample 2 excited with the 4880Å of an Argon-ion laser; (a) Laser power P_L = 5mW; (b) P_1 = 0.5 mW (c) P_L = 0.1 mW

(ii) at 1523.7 meV are resolved only at very low laser power levels. We provide a new interpretation for the photoluminescence spectra. Feature (iv) is identified as the Si(c)→VB transition. Features (ii) and (iii) are interpreted as the neutral donor bound exciton (Si/X) and ionized donor bound exciton (Si$^+$/X), respectively. In the luminescence work of Nomura et al (1985) features (ii) and (iii) were not resolved, and no Si(c)→VB transition was observed. A feature identified as the Si(c)→VB transition was reported in a photoluminescence study of not-intentionally-doped wells (Reynolds, 1984). The binding energies of donors can be determined from the luminescence spectra using the equation: $E_{BD} = E_{hhX} + E_{BX} - E_{Si \to VB}$ where E_{hhX}, E_{BX} and $E_{Si \to VB}$ are the

energies of the heavy hole exciton, the exciton binding energy and the Si(c)→VB transition energy, respectively. The binding energies for donors at the center of the wells determined with the present assignment and using the calculated exciton binding energies of Sanders and Chang (1985) are listed in Table I. In Fig. 2 the donor binding energies obtained in this study are plotted as function of well width (solid circles). The triangles indicate values for E_{BD} obtained with the assumption of Shanabrook and Comas (1984a) that feature (iii) is the Si(c)→VB transition. The solid line represents theoretical values of

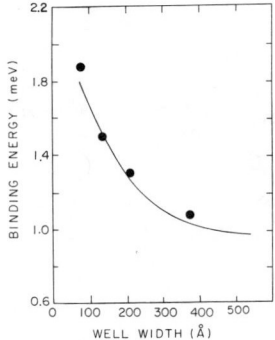

Fig. 3 Binding energies of excitons on neutral donors vs well width. Circles: Experimental values. Curve: Kleinman, 1983.

Greene and Bajaj (1985). The agreement between the photoluminescence results under the present interpretation and the theoretical values is quite good. It is noted that the theoretical values for E_{BD} (Greene 1985, Mailhiot 1982) agree very well with values obtained from FIR experiments performed on the <u>same</u> samples (Jarosik 1985, Mercy 1988).

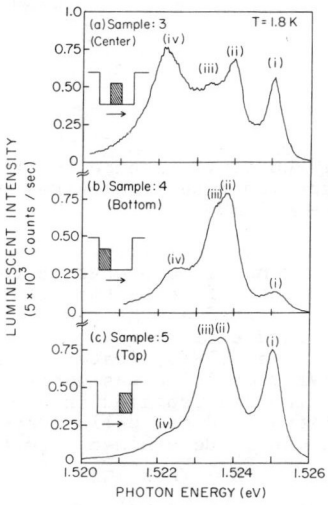

Fig. 4 Photoluminescence spectra recorded at T = 1.8K with the 4880Å line of an Argon-ion laser. (a): Sample 3 (center doped); (b); Sample 4 (bottom edge doped); (c): Sample 5 (top edge doped). The insets represent the nominal doping profile. The arrows indicate the growth direction.

The binding energy of the exciton on the Si donors ($E_{Si/X}$) is given by the difference in energies of the (hhX) and the (Si/X), i.e. features (i) and (ii) in Fig. 1(c). The experimental values for $E_{Si/X}$ obtained in this study are listed in Table I. They are plotted as a function of well width in Fig. 3. The solid line represents calculated values of Kleinman (1983). The agreement between theoretical and experimental values is in this case also rather good. Thus the present assignment of the impurity related photoluminescence features provides good agreement with theory for the confined donor binding energy (and also with the FIR

experiment) as well as the binding energy of the exciton on the Si donor.

The photoluminescence spectra were studied as function of temperature to test the validity of the interpretation. As the sample temperature is raised, features (ii) and (iv) become weaker and disappear. In the case of sample 2, for example, features (ii) and (iv) disappear around 30K. Feature (iii) persists up to 50K. At T=30K, almost 50% of the donors are ionized. As a consequence the Si/X feature becomes very weak. The binding energy of the Si$^+$/X, on the other hand, is slighty larger and since the number of ionized donors increases with T feature (iii) persists up to 50K. The Si(c)→VB feature is a discrete-to-continuum transition and therefore has a linewidth that is strongly dependent on the photoinjected hole distribution. Assuming Boltzman statistics the width of the hole distribution at T = 30K is ~ 6 times larger than that at T=5K. Thus at T=30K the Si(c)→VB transition becomes too broad to be observed.

We also compared the photoluminescence spectra from three quantum well structures which have identical well and barrier dimensions but are doped in different parts of the wells (samples 3,4 and 5). The luminescence spectra from these three samples are shown in Fig. 4. Slightly different laser power was used in this case resulting in the change in the relative intensity of the heavy hole exciton. The three spectra in Fig. 4 show a dramatic variation in the intensity of feature (iv). In the center-doped sample (Fig.4(a)) the Si(c)→VB transition is strong with respect to the bound excitons. In the bottom-edge-doped sample (Fig 4(b)) it becomes quite weak, and finally it has practically disappeared in the spectrum of the top-edge- doped structure (Fig. 4(c)). The variation in the intensity of the Si(c)→VB feature can be understood in terms of the donor distribution in the wells. In sample 3, doped over the central one third of the wells, most of the donors are situated at or near the well centers. Thus the Si(c)→VB transition is strong, with respect to the excitons. In sample 4, doped over the bottom one third of the wells some of the donors have been carried along the growth front in the growth direction towards the well center and contribute to the weak Si(c)→VB transition observed. The energy of the transitions between electrons on donors at the edge of the wells and the top of the highest valence subband (Si(e)→VB) is larger than the energy of the heavy hole exciton, and thus is strongly reabsorbed. In sample 4 the Si(e)→VB transition was observed as a weak feature at 1527 meV. The corresponding binding energy of the donors is 5.0 meV (the theoretical value for E_{BD} calculated by Greene and Bajaj is 5.2 meV). In sample 5, doped over the top one third of the wells, redistribution of the donors during growth is *away* from the well center, into the AℓGaAs barrier. Thus no Si(c)→VB transition is expected, and as can be seen in Fig. 3(c), this feature is practically absent. These results and their interpretation are in agreement with FIR magneto-absorption profile measurements of the hydrogenic impurity transitions on the same samples (Mercy 1988)

3. CONCLUSIONS

We have presented a photoluminescence study of n-type, Si-doped GaAs/AℓGaAs multiple quantum wells. The choice of experimental conditions allowed the observation of two new impurity-related features and led to a new interpretation of the spectra. Based on the new

interpretation the deduced donor binding energies agree with FIR experimental as well as with theoretical values. In addition, experimental values of the exciton binding energy on silicon donors are in good agreement with calculated values. This study has removed uncertainties of earlier work and permitted a consistent explanation of the observed spectra.

ACKNOWLEDGEMENTS

This work was supported in part by the Office of Naval Research under contract #'s N0001483K0219 and N0001486K0730 and by the National Science Foundation through grant #ECS-8200312 to the National Research and Resource Facility for Submicron Structures (now the National Nanostructure Facility) at Cornell University.

REFERENCES

Greene,R.L.and Bajaj, K.K., Phys. Rev. B $\underline{31}$, 913 (1985).

Jarosik, N.C. McCombe, B.D., Shanabrook, B.V. Comas, J. Ralston J., and Wicks, G.Phys. Rev. Lett. $\underline{54}$, 1283 (1985).

Kleinman, D. A., Phys. Rev. B $\underline{28}$, 781 (1983).

Mailhiot, C. Chang, Y-C and McGill, T.C., Phys. Rev. B $\underline{26}$, 4449 (1982).

Mercy, J.M. McCombe, B.D. Beard, W. Ralston, J. and Wicks, G., Surf. Sci. $\underline{196}$, 334 (1988).

Nomura, Y. Shinozaki K. and Ishii, M., J. Appl. Phys. $\underline{58}$, 1864 (1985).

Reynolds, D.C. Bajaj, K.K. Litton, C.W. Yu, P.W. Masselink, W.T. Fischer, R. and Morkoc, H., Phys. Rev. B $\underline{29}$, 7038 (1984).

Sanders G.D. and Chang, Y-C, Phys. Rev. B $\underline{32}$, 5517 (1985).

Shanabrook B.V. and Comas, J. Surf. Sci. $\underline{142}$, 504 (1984a).

Shanabrook, B.V. Comas, J. Perry T.A. and Merlin, R., Phys. Rev. B $\underline{29}$, 7096 (1984b).

Shanabrook, B.V., Surf. Sci. $\underline{170}$, 449 (1986).

/ Inst. Phys. Conf. Ser. No 95: Chapter 1
Paper presented at Int. Conf. Shallow Impurities in Semiconductors, Linköping, Sweden, 1988 /

Circular polarization of impurity luminescence in GaAs/AlGaAs quantum wells

P S Kop'ev, V P Kochereshko, I N Uraltsev and D R Yakovlev

A F Ioffe Physico-Technical Institute, USSR Academy of Sciences, 194021, Leningrad, USSR

ABSTRACT: Using two photoluminescence (PL) techniques, we have probed the impurity profiles in GaAs/AlGaAs quantum well structures (QWS). We have studied for the first time both the oscillator strength of confined electron transitions to residual carbon acceptors and acceptor g-factor as a function of impurity location in QW analysing the spectral dependence of PL circular polarization (CPPL) in the presence of magnetic field. By use of optical orientation technique we have measured the relative electron-acceptor (e-A) recombination rate for the different impurity positions. We have found in short-period superlattices (SL) the absence of the CPPL at low magnetic fields. The effect is explained by the suppression of hole spin relaxation due to the strong splitting of the heavy- and light-hole acceptor states in SL.

1. INTRODUCTION

The main qualitative feature of the confinement is to increase the binding energy of an hydrogenic impurity for decreasing QW width, and to decrease the binding energy, for fixed QW width, as impurity is moved from the center toward the interface (Bastard 1981, Masselink et al 1985). Previous experimental investigations are focused on the examination of these predictions and the strong sensitivity of impurity transition energies to the location in the well is documented, now for GaAs/AlGaAs QW of different thickness.

The magnetic field induced CPPL technique has been suggested by Dyakonov et al (1972) to study the bulk acceptor states. We use this very specific tool to further examine the acceptor wavefunction dependence on location in GaAs/Al$_{.4}$Ga$_{.6}$As MQWS (Kop'ev 1988a). Heterointerface effect on acceptor wavefunction is expected to change both the acceptor g-factor and the oscillator strength of e-A transition. Since the PL transition energies correspond to acceptor location in QW, our finding of the CPPL spectral dependence evidences these changes for the first time.

The e-A recombination rate is governed by the impurity dis-

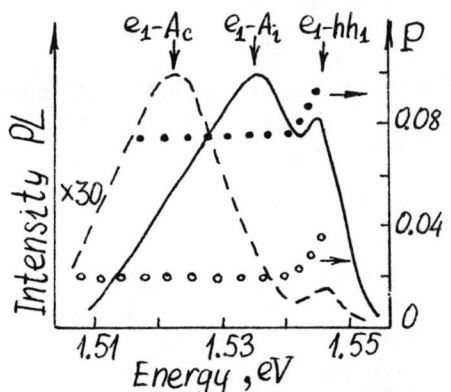

Fig. 1. The carbon acceptor related PL and CPPL degree taken from GaAs/Al$_{.4}$Ga$_{.6}$As MQWS with the 120 Å -thick wells at 6^+ excitation with $\hbar\omega$=1.647$_2$ eV and density 0.01 W/cm^2 are shown, respectively, by dashed line and open circles, and that 0.25 W/cm^2 – solid line and closed circles

tribution function in the well and by the optical response (oscillator strength) of each individual impurity. To evaluate carbon acceptor concentration profile through QW we have measured the relative e-A recombination rate using the optical orientation of electron spins (Kop'ev et al 1988b).

The preliminary experiments in short-period SL indicate that the CPPL techniques can be used to learn about the splitting of the heavy- and light-hole acceptor states.

2. MAGNETIC FIELD INDUCED CPPL

Impurity PL, which is related to confined electron recombination with residual carbon acceptor, is observed at low excitation densities. As excitation density increases, the saturation of impurity PL intensity is observed and intrinsic e_1-hh_1 exciton PL dominates. Two distinct maxima associated with recombination on the well acceptor A_c and the heterointerface acceptor A_i are revealed in PL spectra, shown in Figure 1. These maxima have been observed in PL intensity dependence on both temperature and excitation density.

The magnetic field induced circular polarization degree is found to depend on acceptor location in QW. The values of polarization degree ρ(B) at the A_i and A_c energy positions are shown as a function of magnetic field B in Figure 2. To analyse these dependences we use the expression for bulk e-A recombination at magnetic field (Dyakonov and Perel 1972):

$$\rho(B) = \frac{\tau_h}{\tau_h + \tau_s} \phi(x) \qquad (1)$$

$$\phi(x) = \begin{cases} \frac{5}{4}x, & \text{for } x \to 0 \\ 1, & \text{for } x \to \infty \end{cases}$$

Here $x = \beta_o g_A B/kT$, g_A - acceptor g-factor, β_o = eh/2mc, τ_h - hole life time and τ_s - hole spin relaxation time on acceptor states. When Zeeman splitting of acceptor states becomes higher than kT the polarization degree is governed only by holes spin relaxation in their lifetime $\rho_o = \tau_h/(\tau_h + \tau_s)$. Since the weak interface effect on τ_s is expected and since τ_h under our experimental conditions is determined by the

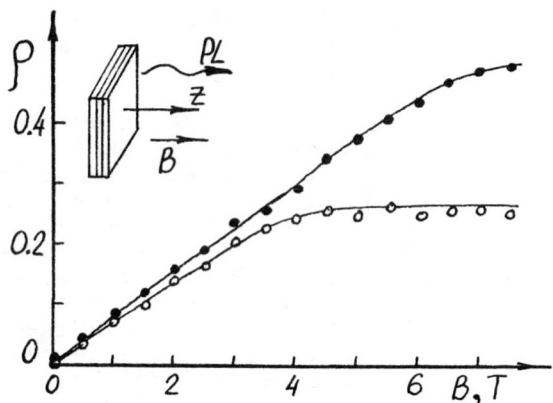

Fig. 2. The degree of CPPL as a function of magnetic field taken at PL energies of the e-A_c and e-A_i transitions is shown by open and closed circles, respectively. The solid lines are the result of fitting with Equation 1. Inset shows the experimental geometry

e-A radiative recombination (Kop'ev et al 1988b), the relative oscillator strength of e-A transitions as a function of acceptor location in QW is deduced from the found ϱ_0 spectral dependence. The oscillator strength of e-A_c transition turnes out to be about 3 times stronger than that of e-A_i transition.

Analysing at low magnetic fields the linear ϱ (B) dependence
$$\varrho(B) = 5\varrho_0 \beta_0 g_A B / 4kT$$
for each photon energy of impurity PL we evaluate the g_A-factor as a function of acceptor location in QW. As expected (Masselink et al 1985), the value $g_{A_c} = 0.52\pm0.05$ coincides with that for bulk acceptor (Bimberg 1978). The reduction of interface acceptor g-factor to the value $g_{A_i} = 0.26\pm0.05$ is found for the first time. The acceptor g value expects to depend in such delicate manner on details of the valence band structure that either reduction or enhancement of g_A-factor is difficult to predict with certainty. But the dramatic change of g_A value one should believe because of the splitting of the heavy- and light-hole states is predicted to increase for interface acceptor (Masselink et al 1985).

It worth notes that the numerical coefficient 5/4 in equation 1 corresponds to degenerate acceptor states. Since the splitting of the 3/2 and 1/2 states is weak to compare with inhomogeneous broadening of impurity PL line, this approachment is reasonable for acceptor located in QW centre. To analyse the most splitted interface acceptor states one would expect to use the coefficient 5/2, as for nondegenerate case, and to evaluate the further 50% reduction of g_{A_i} value. But the lifting of degerancy would results in the absence of CPPL at low magnetic field due to suppression of hole spin relaxation on splitted 3/2 and 1/2 acceptor states, as it will be discussed below. Our finding of CPPL at low magnetic fields evidences that the splitting of interface acceptor states is not more than some millielectronvolts in the 100Å-thick QW. This fact is in quantitative agreement with the numerical calculation of Masselink et al (1985).

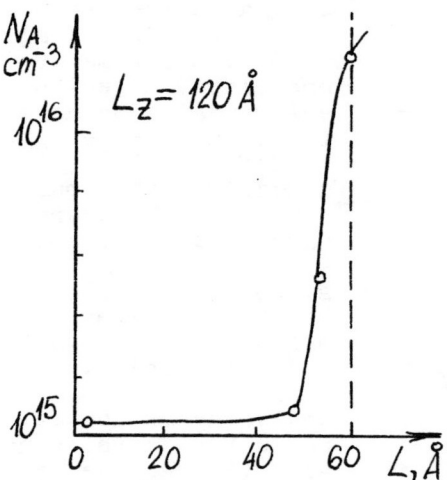

Fig. 3. Carbon acceptor concentration profile through GaAs QW in normalized to residual impurity background $1.5 \cdot 10^{15}$ cm^{-3}. GaAs/AlGaAs interface is shown by dashed line

3. CPPL INDUCED BY THE OPTICAL ORIENTATION OF ELECTRON SPINS

Producing by the circular polarized band-to-band excitation, the electron spin orientation is revealed in CPPL. The degree of CPPL is determined by spin orientation degree, Po, of electrons thermalized to the bottom of the lowest confined subband. So, P, the degree of CPPL is governed by electrons spin relaxation in their lifetime τ_e (Dyakonov and Perel 1971):

$$P = P_o \tau_{es}/(\tau_{es} + \tau_e)$$

We measure the degree of CPPL at two excitation density ranges, when the e-A_c or the e-A_i recombination dominates,[1] as shown in Figure 1. For the e-A_i transitions the enhancement of CPPL degree is found more than 4 times. Since the e-A radiative recombination process contributes mainly to τ_e we connect this effect with electron lifetime shortening.

To examine carefully the electron lifetime shortening we measure the halfwidth of CP decay curve at transverse magnetic field (Hanle effect). When the e-A_i recombination dominates, the recombination rate is found to be 4 times higher than that for the e-A_c transitions.

The evaluated recombination rate dependence on acceptor location in QW is governed by both the acceptor distribution function through QW and the oscillator strength of the e-A transitions. Using the calculations of the acceptor binding energy as a function of impurity location (Masselink et al 1985) we compare the spectral dependences of oscillator strength and recombination rate to carry out the acceptor concentration profile in QW. The evaluated acceptor profile is shown in Figure 3, where residual carbon concentration in the centre of GaAs QW is taken equal to background shallow acceptor concentration in GaAs buffer layer of MQWS under investigation or that in GaAs layers grown in the same MBE cycle.

To treat the found concentration "jump" on the heterointerface as a result of either interface segregation of shallow

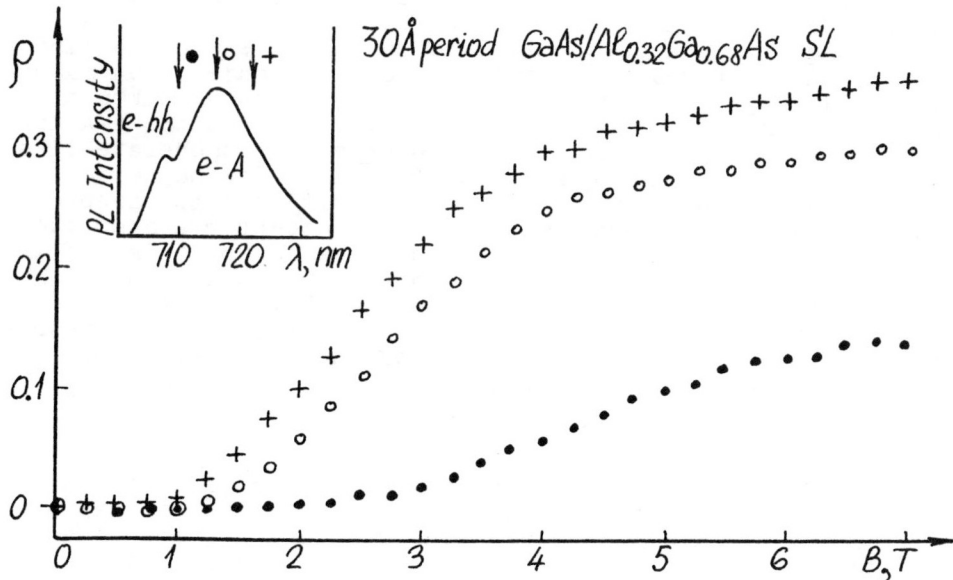

Fig. 4. The degree of CPPL dependence on magnetic field in 30 Å-period SL taken at three spectral positions, shown on inset, of the acceptor related PL

impurities or high residual carbon concentration in AlGaAs barrier layers there are no additional data obtained by PL techniques. One should expect to learn about the electron recombination with a hole which is located in the well and bounded to acceptor located in AlGaAs layer away from interface. This process is documented in GaAs/AlGaAs single heterojunction structure (Kop'ev et al 1988c). The CPPL study of this structure are now in progress.

The obtained acceptor profile allows us to explain the exhibition of two distinct peaks in the PL spectrum of 100Å-thick QW. The longwavelength peak is connected with the maximum of acceptor density of states in the centre of QW. No singularities in shallow impurities sensity of states are expected on the interfaces of thick QW (Bastard 1981) we have to conclude that the $e-A_i$ peak is due to the concentration "jump" on the heterointerfaces. So, we suppose the finding of the $e-A_i$ peak in PL spectra to be the simple probe of shallow acceptor profile in heterostructure.

4. CPPL IN SHORT-PERIOD SUPERLATTICES

When the well width is an order or less of 3D-hole radius of acceptor, the confinement effect is larger than the acceptor binding energy. As a result of this effect the splitting of the 3/2 and 1/2 acceptor states is expected to be an order of the binding energy. We believe it is the case in the

30 Å-period SL, which PL spectrum is shown in inset of Figure 4. Since the splitting is compared with inhomogeneous broadening of the acceptor PL line we should expect to find the revealing of this effect in the CPPL dependence on magnetic field. Figure 4 shows three dependences $\varrho(B)$ taken at different photon energies indicated by arrows on the e-A line. The main feature, which does not observe for degenerate-like acceptor spectra, is the absence of CPPL at low magnetic fields. This effect is suppoused to result from the suppression of hole spin relaxation on the 3/2 acceptor states due to splitting off the 1/2 states. When Zeeman splitting of the $\pm 3/2$ states is compared with the confinement splitting of acceptor states, the CPPL is observed. The CPPL spectral dependence supports this explanation. The effect is more pronounced on the short wavelength wing of the PL line, where the more perturbed acceptor contributes.

5. CONCLUSION

Circular polarization analysis of the shallow impurity PL is found to be both a tool for nondestructive characterization of impurity profiles in heterostructures and a probe of an individual impurity wavefunction in dependence on the location in the quantum well.

REFERENCES

Bastard G 1981 Phys.Rev. B **24** 4714
Bimberg D 1978 Phys.Rev. B **18** 1794
Dyakonov M I and Perel V I 1972 Sov.Fiz.Tverd.Tela **14** 1452
Dyakonov M I and Perel V I 1971 Sov.Phys.JETP **60** 1954
Kop'ev P S, Kochereshko V P, Uraltsev I N and Yakovlev D R 1988a Laser Optics of Condensed Matter ed J L Birman, H Z Cumminz and A A Kaplyanskii (New York, London: Plenum Press) pp 87-93
Kop'ev P S, Kochereshko V P, Uraltsev I N and Yakovlev D R 1988b Sov.Phys.Semicond. **22** 597
Kop'ev P S, Kochereshko V P, Uraltsev I N, Al L Efros and Yakovlev D R 1988c J.Lumines. **40-41** 747
Masselink W T, Chang Y-C, Morkoc H 1985 Phys.Rev. B **32** 5190

Inst. Phys. Conf. Ser. No 95: Chapter 1
Paper presented at Int. Conf. Shallow Impurities in Semiconductors, Linköping, Sweden, 1988

The investigation of impurity distributions around an oval defect in molecular beam epitaxy AlGaAs/GaAs single quantum wells by transmission electron microscopy and cathodoluminescence

J Wang*, J W Steeds* & C W Tu**

*H H Wills Physics Laboratory, University of Bristol, Bristol BS8 1TL, UK
**Bell Laboratories, Room MH 7B-410, Murray Hill, NJ 07974, USA

ABSTRACT: TEM and TEM-CL have been used to study the impurity distributions along lateral and growth directions around a particular type of oval defect in MBE AlGaAs/GaAs SQWs sample with four single wells of different well width. Some twins, dislocations and stacking faults were observed around the core of the oval defect, which was found to consist of misoriented crystal. The results of TEM-CL experiments show that the impurities are trapped mainly in the first grown well. The impurities sitting near the middle of the well are concentrated in a region very near the core of the defect, while those sitting at the interfaces are more widely dispersed around the defect.

1. INTRODUCTION

Microscopic surface defects, so called "oval defects" are commonly observed in GaAs and AlGaAs layers, grown by molecular beam epitaxy (MBE). So far, studies have been focussed on the characterisation of such defects, and the determination of their origins. The observations show that oval defects are generally elongated along the <110> directions. It is known that the origins of these defects are related to a number of factors arising out of the systems employed and the growth conditions used, for example see S Matteson et al (1986) and K Nanbu et al (1986). As a result, there are a number of different oval defects, and many of these have now been identified. A recent thorough classification has been given by K Fujiwara et al (1987). However, in this work, by using a transmission electron microscope (TEM), and TEM cathodoluminescence (TEM-CL), we concentrate on the impurity distributions around a particular type of oval defect, which we observed in a sample where many of the possible causes of oval defects had been eliminated. Our results lead to information about the nature of impurity trapping and diffusion during growth.

2. EXPERIMENT

The GaAs/Al Ga As SQW structures were grown by MBE on (001) oriented semi-insulating GaAs substrates. The sample studied has the detailed structure shown in Fig 1. As we can see, the four single quantum wells have very different well widths, and hence, well distinguished emissions, which allows us to study them separately by CL.
The electron-transparent, thin, plan-view specimen used for TEM and TEM-CL studies was prepared by using selective chemical etching to remove the GaAs substrate and buffer layer.

© 1989 IOP Publishing Ltd

The CL measurements were performed on a Philips EM400 electron microscope in STEM mode at 100kV at liquid He temperatures (approximately 30K). The acquisition of CL spectra and monochromatic CL images was controlled by a Link 860 system.

The TEM investigations were carried out on a Philips EM430 electron microscope, operating at 250kV after all the CL characterisation had been completed (displacement damage of the material occurs at this voltage).

3. RESULT AND DISCUSSION

TEM-CL and TEM studies have been performed on a thin plan-view sample with the structure shown in Fig 1.

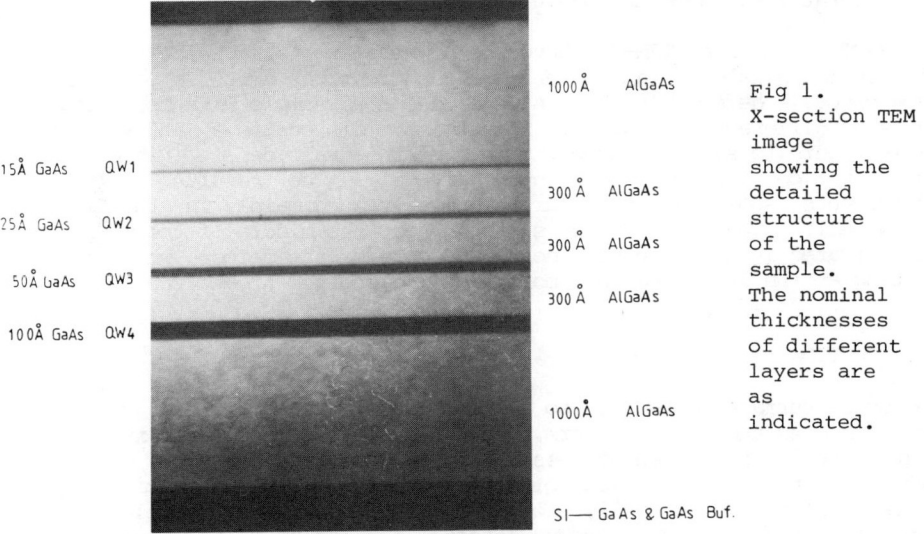

Fig 1. X-section TEM image showing the detailed structure of the sample. The nominal thicknesses of different layers are as indicated.

Fig 2 shows the CL emission from an unfaulted area. The four peaks in the spectrum are attributed to four single quantum wells with nominal widths of 15A, 25A, 50A and 100A respectively, as indicated in the figure. The line widths for the narrower wells are greater than for the wider wells. This is the expected result since the effects of interface roughness and composition fluctuation in the barrier layers are larger for narrower wells. Other techniques (large-angle convergent beam electron diffraction, high resolution electron microscopy) have been employed to investigate the interface roughness of the same sample. However, in this paper we confine our interests to the study of the oval defects in the sample.

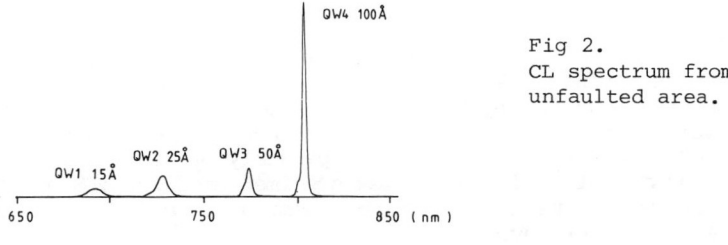

Fig 2. CL spectrum from unfaulted area.

The TEM bright-field image of a typical oval defect is shown in Fig 3. Detailed study shows that the core area of the defect is misoriented crystal and the longer oval axis is along the <100> direction. This is not the usual sense of elongation for oval defects found by other researchers (1,3). As we can see, some twins and dislocations surround the core of the defect.

Fig 3.
TEM image of a typical oval defect.

At the core of the oval defect, the CL emission is totally quenched. The CL emission from the area around the core of the defect (faulted area) is shown in Fig 4. The relative intensities of the spectra in Figs 2 and 4 have been adjusted for easy comparison. Comparing the QW4 (well width

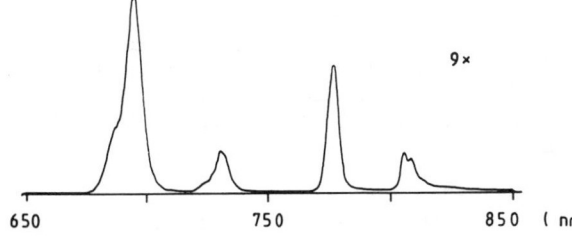

Fig 4.
CL spectrum from faulted area.

100A) emissions in these two spectra, we can easily see that significant changes have occurred.

The dramatic decrease of intensity of emission in the defective area means that there was a high rate of non-radiative recombination there. The lineshape of the emission was also changed greatly, as is shown in Fig 5. Spectrum A is from an unfaulted area. The main peak at 804nm corresponds to the quantum well emission of (hh-e); the smaller peak on the short wavelength side of the main emission is from the recombination of (lh-e). Spectrum B is from the faulted area. Some extra structure appears on the long wavelength side of the main quantum well emission. The lack of

spectral detail and the continuity of the emission band with the main quantum well emission of the unfaulted area is possibly because of the relatively high temperature of observation and the relatively low spectral resolution of the CL system. The broad continuous lineshape of extra structure with a FWHM of about 10MeV is considered to be related to the distribution of impurities along the growth axis rather than different species of impurities. The positions of the long wavelength end and short

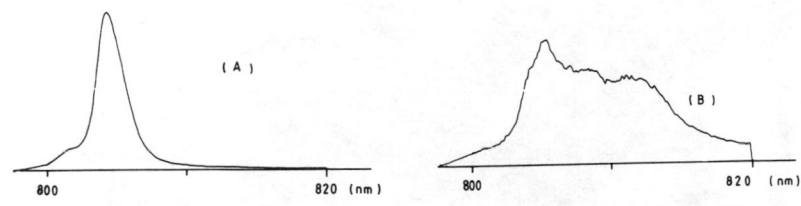

Fig 5. CL emission of QW4 (well width 100Å)
Spectrum (A) from unfaulted area
Spectrum (B) from faulted area

wavelength end of the broad structure correspond to impurity levels of ionization energy, ie impurity ionization energy levels of about 24.8MeV and 13.8MeV respectively, by using the binding energy of the heavy hole exciton (4.8MeV) calculated by Chu and Chang (1987). It is well known that the emission from excitons bound to impurities in quantum wells is dependent on the position of the impurities in these wells. Higher energy emission is generally associated with interfacial impurities and lower energy emission with impurities at the well centres. By comparison with the theoretical calculations done by Delalande (1987), we believe that those impurities with 13.8MeV ionization energy are C acceptors sitting at interfaces, and those with 24.8MeV are C acceptors at the middle of the well. This implies a considerable accumulation of residual C acceptors around the core of the defect in QW4.

Comparing the emissions of the other wells, we notice that the accumulation effect of the C impurities on CL emissions decreases gradually from QW3 (well with 50A) to QW1 (well width 15A) and, in particular, that the emissions of QW1 are almost unchanged.

As QW4 was grown first, and QW1 was grown last, we conclude that residual impurity C acceptors in the growth system were trapped mainly in QW4. This is in agreement with the assumption of Delalande (5). We also notice that a slight shift of the main quantum well emissions towards longer wavelengths occurs for all wells. This may be the result of the strain, distortion of the wells, change of well width or change in composition of the barrier layers near the defect.

By applying the CL monochromatic image technique, we studied the distribution of impurities accumulated by the defects in the wells. Fig 6 presents the results of studying QW4. The TEM image of the same area is shown in Fig 6(d). The CL image in Fig 6(a) is formed by choosing the

main quantum well emission peak. There is only emission from regions (bottom left) remote from the defect. Fig 6(b) and Fig 6(c) show respectively the CL images formed by emissions from C acceptors sitting either at the interfaces or at the middle of the wells.

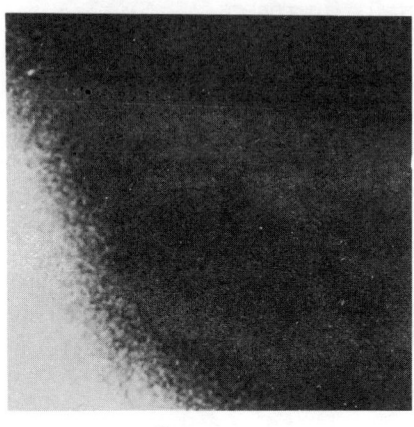

Fig 6a. CL monochromatic image choosing quantum well emission at 804nm.

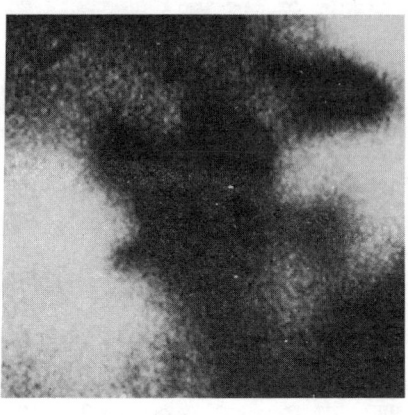

Fig 6b. CL monochromatic image choosing the emission of C acceptor at 808nm.

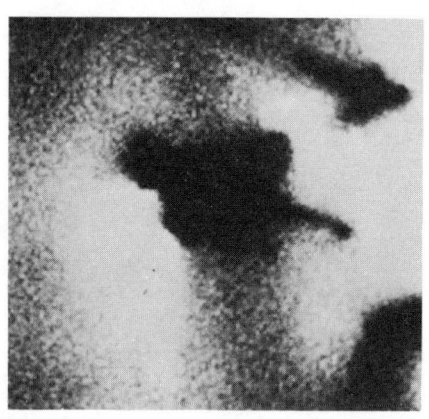

Fig 6c. CL monochromatic image choosing the emission of C acceptor at 812nm.

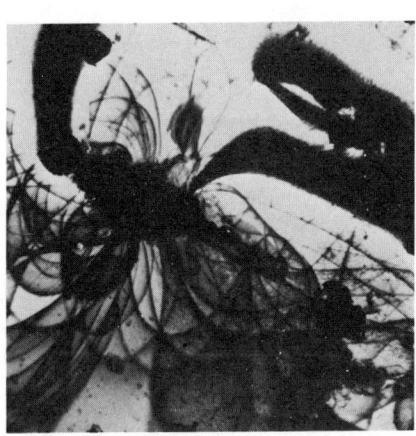

Fig 6d. TEM image of the area of CL acquisition.

Very interestingly, this new result implies that the C acceptors sitting in the middle of the well are restricted to the area very near the centre of the defect, but those sitting at the interfaces have a much wider distribution around the defect. It follows that when impurities diffuse towards defects, they are first trapped at the interfaces, and then occupy the middle of the well as their concentration increases.

4. CONCLUSION

By study of the distribution of impurities around an oval defect, we conclude that the residual impurities in the growth system are trapped mainly in the first well to be grown, and when they diffuse towards the defects they first sit at the interfaces.

5. REFERENCES

Chu H and Chang Y C, 1987, Phys. Rev. B36 p 2946

Delalande C, 1987, Physica 146B p 112

Fujiwara K, Kanamoto K, Ohta Y N, Tokuda Y and Nakayama T, 1987, J. Crystal Growth 80 p 104

Matteson S and Shih H D, 1986, Appl. Phys. Letters 48 p 47

Nanbu K, Saito J, Ishi Kawa T, Kondo K and Shibatomi A, 1986, J. Electrochem. Soc. 133 p 601.

Shallow donor impurities and far-infrared absorption in GaAs-Ga$_{1-x}$Al$_x$As quantum well structures

F Bassani and S Fraizzoli

Scuola Normale Superiore I-56100 Pisa Italy

R Buczko

Institute of Physics, Polish Academy of Sciences 02-668 Warsaw Poland

ABSTRACT: We present a variational method to compute donor eigenstates in GaAs-Ga$_{1-x}$Al$_x$As quantum wells, using the effective mass approximation and expanding the envelope function in the states at $k_\parallel = 0$ of the quantum well without the impurity, including the continuum states. Results are obtained for all impurity positions and for various values of the well thickness, showing the effects of the mass and dielectric constant difference in the two materials. The absorption coefficient is computed for any given impurity distribution and polarization. Good agreement is found with available experimental data.

1. INTRODUCTION

In the last decade semiconductor heterostructures have received a great deal of attention because of their intrinsic physical interest and of their technological applications in electronic devices. In these new materials impurities play the same important role as in traditional semiconductors. The energy levels of impurities in 'quantum wells' (QW) have been studied by means of the effective mass (EM) approximation by a large number of authors: among others Bastard (1981), Mailhiot et al. (1982), Priester et al. (1984) and Greene and Bajaj (1985). The extension of this method to heterostructures requires appropriate boundary conditions at the interfaces; these involve only the envelope functions (Altarelli 1986) provided that constituent materials are chemically similar, like in the GaAs-Ga$_{1-x}$Al$_x$As QW.

In this paper we present a study of far-infrared absorption for shallow donors in an isolated GaAs-Ga$_{1-x}$Al$_x$As QW. The wave functions and eigenvalues of the donor states are obtained by solving the EM equation by means of the variational method. For the sake of accuracy, the envelope function is expanded in the eigenstates of the QW without impurities at $k_\parallel = 0$. Boundary conditions for the envelope function are easily introduced and the different values of the dielectric constants in the two materials are taken into account exactly. The continuum states of the QW are included in our expansion set in order to make it complete and to test the convergence of the results.

In Sec. II we calculate the eigenstates of shallow impurity centres. In Sec. III we calculate the IR-absorption for various impurity distributions, for both polarizations of the incident radiation ($\parallel$ and $\perp$ to the QW axis). We also compare our results with previous theoretical work and with currently available experimental data.

© 1989 IOP Publishing Ltd

2. SHALLOW DONOR STATES

The system we consider is a single QW grown in a $\langle 100 \rangle$ direction, which we take as the quantization axis z. The well extends from $-L/2$ to $L/2$. If the doping density is not too high we can consider an isolated impurity, located at the point z_0 inside the well. The EM Hamiltonian of the system is

$$H = \frac{\hbar^2 k^2}{2m^*} + V_{QW}(z) + V_{imp}(\mathbf{r}), \qquad (1)$$

where $\mathbf{k} = -i\nabla$ and the effective mass m^* is that appropriate to the well and the barrier material for $|z| < L/2$ and $|z| > L/2$, respectively. V_{QW} is the square-well potential, which vanishes inside the well and equals V_0 in the barriers. The impurity potential V_{imp} is the screened Coulomb potential with Maxwell boundary conditions at both interfaces. Summing a series of image charges we obtain

$$V_{imp} = \begin{cases} -\dfrac{(1+\beta)e^2}{\varepsilon_1}\left(\dfrac{1}{R_0} + \sum_{i=1}^{\infty}\beta^i \dfrac{1}{R_i^+}\right) & \text{for } z < -L/2 \\ -\dfrac{e^2}{\varepsilon_1}\left(\dfrac{1}{R_0} + \sum_{i=1}^{\infty}\beta^i \left(\dfrac{1}{R_i^+} + \dfrac{1}{R_i^-}\right)\right) & \text{for } |z| < L/2 \\ -\dfrac{(1+\beta)e^2}{\varepsilon_1}\left(\dfrac{1}{R_0} + \sum_{i=1}^{\infty}\beta^i \dfrac{1}{R_i^-}\right) & \text{for } z > L/2, \end{cases} \qquad (2)$$

where $\beta = (\varepsilon_1 - \varepsilon_2)/(\varepsilon_1 + \varepsilon_2)$, ε_1 and ε_2 being the dielectric constants inside and outside the well, respectively, $R_0 = \sqrt{\rho^2 + (z - z_0)^2}$, and $R_i^\pm = \sqrt{\rho^2 + (z - z_i^\pm)^2}$ are the distances from the image charges, situated in the two barriers at

$$z_i^\pm = \begin{cases} z_0 \pm iL & \text{for even } i, \\ -z_0 \pm iL & \text{for odd } i. \end{cases} \qquad (3)$$

The EM equation for the envelope function $F(\mathbf{r})$, $HF = EF$, is explicitly obtained from (1), (2) and (3). The boundary conditions at each interface require the continuity of $F(\mathbf{r})$ and of the z-component of the current $(1/m^*)\nabla_z F(\mathbf{r})$, with the effective mass appropriate to each material. Since the EM equations do not admit of an analytical solution, we adopt a variational approach. We find it convenient to expand the envelope function in all states of the QW without impurities $\psi_{n\mathbf{k}_\|} = \exp(i\mathbf{k}_\| \mathbf{r}_\|)f_{n\mathbf{k}_\|}(z)$, for all values of the sub-band index n and of the in-plane wave vector $\mathbf{k}_\|$. Allowing for the completeness of the $\psi_{n\mathbf{k}_\|}$ at a fixed $\mathbf{k}_\|$, we can put this expansion in the form

$$F(\mathbf{r}) = \sum_n g_n(\mathbf{r}_\|) f_{n\mathbf{k}_\|=0}(z), \qquad (4)$$

where n runs over the discrete and the continuous spectrum, and the functions $f_{n\mathbf{k}_\|=0}$ are the zone centre QW states, obtained with the generalized boundary conditions stated above. Axial symmetry requires us to choose the function $g_n(\mathbf{r}_\|)$ of the form $e^{im\theta}\rho^{|m|}h_n(\rho)$, and it can be proved that $h_n(\rho) \simeq const.$ for $\rho \ll a^*$ (effective Bohr radius). The whole problem is now reduced to the choice of a suitable expansion set

Shallow Impurities in Quantum Structures

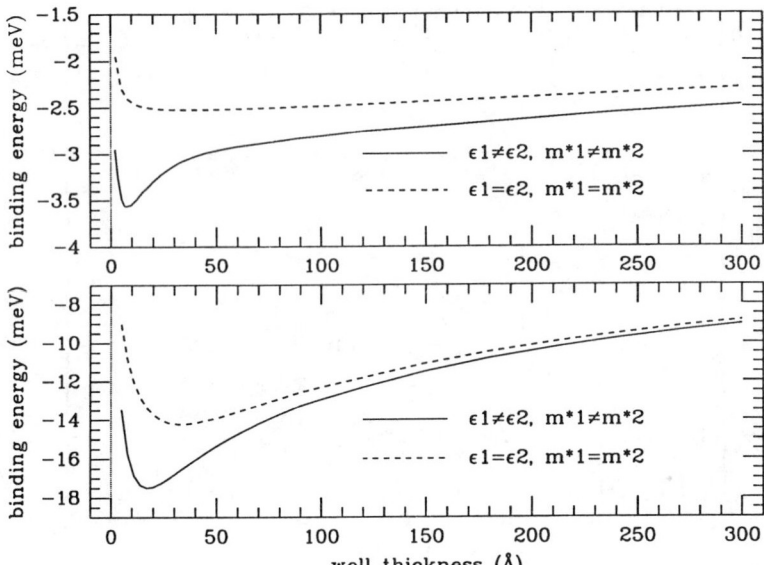

Fig.1. Binding energies of the two lowest even states (upper and lower graphs corresponding to the $2p^\pm$ and $1s$ bulk states, respectively), as a function of the well thickness L, for an on-centre impurity, with $x = .3$.

for $h_n(\rho)$. Since, for $L \to 0$ and $V_0 = \infty$, our problem reduces to the two-dimensional hydrogen problem, where the long-range behaviour of h_n is $\exp(-\alpha\rho)$, we adopt for $F(\mathbf{r})$ the following expansion

$$F(\mathbf{r}) = \sum_{ni} C_{ni} e^{im\theta} \rho^{|m|} e^{-\alpha_i \rho} f_{n\mathbf{k}_\parallel=0}(z), \tag{5}$$

where the α_i are fixed "a priori" so as to cover the physical range, and the C_{ni} are treated as variational parameters, which are determined by solving the eigenvalue problem

$$\langle ni|H|n'i'\rangle C_{n'i'} = E \langle ni|n'i'\rangle C_{n'i'}. \tag{6}$$

We remark that the boundary conditions for $F(\mathbf{r})$ are automatically satisfied. The matrix elements of eq. (6) can easily be computed. The contributions from the kinetic and the V_{QW} parts are straightforward. The contribution from the impurity potential may be reduced to a one-dimensional integral on an auxiliary variable s, by using in the summations (2) the well-known transformation $1/\sqrt{\rho^2 + z^2} = \int ds \exp(-|z|s) J_0(\rho s)$, in terms of the Bessel function J_0; this allows to sum the image charges and to perform the ρ and z integrals analytically.

The convergence of the method requires no more than ten exponentials to obtain ground and excited states. The sum on n has a convergence rate which depends on the well thickness. In general the contribution of the continuum states must be considered, particularly in the case of narrow QW, where, in the absence of donors, there is only one electron bound state at $\mathbf{k}_\parallel = 0$. In order to maintain the discreteness of the basis functions we put the QW in a large box ($V_{box} = 0$ for $|z| < L_0/2$ and

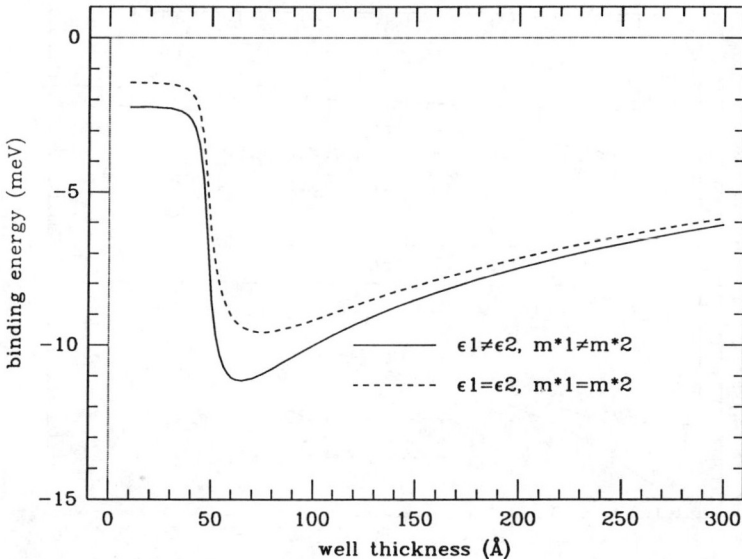

Fig.2. Binding energy of the lowest odd state (corresponding to the $2p^0$ bulk state), as a function of the well thickness L, for an on-centre impurity, with $x = .3$.

$V_{box} = \infty$ for $|z| > L_0/2$, with $L_0 \gg a^*$). When there is at least one discrete bound state at $k_\parallel = 0$ in the QW without donors, the contribution of the continuum is very small (0.7% for the ground state for $L = 30$Å). In the two-dimensional limit (very thin GaAs slab and infinitely high barriers) the exact solutions of the problem are obtained with our expansion procedure. We have also verified the opposite limit: the bulk donor spectrum in GaAs is obtained with better than 1% accuracy when L is very large. We had to extend our calculations to $L \simeq 30a^*$ to obtain convergence for the whole spectrum.

3. RESULTS AND DISCUSSION

We have performed calculations for different QW using a band mismatch $V_0 = .65\Delta E_g$ (Miller et al. 1985 and Menéndez et al. 1986), where $\Delta E_g = 1.247x$ eV (Casey and Panish 1978). The electron effective mass is taken to be $m_1^* = .067m_e$ in the well and $m_2^* = (.067 + .083x)m_e$ in the barriers (Casey and Panish 1978); dielectric constant is taken to be $\varepsilon_1 = 12.53$ in the well and $\varepsilon_2 = 12.53 - 2.73x$ in the barriers (*Landolt-Börstein tables* 1982). By way of example we present in Fig.1 the values of the binding energies (relative to the lowest sub-band at $k_\parallel = 0$) of the two lowest states of even parity with respect to the reflection $z \to -z$, for an on-centre impurity in QW of different thickness and standard composition ($x = .3$). The general behaviour agrees with that already pointed out by many authors (see above). We can observe the difference between the binding energies obtained by using the effective mass value and the dielectric constant of GaAs for both materials (dashed line) and those obtained by using the appropriate parameters (solid line). In the second case the binding energy is considerably increased, and this effect is more relevant the lower the values of L ($\sim 6\%$ at $L \sim 100$Å, $\sim 20\%$ at the minimum of the curve). Our results also give

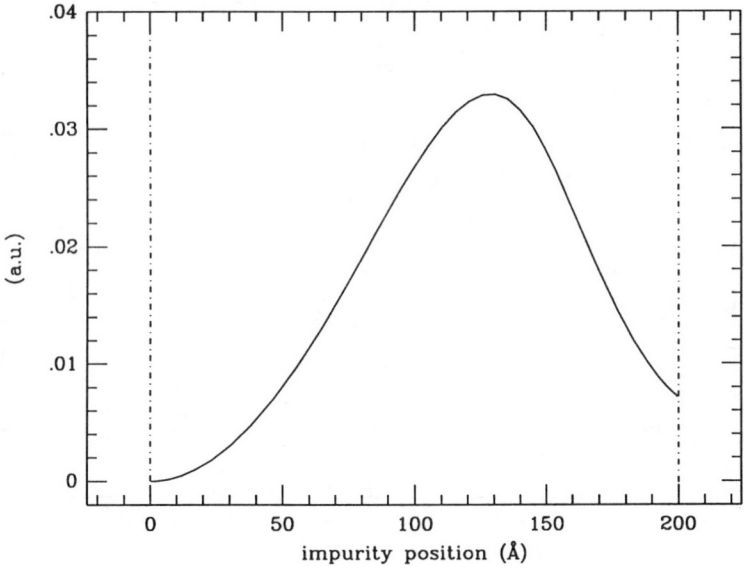

Fig.3. Computed value of $|\langle 1\Sigma|z|2\Sigma\rangle|^2$ as a function of the impurity position z_0 for $L = 400\text{Å}$ and $x = .3$.

larger binding energies for the ground state than those obtained by Mailhiot et al. (1982) with a different variational procedure (for $L = 60\text{Å}$ the difference amounts to $\simeq 1.6 meV$ when the same parameters are used). In Fig.2 we also present the binding energies of the lowest odd parity state, referred to the lowest odd sub-band at $\mathbf{k}_\parallel = 0$, for different values of L. The values for $L < 50\text{Å}$ are obtained by using only the states of the continuum in the expansion, since no bound odd parity state exists without the impurity, and the binding energy is referred to the onset of the continuum, i.e. to V_0. The limit for $L \to 0$ tends to the same value as the $2p^0$ state in $Ga_{1-x}Al_xAs$.

The above results allow for a computation of the optical absorption of donor impurities in QW. General considerations, with a discussion of the line broadening produced by the distribution of the impurities inside the well, are given by Greene and Lane (1986). In Fig.3 we show the square of the dipole matrix element between the states 1Σ and 2Σ, for all positions of the impurity inside the well and with polarization in the z direction. At the centre of the well, parity forces the matrix element to vanish, and away from the centre the matrix element remains small, with a maximum near the interface. The analogous term for polarization in the direction of the plane of growth ($\perp z$) is about ten times larger on the average, because the transition is parity-allowed at the centre of the well; its dependence on the impurity position agrees with that reported by Greene and Lane (1986). Given an impurity distribution in the well, the absorption coefficient can be computed, for different polarizations, for all transitions between the impurity states. In order to display the good agreement with the experiment of Jarosik et al. (1985), in Fig.4 we show the calculated line-shape due to the $1\Sigma \to n\Pi$ transitions with x polarization for a QW with $L = 210\text{Å}$, $x = .31$ and an impurity distribution (a gaussian about the centre with half-width$= 1/6L$) close to the declared one. We also indicate the observed position of the peak.

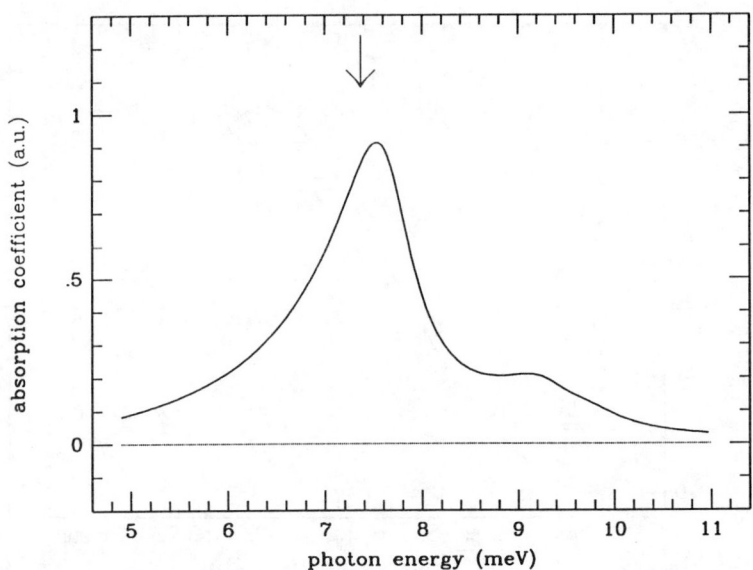

Fig.4. Computed line shape of the absorption coefficient for x-polarized radiation, in a QW with $L = 210$Å and $x = .31$. The arrow marks the experimental position of the absorption peak.

ACKNOWLEDGEMENTS

This research has been partly supported by the European Community under the CODEST Stimulation Program (Contract ST2J-0254-1-I(EDB)). We wish to thank Dr. Zdenka Barticevic for helping to control the accuracy of our results in the course of her work on magnetic field effects.

REFERENCES

Altarelli M 1986 in *Heterojunctions and Semiconductor Superlattices* ed G Allan, G Bastard, N Boccora and M Voos (Berlin: Springer)
Bastard G 1981 *Phys. Rev. B* **24** 4714
Casey H C and Panish M B 1978 *Heterostructure Lasers* (New York: Academic)
Greene R L and Bajaj K K 1985 *Phys. Rev. B* **31** 913
Greene R L and Lane P 1986 *Phys. Rev. B* **34** 8639
Jarosik N C, McCombe B D, Shanabrook B V, Comas J, Ralston J and Wicks G 1985 *Phys. Rev. Lett* **54** 1283
Landolt-Börstein Numerical Data and Functional Relationships in Science and Technology GroupIII vol.17 1982 ed O Madelung (Berlin: Springer)
Mailhiot C, Yia-Chung Chang and McGill T C 1982 *Phys. Rev. B* **26** 4449
Menéndez J, Pinczuk A, Werder D J, Gossard A C and English J H 1986 *Phys. Rev. B* **33** 8863
Miller R C, Gossard A C and Kleinman D A 1985 *Phys. Rev. B* **32** 5443
Priester C, Bastard G, Allan G and Lannoo M 1984 *Phys. Rev. B* **30** 6029

Heterointerface states of shallow impurities

P S Kop'ev, I N Uraltsev, V M Ustinov and A M Vasil'ev

A F Ioffe Physico-Technical Institute, USSR Academy of Sciences, 194021, Leningrad, USSR

ABSTRACT: Measuring the magnetophotoluminescence (MPL) spectra of p-type GaAs/AlGaAs single heterojunction both unintentionally doped and modulation Be-doped we have studied the interface states of shallow impurities for the first time. The interface donor-acceptor (D-A) pair recombination analysis shows that the bound hole states to shallow acceptors located in the barrier material at some distance to the heterojunction interface produce the "impurity band" with the upper energy limited by the interface acceptor binding energy. Measuring the diamagnetic shift anisotropy of the photoluminescence (PL) line, which is attributed to bound electron recombination with 2D-hole, we find the P_z-like character of the bound electron wave-function for donor located near the interface.

1. INTRODUCTION

Since the finding of the new PL line associated with single GaAs/Al$_x$Ga$_{1-x}$As heterojunction has been reported by Yuan et al (1985), a great deal of interest has been devoted to the study of this recombination process as a function of the aluminium content x and doping concentration in both undoped and p- and n-modulation doped structures. The main PL features found previous experimental investigations can be summarized as follows: the PL line, which is shown by step-etching technique to originate from the interface of GaAs active layer and AlGaAs clading layer, reveales both a blue shift with an excitation intensity increase and the strong intensity dependence on temperature up to the line disappearing at 15-20K. The PL peak position ranges from that of the bound excitons to that of D-A pair emission of bulk GaAs. The range width is found by Hooft et al (1987) to be 1 meV in pure samples with background impurity concentration, N=2·10^{14}cm^{-3} to enlarge up to 14 meV (Vasil'ev et al 1986 and Aytieva et al 1987) in unintentionally doped samples with N≈10^{16} cm^{-3} and to reduce to 4 or 5 meV in light doped samples with N not higher than 2·10^{17} cm^{-3} (Yuan et al 1986, Aytieva et al 1987, Altukhov et al 1987). No observation of the interface PL line has been reported in heavily doped samples.

To explain the main properties of the PL line some heterojunction energy band models have been proposed by Yuan et al (1986) and Altukhov (1987), which involve the effects of confinement and tunable transitions. However, using the MPL technique Alferov et al (1986) have documented the impurity related origin of the interface PL in undoped samples. On the other hand, basing on MPL data Ossau et al (1987) has identified this PL to be intrinsic in doping structures with 2D-hole gas.

To learn about the main features of the shallow impurity energy states at (or near) the heterointerface we study the MPL spectra of both unintentionally doped and selectively Be-doped p-type GaAs/Al$_3$Ga$_7$As heterostructures, grown by MBE. The MPL decay kinetics and peak position shift analysis of the PL line, which originates in undoped samples from the interface D-A pair recombination, reveals a broad energy band of shallow acceptor states at the heterointerface. The doublet structure of the interface PL, which is observed in lightly doped samples, we identify to be due to 2D-hole recombination with conduction or donor electrons from nearly flat-band region. The donor related diamagnetic shift analysis elucidates the P_z-like character of bound electron wavefunction for donor located not long distance to the interface.

2. DONOR-INTERFACE ACCEPTOR PAIR RECOMBINATION

There are clear evidences of heterointerface PL to be impurity related in undoped samples. The finding of both the mcs-kinetics of the PL decay and strong anisotropic quenching of these kinetics in the presence of magnetic field normal or parallel to the interface allows Kop'ev et al (1987) to identify this transition to interface acceptor recombination with GaAs donor located not long distance to the interface. Moreover, the donor assignment is supported by the low value of the PL activation energy and the obtained value of the line diamagnetic shift.

Since for D-A pair recombination the Coulomb interaction can never exceed the binding energy of shallower impurity we should consider the interface spectral range of 14 meV to result at large extent from the broad energy band of interface acceptor states. The acceptor binding energy has been shown by Bastard (1981) to reduce for impurity located at the interface. Furthermore, the donor binding energy calculations of Masselink et al (1984) predict the GaAs electron to bound to donor located in wide-gap material with the binding energy diminishing as impurity distance to the heterojunction interface becomes larger than the Bohr radius. We suppose the energy states of acceptors located near the interface in barrier domain to produce the "impurity band". The upper energy of "impurity band" is limited by the interface acceptor binding energy, 15 meV. To estimate the low limit of the impurity band exhibition one should compare this energy to the exciton binding energy in bulk GaAs. The deduced band width is consistent with the acceptor related part of the PL spectral

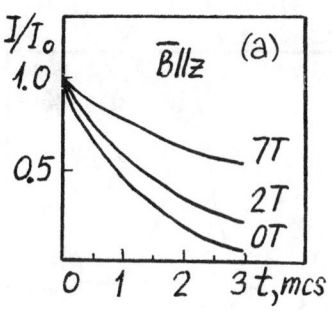
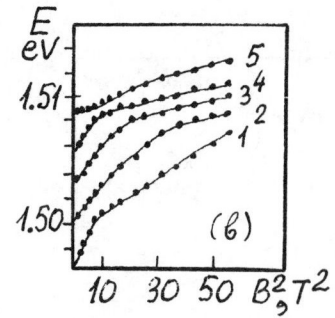

Fig. 1. Magnetic field dependence of a – decay kinetics taken at the PL energy 1.504 eV, b – peak position of the interface line at excitation density, mW/cm^2: 1 –0.06, 2 – 0.2, 3 – 2, 4 – 10, 5 – 80.

range, 10 meV, evaluated by Alferov et al (1986).

The filling of the interface acceptor states at excitation intensity increase results in recombination energy increase mainly due to the shallower acceptor contribution. The PL decay enhancement is simultaneously observed due to decreasing of distance in the D–A pairs. When the PL energy increases from 1496 to 1507 meV the PL decay time shortening is found from 10 to 0.8 mcs. The reversal process of the acceptor states emptying due to recombination is revealed in the PL peak shift to lower energy during long persistent PL decay.

Measuring the MPL decay kinetics we find the effect of the acceptor states filling induced by magnetic field. The PL recombination rate, which is governed by the wavefunction overlap, determines the filling of acceptor states at fixed excitation density. Since in the presence of magnetic field normal to the interface the recombination rate falls down due to shrinkage of the impurity wavefunction in the plane, and concomitant reduction in wavefunctions overlap, the filling increases to produce the PL peak blue shift. The verification of this behaviour is shown on Figure 1, which displays (a) PL decay together with (b) the peak positions taken at different excitation densities as a function of magnetic field. Since no change of the MPL decay kinetics taken at the PL peak is observed we conclude that the peak shift is connected with the filling of impurity states induced by magnetic field. If the PL spectral range to compare with shift reduced on the bulk donor diamagnetic shift we have to conclude the magnetic field induced shift is additional to that under excitation intensity increase. Actually, if the photoexcitation filling is minimal, the "magnetic" shift covers nearly the whole PL spectral range. This giant shift exceed considerably the conduction electron Landau shift. While at high excitation intensity it reduces to the donor related diamagnetic shift. This effect is clear evidence of the magnetic field induced filling of the "impurity band" on heterointer-

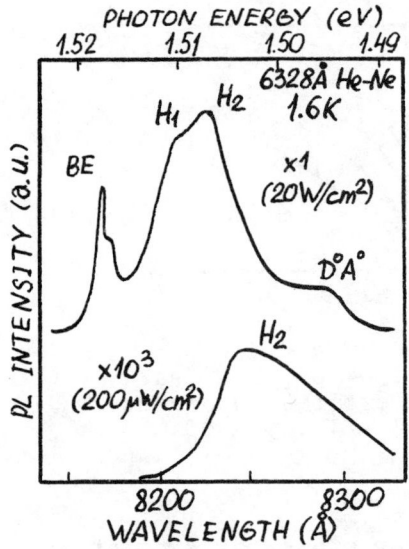

Fig. 2. PL spectra of modulation Be-doped GaAs/AlGaAs heterojunction for three different laser excitation densities

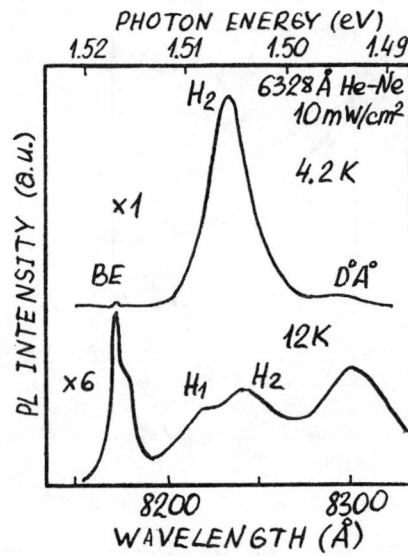

Fig. 3. Temparature dependence of the PL spectra of p-type GaAs/AlGaAs heterojunction

face. It should note, that some hint of the magnetic field effect on the impurity filling has been observed by Willman et al (1973) on the bulk D-A pair recombination in GaAs.

Since the penetration depth of a hole wavefunction in barrier material is about 20 Å, so that an acceptor have to be close to the interface to be involved in the luminescence process. But our findings evidence that a donor, which contributes to the interface D-A pair recombination, is located on the other side of the heterojunction interface. To estimate the donor distance to interface we carefully analyse the MPL shift by use of the D-A pair model taking into account the magnetic field induced shrinkage of the donor wavefunction (Alferov et al 1986). The evaluated distance appears to vary from 700 to 350 Å for the lowest and highest PL energy, respectively, of the interface D-A pair recombination. To examine the modification of the near interface donor wavefunction we study the MPL spectra of p-modulation doped GaAs/AlGaAs heterojunction.

3. RADIATIVE RECOMBINATION OF 2D-HOLE WITH ELECTRON BOUNDED TO NEAR INTERFACE DONOR

Figure 2 displays the PL spectra of p-type modulation doped GaAs/Al$_{0.27}$Ga$_{0.73}$As heterojunction with 2D-hole concentrati-
$8 \cdot 10^{10}$ cm^{-2}. The interface PL, which peak position shifts to the high energies as excitation intensity increases, is found clearly distinguish to have doublet structure. At low

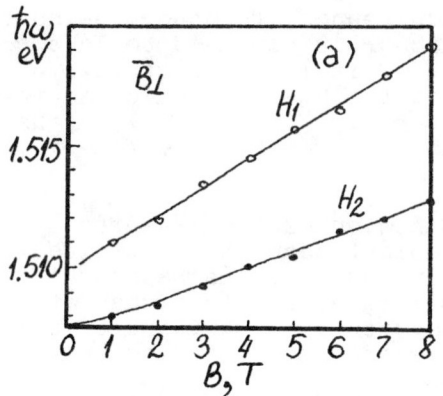 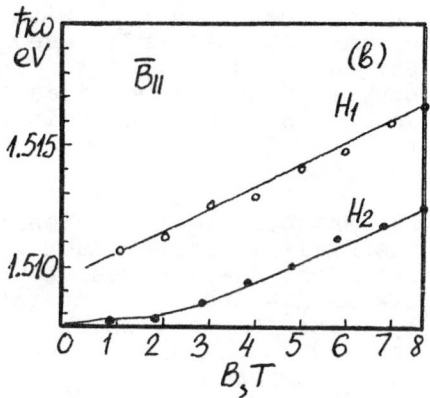

Fig. 4. The H_1- and H_2-peak positions as a function of magnetic field a - normal and b - parallel to the heterointerface

excitation intensities the relatively broad, dominant H_2-line centeres at 1.504 eV. At high tail of the H_2-line the new H_1-line is revealed while the H_2-line intensity saturates with increase of excitation density.

The interface PL is evaluated further by measuring the temperature dependence. The results are given in Figure 3 for two different temperatures and an excitation density of 10 mW/cm². Below 4K the PL intensity remains almost constant, while between 4 and 12K the H_2-line intensity decreases with activation energy of 4.5 meV and the H_1-line becomes observable similar it does with excitation intensity increase. Above 12K the relative intensity of the H_1-line droppes by only one oder of magnitude before the line becomes unnoticeable on the background emission. Since the small intensity range we can not determine the H_1-line activation energy.

To attribute the interface PL to 2D-hole recombination we study, following to Ossau et al (1987), the lines behaviour in the presence of magnetic field, $B_\perp$, normal and $B_\parallel$, parallel to the heterojunction plane, which is shown on Figure 4 a and b; respectively. The linear shift of the H_1-line position at $B_\parallel$, which is found 0.92 meV/T close to the electron lowest Landau level shift 0.86 meV/T, evidences that the H_1-line is due to free electron recombination. As expected at $B_\perp$ there are contribution of 2D-hole lowest Landau level, which is evaluated from the 1.17 meV/T linear shift. The obtained hole mass 0.2 m_0 is closed to predicted by Sanders and Chang (1985) for that of heavy hole along the interface. So we attributes the H_1-line to the 2D-hole recombination with free electron.

Since the H_1-line reveales in the interface PL as the H_2-line satirates with excitation intensity or disappeares with temperature increase, we conclude that the H_2-line is due to

2D-hole recombination with electron bound to donor. Because of the low activation energy the donor is supposed to locate not long distance to the interface. This model is strongly supported by the nonlinear shift of the H_2-line, which is found to enhance with excitation intensity increase, when the filling of donors with smaller distance to the interface is expected.

The anisotropy of the donor wavefunction can be measured by the H_2-line shift reducing on 2D-hole Landau shift obtained from the H_1-line behaviour. The evaluated diamagnetic shift of the H_2-line is about two times more at $B_\parallel$ than that at $B_\perp$. This effect, which is expected for P_z-like bound electron wavefunction, is clear evidence of donor location near the interface.

4. CONCLUSION

In contrast to the energetic models, where the strong band bending sensitivity to the photoexcitation is proposed to explain the blue shift of the interface related PL, we conclude that this PL main feature is at large extent due to extrinsic origin. The acceptor impurity band and donor P_z-like states are identified on the heterojunction.

We gratefully acknowledge the colloboration of V.P.Kochereshko, D.R. Yakovlev and Al.L. Efros on various aspects of the research described in this paper.

REFERENCES

Alferov Zh I, Vasil'ev A M, Kop'ev P S et al 1986 Sov.Phys. JETP Lett. 43 569
Altukhov P D et al 1987 Sov.Phys.Semicond. 21 290
Aytieva G T et al 1986 Sov.Phys.Semicond. 20 828
Bastard G 1981 Phys.Rev. B 24 4714
Hooft G W't, van der Poel W A, Molenkamp L W, Foxon C T 1987 Appl.Phys.Lett 50 1388
Kop'ev P S, Kochereshko V P, Uraltsev I.N et al 1988 J.Lumines. 40-41 747
Masselink W T, Chang Y-C, Morkoc H 1984 J.Vac.Sci.Technol. B 2 376
Ossau W, Bangert E, Weimann G 1987 Sol.St.Commun. 64 711
Sanders G D, Chang Y-C 1985 Phys.Rev. B 31 6892
Vasil'ev A M, Kop'ev P S et al 1986 Sov.Phys.Semicond. 20 220
Willman F et al 1973 Phys.Stat.Sol. b 60 751
Yuan Y R, Mohammed K, Pudensi M A A and Merz J L 1984 Appl. Phys. Lett. 45 739
Yuan Y R et al 1985 J.Appl.Phys. 58 397

Secondary ion mass spectrometry of Si in δ-doped GaAs

C. Grattepain, A.M. Huber, E. Barbier, *G. Gillmann

THOMSON-CSF/LCR, B.P. 10, Domaine de Corbeville, 91401 ORSAY France.
*PERKIN ELMER, Division Instruments B.P. 304,78504 St QUENTIN EN YVELINES France.

ABSTRACT : Secondary ion mass spectrometry (SIMS) is used to quantify the hyper-abrupt doping profile of Si in δ-doped GaAs. The beam-induced broadening artefacts are reduced by performing SIMS measurements at low beam energies. We also report the full width at half maximum (FWHM) of Si-peak doping concentration as a function of primary ion beam energy and it is extrapolated to zero energy. The 17Å FWHM at zero energy indicates that Si is not confined to a single atomic plane, although this result is very close to the surface roughness (2-3 monolayers). Crystal quality observation by metallography completes this study.

1. INTRODUCTION

An advanced epitaxial process such as molecular beam epitaxy (MBE) has shown a potential for growing semiconductor structures with a hyper-abrupt doping profile. This flexibility of MBE allows the growth of new types of devices. Using δ-doping i) new homostructure FET's ii) non alloyed ohmic contacts and iii) a decrease in parasitic resistance have been obtained (Schubert et al 1986a). Atomic planar doping by Ge of GaAs by MBE has been proposed by Wood et al (1980) and Si atomic planar doping of GaAs was investigated by several authors : Sasa et al (1985), Schubert et al (1986b) Ploog et al (1987), Zrenner et al (1987). A two-dimensional electron gas has been produced using this δ-doping technique which theoretically allows impurity atoms to be localized in one atomic monolayer of the host GaAs material, thereby generating a V-shaped quantum well.

Schubert et al (1987) and Gillman et al (1988) have reported the Hall measurements of electron density and mobility in planar-doped GaAs layers and found for Si a saturation value of $2 \times 10^{13} cm^{-2}$ and $2.7 \times 10^{13} cm^{-2}$ respectively.

Recently Zrenner et al (1987), from oscillatory magnetotransport data, demonstrated that the so-called δ-doping layers described in the literature do not represent a system of donor ions confined to a single atomic layer. The existing data from their measurements

© 1989 IOP Publishing Ltd

on all samples can be explained in terms of finite ion spreading. The questions relating to the observed saturation and the mechanism of the Si spreading requires the use of a characterization method which permits(1) the analysis of possible compensating impurities such as C or complex-forming H and O, and provides (2) the real extent of Si from the basic δ-doping plane.

Secondary ion mass spectrometry (SIMS) is one of the most powerful methods which allows us to determine the doping element and parasitic impurity profiles. However it suffers a loss of interfacial information due largely to beam induced broadening. Nevertheless a number of researchers have demonstrated that this phenomenon can be reduced by performing SIMS measurements at low energies (Wittmaack 1985, Turner et al 1987 and Schultz et al 1973). Beall et al (1988) studied the Si migration in δ-doped GaAs by a 2 keV O_2 optimized ion beam.

This paper will present the study of δ-doped GaAs Si depth profiles at various low Cs^+ beam energies. The FWHM of the Si-peak concentration is reported as a function of the primary ion beam energy extrapolated to zero energy. In this way the sputter-induced artefacts of SIMS measurements can be projected and the atom spreading obtained with good approximation. Before the presentation of the results, we shall review briefly the problems associated with the depth resolution in SIMS.

2. DEPTH RESOLUTION

The quality of a depth profile, called depth resolution, is described by the measured width of an interface between two layers. The interface width z is commonly defined as the interval where the intensity drops from 84% to 16% of maximum signal, equivalent to two standard deviation (2σ) of the error curve (Werner 1982). In the case of δ-doped layers the depth resolution is expressed as the full width at half maximum (FWHM) of the peak doping concentration measured in optimized analytical conditions (see paragraph 3).

A number of researchers have investigated the cause of broadening of the secondary ion signal (Magee et al 1982, Wittmaack 1985, Gronet et al 1986, Turner et al 1987, Vandervorst el al 1987, Armour et al 1987, Hoercher et al 1987, Treichler et al 1987, Breuer 1987, McPhail et al 1988).

They have provided parameters enabling the analytical artefacts to be separated from the material effects in SIMS data. In each case a rigorous examination is needed of the major factors which determine the depth resolution presented in table 1 (McPhail et al 1988).

Observable loss of depth resolution occurs over a depth range more than 4000Å (Beall 1988), in this case, and for low beam energies, the characteristic width and shape of a concentration peak will be independent of the depth beneath the initial surface (deeper than 150Å which is the near-surface pre-equilibrum region) Beall (1988).

TABLE 1.
Factors that may affect the depth resolution in SIMS.

1) Fundamental effects
   atomic mixing
   recoil implantation
   segregation
   radiation enhanced diffusion
   sputter induced microtopography

2) Instrumental effects
   uneven etching
   charging

3) Sample specific effects
   particles on or in material
   defects
   surface waviness

4) Calibration errors
   different sputter yields

3. EXPERIMENTAL TECHNIQUE

The Si δ-doped GaAs sample was grown by MBE in a VARIAN GEN II system on a semi-insulating GaAs (100) substrate. Particular care was taken with source outgassing and residual pressure of the growth chamber. The substrate temperature, As_4/Ga flux beam equivalent pressure ratio and the growth rate were respectively 590°C, 11 and 1 μm/h. A GaAs buffer layer 1 μm thick (with a residual hole concentration in the $10^{14} cm^{-3}$ range) was first grown. Then the planar doping was obtained by closing the Ga shutter and at the same time opening the Si shutter. The As shutters were kept open so that the Si atoms were deposited preferentially in Ga sites. The Si flux was calibrated by van der Pauw Hall measurement on a thick homogeneously Si doped GaAs layer. Substrate temperature was measured with an infra-red radiation pyrometer.

The δ-doped layer was deposited at beneath 300Å depth (Sample VAR 135) and 4800Å depth (Sample VAR 101) beneath the surface for intended Si concentrations of $1 \times 10^{13} cm^{-2}$ and $6 \times 10^{12} cm^{-2}$ respectively. Sample VAR 135 was used to perform the quasi-static SIMS Si analyses, and VAR 101 to determine the C, H and O concentration by dynamic SIMS.

SIMS depth profiling was carried out using a Cs^+ primary ion bombardment in a CAMECA IMS 4F. With Cs^+ bombardment we used positive secondary ion detection. Positive secondary ion detection lowers the Cs^+ impact energy which allows analyses to be performed with 2,2.5 and 3 keV. The dimension of the square eroded and analysed area, and the primary ion current were optimized to achieve high secondary ion yields without introducing edge effects. $^{161}CsSi^+$ as the secondary ion analyzed without offset and $^{75}As^+$ was used as the reference mass with -110V offset. $^{161}CsSi_2^+$ ions give better reproducibility than $^{28}Si^+$ in a GaAs matrix. $^1H^-$, $^{12}C^-$, $^{16}O^-$ with $^{69}Ga^-$ reference ions were also analyzed (Cs^+ 14.5 keV, 110V offset).
Calibration of the quantitative Si determination has been obtained by a homogeneously Si doped epitaxial layer (Si $1 \times 10^{18} at.cm^{-3}$) and by Si, C, H and O implanted samples.

Finally the crystal perfection of the δ-doped layer was investigated using the etching technique on a low angle polished sample (ETOCAPS) and optical microscopy observation. This method (Huber et al 1983) achieved

an angle of 0.01-0.06° to the GaAs (100) surface by dissolving with 15% Br in methanol. This small angle gives a 10^3-5x10^3 magnification of the layer and interface thicknesses. Figure 1 is the schematic representation of a chemically angle polished sample. The angle polished sample etched by modified AB bath (25° 2 min, Abrahams et al 1965) is a sensitive means for observation of crystal perfection.

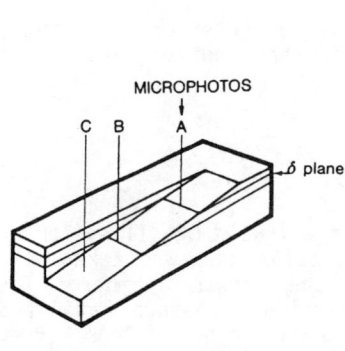

Figure 1. Schematic representation of a chemically angle polished specimen A,B and C show the location of δ-doped layer, layer-substrate interface and substrate respectively.

Figure 2. Si depth profile of a δ-doped GaAs MBE layer. Sample VAR 135. Analysed by quasi-static SIMS (Cs^+, 2.5 keV secondary positive ions).

4. RESULTS

A SIMS profile of Si -doped GaAs (sample VAR 135) is shown in figure 2. (primary beam energy is 2.5 keV in this case). The Si peak is almost symmetrical and the mean value of FWHM is given in table 2.

The Si doping peak FWHM extracted from SIMS depth profiles as a function of the Cs^+ primary beam energy for our Si -doped structure is shown in figure 3. The mean FWHM values extrapolated to zero energy give 17Å with +10Å error bar. This value indicates that the Si-donors are indeed distributed in more than one lattice plane in the host crystal, although the result is very close to the surface roughness step (2-3 monolayers X 2.8Å).

Table 2.
The mean FWHM (Å) versus beam energy.

Primary Beam Energy (keV)	FWHM (Å) mean value
2	56
2.5	66
3	79

No noticeable C, H and O concentration was detected at the $_{16}$-site. It is less than the instrument background level which is 1x10^{16}, 7x10^{17} and 2x10^{17} cm^{-3} respectively in this analytical condition.

Figure 3. Full width at half maximum (FWHM) of CsSi signal versus primary beam energy in a δ-doped GaAs MBE layer.

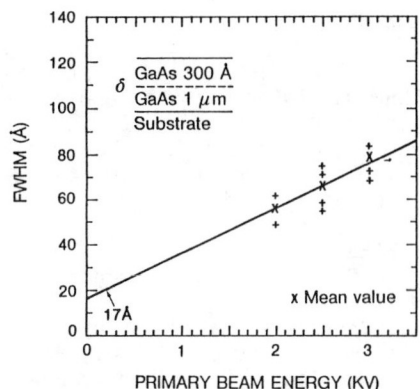

We presented here the Si distribution in optimized analytical conditions and determined that the only atom located in the host crystal is Si, so the main question still remains : why the Hall saturation is produced at a relatively low Si level ($2.7 \times 10^{13} cm^{-2}$) compared to the available sites for group III atoms ($6.25 \times 10^{14} cm^{-2}$ Ga) on the (100) face of a GaAs crystal.

In fact the sensitive ETOCAPS metallographical observation shows, in figure 4, the loss of crystal perfection by Si doping. The interwoven " S-pits" originating from the δ-doping zone are connected with some lattice disorder. This disorder probably results from the high impurity concentration, where Si atoms no longer act as donors but form a Si lattice (Schubert 1986a) and in consequence a strained, disordered crystal (not necessarily dislocated).
Similar etch figures were observed on VAR 101 and other δ-doped samples (δ-plane at 4800Å depth). Etch figures seem to arise from the zone of the doping plane and extend to the surface.

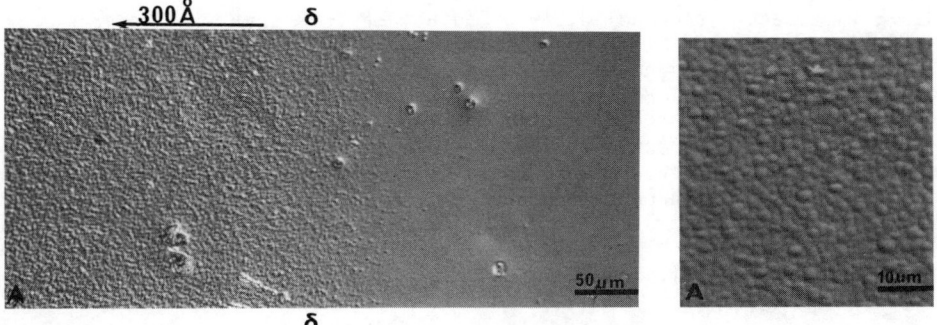

Figure 4. High density of S-pits on the epitaxial layer arising from the δ-doped plane VAR 135. ETOCAPS.

CONCLUSION

With optimized SIMS analysis, the Si δ-doping peak was determined and a 17Å FWHM was found for the zero-beam-energy extrapolated value. No other impurities in particular C, H or O were found on the δ-plane in noticeble concentration. The possible local Si lattice formation and crystal disorder arising from the δ-doping zone is probably one of the causes of Hall density saturation.

REFERENCES.
Armour D G, Wadsworth M, Badheka R, Van den Berg, Blackmore G, Courtney S, Whitehouse C R, Clark E A, Sukes D E and Collins R, 1987 Proc. of SIMS VI Versailles, (John Wiley et Sons) 399.
Abrahams M S and Buiocchi C J, 1965 J. Appl. Phys. 36 2855.
Beall R B, Clegg J B and Harris J J 1988 Semiconductor Science and Tech. 3 612.
Breuer U 1987 Proc. of SIMS VI Conf. Versailles, (John Wiley et Sons) 481.
Gillmann G, Vinter B, Barbier E and Tardella A 1988 Appl. Phys. Lett. 52 972.
Gronet C M, Sturm J C, Williams K E and Gibbon J F 1986 Appl. Phys. Lett 48 1012.
Hoercher G, Forchel A, Steiner S and Germann R, 1987 Proc of SIMS VI Conf. Versailles, (John Wiley and Sons) 457.
Huber A M, Laurencin G and Razeghi M, 1983 Journ. de Phys. C 4 409.
Kirchner P D, Jackson T N, Petit G D and Woodal J M, 1985 Appl. Phys. Lett 47 26.
Magee C W and Honing R E, 1982 Surf. and Interf. Anal. 4 35.
McPhail D S, Dowsett M G, Fox H, Houghton R, Leong W Y, Parker E H C and Parker G K, 1988 Surf. and Interf. Anal.11 80.
Ploog K, 1987 J. Cryst. Growth 81 972.
Sasa S. Muto S, Kondo K, Ishikawa H and Hiyamizu S, 1985 Jap. J. of Appl. Phys. 24 L 602.
Schubert E F, Cunningham J E and Tsang W T 1986a Appl. Phys. Lett. 49 1729.
Schubert E F, Ploog K 1985b JAP. Jap. Lett 24 L 603.
Shulz F, Wittmaack K and Maul J, 1973 Radiat. Eff. 18 211.
Treichler R, Korte L and Von Criegern R 1987 Proc. of SIMS VI Versailles, (John Wiley and Sons) 469.
Turner J A and Amano J 1987 Appl. Phys. Lett. 50 1601.
Vandervorst W and Shepherd F R 1987 J. Vac. Sci. Technol. A 5 313.
Werner H A, 1976 Acta Electron. 19 53
Wittmaack K, 1985 J. Vacv. Sci. Technol. A 3 1350.
Wood C E C, Metze G, Berry J and Eastman L F, 1988 Apl. Phys. Lett. 52 972.
Zrenner A, Koch F and Ploog K, 1987 Proc. of 14th GaAs and Related Compounds, Heraklion 341.

Inst. Phys. Conf. Ser. No 95: Chapter 2
Paper presented at Int. Conf. Shallow Impurities in Semiconductors, Linköping, Sweden, 1988

Metal-insulator transition in Si:P

M.A.Paalanen and R.N.Bhatt

AT&T Bell Laboratories, Murray Hill, NJ 07974, USA

> Abstract: This paper presents a review of recent experimental work near the metal-insulator transition in phosphorus doped silicon (Si:P). It contains electrical conductivity and thermodynamic studies and both electron and nuclear spin resonance measurements for uncompensated and compensated samples. In the insulating phase the data is compared with the predictions for random Heisenberg antiferromagnet modelled by Bhatt and Lee. On the metallic side, a number of models are tested and the available data is found to support the existence of localized magnetic moments near the transition.

1. INTRODUCTION

During the last ten years our understanding of the electron transport in disordered metals has been revolutionized by the scaling theories (Abrahams et al 1979, Finkelstein 1983, Castellani et al 1984). These theories have succesfully described the metal-insulator (MI) transition in most disordered systems like metal-semiconductor alloys (Hertel et al 1983), magnetic semiconductors (Washburn et al 1985) as well as compensated semiconductors (Thomas et al 1982). However, applying the scaling ideas to the MI transition of uncompensated semiconductors like Si:P or Si:As has proven to be more difficult (Milligan and Thomas 1985). For example, the onset of the metallic conductivity in uncompensated materials is steeper than predicted by the lowest order scaling calculation. This is somewhat surprising because a doped semiconductor, a collection of hydrogenic "atoms", is one of the simplest systems for studying the MI transition. The interactions between the donor electrons are well known and these materials are better characterized than other systems (see e.g. Bhatt 1987) in terms of their optical, dielectric and magnetic behavior on both the dilute doping limit (insulating phase) and the heavy doping limit (metallic phase). However, in the intermediate doping regime near the MI transition, the theoretical modelling combining the effects of disorder and interactions has turned out to be a difficult task.

We begin this paper in section 2 with a discussion of thermodynamic behavior, e.g. susceptibility and specific heat measurements in uncompensated Si:P. The Wilson ratio, which is equal to the temperature multiplied by the ratio of susceptibility to specific heat, is shown to be a powerful tool in testing different theoretical models. In section 3 we review recent measurents in compensated Si:P;B and compare them with corresponding results in uncompensated Si:P. Based on sections 2 and 3 we argue that electron spin dynamics appears to play an important role at the MI transition and in section 4 we will review the recent ESR and NMR measurements in Si:P near the MI transition. Both electron and nuclear spins are local probes and complement the global thermodynamic measurements. Finally in section 5 we conclude by presenting our current understanding of the MI transition in semiconductors and future directions.

© 1989 IOP Publishing Ltd

2. THERMODYNAMIC MEASUREMENTS

2.1 Experimental

Electron spin susceptibility of Si:P was first measured by Quirt and Marko (1971) and later by Ue and Maekawa (1971), Murayama et al (1984), Ikehata et al (1985), Paalanen et al (1986a and 1986b) and by Sarachick et al (1986). Specific heat has been similarly measured by several groups (Marko et al 1974, Kobayashi et al 1977 and Paalanen et al 1988a). We have limited our recent studies of these quantities closer to the MI transition but extended them further down in temperature than the previous ones. Furthermore, we measured both quantities for the same samples in order to obtain accurate values for the Wilson ratio.

In Figure 1 we show the paramagnetic susceptibility χ for three samples as a function of temperature down to $T = 30$ mK. These χ values were obtained by integration from ESR lines and by calibrating our ESR spectrometer against the known Curie susceptibility of ^{29}Si nuclei. The susceptibilities in Figure 1 have been scaled by $\chi_0 = 3n\mu_B^2/2k_BT_F$, the Pauli susceptibility of a degenerate electron gas with density n. The χ of the heavily doped Si:P agrees with the Pauli value χ_0. However the susceptibilities near the MI transition are larger than χ_0 and increase without saturation towards lower temperatures. On the other hand the susceptibilities of both the metallic and the insulating samples are less than the Curie value $\chi_{Curie} = n\mu_B^2/k_BT$, which describes well the susceptibility of noninteracting localized spins in the dilute doping limit. In the insulating phase χ seems to fit a powerlaw behavior $T^{-\alpha}$ with $\alpha \approx .65$. This was first observed by Andres et al (1981) at low doping densities and explained by Bhatt and Lee (1982) as being due to the antiferromagnetic interaction between the donor electrons.

In Figure 2 we show the specific heat C for the above three samples. In order to emphasize the electronic contribution to the specific heat we have subtracted from the measured values the known phonon specific heat AT^3 which is shown as broken lines in Figure 2. For comparison we also present for each sample the expected itinerant electron specific heat $C_0 = \gamma_0 T$ as solid lines. These values are

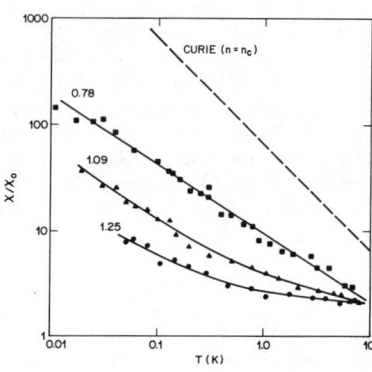

Fig. 1. Normalized susceptibility of Si:P as a function of temperature for three samples near n_c ($n/n_c = 0.78$, 1.09 and 1.25)

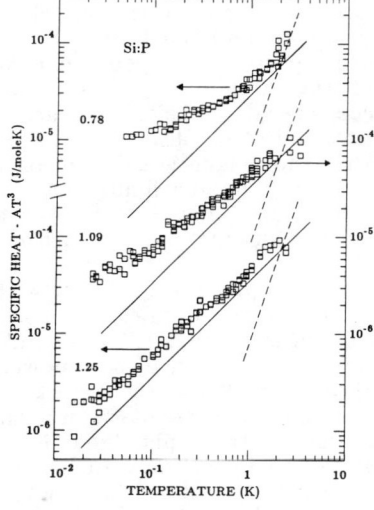

Fig. 2. Electronic part of the specific heat for the same samples

estimated from the Si conduction band mass $m_0^* = 0.34 m_0$ and from the known electron density n. The measured specific heat is enhanced over the itinerant electron prediction and the enhancement seems to increase toward lower temperatures much like the low temperature susceptibility enhancement.

2.2 Theories

The enhanced susceptibility of Si:P was first described by Curie type localized spins in both the insulating and metallic phases (Quirt and Marko 1971). However, for noninteracting Curie spins, one wuold expect a 1/T temperature dependence in susceptibility and no specific heat at all contrary to the observations of Figures 1 and 2. Consequently better models have been constructed by taking into account the e-e interactions. In the insulating phase the interactions are well described by the Heisenberg antiferromagnetic exchange coupling J. In the BL model the spatial randomness of the donors gives rise to an effective pairwise coupling distribution $P(J) = J^{-\alpha}$. The quantity α was calculated to be $\approx .6$, relatively temperature independent and only weakly n-dependent. Bhatt and Lee argued further that at temperature T the pairs with $J > k_B T$ are locked into an inert singlet ground state and only pairs with $J \leq k_B T$ contribute to the thermodynamic properties. This model leads to $\chi \propto T^{-\alpha}$ and $C \propto T^{1-\alpha}$ temperature dependences for susceptibility and specific heat. This prediction agrees well with the susceptibility of the insulating sample in Figure 1. The specific heat of the same sample is consistent with the BL model only below .7 K temperature. We believe that our insulating sample with $n \approx n_c$ has also nonmagnetic excited states (upper Hubbard band) within few Kelvin above the Fermi level and these states are not accounted for in the BL model. In the metallic phase the Fermi level finally reaches these states and their contribution to C and χ is expected to dominate. Overall we find, however, that the BL model adequately describes the low temperature thermodynamic properties of the uncompensated semiconductors in the insulating phase.

In the metallic phase the susceptibility and specific heat are harder to calculate because the e-e interactions are difficult to handle in a disordered medium. The Brinkman-Rice model (1970) takes into account only the interactions without disorder. However, it demonstrates that for the half filled band (uncompensated case) the MI transition can be driven by the e-e interactions which give rise to a diverging density of states and consequently to diverging C and χ when n approaches n_c. The BR calculation is in qualitative agreement with the experimental results but being only a zero temperature calculation a direct comparison with the data is not possible.

The combined effect of disorder and e-e interactions on the finite temperature thermodynamic properties as well as on the conductivity was recently addressed by the scaling theory for disordered interacting electrons (Finkelstein 1983, Castellani et al 1984 and 1987). This theory predicts that the MI transition is not characterized by a unique critical exponent and depending on the strength of the spin-flip or spin-orbit scattering the disordered materials belong to different universality classes. Materials with weak spin-orbit or spin-flip scattering rates $\hbar/\tau_s < k_B T$ are predicted to have a diverging $\gamma = C/T$ and χ toward zero temperature. In contrast, such a divergence should not take place in materials with high spin scattering rates $> k_B T$. We expect that the spin-orbit scattering is weak in Si:P. The P impurity is light, the nearest neighbour to Si in the periodic table and substitutional in the Si host lattice and therefore causes very little lattice distortion. Similarly the spin-flip scattering due to the hyperfine interaction between the P nuclei and the electrons is weak. Consequently we expect Si:P to belong to the universality class with diverging χ and γ in qualitative agreement with the data in Figures 1 and 2. The ESR linewidth should give more information about the different spin-flip rates and we will return to this question in section 4.

In the scaling theory the susceptibility and specific heat enhancements are due to the low lying long wavelength excitations. However, these enhancements could also be explained by the existence of localized magnetic moments. As we argued above, we have to take into account their interaction with other localized moments and possibly with the itinerant electrons, which was not done in the earlier model by Quirt and Marko (1971). In the following we will assume that the interaction between the localized moments dominate and can be treated by the BL model in analogy with the insulating phase. We will also assume that the itinerant electrons can be described by the standard Fermi liquid theory. In this two fluid picture (BL+FL) the specific heat and susceptibility will be the sum of two contributions

$$\gamma/\gamma_0 = m^*/m_0^* + (T/T_0)^{-\alpha} \qquad (1)$$

and

$$\chi/\chi_0 = m^*/m_0^* + \beta(T/T_0)^{-\alpha} \qquad (2)$$

where m^*/m_0^* is the effective mass of the itinerant electrons and β depends only on α and is about 10.5 for our averaged experimental value of $\alpha = .62$. The parameter T_0 measures the fraction of electrons which can be described as localized spins. Eqs. (1) and (2) agree reasonably well with the metallic phase data in Figures 1 and 2. Especially the low temperature susceptibility increase resembles the powerlaw behavior of the insulating sample. From the high temperature specific heat we estimate that m^*/m_0^* is about 1.3 in reasonable agreement with the earlier specific heat measurements by Marko et al (1974) and Kobayashi et al (1977).

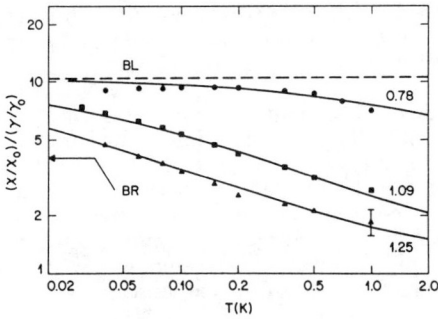

Fig. 3. Wilson ratio as a function of temperature for the above three samples

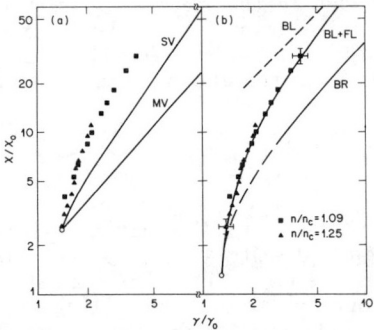

Fig. 4. Susceptibility versus specific heat γ. In 4a. the data is compared with single valley (SV) and multi valley (MV) scaling theories and in 4b. with Brinkman-Rice (BR) and extended Bhatt-Lee (BL+FL) models

The above three models are hard to test with the data in Figures 1 and 2 because of too many adjustable parameters. We can improve the testing by forming the Wilson ratio $(\chi/\chi_0)/(\gamma/\gamma_0)$ which is independent of the density of states at the Fermi level and has fewer parameters. However it is easyly obtained only for the extended BL model. In Figure 3 we have plotted the Wilson ratio as a function of temperature and solid lines through the data are the best fits using Eqs. (1) and (2). The only parameter in these fits is the characteristic temperature T_0. The fits look good and the values of T_0 correspond to reasonable values of 6% and 2% for the fraction of localized moments with interactions J less than 2 K in the 1.09 and 1.25 samples respectively. (These fractions depend on the cutoff parameter J_0 of the BL model). We also notice from Figure 3 that the low temperature values of the Wilson ratio are larger than the zero temperature value 4 of the BR theory.

We can test the scaling theory more accurately by integrating the scaling equations so that the susceptibility is directly a function of specific heat. This is done in Figure 4 where we have plotted χ/χ_0 as a function of γ/γ_0. In Figure 4a the SV and MV curves are the somewhat different scaling theory predictions for a single valley and a multivalley semiconductor correspondingly. The only adjustable parameter in this plot is the starting point of the scaling, the highest temperature data point. The fit is not good and can not be improved by moving the starting point around. The first order scaling equations fail to generate enough magnetism, but we have not ruled out the possibility that higher order scaling terms, which are not available at the moment, can correct this disagreement. For comparison we also show in Figure 4b the extended BL model (BL+FL). As expected from the previous fits the agreement is excellent. The only adjustable parameter in this case has been m^*/m_0^*, which we selected to be 1.3 as previously determined from the high temperature specific heat. We also test in Figure 4b the zero temperature BR model. Its prediction disagrees again with the experimental results.

The thermodynamic measurements seem to imply that there are localized moments in the metallic phase of Si:P, but presumably only near the MI transition. Recently Alloul and Dellouve (1987) obtained more evidence about the localized moments by studing NMR of ^{31}P nuclei. They noticed a loss of the signal for insulating samples and partial loss even in the metallic phase. They argued that the NMR signal comes from the nuclei in contact with itinerant electrons and the nuclei close to localized electrons were in so large hyperfine field that their signal was shifted away from the frequency range covared in this experiment. The fraction of localized moments measured in this experiment is somewhat larger than our values.

3. COMPENSATED SEMICONDUCTORS

Compensated semiconductors contain both donors and acceptors. In a n-type material each acceptors binds one electron and appears as a scattering center. The remaining $n = N_D - N_A$ electrons are shared among the N_D donors and constitute the system where we want to study the MI transition. Compared to an uncompensated semiconductor a compensated one has more disorder and less e-e interaction because there are several donor sites available for each free electron. By varying the compensation ratio $K = N_A/N_D$ one can in principle study the relative effects of disorder and interactions on the MI transition.

In the early studies of the MI transition the critical doping density n_c was found to increase with increased compensation K. This is an expected result if one only considers the effects of increased disorder. Thomas et al (1982) also noticed in Ge:Sb that the critical conductivity exponent μ [$\sigma = \sigma_0(n-n_c)^\mu$] increases with increasing K. Similar studies were recently conducted in Si:P;B by Hirsch et al (1988b) and some of their results are presented in Figure 5 along with the uncompensated Si:P results. The critical conductivity exponent of compensated Si:P;B is close to 1 and similar to what is found in most materials like metal-semiconductor alloys. In contrast the onset of critical conductivity of uncompensated Si:P is much steeper with exponent $\mu \approx .5$. This result for Si:P is still unexplained within the first order scaling expansion. However as shown by Castellani et al (1987) the universality class with low spin scattering rate might yield an exponent less than one when higher order terms are included. In this light the difference between the compensated and uncompensated materials is interesting and might provide an important clue for solving the exponent problem.

In Figure 6 we have compared the susceptibilities of Si:P and Si:P;B in the metallic phase (Hirsch and Paalanen 1988a). Both of these samples have approximately the same carrier density n but somewhat different normalized densities n/n_c =1.25 (Si:P) and 1.1 (Si:P;B). In this plot we have also normalized the susceptibilities by the Curie value χ_{Curie} in order to convert χ into the fraction of localized moments.

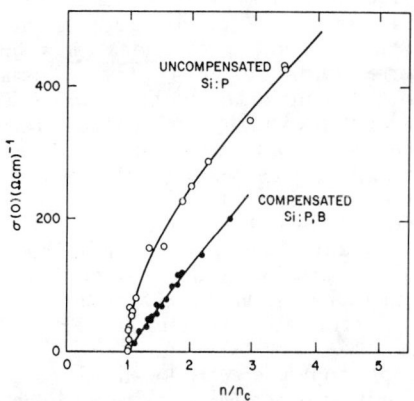
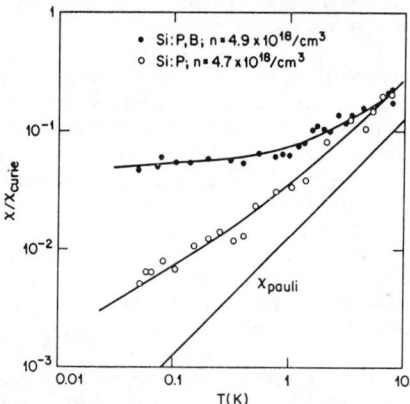

Fig. 5. Zero temperature conductivity versus normalized doping density for uncompensated Si:P and compensated Si:P;B

Fig. 6. Susceptibility comparison between compensated and uncompensated samples with similar free electron densities n

The normalized susceptibility of compensated Si:P;B is larger than in uncompensated Si:P at low temperatures. This is an unexpected result and seems to indicate that the compensated samples have more local moments than uncompensated ones. At the moment we do not have a quantitative theory tying up the conductivity and susceptibility results in Figures 5 and 6. However we can argue that localized moments can cause spin-flip scattering among the itinerant electrons and there is more spin scattering in compensated materials. Within the scaling theory the compensated materials presumably belong to the universality class with high spin-flip rate and critical exponent 1. The uncompensated materials on the other hand have less spin-flip scattering and might belong to the universality class with $h/\tau_s < k_B T$.

4. SPINDYNAMICS IN Si:P

In Figure 7 we introduce the different spin subsystems of Si:P and the couplings between them. We will assume that localized moments exist in the metallic phase and they interact weakly with the delocalized electrons. Our main interest is the spin-flip scattering of the delocalized electrons. However this scattering rate is determined indirectly and we have to test our modelling of spin dynamics with other quantities like T_1 of ^{29}Si nuclei and the ESR linewidth in both the insulating and metallic phases.

The temperature dependence of spin-lattice relaxation time T_1 of ^{29}Si nuclei is shown in Figure 8 for several insulating samples in 0.2 $n_c <$ n $< n_c$ regime (Zadrozny et al 1987). The data from different sources and magnetic fields have been scaled to 1 T magnetic field with a linear scaling law. From Figure 8 we find T_1 independent of electron density n, roughly linear in magnetic field B and proportianal to $T^{-.3}$. The T_1 of ^{29}Si nuclei is due to the hyperfine and dipolar couplings to the electrons and has been estimated by Gan and Lee (1986) and by Hoch and Holcomb (1988). In the insulating phase the electrons are exchange coupled, which gives rise to fluctuating interaction fields. After Hoch and Holcomb T_1^{-1} is proportional to the density of free Curie like magnetic moments $n\chi/\chi_{Curie}$. The closely coupled pairs in singlet state $(J > k_B T)$ are not contributing to T_1,

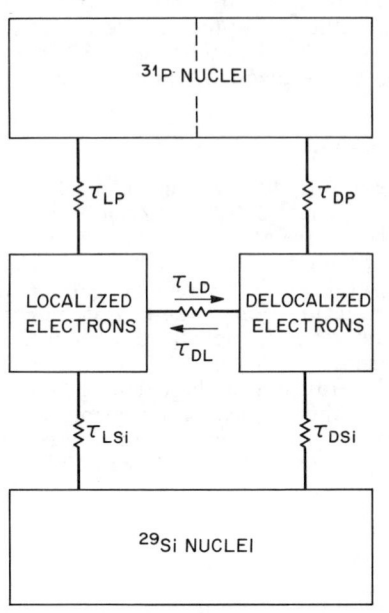

Fig. 7. Schematic diagram of different spin subsystems, their interactions and corresponding relaxation times in doped Si

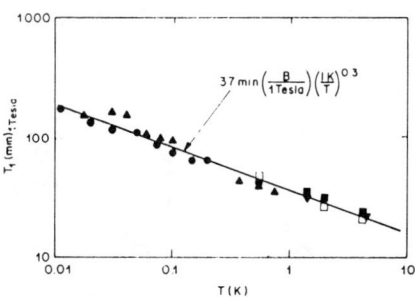

Fig. 8. ^{29}Si spin-lattice relaxation time T_1 versus temperature in the insulating phase near n_c

because the dominating dipolar relaxation is proportional to S_z and therefore zero for the singlet state. From the susceptibility measurements by Andres et al (1982) and subsequent analyses by Bhatt and Lee (1982) we find that $n\chi/\chi_{Curie}$ is nearly independent of n in this region and that the temperature dependence $T^{1-\alpha}$ in reasonable agreement with Figure 8. The linear B dependence is harder to explain. Hoch and Holcomb suggest that the fluctuation correlation time has a wide distribution due to the spatial randomnes, but only a specific shape of the distribution will give the observed B dependence. The Si T_1 in the metallic phase near n_c is similar to T_1 in Figure 8 and much shorter than the T_1 predicted by Koringa (1950) for metals and observed by Sundfors and Holcomb (1964) in heavily doped Si:P. Therefore we believe that the T_1 results support the BL picture in the insulating phase and the existence of localized spins in the metallic phase.

The ESR linewidth $\Delta H_{\frac{1}{2}}$ has been recently measured by Murayama et al (1984) and by Paalanen et al (1986) for uncompensated Si:P and Hirsch and Paalanen (1988) for compensated Si:P;B. In all these measurements $\Delta H_{\frac{1}{2}}$ was found to increase toward low temperatures. In Figure 9 the normalized linewidth $\Delta H_{\frac{1}{2}}/(\Delta H_{\frac{1}{2}})_{3K}$ has been plotted against the normalized susceptibility $\chi/(\chi)_{3K}$ for three samples near the MI transition. We find several interesting features in the data. First of all there is a large difference between the uncompensated and compensated materials. Secondly insulating and metallic samples seem to behave approximately the same way pointing to a similar mechanism in both phases. A third interesting feature in the data is that for the uncompensated samples $\Delta H_{\frac{1}{2}}$ is proportional to χ just below the 3 K temperature but grows slowlier than χ at the lowest temperatures.

This linewidth behavior has been analyzed by Sachdev and Bhatt (1986), Sachdev (1986) and by Finkelstein (1987). After Sachdev the e-e interactions lead in the scaling theory aproach to both susceptibility and linewidth increases at low temperatures. However only in first order calculation $\Delta H_{\frac{1}{2}}$ is proportional to χ and in higher order expansion χ increases faster than $\Delta H_{\frac{1}{2}}$. This is in qualitative

agreement with the linewidth behavior of uncompensated samples. However the similarity between the insulating and metallic samples suggest that the linebroadening is due to the localized moments. After Finkelstein (1987) the ESR linewidth can be written in the presence of both localized (L) and delocalized (D) electrons in the form

$$\Delta H_{\frac{1}{2}} = (\chi_L/\tau_{LP} + \chi_D/\tau_{DP})/(\chi_L + \chi_D) \tag{3}$$

where the broadening is due to the hyperfine coupling to phosphorus nuclei. At the lowest temperatures $\chi_L \gg \chi_D$ and we can write $\Delta H_{\frac{1}{2}} = 1/\tau_{LP} \approx \frac{1}{2}\gamma_e H_{hf}^2 \tau_c$, where the hyperfine field $H_{hf} = 21\,G$ and τ_c is the correlation time for the hyperfine interaction when a spin excitation diffuses within a given cluster. Finkelstein's result for τ_c takes into account only the interaction with delocalized electrons. Motivated by the success of the two fluid picture for χ and γ it might be appropriate to include also the interactions between the localized moments. This has been done numerically by Sachdev and Bhatt (1986) in the insulating phase, where they emphasized the role of the broad distribution of the exchange constants in localizing the spin excitation. If we make the drastic assumption that the distribution of exchanges (including further neighbor interactions) can be described by a single parameter α then we may obtain τ_c by averaging J^{-1} over $P(J) = J^{-\alpha}$ distribution from $\hbar\gamma_e H_{hf}$ to $k_B T$. This will give a linewidth

$$\Delta H_{\frac{1}{2}} = cH_{hf}\left(\frac{\bar{h}\gamma_e H_{hf}}{k_B T}\right)^{1-\alpha} \tag{4}$$

where the parameter $c \approx 1$. This result is in qualitative agreement with the slower variation of the linewidth in the compensated samples compared to the uncompensated samples in Figure 9. Clearly a more sophisticated calculation is necessary for obtaining a quantitative result.

In the scaling theory the important quantity is the spin-flip rate $1/\tau_s = 1/\tau_{DL}$ of the delocalized electrons which we believe to be mainly due to the localized moments. This quatity is hard to measure directly but according to the single electron scaling theory the negative magnetoresistance contains information about

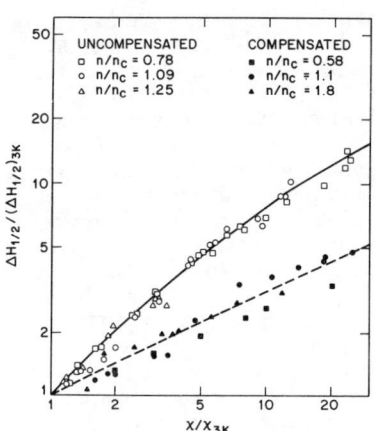

Fig. 9. Normalized ESR linewidth versus normalized susceptibility for compensated and uncompensated samples

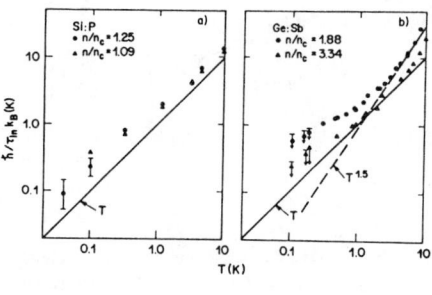

Fig. 10. Dephasing scattering rate versus temperature for uncompensated Si:P and compensated Ge:Sb

it (Kawabata 1980). Paalanen *et al* (1988b) have recently analyzed the dephasing rate $\hbar/\tau_{in}$ from the observed negative magnetoresistance. The results are presented in Figure 10 as a function of temperature along with similar measuments in compensated Ge:Sb (Ootuka *et al* 1987) . The dephasing rate has at least two contributions, one from e-e inelastic scattering and the other from spin-flip scattering. After Schmid (1974) the e-e scattering has $T^{1.5}$ temperature dependence at higher temperatures and $T^{1.0}$ dependence at lower temperatures [Belitz and Wysokinski 1987)]. The spin-flip rate due to localized moments is expected to be proportianal to the number of localized moments $\approx \chi/\chi_{Curie} \approx T^{1-\alpha}$. The low temperature dephasing rate in Figure 10 is larger than the e-e scattering estimates and we believe that it is due to the spin-flip scattering. If this is the case we make note of the fact that $\hbar/\tau_s \approx k_B T$ for Si:P, but the spin-flip rate of compensated Ge:Sb is somewhat larger. This agrees with our earlier speculation that the compensated materials have more local moments, higher spin-flip rate and therefore belong to a different universality class than the uncompensated materials.

5. CONCLUSIONS

We have presented a review of recent experimental work and associated theoretical modelling of the metal-insulator transition in uncompensated Si:P and compensated Si:P;B. The puzzling difference in the critical conductivity exponent of these two groups of materials is found to be accompanied by differences in low temperature susceptibility, ESR linewidth and in spin-flip rate. The available data suggests that the disordered interacting electron gas has strong tails in the distribution (presumably due to disorder). Because of this, it behaves like a two fluid system with itinerant electrons and localized magnetic moments in just metallic phase. The exact mechanism for more local moments in compensated than in uncompensated materials is not compeletely clear, though it may be due to coulomb interaction effects (Bhatt 1988). However, we argue that this difference leads to different spin-flip rates and critical conductivity exponents, because the delocalized electrons undergo spin-flip scattering, which is proportional to the density of local moments. While the above scenario is qualitatively plausible more theoretical work is needed to explain the existence of localized magnetic moments in a dirty metal and to estimate their interaction with the delocalized electrons. Experimentally it would be nice to check how universal the existence of localized moments is in other nonmagnetic dirty metals like amorphous alloys.

6. ACKNOWLEDGEMENTS

Portions of this research was done in collaboration with various persons. We are especially indebted to Subir Sachdev, John Graebner, Mark Hirsch and Don Holcomb for their help during this work. One of us (R.N.B) would like to thank Aspen Center of Physics for the hospitality while this manuscript was typed.

REFERENCES

Abrahams E, Anderson P W, Licciardello D C and Ramakrishnan T V 1979 Phys. Rev. Lett. *42* 673
Alloul H and Dellouve P 1987 Phys. Rev. Lett. *59* 578
Andres K, Bhatt R N, Goalwin P, Rice T M and Walstedt R E 1981 Phys. Rev. *B24* 244
Belitz D and Wysokinski K I 1987 Phys. Rev. *B36* 9333
Bhatt R N and Lee P A 1982 Phys. Rev. Lett. *48* 344
Bhatt R N 1987 Physica *146B* 99
Bhatt R N 1988 to be published
Brinkman W F and Rice T M 1970 Phys. Rev. *B2* 4302

Castellani C, DiCastro C, Lee P A and Ma M 1984 Phys. Rev. *B30* 527; Castellani C, DiCastro C, Lee P A, Ma M, Sorella S and Tabet E 1984 Phys. Rev. *B30* 1596 and 1986 Phys.Rev. *B33* 6169
Castellani C, Kotliar G and Lee P A 1987 Phys. Rev. Lett. *59* 323
Finkelstein A M 1983 Zh. Eksp. Teor. Fiz. *84* 168 [Sov. Phys. JETP *57* 97 (1983)] and 1984 Z. Phys. *B56* 189
Finkelstein A M 1987 JETP Lett. *46* 513 [Pis'ma Zh. Eksp. Teor. Fiz. *46* 407 (1987)]
Gan Z Z and Lee P A 1986 Phys. Rev. *B33* 3595
Hertel G, Bishop D J, Spencer E G, Rowell J M and Dynes R C 1983 Phys. Rev. Lett. *50* 743
Hirsch M J and Paalanen M A 1988a Bull. Am. Phys. Soc. *33* 385
Hirsch M J, Thomanschefsky U and Holcomb D F 1988b Phys. Rev. *B37* 8257
Hoch M J R and Holcomb D F 1988 submitted to Phys. Rev. *B*
Ikehata S and Kobayashi S 1985 Solid State Commun. *56* 607
Kawabata A 1980 J. Phys. Soc. Jpn *49* 375, J. Phys. Soc. Jpn *49* 628 and Solid State Commun. *34* 431
Kobayashi N, Ikehata S, Kobayshi S and Sasaki W 1977 Solid State Commun. *24* 67; 1979 ibid *32* 1147; Thomas G A, Ootuka Y, Kobayashi S and Sasaki W 1981 Phys. Rev. *B24* 4886
Korringa J 1950 Physica (Utrecht) *16* 601
Marko J R, Harrison J P and Quirt J P 1974 Phys. Rev. *B10* 2448
Milligan R F and Thomas G A 1985 Am. Rev. Phys. Chem. *36* 139
Murayama C T, Clark W G and Sanny J 1984 Phys. Rev. *B29* 6063
Ootuka Y, Matsuoka H and Kobayashi S 1987 Disordered Semiconductors eds M A Kastner, G A Thomas and S C Ovshinsky (NY: Plenum Press)
Paalanen M A, Sachdev S, Bhatt R N and Ruckenstein A E 1986a Phys. Rev. Lett. *57* 2061
Paalanen M A, Sachdev S and Bhatt R N 1986b Proc. 18th Int. Conf. Physics of Semiconductors ed O Engstrom (Singapore: World Scientific) pp 1249-52
Paalanen M A, Graebner J E, Bhatt R N and Sachdev S 1988a Phys. Rev. Lett. *61* 597
Paalanen M A, Bhatt R N and Sachdev S 1988b, to be published
Quirt J P and Marko J R 1971 Phys. Rev. Lett. *26* 318 and 1973 Phys. Rev. *B7* 3842
Sachdev S and Bhatt R N 1986 Phys. Rev. *B34* 4898
Sachdev S 1986 Phys. Rev. *B34* 6049 and 1987 Phys. Rev. *B35* 7558
Sarachick M P, Roy A, Turner M, Levy M, He D, Isaacs L L and Bhatt R N 1986 Phys. Rev. *B34* 387
Schmid A 1974 Z. Phys. *271* 251
Sundfors R K and Holcomb D F 1964 Phys. Rev. *136* 810
Thomas G A, Ootuka Y, Katsumoto S, Kobayashi S and Sasaki W 1982 Phys. Rev. *B25* 4288
Ue H and Maekawa S 1971 Phys. Rev. *B3* 4232
Washburn S, Webb R A, von Molnar S, Holtzberg F, Flouquet J and Remenyi G 1985 Phys. Rev. *B30* 6224
Zadrozny J, Sachdev S and Paalanen M A 1987 Bull. Am. Phys. Soc. *32* 815

Excited states of deep defects in silicon

M Kleverman, J Olajos, G Grossmann, and H G Grimmeiss

Department of Solid State Physics, University of Lund, Box 118, S-221 00 Lund, Sweden

ABSTRACT: We briefly review line spectra of deep centers in silicon which are due to excitations from the ground state to shallow excited states. The well known optical properties of shallow centers and chalcogen double donors enable us to interpret the transition-metal spectra when taking into account the deep character of their ground states. The donor or acceptor character of the centers is revealed by the phonons involved in the phonon-assisted Fano resonances. Many-particle effects and spin-orbit interactions are observed and the interaction between the $P_{1/2}$ acceptor states and the valence-band continuum, finally, leads to typical Fano line shapes for deep acceptors.

1. INTRODUCTION

Shallow centers in silicon are among the most investigated systems in solid state physics. Already a long time ago their technological importance initiated a great interest both experimentally and theoretically, and the study of their well resolved line spectra is one of the main reasons for our detailed understanding of their electronic structure. Deep centers also have great technological importance but detailed information on their electronic structure has to a large extent only been available for their ground state which is probed in electron paramagnetic resonance (EPR) studies. What are the general differences between shallow and deep centers? The electronic structure of shallow centers is essentially determined by the band structure of the host crystal and their wavefunctions are delocalized in $\mathbf{r}$-space. For deep centers, many parts of the Brillouin zone contribute to the wavefunction and, hence, deep levels tend to be localized. Deep centers with their in general larger binding energies usually show a much stronger electron-phonon coupling and their optical spectra exhibit properties which may significantly differ from those of shallow centers e.g. in their intensity distribution.

Shallow centers have bound states due to the screened Coulomb potential and the electronic structure is well described by the effective-mass approximation (EMA) leading to a series of hydrogenic bound states. Extensions of EMA, especially for the isocoric impurities, account to a large degree for the difference in the observed binding energy and that predicted by EMA. The deeper a states penetrates the central cell, the more will it be shifted by the localized potential of the impurity. For deep centers, this localized potential is even more important in determining the properties of the ground state. Nevertheless, the Coulomb potential of a charged impurity core can still bind less localized states in accordance with shallow centers. The first observation of shallow excited states of a deep impurity was reported as early as 1962 by Krag and Zeiger (1962) who assigned the observed line spectra of sulfur-related centers in silicon to excitations from the deep ground state to excited states due to the Coulomb tail of the impurity potential. Surprisingly, this observation remained more or less unnoticed by the physics community interested in deep levels until the late 1970´s when detailed optical studies were performed on the chalcogen double donors in silicon.

The majority of the line spectra presented in this paper are dominated by excitations to p like excited states. The chemical shift of the shallow states may be quite marked for the s states, since

only these have a non-vanishing amplitude at the impurity site. On the other hand, the p like excited states, whose amplitude vanishes at the impurity site and who thus closely follow EMA, are virtually independent of the impurity studied. This ensures that all centers with strong transitions to p like final states will have closely similar spectra, both concerning the line spacing and, in most cases, the relative intensities. However, as the oscillator strength is sensitive to the ground state properties, we will find some centers to show totally different relative intensities.

In this paper we will focus our attention on some recent results on optical properties of deep centers in silicon in general and on transition metals (TM) in particular with emphasis on the differences between deep and shallow-center spectra. It will be shown that the study of phonon-assisted Fano resonances is a valuable tool for the identification of the character of the center. Furthermore, the important task of chemical identification of impurities is stressed and exemplified by combining results from photo-EPR and high resolution optical methods. Finally, the importance of studying optical spectra by different methods is emphasized since deeper insight in the optical processes can be gained.

2. EXPERIMENTAL DETAILS

The majority of the spectra presented in this paper are obtained from absorption or photo-thermal ionization spectroscopy (PTIS). As light source, a Bomem DA3.01 Fourier-transform spectrometer was employed, equipped with suitable beamsplitters, detectors, and optical filters. In PTIS, the current measured in the external circuit is due to optical excitation of a charge carrier from the ground state to an excited state from which it is subsequently thermally emitted to the corresponding band. Of course, direct ionization of defects also contributes to the current. The Boltzmann factor favors those excited states which have small binding energy and, consequently, shallower excited states are observed with higher relative intensity in PTIS than when observed in transmission. The dominant part of the PTIS line spectra, presented in this paper, are superimposed on a much more intense background signal due to e.g. direct excitation of other impurities having smaller binding energy. For a given photon energy all possible excitations compete for the photons available and, hence, less efficient processes, with regard to free charge-carrier production, may decrease the total current. As a consequence, line spectra may be observed as negative dips on a positive background. In border-line cases, due to the Boltzmann factor, deeper excited states may appear as negative dips whereas shallower states are observed as positive peaks. It is an advantage to be able to use both methods, PTIS and absorption, since i) the direct ionization is observed more strongly in PTIS and the ionization cross section may be determined, ii) phonon-assisted Fano resonances have different line shapes in absorption and in PTIS, and iii) in transmission, a weak absorption signal may be difficult to observe against the intense background, whereas in PTIS only the free charge carriers produced by the photon absorption contribute to the net current. The doping procedures of the samples are described in the references.

Line spectra of the transition metal impurities in silicon are observed, both in absorption and PTIS, superimposed on a high intensity background signal and a correction procedure is employed to increase the transparency of the TM line spectra (for details see Kleverman et al. 1987a, b).

3. SHALLOW CENTER SPECTRA

Silicon has six equivalent conduction-band minima along [001] and the wavefunctions of a shallow donor are therefore written as symmetry-adapted linear combinations

$$\Psi(\mathbf{r}) = \sum_{j=1}^{6} \alpha_j F_j(\mathbf{r}) \phi_j(\mathbf{r}) \qquad (1)$$

where the coefficients α_j are determined by the point-group symmetry of the defect, F_j is the

envelope function and ϕ_j is the Bloch wave from the jth minimum. The valley-orbit interaction splits the six-fold degenerate ns states in tetrahedral symmetry (T_d) into ns(A_1), ns(E), and ns(T_2) states of one-, two-, and three-fold degeneracy, respectively. The ground state of shallow donors has experimentally been verified to be 1s(A_1) except for the interstitial Li donor which has a 1s(E+T_2) ground state (Ramdas and Rodriguez 1981) and for the thermal donors for which Stavola et al. (1985) showed that one of the T_2 components is the ground state. The np states have a magnetic quantum number m=±1,0 and, due to the anisotropic effective-mass tensor in silicon, the $np_\pm$ and np_o states are not degenerate. For shallow centers, the absorption spectra are dominated by 1s(A_1) - $np_\pm$ and 1s(A_1) - np_o transitions which are the only ones allowed by EMA. An example of shallow-donor absorption spectra is that of phosphorus presented in Figure 1. A series of sharp lines is seen which is almost identical for all shallow donors. However, the spectra appear at different photon energies due to the chemical shift of the ground state.

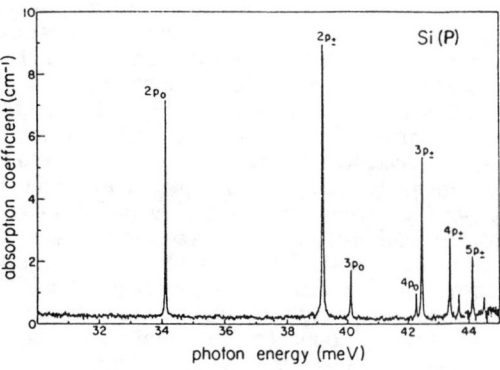

Figure 1. Absorption spectrum of phosphorus donors in silicon from Jagannath et al. (1981).

The shallow acceptor states are derived from the top of the valence band which in silicon is split into an upper $P_{3/2}$ and a lower $P_{1/2}$ band by spin-orbit interaction. Due to the screened Coulomb potential, two series of excited states are formed, mainly derived from the $P_{3/2}$ and $P_{1/2}$ valence bands, respectively. Within EMA, excitations are allowed only from the s like ground state to p like excited states. The absorption spectrum of boron is shown in Figure 2. Both $P_{3/2}$ and $P_{1/2}$ lines are detected. The shallow acceptor ground state transforms as Γ_8 of the T_d double group and is four-fold degenerate. The orbital degeneracy makes the shallow acceptor sensitive to Jahn-Teller distortions. However, since the ground state wavefunction is delocalized, the coupling to the ligands' motion is expected to be small for all group III acceptors except those with larger chemical shifts such as In and Tl. Ultrasonic attenuation experiments on indium doped silicon indeed show that In has an excited state 4.1 meV above the ground state (Schad and Lassmann 1976) which is interpreted in terms of a tunneling state due to the dynamical Jahn-Teller effect.

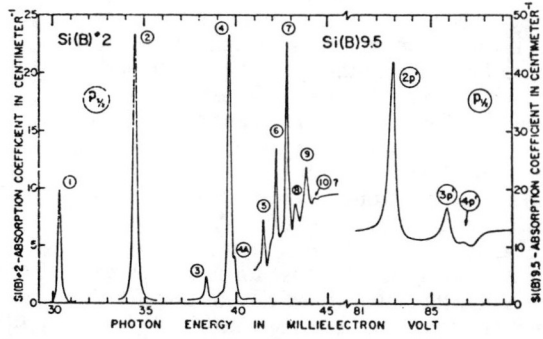

Figure 2. Absorption spectrum of boron acceptors in silicon from Onton et al. (1967a).

In silicon, direct electron-phonon interactions as e.g. phonon replicas are not observed in connection with shallow donor or acceptor spectra and only rarely in deep-level spectra. However, a higher order process involving optical phonons has been observed for both shallow and deep centers in silicon. The excitation of a bound charge carrier from the ground state into the continuum may interact resonantly with the transition to a shallow excited state accompanied by emission of an optical phonon. Selection rules restrict the phonons involved to a very narrow en-

ergy range, and the resulting interaction between a quasi-discrete state and a continuum, similar to the Fano process studied in atomic spectra, is described as a phonon-assisted Fano resonance (Watkins and Fowler 1977, Janzén et al. 1985). The study of these resonances has proven to be a valuable tool for investigating symmetry-forbidden transitions not visible in ordinary absorption spectra. In this way, for example, the binding energy of some ns(E) lines have been determined for the chalcogen donors. Different phonons are involved in the phonon-assisted Fano resonances of acceptors and donors. For donors, both the g LO and f TO intervalley phonons have experimentally been verified to participate in the resonances whereas for acceptors only the Γ zone-center phonon has been observed. This makes it possible to decide whether a spectrum is due to excitation of a donor or an acceptor by analyzing its phonon-assisted Fano resonances. In addition to these resonances another resonant process in Si is responsible for broadening of shallow level lines and is due to the coincidence of the excitation energy of the bound-to-bound transition with an optical bulk phonon energy and has e.g. been observed in Bi (Onton et al. 1967b).

4. DEEP DONOR SPECTRA

Electric-dipole selection rules for shallow donors follow from EMA and from symmetry. EMA requires the envelope functions of the initial and final states to have opposite parity. This makes the $1s(A_1) - 1s(T_2)$ transition forbidden, although, for a T_d center, it is symmetry allowed. However, it is expected that the EMA selection rules should weaken as the deep-level character of the ground state increases. Figure 3 shows the line spectrum of the neutral substitutional Se double donor (Se^0). At a first glance the spectrum is very similar to those of group V donors. However, here the $1s(A_1) - 1s(T_2)$ transition gives rise to one of the most intense lines due to the ground state deviation from EMA. No $1s(A_1) - ns(E)$ transitions are revealed. The strong resonance structures observed in the continuum part of the spectrum have been shown to be due to phonon-assisted Fano resonances from which the binding energy of 1s(E) and 2s(A$_1$) may be inferred. In the case of the chalcogen double donors, the intensity of the f TO resonances exceeds those for the g LO resonances and this intensity rule holds also for the transition-metal donors investigated.

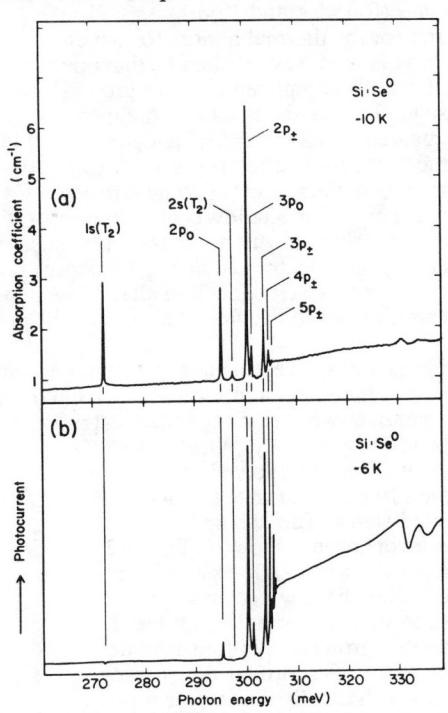

Figure 3. Absorption and PTIS spectra of neutral selenium double donors in silicon from Janzén et al. (1984).

The binding energies of the p states are very close to those predicted by EMA for a hydrogenic center. It should be remembered that a double donor is the solid state analogue of the helium atom and that all two-electron configurations should by exchange interaction give rise to spin singlet and spin triplet terms, unless prohibited by the Pauli exclusion principle, as for the ground state. However, as the agreement between hydrogenic EMA and the experimental binding energies of p states are close to perfect, the inner electron, which is left in a 1s(A$_1$) orbital, must completely shield the outer one in a p state. The exchange interaction is strongest for those states that have the largest overlap with the inner electron. At somewhat lower energy than the 1s(T$_2$) line, an additional line is revealed (Figure 4) which cannot be accounted for by using hydrogenic EMA. It has been shown that this is the spin triplet partner to the strong 1s(T$_2$) singlet line. As long as spin is a good quantum number, transitions from the singlet ground state to the triplet 1s(T$_2$) state, i.e. transitions from the $1s(A_1)^2$ ground state configuration to the spin-triplet terms of the 1s(A$_1$)1s(T$_2$) configuration, are forbidden. Spin-orbit interaction splits the triplet T$_2$ state, as for the 1s2p triplet term of helium, into levels with J = 2, 1, and 0 as well as mixes the J=1 level with

the singlet T_2 level. Since the oscillator strength is carried by the spin-singlet state, the mixing makes the spin triplet T_2 line optically visible. This interpretation was first reported by Bergman et al. (1986) who observed an avoided crossing behavior between spin singlet and triplet absorption lines under uniaxial-stress. The assignment was later confirmed in Zeeman spectroscopy by Peale et al. (1988).

In silicon, interstitial impurities from group II of the periodic table are expected to form double donors. Two EMA-donor series have in fact been observed in Mg doped silicon due to bound-to-bound transitions at the neutral and singly positively charged center, Mg° and Mg⁺, with ionization energies of about 107 meV and 254 meV, respectively. Both these values are considerably larger than expected from hydrogenic EMA, which may at first sight indicate a chemical shift of the ground state due to a localized potential. However, a more appropriate estimate of the Mg donor ionization energy is obtained within EMA by treating Mg as a helium like defect, i.e. scaling atomic helium to yield 56 meV for Mg°. Numerical calculations indicate a rather weaker central cell potential than for the chalcogens which would let Mg retain a stronger helium like character. This may also explain why no EMA forbidden $1s(A_1)$-$1s(T_2)$ transitions have been observed.

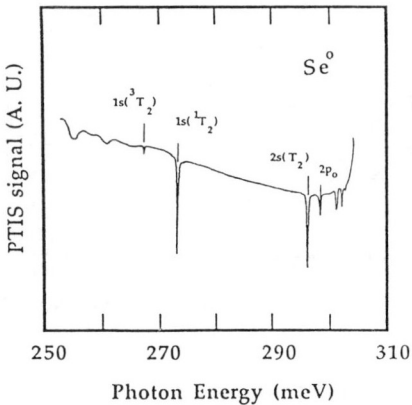

Figure 4. PTIS spectrum of neutral selenium double donors in silicon.

For donors in T_d symmetry, p_o states transform as A_1+E+T_2, $p_\pm$ as $2T_1+2T_2$, and the electric-dipole operator transforms as T_2. Transitions are thus only allowed between the $1s(A_1)$ ground state and excited states transforming as T_2. The $1s(A_1) - 2p_\pm$ line for the Mg⁺ center is split into two components indicating a small split of the two T_2 components. We have performed high resolution studies of the Mg⁺ donor which show that also the $3p_\pm$ line is split although less than the $2p_\pm$ line (see Figure 5). The observed splitting for $2p_\pm$ is 1.87 cm⁻¹ and that for $3p_\pm$ 0.65 cm⁻¹ at about 10 K. The ratio $\Delta E(2p_\pm)/\Delta E(3p_\pm)$ is then 2.88. Assuming that this splitting is due to penetration of the central cell by the p electrons, the observed splitting can be estimated by simply assuming that it is proportional to the probability of finding the electron inside a sphere of radius R. It is found numerically, using simple hydrogenic wavefunctions, that the ratio is 2.85 when $R \rightarrow 0$ and that it only slowly increases for increasing R. It remains uncertain what causes this splitting in detail. No such splitting has been observed for the chalcogen donors with their strong central cell potential, on the other hand, the Mg impurity may be less effectively screened at its interstitial site (Weiler et al. 1984).

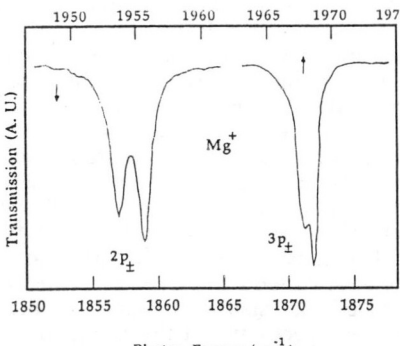

Figure 5. Transmission spectrum of Mg⁺ showing the splitting of the $p_\pm$ lines.

Both the chalcogen and Mg donors show line spectra that are very similar to those for group V donors. It thus becomes very simple to identify these impurities as donors. On the other hand, one sometimes finds impurity spectra which neither resemble ordinary donor spectra nor acceptor spectra. An example is found in Si doped with the 4d metal Ag. A series of lines is observed (see Figure 6) at an energy which is close to the binding energy previously deduced for a Ag related level, Ag(D), using junction space charge techniques (JSCT) (Baber et al. 1987). The Ag(D) center has been interpreted as a single substitutional Ag donor. The understanding of phonon-assisted Fano reso-

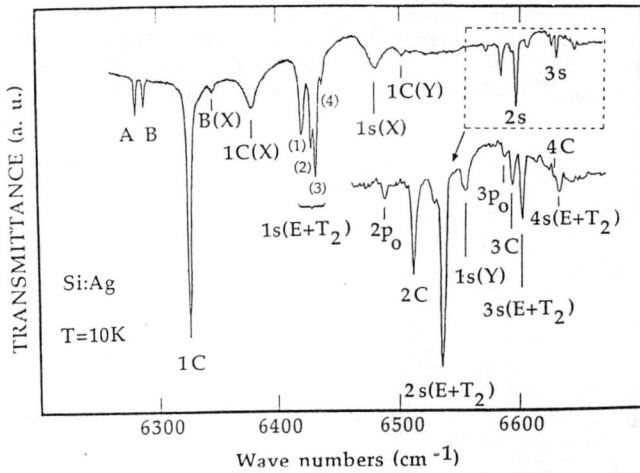

Figure 6. Line spectrum of the Ag donor showing strong s lines.

nances makes it possible to identify the donor character of this center. It has already been mentioned that different phonons participate in the phonon-assisted Fano resonances of donors and acceptors. It hereby becomes possible to distinguish between spectra of donors and acceptors. However, the g LO and the Γ phonon energies are nearly the same and since the resonances are rather broad features both g LO and f TO resonances have to be identified for a donor assignment. In the case of Ag, both the g LO and f TO phonons were in fact observed and thus the Ag spectrum is due to donor transitions. A careful analysis shows that the EMA forbidden s - s transitions dominate the spectrum which may be understood if one assumes that the ground state is very spread in k space due to its deep level character. Further details about the Ag spectrum are given in another paper at this conference.

3d-metal impurities in Si have rendered a great deal of interest both theoretically (Zunger 1987) and experimentally but line spectra have only recently been revealed. The main experimental techniques for studying 3d impurities have been JSCT and EPR. JSCT have high sensitivity but rather poor optical resolution and EPR only probes the ground state. By using transmission spectroscopy and PTIS on iron doped silicon Olajos et al. (1988) observed a line spectrum close to 6300 cm^{-1} which is presented in Figure 7. As seen in the PTIS raw spectrum, the ionization limit can be estimated to be about 6300 cm^{-1} which is close to the value generally accepted for the neutral interstitial iron impurity, Fe_i^0. A closer examination of both the raw and corrected PTIS spectrum strongly suggests that at least two equivalent series of lines are recorded since two similar spectra, somewhat displaced in transition energy, can be distinguished. Above the ionization limit phonon-assisted Fano resonances are clearly revealed and establish a connection between the line spectrum and the direct ionization signal. Other lines were also detected in the samples and complementary experimental techniques have to be employed in order to identify the constituents of the center. Therefore, photo-EPR measurements were carried out on the same samples. In EPR, both tetrahedral Fe_i^0 and Fe_i^+ centers were observed, and the Fe_i^0 electron photo-ionization cross section was determined by observing the

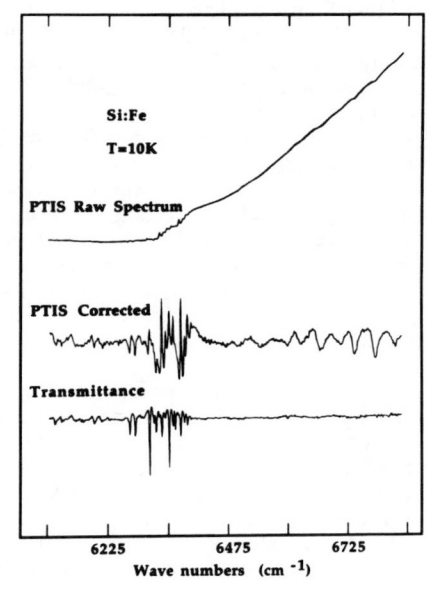

Figure 7. Line spectra of Si:Fe.

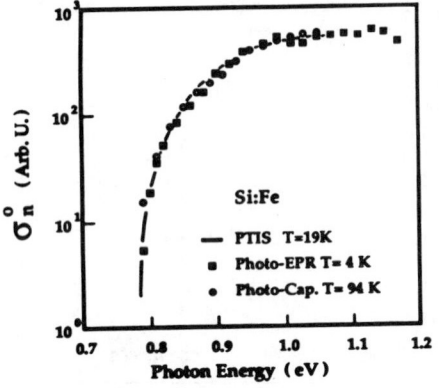

Figure 8. Electron photoionization cross section of Si: Fe_i^0 measured by PTIS, photo-EPR, and JSCT (photo-capacitance).

production rate of Fe_i^+. A comparison of the photo-EPR result with that obtained from PTIS is presented in Figure 8 and clearly shows that the line spectrum is indeed due to excitation of the Fe_i^0 center. In order to further confirm this identification, an isocronal annealing experiment was carried out and it was found that the annealing behavior was identical for the Fe_i^0 EPR signal and for the optical line spectrum. The annealing behavior also show that the other lines observed in Fe doped silicon are due to other centers.

All results so far show that the line spectrum is due to excitations of the Fe_i^0 donor. What is the origin of these lines? Are they due to intra-d transitions or excitations to shallow donor states? The fact that the lines appear close to the ionization limit provides strong evidence that the lines involve shallow excited states. Comparing the Fe_i^0 spectrum presented in Figure 9 with shallow donor spectra shows that more lines are revealed in the case of Fe_i^0. A closer inspection shows that the spectrum can be interpreted in terms of three superimposed shallow-donor spectra, having ionization energies of about 6361.5 cm^{-1}, 6403.6 cm^{-1}, and 6417.3 cm^{-1}. The spectrum is dominated by transitions to p states but the possibility of transitions to shallow s states has to be considered. It could be ruled out that three different Fe centers were causing the three line spectra since only one center was observed in EPR and since all line spectra annealed out together. Furthermore, the relative intensity of the spectra did not vary from sample to sample. It was also possible to rule out that the three spectra were caused by a split ground state since no thermalization was observed when the sample temperature was varied. The possibility that the extra lines were phonon replicas was excluded by the fact that the FWHM was the same for all three series. All together, these observations show that the cause of the three displaced line spectra is to be sought in the final states of the transitions. An electron excited from the deep ground state of Fe_i^0 enters a delocalized orbital where it only weakly interacts with the impurity core. The final state can therefore be viewed as a slightly perturbed Fe_i^+ center with a loosely bound, delocalized electron. It is known that the Fe_i^+ center has a $^4T_1$ ground state which is split by the spin-orbit interaction into three levels with different J values: 5/2, 3/2, and 1/2. It was therefore suggested by Olajos et al. (1988) that the origin of the three donor series of Fe_i^0 is the spin-orbit split Fe_i^+

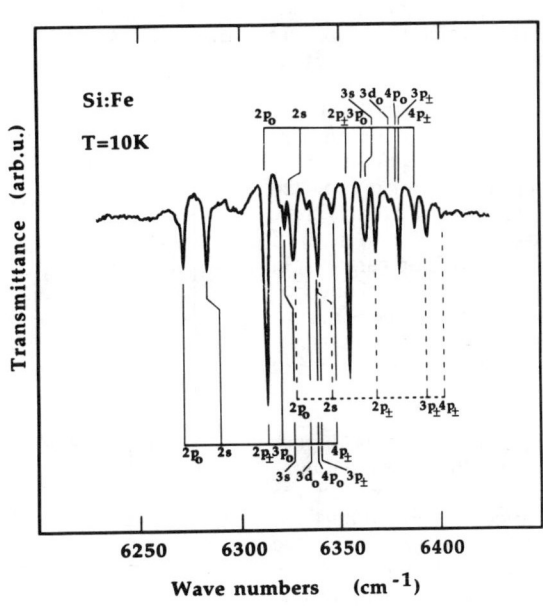

Figure 9. Identification of the three different EMA series contained in the Si:Fe_i^0 spectrum.

ground state. The observed spin-orbit splitting also is in agreement with the expected values.

5. TRANSITION METAL ACCEPTORS.

In silicon, only two TM acceptors have been studied in detail using high resolution optical techniques: Au and Pt (Armelles et al. 1985, Kleverman et al. 1987a, b, and 1988). From JSCT the acceptor has been found to lie 545 meV below the conduction band for Au (Lang et al. 1980) and 230 meV for Pt (Braun and Grimmeiss 1977). In Au doped silicon, several series of lines have been detected of which one is close to the binding energy previously found for the Au acceptor. Unfortunately, no EPR signal has hitherto been attributed to the Au center in silicon. The Au acceptor spectrum is presented in Figure 10 where a series of sharp lines is observed close to 4900 cm^{-1}. A comparison with shallow acceptor spectra

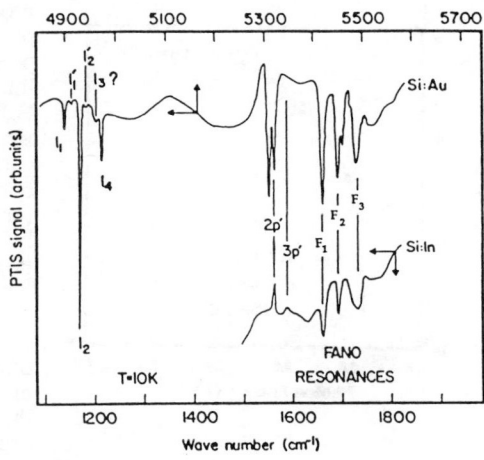

Figure 10. PTIS spectrum of the Au and In acceptors in silicon

shows that the Au spectrum is very similar and that it is caused by excitations from the deep ground state to shallow excited $P_{3/2}$ states. It is interesting to note that no strong lines are detected for Au that have not also been observed for group III acceptors, whereas for deep donors, the EMA forbidden $1s(A_1)$-$1s(T_2)$ line has an intensity comparable to that of the $2p_\pm$ line. At somewhat higher energies a doublet line is found close to 5300 cm^{-1}. A comparison with shallow acceptor spectra shows that these lines appear close to the expected position of the lowest p like $P_{1/2}$ line ($2p'$). It may be ruled out that the doublet is due to $2p'$ and $3p'$ since the delocalization of the $P_{1/2}$ states ensures that they have binding energies in close agreement with shallow acceptors. The doublet has therefore been attributed to excitations to the same final state as for the $2p'$ line of group III acceptors. It is known from uniaxial-stress experiments that the final state of the $2p'$ line transforms as Γ_6, an orbital singlet, and, hence, no splitting is expected for the shallow excited state. However, the initial state of the $2p'$ transition can be split or, as in the case of the Fe_i^0 donor, the final state of the impurity core. At present, the origin of the split doublet is not clear.

At photon energies higher than the $2p'$ lines phonon-assisted Fano resonances are observed. By subtracting the Γ phonon energy from the energy positions of the resonances it is established that they involve the shallow excited states responsible for the I_1, I_2, and I_4 lines. It is interesting to note that the F_2

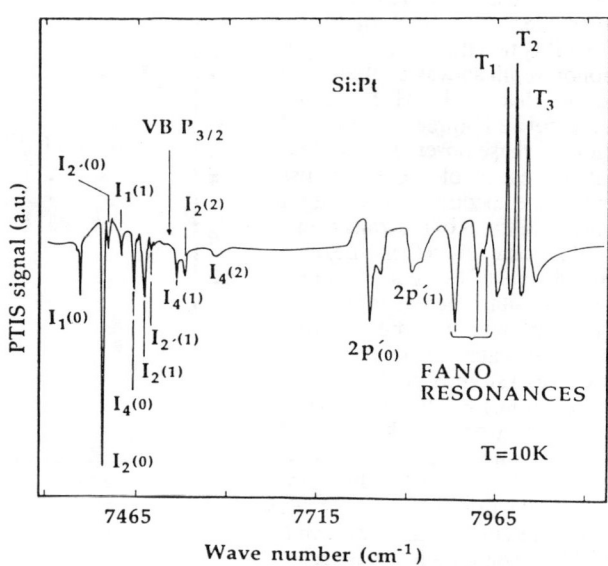

Figure 11. PTIS spectrum of the Pt acceptor in silicon.

resonance shows a doublet structure and that the energy separation is similar to that found for the 2p´ doublet. A closer inspection of the Au PTIS spectrum shows that the I_2 line has a weak partner $I_{2'}$ at somewhat higher energy. The energy difference between I_2 and $I_{2'}$ is about the same as the splitting of the 2p´ line and of the F_2 Fano resonance.

PTIS and transmission measurements on Pt doped silicon show a line spectrum similar to that of the Au acceptor although the number of lines exceeds that for Au as seen in Figures 9 and 10. Moreover, the majority of the lines are observed as negative dips whereas the lines T_1, T_2, and $T_{2'}$ are observed as positive peaks. The relative intensities of all lines are the same for all samples irrespective of shallow dopant concentration and diffusion temperatures. This indicates that all lines in Figure 11 belong to the same Pt related center or that different centers have similar solubility in that temperature range where the diffusion was carried out. The group of lines at lower energy is similar to the Au $P_{3/2}$ line spectrum and Kleverman et al. (1988) showed that all lines may be accounted for by assuming three overlapping $P_{3/2}$ series: a zero-phonon series, as in the case of Au, and two phonon replicas. The phonon giving rise to the

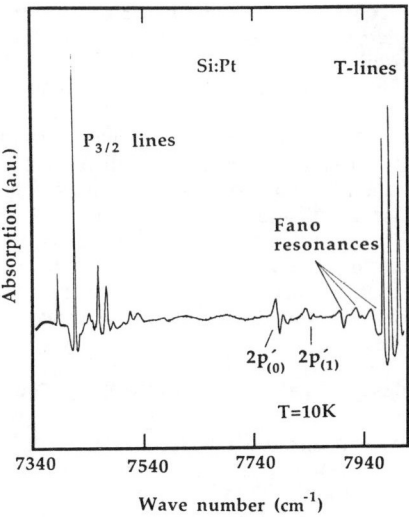

Figure 12. Absorption spectrum of the Pt acceptor in silicon showing the resonance line shapes of the 2p´ line.

replicas has an energy of about 57.5 cm^{-1} (7.3 meV) and it is resonant with the acoustic phonon band. The interaction with the band phonons is weak since the FWHM of the one-phonon replicas is only about 50% larger than for the corresponding no-phonon lines. Such pseudo-localized phonons have been observed previously for other defect systems in silicon, e.g. Cu-Cu pairs (Weber et al. 1982). Using the expression $(S^n/n!)\exp(-S)$ for the relative intensities of the nth phonon replica, the Huang-Rhys factor S is estimated to be about 0.4, such a small value indicating a weak electron-phonon coupling. The three T lines have so far not been identified.

As for the Au acceptor, the Pt 2p´ line is split into two components whereas the splitting energy is somewhat smaller but similar to the energy difference between the I_2 and $I_{2'}$. Since these splittings are impurity dependent they provide additional support for assuming a split ground state for the Au and Pt acceptors.

For both Au and Pt acceptors, the $P_{3/2}$ and the $P_{1/2}$ lines showed similar line shapes in PTIS. Surprisingly, a different result is obtained in absorption (Figure 12). The $P_{3/2}$ line shapes are similar to those observed in PTIS but the $P_{1/2}$ lines now show a resonance form like the phonon-assisted Fano resonances. Janzén et al. (1985) showed that for phonon-assisted Fano resonances different line shapes are expected in PTIS and absorption. This is due to the fact the all final states of the optical transitions contribute to the absorption signal whereas the PTIS spectrum only reflects those final states that actually contain free charge carriers. The line shape of the 2p´ line can be explained by a similar process. The shallow Γ_6 final state is only discrete in zeroth-order approximation and is resonant with the $P_{3/2}$ valence-band continuum, a situation very similar to the atomic Fano resonances. No Fano profiles have been revealed for shallow acceptors and the origin of the Au and Pt 2p´ line shape must be related to the

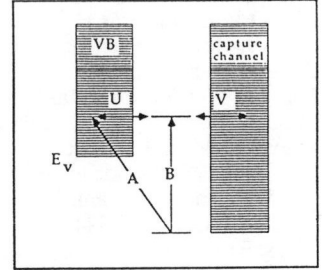

Figure 13. Schematic view of Fano resonances.

electronic properties of the deep ground state. As it is unlikely that the localized potential increases the coupling between the Γ_6 state and the valence band, it is rather due to the larger oscillator strength for direct ionization deep into the valence band that the Fano profile of the $2p'$ line is revealed for these deep acceptors. We model the $2p'$ transition in accord with Figure 13, where the interactions U and V describe the mixing between the discrete state and the valence-band continuum and the recapture to the ground state, respectively. A and B are the electric-dipole matrix elements for direct ionization and transition to the discrete state, respectively. When A is zero no direct transitions to the continuum are possible and the interaction U only gives rise to a life-time broadened discrete line. To obtain a Fano-resonance line shape, both the A and B excitations must be allowed. For shallow levels, the ground state has only a appreciable amplitude at $k = 0$ in k-space and since no-phonon lines imply $\Delta k = 0$, A is expected to be much smaller than B and no Fano line shape is observed. Deep levels, on the other hand, contain contributions from many parts of k-space, A is no longer negligible, and the Fano line shape is observed.

6. SUMMARY AND CONCLUSION

It has been shown that several deep donors and acceptors in Si have shallow excited states. The corresponding line spectra may be interpreted on the basis of the EMA if the analysis includes detailed considerations concerning e.g. electronic structure of initial and final states, electron-phonon coupling, resonant interaction with band continua, and many-particle effects. Furthermore, the wealth of information gained from these spectra, obtained from PTIS and transmission spectroscopy, brings new insight into the electronic structure of deep levels.

7. ACKNOWLEDGEMENT

The authors acknowledge financial support from the Swedish Natural Science Research Council

8. REFERENCES

Armelles G, Barrau J, Brosseau M, Pajot B, and Naud C 1985 Solid State Commun. **56** 303
Baber N, Grimmeiss H G, Kleverman M, Omling P, and Zafar N 1987
  J. Appl. Phys. **62** 2855 and references therein.
Bergman K, Grossmann G, Grimmeiss H G, and Stavola M 1986 Phys. Rev. Lett. **56** 2827
Braun S and Grimmeiss H G 1977 J. Appl. Phys. **48** 3883
Janzén E, Stedman R, Grossmann G, and Grimmeiss H G 1984 Phys. Rev. **B29** 1907
Janzén E, Grossmann G, Stedman R, and Grimmeiss H G 1985 Phys. Rev. **B31** 8000
Jagannath C, Grabowski Z W, and Ramdas A K 1981 Phys. Rev. **B23** 2082
Kleverman M, Olajos J, and Grimmeiss H G 1987a Phys. Rev. **B35** 4093
Kleverman M, Olajos J, Grossmann G, and Grimmeiss H G 1987b
  Mat. Res. Soc. Symp. Proc. **104** 141
Kleverman M, Olajos J, and Grimmeiss H G, 1988 Phys. Rev. **B37** 2613
Krag W E and Zeiger H J 1962 Phys. Rev. Lett. **8** 458
Lang D V, Grimmeiss H G, Meijer E, and Jaros M 1980 Phys. Rev. **B22** 3917
Olajos J, Bech Nielsen B, Kleverman M, Omling P, Emanuelsson P, and Grimmeiss H G
  1988 to be published
Onton A, Fisher P, and Ramdas A K 1967a Phys. Rev. **163** 686
Onton A, Fisher P, and Ramdas A K 1967b Phys. Rev. Lett. **19** 781
Peale R E, Muro K, Sievers A J, and Ham F S 1988 (to be published in Phys. Rev. B)
Ramdas A K and Rodriguez S 1981 Rep. Prog. Phys. **44** 1297
Schad Hp and Lassmann K 1976 Phys. Lett. **56A** 409
Stavola M, Lee K M, Nabity J C, Freeland P E, and Kimerling L C 1985
  Phys. Rev. Lett. **54** 2639
Watkins G D and Fowler W D 1977 Phys. Rev. **B16** 4524
Weber J, Bauch H, and Sauer R 1982 Phys. Rev. **B25** 7678
Weiler H, Meskini N, Hanke W, and Altarelli M 1984 Phys. Rev. **B30** 2266 and ref. therein
Zunger A 1987 Solid State Phys. **39** 275

Zeeman splitting of double-donor spin-triplet levels in silicon

R.E.Peale, R.M.Hart and A.J.Sievers
Laboratory of Atomic and Solid State Physics, Cornell University, Ithaca N.Y.14853.2501
F.S.Ham
Department of Physics and Sherman Fairchild Laboratory, Lehigh University, Bethlehem, PA 18015

ABSTRACT: Observation of the Zeeman effect confirms the identification of spin-triplet terms for double donors in silicon. The theory of Landé g-factors fits the data well when the orbital magnetic moment of the $1s(A_1)1s(T_2)$ configuration is zero as predicted by effective-mass theory. The spin-orbit interaction parameter is determined from the non-linear part of this splitting, and the Zeeman splitting of the singly-ionized double-donor, spin-orbit-split $^2T_2$ line is studied in order to confirm the validity of this method.

1. INTRODUCTION

New infra-red absorption lines, attributed to spin-forbidden transitions to spin-triplet states of the $1s(A_1)1s(T_2)$ configuration of the double donors Se^0 and Te^0 in silicon, have been observed (Bergman et al. 1986, 1988) in experiments employing uniaxial stress to tune these lines and thereby enhance their intensities. In previous work (Peale et al. 1988) we have observed the Zeeman effect of the corresponding lines in zero stress in fields up to 4 T, and from this Zeeman splitting we have confirmed the identification of these lines with the spin-triplet states. We have shown that these lines result from transitions to components of the $J=1$ spin-orbit level of the $^3T_2$ term and that the Zeeman splitting of this level is described well by effective-mass theory.

The purpose of the present paper is to report the results of new experiments that extend the earlier Zeeman measurements on Si:Se^0 and Si:Te^0 to fields of 9 T in order to provide a more exacting test of the theoretical model. The lines are not linear in the field but instead show a curvature which we attribute to magnetic coupling between the $J=0,1,2$ spin-orbit levels of the $^3T_2$ term and which we use to obtain the value of the spin-orbit parameter λ that gives the separation of these levels. We have been unable to check the validity of our procedure by direct observation of the transitions to the $J=0$ and $J=2$ levels, since these transitions are strictly forbidden at low fields. We have therefore applied the same theoretical procedure to interpret the Zeeman splitting of the $1s(A_1) \rightarrow 1s(T_2)$ transition of the singly ionized donors S^+ and Se^+ in silicon, the spin-orbit splitting of which is directly observed. The Zeeman splitting of S^+ and Se^+ has not been reported previously and is described in this paper. The strength of the spin-orbit coupling obtained from our results for the $^3T_2$ term of Se^0 and Te^0

© 1989 IOP Publishing Ltd

is found to be larger than that obtained by Bergman et al. (1986) from the interaction of the $^3T_2$ and $^1T_2$ terms.

2. THEORY

The theory of the spin-orbit and Zeeman splitting of the $^3T_2$ term of the $1s(A_1)1s(T_2)$ configuration has been developed (Peale et al 1988) using the analogy of a ^3P term of a free atom and the formalism of the theory of the Landé g-factor. Only the results are given here.

Within the $^3T_2$ term, the spin-orbit interaction is described by the parameter λ. As in the Landé interval rule for the relative energies of the spin-orbit levels of a ^3P term there are three levels with $J=0$, 1 and 2 and energies -2λ, $-\lambda$ and $+\lambda$, respectively. The Zeeman energies are found by solving the secular equation for each value M_J of the component of the total angular momentum J along the field direction. The relative energies of the two states with $M_J = +1$ are given exactly by

$$E_{+1} = \tfrac{1}{2}(g_S + g_L)\mu_B B \pm \left[\lambda^2 + \tfrac{1}{4}(g_S - g_L)^2 (\mu_B B)^2\right]^{1/2}, \qquad (1)$$

and those for $M_J = -1$ by

$$E_{-1} = -\tfrac{1}{2}(g_S + g_L)\mu_B B \pm \left[\lambda^2 + \tfrac{1}{4}(g_S - g_L)^2 (\mu_B B)^2\right]^{1/2}. \qquad (2)$$

The lower sign in both Eqs. (1) and (2) corresponds to the states originating in the $J = 1$ level, the only level of $^3T_2$ to which optical excitations from the $^1A_1$ ground state are allowed in low fields. States with $M_J=0$ are given by the three roots of the equation

$$E^3 + 2\lambda E^2 - [\lambda^2 + (g_S - g_L)^2 (\mu_B B)^2] E - 2\lambda^3 = 0. \qquad (3)$$

If ξ denotes the one-electron spin-orbit parameter in the $1s(T_2)$ state, we should have

$$\lambda = \xi/2. \qquad (4)$$

For the singly-ionized double donor, the Zeeman splitting of the $1s(A_1) \rightarrow 1s(T_2)$ transition $[^2A_1 \rightarrow {}^2T_2]$ is found by the same method. The $^2T_2$ term comprises two spin-orbit levels Γ_7 and Γ_8, with $J = 1/2$ and $3/2$, respectively. The Γ_7 line splits into two lines for Faraday geometry in a magnetic field, with transition energies given by

$$E_{\pm 1/2} = -\frac{\lambda}{4} \pm \tfrac{1}{2}(g_S+g_L)\mu_B B - \tfrac{3}{4}\lambda\left[1 \pm \frac{4(g_S-g_L)}{9\lambda}\mu_B B + \frac{4(g_S-g_L)^2}{9\lambda^2}\mu_B^2 B^2\right]^{1/2}, \qquad (5)$$

where the notation $E_{\pm 1/2}$ indicates the M_J value of the final state. The Γ_8 line has in general four components in Faraday geometry, given by

$$E_{\pm 1/2} = -\frac{\lambda}{4} \pm \tfrac{1}{2}(g_S+g_L)\mu_B B + \tfrac{3}{4}\lambda\left[1 \pm \frac{4(g_S-g_L)}{9\lambda}\mu_B B + \frac{4(g_S-g_L)^2}{9\lambda^2}\mu_B^2 B^2\right]^{1/2}, \qquad (6)$$

$$E_{\pm 3/2} = \lambda/2 \pm g_L \mu_B B. \qquad (7)$$

We note that the $\pm 3/2$ transitions coincide if the orbital g-factor g_L is zero as predicted by

Shallow Impurities in Si and Ge

effective-mass theory. In Eqs. (5)-(7) we have $\lambda=\xi$, and we assume that to a sufficient approximation we may take the spin g-factor g_S to be the same in the initial and final states.

3. MEASUREMENTS

All spectra were taken with an IBM Fourier transform infrared interferometer with the sample immersed in pumped liquid helium at a temperature of 1.7K. A wire-grid polarizer was used to polarize the beam when necessary. Magnetic fields were applied by placing the sample in the bore of 9T solenoid. Voigt geometry was achieved by placing a reflection device inside the magnet bore.

The center frequencies vs. magnetic field of the $^3T_2$ (J=1) spin-triplet Zeeman components for Si:Se^0 are plotted in Figure 1a, and the same for Si:Te^0 appears in Figure 1b. The data for the component of the split absorption line with π polarization are denoted by triangles and those for the σ polarized components are plotted as circles. We also plot $E_{\pm 1}$ from Eqs. (1) and (2), and the solution E_0 to Eq. (3) corresponding to J=1. For Figure 1a, Eq. (3) was solved numerically for discrete values of the magnetic field B. For Figure 1b, E_0 was assumed to have a series form and the first three terms were found. For all curves in Figs. 1a and 1b, we took $g_L = 0$ and used the experimentally determined spin-orbit parameter λ extracted from the curvature of the data (described below) and appearing in the Table.

We obtain the spectroscopic splitting factor $g_J = (g_S+g_L)/2$ for the J=1 level from the linear part of the Zeeman effect by fitting $(E_{+1} - E_{-1})$

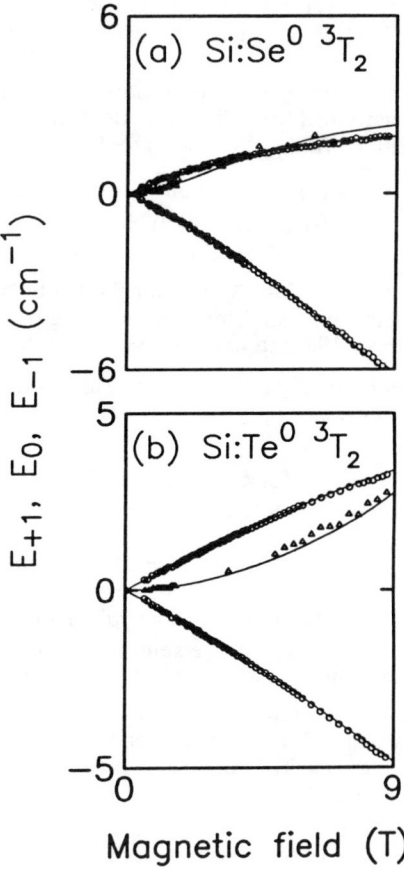

Figure 1. Theory curves (solid lines) and center frequencies of the π (Δ data) and σ (O data) Zeeman components vs. magnetic field for the J=1 spin-orbit component from the $^3T_2$ term of (a) Si:Se^0 and (b) Si:Te^0.

Table	Present work		Bergman et al (1986)	
	Se^0	Te^0	Se^0	Te^0
λ(cm^{-1})	2.93 ± 0.02	11.18 ± 0.06		
ξ(cm^{-1})	5.86	22.36	3.2	12.4
g_J (J=1)	0.983 ± 0.002	0.980 ± 0.001		

vs. magnetic field. The slope of the fit equals $2g_J\mu_B$. These data yield g_J-values of 0.983±0.002 (Se⁰) and 0.980±0.001 (Te⁰).

The spin-orbit interaction parameter is determined from the curvature of the Zeeman data by fitting ($E_{+1} + E_{-1}$) to the sum of Eqs. (1) and (2). We obtain the values for λ of 2.93 ± 0.02 cm^{-1} (Se⁰) and 11.18 ± 0.06 cm^{-1} (Te⁰). From Eq. (4) and our values for λ we determine values for the parameter ξ, which are also determined by Bergman *et al.* (1986, 1988) from the stress-tuned interaction between $^3T_2$ and $^1T_2$ terms. These values are given in the Table.

The values for ξ that we obtain for Se⁰ and Te⁰ from the spin-orbit splitting of the $^3T_2$ term are nearly twice as large as those obtained by Bergman *et al.* (1986, 1988) from the avoided crossing of the components of the $^1T_2$ and $^3T_2$ terms under applied stress. The simplest model of the spin-orbit coupling, on the other hand, predicts that these two determinations of ξ should agree. Further evidence that such a difference is real is found for Se⁰ and Te⁰ from the relative intensities of the transitions to the $^3T_2$ and $^1T_2$ states in zero stress (Peale *et al.* 1988) and, for Te⁰, from the size of the stress-induced splitting of the $^3T_2$ term at high stress for uniaxial compression along the [110] crystal axis (Bergman et al. 1988).

In Figure 2a we plot the Zeeman data for the Si:S$^+$ 1s $^2A_1 \rightarrow$ 1s $^2T_2$ (Γ_7, Γ_8) line along with Eqs. (5)-(7). Here g_L is assumed to be zero and the value of λ used was determined from the Γ_7, Γ_8 zero-field splitting to be 2.0 cm^{-1}. For g_S we used the value 2.0054 (Ludwig 1965). In Figure 2b we plot the same for Si:Se$^+$ using g_S=2.0057 (Grimmeiss *et al.* 1981) and λ=11.8 cm^{-1}. For Si:Se$^+$, the weak coupling between the Γ_7 and Γ_8 components and knowledge of g_S allow us to extract a value for g_L of 0.008±0.002, which

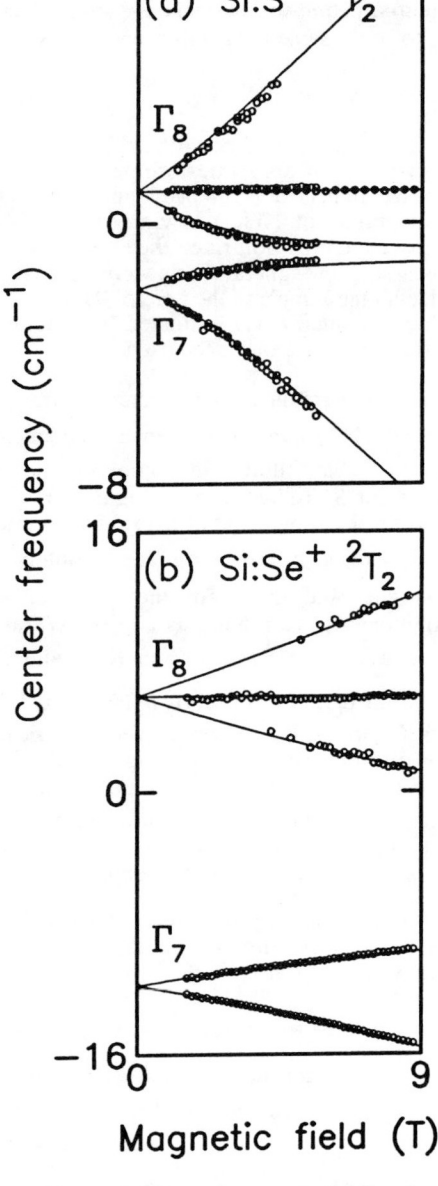

Figure 2. Theory curves (solid lines) and center frequencies (points) of the Zeeman-split $1s(^2A_1) \rightarrow 1s(^2T_2(\Gamma_7))$, and $1s(^2A_1) \rightarrow 1s(^2T_2(\Gamma_8))$ lines for (a) Si:S$^+$ and (b) Si:Se$^+$.

has been used in evaluation Eqs. (5), (6) and (7) for use in Figure 2b. The good agreement of data and theory evident in Figures 2a and 2b leads us to the conclusion that there is no additional source of nonlinearity in these data beyond that described by Eqs. (5) and (6). These data therefore confirm the validity of our method of extracting the spin-orbit interaction parameter λ within the neutral double-donor $^3T_2$ term from the curvature of its Zeeman splitting.

4. DISCUSSION

The new data reported in this paper more than double the range of magnetic field over which the Zeeman splitting of the transition into the $^3T_2$ term of the $1s(A_1)1s(T_2)$ configuration of Si:Se^0 and Si:Te^0 have been observed, as compared to our previous work (Peale et al. 1988). These data continue to support the identification of these lines (Bergman et al. 1986) with transitions into the spin-triplet states of the neutral double donor, and they increase the accuracy with which we are able to determine the value of the spin-orbit coupling parameter λ effective within the $^3T_2$ term. Indeed our new value of λ for Se^0 is nearly 6% smaller than that obtained in the earlier work, but it remains still nearly twice as large as that inferred from the strength of the spin-orbit coupling between the $^3T_2$ and $^1T_2$ terms as determined in the stress experiments of Bergman et al. (1986, 1988). A similar discrepancy occurs for Te^0 as well. The theory we have presented for the Zeeman splitting of the neutral donors gives an excellent fit with our value for λ to all the observed lines over the full range of field, as seen in Fig. 1. Despite the higher fields, however, we have not been able to observe transitions into any of the states derived from the $J=0$ and $J=2$ spin-orbit levels of $^3T_2$, which should borrow intensity from the $J=1$ states with increasing field.

Our new Zeeman spectra for the singly-ionized donors S$^+$ and Se$^+$, in which both the $J=1/2$ and $J=3/2$ levels are observed, also give an excellent fit to the theory, as seen from Fig. 2. The near-vanishing of the orbital g-factor $g_L \approx 0$ in these spectra shows that effective-mass theory remains an excellent approximation for this charge state of the donor as well, despite the increased binding energy of the $1s(T_2)$ state in the presence of the doubly charged core.

We conclude from the success of the theory in fitting the S$^+$ and Se$^+$ Zeeman data that our use of this model to obtain the value of λ for the neutral donor should be valid. We conclude that a real discrepancy occurs between the strength of the spin-orbit coupling inferred in this way and that obtained from the stress data of Bergman et al. (1986, 1988). This difference may reflect a difference in the $1s(T_2)$ wave function between the $^1T_2$ and $^3T_2$ states, as suggested by King and Van Vleck (1939) in accounting for similar, though smaller, discrepancies in the excitation spectra of atoms such as mercury, cadmium and zinc with the electronic ground-state configuration $(ns)^2$.

5. ACKNOWLEDGMENTS

We thank Dr. Peter Wagner and Dr. C. Holm of Heliotronic GmbH for kindly providing us with the tellerium-doped silicon samples used in this research, and K. Muro of Osaka University for his early guidance and for preparing most of the sulfur- and selenium-doped samples used in this work. The work by three of us (R.E.P., R.M.H. and A.J.S.) was supported by the U.S. National Science Foundation under Grant No. DMR-84-03597 and by the U.S. Army Research Office under Grant No. DAAL03-86-K0103. The portion of this research contributed by F.S.H. was supported by the U.S. Office of Naval Research (Electronics and Solid State Science Program) under Contract No. N00014-84K-0025.

6. REFERENCES

Bergman K, Grossmann G, Grimmeiss H G, and Stavola M 1986 *Phys. Rev. Lett.* **56** 2827

Bergman K, Grossmann G, Grimmeiss H G, Stavola M, Holm C and Wagner P 1988 *Phys. Rev. B* **37** 10738

Grimmeiss H G, Janzén E, Ennen E, Schirmer O, Schneider J, Wörner R, Holm C, Sirtl E, and Wagner P 1981 *Phys. Rev. B* **24** 4571

King G W and Van Vleck J H 1939 *Phys. Rev.* **56** 464

Ludwig G W 1965 *Phys. Rev.* **137** A1520

Peale R E, Muro K, Sievers A J and Ham F S 1988 *Phys. Rev. B* **37**, 10829

Piezospectroscopy of boron impurity in silicon

R A Lewis, P Fisher and N A McLean

Department of Physics, The University of Wollongong, Wollongong, NSW, 2500. Australia

ABSTRACT: The infrared absorption of boron impurity in silicon has been studied with and without applied uniaxial compression. Excellent agreement is found in comparing the results with the recent theoretical work of Buczko and of Binggeli and Baldereschi.

1. INTRODUCTION

Recently, important advances have been made in the study of shallow impurities in silicon and germanium. The technique of Freeth *et al* (1986; 1987) in which a uniaxial force and magnetic field are applied simultaneously (the piezo-Zeeman configuration) has proved to be a valuable way to obtain additional information and clarify the absorption spectra of group III acceptors in germanium. This experimental work is now complemented by the exact theoretical determination by Duff *et al* (1988) of the behaviour of the acceptor states under these dual perturbations; this work supersedes the approximations of Bir and Pikus (1974). Piezospectroscopic theory has been advanced by Buczko (1987) who has used wave functions of the type developed by Baldereschi and Lipari (1976) to calculate the behaviour of shallow acceptor states in silicon and germanium under applied uniaxial force; his calculations extend to substantial stresses and include interaction effects between states of the same symmetry. Buczko (1987) gives theoretical values for the deformation potential constants, values for the intensity parameters u and v of Rodriguez *et al* (1972) and the energies of a larger number of acceptor states than calculated by Baldereschi and Lipari (1976). Very recently, such calculations have been further refined by Binggeli and Baldereschi (1988) and extended to give the first detailed results for the absolute intensities of the lines in the Lyman series of shallow acceptors in germanium and silicon.

In view of the above, an experimental piezospectroscopic study has been made on boron in silicon to determine the nature of some of the higher lying states and to make a comparison with the new theoretical results, to re-examine and extend some of the earlier work of Chandrasekhar *et al* (1973) and to undertake piezo-Zeeman studies on the more dominant transitions. Some of the results of this investigation are presented, although only preliminary information is available on the piezo-Zeeman aspect.

2. EXPERIMENTAL

Almost all the spectroscopy to be reported has been carried out using a SPEX 1402 double monochromator equipped with gratings blazed at 30μm. The detector was a cooled germanium bolometer. A "wire" grid polariser on a polyethylene substrate was used to orient the electric field of the radiation, $\mathbf{E}$, either parallel ($E_{\parallel}$) or perpendicular ($E_{\perp}$) to the applied, compressive, uniaxial force, $\mathbf{F}$. The force is applied by calibrated lead weights.

© 1989 IOP Publishing Ltd

2.1 Unperturbed Spectrum

The unperturbed spectrum of part of the Lyman series of boron in silicon is shown in Figure 1; the rest of this spectrum, lines 1 - 4A, is well known (see Chandrasekhar *et al* (1973)). Many of the features shown in Figure 1 have been observed before but little attention has been given to them until now. In view of the richness of the spectrum in this region and the fact that the more prominent lines have already been assigned a number, it is somewhat cumbersome to use the technique of attaching a letter to an existing number to identify a previously unlabelled weaker feature. The authors suggest the **revised numbering system** used in Figure 1. The relationship between the old scheme and the new is given in Table 1, with the old labels shown in parentheses. The present measurements of the transition energies, $h\upsilon_x$, are also listed; here x = 1,2,3, ... Thus, for example, the spectral lines 4, 4B and 4A are renumbered as the lines 4, 5 and 6, respectively. Lines 8, 11 and 14 have been observed by Skolnick *et al* (1974) and Jagannath *et al* (1981), who also observed line 9.

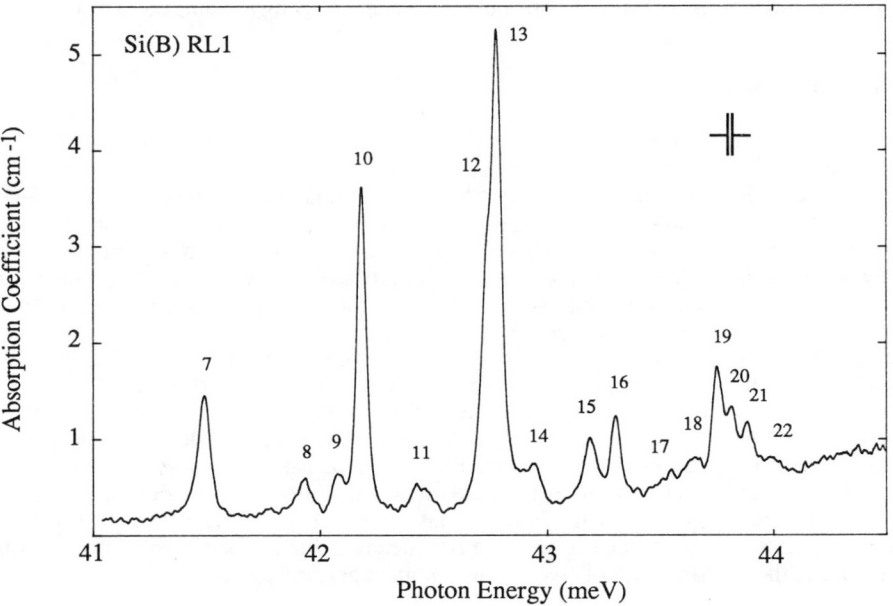

Figure 1. Absorption spectrum of the higher energy transitions of the Lyman series of boron in silicon. Room temperature resistivity is 20Ω-cm. Coolant is liquid helium.

Symmetry labels for the final state of some transitions are shown in Table 1. These follow Binggeli and Baldereschi (1988), with the conventional parity superscript added; Buczko (1987) labels fewer states and transposes $1\Gamma_6^-$ and $1\Gamma_7^-$. Also shown are the binding energies of each final state as given by these authors. As well, Table 1 contains the oscillator strength, f, of the transitions calculated by Binggeli and Baldereschi (1988) and the relative intensities deduced from Figure 1 by a least squares fit using Lorentzian profiles. Since numerical values have been given for only a limited number of the calculated binding energies and intensities, the present authors have extracted additional values from the graphed theoretical results. These are given in italics in Table 1. The experimental relative intensities are those obtained if line 7(5) is assigned the theoretical value of $f = 7.77 \times 10^{-4}$.

Table 1. Experimental transition and binding energies and relative intensities of spectral lines of boron in silicon. Theoretical binding energies and intensities of acceptors in silicon.

Line	$h\upsilon_x$ (meV)	Final State	$\varepsilon_b{}^a$ (meV)	$\varepsilon_b{}^b$ (meV)	$\varepsilon_b{}^b$ (meV)	ε_b(exp) (meV)	f(theory)^b	f(theory)^b	f(exp)
1(1)	30.37	$1\Gamma_8^-$	15.65	15.8	15.81	15.81	.0051	.0053	
2(2)	34.51	$2\Gamma_8^-$	11.47	11.7	11.72	11.50	.0283	.0283	
3(3)	38.38	$3\Gamma_8^-$	7.25	7.5	7.49	7.46	.0021	.0021	
4(4)	39.67	$1\Gamma_6^-$	6.15	6.1	6.17	6.12	.0100	.0100	
5(4B)		$1\Gamma_7^-$	5.91	6.1	6.07		.0092	.0092	
6(4A)	39.93	$4\Gamma_8^-$	5.61	6.0	6.00	5.85	.0033	.0033	
7(5)	41.47	$5\Gamma_8^-$	4.02	4.2	4.26	4.23	.0008	.0008	.0008
8(-)	41.92	$2\Gamma_6^-$	3.61	3.8	3.83	3.77	.0002	.0002	.0001
9(-)	42.07	$6\Gamma_8^-$	3.47	3.7	3.72	3.62	.0005	.0005	.0001
10(6)	42.17	$2\Gamma_7^-$	3.39	3.5	3.55	3.50	.0010	.0010	.0015
11(-)	42.43		3.07		3.29	3.24		.0003	.0001
12(-)	42.73		2.80		2.97	2.93		.0008	.0009
13(7)	42.77		2.74		2.91	2.88		.0015	.0021
14(-)	42.94				2.72	2.71		.0004	.0000
15(8)	43.18					2.45			.0003
16(-)	43.29					2.34			.0003
17(-)	43.49					2.13			
18(-)	43.63					1.98			
19(9)	43.75					1.86			.0003
20(-)	43.81					1.80			.0003
21(-)	43.88					1.72			.0001
22(-)	43.99					1.61			.0000

^{a}Buczko (1987). ^{b}Binggeli and Baldereschi (1988).

In order to compare the calculated energies with those of the observed transitions, we have considered the binding energy, ε_b, of the final states of the set of lines 1(1), 2(2), 3(3), 7(5) and 10(6). These span much of the spectrum and are transitions whose final states appear to be simple. When $h\upsilon_x$ vs ε_b was plotted for this set, using Binggeli and Baldereschi's (1988) values of ε_b, it was found that line 2(2) did not lie on the same straight line as the rest and so was discarded for this correlation. The result of a linear fit to the other four transitions gave $\varepsilon_b(\exp) = (47.49 \pm 0.15) - (1.043 \pm 0.004) h\upsilon_x$ (where $h\upsilon_x$ and ε_b are expressed in meV), from which the values of $\varepsilon_b(\exp)$ given in Table 1 were determined.

It is seen that the recent results of Binggeli and Baldereschi (1988), for both the energies and the intensities, are in very good agreement with experiment. Surprisingly, it was found that Buczko's (1987) value of ε_b for the $2\Gamma_8^-$ state (line 2(2)) lies closer to the above straight line than does that of Binggeli and Baldereschi (1988).

2.2 Piezospectroscopy

The behaviour of several spectral lines of boron in silicon with **F**∥<100> and <111> has been studied. The results for lines 1(1) and 2(2) reported previously by Chandrasekhar et al (1973) have been confirmed except that the present measurements produce lower deformation potential constants for the ground state and first two excited states. The origin of these differences is not clear. Table 2 summarises the present and previous experimental results and the calculated values of Buczko (1987). An example of the present data is shown

Table 2. Deformation potential constants for some of the states of boron in silicon.

State	b(eV)			d(eV)		
	This work	Experiment[a]	Theory[b]	This work	Experiment[a]	Theory[b]
$1\Gamma_8^+$	-1.38±0.04	-1.61±0.07	-1.43	-4.02±0.09	-4.50±0.15	-3.84
$1\Gamma_8^-$	-0.055±0.10	0.20±0.15	-0.026	-1.76±0.05	-2.31±0.25	-1.69
$2\Gamma_8^-$	-	1.61	1.13	1.9±0.6	2.64±0.25	1.87
$3\Gamma_8^-$	-0.13±0.02	-	0	-1.43±0.07	-	-1.56
$5\Gamma_8^-$	0.09±0.03	-	0	-0.79±0.09	-	-1.52

[a]Chandrasekhar *et al* (1973). [b]Buczko (1987).

in Figure 2 where the stress-splitting of the ground state ($1\Gamma_8^+$) and the first excited state ($1\Gamma_8^-$), as determined from line 1 with F||<100>, is given. Also shown is the splitting predicted by Buczko (1987). It should be noted that in the present measurements, the two stress-induced components allowed for $E_{||}$ were consistently the lowest and highest energy components which, from the selection rules (see Rodriguez *et al* (1972)), indicates that the signs of the deformation potential constants of $1\Gamma_8^-$ and $1\Gamma_8^+$ are the same, as predicted by Buczko (1987). The values of b and d for $1\Gamma_8^+$ have been determined from lines 1(1) and 3(3) for F||<100> and from lines 1(1), 2(2) and 3(3) for F||<111>, respectively. The present results and Buczko's suggest that the stress-isotropy of the ground state as discussed by Chandrasekhar *et al* (1973) does not occur.

Figures 3a and 3b show typical spectra of line 3(3) with F||<100> and F||<111>, respectively. For the latter direction (Figure 3b), the splitting pattern is characteristic of the behaviour of a line due to a $\Gamma_8 \rightarrow \Gamma_8$ transition with two allowed $E_{||}$ components and three allowed $E_\perp$ components, with one of the $E_{||}$ components completely polarised. For F||<100> (Figure 3a),

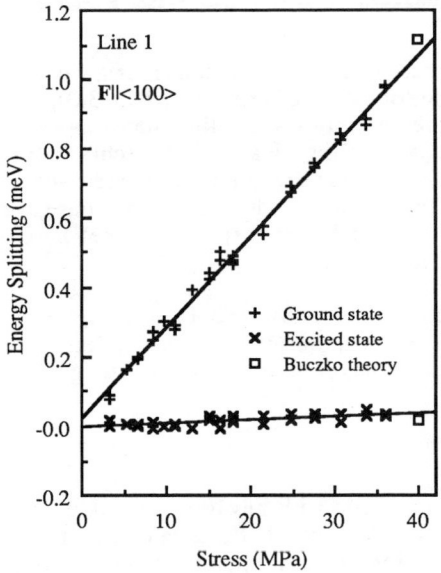

Figure 2. Splitting of the ground state and the excited state of line 1, F||<100>.

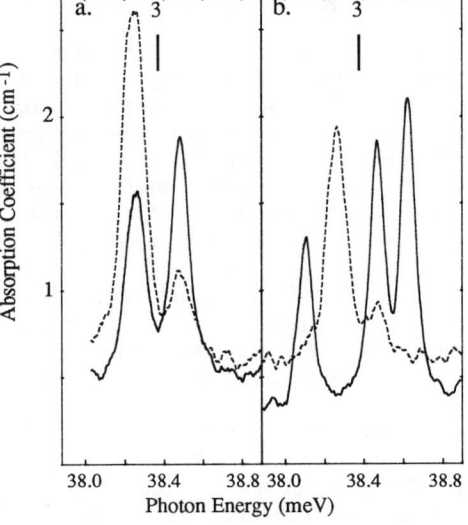

Figure 3. Behaviour of line 3 under uniaxial compressive forces. $E_{||}$ – – –; $E_\perp$ ——. a. F||<100>, 10.4MPa; b. F||<111>, 15.8MPa.

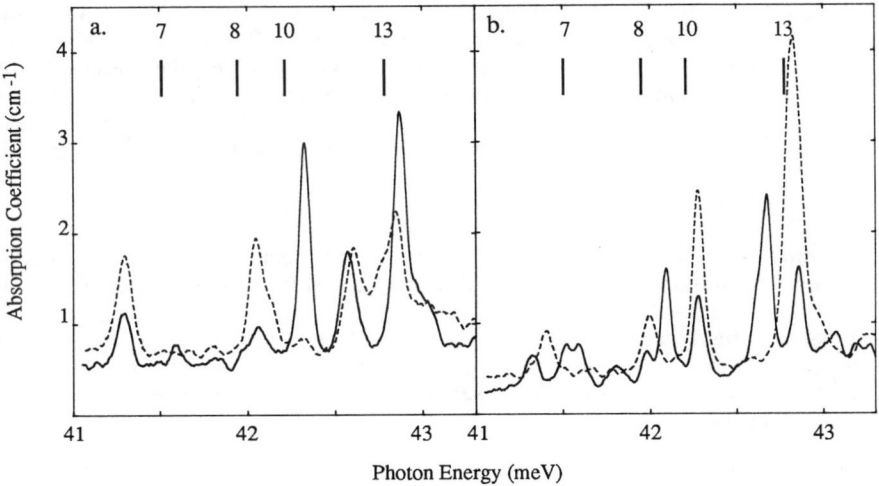

Figure 4. Behaviour of lines 7(5)-13(7) under uniaxial compressive forces. $E_\parallel$ – – –; $E_\perp$ ——. a. F||<100>, 11.0MPa ; b. F||<111>, 9.6MPa.

the result is again that of a $\Gamma_8 \to \Gamma_8$ transition but in this case for an excited state undergoing very little splitting. Figures 4a and 4b give spectra for lines 7(5) - 13(7) while Figures 5a and 5b show the stress dependence of some of these lines. In Figure 5, the data for some of the very weak transitions are not included. These results and the selection rules establish unambiguously that line 7(5) is due to a $\Gamma_8 \to \Gamma_8$ transition and lines 8 and 10(6) are due to

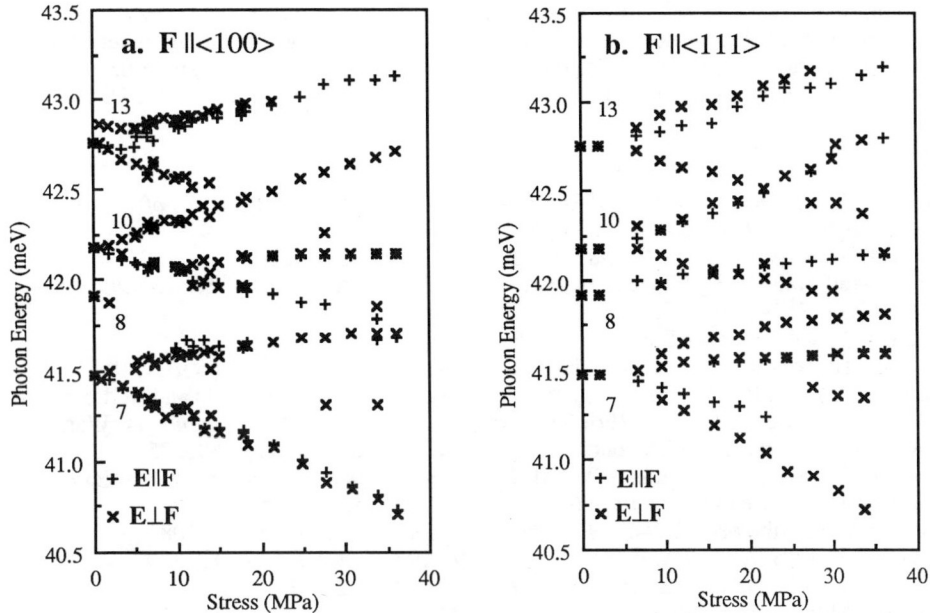

Figure 5. Stress dependence of the components of lines 7(5), 8, 10(6) and 13(7).

$\Gamma_8 \to \Gamma_7$ transitions while line 13(7) is either a $\Gamma_8 \to \Gamma_6$ or a $\Gamma_8 \to \Gamma_7$ transition. The distinction between these possibilities can only be made with F||<100> (Rodriguez et al (1972)). Line 13(7) lies close to line 12, the latter being a low energy shoulder; the present data is not sufficient to separate lines 12 and 13 under stress. The deformation potential constants for the excited states of lines 3(3) and 7(5) are given in Table 2. The predictions made by Buczko (1987) are in agreement with these results, as can be seen from the data in Table 2.

2.3 Piezo-Zeeman Effect

Piezo-Zeeman studies have been made of some of the spectral lines of boron in silicon. These measurements were made with a Fourier transform spectrometer using various piezo-Zeeman configurations and magnetic field strengths up to a maximum of 1.1T. It was found that this field was insufficient to produce well resolved Zeeman components of the stress-induced line components, in agreement with that predictable from the direct Zeeman studies of Merlet et al (1975). The observations are being extended to larger values of magnetic field strengths.

3. CONCLUSION

The excellent agreement between theory and the present quantitative piezospectroscopic results for lines 1(1), 2(2), 3(3), 7(5), 8, 10(6) and 13(7), the binding energies of the states and the relative intensities of the transitions illustrates the exceptional progress that the theoretical description of acceptors in silicon has made in approximately the last decade. The quality of the theoretical model is clearly demonstrated by its ability to explain such a rich and complex spectrum so well, particularly near the crowded end of the series.

ACKNOWLEDGEMENTS

The authors wish to thank Dr J A Campbell of the University of Canterbury for assistance with some of the measurements. The work was supported by the Australian Research Grants Council and the University of Wollongong Board of Research and Postgraduate Studies. The silicon samples were kindly provided by Professor A K Ramdas of Purdue University.

REFERENCES

Baldereschi A and Lipari N O 1976 *Proc. 13th Int. Conf. on the Physics of Semiconductors* (Rome:Marves) pp 595-8
Binggeli N and Baldereschi A 1988 *Solid State Commun.* (in press)
Bir G L and Pikus G E 1974 *Symmetry and Strain-Induced Effects in Semiconductors* (New York:Wiley) pp 438-43
Buczko R 1987 *Nuovo Cimento* **9D** 669
Chandrasekhar H R, Fisher P, Ramdas A K and Rodriguez S 1973 *Phys. Rev. B* **8** 3836
Duff K J, Vickers R E M, Fisher P, Freeth C A, Takacs G J, Warner A D and McLean N A 1988 (to be presented at the *19th Int. Conf. on the Physics of Semiconductors*, Warsaw)
Freeth C A, Fisher P and Simmonds P E 1986 *Solid State Commun.* **60** 175
Freeth C A, Fisher P and Vickers R E M 1987 *Proc. 18th Int. Conf. on the Physics of Semiconductors* ed O Engström (Singapore:World Scientific) pp 841-4
Jagannath C, Grabowski Z W and Ramdas A K 1981 *Phys. Rev. B* **23** 2082
Merlet F, Pajot B, Arcas Ph and Jean-Louis A M 1975 *Phys. Rev. B* **12** 3297
Rodriguez S, Fisher P and Barra F 1972 *Phys. Rev. B* **5** 2219
Skolnick M S, Eaves L, Stradling R A, Portal J C and Askenazy S 1974 *Solid State Commun.* **15** 1403

Shallow excited states of the deep silver donor in silicon

J Olajos, M Kleverman, B Bech Nielsen, and H G Grimmeiss

Department of Solid State Physics, University of Lund, Box 118, S-221 00 Lund, Sweden

ABSTRACT: Silver-doped silicon was investigated by transmission and photothermal ionization spectroscopy (PTIS). The spectra revealed a number of sharp lines between 6200 and 6700 cm^{-1}. The donor character of the center was determined from an analysis of the phonon-assisted Fano resonances and uniaxial stress measurements. By comparing spectra of phosphorus and neutral tellurium in silicon with the spectrum obtained for silver it is shown that the silver spectrum is dominated by excitations to s states.

1. INTRODUCTION

Silver gives rise to several deep centers in silicon. The two most commonly observed centers have binding energies of about $E_a = E_v + 0.54$ eV (4355 cm^{-1}), Ag(A), and $E_d = E_v + 0.34$ eV (2742 cm^{-1}), Ag(D) (Baber et al. 1987). One of them, Ag(A), is assumed to be an acceptor whereas the other Ag(D) is believed to be a donor. It is not known whether the Ag(A) and Ag(D) levels really belong to the same center. No evidence has yet been presented in the literature yet for the presumed acceptor- and donor-like behavior of Ag(A) and Ag(D) and no line spectrum has so far been reported for silver-doped silicon. Silver, as well as gold and copper, belongs to group 1B of the periodic table. The free atoms of this group have one ns electron outside a filled (n-1)d shell, n=4,5 and 6. It is therefore not unreasonable to assume that single substitutional Ag and Au centers in silicon should have similar electrical and optical properties, provided that the two centers have similar microscopic structures. No data have yet been published revealing the lattice position of the centers, but a comparison of the binding energies of the gold-related levels ($E_a = E_v + 0.63$ eV and $E_d = E_v + 0.35$ eV) (Braun and Grimmeiss 1974) with the values mentioned above for silver points to a similarity.

High-resolution spectroscopic studies in silicon have recently revealed detailed information on excited Coulomb states of both deep donor and acceptor centers (e.g. Janzén et al. 1984 and Armelles et al. 1985) in addition to the earlier investigated shallow centers. The deep acceptor spectra showed overall features that closely resembled those for shallow acceptors. The deep donor spectra, such as the spectra obtained from chalcogen double donors showed, in addition to the commonly observed excitations to p states, also excitations to valley-orbit split ns states. The transition 1s(A_1)-1s(T_2) is forbidden in the Effective Mass Theory (EMT) but is symmetry allowed. The intensity of this absorption line in the deep donor spectra is comparable to the strongest p lines, while it is absent in the spectra for shallow donors. Within EMT, optical transitions from the ground state to s-like states are not allowed. The deeper the ground state, the more this selection rule is expected to be relaxed. The selection rules will instead be governed by the symmetry of the center. There is reason to believe that for donors still deeper than the chalcogens, the intensity of the ns lines will increase and may even exceed the strength of the p lines. A survey of the EMT and symmetry-allowed transitions is given in Fig 1. The ground state has been assumed to be 1s(A_1) because most of the donors in silicon have a ground state of this symmetry. It should be noted that the 1s(A_1) state is the only state which has a non-vanishing amplitude at the impurity site.

© 1989 IOP Publishing Ltd

In this paper we report for the first time on the observation of sharp line spectra in Ag-doped silicon in the energy range between 6200 cm^{-1} and 6700 cm^{-1}. This energy is close to the

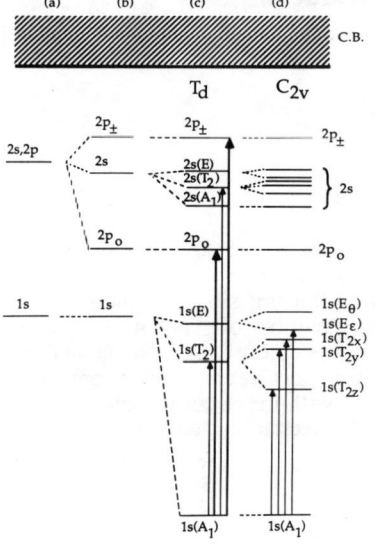

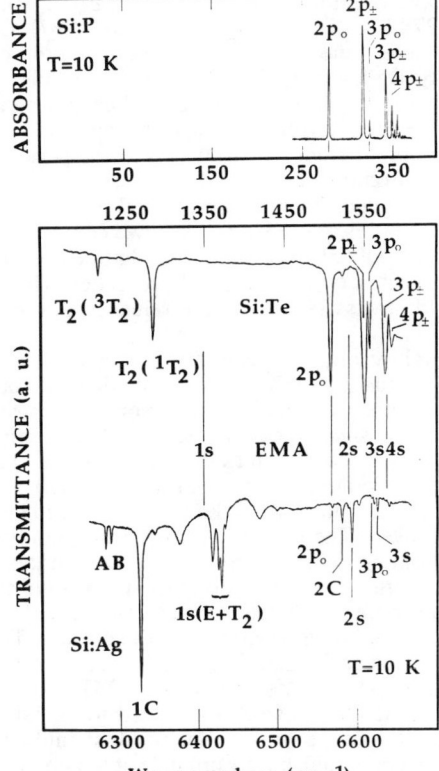

Fig. 2 Transmission spectrum of Ag doped silicon. For the assignment of the different lines see text.

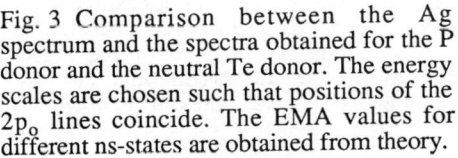

Fig. 1 Schematic figure showing the energy positions of ns- and np-states for different approximations and point group symmetries for donors in Si. The EMA and symmetry-allowed transitions are indicated by the thick and thin arrows respectively.
a) Isotropic effective mass.
b) Anisotropic effective mass
c) Valley-orbit interaction included for a center with T_d symmetry.
d) Valley-orbit interaction for center in C_{2v} symmetry.(only transitions to 1s states are indicated)

Fig. 3 Comparison between the Ag spectrum and the spectra obtained for the P donor and the neutral Te donor. The energy scales are chosen such that positions of the $2p_o$ lines coincide. The EMA values for different ns-states are obtained from theory.

binding energy previously reported for the Ag(D) center. The sharp lines are seen in transmission and PTIS. The excitation spectra differ from previously studied donor and acceptor spectra in the sense that the Ag-spectrum is dominated by excitations to ns states.

2. RESULTS AND DISCUSSION

A typical transmission spectrum, obtained from Ag-doped, silicon is shown in Figure 2. The sharp lines in the energy range 6200 cm^{-1}- 6700 cm^{-1} have the same relative intensities in all our samples. It is therefore believed that the lines originate from the same center. The reported energy difference between the conduction band and the Ag(D) center is 6690 cm^{-1}. This value is very close to the high-energy limit of the spectrum in Figure 2. Furthermore, the lines converge when going from lower to higher energy. It is therefore reasonable to assume that the spectrum observed is caused by the Ag(D) center and the lines are due to excitations from the ground state to delocalized excited states just below the conduction band caused by the long-range, screened, Coulomb potential. The spectrum does not, however, show close resemblance to previously studied donor spectra in silicon.

In Figure 3 the Ag(D) spectrum is compared with the donor spectra of phosphorus and neutral tellurium. The EMT values for s-states are also shown for comparision. The Ag spectrum is adjusted to these values in such a way that the energy position for the $2p_0$ lines coincide.(For information how the $2p_0$ line is identified in the the Ag spectrum, see below.) Considering the 1s EMT value as a rough estimate for the $1s(T_2)$ and $1s(E)$ states of the Ag(D) center, which is approximately correct for the chalcogen donors, it is interesting to note that a sharp line structure is observed very close to this energy. We therefore tentatively assign this structure as originating from $1s(A_1)$-$1s(E+T_2)$ transitions of the Ag(D) center, assuming the ground state to be $1s(A_1)$. The $1s(E+T_2)$ structure observed consists of four closely spaced lines indicating a lower symmetry than T_d for the Ag(D) center (Figure 1d) since in T_d symmetry only one line is expected.

Recent photo-luminescence studies of Au-doped silicon, by e.g. Thebault et al. (1984), revealed a sharp line structure which was attributed to the $1s(E+T_2)$-$1s(A_1)$ transition of the Au donor. This assignment was confirmed by Zeeman and uniaxial stress measurements. Corresponding transitions have been observed in transmission. In Figure 4 the Ag $1s(E+T_2)$ lines are compared with the corresponding lines of the Au donor. A remarkable similarity is observed which gives further support to our assignment of the structure at about 6430 cm^{-1} as

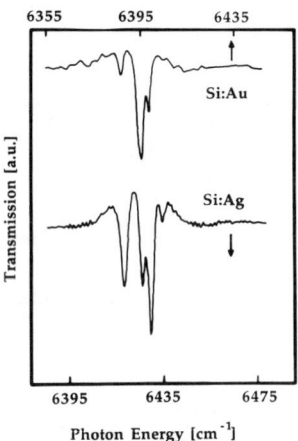

Fig. 4 Comparison of the $1s(A_1)$-$1s(E+T_2)$ lines for the Ag and Au donors.

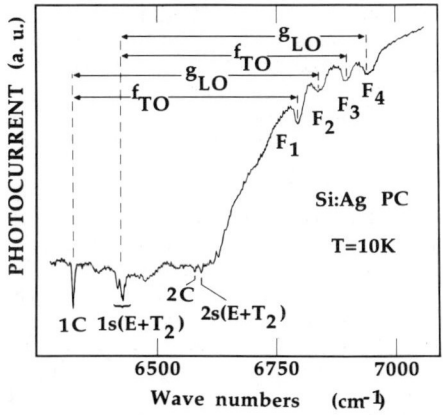

Fig. 5 The PTIS spectrum of the Ag donor showing bound-to-bound excitation lines as well as direct excitations to the electronic continuum and phonon-assisted Fano resonances. The identification of the phonons involved in the resonances is indicated.

lines due to $1s(A_1)$-$1s(E+T_2)$ transitions of the silver donor.

Additional information on the Ag spectrum was obtained from PTIS measurements, the results of which are presented in Figure 5. Besides the sharp lines caused by bound-to-bound transitions, the onset of the excitation to the conduction band (continuum part of the spectrum) is clearly seen. Transitions which ionize the center directly give a positive contribution to the photo-current whereas the bound-to-bound transitions are seen as dips, caused by the absorption of photons from other, more efficient excitation processes in the production of charge carriers, unless the carriers are are thermally excited into the conduction band. This is the reason why shallower excited states are seen as positive peaks. The threshold energy for the onset of transitions from the ground state directly into the conduction band is deduced to be about 6620 cm^{-1} (0.82 eV) (Figure 5). This result further supports our assignment of the spectrum, namely that it is caused by the Ag(D) center. The data also confirm that the lines close to the onset of the continuum part in Figure 5 are most probably caused by excited Coulomb states.

An interesting spectral feature was observed in the continuum part of the PTIS spectrum of Figure 5 which we identify as phonon-assisted Fano resonances. These structures appear at energies where the direct ionization is resonant with an optical transition to a discrete bound state followed by emission of certain bulk phonons. The phonons are characteristic for the type of center i.e. whether the center is a donor or an acceptor. For donors, in principle three different inter-valley phonons are allowed (Janzén et al. 1985), namely the f LA (387.9 cm^{-1}), g LO (515.4 cm^{-1}) and f TO (476.6 cm^{-1}) phonons, but it has been shown experimentally that only the g LO and f TO phonons couple strongly enough to be observed in phonon-assisted Fano resonances. Since in the case of acceptors, only the zone-center Γ phonon (519 cm^{-1}) has been found to couple, a study of Fano resonances offers the unique possibility of identifying the type of center. The Γ and the g LO phonons are very close in energy, and it is therefore difficult to determine the nature of the center based on these two phonons alone. However, if an f TO phonon is involved, it is highly probable that the center is a donor. In Figure 5, arrows have been inserted with the corresponding phonon energy which show that both the f TO and gLO phonons are involved.

Uniaxial stress measurements were carried out in order to gain further information on the center observed. Figure 6 shows the behavior of the strongest lines at about 6300 cm^{-1} - 6400 cm^{-1} (A, B, 1C and $1s(E+T_2)$ in Figure 2) under uniaxial stress, applied in the <100> direction. Without analyzing the data in detail, it is clearly seen that the stress response of the two most intense lines, namely 1C and $1s(E+T_2)$, is very similar, with respect to energy shifts and splittings per stress unit. While the total splitting at 200 MPa for each of the lines is approximately 150 cm^{-1}, the components moving towards lower energies have shifted 100 cm^{-1} and the others 50 cm^{-1}. This behavior, and the absolute value of the splitting per stress unit are characteristic for the six conduction band minima in silicon, these being oriented along <100> and equivalent directions. The shear deformation potential constant calculated for the Ag-related lines is in very good agreement with values obtained for other deep donors, such as the chalcogens in silicon. These

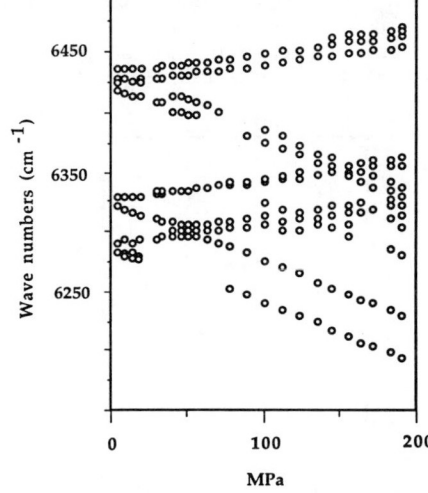

Fig.6. Uniaxial stress behavior of the two major absorption line complexes, 1C and $1s(E+T_2)$ for stress in [001] direction. The splitting pattern is characteristic for shallow donor states pinned to the conduction band.

results clearly show that the two major Ag lines, (1C and $1s(E+T_2)$) are of common origin and that they are due to transitions from a deep ground state to shallow excited states which are pinned to the conduction band minima.

The Ag spectrum, shown together with the Te and P spectra, in Figure 3 seems to be very different from the other two spectra. If lines within the 1s multiplet are excluded, the $2p_0$ line is known to be the one with the lowest energy for all donors in silicon. When analyzing the Ag spectrum it was assumed that the next line above the 1s multiplet was caused by the $2p_0$ state. Assuming that the $2p_0$ state shows only a minor central-cell correction, the binding energy of the Ag(D) center is calculated to be 826 meV (6661.2 cm^{-1}) by adding the theoretical EMT binding energy of $2p_0$ (92.68 cm^{-1}) to the measured transition energy (6568.53 cm^{-1}). Excluding the 1s multiplet, the $2p_0$ line of a donor spectrum in silicon is normally the one with the largest binding energy, while the $2p_\pm$ line has the highest intensity. No $2p_\pm$ line has been observed in the Ag spectrum. Shifting the energy scales of the spectra in Figure 3 such that the $2p_0$ lines coincide, it is seen that the relative intensity of the p lines in the Ag spectrum is very low compared with the other two spectra.

The energy positions of different ns states can be estimated from their EMT values which are indicated in Figure 3. The energy position of the 1s EMT state is seen just below the previously discussed $1s(E+T_2)$ multiplet, and the EMT positions of additional s lines are close to other groups in the Ag spectrum. It is reasonable to assume that each group of lines is caused by corresponding transitions from the assumed $1s(A_1)$ ground state to split ns states. We therefore denote these lines the $ns(E+T_2)$ lines.

A closer inspection of the tellurium spectrum (Figure 3) shows that there is a weak line just below the $1s(T_2)$ line which has previously been proven to be the spin-triplet state of the $1s(T_2)$ state. Neutral tellurium binds two electrons and via exchange interaction one spin singlet and one spin triplet series are formed (Bergman et al. 1986). The ground state of the single substitutional tellurium donor is a spin singlet, and electrical dipole transitions from the ground state to the triplet series are therefore, in principle, forbidden. In the case of tellurium, however, a strong spin-orbit interaction mixes the T_2 levels originating from the singlet T_2 ($T_2(^1T_2)$) and triplet T_2 ($T_2(^3T_2)$) states so that both the singlet and triplet T_2 levels become visible. Comparing the tellurium spectrum with the silver spectrum (Figure 3) and assuming that there is an exchange interaction between the loosely bound electron and the electrons of the donor core, it is tempting to suggest that the A, B, and 1C lines may originate from similar many-particle and spin-orbit effects.

Each $ns(E+T_2)$ line is accompanied by another line at somewhat lower energy (Figure 2) which we belive is the corresponding C line for various values of n. In the limit of a strong localization of the electrons in the defect core the exchange splitting is proportional to the probability of finding the delocalized excited electron at the impurity site. This probability is given by $|\Psi_n(0)|^2$ where $\Psi_n(0)$ is the value of the wavefunction of the excited electron at the origin. For hydrogenic ns states, $\Psi_n(0) \sim 1/n^{3/2}$ and, hence, the splitting is expected to scale approximately as 1: 1/8 : 1/27 for n = 1, 2, and 3. In the case of the deep Ag donor, the exchange interaction between an electron in the delocalized $1s(T_2)$ state and the strongly bound electron(s) is expected to follow roughly the same scaling. As already mentioned, the energy difference between the two dominating line structures of the Ag

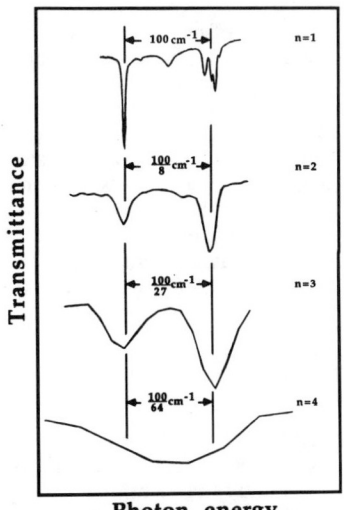

Fig. 7 The splitting of various s-states. The energy scales are multiplied with 1, 8, 27, and 64 for n=1, 2, 3, and 4 in order to demonstrate the scaling procedure discussed in the text.

spectrum, i.e. the C line and the 1s(E+T$_2$) multiplet, is about 100 cm^{-1}. If it is assumed that both structures involve 1s states split by an exchange interaction, then the corresponding splitting of the 2s, 3s, and 4s states should be about 13 cm^{-1}, 4 cm^{-1}, and 2 cm^{-1}, respectively. This is indeed observed, as is readily seen from Figure 2 where the energy differences between the ns(E+T$_2$) and nC lines are found to be about 12 cm^{-1}, 4 cm^{-1}, and 2 cm^{-1} for n=2, 3, and 4, respectively. To show this interesting agreement in more detail we have replotted the different ns multiplets in Figure 7. using energy scales which are scaled in accordance with the ns splitting i. e. 1 : 1/8 : 1/27 : 1/64.

To the best of our knowledge, the Ag(D) spectra presented here ar the first donor spectra observed in silicon which are not dominated by transitions to p states. Moreover, the 2p$_\pm$ line, normally observed as the strongest line for other donors with excited Coulomb states, is not observed at all. These particular properties of the silver spectrum may result from the very large binding energy of the silver donor, which is almost equal to the band gap. Excluding 1s states, the electronic structure of shallow or moderately deep donors, such as the chalcogens, is well described by EMT. In EMT, the wavefunction of a donor is obtained as the product of the envelope function and the Bloch waves close to the conduction band minima. With increasing strength of the central-cell potential the binding energy of the 1s (A$_1$) ground state increases, but simultaneously the probability of previously EMT-forbidden transitions taking place may increase. A consequence of the increased binding energy, and hence, degree of localization in real space, is that a larger part of the conduction band minima in **k** space contributes to the wavefunction of the 1s (A$_1$) ground state. Simultaneously, the overlap between the wavefunctions of the ground state and excited Coulomb states decreases and so do the electric-dipole matrix elements. Since no-phonon lines are observed as direct transitions in **k** space, this means that the relative intensity of the lines due to transitions to excited Coulomb states may decrease. This effect is strongest for p states since s states are more affected by the central-cell potential than p states and are therefore more delocalized in **k** space. One would therefore expect that for a donor with increasing deviation from EMT the *relative* intensity of transitions to s states would become so strong, compared with transitions to p states, that they would dominate the excitation spectrum.

The absence of np$_\pm$ lines may be explained by recalling that the effective mass is anisotropic in silicon and that the constant-energy surfaces are ellipsoids along equivalent <100> directions. The principal components of the effective-mass tensor are the longitudinal (m_l) and transverse (m_t) effective masses. Since $m_l > m_t$, the curvature of the conduction band is smaller along the <100> direction than in the perpendicular directions. A p$_0$ state is oriented in the <100> direction as m_l is, whereas a p$_\pm$ state is oriented along the m_t principal axis. This implies that it is easier for a p$_0$ state to localize in real space than for a p$_\pm$ state since more energy is needed for a p$_\pm$ state to include more Bloch waves into the wavefunction. For the same reason, the electric dipole matrix elements of P$_0$ states for a very deep donor are then expected to be larger than those of p$_\pm$ states, which means that excitations to p$_0$ states are still seen when excitations to p$_\pm$ states are already too weak to be observed.

REFERENCES

Armelles G, Barrau J, Brousseu M, Pajot B and Naud C 1985 *Solid St. Commun.* **56** 303
Baber N, Grimmeiss H G, Kleverman M, Omling P and Zafar N 1987 *J. Appl. Phys.* **62** 2855 and references therein
Bergman K, Grossmann G, Grimmeiss H G and Stavola M 1986 *Phys. Rev. Lett.* **56** 2827
Braun S and Grimmeiss H G 1974 *J. Appl. Phys.* **45** 2658
Janzén E, Stedman R, Grossmann G and Grimmeiss H G 1984 *Phys. Rev.* **B29** 1907
Janzén E, Grossmann G, Stedman R and Grimmeiss H G 1985 *Phys. Rev.* **B31** 8000
Thebault D, Barrau J, Armelles G, Lauret N, and Noguier J P 1984 *Phys.Stat.Sol.(b)* **125** 357

Shallow acceptor bound and resonant states in Si

Ryszard Buczko[a),b)] and Franco Bassani[a)]

[a)] Scuola Normale Superiore, 56100 Pisa, Italy
[b)] Institute of Physics, Polish Academy of Sciences, 02-668 Warsaw, Poland

ABSTRACT: The energy spectrum and the wave functions of shallow acceptors in Si are computed within the effective-mass approximation using the full cubic symmetry and including the spin-orbit split-off band Γ_7^+. In addition to the usual bound states, resonant states are found, which are strictly connected with the Γ_7^+ band. The relative intensities of the optical transitions from the ground to the odd parity states allow an interpretation of the experimental spectra in terms of bound and resonant states

1. INTRODUCTION.

The study of shallow acceptor states within the effective mass approximation has ben carried out by Baldereschi and Lipari (1973) in the spherical approximation. It has been shown (Baldereschi and Lipari 1974, Kogan and Polupanov 1978, Lipari et al 1980, Beinikhes et al 1985) that consideration of warping effects, split-off band contributions and dispersive screening lead to a much better agreement with experimental data. These consist principally of the optical absorption spectra observed in Si by Onton et al (1967) and Jagannath et al (1981), and in Ge by Butler and Fisher (1976), whose characteristics have been amply discussed (Ramdas and Rodriguez 1981, Altarelli and Bassani 1980). An analysis of the line intensities has been given for the bound states of Ge (Polupanov and Kogan 1979). More recently Binggeli and Baldereschi (1988) have shown a systematic dependence of the oscillator strengths on the parameter μ of the band structure, which gives a measure of the difference between the heavy and the light hole mass, and also a strong dependence on dispersive screening and nonspherical effects.

In this work we study the energies and oscillator strengths of shallow acceptor states in Si, using the full Luttinger matrix and spacial dependent impurity screening. The general features of our results are in agreement with those of Binggeli and Baldereschi (1988) for the bound states. In addition, we identify some states above the Γ_8^+ ionization limit which have the characteristics of resonant states associated with the split-off Γ_7^+ band. Their energy values however differ from those expected from the effective mass approximation on the Γ_7^+ band alone, because the admixture of the Γ_8^+ band is non-negligible. We also find that the hydrogenic degeneracies of the resonant states are removed, according to the requirements of cubic symmetry, the splitting being larger the more localized is the state. The oscillator strengths of the optical transitions to resonant states are smaller than those to the bound states, and are nearly hydrogen-like, as indicated by the experiments.

2. THEORY.

To compute the energies and the wave functions of shallow acceptor centers in Si, we follow the same method already adopted for studying the effects of uniaxial stress (Buczko 1987). We find the variational solutions of the matrix equation for the envelope functions with the full six by six Luttinger Hamiltonian. The impurity potential is given by

$$V(r) = - \frac{e^2}{\varepsilon_0 r} \left[1 + (\varepsilon_0 - 1) e^{-\alpha r} \right] \quad , \qquad (1)$$

with $\varepsilon_0 = 11.4$ and $\alpha = 1.01$ a.u. .
The six envelope functions (each applied to a Bloch function at the band extrema) are expressed as the products of radial functions and spherical harmonics.

$$\Phi_i^{(\nu)}(\vec{r}) = \sum f_{ilm}^{(\nu)}(r) \; | \; l,m > \qquad (2)$$

The radial functions are expanded in a series of exponential functions of the type described in the previous paper (Buczko 1987)To optimize the choice of our set, we adopt here a total of 20 exponential terms in the expansion of each radial function.

The values of the energies and the expressions for the envelope functions are obtained for a large number of even and odd states. The discreteness of the set implies that the continuum threshold of the valence band is not identified numerically, but is fixed as the conventional zero by the choice of the Hamiltonian. Many of the numerically obtained states lie above the continuum,and among them some are mostly composed of the split-off band Γ_7^+, with respect to which they are bound states. This is precisely the condition required for resonant states (Bassani et al 1974), and therefore we identify them with the resonant acceptor states experimentally observed (Onton et al 1967, Rome et al 1982, Fisher and Rome 1982).

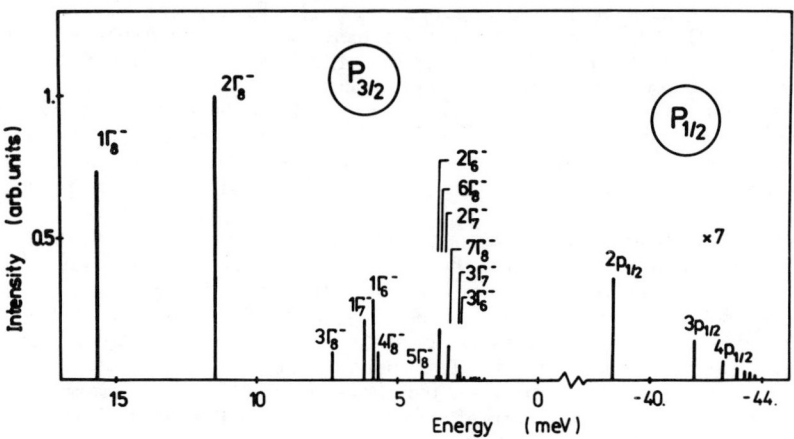

Fig.1 Oscillator strengths of the optical transitions between ground and excited states. The resonant state transitions ($P_{1/2}$ series) have been multiplied by 7 to allow visualization. The sum of the oscillator strengths of different components has been reported for every np state.

Shallow Impurities in Si and Ge 109

The optical transitions which take place between acceptor states can be computed from the energies and wavefunctions. Since the ground state is even, only the odd excited states need to be considered. From the impurity envelope functions $\Phi_{\nu,i}$ we can compute the oscillator strengths

$$f_{o\nu} = \frac{2m_0}{\gamma_1 \hbar^2}\left[E_\nu - E_0\right]\sum_{i,s,s'} |<\Phi_i^{(\nu,s)}|z|\Phi_i^{(0,s')}>|^2 \quad , \qquad (3)$$

where γ_1 is the first Luttinger parameter and the summation over s and s' is due to degeneracy of the states.

Table I. Energies and oscillator strengths of odd resonant states. The total and angular quantum numbers are given in the first and second columns. The crystal symmetry states are labeled in the third column and their computed energy values are given in the forth column. For comparison the Rydberg series and the experimental values are also given. The values of the oscillator strengths, relative to $2\Gamma_8^-$ are given in the last column.

n	l	Cubic symmetry	Energy [meV]	$\frac{R^*}{n^2}$ [meV]	E_n exp. [meV]	transition intensity
2	1	Γ_6^-	5.41	6.20	5.49	0.0444
		Γ_8^-	5.30			0.0079
3	1	Γ_6^-	2.44	2.76	2.48	0.0146
		Γ_8^-	2.41			0.0025
4	3	$2\Gamma_8^-+2\Gamma_7^-+\Gamma_6^-$	1.50±0.01	1.55	1.43	
	1	Γ_6^-	1.40			0.0071
		Γ_8^-	1.39			0.0022
5	3	$2\Gamma_8^-+2\Gamma_7^-+\Gamma_6^-$	0.96±0.01	0.99	0.93	
	1	Γ_6^-	0.90			0.0045
		Γ_8^-	0.89			0.0012
6	5	$4\Gamma_8^-+2\Gamma_6^-+\Gamma_7^-$	0.68	0.69		
	3	$2\Gamma_8^-+2\Gamma_7^-+\Gamma_6^-$	0.67			
	1	Γ_6^-	0.63			0.0039
		Γ_8^-	0.62			0.0008

Effective Rydberg R^*=24.81 meV.
(a) Fisher and Rome (1982)

3. RESULTS.

The computed values of the energy of the ground state and of the bound states are in agreement with the results computed by other authors. We report in Table I the results for the odd parity resonant states. It can be observed that their energy values are different from those which would be obtained from considering only the Γ_7^+ split-off band; furthermore, a small separation between different crystal symmetries results because the full Hamiltonian has been used. The resonant state energies are in good agreement with the experimental values. Our results are also in better agreement with experiments then those of Zwerdling et al (1959), who included a contribution from the Γ_8^+ band by perturbation theory.

As found by Binggeli and Baldereschi (1988), the absolute values of the oscillator strengths are very sensitive to the short part of the impurity potential (q dependence of dielectric screening), but the relative values are not. Therefore we report in fig.1 the relative values of the oscillator strengths for transitions to bound and resonant states. The strongest absorption is the $2\Gamma_8^-$, as observed in the experiments (Ramdas and Rodriguez 1981). The other absorption intensities are also in fair agreement with experiment. In particular, the intensities of the resonant states are nearly hydrogen-like, the strongest being the lowest one ($1\Gamma_6^-$). The splittings between the different symmetry states belonging to the same hydrogen-like resonant state are to small to have been observed

It is also of interest to observe that the probability of finding a hole at a given distance from the impurity site is quite different from what expected from the hydrogen-like spherical model. We can display this by plotting the moduli of the wave functions

$$| \Psi^{(\nu)} | = \sqrt{ \sum_i | \Phi_i^\nu(\bar{r}) \, \psi_i(0,\bar{r}) |^2 } \quad , \qquad (4)$$

where the radial dependence of the Bloch functions can be neglected.

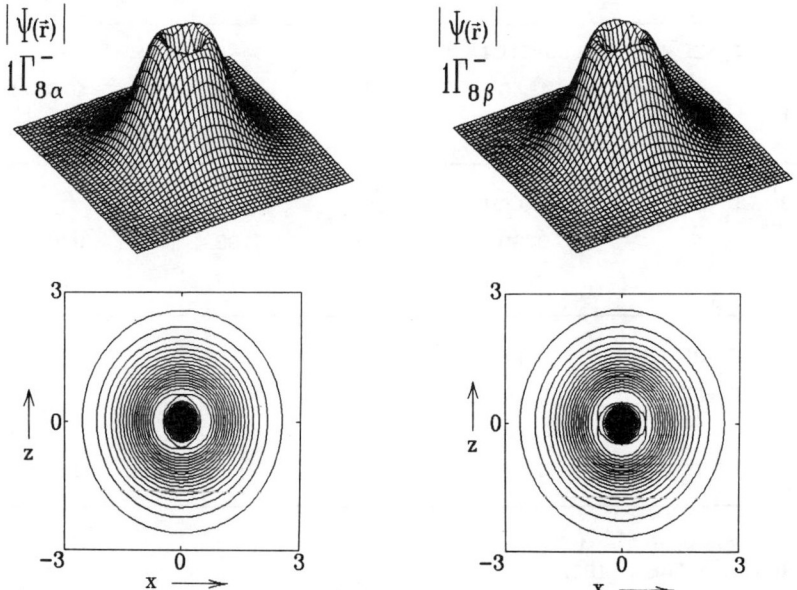

fig.2. The modules shapes of $1\Gamma_8^-$ excited state components. The distances are given in effective Bohr radius units.

We display in fig.2 the wave function modulus of the Γ_8^- bound state in the zx plane, z being the quantization axis. Since the Γ_8^- state is fourfold degenerate and transforms as the 3/2 spinor, we give in fig.2 the ±3/2 (denoted as α) and the ±1/2 (β) components. It can be noticed that the hole density is nearly spherical for both spinor states with a maximum at about one effective Bohr radius.

We report the modulus wave functions for the two resonant states $1\Gamma_6^-$ and $1\Gamma_8^-$ in fig.3 and fig.4, respectively. We can observe that the spherical symmetry is completely lost, and the p-like character of the wave functions is more evident. In particular, the $\Gamma_8^- \alpha$ state of fig.4 displays the main contribution from the 2p envelope functions connected to the Γ_7^+ valence band, but contains also some contributions from the Γ_8^+ states which disturb the symmetry.

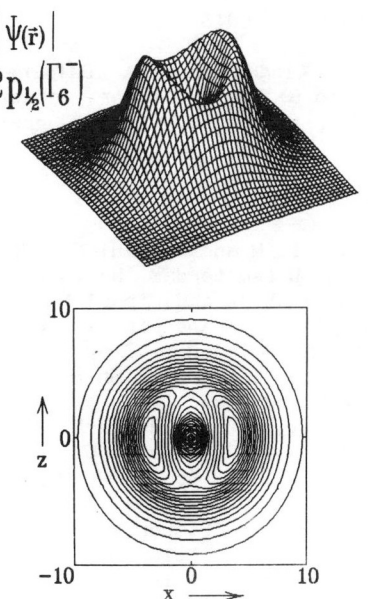

fig 3. The same as for fig.2 for $2p_{1/2}$ (Γ_6^-) state.

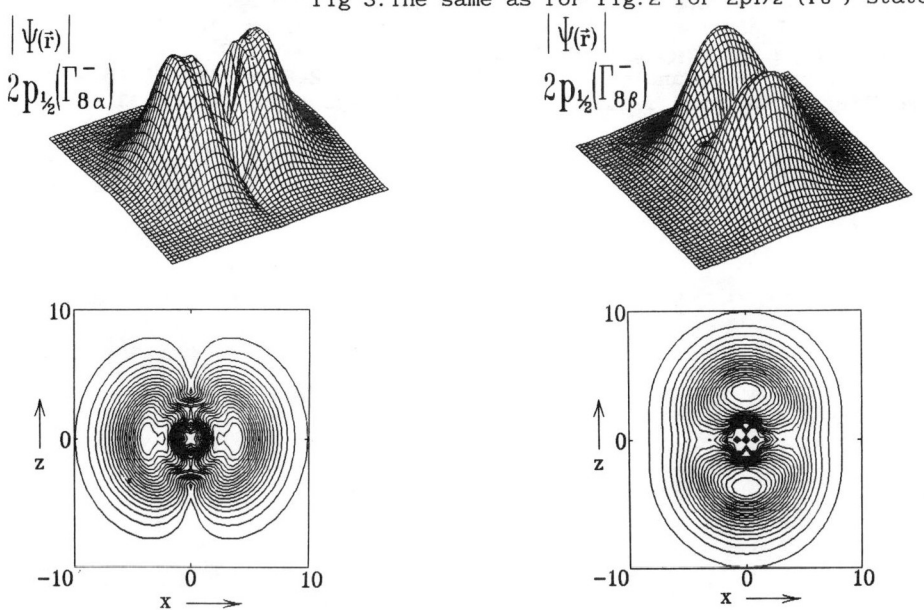

fig.4. The same for $2p_{1/2}(\Gamma_8^-)$ state.

4. CONCLUSIONS.

We have calculated the optical transitions to the excited and resonant states of shallow acceptors in silicon. We have obtained new numerical data about energy positions and splitting of the resonant states. The computed oscillator strengths allow an explanation of the intensity features of the optical absorption spectrum.

ACKNOWLEDGMENTS

We are indebted to A. Baldereschi and N. Binggeli for useful discussions and to S. Fraizzoli for a stimulating collaboration. Financial support from C.N.R. through G.N.S.M. is gratefully acknowledged.

REFERENCES

Altarelli M and Bassani F 1980 in *Handbook of Semiconductors* **1**, ed. by Paul W (Amsterdam, North Holland)
Baldereschi A and Lipari N O 1973 *Phys. Rev.* **B8** 2697
Baldereschi A and Lipari N O 1974 *Phys. Rev.* **B9** 1525
Bassani F, Iadonisi G and Preziosi B 1974 *Rep. Prog. Phys.* **37** 1099
Beinikhes I L, Kogan Sh M, Polupanov A F and Taskinboev R 1985 *Solid State Comm.* **53** 1083
Binggeli N and Baldereschi A 1988 *Solid State Comm.* **66 323**
Buczko R 1987 *Il Nuovo Cimento* **9D** 669
Fisher D W and Rome J J 1982 *Phys. Rev.* **B27** 4826
Jagannath C, Grabowski Z W and Ramdas A K 1981 *Phys. Rev.* **B23** 2082
Jones R L and Fisher P 1970 *Phys. Rev.* **B2** 2016
Kogan Sh M and Polupanov A F 1978 *Soviet Phys. Semicon.* **12** 1094
Kogan Sh M and Polupanov A F 1978 *Solid State Comm.* **27** 1281
Lipari N O, Baldereschi A and Thewalt M L W 1980 *Solid State Commun.* **33** 277
Onton A, Fisher P and Ramdas A K 1967 *Phys. Rev.* **163** 689
Polupanov A F and Kogan Sh M 1979 *Soviet Phys. Semicon.* **13** 1368
Ramdas A K and Rodriguez S 1981 *Rep. Prog. Phys.* **44** 1297
Rome J J, Spry R J, Candler T C and Brown G J 1982 *Phys. Rev.* B **25** 3615
Zwerdling S, Button K J, Lax B and Roth L M 1959 *Phys. Rev. Lett.* **4** 173

Stability of bound exciton state in highly stressed Si:B

H Nakata, A Uddin and E Otsuka

Department of Physics, College of General Education, Osaka University, Toyonaka Osaka 560 Japan

ABSTRACT: The shell model of bound multi-exciton complexes is examined in uniaxially deformed Si:B samples by photoluminescence measurements. The lifetime of bound exciton is shortened from 0.97 to 0.37 µs by stress up to 124MPa along [100] direction. Lifetime shortening is partly responsible for decrease in the luminescence intensity of bound excitons under stress. The idea of bound exciton-electron scattering is consistent with the shell model.

1. INTRODUCTION

The shell model of bound multi-exciton complexes (BMEC) has been well established after detailed study by Thewalt (1977) and Kirczenow (1977). Recently we observed phenomenon appearently inconsistent to the shell model in Si:B (Otsuka et al. 1986). Here we have solved appearent inconsistency by the idea of lifetime shortening of bound excitons (BE) under stress.

Phononless Auger recombination is the main recombination process of an electron and a hole in an acceptor BE in Si (Osbourn and Smith 1977). According to this process, the lifetime of BE depends on the ionization energy of the acceptor, since the hole wave function of a deep acceptor spreads in a momentum space and has a big matrix element with an electron. The lifetime of BE is expected to depend on stress, since phononless Auger recombination is sensitive to the band structure deformed by stress. We first observed the lifetime shortening of BE in Si:B under stress.

The lifetime shortening of BE reduces the number of BE in a stationary condition. The BE seems unstable in a high stress limit. This fact looks contradictory to the shell model of BMEC. According to the shell model, electrons and holes in BMEC form electron and hole shells, respectively. If the electron shell is filled up with electrons or the hole shell with holes, BMEC is stable like an inert gas such as He, Ne etc.. The atom with an unfilled shell like a hydrogen atom can not have a short lifetime but a large capture rate of electrons. As a consequence, shell structure is responsible for the electron capture rate but not for the lifetime.

Two decades ago, Otsuka et al. (1966) observed narrowing of electron cyclotron resonance lines with stress in Si:B. This phenomenon was first interpreted by desappearance of trapping centers which determine the lifetime of electrons. After they observed long lifetime of electrons, Otsuka (1981) reinterpreted their data by the model of electron-BE

scattering. According to his interpretation, the linewidth of electron cyclotron resonance in Si:B at 2 K is dominated by electron-BE scattering and the density of BE decreases with increasing stress. The absorption line of electron cyclotron resonance becomes narrower because of decrease in the number of scattered centers. Our measurements on lifetime shortening and decrease of luminescence intensity of BE with stress supports his idea.

2. EXPERIMENTAL

The samples which we used in this study were Si:B ($N_A = 4.3 \times 10^{14}$ cm^{-3}) and pure Si ($N_A = 3 \times 10^{11}$ cm^{-3}) denoted Si/B-1 and UHP-Si, respectively. The sample with size of 1 x 1 x 4 mm^3 was stressed by a spring to monitor the stress. Excitation source was a CW Ar$^+$ laser and its beam was chopped by a mechanical chopper (200Hz) or an A/O modulator(1kHz). Luminescence was dispersed by a monchromator (SPEX 1269) and detected by a Ge PIN detector (North Coast EO 817 P). Time resolution was limited to 0.1μs by a pre-amplifier.

Figure 1 shows time-resolved spectra of BE TO/LO phonon replica at zero stress. The peak has low energy tail consisting of several BMEC lines because of low resolution but the main peak is due to BE. The BMEC tail has a short lifetime as reported previously(Thewalt 1978). Under stress, the BMEC tails disappear and the BE peak decays faster than that without stress as shown in Fig. 2. The peak intensity is monitored to estimate the lifetime of BE as shown in Fig. 3. We observed the exponential decay

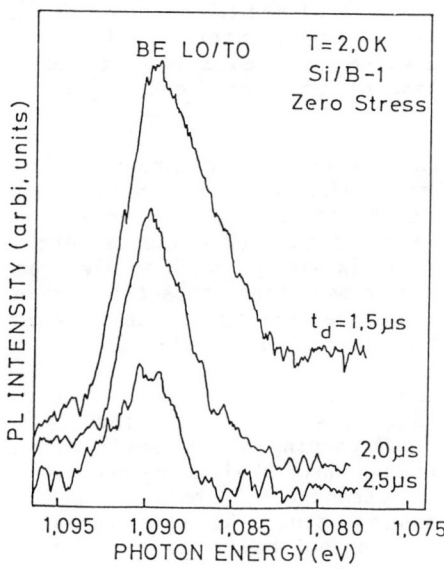
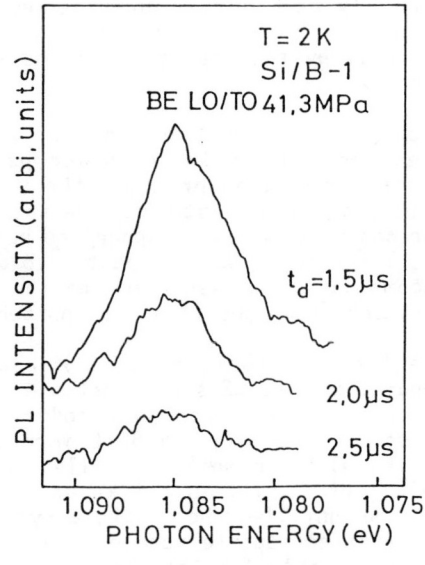

Fig. 1. Time-resolved spectra of bound exciton LO/TO phonon line at zero stress in Si:B.

Fig. 2. Time-resolved spectra of bound exciton LO/TO phonon line under stress in Si:B.

of BE over the whole range of stress.

The density of BE n_{BE} decays after the shutting-off of the excitation light obeying the following rate equation

$$\frac{dn_{BE}}{dt} = -\frac{n_{BE}}{\tau_{BE}} + Cn_{FE}(N_A - n_{BE}) - C'n_{FE}n_{BE} - An_{BE} \qquad (1)$$

Here τ_{BE} is the lifetime of BE, C and C' are capture rate of excitons by neutral acceptors and BE, respectively, N_A is an acceptor concentration, n_{FE} is the density of free excitons and A is the dissociation rate of free excitons from BE. As the binding energy of an exciton to a neutral boron is 4meV, excitons cannot dissociate at 2K. We can neglect the last term. The second and third term influence the decay of BE little since the density of free excitons is small. In fact the previously reported BE lifetime of 1.1μs in Si:B (Lyon et al. 1977) is almost the same as ours. The lifetime obtained by such a procedure of BE is shortened from 0.97 to 0.37μs under stress up to 124MPa, as shown in Fig.4.

Another experiment was performed on luminescence intensity of BE under stress. Figure 5 shows TO/LO phonon replica of BE luminescence. The luminescence intensity measured at 4.2 and 2K decreases with increasing stress, as shown in Fig.4. The other phonon replicas-TA phonon and no phonon lines also behave like TO/LO phonon replica. The broadening of each lines were observed because of splitting of BE and BMEC lines with stress.

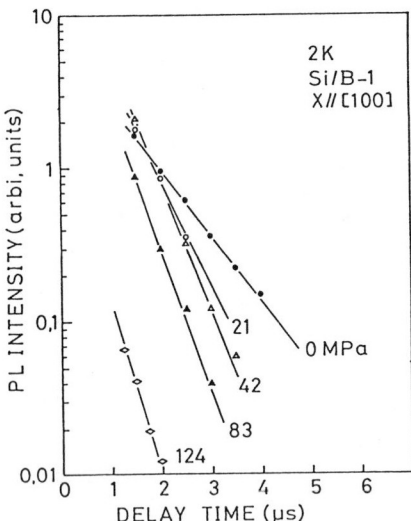

Fig. 3. Decay profile of peak intensity of TO/LO phonon replica of bound exciton under stress.

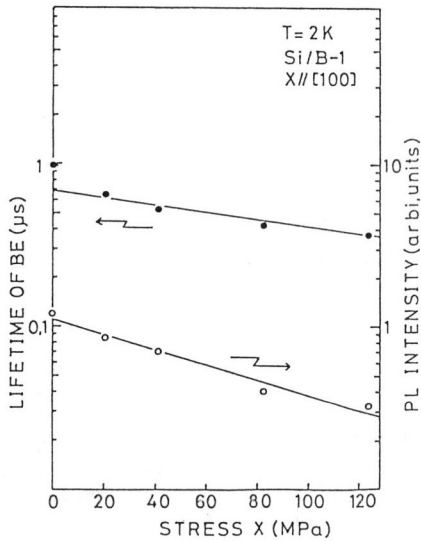

Fig. 4. Stress dependence of lifetime and density of bound exciton in Si:B.

The other interesting feature is that the free exciton peak intensity does not increase with increasing stress in contrast with Ge:Zn, where free exciton density increases with increasing stress as shown in Fig.5.

3. DISCUSSIONS

We observed the lifetime shortening of BE with stress in Si:B. In Si, phononless Auger recombination is responsible for the lifetime of BE. An electron in an acceptor BE is loosely bound to an A^+ center and the wavefunction does not spread in the momentum space. On the contrary, holes are tightly bound and wavefunctions spread in the momentum space. In the phononless Auger precess, the expansion of the hole wave function in the momentum space induces a large transition rate. In our case, the hole wave function may expand in the momentum space due to stress or the density of final states increases with increasing stress.

Decrease of BE density can be partly explained by lifetime shortening of BE through following simplified rate equations in a stationary condition.

$$\frac{dn_{FE}}{dt} = G - Cn_{FE}N_A = 0 \quad (2)$$

$$\frac{dn_{BE}}{dt} = -\frac{n_{BE}}{\tau_{BE}} + Cn_{FE}N_A = 0 \quad (3)$$

, where we neglect BMEC for simplification and obtain the density of BE

$$n_{BE} = \tau_{BE} G \quad (4)$$

If the generation rate G does not change under stress, density of BE n_{BE} is directly proportional to the lifetime of BE τ_{BE}.

In Fig.5, both lifetime and photoluminescence intensity depends on stress X as

$$\tau_{BE} = \tau_{BE}^0 \exp(-\gamma x) \quad (5)$$

$$I_{PL} = I_{PL}^0 \exp(-\gamma' x) \quad (6)$$

Parameters and are estimated to be 4.9×10^{-3} and $1.08 \times 10^{-2} (MPa^{-1})$, respectively. To satisfy equation (4), we must assume stress dependence of generation rate as follows

$$G = G^0 \exp(\gamma' - \gamma)x \quad (7)$$

, where $\gamma' - \gamma = 5.9 \times 10^{-3} (MPa^{-1})$. We observed stress dependence of free exciton photoluminescence intensity in pure Si. The intensity decreases with increasing stress as shown in Fig. 6. In Si, electrons populate in

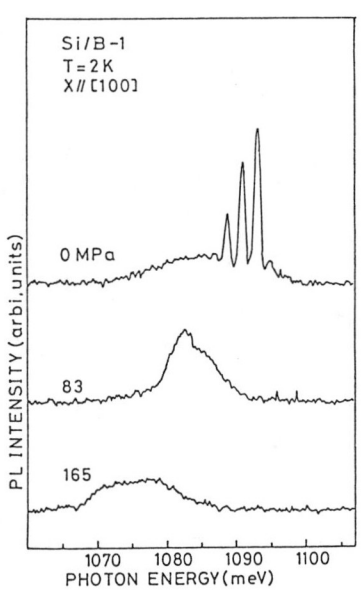

Fig. 5. Photoluminescence spectra of TO/LO phonon replica of bound multi-exciton complex under stress in Si:B.

6 valleys of conduction minima. These valleys are split into two kinds of valleys under stress. The cold exciton denoted in Fig.6 is the complex with down valley electron and the hot exciton is one with up valley electron. The decrease of the photoluminescence intensity supports decrease of generation rate. As a concequence, we cannot explain the decrease of BE density under stress only by lifetime shortening. But considerable part of the decrease is induced by the lifetime shortenig.

One of the main purpose of this paper is to solve the apparent contradiction of our experimental results to the shell model of BMEC. Here we insist that the shell model itself explains the activity of BMEC but not its stability. BMEC in a closed shell structure does not tend to capture an exciton, whereas one in an unclosed shell structure tends to capture and exciton and becomes a closed shell structure. It means that the shell model explains capture rate of excitons to neutral acceptors or BMEC.

Zinc is a double acceptor in Ge and it has two holes in a neutral state. The BE in Ge:Zn has three holes and one electron. The top of the valence band is four fold degenerate, so the first hole shell can contain four holes without stress. When stress is applied, the top of the valence band is split into two bands and the first hole shell can contain only two holes.

At zero stress, a neutral Zn acceptor has an unclosed shell structure and it captures an exciton to form BE. In a high stress limit, a neutral Zn acceptor has a closed shell structure and it cannot capture an exciton.

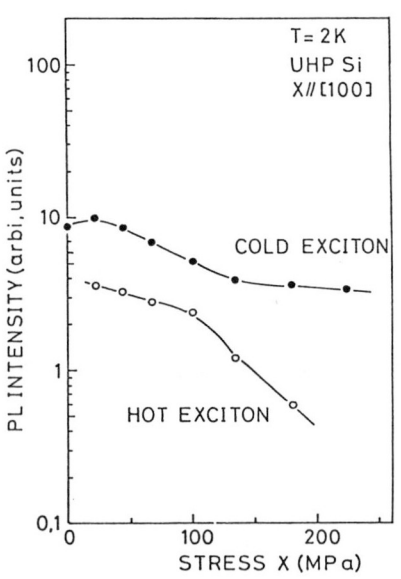
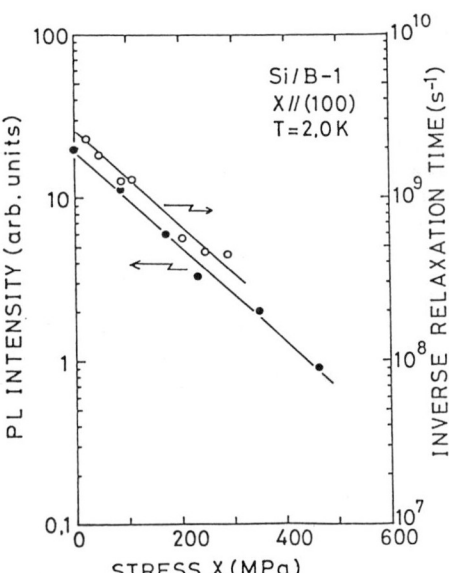

Fig. 6. Photolumincence intensity of TO/LO phonon replica of free exciton with stress in pure Si.

Fig. 7. Stress dependence of photoluminescence intensity of bound exciton and inverse relaxation time of bound exciton-electron scattering in Si:B.

It explains well decrease of BE density and increase of free exciton density under stress (Thewalt et al. 1985, Otsuka et al. 1986).

In Si:B, a neutral acceptor does not have a closed hole shell structure regardless of stress. The capture rate of an exciton to a neutral acceptor does not depend strongly on stress . We cannot expect sudden increase of free exciton density in Si:B with stress. The simultaneous decrease of free excitons and BE cannot be explained by the change of the capture rate. We conclude that the decrease of free exciton density and BE under stress is caused by the lifetime shortening of BE and decrease of generation rate of free excitons, etc.. This phenomenon does not contradict the shell model.

Another important conclusion is that our results support the idea of BE-electron scattering. Otsuka (1981) proposed the scattering mechanism to explain line narrowing of the electron cyclotron resonance in Si:B and Ge:Zn under stress. According to his proposal, decrease of BE density under stress is expected. The linewidth of electron cyclotron resonance is proportional to BE density under stress as shown in Fig.7. In conclusion, we suggest that the BE-electron scattering is dominant process in both materials and explained in accordance with the shell model.

4. ACKNOWLEDGEMENT

We are indebted to T. Ohyama for fruitful discussions and Y. Ichikawa for technical assistance.

REFERENCES

Kirczenow G 1977 Can. J. Phys. **55** 1787
Lyon S A, Osbourn G C, Smith D L and McGill T C 1977 Solid State Commun.**23** 425
Osbourn G C and Smith D L 1977 Phys. Rev. **B16** 5426
Otsuka E, Murase K and Ohyama T 1966 Proc. 8th Int. Conf. on Physics of Semiconductors, Kyoto 1966, J. Phys. Soc. Jpn **21** (1966), Suppl. pp 327-330
Otsuka E 1981 J. Phys. Soc. Jpn **50** 2631
Otsuka E, Nakata H and Ichikawa Y 1986 Proc. 18th Int. Conf. on Physics of Semiconductors ed O. Engstrom (Singapore: World Scientific) pp 1433-1436
Thewalt M L W 1977 Can. J. Phys. **55** 1463
Thewalt M L W 1978 Solid State Commun. **25** 513
Thewalt M L W, Lightowlers E C and Haller E E 1985 Solid State Commun. **55** 1043

Phonoionization of A^+ states in Si under uniaxial stress: influence of the valence-band structure

R. Haug, and E. Sigmund

Institut für Theoretische Physik, Universität Stuttgart,
Pfaffenwaldring 57, D- 7000 Stuttgart 80, FRG

ABSTRACT: Scattering of phonons at shallow bound A^+ states in Si can lead to the creation of mobile charge carriers through excitation of the second hole. This leads to an increase of the electrical conductivity. We investigate the coupling of phonons to A^+ states in Si and calculate the phonon-induced conductivity. We obtain a strong dependence on the phonon energy due to the reduced electron-phonon coupling of short-wavelength phonons. Uniaxial stress changes the valence band structure leading to striking effects on the phonoconductivity response signal which has been observed experimentally.

1. INTRODUCTION

Shallow impurities in semiconductors at low temperatures can bind a second carrier to form H^--like states. Then phonon scattering at these states can create mobile charge carriers through excitation. This induces a change in the electrical conductivity which can be used as detecting mechanism for this effect. An estimate of the binding energies of D^- states of donors and A^+ states of acceptors can be obtained from thresholds in far infrared (FIR) photoconductivity and phonon spectroscopy. Both experiments lead to the same threshold energies e.g. in the case of B^+ or P^- in Si and thereby prove that the neutralization by phonons is mainly a one-phonon process (Burger and Laßmann 1984).

In this paper we study the scattering of phonons at A^+ states in Si. By means of phonon spectroscopy the binding energies of A^+ states in Si have been found to be about 1.8 meV for B^+, Al^+ and Ga^+ and 5.9 meV for In^+ (Burger and Laßmann 1986). The neutralization of these states by phonons leads to an increased electrical conductivity which is called **phonoconductivity** in contrast to the **photoconductivity** where the mobile charge carriers are created by optical methods. In Sec. 2. the wave functions for A^+ states in Si based on the effective-mass approximation are discussed. Furtheron, the Hamiltonian for the electron-phonon interaction is presented. In Sections 2.1 to 2.4 we investigate the influence of uniaxial stress on the phonoconductivity. It turns out that externally applied stress leads to a splitting of the A^+ state as well as of the valence band states. Additionally, we show the stress-dependent mixing and coupling of the valence bands and their influence on the electric conductivity. In Sec. 2.5 matrix elements are calculated describing the dependence of the electron-phonon interaction on the phonon energy.

Then, in Sec. 2.6 we develop a microscopic single particle theory which allows to calculate the phonon-induced electrical conductivity starting from the Kubo formula (Kubo 1957) and applying the Mori formalism (Mori 1965). Finally, in Sec. 3 the results are presented and compared with experimental measurements.

2. THEORY

The top of the valence band in Si is at the center of the Brillouin zone at the Γ-point and has sixfold degeneracy, resulting from a threefold orbital (L=1) and a twofold spin degeneracy (S=1/2). Spin-orbit interaction partially splits these degeneracies leading to a higher lying quadruplet (J=3/2) and a lower lying doublet (J=1/2) state. This structure is reflected in the bound states of shallow acceptor impurities, which in an effective mass approximation can be represented by wave packets made up largely of the six Bloch waves chosen from the top of the valence band with appropriate envelope functions describing the localization of the defect. The envelope functions are equivalents to the hydrogenic eigenfunctions and can be calculated by solving the set of effective mass equations (Fjeldly et al. 1973). For this the envelope functions are expanded in spherical harmonics characterized by the quantum numbers l and m. The effective mass equations have inversion symmetry, and therefore either even l- or odd l-terms contribute to the solution only. Since the ground state wave functions are mainly s-like (l=0), the even l-expansion is chosen. In the following we consider the l=0 (s-like) and l=2 (d-like) contributions only.

The A^+ state is fourfold degenerate due to the tetrahedral symmetry of the surrounding crystal. We neglect correlation effects of the two holes bound by the acceptor. This is justified since the hole-hole coupling would lead to a multiplet structure which, however, has not yet been found in measurements of the A^+ binding energies in Si (Burger and Laßmann 1986). Therefore as wave function for the second hole we can adopt the form of the wave function of the acceptor ground state quartet (J=3/2), specified by M_J=3/2, 1/2, -1/2 and -3/2.

This function has been determined by Suzuki, Okazaki and Hasegawa (1964), within the framework of the effective mass approximation. Including s- and d-like envelope parts they used a variational procedure to calculate the explicit structure of the wave functions Ψ^{M_J} of the acceptor ground state quartet. Each wave function can be represented as a linear combination of six orthogonal component vectors $\Phi_i^{M_J}$ whereby one of these includes only s-like terms, and the remaining five have only d-like character. Thus the wave function can be written as

$$\Psi^{M_J} = \sum_{i=0}^{5} C_i \, \Phi_i^{M_J} \qquad (2.1)$$

where in the first step of the variational procedure the coefficients C_i have been determined and then in the second one the corresponding effective Bohr radii describing the extent of the wave function.

We approximate the wave function of the second hole by the single particle function of Suzuki et al. with effective Bohr radii of about 1.5 times the values of the neutral acceptor states. This is suggested by variational calculations for two-electron systems (Bethe and Salpeter 1957). The expression for the wave function $\Phi_i^{M_J}$ is derived for the axis

of quantization along [001]. Fjeldly et. al (1973) worked out the corresponding expression for the axis of quantization along [111].

The strain Hamiltonian describing the coupling of phonons to the A^+ state and the splitting of the ground state quartet by external stress can be derived from symmetry considerations (Hasegawa 1963)

$$H_I^v = D_d^v (e_{xx}+e_{yy}+e_{zz}) + 2D_u [(L_x^2-\frac{1}{3}L^2) e_{xx} + C.P.]$$
$$+ D_{u'} [(L_xL_y + L_yL_x) e_{xy} + C.P.] \qquad (2.2)$$

where L_α is the α component of the angular momentum operator $\vec{L}$ (L=1). α = x, y or z refer to the fourfold symmetry axes. D_d^v, D_u and $D_{u'}$ are the valence band deformation-potential constants. C.P. denotes cyclic permutation of the indices x, y and z. $e_{\alpha\beta}$ are the conventional strain components (Hasegawa 1963). The coupling of the totally symmetric part of the elastic distortion is neglected, since it yields only a constant energy shift without splitting of the ground state. D_u and $D_{u'}$ define the valence band splitting for uniaxial stress along the [001] and [111] directions, respectively.

The strain Hamiltonian (2.2) describes the splitting of the energy in the A^+ state quartet caused by static strain and the coupling of the A^+ state to phonons. In both cases the Hamiltonian operates on the A^+ wave function (2.1). The calculations can now be carried out by representing H_I^v as a 6x6 matrix within the same basis we used to build up Ψ^{MJ} (Fjeldly 1973). It is convenient to choose the direction of the external stress as the axis of quantization. Thus for uniaxial stress along [001] α corresponds to x, y and z of Eq. (2.2). In the case of [111] stress, α refers to the axes of a coordinate system with the z' axes (α=3) along the [111] direction. The axes x' and y' are chosen along [1$\bar{1}$0] and [11$\bar{2}$], respectively. L_x, L_y and L_z must then be decomposed along the new coordinate axes (Fjeldly 1973).

2.1 Stress-induced energy splitting

For stress along a three or fourfold symmetry axis, the A^+ quartet splits into two doublets $M_J=\pm 3/2$ and $M_J=\pm 1/2$ with an energy separation Δ. D_u and $D_{u'}$ are the A^+ state deformation potential constants depending on the coefficients C_i of the wavefunction (2.1) (Fjeldly et al. 1973).

2.2 Valence band structure in the presence of strain

In the absence of stress the interaction between the degenerate valence bands at k=0 disturbs the energy surfaces and leads to quartic terms in the wavevector expansion. Uniaxial stress lifts the cubic symmetry of the crystal and removes the degeneracy at k=0. For stress along the [001] and [111] directions the decoupled states are degenerate Kramers' doublets identified by the quantum number $\pm M_J$. In the case that the applied stress is strong enough to decouple the two bands with J=3/2 completely their energy surfaces are ellipsoidal (Hensel and Feher 1963).

2.3 Linear mass shift due to valence band mixing

The existence of the degeneracy with J=3/2 in the case of X=0 is a consequence of the spherical isotropy of the spin-orbit interaction (Hasegawa 1963). This isotropy is violated by uniaxial stress resulting in a splitting of the terms and a mixing of the eigenstates of J=3/2 with J=1/2. Since M_J is still a good quantum number under the uniaxial stress

the mixing takes place only between the two doublets with $M_J = \pm 1/2$. The degree of the mixing depends on the relative strength of the energy separation of the A^+ doublets to the spin-orbit separation of the valence bands with J=3/2 and J=1/2 (Hasegawa 1963). If one calculates the energy bands by omitting the off-diagonal elements of the Hamiltonian for the valence band states the energy surfaces described in the last section are obtained. When including the off-diagonal elements the admixture of the spin-orbit split-off state (J=1/2) is taken into account. In low order the inverse effective mass of the states with $M_J = \pm 1/2$ depends linearly on the stress.

2.4 Effects of fourth-order terms

Second order perturbation theory leads to a correction of the energy surfaces discussed in the last two sections which is given by fourth-order terms in the three variables k_x, k_y, and k_z. These terms are important when the applied stress is not strong enough to decouple the upper two valence bands completely. In this case their nonparabolic structure becomes significant. This second order correction for the energy always leads to an increased effective mass and is directly proportional to the applied uniaxial stress.

2.5 Phonon matrix elements

For the description of the dynamics of the system the coupling of the A^+ state to acoustic phonons has to be considered instead of the coupling to static deformations discussed in Sec. 2.1 . The elastic constants $e_{\alpha\beta}$ can be expressed by the displacement vectors **u** of the atoms. The displacement **u** can be expanded in terms of the vibrational modes of the crystal. The matrix elements describing the phonon-induced coupling between the stress-split doublets of the A^+ state and the valence band states are obtained by substituting the expansion in normal modes for the strain components into the strain Hamiltonian (2.2). We use the wave function as discussed in Sec.2 for the A^+ state and Bloch waves in the same representation for the valence band states. The axis of quantization is chosen along the direction of uniaxial stress.

The matrix elements contain integrals over the space variables which consist of products between the various s- and d-like angular parts of the A^+ state (see Fjeldly et al. 1973), their radial parts, a phase factor $e^{i\mathbf{q}\mathbf{r}}$ from the expansion of the strain components into normal modes and a factor $e^{i\mathbf{k}\mathbf{r}}$ from the Bloch function of the valence band state. These integrals are rather complicated, but assuming that $\mathbf{k} \ll \mathbf{q}$ the integrations can be performed analytically. As a result one is left with only two form-factors which appear in different linear combinations in the matrix elements for the various transitions between the states of the A^+ quartet and the upper valence bands. The s-like part of the A^+ state gives a formfactor f_0 which reads

$$f_0 = (1 + r_1^2 q^2)^{-2} \tag{2.3}$$

and the d-like part

$$f_2 = r_2^2 q^2 (1 + r_2^2 q^2)^{-4}. \tag{2.4}$$

r_1 and r_2 are the effective Bohr radii for the s- and d- like parts, respectively. These formfactors reflect the strong dependence of the electron-phonon interaction on the phonon energy since the wavelengths of the phonons which are important for the scattering process are comparable to the effective Bohr radii (Fig. 1).

As a result the matrix elements M can then be written as sums of products between the amplitude factors C_i and the formfactors f_0 and f_2. We have calculated the explicit expressions for the coupling constants for two cases of configurations with different stress and phonon propagation directions: (i.) first, uniaxial stress along [001] with $\mathbf{q} \parallel$ [100] and (ii.) second, stress along [111] with $\mathbf{q} \parallel$ [110]. In order to obtain analytic expressions a quasi-isotropic model for the elastic properties of the crystal has been adopted (Fjeldly et al. 1973). This results in simple expressions for the polarization vectors $\mathbf{e}_{\mathbf{q}\lambda}$ for the three acoustical modes.

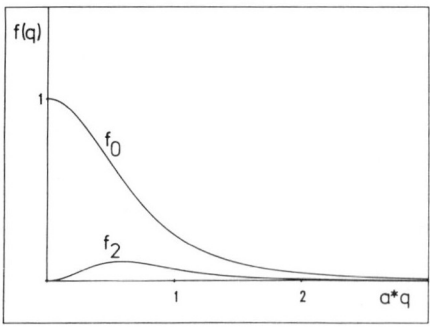

Fig. 1 : Formfactors f_0 and f_2 in dependence of the wavevector q of the phonons. a^* corresponds to the Bohr radii r_1 and r_2, respectively.

2.6 Phonoconductivity

Phonons with energy greater than the binding energy of an A^+ state can neutralize this state and create mobile charge carriers which causes an increase of the electrical conductivity. We use the Kubo formula (Kubo 1957) to derive an expression for the phonon-induced change of conductivity. Therefore a current-current correlation function has to be calculated for which we derive an equation of motion by means of Mori's formalism (Mori 1965). We introduce a relaxation time ansatz and apply Born-Markov approximation. Furtheron one has to take into account that the temperature dependent formula for the conductivity is the average of an energy dependent expression over a phonon distribution in equilibrium. Performing the explicit calculations for the energy dependent phonoconductivity one obtains an expression which contains the matrix elements discussed in Sec. 2.5, a power of the wavevector k of the valence band state, which comes from the current operator in the correlation function, and finally a distribution function which describes the possibility of exciting a hole from the A^+ state into the valence band taking the occupation of the A^+ state and the valence band states into account. The effect of applied uniaxial stress enters the expression for the phonoconductivity via the dependence on the formfactors which we have discussed in Sec. 2.5 and the dependence on the structure of the valence band. Finally the phonoconductivity is very sensitive to the mass shift of the states with $M_J = \pm 1/2$.

3. RESULTS

Taking the k^4-correction for the energy surface of the valence band and the valence band mixing into account our theoretical model is able to describe qualitatively the behavior of the experimental observations (Groß et al. 1987). To measure the phonoconductivity thin Al junctions as tunable phonon generators are evaporated on one side of the crystal sample. For the measurement of the phonon-induced conductivity Al contacts are evaporated on the opposite face which can be illuminated with light for generating a sufficiently large number of carriers, necessary for the production of A^+ states. In contrast to FIR-photoconductivity

results (Sugimoto et al. 1979), the behavior of the phonoconductivity response is much more complicated which is due to the reduced electron-phonon coupling for short wavelength phonons.

Fig. 2 shows the phonoconductivity as a function of the phonon energy in the case of In^+ in Si. The calculations have been carried out for various values of uniaxial stress along the [001] direction with the k^4-correction for the energy surface and the valence band mixing discussed above. The deformation potential constant of In^+ in Si is smaller than the deformation potential constant of the valence band. Thus the threshold energy shifts to smaller values under uniaxial stress and the formfactors f_0 and f_2 lead to an increased response signal. The k^4-correction for the energy surface in the case of 300 bar in fig. 2 causes the stronger decrease of the conductivity with increasing

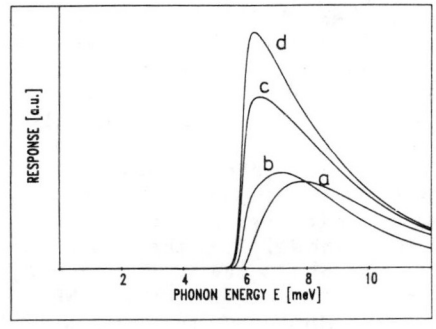

Fig. 2: Phonoconductivity response signal of In^+ in Si for uniaxial stress along [001]. a: 0 bar, b: 300 bar, c: 600 bar, d: 800 bar.
(r_1= 20 Å, r_2= 13 Å, D_u= 3.0 eV)

phonon energy compared with the case of zero stress. Uniaxial stress of 600 and 900 bar is strong enough to decouple the valence bands completely. In these two cases no influence of the quartic terms in the wavevector expansion of the energy surfaces can be seen. Thus we obtain the same behavior of the response signal as in the experimental observations (Groß et al. 1987).

We have also calculated the phonoconductivity of B^+ in Si with the k^4-correction and the valence band mixing (Haug and Sigmund 1988).

We gratefully acknowledge valuable discussions with P. Groß and K. Laßmann and we want to thank the Deutsche Forschungsgemeinschaft (DFG) for financial support.

REFERENCES

Bethe H A and Salpeter E E 1957 *Quantum Mechanics of One- and Two-Electron Systems,* Handbuch der Physik, ed Flügge S (Berlin: Springer)
Burger W and Laßmann K 1984 *Phys. Rev. Lett.* **53** 2035
Burger W and Laßmann K 1986 *Phys. Rev.* **B 33** 5868
Fjeldly T, Ishiguro T and Elbaum C 1973 *Phys. Rev.* **B 7** 1392
Groß P, Gienger M and Laßmann K 1987 *Jap. J. Appl. Phys. Suppl.* **26/3**, 673
Hasegawa H 1963 *Phys. Rev.* **102** 1029
Haug R and Sigmund E 1988 *Physica Scripta* **38** 114
Hensel J C and Feher G 1963 *Phys. Rev.* **129** 1041
Kubo R 1957 *J. Phys. Soc. Jap.* **12** 570
Mori M 1965 *Prog. Theor. Phys.* **33** 423
Sugimoto N, Narita S, Taniguchi M and Kobayashi M 1979 *Sol. St. Comm.* **30** 395
Suzuki K, Okazaki M and Hasegawa H 1964 *J. Phys. Soc. Jap.* **19** 930

Inst. Phys. Conf. Ser. No 95: Chapter 2
Paper presented at Int. Conf. Shallow Impurities in Semiconductors, Linköping, Sweden, 1988

Vibronic coupling in shallow excited states of optical centres in silicon

Gordon Davies+ and M.C. do Carmo*

+ Physics Department, King's College London, Strand, London WC2R 2LS, UK
* Departamento de Fisica, Universidade de Aveiro, 3800 Aveiro, Portugal

ABSTRACT : The 1067 meV vibronic band is observed in silicon after in-diffusing Fe. We show that its unusual bandshape is the result of a vibronic interaction between the excited states of the centre. The magnitude of the interaction is consistent with the stress-induced interaction observed at other centres. This is the first identification of a vibronically induced bandshape in silicon.

1. INTRODUCTION

Transition metals are readily introduced into crystalline silicon, producing many photoluminescence bands (Conzelman 1987). In some cases the vibronic bandshapes consist of simple progressions of vibrational modes. Simple progressions occur, for example, at the well-known 1014 meV Cu-related band (Sauer and Weber 1984) and the similar 944 meV band (McGuigan et al 1988). In contrast, the 1067 meV band has a very unusual shape (figure 1a; Mohring et al 1984). At low temperature it consists mainly of a one-phonon sideband (labelled B0 and B1 in figure 2). To emphasise this unusual bandshape, we present in figure 1b the luminescence spectrum of the 735 meV band, which, like the 1067 meV band, is observable after in-diffusing Fe to crystalline silicon. (The precise chemical natures of the two optical centres are uncertain (Schlessinger et al 1983), but this is of no concern here). The 1067 meV (Mohring et al 1984) and 735 meV bands have several similarities :
1) their vibronic sidebands involve two low energy resonance phonons. The band at 1067 meV has resonances of quanta
$$\hbar w_A = 7.2 \text{ and } \hbar w_B = 9.7 \text{ meV} \qquad (1)$$
and the 735 meV band has quanta of
$$\hbar w_A = 7.3 \text{ and } \hbar w_B = 9.3 \text{ meV}. \qquad (2)$$
2) the bands have three zero-phonon lines (labelled 0,1,2 on figures 2 and 3). The separations are $E_1 - E_0 = 0.36$ meV and $E_2 - E_0 = 2.85$ meV at the 1067 meV band, and 0.45 and 2.9 meV at the 735 meV band.
3) the relative transition probabilities in these zero-phonon lines are of the same general form with the lowest energy pair are relatively weak compared to the highest energy line. At the 1067 meV band the transition probabilities in the three lines are (in order of increasing energy) 0:1:25, and at the 735 meV band they are 1:1:8.

But despite these similarities the one-phonon regions of the bands are very different. The ratios of the intensities of the one-phonon sidebands to the zero-phonon lines for the 735 meV band give Huang Rhys factors for the two modes of equation 2 of

© 1989 IOP Publishing Ltd

$$SA = 0.8, \quad SB = 1.0 \tag{3}$$

In a simple vibronic progression the relative intensity of the nth phonon sideband is given by the well-known distribution $(S)^n \exp(-S)/n!$ (Maradudin 1966). The two-phonon sidebands of the 735 meV band are closely as expected for a simple vibronic progression of this form, as shown by the labelling of the combination modes in figure 3. In contrast, at low temperature the high energy part of the 1067 meV band consists of the very weak zero-phonon lines 0 and 1, and the far stronger one-phonon sidebands (figure 2). From the ratio of the intensities of the 9.7 meV phonon and the 0 and 1 zero phonon lines, the Huang Rhys factor is apparently "SB" = 15, but succeeding phonon sidebands are weak : the vibronic sidebands do not follow a simple vibronic progression. In the next section we explain this bandshape in terms of a strong vibronic interaction between the nearly-degenerate excited states of the centre.

2. VIBRONIC INTERACTIONS BETWEEN THE EXCITED STATES OF THE 1067 MEV BAND

The 1067 meV band has two resonance modes (equation 1) but the 9.7 meV mode is by far the stronger in the sideband (figure 2). To reduce the number of arbitrary parameters, we will consider only one vibrational mode, with a quantum $\hbar w$ = 9.7 meV. Similarly, the small splitting of lines 0 and 1 will be neglected, and for simplicity we will represent the excited states by two non-degenerate states ϕ_1 (for the excited states of lines 0 and 1) and ϕ_2 for the excited state of "2". If there was no interaction between ϕ_1 and ϕ_2, the vibronic states of the optical centre would be the Born Oppenheimer states

$$\psi_{1n}(r,Q) = \phi_1(r)X_n(Q); \quad \psi_{2n}(r,Q) = \phi_2(r)X_n(Q). \tag{4}$$

A linear electron phonon term aQ is assumed to mix ϕ_1 and ϕ_2:

$$\int dr \phi_1 a \phi_2 = A, \tag{5}$$

At high temperature there is little evidence in figure 2 for one-phonon sidebands of the strongly allowed line 2. The equilibrium positions of the centre are therefore the same in the ground and excited states. Hence

$$\int dr \phi_1 a \phi_1 = \int dr \phi_2 a \phi_2 = 0. \tag{6}$$

Because the coupling is linear in Q it can only mix those Born Oppenheimer states which differ by ±1 in the vibrational quantum number :

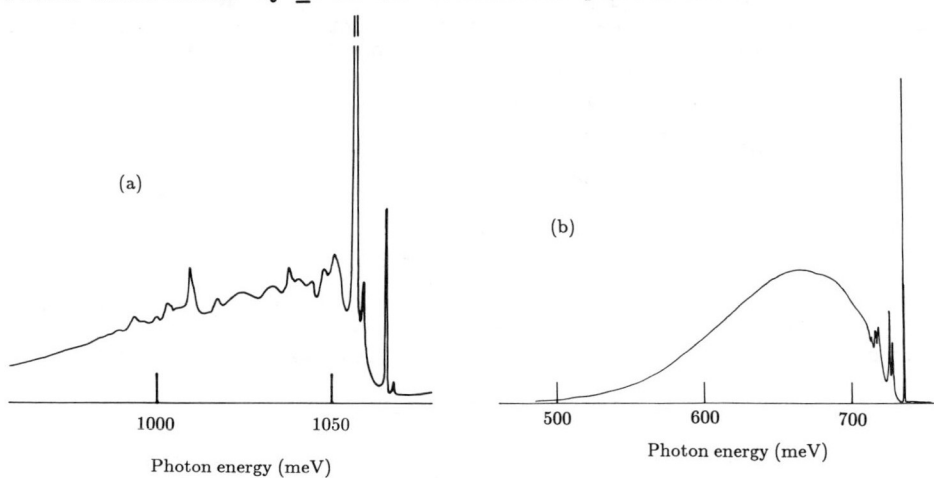

Figure 1. (a) The 1067 meV band at 4.2K (Mohring et al 1984). (b) The 735 meV band. The one-phonon peak in (a) goes to twice the height shown.

Shallow Impurities in Si and Ge

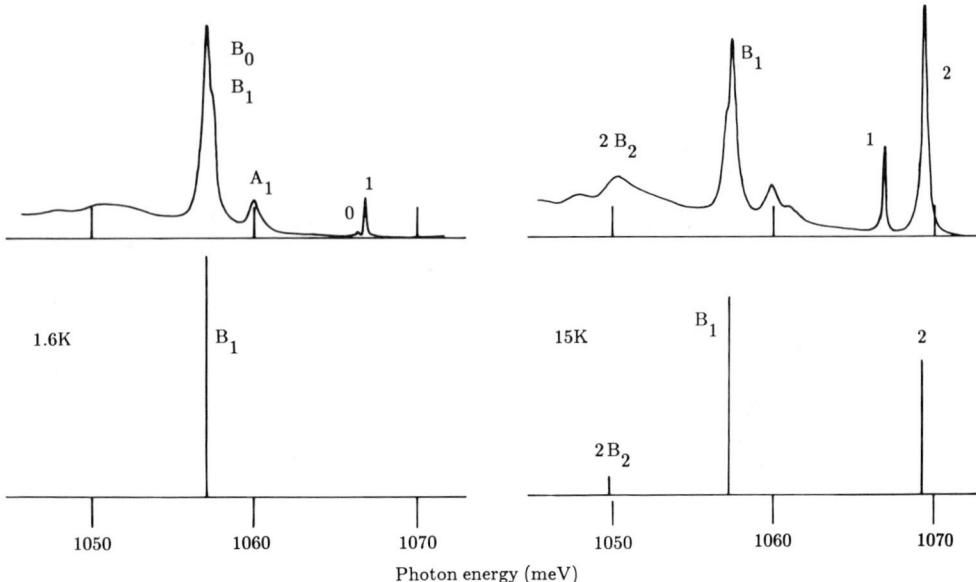

Figure 2. Top row : the zero- and one-phonon regions of the 1067 meV band at 1.6K (left) and 15K (right) recorded by Mohring et al 1984. Zero-phonon lines are labelled 0,1,2. "Xi" denotes the phonon sideband involving mode X of zero-phonon line i. Bottom row : spectra at 1.6 and 15K calculated as in section 2.

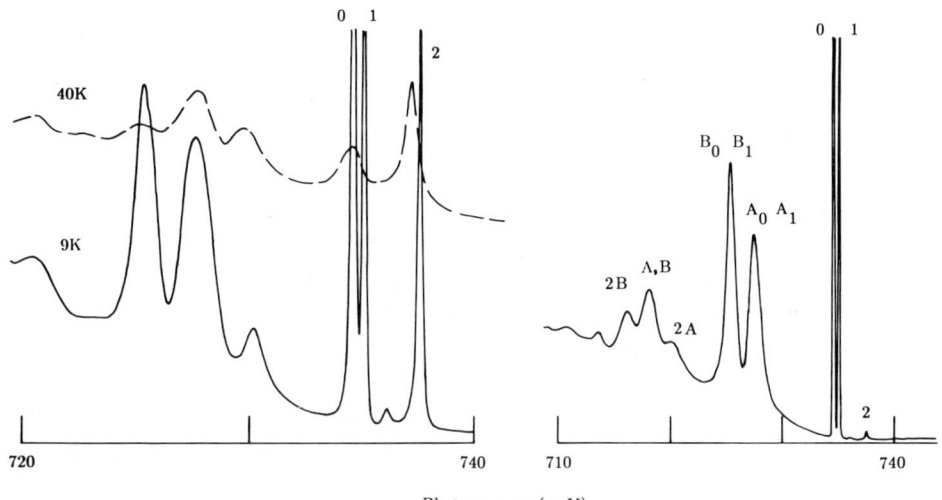

Figure 3. Left : zero- and one-phonon regions of the 735 meV band at 9 and 40K; right : zero-, one- and two-phonon regions at 4.2K. Notation is as figure 2. The two-phonon sidebands of lines 0 and 1 are labelled 2A, AB and 2B.

and
$$\int dr dQ \psi_{1n} aQ \psi_{2,n+1} = A\,[(\hbar/mw)(1+n)]^{1/2},$$
$$\int dr dQ \psi_{1n} aQ \psi_{2,n-1} = A\,[(\hbar/mw)(n)]^{1/2} \qquad (7)$$

The effect on the Born Oppenheimer states (equation 4) whose electronic states ϕ_1 and ϕ_2 are separated by energy E (at Q = 0) can be calculated from the secular matrix whose off-diagonal elements are in equation 7. The calculated bandshape is shown in figure 2 when the energy separation E and the electron phonon coupling A are

$$E = 4 \text{ meV}, \quad A[\hbar/mw]^{1/2} = 8.5 \text{ meV} \qquad (8)$$

The calculation clearly reproduces the observed optical properties within the simplifying limitations used. For example, although there is no detectable one-phonon sideband of the zero-phonon line "2", a two-phonon sideband "2B2" is produced by the vibronic coupling, as observed in the high temperature spectrum of figure 2. We must now check whether the parameter A has a realistic value.

3. THE SIZE OF THE PARAMETER A

The vibronic coupling parameter A describes the interaction of the excited states under a unit strain of the appropriate symmetry (equation 5). It can be measured from the <u>stress-induced</u> interaction of the excited states if the elastic constants near the optical centre are known and if the effective mass of the vibration is known (for making the comparison with equation 8). Unfortunately relevant data are not yet available for the 1067 meV line. However, as Conzelman (1987) has observed, many of the

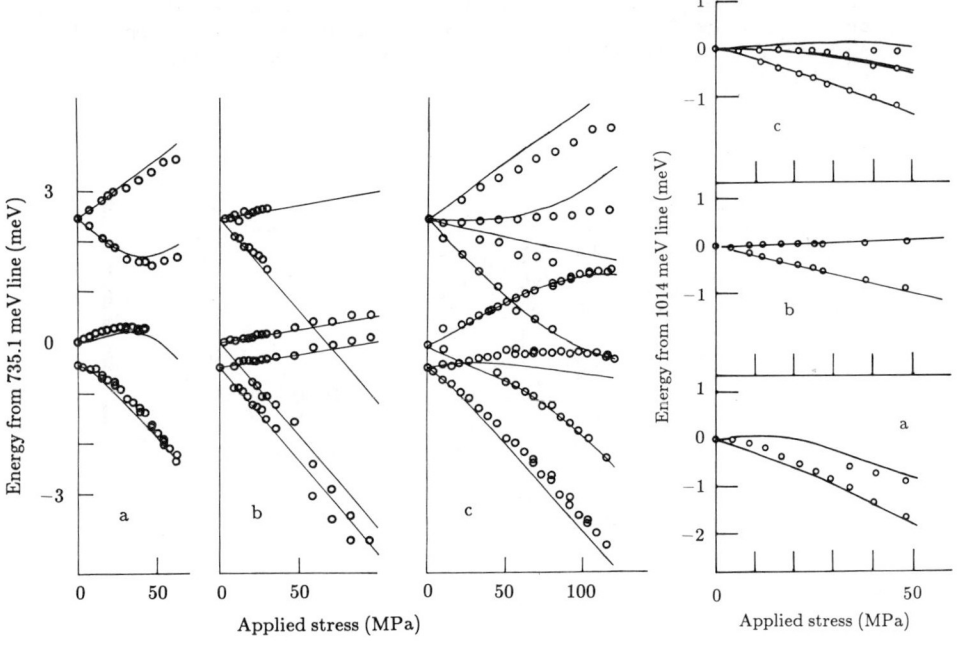

Figure 4. Left : points show uniaxial stress data for the three 735 meV zero-phonon lines. Lines calculated as in section 3. Right : points show data for the unresolved 1014 meV low-energy doublet (Weber et al 1982), lines calculated as in section 3. For both centres, the stress axes are (a) ⟨001⟩, (b) ⟨111⟩, (c) ⟨110⟩.

optical centres produced by transition metals appear to have similar properties. We have been able to measure the effects of uniaxial stresses on the 735 meV zero-phonon lines, which are very similar to the 1067 meV lines (section 1) and are also produced by in-diffusing Fe. We show below that by using the strength of the one-phonon sideband (equation 3), we can avoid having to guess the local elastic constants and the effective mass of the mode.

The uniaxial stress data for the 735 meV lines are shown in figure 4. The curves have been calculated for the effects of stress on transitions between a non-degenerate (A1) orbital ground state and a pair of doubly degenerate (E) orbital excited states at a trigonal centre (do Carmo et al 1988). The 0.45 meV splitting of the lines 0 and 1 is consistent with a spin-orbit interaction in the lower E excited state. In the notation of Hughes and Runciman (1967) a hydrostatic compression perturbs the transitions by A1 = -5.4 meV/GPa; a [111] compression perturbs the transitions at the [111] oriented centre by (A1+2A2) = -36.8 meV/GPa; a ⟨001⟩ compression splits both E excited states by 4B = 49 meV/GPa; and a ⟨111⟩ compression gives no splitting of the E states (ie C=0). With increasing ⟨001⟩ stress, components split from the E states repel each other. The magnitude of the matrix element producing the repulsion is (defined analogously to B)
$$B12 = \pm 8.7 \text{ meV/GPa}, \quad (9)$$

Uniaxial stress perturbations have been reported by Weber et al (1982) for the 1014 meV Cu-related band, but these data have not previously been analysed quantitatively. The 1014 meV band has two lowest energy lines separated by 0.15 meV and another line 1.9 meV higher. The reported stress data are only for the lower pair of lines, which were not resolved. Figure 4 shows that they can be fitted using the same theory as for the 735 meV band, and with parameters of very similar magnitude :
$$A1 = -4.6, \; A2 = -7.2, \; B = 10.0, \; C = 0, \; B12 = 15 \text{ meV/GPa} \quad (10)$$

The striking similarity in the parameters confirms the suggestion by Conzelman (1987) that several of the luminescence centres introduced into silicon by transition metals have very similar excited states. The excited states consist of a bound exciton with a shallow electron and a deeply bound hole. This situation also occurs at other centres in silicon : for example the 1045 meV Li-related centre (Davies et al 1984), has very similar uniaxial stress parameters to those derived here. The failure of a ⟨111⟩ compression to split the bound exciton shows that the hole component is in an orbital singlet state. Its orbital angular momentum is then quenched, producing the nearly isotropic magnetic reponse reported by Weber et al (1982) for the 1014 meV line and by Mohring et al (1984) for the 1067 meV line. The E orbital states derive from the shallow electron, so that the parameter C is zero, reflecting the lack of splitting of the conduction band minima under a ⟨111⟩ uniaxial stress (Laude et al 1971). Repulsion effects are not observed under ⟨111⟩ compression, showing that the excited states do not interact under totally symmetric perturbations. E-symmetry deformations are necessary for the coupling, which therefore takes place in the relatively shallow electron states.

The parameter A (describing the interaction of the excited states) is proportional to the corresponding uniaxial stress parameter B12. We can relate them without having to make any assumptions about the elastic constants near the optical centre and also without explicit use of an effective mass for the vibration as follows. The Huang Rhys factor S for

coupling to a single mode of quantum hw is
$$S = (A')^2/(2m\hbar w^3)$$
where the vibronic coupling within one state is $V' = a'Q$ (in analogy with equation 5). A' is proportional to the stress parameter B, the proportionality involving the same elastic constants as the relationship between A and B12 :
$$A/B12 = A'/B \; ; \; \text{giving} \; A[\hbar/mw]^{1/2} = [2S]^{1/2}\hbar w \, B12/B.$$
For the 9.7 meV mode of the 735 meV band, $S = 1$ (equation 3) and hence
$$A[\hbar/mw]^{1/2} = 10 \text{ meV}$$
in sufficient agreement with equation 8. With the assumption that the 735 and 1067 meV bands are similar, we have shown that the strength of vibronic coupling necessary to produce the observed pseudo-Jahn-Teller effect is as expected from the existing uniaxial stress data.

4. SUMMARY

We have shown that the unusual shape of the 1067 meV band is produced by a vibronic interaction of the nearly degenerate excited states. Vibronic interactions are not usually observed at optical centres in silicon, despite their frequently having multiple excited states (Davies 1988). A vibronic interaction is a second order perturbation effect : the luminescence intensity transfered to the forbidden state is proportional to the square of the coupling between the two interacting states (ie to A^2), and is inversely proportional to the square of the energy separation $(E + \hbar w)$ of the two uncoupled states. The transition metal centres have appreciable electron-phonon coupling to well-defined resonance modes (in contrast to the coupling being spread over a continuum of lattice modes, as happens at most optical centres in silicon). The resonance modes also have low energies. Both these features enhance the possibility of vibronic interactions at transition metal centres. A related effect is the frequent occurence of bi-stability at transition metal centres (Chantre 1988). Uniaxial stress data on the 1067 meV band are clearly urgently required.

ACKNOWLEDGEMENTS

We are grateful for support from our respective national funding agencies (SERC, and INIC and JNICT).

REFERENCES

Chantre A 1988 Mat. Res. Soc. Symp. Proc. 104 37
Conzelman H 1987 Appl. Phys. A42 1
Davies G 1988 Physics Reports, in press.
Davies G, Canham L and Lightowlers EC 1984 J. Phys. C17 L173
do Carmo MC, Calao MI, Davies G and Lightowlers EC 1988 Proc. 15th Int.
  Conf. Defects in Semiconductors, Budapest, in press
Hughes AE and Runciman WA 1967 Proc. Phys. Soc. 80 827
Laude LD, Pollack FH and Cardona M 1971 Phys. Rev. B3 2623
Maradudin AA 1966 Solid State Phys. 18 273
McGuigan KG, Henry MO, Lightowlers EC, Steele AG and Thewalt MLW 1988
  Solid State Commun. 1988 in press
Mohring HD, Weber J and Sauer R 1984 Phys. Rev. B30 894
Sauer R and Weber J 1984 Solid State Commun. 49 833
Schlessinger TE, Hauenstein RJ, Feenstra RM and McGill TC 1983
  Solid State Commun. 46 321
Weber J, Bauch H and Sauer R 1982 Phys. Rev. B25 7688

Linear stark coupling to the ground state of effective mass acceptors

Andreas Köpf, Anton Ambrosy, Kurt Laßmann

1. Physikalisches Institut, Univ. Stuttgart
Pfaffenwaldring 57, D-7000 Stuttgart 80, FRG

ABSTRACT: It is shown by dielectric resonance absorption at 24 GHz and by ultrasonic resonance spectroscopy between .5 and 10 GHz that linear coupling of the electric field to the ground state of effective mass acceptors in Si exists and has a distinct chemical shift from B to In.

1. INTRODUCTION

Linear coupling of an electric field to the Γ_8 acceptor ground state is forbidden by inversion symmetry within the Effective Mass Approximation. It becomes, however, possible by the local T_d symmetry of the central cell (Kohn 1957, Bir et al 1963) and should therefore be stronger for deeper acceptors with more localized envelope functions. We have shown by dielectric resonance absorption at 24 GHz (corresponding to electric dipole transitions between ground state levels split by a magnetic field) that there is indeed a chemical shift of the coupling strength for the acceptors B, Al, Ga, In by about a factor of 5, which is larger than the shift of the binding energies. For all the dopants the ground state splittings from internal strain fields are small enough as compared to the magnetic field splitting i.e. small overlap of the inhomogeneously broadened lines. Nevertheless, higher resonant frequencies or smaller linewidths would improve the precision of the evaluation.
The residual random splitting existing even in high quality samples means likewise an experimental difficulty if the (additional) splitting of the ground state by a <u>static</u> external electric field is to be observed: It is too small to be easily measurable before the breakdown field is reached. We have made use of the larger random internal electric fields in Si:B compensated by neutron transmutation to generate a broadening of the ground state splittings by these electric fields additional to the random strain fields. The electric fields can be switched off by bandgap illumination neutralizing the donor-acceptor ion pairs persistently at low temperatures. This change in the width of the splitting distribution is probed in the frequency range from 0.5 to 12 GHz by ultrasonic resonance absorption. Comparison of the measured distributions is made with Monte Carlo (MC) calculations. The analysis of the angular dependence of the linewidth in acoustic paramagnetic resonance shows that only the linear Stark coupling is effective in line broadening whereas a quadratic contribution could not be observed. The measured value for the linear coupling in the case of Si:B compares well with the value obtained by the resonance method. Other dopants could not be measured this way because appropriately compensated samples are difficult to obtain.

2. THEORETICAL CONSIDERATIONS

The interaction Hamiltonians for elastic, electric, and magnetic perturbations, have been summarized by Bir et al (1963) on the basis of group theoretical considerations together with estimates for the respective coupling constants with EMA-wavefunctions. As discussed by Zeile and Laßmann (1982) the introduction of a five-dimensional state vector Δ is useful for an evaluation not only of the splitting but also of the coupling to the elastic or electric alternating fields. The matrix elements for the elastic and electric perturbations of the Γ_8-state with its substates with magnetic quantum numbers m = 3/2, 1/2, -1/2, -3/2 (indexed 1,2,3,4, respectively) are then:

$$H_{11} = -H_{22} = -H_{33} = H_{44} = \Delta_1, \quad H_{14} = H_{23} = 0, \quad H_{ik} = H_{ki}^*,$$

$$H_{12} = -H_{34} = \Delta_4 + i\Delta_5, \quad H_{13} = H_{24} = \Delta_2 + i\Delta_3,$$

where the Δ_i are given by the elastic (ϵ) and electric (E) fields and the corresponding coupling constants as follows:

$\Delta_{1\epsilon} = (2\epsilon_{zz} - \epsilon_{xx} - \epsilon_{yy})b'/2$ $\quad\quad \Delta_{1E} = (2E_zE_z - E_xE_x - E_yE_y)\beta'/2$

$\Delta_{2\epsilon} = (\epsilon_{yy} - \epsilon_{xx})\sqrt{3}b'/2$ $\quad\quad \Delta_{2E} = (E_yE_y - E_xE_x)\sqrt{3}\beta'/2$

$\Delta_{3\epsilon} = \epsilon_{xy}d'$ $\quad\quad \Delta_{3E} = E_xE_y\delta' + E_z p_\chi$

$\Delta_{4\epsilon} = \epsilon_{yz}d'$ $\quad\quad \Delta_{4E} = E_yE_z\delta' + E_x p_\chi$

$\Delta_{5\epsilon} = \epsilon_{xz}d'$ $\quad\quad \Delta_{5E} = E_xE_z\delta' + E_y p_\chi$

b' and d' are the two deformation potential constants, $\beta' = -(p_0^2/E_B)\beta$ and $\delta' = -(p_0^2/E_B)\delta$ the two coupling constants for the quadratic, p_χ for the linear Stark effect with dipole moments $p_0 = e\sqrt{\langle r^2 \rangle}$ and $p = ea_B$ introduced to have the dimensionless constants β, δ, and χ. The definition $a_B = \hbar/\sqrt{2 m_h E_B}$ (m_h is the hole effective mass and E_B the experimentally determined binding energy.) for the Bohr-radius a_B is common to the EMA-, quantum defect, and δ-potential models (see e.g. Rynne et al 1976). The constant χ may be regarded as a characteristic of the wavefunction in the central cell and will be different for the three models. However, the wavefunctions corresponding to these models are centro-symmetric (i.e. do not reflect the local T_d-symmetry at the center) and so χ should be zero since the dipole moment $e\langle r \rangle$ for these functions is zero in contrast to the constants β and δ for the quadratic Stark effect.
For any splitting due to static elastic or electric fields characterized by a splitting state vector Δ the coupling to a phonon with unit state vector Δ_ϵ the coupling is given by $\Delta_\epsilon \sin\psi$ (ψ is the angle between the five-dimensional vectors). This relation simplifies MC calculations of the resonant phonon coupling to random internal elastic or electric fields. It has been done for the strain fields from point defects or dislocations and for the electric fields from correlated and noncorrelated ion pairs (Zeile and Laßmann 1982). At 4.2 K and for the ion concentrations achieved nearest (= correlated) ion pairs will render the thermodynamically favourable configuration which therefore has been used for estimating the constant χ by comparison of MC calculations with the measured additional broadening.
The combined effect of a large magnetic field B and small electric or elastic fields on the energy of the substates E_m of Γ_8 is (Yafet 1965):

$E_m = \mu_0(g_1' + g_2'<(41/20)+(m^2-(41/20))p(\theta)>)mB + m^2 E_{\epsilon,E}(\theta)/2$

with $p(\theta) = 1 - 5\sin^2(\theta) + (15/4)\sin^4(\theta)$

and $E_{\epsilon,E}(\theta) = -\Delta_1(1-3\cos^2(\theta)) + \sqrt{3}<\Delta_3\sin^2(\theta)+(\Delta_4+\Delta_5)\sin(2\theta)/\sqrt{2}>$

The Δ_i are the sum of the state vector components of the elastic and electric perturbations. In the case of random distributions of these fields the various components will add to the inhomogeneous broadening of the line depending on the orientation of the magnetic field. θ is the angle between the <001>- or z-direction and the magnetic field in the (1$\bar{1}$0)-plane. The influence of the linear and the quadratic Stark effect on the linewidth can be separated for $\theta = 0°$ and $\theta = 55°$ since

$E(\theta = 0°) = 2\Delta_1$ and $E(\theta = 55°) = 2(\Delta_3+\Delta_4+\Delta_5)/\sqrt{3}$.

Phonon or electric dipole transitions are allowed for $m = \pm 1/2 \leftrightarrow \pm 3/2$ ($\Delta m = 1$) or $\mp 1/2 \leftrightarrow \pm 3/2$ ($\Delta m = 2$) depending on orientation whereas for $m = -1/2 \leftrightarrow +1/2$ and $-3/2 \leftrightarrow +3/2$ they are always forbidden. The matrix elements for phonon transitions have been calculated by Yafet (1965). With the unit state vector $\Delta_{\epsilon,E}$ for the elastic or electric alternating field they can be written as follows:

$2H_{12} = -\Delta_1\sqrt{3}\sin 2\theta + \Delta_3\sin 2\theta + (\Delta_4+\Delta_5)\sqrt{2}\cos 2\theta - i<2\Delta_2\sin\theta - (\Delta_5-\Delta_4)\sqrt{2}\cos\theta>$

$2H_{13} = +\Delta_1\sqrt{3}\sin 2\theta + \Delta_3(1+\cos^2\theta) - (\Delta_4+\Delta_5)(1/\sqrt{2})\sin 2\theta - i<2\Delta_2\cos\theta+(\Delta_5-\Delta_4)\sqrt{2}\sin\theta>$

$H_{34} = -H_{12}$, $H_{24} = H_{13}$, $H_{14} = H_{23} = 0$.

For electric dipole transitions the components Δ_1 and Δ_2 of the state vector are zero. This allows to separate the $\Delta m = 1$ and $\Delta m = 2$ electric dipole transitions by orientation of the sample in the magnetic field: For the microwave electric field E // to the <1$\bar{1}$0>-direction and the magnetic field B // to the <001> or the <110>-direction we get

$H_{12} = ip_XE/2$, $H_{13} = H_{14} = H_{23} = 0$, or $H_{13} = -ip_XE/2$, $H_{12} = H_{14} = H_{23} = 0$,

respectively. The vanishing of the $\Delta m = 1$ signal for <110> then means that the alternating magnetic field is negligible within the sample.

3. EXPERIMENTAL RESULTS AND DISCUSSION

The dielectric absorption was measured by positioning the samples (<110>-oriented cylinders of 2 mm diam., 1 mm thickness) in the capacitive part of a 24 GHz coaxial cavity in the center of a 4.5 T superconducting magnet at 4.2 K and below. Care was taken to keep extra strains and line broadening due to shaping and mounting to a minimum. The change in reflection of the weakly coupled cavity tuned to resonance was detected by a diode. To improve the signal to noise ratio we applied amplitude modulation at 1 kHz by switching the frequency of the 180 mW Gunn source from cavity resonance to a frequency far off the resonance. The modulated detector signal was integrated by a standard VSWR-meter. We do not apply magnetic field modulation i.e. we obtain the absorption curves instead of their derivatives. Thus we get the absorption strength by integration of the lines with the simplifying assumption that the electric energy of the resonator is completely stored within the sample. The magnetic field was measured with a Hall probe fixed to the outside of the cryostat. It was

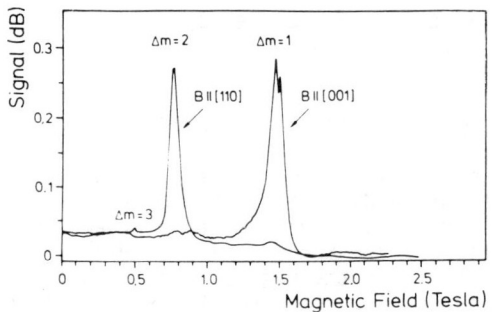

Fig.1 Dielectric resonance absorption for two orientations of a Si:B sample where either the $\Delta m = 1$ or $=2$ transitions are allowed. $\Delta m = 3$ appears since the high field case is not attained. The central dip of the $\Delta m = 1$ line is discussed in the text.
Boron concentration $c_B = 6 \times 10^{22} m^{-3}$.

calibrated at constant currents with the current/field conversion factor supplied by the manufacturer.

Fig. 1 shows the magnetic field dependence of the absorption for a Si:B sample with orientations appropriate for either the $\Delta m = 1$ or $= 2$ allowed transitions. The dip in the centre of the $\Delta m = 1$ line we find for all acceptors; it corresponds to the sharp central line in the EPR of Si:B reported by Neubrand (1978). It may be due to the additional paths in the relaxing Γ_8 - quartet when the upper and lower $\Delta m = 1$ transitions become equal in the centre of the line. Such a dip was not found in (pulsed) acoustic paramagnetic resonance (APR) up to 12 GHz (Zeile and Laßmann 1982 and this work) and at 24 GHz (Schulze 1984). On top of the $\Delta m = 2$ line of Fig. 2 (as well as in other cases) a weak extra peak is discernible also consistent with Neubrand's finding for the centre of the $\Delta m = 2$ line. For the orientation allowing a $\Delta m = 2$ transition the forbidden $\Delta m = 3$ transition is also seen because the high field case is not attained and has similarly been found as an asymmetric line in EPR and APR. The occurrence and the asymmetry of this line means that any transition probabilities as evaluated for the high field case must be corrected for the zero field splitting. An experimental demonstration of this has been given by Zeile and Laßmann (1982). Also the higher field $\Delta m = 1$ line contains some contribution of the forbidden $+1/2 \leftrightarrow -1/2$ transition which is resolved for an intermediate orientation as shown in Fig. 2. Thus, for an absolute evaluation of the coupling strength the resonance splitting should be larger. The relative change, however, for the different acceptor species may be estimated more reliably. (There was no systematic variation of the integrated transition probability on the small variation of linewidth of different samples with same dopant.) Some scatter in our data is due to the difficulty to mount the small samples exactly in the required orientation. Another problem was due to the fact that the cavity has some moveable parts for adapting to the sample and for tuning: This apparently lead to some mechanical instability at higher field

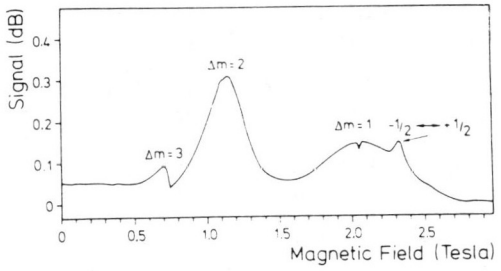

Fig.2 Dielectric resonance absorption for an intermediate orientation of a Si:In sample where in addition two both allowed the two forbidden $\Delta m = 1$ and $= 3$ transitions appear. Indium concentration $c_{In} = 3.9 \times 10^{22} m^{-3}$.

strengths with some signal drift as a consequence. With the provisos just mentioned we get for the coupling constants (dipole moments) p_χ:

dopant / E_B (meV)	dipole moments p_χ (Debye)	
In / 156	1.0±.5	3 samples Si:In, C_{In} = 4 to 8×10^{22} m^{-3}
Ga / 73	.5±.35	1 sample Si:Ga, C_{Ga} = 1.2×10^{22} m^{-3}
Al / 69	.7±.4	1 sample Si:Al, C_{Al} = 6.9×10^{22} m^{-3}
B / 44	.2±.15	5 samples Si:B, C_B = .1 to 6×10^{22} m^{-3}

The p_χ are average values of several runs for the two orientations of

Our g-factors compare well with those derived from the data of Feher et al (1960). The absolute value of p_χ can only roughly be estimated since the cavity/sample geometry was not well enough defined. Assuming that the capacitive part of the cavity is identical with the sample volume p_χ was obtained from the integrated <u>change in Q-factor</u> in the case of In where the signal was strong enough. For the weaker signals of Ga, Al, and B the values given are from comparison of the integrated <u>change in cavity reflection</u> signal to that for In. The error bars indicated contain the scatter of these latter relative values for each dopant (which were only about 20%) plus the scatter in the determination of the Q-factor in the case of In. (These absolute values have been evaluated without the consideration of the internal field due to the polarization of the silicon lattice as discussed by Mims (1976).) The scatter in the relative values is smaller: We obtain a factor of 5.1±1.6 for the ratio of p_χ between In and B. Since the coupling is determined by the innermost part of the wave function in the central cell it is difficult to estimate. As a guideline we have calculated $\langle |r| \rangle$ (instead of $\langle r \rangle$ which is zero for centro-symmetric wavefunctions) for the EMA and δ-potential wavefunctions (with a_B defined by the experimental binding energy as given above) over spheres of the lattice constant (as well as half the lattice constant) and obtained a ratio of 3.8 (4.9) between In and B for EMA and a factor of 1.2 (1.5) for δ-potential. In the case of a crossover (EMA for B, δ-potential for In) the ratios would be larger: 8.8 (40).

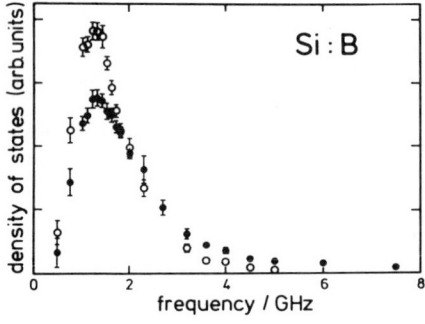

Fig.3 Splitting distributions of a Si:B sample partly compensated with P by neutron transmutation. In the neutralized (illuminated) state (o) the distribution is narrower. Electric fields from donor-acceptor pairs (C_{da} = 1×10^{21}m^{-3}) are responsible for the broadening in the ionized (nonilluminated) state (●). Boron concentration C_B = 5.5×10^{21}m^{-3}.

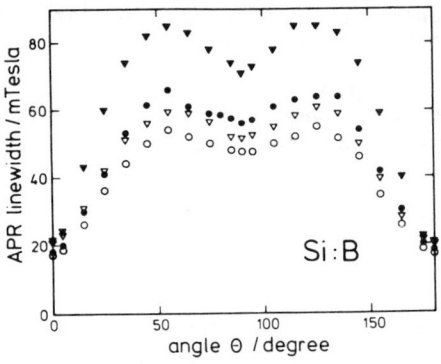

Fig.4 Angular dependence of the width of the $\Delta m = 1$ line in acoustic paramagnetic resonance at 7.5 GHz of the sample of Fig. 3 after partial ($\triangledown$,▼) and after complete (o,●) annealing of the radiation defects (i.e. more or less ion-pairs). Since the difference between illuminated (open) and nonilluminated (filled) state is zero at 0° there is negligible contribution due to the quadratic Stark effect.

The B-doped samples (c_B = 5.5 and 20×10^{21} m^{-3}) for the measurement of the influence of internal electric fields were irradiated with thermal neutrons at Harwell resulting in donor concentrations .5 and 1×10^{21} m^{-3}. They were then annealed at several temperatures. After the final step at 750 °C the splitting distribution was close to that before n-transmutation if the sample was illuminated with bandgap light at low temperatures to neutralize the donor-acceptor ion pairs, i.e. the defects were essentially healed. Without illumination, however, the distribution is broader due to the random electric fields from the ions as seen in Fig. 3. As mentioned above and described by Zeile and Laßmann (1982) these curves are obtained by measuring the resonance attenuation of ultrasound in the corresponding frequency range. To distinguish between quadratic and linear Stark effect measurement of the APR is helpful: The linewidth for certain directions of the magnetic field does not contain a contribution of the linear coupling. In Fig.4 this is the 0° direction. From comparison of the angular dependence of the linewidth with and without illumination we find that the quadratic coupling has no measurable effect so that the broadening of the distribution is solely due to the linear coupling. From comparison with MC-calculations we derive $p_X = (0.26 \pm 0.13)$ Debye which compares quite well with the result of the resonance method.

Financial support by the Deutsche Forschungsgemeinschaft is gratefully acknowledged. We are obliged to many colleages for the supply of samples: W. Zulehner, Wacker Chemitronic (Si:B,Al,Ga); J.S. Blakemore, then at Florida Atlantic Univ. (Si:In); M. Vilain, LETI, Grenoble (Si:In); N.W. Crick, Harwell (neutron transmutation) and L. Challis, Nottingham, for arranging this possibility.

References

Bir G L, Butikov E I and Pikus G E 1963 J. Phys. Chem. Sol. 24 1467, 1475
Feher G, Hensel J C and Gere E A 1960 Phys. Rev. Lett. 5 309
Kohn W 1957 Solid State Physics 5 258, ed. F. Seitz et al, Acad. Press.
Mims W B 1976 The linear electric field effect in paramagnetic resonance
  Clarendon Press, Oxford
Neubrand H 1978 phys. stat. sol. (b) 86 269 and 90 301
Rynne E F, Cox J R, McGuire J B and Blakemore J S 1976
  Phys. Rev. Lett. 36 155
Schulze H 1984 Thesis, Stuttgart
Yafet Y 1965 J. Phys. Chem. Sol. 26 647
Zeile H and Laßmann K 1982 phys. stat. sol.(b) 111 555

Position of the chemical potential, near the melting temperature, in heavily shallow-impurity doped silicon

B E Sernelius

Department of Physics and Measurement Technology, Linköping University,
S-58183 Linköping, Sweden.

ABSTRACT: The chemical potential, near the melting temperature, of n-type doped silicon is presented for doping levels ranging from moderate to extremely high. The derivation is within the finite-temperature formalism and is based on the thermodynamic potential. The effects from electron- electron interactions, i.e. exchange and correlation, as well as from electron-donor-ion interactions are included. The results are compared to those from a simpler calculation in which the non-interacting chemical potential is determined and to which is added the zero-temperature shifts caused by the interactions. Both derivations are within RPA (the Random Phase Approximation).

1. INTRODUCTION

In laser annealing experiments on Si one can reach much higher, homogeneous doping levels than with other doping methods. In these experiments, and in the related experiments on rapid solidification, the thermodynamics of the different phases appearing in the processes is not fully understood (Aziz 1986, Lowndes 1986). One key entity in the theory is the chemical potential for the carriers, which depends on both the local doping level and the local temperature. At low temperatures it is a well-established fact that high doping levels mean important shifts of the conduction- and valence bands, leading to a reduction in the band-gap value (Abram et al 1978). The chemical potential is shifted in a similar way. This shift might be of fundamental importance in the processes. It is, e.g., found to be much easier to melt, with the laser, a heavily doped region as compared to an undoped region (Lowndes 1986).

Let us discuss n-type doping. At low temperatures the conduction-band edge and the chemical potential are both shifted downwards in energy. In the case of silicon these shifts are of almost the same size (Berggren and Sernelius 1981) but this is not true in general (Sernelius 1986a,1986b,1987). The shifts can be divided into contributions from electron-electron- and electron-donor-ion interactions. The electron-electron contribution can further be separated into an exchange part and a correlation part. The exchange part dominates, and comes from the blocking (according to the Pauli principle) of channels for interaction processes, involving the occupied conduction-band states. This effect is reduced with increasing temperature since the state occupancy is decreasing. Thus, it is obvious that the exchange contribution to the energy shift is reduced with increasing temperature but it is difficult to predict the temperature dependencies of the other two contributions.

We use the finite-temperature formalism to determine the chemical potential and use as a starting point the thermodynamic potential. These derivations as well as the numerical results are presented in Section 3. The numerical effort needed here is more involved as compared to

© 1989 IOP Publishing Ltd

the zero-temperature formalism. This is because the finite-temperature form of the dielectric function for the donor electrons is here needed and the RPA expression cannot be obtained in analytical form. The imaginary part, however, can be reduced to analytical form (Sirko and Mills 1978) and the real part is found numerically from it with the use of Kramers Kronig dispersion relations. We do not need, however, the real part here. We need the dielectric function for a discreete set of frequencies on the imaginary frequency axes. This is found numerically from the imaginary part through similar dispersion relations.

Before the full finite-temperature calculation is presented we give in Section 2 the general results for the non-interacting chemical potential for parabolic bands. Figure 1 displays the result from extremely low to extremely high effective temperatures together with the high and low temperature limits. We further show how to overcome the numerical problems with an integrand containing the Fermi-Dirac distribution function.

The numerical results from the full finite-temperature calculation are presented in Section 3 and compared to simpler results obtained by adding the zero-temperature interaction-shifts to the non-interacting chemical potential from Section 2.

Finally, in Section 4 we make a brief summary.

2. NON-INTERACTING CHEMICAL POTENTIAL

In this section we study the non-interacting chemical potential at finite temperatures. It is in the general case implicitely found from the standard relation

$$n = \int_0^\infty d\varepsilon \, g(\varepsilon) \frac{1}{e^{(\varepsilon-\mu)\beta} + 1} \tag{1}$$

where **n** represents the carrier density, $g(\varepsilon)$ the density-of-states and the last factor in the integrand is the Fermi-Dirac distribution function. By introducing the unitless quantities $B=\beta \cdot E_F$, $M=\mu/E_F$, $x=\beta \cdot \varepsilon$ and $A=B \cdot M=\beta \cdot \mu$ Equation 1 can, in the case of parabolic bands, be transformed into the following universal, implicit relation between M and B:

$$B^{3/2} = \frac{3}{2} \int_0^\infty dx \, x^{1/2} \frac{1}{e^{x-A} + 1} \tag{2}$$

The parameters of the system are contained in E_F. Silicon has $\nu=6$ equivalent anisotropic conduction-band minima. The energy dispersion around each minimum is parabolic in all directions (which is enough for Equation 2 to be valid) and the density-of-states effective mass, m_{de}, has the value 0.322.

It is difficult to solve this integral in Equation 2 numerically as the integrand contains the Fermi-Dirac distribution function. To circumvent this problem we make the following substitution:

$$y^2 = \frac{1}{e^{x-A} + 1} \tag{3}$$

which reduces Equation 2 to the form

$$B^{3/2} = \int_0^{(1+e^{-A})^{-1/2}} dy \frac{3y}{1-y^2} (A+\ln(\frac{1-y^2}{y^2}))^{1/2} \quad (4)$$

This integrand is well-behaved and vanishes at the two limits of integration. For a given A one obtains B. M is then obtained as A divided by B. This means that a relation between M and B is found.

The result from Equation 4 is presented in Figure 1 (solid curve) in the form of 1-M as a function of 1/B, i.e., the distance of μ, in units of E_F, below E_F as a function of T/T_F. T and T_F are the temperature and Fermi-temperature, respectively. The dashed curves are the low- and high temperature limits of the relation, viz.

$$M = 1 - \frac{\pi^2}{12} \frac{1}{B^2} \quad ; \quad T \rightarrow 0 \quad (5)$$

and

$$M = \frac{1}{B} \ln(\frac{4 B^{3/2}}{3 \pi^{1/2}}) \quad ; \quad T \rightarrow \infty \quad (6)$$

respectively

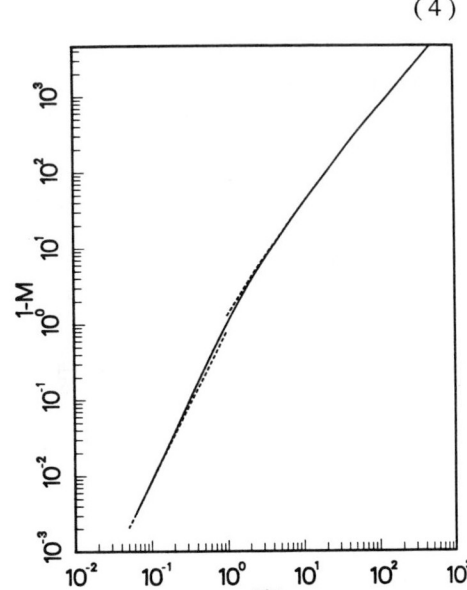

Fig. 1 The solid curve shows the position below E_F, in units of E_F, of the non-interacting chemical potential as a function of T/T_F. The dashed curves are the low-and high temperature limits, given in Equations 5 and 6. The results are general for parabolic bands.

3. FINITE-TEMPERATURE CHEMICAL POTENTIAL

We determine the interacting chemical potential from the thermodynamic potential Ω which can be separated into a non-interacting part, Ω_o, and interacting parts according to

$$\Omega(T,V,\mu) = \Omega_o + \Omega_1 + \Omega_r + \Omega_{ion} + \ldots \quad (7)$$

where Ω_1, Ω_r and Ω_{ion} are the exchange part the ring-diagram part and the contribution from the interaction with the donor ions, respectively. The dots represent all contributions not included here. We want μ as a function of T and **n**, but the formalism gives us **n** as a

function of T and μ from the relation:

$$N(T,V,\mu) = -\left(\frac{\partial \Omega}{\partial \mu}\right)_{T,V} \tag{8}$$

Dividing both sides of the equation by V we get **n** on the left side and Ω on the right hand side is changed to the thermodynamic potential per unit volume.

Before we continue we must specify our approximations. We replace the ν anisotropic conduction band valleys with isotropic ones, assume that the donor ions are distributed randomly and that the potentials to a good approximation can be replaced by pure Coulomb potentials, $v(q)$. The interaction with the donor ions is treated to second order in the impurity potential. We let κ denote the back-ground dielectric constant, which for silicon has the value 11.4.

The expressions for the thermodynamic potentials are

$$\Omega_1 = \frac{1}{2}\sum_{\vec{q}}\left[\frac{\hbar}{\beta}\sum_n\{\alpha_o(q,i\omega_n)\} - v(q)\cdot n_o\right], \tag{9}$$

$$\Omega_r = \frac{1}{2}\sum_{\vec{q}}\left[\frac{\hbar}{\beta}\sum_n\{\ln(1+\alpha_o(q,i\omega_n)) - \alpha_o(q,i\omega_n)\}\right] \tag{10}$$

and

$$\Omega_{ion} = -n_{ion}\frac{2\pi e^2}{\kappa}\sum_{\vec{q}}\frac{\alpha_o(q,0)}{q^2\varepsilon(q,0)} \tag{11}$$

where n_{ion} is the density of donor ions, n_o is the density of electrons at the particular μ and T in the absence of interactions. The function α_o is the polarizability for the electrons for the given T, μ and electron density n_o. The function ε is the corresponding dielectric function, i.e., $\varepsilon = 1 + \alpha_o$. As can be seen in Equations 9 and 10 the polarizability is needed in the discrete set of frequencies $i\omega_n$, where

$$\omega_n = \frac{2n\pi}{\beta\hbar} \quad ; \quad n = 0,\pm 1,\pm 2,\pm 3,\dots\dots \tag{12}$$

The polarizabilities are real-valued on the imaginary frequency axes and can be obtained from the knowledge of the imaginary part of the retarded polarizability on the real axes through the following relation:

$$\alpha_o(q,i\omega_n) = \frac{1}{\pi} \int_0^\infty d\omega \, \varepsilon_2^R(q,\omega) \frac{2\omega}{\omega_n^2 + \omega^2} \qquad (13)$$

Much could be said about the numerical calculation and about tricks to improve the convergence, but there is no space available here. Therefore, we directly present the numerical results in the form of Figure 2.

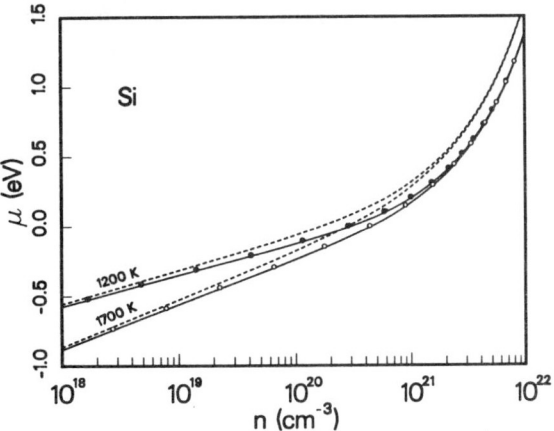

Fig. 2. The results of the calculation for two temperatures, bracketing the melting temperature. The circles, filled for 1200 K and open for 1700 K, are the results from the full finite-temperature calculation. The dashed curves are the corresponding non-interacting results from Section 2. The solid curves are obtained by adding, to the dashed curves, the zero-temperature interaction shifts in the chemical potential. The reader is referred to the works of Berggren and Sernelius (1981) and of Sernelius (1986a,b,1987) for details in how these shifts are calculated. The reference energy is the bottom of the conduction band in the undoped semiconductor at zero temperature. The curves are only valid in regions where the chemical potential is well above -0.5 eV. Outside those regions thermally excited electrons and holes will give important contributions, not included here.

4. SUMMARY AND CONCLUSION

We have determined the position of the chemical potential for n-type doped silicon as a function of doping concentration at finite temperatures. The interactions within the system of donor electrons and donor ions were included but not the effects from interactions with phonons. The calculation was performed in the finite-temperature formalism. A second, approximate calculation was performed in which the non-interacting chemical potential was determined and to which was added the zero-temperature shifts. As is found in Figure 2 this simplified approach gives results, solid curves, that are surprisingly close to the results from the full calculation, circles. The results cover the range in effective temperatures, T/T_F, from 0.1 to 24. This means that the reduction in the contribution to the energy-shifts from exchange with temperature is dramatically compensated by the two other contributions, viz., correlation and ion-interactions. This is in line with the results found by Rösler et al (1984) and Thuselt and Rösler (1985).

ACKNOWLEDGMENTS

Research support is acknowledged from the Swedish Natural Science Research Council. Part of this work was carried out during a stay at the University of Tennessee/Oak Ridge National Laboratory.

REFERENCES

Abram R A, Rees G J and Wilson L H 1978 *Adv. Phys.* **27** 799
Aziz M J 1986 Private communications
Berggren K-F and Sernelius B E 1981 *Phys. Rev.* B **24** 1971
Lowndes D H 1986 Private communications
Rösler M, Zimmermann R and Richert W 1984 *Phys. Stat. Sol. (b)* **121**
Sernelius B E 1986a *Phys. Rev.* B **33** 8582
Sernelius B E 1986b *Phys. Rev.* B **34** 5610
Sernelius B E 1987 *Phys. Rev.* B **36** 4878
Sirko R and Mills D L 1978 *Phys. Rev.* B **18** 4373
Thuselt F and Rösler M 1985 *Phys. Stat. Sol. (b)* **130** 661

Uniaxial stress by substrate bending: application in a study of boron acceptors in Si MBE layers

J.A. Chroboczek
Centre National d'Etudes des Télécommunications, 38240 Meylan, France,

and A. Briggs
CRTBT-CNRS, Avenue des Martyrs, 38042 Grenoble Cedex, France.

ABSTRACT : Thin Si:B layers grown by molecular beam epitaxy (MBE) on Si have been shown to possess the transport properties of bulk Si:B. Because the active layer/substrate thickness ratio is small (1/300) the active layer can be uniformly stressed by bending. We show that the values of stress thus obtained (Stress By Substrate Bending, 2SB) are well accounted for by classical theory. The main advantage of the 2SB method, apart from its simplicity, is that it allows both compressive and tensile stresses to be applied. We demonstrate the correct functioning of the method by studying piezoresistance in Si:B in conditions of strong localisation. The results compare favorably with appropriate bulk data, where stress can be directly calibrated. The 2SB method can be used in studies of piezo effects in other thin layer systems and heterostructures in particular.

1. ELECTRICAL CURRENT TRANSPORT IN MBE Si/B LAYERS

The MBE Si:B layers were fabricated by coevaporation of Si and B, using a novel high-temperature effusion cell. The fabrication method and sample characterisation have previously been discussed by Andrieu et al., (1988). At room temperature (RT) the mobility, μ, vs. impurity concentration, N, and the resistivity, ρ vs. N dependencies were found to be very close to those of bulk Si:B (Thurber et al., 1980). At 77K<T<300K the values of μ were found to be slightly inferior to those of the bulk Si:B, suggesting the presence of scatterers other than ionized B atoms. In this work we studied transport at low T in lightly doped samples, and the resistivity vs. 1/T plot is shown in Fig. 1, for two samples of Si with B concentration of $1.67 \times 10^{17}/cm^3$. The plot is typical for impurity conduction in Si (cf. e.g. Pollak, 1987). The data of Fig. 1 were taken using a collinear, 4-probe, equidistant contact array, with spacings for each sample differing by about 20%. The left hand scale of Fig. 1 gives the V/I values, in Ω and the right-hand scale the resistivity, in Ω cm. Note that the V/I values for both samples coincide almost perfectly. This feature is characteristic of thin layers alone; when the probe spacing exceeds by more than 10 times the film thickness, d, the resistance (called then Sheet Resistance, $R_\Box$), does not depend on the probe spacing and can be expressed as,

$$R_\Box = (\pi/\ln 2)(V/I) \text{ and } \rho = R_\Box \times d, \qquad (1)$$

the latter corresponding to the right-hand scale in Fig. 1, where we used d = 0.95 μm, determined separately by spreading resistance measurements at RT. The $R_\Box$ concept is commonly used in thin film physics, but relatively

© 1989 IOP Publishing Ltd

unknown in impurity conduction studies, hence this discussion. It should also be emphasized that the correct determination of $R_\square$ with different probe arrays proves that the current is restricted to the B-doped layer, which is essential for the interpretation of the data we present.

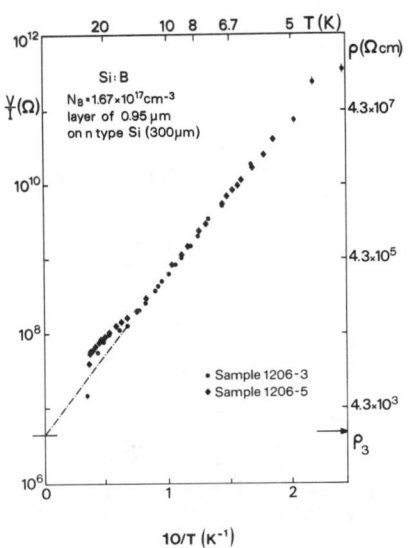

Fig. 1. Temperature dependence of V/I ($R_\square$ × ln2/π) and resistivity measured in thin Si:B layers. The dependence is typical for impurity conduction involving shallow states.

2. STRESS BY SUBSTRATE BENDING (2SB)

Having established that the current is restricted to the Si:B active layer, which is about 300 times thinner than the Si substrate, we realized that uniform uniaxial stresses can be obtained in the active layer by substrate bending and be calculated using the classical solution of Bernoulli for a bent beam problem. Let us define the system of coordinates by positioning the undeformed beam (length L, thickness t, and width W) in the plane z=0, along the x axis, with the point of support at x=0, and the force F, parallel to the z-axis, applied at x=L. When the deformation z(x)≪L, the stress in the infinitely thin layer at the surface of the beam is,

$$S(x) = (6E(L-x)/Wt^2)F \quad \text{or} \quad S(x) = 2/3(Et(L-x)/L^3)z(L) \quad (2)$$

where E is Young's modulus. Because the stress value depends on the distance from the point of support, x, stress is uniform in infinitesimally narrow strips across the beam.

2.1 Experimental details

Samples. We used narrow (L=26mm, W=5mm, t=0.3 mm), [1$\bar{1}$0] oriented, slabs cut from the (111) Si wafers, with the MBE Si:B active layer of the thickness d = 0.95 μm. The orientation of the slabs is not without importance for mechanical properties of the system ; orientation of the cleavage planes normal to the beam axis should be avoided. In our case the cleavage planes are oriented at 35° from the axis and whenever the samples broke, the fracture lines followed the <110> orientations. The bending force was applied normal to the surface, thus we used the value of E(111) = 1.9x10^{12} dynes/cm^2. Tests performed at RT showed that a strip of Si of the dimensions

specified above can be bent by about 3.5 mm. For x<<L, Eq.(2) gives the stress values of the order of 10^8 Pa (1kbar) per 1mm deflection of the slab, thus the maximum value of the stress should exceed 3×10^8 Pa.

Electrical Contacts. In this work we used a novel method of contact formation to Si, developed at the C.N.E.T. laboratories and described in part by Chroboczek and Arnaud d'Avitaya (1986). The principle of the method lies in forming the Si/Au or Si/Al eutectics directly on the Si surface; the former proved to be suitable for undoped or n-type Si, the latter for p-type Si. Both Si/Au and Si/Al contacts have an interesting morphology, reflecting the surface orientation ; on the (100) surfaces the contact spots are usually quadratic and on (111) surfaces they are hexagonal. The Si/Au eutectic mass penetrates into the Si and is delimited by crystalographic planes of high symmetry (equilateral pyramid in the case of the (001) face). The Si/Al eutectic contacts to p-type Si have excellent ohmic properties down to 4.2K. Their penetration depth is less extensive than that of Si/Au and is of the order of 1μm. The Si/Al contacts are, therefore, suitable for studies of thin layers of p-type Si, deposited on resistive or n-type Si (where a barrier formation is likely to occur). A typical I/V characteristic and a microphotograph of a Si/Al contact is presented in Fig. 2. The method of contact formation will be discussed in a forthcoming publication.

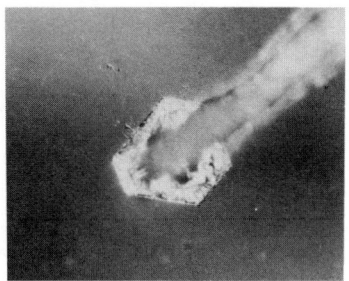

A   B

Fig. 2 : A) Microphotograph of a Si/Al eutectic contact formed on a (111) surface of Si, using 50μm dia. Al wire; B) An I-V characteristic, current passing through two contacts at RT. Note a hexagonal form of the contact spot (A) and the linear (ohmic) characteristic (B), 50μA/cm vs. 2V/cm.

Stress Rig. Because the force needed for bending the Si strip up to the breaking point is small, the stress can be applied by fixing one end of the sample and moving the other with a fork-like device, displaced by a screw or some other mechanical system. We used an L-shaped copper block, both for supporting the sample and housing a screw for displacing the sample's free end. The longer arm of the L (horizontal) was in thermal contact with a cold finger of a He cryostat. The shorter arm of the L pointed down and the sample was mounted on its end, parallel to the longer arm. A fine-thread copper screw was passed vertically through the extremity of the longer arm of the L and its end was fitted with a grooved (0.5mm) Teflon wheel, to house the sample's free end. The screw could be turned from the top of the cryostat, however, its actual displacement in the cryostat was directly monitored using a device consisting of a minature drum, resembling a collector of an electric motor (with parallel Cu and insulator strips), which was rigidly attached to the screw's axis. A passage of each Cu strip

by a slider pressed against the drum was registered by a telephone counter at the cryostat's top and thus the real displacement of the sample's end could be measured. Knowing z(L) and the position of the contacts, x, the value of S(x) could be obtained from Eq. (2).

Electrical measurements. As mentioned above, we used a collinear system of four equidistant contacts made across the sample's width and determined the sheet resistance of the active layer, as discussed in Sec. 1. Resistance values between contacts at low temperatures could be as high as 10^{11} Ω, (cf. Fig. 1), which required employing high-input impedance instruments. We used for this purpose a four-probe electrometer system with driven shields (Chroboczek and Link, 1985) which allows resistances up to 10^{11} Ω to be measured, with a time constant of the order of 1ms. Connections of the sample contacts to the electrical system were made by floating coaxial cables inside the cryostat and by triaxial cables outside (external shield grounded). The entire system showed perfect linear response at the voltages developed between central contacts, which were selected not to exceed 0.2 V.

2. DISCUSSION OF THE DATA

2.1 Impurity conduction in MBE Si:B layers

The ρ vs. 1/T dependence at low T, shown in Fig. 1, is typical for nearest neighbor hopping (NNH), where ρ can be expressed as a product $\rho_3 \exp(\varepsilon_3/kT)$.

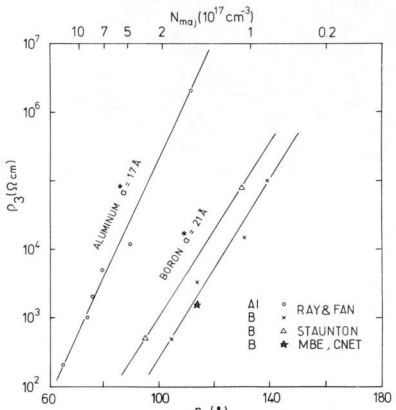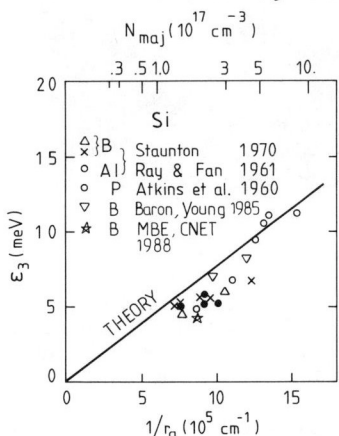

Fig.3. The preexponential and the activation energy for NNH in bulk Si, compared with the values that we obtained on MBE Si:B thin layer samples.

The pre-exponential, ρ_3, and the activation energy, ε_3, obtained in the MBE Si : B are compared in Fig. 3, with other available data for bulk Si samples, indicating that thin layer and bulk Si:B behave similarly. The value of ε_3 is determined from the low temperature part of log ρ vs. 1/T plot. At higher T, when kT becomes comparable with ε_3, NNH becomes less efficient (saturation of NNH) because the final states for tunneling transitions become heavily populated. The overshoot of the values of ρ above the extrapolated $\exp(\varepsilon_3/kT)$ line (dashed) results from this effect. This phenomenon, which has been reported in bulk samples, however, has not yet been analysed in detail, because the latter requires the knowledge of the

density of states in the impurity band. The MBE samples, where N can be more easily controlled, seem to provide a more manageable system for such analysis. (For definitions of ρ_3, ε_3, their physical meaning, and the NNH saturation effect, see, for example Chap. 3, Vol.4, Pollak, 1987).

2.2 Piezoresistance in percolative transport via B ground states

The NNH piezoresistance (PR) in bulk Si:B has been studied for $S||[001]>0$ (compression) for S up to 1.6×10^9 Pa (Staunton, 1970; Chroboczek, Pollak, Staunton, 1980). Some of our results (obtained using 2SB) are compared, in Fig. 4, with former results obtained on two bulk samples with similar B concentrations. The comparison should be considered as qualitative because the bulk Si data were taken for $S||[001]$. However, recent calculations by Buczko 1987, show that the splitting of the ground state of the acceptor does not depend crucially on the stress orientation. It should be noted that the magnitude of PR for all three samples scales correctly with the B concentration. The 2SB data supplement the bulk data providing a set of points for S<0. Note that for the bulk samples the PR passes through a maximum at S near 3×10^8 Pa, the value close to the maximal stress that could be attained by the 2SB method, in its present state. This rather unfortunate coincidence did not permit us to reach stress values big enough to reproduce the maximum in the NNH PR, the feature important for stress calibration. It should be noted, however, that the PR peaks for $S||[110]$ might lie at somewhat higher values of S. Measurements involving such oriented bulk samples are, therefore, envisaged.

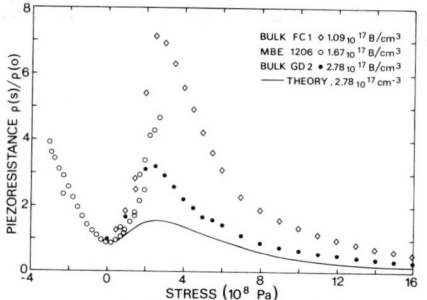

Fig.4. Piezorestance of NNH conduction in Si:B. The data for two bulk samples and one MBE Si:B samples are compared. Note that the 2SB method permits both, tensile and compressive stresses to be applied to the same sample.

The model of the PR in group IV, p-type semiconductors in NNH has been discussed in detail elsewhere (see e.g. Buczko and Chroboczek, 1984, and references therein), however, to make our results comprehensible, we shall very briefly describe its main features. The acceptor ground state (4-fold degenerate) is split by a uniaxial deformation into two, 2-fold degenerate states. The amplitudes of the wave functions of each pair are spatially complementary to each other, in the sense that they form a spherical surface when superimposed. If the states are split by more than kT, some transitions in the solid angle, where the wave function associated with the more populated states extends less, are inhibited. Consequently, when a contribution of one of the wave functions in NNH transport diminishes with increasing splitting, resistance should increase, symmetrically around S=0, as observed. At higher values of S the states that remain populated approach the continuum (the ground state binding energy diminishes with increasing S). The consequence of this is the overall swelling of the wave function, and a consecutive drop of resistance, as seen for S>0 in the bulk

samples. Results of a calculation of the PR effect for $S||[100]$ for $N = 2.78 \times 10^{17}/cm^3$ (Buczko and Chroboczek, 1984) are shown by the solid curve in Fig. 3. The calculation was based on the effective mass approach with no central cell corrections taken into account. This might be a reason for rather severe discrepancy between experiment and theory.

3. CONCLUSIONS

We demonstrated that uniform uniaxial stresses, both tensile and compressive, up to $3 \times 10^8 Pa$ (3kbar) can be obtained by the 2SB method, in thin films of doped Si deposited by MBE on standard Si substrates. We showed that the values of stress thus obtained are well accounted for by the Bernoulli solution of the bent beam problem. We further demonstrated that existing piezoresistance data (for compression) obtained on bulk Si:B in the strong localisation regime compare favorably with the results obtained by substrate bending and extended piezoresistance data for tensile stresses. It should be noted that the top of the valence band (VB) of common semiconductors (Si, Ge, GaAs, etc.) can be split by about $6meV/10^8 Pa$, thus the 2SB method should permit one to tune transitions involving VB states by about +/-20meV. The systems that seem to be particularly well-adapted for studies involving the 2SB method are the Si/III-V MBE buffered heterostructures which are known to be strained due to the difference in the thermal dilatation coefficients, with S values of a few $10^8 Pa$. For completeness we should mention two papers, relevant to this study, that we found in the course of this work. Dorda and Eisele (1973) used bending of Si samples in a study of inversion layers in MOSFETS and Pinczuk et al. (1986) in their study of III-V heterostructures used uniaxial stress up to $1.5 \times 10^8 Pa$, without mentioning the method of stress application, however a method involving bending was probably used.

ACKNOWLEDGMENTS

Our thanks are due to our colleagues from the MBE group of the CNET for their help and assistance at various stages of this work.

REFERENCES

Andrieu A, Chroboczek J A, Campidelli Y, André E and Arnaud d'Avitaya F, 1988, *J. Vac. Sci. Techn.*, in print
Atkins K R, Donovan R and Walmsley R H 1960 *Phys. Rev.* **118** 411
Baron R and Young M H 1985 *Solid State Electron.* **28** 204
Buczko R 1987 *Il Nuovo Cim.* **9** 669
Buczko R and Chroboczek J A 1984 *Phil. Mag. B* **50** 429
Chroboczek J A and Arnaud d'Avitaya 1986 Patent No 8616030 (France) *Procédé de formation de contacts ohmiques sur du silicium*
Chroboczek J A and Link J 1985 *J. Phys. E, Sci. Instr.* **18** 568
Chroboczek J A, Pollak F H and Staunton H F 1984 *Phil. Mag. B* **50** 113
Dorda G and Eisele I 1973 *phys. stat. sol.(a)* **20** 263
Pinczuk A, Heinman D, Sooryakumar R, Gossard A C and Wiegmann W 1986 *Surf. Sci.* **170** 573
Pollak M 1987 *Noncrystalline Semiconductors* (Boca Raton: CRC)
Ray R K and Fan H Y 1961 *Phys. Rev.* **121** 768
Staunton H F 1970 *Thesis, Brown University*, U S A (unpublished)
Thurber W R, Mattis R L, Liu Y M and Filliben J J 1980 *J. Electrochem. Soc.* **127** 2291

The stability of In-As pairs in silicon observed by the PAC-method

S Deubler, N Achtziger, P Dohlus, D Forkel, H Wolf, and W Witthuhn

Physikalisches Institut der Universität Erlangen-Nürnberg,
D-8520 Erlangen, Germany

Abstract: The formation and dissociation of In-As pairs in Silicon are studied by the Perturbed Angular Correlation method (PAC) using radioactive ^{111}In(--> ^{111}Cd) probe atoms. The In-As pairs are stable at temperatures below 1400 K and characterized by a strongly temperature dependent electric field gradient (EFG). The data are explained quantitatively in the framework of a model based on the thermally activated population of the Cd-As acceptor level.

1. INTRODUCTION

The electronic properties of semiconductors are influenced by the mutual compensation and passivation of acceptors and donors, even small impurity or defect concentrations are able to alter drastically the charge carrier density. Therefore the acceptor-donor interaction has been studied intensively, applying numerous techniques like ESR, IR-spectroscopy or DLTS. Despite these great efforts the chemical identity, the electronic properties and the local structure of many acceptor-donor configurations are still discussed controversely. Here the PAC-method offers a direct experimental access to the microscopic investigation of next-nearest acceptor-donor pairs, since the local charge distribution is measured at the acceptor or donor atom itself.

The first results about the acceptor-donor interaction in Silicon studied by the PAC-method are published by Forkel et al. (1985), stimulating further PAC investigations (Forkel et al. 1986, Wichert et al. 1986, Swanson et al. 1986). Here we report on the formation and stability of In-As pairs in Si and the electronic properties of the Cd-As acceptor comlex.

2. EXPERIMENT

In the present experiment the PAC-measurements were performed at the isomeric $5/2^+$ level of ^{111}Cd ($T_{1/2}$ = 84 ns) which was populated by the EC-decay of ^{111}In ($T_{1/2}$ = 2.8 d). Therefore the trapping and detrapping of impurities or defects is governed by the electronic properties of the

© 1989 IOP Publishing Ltd

In atom whereas the EFG is measured at the Cd nucleus, thus yielding information about the electronic structures of the In- and the Cd-impurity complexes.

Samples of n-Si ($[As] = 8 \cdot 10^{18}/cm^3$) were doped with ^{111}In atoms by the Silicon Direct Bonding process, which was adapted to the specific requirements of the PAC-spectroscopy (Deubler et al. 1988): after hydrophilizing the surfaces of two Si wafers, ^{111}In was applied to one of them. The bonding was performed by keeping the samples several hours at temperatures between 1200 K and 1500 K. During this process the In atoms were diffused into the Si bulk. Based on thermal diffusion data an In concentration of about 10^{15} at/cm^3 could be estimated. After this preparation PAC spectra were taken at temperatures between 4.2 K and 1450 K.

For the time differential detection of the γ-γ-cascade emitted from the excited Cd isotopes a standard four-detector set-up was used. An appropriate combination of the 12 coincidence time spectra leads to an intensity ratio R(t), which is proportional to the product of the angular correlation coefficient A_{22}^{eff} and the perturbation factor $G_{22}(t)$:

$$R(t) = A_{22}^{eff} \sum_i f^{(i)} G_{22}^{(i)}(t) \qquad (1)$$

with

$$G_{22}^{(i)}(t) = \sum_{n=0}^{3} s_{2n}(\eta^{(i)}) \cos(a_n(\eta^{(i)}) \omega_0^{(i)} t) \exp(-a_n(\eta^{(i)}) \delta_0^{(i)} t) \qquad (2)$$

Here $f^{(i)}$ denotes the fraction i of probe atoms exposed to a distinct $EFG^{(i)}$, which is characterized by the basic frequency $\omega_0^{(i)}$. This frequency is related to the quadrupole coupling constant ν_Q, which in turn depends on the largest component V_{zz} of the EFG-tensor:

$$\omega_0 = (3\pi/10) \nu_Q \quad \text{with} \quad \nu_Q = eQV_{zz}/h \qquad (3)$$

The parameter η describes the asymmetry of the EFG tensor:

$$\eta = (V_{xx} - V_{yy})/V_{zz} \qquad (4)$$

The amplitudes s_{2n} as well as the factor a_n depend on η. In the case of an axially symmetric EFG ($\eta = 0$) a_n is equal to n. The quantity δ_0 denotes the width of the EFG distribution, which yields information about remote defects or dynamical processes.

More details concerning the PAC-method applied to semiconductors are given by Forkel et al. (1988).

3. RESULTS AND DISCUSSION

A typical PAC-spectrum for ^{111}Cd in As-doped Si is given in fig. 1. One distinct EFG characterized by ν_Q = 230.9(3)MHz (T=293K), η = 0 and a <111> orientation is observed. The results are in agreement with those obtained by Wichert et al. (1986) for isolated In-As pairs in ^{111}In implanted n-Si.

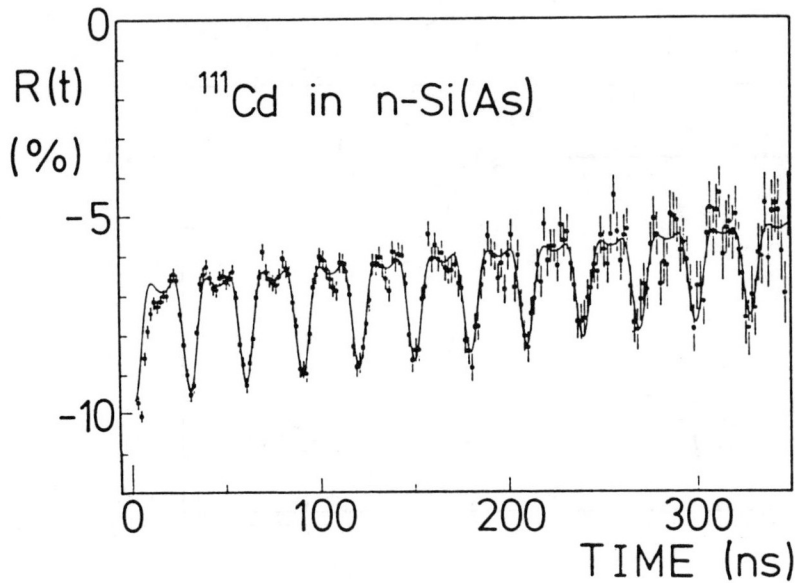

fig. 1: typical PAC-spectra for ^{111}Cd in As doped Si (T_m = 300 K)

The As-induced EFG is nearly constant at temperatures below 800 K, between 800 K and 1300 K it decreases drastically from 230 MHz to 120 MHz (see fig. 2a). Within the same temperature interval a strong damping of the PAC time spectra is observed (see fig. 2b). At temperatures above 1400 K the PAC signal vanishes, here the In-As pairs dissolve.

The temperature dependence of the As-induced EFG can be explained quantitatively by a general model, which was first applied to the compound semiconductor In_2Te_5 (Forkel et al. 1983).

The radioactive EC-decay of ^{111}In transforms the electrically inactive In-As pairs into an electrically active Cd-As complex. The charge states $(Cd-As)^0$ and $(Cd-As)^-$ result in two different electric field gradients $V_{zz}(0)$ and $V_{zz}(-)$ at the Cd nucleus, provided the Cd-As pair represents a deep acceptor in Si. At low temperatures all Cd-As pairs are ionized due to the high As donor concentration and therefore all probe atoms are exposed to the EFG $V_{zz}(-)$. With increasing temperature the Fermi level

shifts to the center of the energy-gap and the ionization probability P(T) of the Cd-As impurity becomes temperature dependent:

$$P(T) = 1/[1 + g \exp((E_A - E_F)/kT)] \tag{5}$$

Here E_A is the acceptor energy level generated by the Cd-As pair, g the corresponding degeneracy factor and E_F the fermi level.

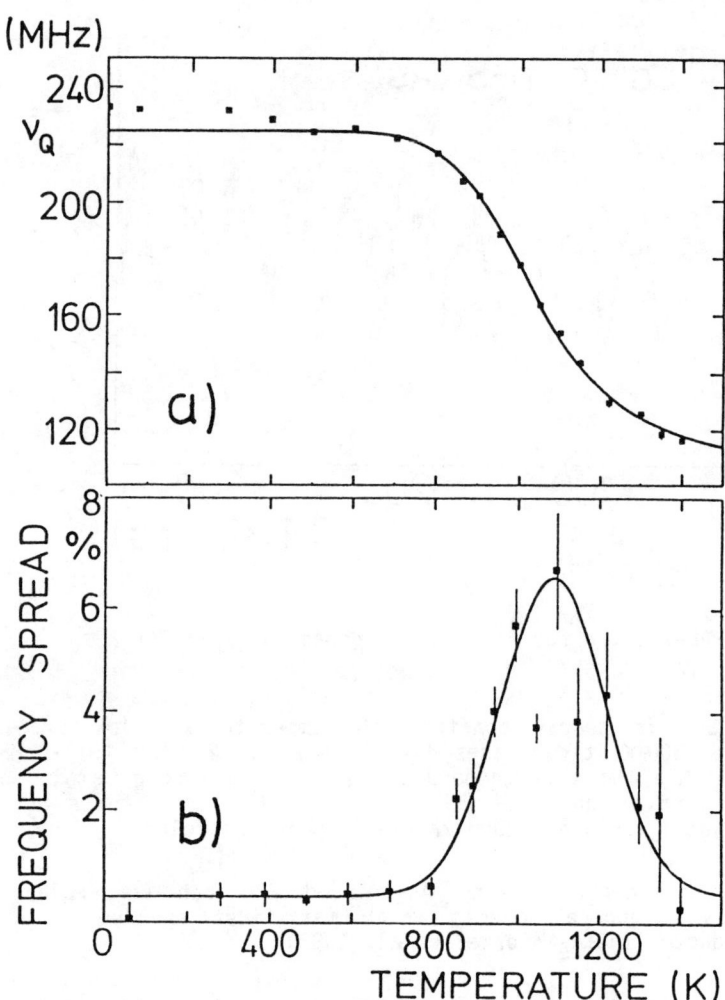

fig. 2: a) temperature dependence of the As-induced EFG
       b) temperature dependence of the frequency spread of the As-induced EFG

The average charge state of the acceptor is given by a dynamical equilibrium between $(Cd-As)^0$ and $(Cd-As)^-$. This dynamic population of the Cd-As acceptor levels induces fluctuating field gradients and the effective EFG at the Cd nucleus can be expressed by [for details see (Forkel et al. 1983)]:

$$V_{zz}(T) = V_{zz}(0)[1 - P(T)] + V_{zz}(-)P(T) \qquad (6)$$

A fit to the data according to this model is represented by the solid line of fig. 2a. The results for the fit parameters are listed in table 1. The fit yields the energy level of the Cd-As acceptor to be $E_A = E_v + 0.64$ eV and an As concentration reduced by a factor of two compared to the bulk value. The latter is supported by spreading resistance data of our samples.

E_A	N_D	$V_{zz}(-)$	$V_{zz}(0)$
$E_v + 0.64$ eV	$4 \cdot 10^{18}$ cm^{-3}	226 MHz	84 MHz

tab. 1: Results of a least square fit of function 6 to the data (a value of g = 4 was used)

The model is further strongly supported by the observed damping of the PAC time spectra. At low temperatures the acceptor charge state is not changed during the time interval of observation (in the order of 1 μs), and thus a static, sharply defined EFG is observed. At temperatures above 800 K the correlation time τ of the thermally induced population of the Cd-As level is in the order of the characteristic times related to the interaction strength, i.e. $\omega_0 \tau \approx 1$, resulting in damped PAC spectra. At higher temperatures the correlation time τ is small compared to ω_0^{-1} and the mean value $V_{zz}(T)$ is sharply defined again ("motional narrowing" region).

4. CONCLUSION

The present model describing the temperature dependence of an EFG related to a defect complex is applicable in all cases where a thermally activated change of the ionization state occurs. The only assumptions for this model are:
  a) the impurity represents a deep acceptor or donor
  b) the EFG at the probe nucleus is different for the neutral and the ionized impurity.

5. ACKNOWLEDGEMENT

We thank Dr. Boit [Siemens, München] for his assistance during the spreading resistance measurements and R. Lindner for helpful cooperation concerning the bonding process. This work has been funded by the German Federal Minister for Research and Technology (BMFT).

Deubler S, Forkel D, Lindner R, and Witthuhn W 1988 to be published
Forkel D, Engel W, Iwatschenko-Borho M, Keitel R and Witthuhn W
   1983 Hyp. Int. 15/16 821
Forkel D, Föttinger H, Iwatschenko-Borho M, Meyer F, Witthuhn W and
   Wolf H 1985 Mat. Res. Soc. Symp. Proc. 46 481
Forkel D, Föttinger H, Iwatschenko-Borho M, Malzer S, Meyer F,
   Witthuhn W and Wolf H 1986 Materials Science Forum, 10 - 12 557
Forkel D, Baurichter A, Deubler S, Wolf H and Witthuhn W 1988 to appear
   in Applied Physics A
Swanson M L, Wichert Th, Quenneville A F 1986 Appl. Phys. Lett. 49 265
Wichert Th, Swanson M L, Quenneville A F 1986 Phys. Rev. Lett. 57 1757

On the nature of defects formed during chemomechanical polishing of silicon

M. Deicher, G. Grübel, R. Keller, E. Recknagel, N. Schulz, H. Skudlik, Th. Wichert, H. Prigge[*] and A. Schnegg[*]

Fakultät für Physik, Universität Konstanz, D-7750 Konstanz, FRG
[*] Wacker-Chemitronic GmbH, D-8263 Burghausen, FRG

Abstract: First experiments using the perturbed angular correlation (PAC) technique on the identification of the defect species leading to an electrically passivation of acceptors in Si after chemomechanical polishing are presented. The formation of close pairs between the acceptor atom and an unknown defect X is shown. The similarities between the behaviour of the X defect and the also neutralizing H and Li atoms will be discussed.

1. Introduction:

During the past few years the passivation of acceptor atoms in Si by H has become well known (Pearton et al 1987), whereby the H atoms can be introduced via a variety of different procedures: plasma or chemical etching, boiling in water, avalanche injection in SiO_2 and other common wafer processing steps. It has also been observed that chemomechanical polishing of p-type Si wafers doped with either B, Al, Ga or In using an alkaline slurry containing ethylendiamine causes an increase of resistivity (Schnegg et al 1986), corresponding to a neutralization of up to $1 \cdot 10^{17}$ cm^{-3} acceptor atoms throughout a mm thick wafer. Until now, the mechanism causing this neutralization is unknown. It has been proposed that the passivating species might be H (Schnegg et al 1988) or an interstitial Si atom (Reichel and Sevcik 1987); however, direct evidence is still lacking.

Using the radioactive acceptor ^{111}In in conjunction with the perturbed angular correlation technique (PAC) it will be shown that close In-defect pairs are formed as a consequence of the polishing procedure and that these pairs are different from the recently observed In-H pairs (Wichert et al 1987).

2. Experimental Details:

Radioactive ^{111}In atoms (lifetime 4 days) are implanted with 350 keV into p-type (100) and (111) Si wafers doped with either $1 \cdot 10^{15}$ or $6 \cdot 10^{16}$ B cm^{-3}. After implantation and annealing at 1170 K the implanted ^{111}In atoms are located within a Gaussian concentration profile centered 1600 Å underneath the surface and do not exceed a local concentration of about $5 \cdot 10^{16}$ cm^{-3}. These samples are chemomechanically polished in the usual way (Schnegg et al 1986) for typically 3 to 5 hours at ambient temperature on the

© 1989 IOP Publishing Ltd

surface opposite to the implanted one which was protected by a lacquer. After polishing, the degree of passivation is characterized with help of a four point probe and subsequently the samples are stored at LN_2 temperature for the PAC measurements in order to prevent the reported reactivation of the acceptor atoms at ambient temperature.

In the case that close In-defect complexes have been formed during the preceeding polishing procedure, a defect specific electric field gradient (efg) would show up. It is measured at the site of the acceptor atom ^{111}In by use of the perturbed angular correlation (PAC) of the emitted γγ cascade when the ^{111}In nucleus decays to an excited ^{111}Cd nucleus. If both γ rays are recorded in coincidence by two γ detectors, forming an angle Θ, the coincidence rate

$$I(t,\Theta) = I_o \cdot \exp(-t/\tau) \cdot (1 + A_{22} \cdot G_2(t,\Theta))$$

can be recorded. Here I_o denotes the coincidence rate at zero time, $\tau = 123$ ns the lifetime of the nuclear state fed by the first γ transition and depopulated by the second one, and $A_{22} = -0.12$ the spatial anisotropy of the γγ coincidence probability. The perturbation function

$$G_2(t,\Theta) = S_o + \sum_{n=1}^{3} S_n \cdot \cos(\omega_n t)$$

contains the information of the defect specific efg tensor. For the fundamental frequency

$$\omega_1 = (3\pi/10) \cdot eQV_{zz}/h = (3\pi/10) \cdot \nu_Q$$

in case of axial symmetry of the tensor, i.e. $\eta = (V_{xx}-V_{yy})/V_{zz} = 0$; this symmetry is visible from the fact that $\omega_n = n \cdot \omega_1$. From the S_n coefficients as determined for different detector-sample geometries the orientation of the efg tensor is extracted in a known way (Deicher et al 1983).

By using four γ detectors eight coincidence spectra are recorded simultaneously, which are combined in such a way that the so-called PAC time spectrum $R(t) = A_{22}G_2(t)$ is obtained.

3. Results:

In Figure 1, the results for a (111) Si:B ($1 \cdot 10^{15}$ cm^{-3}) wafer are shown which has been completely passivated ($\rho = 28$ Ωcm → 1120 Ωcm). The PAC time spectrum R(t) (top), measured at 78 K, indicates through the pronounced oscillations of the γγ coincidence probability the formation of close In-defect complexes, which we will call In-X complexes. Their Fourier transforms F(ω) exhibit the presence of two different efg characterizing two different complexes. Each complex is characterized by the three frequencies ω_n giving rise to the coupling constant $\nu_Q = 237$ MHz for complex In-X1 and 334 MHz for complex In-X2. The frequency ratios, 1:2:3, show in both cases the axial symmetry of the efg tensor. From the dependence of the amplitudes S_n on the detector orientations, being along <100>, <110> (middle) and <111> (bottom), the <111> orientation of the symmetry axis of both complexes is obtained.

Shallow Impurities in Si and Ge

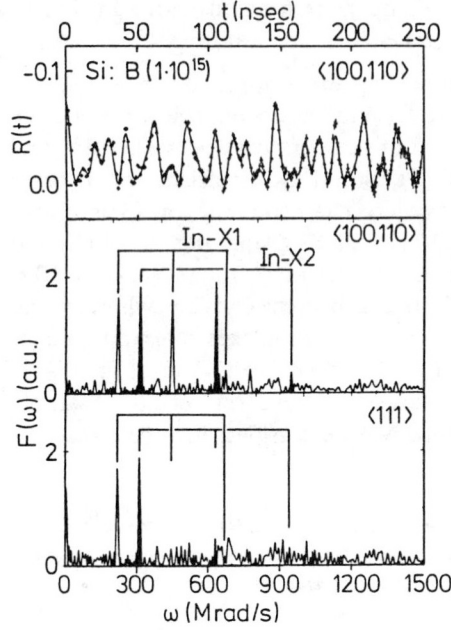

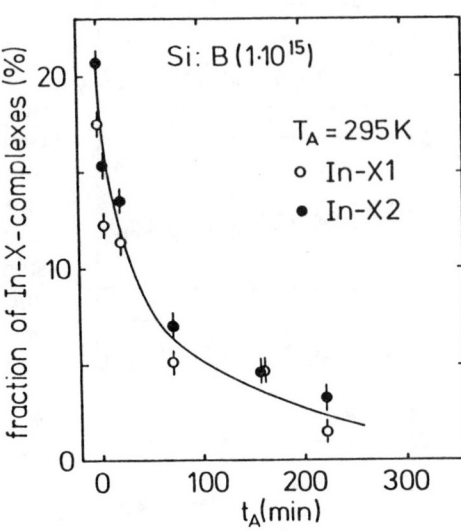

Figure 1: PAC spectrum (top) of a (111) Si:B ($1 \cdot 10^{15}$ cm^{-3}) wafer measured at 78 K after chemomechanical polishing along with Fourier transforms of measurements with detectors along <100> and <110> (middle) and along <111> (bottom) directions.

Figure 2: Influence of isothermal annealing at 295 K on the fractions of In-X complexes in a chemomechanical polished (111) Si:B ($1 \cdot 10^{15}$ cm^{-3}) wafer.

From the amplitudes observed in the PAC spectra, the fraction of In atoms forming the different complexes is obtained. These fractions, observed during isothermal annealing at 295 K, are plotted in Figure 2. It clearly shows the decay of both complexes at ambient temperature within a time of about 300 min. The decay rate of both complexes is identical within the experimental errors. Summing up both fractions, Figure 2 shows that about 40% of the In atoms form the two complexes. This fraction is significantly smaller after polishing (100) Si:B ($1 \cdot 10^{15}$ cm^{-3}) wafers.

If a (111) Si:B ($6 \cdot 10^{16}$ cm^{-3}) wafer is investigated, which differs in B concentration and the degree of passivation ($\rho = 0.38$ Ωcm $\rightarrow 2.0$ Ωcm), a new efg is observed characterizing a third complex In-X3. The PAC spectrum in the top panel of Figure 3 reveals an efg which is axially symmetric around a <111> lattice direction and is characterized by $\nu_Q = 408$ MHz. This spectrum is obtained at 78 K after annealing at 300 K. The following spectrum shows that annealing at 423 K is required to reduce the fraction of In-X3. At the same time a different axially symmetric efg with $\nu_Q = 270$ MHz (at 78 K) is observed. This frequency is known to characterize In-hydrogen pairs in Si (and has been denoted In-H3). Annealing at 460 K effects that all In atoms are in an

environment with zero efg (Figure 3 bottom). In Figure 4, the fractions of the In-X3 and In-H3 complexes are plotted as a function of the annealing temperature.

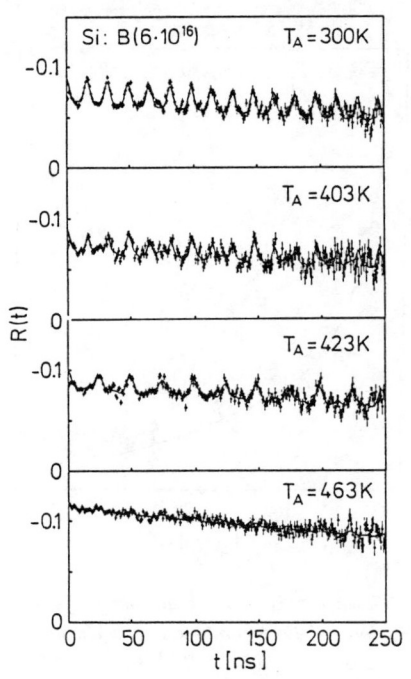

Before annealing this sample above 300 K, different PAC spectra have been measured with the sample held at temperatures between 10 K and 300 K using a closed cycle refrigerator. It turns out that the efg characterizing In-X3 is visible between 10 K and 80 K but not at 140 K and above. In the temperature range between 10 K and 40 K, In-X2 becomes visible whereby the fractions of all In-X complexes sum up to a constant value. The detailed behaviour of the different complexes as function of the sample temperature will be clarified in further experiments.

Figure 3: PAC spectra of a (111) Si:B ($6 \cdot 10^{16}$ cm^{-3}) partially passivated after chemomechanically polishing and annealed at different temperatures.

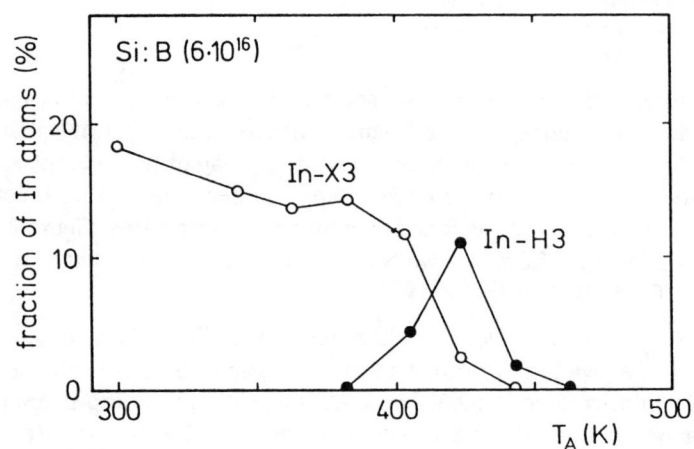

Figure 4: Isochronal annealing of a chemomechanically polished Si:B ($6 \cdot 10^{16}$ cm^{-3}) wafer.

4. Discussion:

In Table I the three different In-X complexes are listed along with their characteristic frequencies. In addition, the frequencies observed for the three In-H complexes are given. A comparision of the frequencies immediately shows that the In-X complexes are

not identical with the In-H complexes, although the symmetry of the complexes (last two columns) is identical. A comparision with the frequencies caused by native, irradiation induced defects at the In atoms (Deicher et al 1986) shows that X is also different from those defects, i.e. a neutral vacancy or a larger defect complex. The fact that depending on the sample conditions three different efg are observed for In-X as well for In-H points to the possibility that the complexes can assume different atomic or electronic configurations. That means all In-X and In-H complexes involve the same number of trapped defects, in the case of In-H a single hydrogen atom. In-X3, like In-H3, is observed in Si wafers doped with at least $6\cdot 10^{16}$ B cm^{-3} and a sample temperature of around 78 K. Therefore, the appearance of both complexes in the Si:B ($6\cdot 10^{16}$ cm^{-3}) sample is consistent.

Table I: In-X complexes observed after chemomechanical polishing of Si wafers and In-H complexes observed after doping of Si wafers with hydrogen. The frequencies are measured at 78 K and <ijk> denotes the orientation of the symmetry axis with respect to the Si lattice.

Complex	eQ V_{zz}/h (MHz)	$(V_{xx}-V_{yy})/V_{zz}$	<ijk>
In-X1	237	0.	<111>
In-X2	334	0.	<111>
In-X3	408	0.	<111>
In-H1	360	0.	<111>
In-H2	480	0.	<111>
In-H3	270	0.	<111>

Since, up to now, the measured efg can not be used to identify the trapped defect on the basis of theoretical calculations, the nature of the X defect still has to be unravelled. However, because of the unique labelling of the In-X complex by its characteristic efg, an identification of the defect should be feasible in future experiments. Forkel et al (1988) observed the identical frequencies in ^{111}In doped Si wafers quenched from about 700 K into water and subsequently transferred into a LN$_2$ bath. Also these observed thermal stabilities are in agreement with the present results.

It should also be noted, that the In-X complexes show close similarities with the In-H and In-Li complexes: All are characterized by three different efg of similar strength and identical symmetry. Because of its high mobility, X is most probably an interstitial-type defect so that its trapping site at the In atom can be the bond center, antibonding or tetrahedral interstitial site.

Comparing the formation of the In-X complexes in Si:B ($1\cdot 10^{15}$ cm^{-3}) samples with the electrical passivation of samples polished under identical conditions (Schnegg et al 1986) it can be stated, that the slower passivation of (100) samples compared to (111) samples is paralleled by the formation of less In-X complexes in the (100) wafers. In addition, the thermal stability of the electrically observed passivation of B and of the In-X1 and In-X2

complexes are comparable; the slightly lower stability of the In-X complexes may reflect the influence of the different acceptor atoms In and B. Using capacitance-voltage measurements, Zundel et al (1988) have shown the stability of the passivation for (111) Si:In ($2 \cdot 10^{15}$ cm^{-3}) wafers; their values are in close agreement with the present results. The fact, that Figure 2 shows only 40% In atoms participating in the formation of the In-X complexes at the beginning of the isothermal annealing sequence can be explained due to annealing during the time elapsed between the end of the polishing procedure and the storage of the samples in LN$_2$.

The experiment with the higher B-doped samples showing the appearance of the In-X3 and the In-H3 complex differs in two parameters from the others, i.e. the B-concentration and the degree of passivation. A consistent interpretation can be given taking into account the different B concentrations: Also in the case of the In-H complex, In-H3 is exclusively observed for higher B concentrations. Thus these efg in both cases seem to be caused by the doping level, i.e. the position of the Fermi level. The efg in both cases become unstable above 78 K. The apparent higher stability of In-X3 compared to In-X1 and In-X2 can be caused by the release of X defects from the B atoms during annealing leading to the formation of In-X3. A similar behaviour was observed for the stability of In-H complexes (Wichert et al 1987). Finally, the appearance of the In-H3 complex at 420 K - the temperature where In-H complexes dissociate - can be caused by the high B concentration: During preceeding processing steps H might have been introduced into the wafer and trapped at B atoms in the surface region. The release of these H atoms at 420 K then causes the formation of the In-H3 complex. However, this interpretation has to be considered as preliminary because of the lack of sufficient data.

Summarizing, the formation of an In-X complex as a consequence of the chemomechanically polishing procedure, leading to an electrically passivation, has been unambigeously shown. This type of passivation seems not to be caused by the trapping of H atoms. Although an identification of X is not possible at this time, the resemblance with H and Li has to be pointed out. Due to its characteristic efg, the identification of X should be possible.

This work was financially supported by the Bundesminister für Forschung und Technologie and the Deutsche Forschungsgemeinschaft (SFB 306)

References:
Deicher M, Minde R, Recknagel E and Wichert Th 1983 *Hyp. Inter.* **15/16** 437
Deicher M, Grübel G, Recknagel E and Wichert Th 1986 *Materials Science Forum* **10-12** 1141
Forkel D et al. 1988 *Appl. Phys. A*, submitted
Pearton S J, Corbett J W and Shi T S 1987 *Appl Phys A* **43** 153
Reichel J and Sevcik S 1987 *Phys. Stat. Sol. A* **103** 413
Schnegg A, Grundner M and Jacob H 1986 *Semiconductor Silicon, Proc 86-4*
  ed H R Fuff, T Abe, B Kolbesen (Pennington: The Electrochemical Society) p 198-205
Schnegg A, Prigge H, Grundner M, Hahn P O and Jacob H 1988 *Mat Res. Soc. Symp. Proc.* **104** 291
Wichert Th, Skudlik H, Deicher M, Grübel G, Keller R, Recknagel E and Song l 1987
  Phys. Rev. Lett. **59** 2087
Zundel T, Weber J, Benson B, Hahn P O, Schnegg A and Prigge H 1988 *Appl. Phys. Lett.*, submitted

Inst. Phys. Conf. Ser. No 95: Chapter 2
Paper presented at Int. Conf. Shallow Impurities in Semiconductors, Linköping, Sweden, 1988

New results in the theory of electron structure and spectra of shallow non-hydrogenlike impurities in semiconductors. I. Donors in germanium and silicon

I.L.Beinikhes and Sh.M.Kogan

Institute of Radioengineering and Electronics of the USSR Academy of Sciences, Moscow, 103907, USSR

ABSTRACT: A number of properties of donors in Ge and Si are calculated: the levels of the odd excited states in the effective mass approximation (EMA), the ground-state wave functins of Sb, P, As and Bi donors in the zero-radius central-cell approximation (ZRCCA), the oscillator strengths of line spectra, the photoionization cross-section spectra, the polarizabilities, and the magnetic change of the valley-orbit splitting for each impurity. Two odd levels previously omitted in the spectra are found.

INTRODUCTION

The shallow impurities of one group and in one and the same semiconductor may have very different properties if these properties are determined by the ground-state wave-function (GSWF) which strongly depends on the ionization energy E_i, i.e. on the chemical shift. We have calculated the GSWF of group-V donors in Ge and Si in ZRCCA (Beinikhes et al. (1985a)). In this approximation the GSWF is the solution of EMA equations which diverges (in the case of nonzero chemical shift) in the vicinity of the central cell ($r \to 0$) as $1/r$ and can be normalized. All calculations have been made using the highly accurate nonvariational method of finiteness conditions transfer (Beinikhes et al. (1985a)).

1. ENERGIES AND INTENSITIES OF SPECTRAL LINES

The EMA energies $E(nPm)$ of the odd excited states Pm (Beinikhes et al. (1986), see also Broeckz et al. (1986)) and the corresponding oscillator strengths (OS) $f(nPm)$ are presented in Tables 1 and 2. For a fully confident interpretation of experimental spectra the energies $E(nP\pm)$ are shown in Fig.1 (solid lines) as functions of effective-mass anisotropy $\gamma = m_t/m_l$. The results of Faulkner's (1969) variational calculation are shown by broken lines. As γ decreases Faulkner's energies deviate from ours. In the case of Si the inaccuracy of Faulkner's results being higher than the experimental one is however smaller than the interlevel spacings and therefore has not brought to misinterpretation of the lines. But at γ = 0.05134(Ge) and $n > 5$ the deviations become of the same order of and even greater then the difference between adjacent levels. This resulted in an erroneous identification of all optical transitions in Ge to levels higher then $5P\pm$. This fact remained unnoticed because the transitions to $6P\pm$ and $9P\pm$ were omitted in experiments and some of Faulkner's levels happened to be close to experimental points corresponding however to quite different levels. The lines $6P\pm$ and $9P\pm$ were not observed for a long time because

© 1989 IOP Publishing Ltd

their OS are significantly smaller then the OS of the adjacent lines $5P\pm$ and $8P\pm$ (Table 1). The existence of the line $6P\pm$ has been quite recently verified in $Ge:Li-O$ by Andreev et al. (1988). The comparison with spectroscopic data (Seccombe et al. (1972), Scolnik et al. (1974), Andreev et al. (1988)) shows that in Ge at $\varkappa = 15.40$ the odd states of the shallow donors coincide with the EMA within the experimental accuracy (≤ 5 μeV). In Si the odd levels higher than $2Po$ are also consistent with EMA with an accuracy of order of 10^- Ei.

TABLE 1. Energy levels E of odd states of shallow donors in Ge and oscillator strengths of optical transitions from the ground level $1s(A_1)$. The parantheses contain the level designation used by Faulkner (1969); the number under the impurity symbol is the ionization energy in meV (Ed = 9.352 meV).

Levels	$|E|/E_d$	$|E|$, meV	$f(nPm)*10^3$			
			EMA 9.783	Sb 10.46	P 12.88	As 14.16
$2Po(2po)$	0.5079	4.750	18.8	17.2	12.4	10.5
$3Po(3po)$	0.2752	2.573	1.91	2.07	2.16	2.05
$2P\pm(2p\pm)$	0.1839	1.720	233	202	126	102
$4Po(4po)$	0.1806	1.689	0.648	0.738	0.852	0.838
$5Po(4fo)$	0.1301	1.217	0.316	0.368	0.446	0.446
$3P\pm(3p\pm)$	0.1109	1.037	40.6	36.4	25.0	20.7
$6Po(5po)$	0.0992	0.928	0.184	0.217	0.270	0.272
$7Po(5fo)$	0.0856	0.800	$1.7*10^{-3}$	$1.6*10^{-3}$	$1.3*10^{-3}$	$1.2*10^{-3}$
$4P\pm(4p\pm)$	0.0802	0.750	21.8	19.8	14.0	11.7
$8Po(6po)$	0.0786	0.735	0.116	0.138	0.176	0.180
$5P\pm(4f\pm)$	0.0649	0.607	20.3	18.2	12.6	10.5
$6P\pm(5p\pm)$	0.0613	0.573	2.26	2.13	1.63	1.40
$7P\pm(5f\pm)$	0.0449	0.467	7.11	6.56	4.80	4.07
$8P\pm(6p\pm)$	0.0426	0.399	7.44	6.72	4.78	4.02
$9P\pm(6f\pm)$	0.0410	0.384	1.40	1.32	1.01	0.88
$10P\pm(6h\pm)$	0.0350	0.328	5.71	5.27	3.86	3.28
$11P\pm(7p\pm)$	0.0335	0.313	2.01	1.86	1.27	1.05
$12P\pm(7f\pm)$	0.0310	0.290	0.16	0.14	0.10	0.083
$13P\pm(7h\pm)$	0.0302	0.282	2.28	2.12	1.16	1.37
$14P\pm(8p\pm)$	0.0267	0.250	2.80	2.56	1.85	1.57
$15P\pm(8f\pm)$	0.0261	0.244	0.79	0.75	0.58	0.50
$16P\pm(8h\pm)$	0.0233	0.217	2.87	2.62	1.87	1.57
$17P\pm(8k\pm)$	0.0221	0.207	1.44	1.36	1.04	0.90

The transitions to nPo levels are induced by the component of the radiation electric field parallel to direction of the maximum effective mass ($m_{\parallel}$). Therefore the OS of the corresponding lines are smaller than the OS of $nP\pm$ lines with the same or even greater energy. This effect is extremely strong in Ge the OS $f(4Po)$ being 3 times smaller then the OS of the highest identified transition ($11P\pm$).

The OS $f(nP\pm)$ change with n quite nonmonotonously (Tables 1 and 2) in qualitative agreement with experimental data. This feature can be related to the quantum chaos in the anisotropic Kepler problem at $\gamma < 8/9$ (Gutzwiller (1973,1977)). The OS of the brightest lines ($2Po$ and $nP\pm$) strongly decrease as Ei increases. This effect is different for different lines therefore the ratios of OS also exhibit a significant "chemical effect".

2. THE PHOTOIONIZATION SPECTRA

A general method of calculation of orthonormalized continous spectrum wave functions of nonhydrogenlike one-particle systems was developed earlier (Beinikhes et al.(1985a,b)). If σ_l and σ_t are respectively the cross-sections for the polarization parallel and perpendicular to the valley axis the cross-section which is measured in an unstrained crystal is σ = = $(2\sigma_t + \sigma_l)/3$. The calculated photoionization cross-sections $\sigma(h\nu)$ of shallow donors in Ge and Si are presented in Fig.2. The larger is E_i the lower is the cross-section at the threshold $\sigma(E_i)$ and the less steeper is the decrease of $\sigma(h\nu)$ as a function of quantum $h\nu$. In Ge at $n = (E_i^{EMA} / E_i)^{1/2} \approx 0,7$ (in Si $n \approx 0,6$) the maximum of $\sigma(h\nu)$ coincides with the threshold. At smaller n the maximum corresponds to $h\nu > E_i$.

TABLE 2. Energy levels E of odd states of shallow donors in Si and oscillator strengths of optical transitions from the ground level $1s(A_1)$. The parantheses contain the level designation used by Faulkner (1969); the number under the impurity symbol is the ionization energy in meV (E_d = 39.88 meV).

Levels	$|E|/E_d$	$|E|$, meV	$f(nP)*10^3$			
			EMA 31.27	Sb 42.74	P 45.59	As 53.76
2Po(2po)	0.28810	11.491	57.9	35.1	31.3	23.0
2P±(2p±)	0.16050	6.401	287	153	133	93.4
3Po(3po)	0.13753	5.485	7.81	6.86	6.44	5.29
4Po(4po)	0.08297	3.309	2.75	2.68	2.55	2.16
3P±(3p±)	0.07822	3.120	53.9	33.9	30.4	22.5
5Po(4fo)	0.05864	2.339	0.057	0.052	0.049	0.042
6Po(5po)	0.05603	2.235	1.27	1.30	1.25	1.08
4P±(4p±)	0.05483	2.187	18.7	12.1	10.8	8.09
5P±(4f±)	0.04749	1.894	6.00	4.09	3.70	2.80
7Po(5fo)	0.04089	1.631	0.74	0.77	0.74	0.65
8Po(6po)	0.03785	1.510	0.014	0.017	0.017	0.015
6P±(5p±)	0.03634	1.449	14.9	10.1	9.09	6.87
7P±(5f±)	0.03158	1.259	0.594	0.37	0.33	0.24
9Po(6fo)	0.03116	1.243	0.48	0.51	0.49	0.43
8P±(6p±)	0.02684	1.071	6.89	4.74	4.30	3.27
9P±(6f±)	0.02513	1.002	$4*10^{-4}$	$<10^{-4}$	$<10^{-4}$	$<10^{-4}$
10P±(6h±)	0.02222	0.886	3.64	2.43	2.20	1.66
11P±(7p±)	0.02064	0.823	2.26	1.44	1.30	0.97
12P±(7f±)	0.01880	0.750	1.41	0.90	0.81	0.61
13P±(7h±)	0.01699	0.678	2.69	1.74	1.57	1.19
14P±(8p±)	0.01597	0.637	$7*10^{-4}$	0.003	0.003	0.003
15P±(8f±)	0.01494	0.596	2.39	1.50	1.35	1.00
16P±(8h±)	0.01419	0.566	0.95	0.63	0.57	0.43

The calculated values of $\sigma(E_i)$ for donors in Si and the measured ones (Picus et al. (1956), Baron et al. (1977)) are respectively (in units 10^{-15} cm^2) 2.55 and 8.5(Sb), 2.34 and 2.5(P), 1.81 and 1.6(As), 1.10 and 0.72(Bi). For the same impurities in Ge (experimental data of Reuszer et al.(1965) have been presented by Bratt (1977)) the calculated and measured $\sigma(E_i)$ are (in units 10^{-14} cm^2) 2.20 and 1.8(Sb), 1.64 and 1.5(P), 1.40 and 1.1(As).

As E_i increases the part of the whole area under the absorption spectrum that corresponds to the continous spectrum ($h\nu > E_i$) becomes greater. The sums of the OS of all lines listed in Table 1(Ge) are 0.37(EMA), 0.33(Sb),

0.22(P), and 0.18(As). The contribution of higher levels estimated by using the known value of $\sigma(Ei)$ is 0.01 - 0.02. The OS corresponding to the true continous spectrum is 0.60(EMA), 0.65(Sb), 0.76(P), 0.79(As). Within the accuracy 1 - 3% the sum of all OS is unity as it has to be. In Si the sums of OS in Table 2 are respectively 0.47 (EMA), 0.27 (Sb), 0.24 (P), 0.18(As), the corresponding OS of continous spectrum are 0.52, 0.71, 0.74, and 0.79. As γ decreases the OS corresponding to the continous spectrum increases. This agrees with the known fact that in a hydrogen atom this OS is 0.435 and in a two-dimensional hydrogen-like atom 0.721.

Under a strong enough and suitably directed uniaxial stress the ground state of the donor is related to only one valley (in Ge) or only two valleys (in Si). In such a strained crystal the photoionization cross - section is anisotropic and depends on the polarization of the radiation the principal values of the tensor being $\sigma_\parallel$ and σ_t. The calculated ratios $\sigma_t/\sigma_\parallel$ are shown in Fig.3 as functions of $h\nu/Ei$. The cross-section anisotropy is extremely high in Ge especially in the case of impurities with small chemical shift. For donors in Si it is also significant. The fact that $\sigma_\parallel < \sigma_t$ is an obvious consequence of $m_\parallel > m_t$.

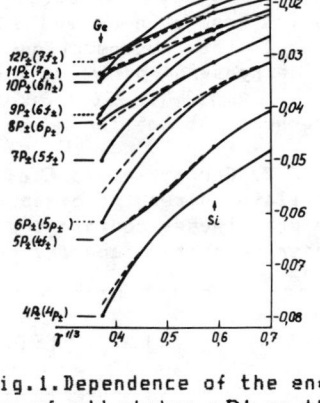

Fig.1. Dependence of the energy of odd states $nP\pm$ on the effective-masses anisotropy parameter γ. Solid lines - present calculation, dashed - result of Faulkner (1969). The arrows correspond to the values of γ for Si and Ge. The solid and dashed horizontal lines mark respectively the experimental (Seccombe et al. (1972), Scolnik et al. (1974) values of the level energy and the levels to which transitions were not observed before present calculation.

3. POLARIZABILITIES

The polarizability tensor of an atomic system being in a state $\psi^{(n)}$ is $\alpha_{ik}^{(n)} = 2\langle \psi^{(n)} | \hat{d}_i | w_k^{(n)} \rangle$ where $\hat{d}_i$ is the component of the dipole moment, $w_k^{(n)}$ is the solution of the inhomogeneous Schrodinger equation $(\hat{H} - E)w_k^{(n)} = d_k \psi^{(n)}$. This equation and homogeneous equation for $\psi^{(n)}$ form a system. By solving it we are able to calculate $\alpha_{ik}^{(n)}$. The results are listed in Table 3 in which $\alpha = (2\alpha_t + \alpha_\parallel)/3$ is the donor polarizability in an unstrained crystal. One can see that α strongly depends on Ei. The agreement with the most accurate experimental data is good. Only in the case of As in Si the calculated value is slightly beyond the experimental errors.

4. THE EFFECT OF MAGNETIC FIELD ON VALLEY-ORBIT SPLITTING.

It is well known that the donor's ground level which is degenerate in EMA is splitted by the valley-orbit interaction into two (Ge) or three (Si) levels. The magnetic field which is symmetric against all valleys shifts these levels and these shifts (mainly diamagnetic) are different because the wave function of the upper level is more extended. The calculated coefficients β of the B^2 shifts of the levels are presented in Table 4. The difference $\Delta\beta = \beta(T_2) - \beta(A_1) = [4\Delta(B) - 4\Delta(0)]/B^2$ which was directly measured in Ge is also given.

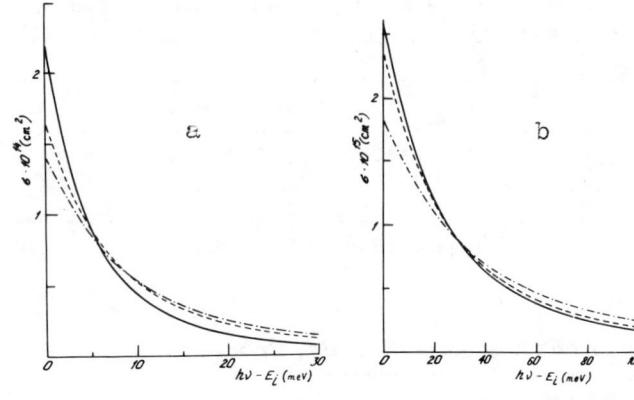

Fig.2. Cross section for photoionization of shallow Sb (solid curve), P(dashed), and As(dash - dot) donors in Ge (a) and Si (b) vs the photon energy.

TABLE 3. Shallow-donor polarizability in Si and Ge (in units of 10^5 Å^3)

Impurity	Si				Ge			
	α_t	α_l	α	α_{exp}	α_t	α_l	α	α_{exp}
EMA	5.35	2.26	4.32	–	108	21.0	79.2	–
Sb	1.96	0.70	1.54	1.9±0.6[a]	88.0	15.7	63.9	68±15[d]
P	1.61	0.56	1.26	{1.1±0.1[b], 1.2±0.2[a]}	45.7	6.50	32.6	–
As	0.96	0.31	0.74	0.52±0.09[c]	34.6	4.55	24.6	–

[a] Tan et al. (1981)
[b] Capizzi et al. (1980)
[c] Deri et al. (1986)
[d] D'Altroy et al. (1956)

5. ACCURACY OF ZRCCA

Among those parameters of donors in Si which are determined by the form of the ground-state wave function the polarizabilities are now perhaps the most accurately measured (error ~10 - 30%). The agreement between the calculated and experimental values (Table 3) indicates that the inaccuracy of ZRCCA in this case is not greater. In the case of donors in Ge the most accurately measured is perhaps $\Delta\beta$ (error ~10%). The computed values are slightly beyond the experimental errors (Table 4). In any case the inaccuracy of ZRCCA in Ge is expected to be not greater than in Si.

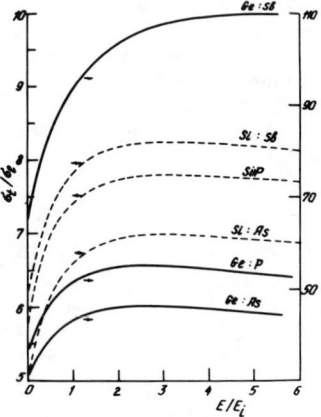

Fig.3. Anisotropy of the cross section for the donor photoionization in Ge (solid curves) and in Si (dashed) vs the photon energy.

TABLE 4. Values of β, which determine the quadratic shifts of the donor levels in a magnetic field (in $\mu eV/T^2$)

Impurity	Ge				Si		
	$\beta(A_1)$	$\beta(T_2)$	$\Delta\beta$	$\Delta\beta^{a,b}_{exp}$	$\beta(A_1)$	$\beta(E)$	$\beta(T_2)$
P	5.10	7.93	2.83	2.32±0.25	0.339	0.614	0.572
As	4.33	8.11	3.78	3.35±0.25	0.256	0.663	0.612
Sb	7.38	7.81	0.43	-	0.378	0.694	0.604

[a] Aggarwal (1981)   [b] Jagannath et al. (1985)

REFERENCES

Aggarwal R L 1981 *Physics of High Magnetic Fields, Proc. Oji Int. Seminar* Hakone Sept.10-13 1980 (Springer) p 105
Andreev B A, Ikonnikov V B, Kozlov E B et al. 1988 *Vysokochistye Veschestva (High-purity Substances)* no.2 180
Baron R, Young M H, Neeland J K et al. 1977 *Appl. Phys. Lett.* **30** 594
Beinikhes I L, Kogan Sh M, Polupanov A F et al. 1985a *Sol. St. Comm.* **53** 1083
Beinikhes I L and Kogan Sh M 1985b *Zh. Eksp. Teor. Fis.* **89** 722 [*Sov. Phys. JETP* **62** 415 (1985)]
Beinikhes I L and Kogan Sh M 1986 *Pis'ma Zh. Eksp. Teor. Fis.* **44** 39 [*JETP Lett.* **44** 48 (1986)]
Bratt P A 1977 *Infrared Detectors, part II. Semiconductors and Semimetals* vol 12 (R K Willardson and A S Beer eds) Academic p 39
Broeckz J, Clauws P, and Vennik J 1986 *J. Phys.* **C19** 511
Capizzi M, Thomas G A, DeRose F et al. 1980 *Phys. Rev. Lett.* **44** 1019
D'Altroy F A and Fan H Y 1956 *Phys. Rev.* **103** 1671
Deri R J and Castner T G 1986 *Phys. Rev.* **B33** 2796
Faulkner R A 1969 *Phys. Rev.* **184** 713
Gutzwiller M S 1973 *J. Math. Phys.* **14** 139
Gutzwiller M S 1977 *J. Math. Phys.* **18** 806
Jagannath C, Aggarwal R L, and Larsen D M 1985 *Sol. St. Comm.* **53** 1089
Pikus G, Burstein E, and Henvis B 1956 *J. Phys. Chem. Sol.* **1** 75
Reuszer J H and Fischer P 1965 *Phys. Rev.* **140** A245
Seccombe S D and Korn D M 1972 *Sol. St. Comm.* **11** 1539
Scolnik M S, Eavens L, Stradling R A et al. 1974 *Sol. St. Comm.* **15** 1403
Tan H S and Castner T G 1981 *Phys. Rev.* **B23** 3983

Impurity impact ionization in indium doped germanium

U Rau, J Parisi, K M Mayer, W Clauß, J Peinke, B Röhricht and R P Huebener

Physikalisches Institut, Lehrstuhl Experimentalphysik II, Universität Tübingen, Morgenstelle 14, D-7400 Tübingen, Fed. Rep. Germany

ABSTRACT: Conductivity and Hall-effect measurements were performed on single-crystalline p-doped germanium, electrically driven into low-temperature avalanche breakdown via impact ionization of the shallow impurities. The electric transport properties (i.e., density and Hall mobility of the charge carriers) were determined in the pre- and post-breakdown regime. The results are further discussed with respect to the negative differential resistance and the formation of current filaments, as observed experimentally.

1. INTRODUCTION

Magnetoresistance and Hall-effect measurements are one of the classical frameworks for the understanding of the electric transport properties in semiconductor materials. Particularly the low-temperature electric breakdown effect has been the subject of intensive researches in the late fifties and the early sixties (Sclar and Burstein 1957; Koenig and Gunther-Mohr 1957; Yamashita 1961; Koenig, Brown and Schillinger 1962). The main mechanism involved in the nondestructive avalanche breakdown was found to be attributed to impact ionization of the shallow impurities by mobile charge carriers.

In recent years, the exemplary system of p-doped germanium (as well as similar semiconductor materials) has become the subject of growing interest with respect to a totally different point of view. Exactly in the strongly nonlinear breakdown regime the current transport in the system displays the self-organized formation of both temporal oscillations and spatially inhomogeneous filament structures (for an overview see: Huebener et al 1987; Mayer et al 1987a; Peinke et al 1988). The present paper intends to be a first step to close the gap between the classical semiconductor physics of the avalanche breakdown and the understanding of the system in terms of its nonlinear dynamics.

2. PHYSICAL MECHANISM AND HALL-EFFECT

At the temperature of liquid helium all charge carriers (holes) of the p-germanium are frozen out at the shallow impurities. So in the range of zero electric field the material nearly becomes an electric insulator. Due to the weak interaction between the holes and the lattice at low temperatures, a relatively small electric field heats the carriers up to energies sufficiently strong to ionize the shallow impurities. At a critical electric field strength ε_c of about 4 V/cm the avalanche breakdown takes place in the bulk of the semiconductor, and the number of charge carriers increases

© 1989 IOP Publishing Ltd

by many orders of magnitude. The autocatalytic character of this impact ionization process is the origin of typical nonlinear effects, so as the spontaneous current oscillations and the formation of current filaments.

In the presence of a transverse magnetic field the mobile charge carriers are deflected from the direction of motion by the Lorentz force. This deflection causes a Hall field which, on its part, compensates the Lorentz force relatively to the mean ensemble velocity (drift velocity). Therefrom, we obtain the well-known formula for the Hall mobility

$$\mu_H = \varepsilon_H / \varepsilon_o B \qquad (1)$$

with the Hall field ε_H, the electric field strength along the sample ε_o, and the transverse magnetic field B. The Hall mobility is proportional to the drift mobility with a proportionality constant r_H, the Hall scattering factor, which, on its part, is a function of the scattering mechanisms involved. r_H can be assumed of the order of unity, if the Hall condition, $\mu_H \cdot B < 1$, is fullfilled. So the error introduced by assuming equal drift and Hall mobility is not too large, and we proceed from (1) to the determination of the charge carrier density by

$$n = j/q\mu_H\varepsilon_o \qquad (2)$$

with the measured current density j and the elementary charge q. Although the Hall field compensates the Lorentz force for carriers moving with the drift velocity, this does not hold for all carriers of the ensemble. So the effect of the transverse magnetic field is to diminish the projection of the mean free path of the carriers in the direction of the electric field (cooling effect of the magnetic field). This results in an increase of the critical electric field and a corresponding shift in the current-voltage characteristic.

3. EXPERIMENTAL

Our experimental studies were performed on single-crystalline p-type germanium material with an indium doping concentration of about 3×10^{14} cm^{-3} (corresponding to a room-temperature resistivity of about 10 Ωcm). The shallow acceptor level of about 12 meV above the valence band edge is small compared to the energy gap between the conduction band and the valence band of about 0.75 eV. Since the measurements to be reported were made at liquid-helium temperature, low enough for the contribution of intrinsic carriers to be neglected, mobile carriers of only one sign (holes) are present (in the valence band). The extrinsic germanium samples investigated were prepared from (111) oriented single-crystal slices in the standard manner by cutting the desired "bridge" shape (rectangular bar with side arms) via a combined masking and chemical etching process, the bar having the dimensions of about $0.25 \times 1.5 \times 11$ mm^3. Properly arranged ohmic aluminum contacts were evaporated upon one of the two largest crystal surfaces, polished and etched previously. The large areas on both ends of the bar were used for the current and voltage leads, the side arms for the Hall measurements.

To provide the outer ohmic contacts of the bar with an electric field ε_o, a d.c. bias voltage V_o was applied to the series combination of the sample and an 1 Ω load resistor. A d.c. magnetic field B perpendicular to the broad sample surfaces could be applied by a superconducting solenoid surrounding the semiconductor sample. The resulting electric current I was either found from the voltage drop at the load resistor or, in the pre-breakdown region

of extremely small sample currents, directly via a sensitive current meter connected in series. As stated above, the longitudinal and transverse voltages V and V_H were measured along and across the sample bridge via the corresponding probe contacts, respectively. During the experiments, the semiconductor sample was always kept at the liquid-helium temperature of 4.2 K and carefully protected against external electromagnetic irradiation (visible, far infrared). A more detailed treatment of the experimental set-up can be found elsewhere (Parisi et al 1988).

4. RESULTS AND DISCUSSION

The autocatalytic process of impurity impact ionization is reflected in a strongly nonlinear curvature of the measured current-voltage characteristic in the breakdown regime. Figure 1 gives an example of a typical I-V curve plotted on a semilogarithmic scale. At the apparent breakdown voltage of 4.7 V (corresponding to a critical electric field of about 4.3 V/cm) the current flow increases several decades from a few nA in the pre-breakdown up to a few mA in the post-breakdown regime for extremely small incremental changes in voltage.

More detailed information on the electronic transport properties can be extracted from the Hall-effect measurements using Equations (1) and (2). Figure 2 displays the so-obtained quantities n and μ_H as a function of the applied sample voltage. As already indicated in the integral current-voltage characteristic (Fig. 1), there is even in the pre-breakdown region a gradual increase of the carrier density with electric field intensity associated with a maximum in the Hall mobility (Fig. 2). Upon further increasing the sample voltage, breakdown is first announced by a decrease of the Hall mobility from about 5×10^5 cm^2/Vs to about 2×10^4 cm^2/Vs in the voltage range 3.5 V < V < 5.0 V. Simultaneously, the density of the charge carriers begins to increase and displays a drastic jump at V = 4.7 V from about 5×10^7 cm^{-3} to a value of about 10^{12} cm^{-3} (quasi at the same sample voltage). So far, our results correspond to quantities found by Koenig et al (1957, 1962) for a somewhat different n-germanium system.

In the region of breakdown the carrier density increases rapidly up to values of about 6×10^{13} cm^{-3} at V = 5.0 V, while the Hall mobility attains a minimum value of about 2×10^4 cm^2/Vs. Thereafter, this quantity reincreases again up to 1.5×10^5 cm^2/Vs at V = 5.3 V. Beyond that voltage, the Hall mobility remains nearly constant up to values of twice the breakdown voltage. Breakdown is thus characterized by a mobility which essentially depends on the density of the mobile charge carriers, such that carrier density and mobility are both involved. This result, which is in contrast to findings of Sclar and Burstein (1957), is in our intention closely connected to the formation of current filaments in p-germanium, as it was observed experimentally by Mayer et al (1987a,b).

Next we turn to first Hall-effect measurements performed simultaneously with the direct visualization of the current filament structures developing during avalanche breakdown. Hereto the well-established imaging method of low-temperature scanning electron microscopy was applied that combines a liquid-helium cryostage with a commercial scanning electron microscope (Huebener 1984). The p-germanium sample with a typical Hall geometry (see Fig. 3a) was glued onto a sapphire disc, the bottom of which was kept in direct contact with the liquid-helium bath. For two-dimensional imaging of the current filaments, the sample was scanned by an electron beam, and the beam-induced changes of the sample conductance were recorded as a function of the coordi-

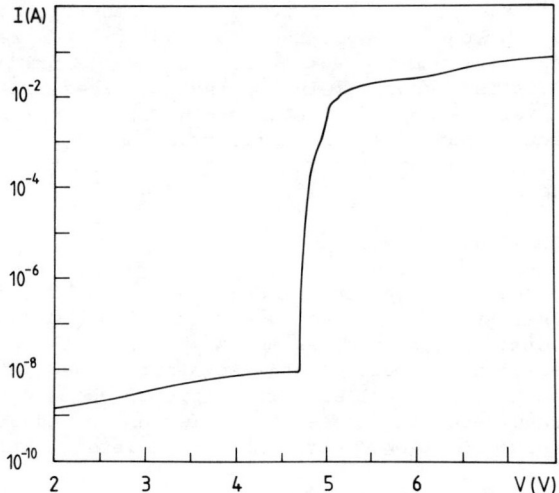

Fig. 1. Current-voltage characteristic obtained for an applied transverse magnetic field B = 30 mT at the temperature T = 4.2 K.

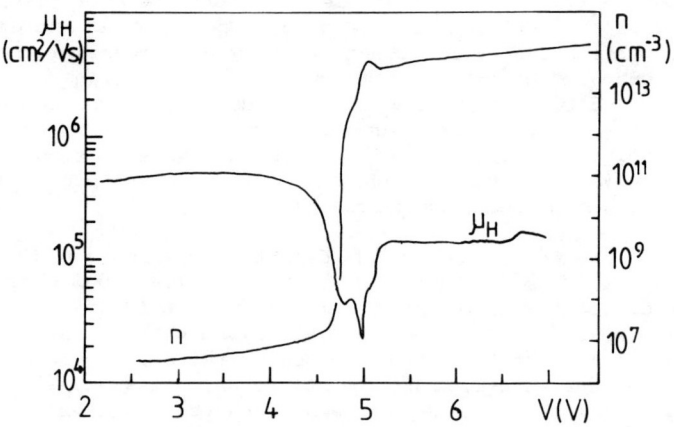

Fig. 2. Breakdown characteristics of the electric transport properties (carrier density and mobility versus voltage) obtained from conductivity and Hall-effect measurements for an applied transverse magnetic field B = 30 mT at the temperature T = 4.2 K.

nates (x,y) of the beam focus. Due to the low beam power of typically 1 µW, a considerable response signal is obtained only along the boundaries of the current filaments, where the electric sample behavior is most sensitive to the beam-injected disturbation (for details see Mayer et al 1987a,b).

With these experiments the onset of lateral structure formation was found at a sample current of about 40 µW, independent of the applied magnetic field. An increasing sample voltage causes a monotonic growth of filament size. In Fig. 3b,c we present a two-filament state at the sample voltages

Shallow Impurities in Si and Ge

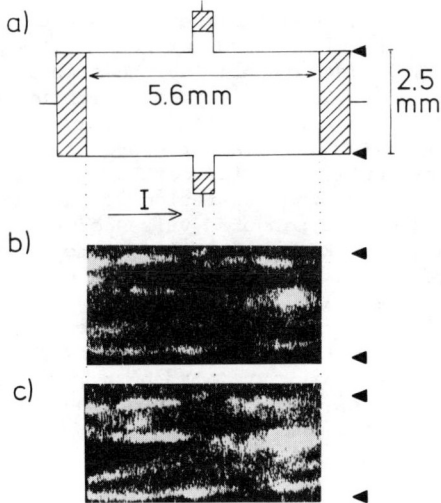

Fig. 3. Combined scanning electron microscopy and Hall-effect experiments performed with a different Hall sample for an applied transverse magnetic field $B = 15$ mT at the temperature $T = 4.2$ K. (a) Sample geometry with dimensions $0.17 \times 2.5 \times 5.6$ mm^3. The hatched areas upon the ends of the bar and the side arms of the Ge sample indicate the evaporated ohmic Al contacts. (b,c) Filament structures of the current flow in the breakdown region observed at the voltages $V = 2.23$ V (b) and $V = 2.27$ V (c) with the corresponding currents $I = 0.64$ mA (b) and $I = 1.13$ mA (c). The bright regions in the two-dimensional images represent the filament boundaries. The noisy underground results from superposed spontaneous current oscillations.

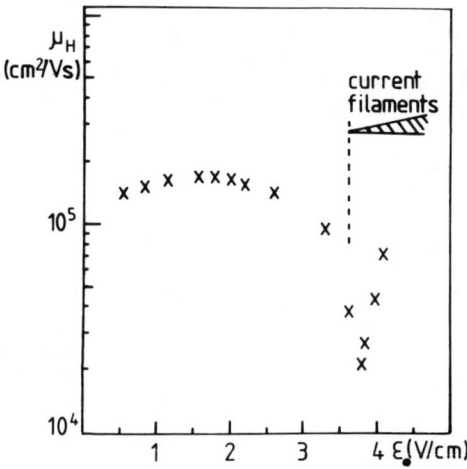

Fig. 4. Hall mobility versus electric field characteristic obtained from combined scanning electron microscopy and Hall-effect experiments performed with the sample shown in Fig. 3a for an applied transverse magnetic field $B = 15$ mT at the temperature $T = 4.2$ K. Note the onset of spatial structure formation at $\varepsilon_0 = 3.6$ V/cm.

V = 2.23 V and V = 2.27 V, covering about 35 % and 50 % of the sample diameter, respectively. On the other hand, we point out that the simultaneously performed Hall-effect measurements display the same features as the corresponding experiments done in the bath cryostat without imaging (results shown in Figs. 1,2), including the characteristic valley of the Hall mobility in the breakdown region (see Fig. 4). The presence of a spatially inhomogeneous state forces a modification for the further interpretation of our experimental data. Taking into account the separation of the sample in a conducting and a nonconducting phase, the measured values of the Hall mobility and density of the charge carriers (Eqs. (1),(2)) have to be understood as spatial averages over the local values (i.e., inside and outside the current filaments), as explained in detail elsewhere (Parisi et al 1988). Moreover, the reincrease of the Hall mobility is closely connected to a considerable growth of the filaments. Thus, it makes sense to identify the two phases not only by a low and a high carrier density, but also by a low and a high mobility, respectively. This fact, on its part, is in accordance with some aspects of convenient breakdown theory (Yamashita 1961).

ACKNOWLEDGMENT

This work was supported by a grant of the Deutsche Forschungsgemeinschaft.

REFERENCES

Huebener R P 1984 Rep. Prog. Phys. **47** 175
Huebener R P, Mayer K M, Parisi J, Peinke J and Röhricht B 1987 Nuclear Physics B (Proc. Suppl.) **2** 3
Koenig S H and Gunther-Mohr G R 1957 J. Phys. Chem. Solids **2** 268
Koenig S H, Brown R D and Schillinger W 1962 Phys. Rev. **128** 1668
Mayer K M, Peinke J, Röhricht B, Parisi J and Huebener R P 1987a Physica Scripta **T 19** 505
Mayer K M, Gross R, Parisi J, Peinke J and Huebener R P 1987b Solid State Commun. **63** 55
Parisi J, Rau U, Peinke J and Mayer K M 1988 Z. Phys. B - Condensed Matter (to be published)
Peinke J, Parisi J, Röhricht B, Mayer K M, Rau U and Huebener R P 1988 Solid State Electronics **31** 817
Sclar N and Burstein E 1957 J. Phys. Chem. Solids **2** 1
Yamashita J 1961 J. Phys. Soc. Japan **16** 720

One-phonon ionization of neutral donors in germanium

Martin Gienger, Peter Groß, Kurt Laßmann

1. Physikalisches Institut, Univ. Stuttgart
Pfaffenwaldring 57, D-7000 Stuttgart 80, FRG

ABSTRACT: By phonon spectroscopy with superconducting Al tunnel-junctions we show that effective mass donors in germanium can be ionized by a one-phonon excitation.

1. INTRODUCTION

In the last decades there has been continuing interest in the problem of dynamics and interactions of carriers excited from and trapped by shallow states of semiconductors. Impact ionization or photons were used to excite the carriers from their bound states at low temperatures. For the reverse process of relaxation phonons in many cases play the dominant rôle, however, specific experimental evidence for the phonons involved as yet is missing. A problem connected with relaxation by phonons is that the deformation potential coupling may be drastically reduced by the condition of momentum conservation because the wavelength of the phonons involved is much smaller than the extent of the wave function of the shallow bound states. In particular, the deexcitation to the EMA ground state by the emission of one phonon was estimated to be small (Gummel and Lax, 1955).

Here we show by phonon spectroscopy with superconducting tunnelling junctions that one-phonon excitations from the ground state to the conduction band (CB) are measurable as phonon induced conductivity (PIC) changes in the special case of donors in Ge. For the conductivity thresholds we find somewhat smaller binding energies than evaluated from optical measurements whereas for the 1s-singlet to 1s-triplet ground state splitting as seen by elastic phonon scattering we obtain the optical values.

2. EXPERIMENTAL

The experimental setup is similar to that described by Burger and Laßmann (1986) and is shown in Fig. 1: An Al-junction as tunable phonon source is evaporated onto one of the 15 mm x 5 mm faces of the 2 mm thick samples. 100 nm thick Al-contact films are evaporated on the side opposite to the junction. It is the bias to the superconducting tunnelling junction (bath temperature typically 1 K) that determines the maximum energy of the phonons emitted by the tunnelling quasiparticles. This maximum phonon frequency is filtered by Lock-In technique if the junction bias is modulated (Eisenmenger 1976).
The side opposite to the junction was irradiated with visible light from an incandescent lamp on top of the cryostat via a glass fibre rod to pro-

© 1989 IOP Publishing Ltd

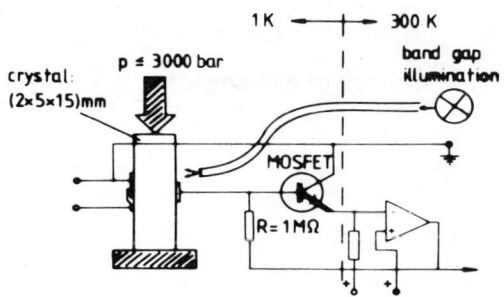

Fig. 1. Experimental setup for PIC measurements: Al-junction as tunable phonon source bandgap illumination for free carrier and D^- production, constant bias across the sample to measure the induced current. The transimpedance amplifier allows higher modulation frequencies.

duce free carriers for a finite resistance of the sample (typically 1 MΩ) to measure the phonon induced resistance changes. (Part of the electrons is trapped by the neutral donors to form D^--states (Taniguchi et al 1975) with binding energy of about 1.5 meV.) The high sample resistance allows modulation frequencies up to about 500 Hz because of the large electronic time constant. Higher modulation frequencies were possible by the installation of a transimpedance amplifier with small heat dissipation close to the sample within the cryostat rendering a better signal to noise ratio (SNR) and, in addition, some information on the carrier dynamics in the sample. A high upper cut-off frequency (given by $1/R_L \cdot C_L$) is conflicting with a high gain which is proportional to R_L. For $R_L = R_{sample}$ we get minimum SNR with a cut-off at about 100 kHz. Similar devices have been applied by several authors in low temperature photoconductivity (see e.g. Haller 1987). The sample was essentially kept at constant voltage. The current change was measured across the sample with the junction being one of the contacts. These evaporated Al-contacts show large contact resistances. The I-V-curves start with a linear part followed by a region of constant current where all the carriers produced by the light arrive at

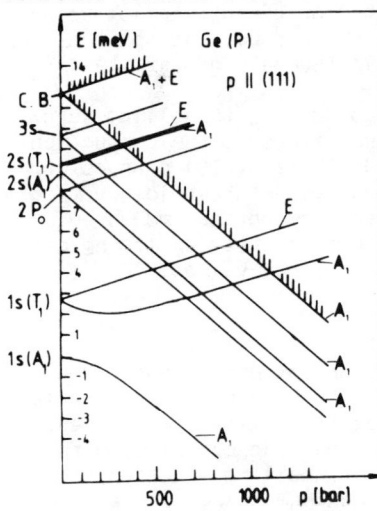

Fig. 2. Stress dependence of the energy of donor states and conduction band in Ge: decrease of binding energy, level crossings.

the contacts. Beyond this plateau the current again increases steeply due to injection from the contacts. In this region the signal becomes noisy. Alloyed ohmic contacts did not improve the SNR as compared to measuring in the plateau region with evaporated contacts.

Uniaxial stress shifts the valleys of the CB and splits the degenerate bound states of the donors as shown in Fig. 2. As a consequence there will be a downshift of the binding energy at lower stresses and level crossings with the lowest valley(s) at higher stresses. To investigate this, stress was applied by tearing a yoke against the sample with a wire strained from top of the cryostat by a screw. It is measured either with resistive strain gauges near the top of the cryostat or with a piezoelectric strain gauge adjacent to the sample. The analysis of our experiments (see below) indicates that "zero" stress may be indefinite up to 50 bar because of the needs of a firm positioning of the narrow samples.

Some inhomogeneous residual stress may be due to the fact that the electrical connection to the junction and contact films is made by indium cones pressed against the sample. Such contacts have the advantage of low induction wiring to the low ohmic junctions.

Another source of residual strains is the surface damage produced by polishing with .25 μm diamond grain. These strains will be more important for the highest phonon frequencies where the mean free path (mfp) is in the μm range. Two types of processes determine the phonon mfp at high frequencies and low temperatures: elastic isotope scattering $l_{is} \sim E_{ph}^{-4}$ and anharmonic decay $l_{an} \sim E_{ph}^{-5}$. Only rough estimates are possible in the high frequency range: l_{is} and l_{an} range from several 1000 μm at 2 meV (- the threshold of the D$^-$-states) to several μm at 12 meV (- the binding energy of the D^o.) Thus the 2 meV phonons may excite the D$^-$ across the whole thickness of the sample increasing the carrier density everywhere across the sample whereas the 12 meV-phonons ionize the D^o only in a thin layer beneath the junction generating a thin space charge layer which will mainly influence the contact resistance.

3. RESULTS AND DISCUSSION

We have found phonon induced conductivity thresholds corresponding to the respective D^o-binding energies for samples containing Sb, P, and As with dopant concentrations of 2 to 6x10^{14} cm^{-3}. This means that Al-junctions emit primary phonons as determined by the junction bias up to the Debye frequencies of the transverse acoustic phonon branches in Ge. The PIC-signal of a Ge:Sb sample is shown in Fig. 3 as a function of phonon energy: The threshold near 2 meV is due to the excitation of the Sb$^-$ and the sharp threshold at 9.9 meV (obtained by extrapolating both the "base line" on the low energy side and the turning point tangent of the threshold to a common foot point) is attributed to the ionization of Sb^o. The feature in front of this threshold is a spectral precursor emitted by the Al-phonon source at an energy $2\Delta = 0.6$ meV before the main line. It can be distinguished only for sharp and prominent spectral structures on the detecting side. It proves that the gap of the Al-junction has not been reduced by the injection of quasiparticles and phonons.

As compared to the P^o and As^o thresholds the SNR for Sb^o is rather good because the binding energy (phonon frequency) is lowest and perhaps also because the threshold coincides with a peak in the slow transverse phonon density of states. These phonons with their small group velocity should have a penetration depth even smaller than indicated above. In this thin surface layer the strains may be larger than 100 bar. As an indication for this we take the fact that the observed threshold does not depend on stress; a shift of about .3 meV would have been expected for stresses up

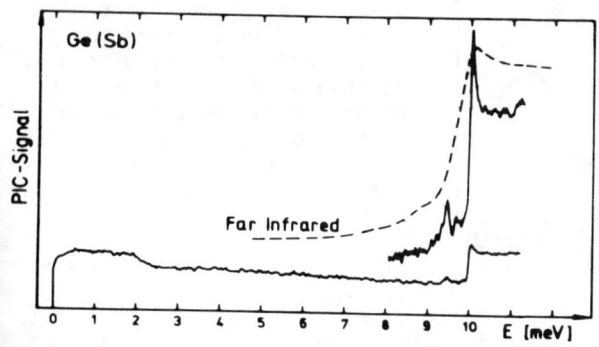

Fig. 3. PIC signal of Ge:Sb. The threshold at 2 meV belongs to Sb$^-$; at 9.8 meV to Sb^o. The precursor $2\Delta_{Al} = .6$ meV in front is a spectral property of the Al-junction The threshold is much sharper for PIC than for FIR. A peak in the phonon density of states may be one reason.

to 100 bar. Thus the threshold value measured here should correspond to the high stress case.

In contrast to the FIR-photoconductivity threshold at nearly the same temperature (Nagasaka and Narita 1973) the PIC thresholds are rather sharp. In the case of Sb a peak in the phonon density of states at 10 meV may be effective. There are also two "cut-off" functions that may be responsible for a steepening: The k-conservation condition reduces the interaction with phonons rapidly with increasing frequency. Also, the decreasing phonon lifetime due to spontaneous anharmonic decay will diminish the effective phonon density and thus the signal to higher energies. The photoconductivity signal below the isolated donor ionization energy <as determined by combining the measured energy differences between ground state to excited bound states from optical transitions with calculations with EMA for the excited states> down to about 9 meV has been interpreted by Nagasaka and Narita (1973) as due to donor complexes with reduced binding energies. This would then mean that the phonon interaction with these D^o-complexes is weak though, on the other hand, the broadening and shift to higher binding energies of A^+-complexes in Si is readily observed with PIC (Burger and Laßmann 1986.)

The high sensitivity for Sb is evident from the PIC signal for a Ge:As sample (Fig. 4) where the Sb concentration should be below 10^{12} cm^{-3}, the detection limit in photoluminescence measurements of the sample (Thonke 1988). The special form of the As^o-signal was reproducibly obtained with various samples. The increase of the signal after the steep step may be a sensitivity increase due to the peak in the density of states of the fast transverse acoustic phonons around 14 meV. The increase in front of the step is perhaps the effect of the strained surface layer reducing the binding energy randomly in part of the detection layer.

Whereas the SNR of the D^--signal increases for higher modulation frequencies the D^o-signal starts to diminish at about 500 Hz and was not measurable above 1.5 kHz. Since this decay shifts to somewhat higher frequencies with higher bias at the sample we believe that delay times of drifting carriers are responsible for the signal reduction, not the trapping time constants of the D^+. In the same range of modulation frequencies the phase of the D^o shifts with respect to the D^--signal and this shift can be compensated by increasing the bias also consistent with a drift delay. Apart from the precursor there is no structure between the Sb^-- and the Sb^o-thresholds which could be attributed to <u>scattering</u> of phonons by transitions from the 1s to e.g. the 2s or 2p excited states since the PIC-signal has reduced spectral sensitivity far beyond the D^--threshold. (Phonothermal detection of these states is not possible at the bath

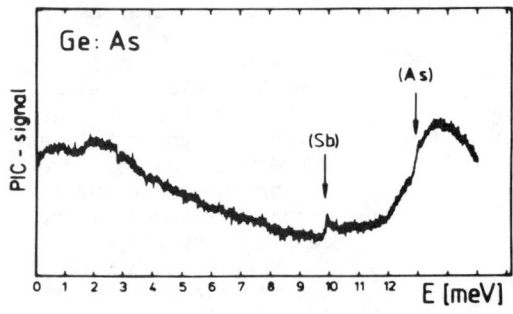

Fig. 4. PIC signal for Ge:As (<As> = 6×10^{14} cm^{-3}) The peak at 9.9 meV is ascribed to residual Sb^o. The form of the As^o-signal differs from that of Sb^o. Peaks in the phonon density of states around 10 and 14 meV may be one reason.

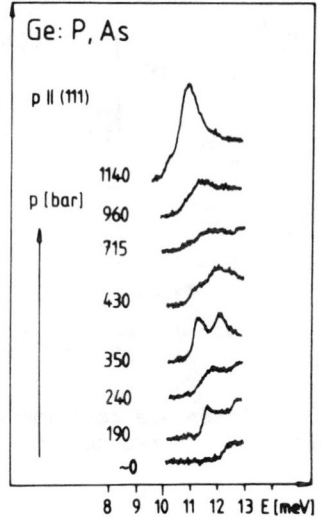

Fig. 5. Effect of level crossings on signal height.
⟨P⟩ = 2.4x10^{14} cm^{-3}
⟨As⟩ ~ 10^{13} cm^{-3}

temperature of 1 K since the distance from the excited states to the band is much larger than the thermal energy).
Interaction with excited ns states is, however, observable by the enhancement of the PIC signal when an excited level crosses the CB of the lowest valley under stress. This is shown in Fig. 5 for a sample containing P (2.4x10^{11} cm^{-3}) and As (~ 10^{13} cm^{-3}): In the curve for zero stress only the threshold of P^0 at about 12 meV is visible. Under stress the threshold shifts down as expected from Fig. 2. Around 190 bar there is a first increase due to the crossing of the A$_1$ level of the 3s-triplet of P and at about 12.5 meV due to the crossing of the same level belonging to As. The next increase of the thresholds due to A$_1$ of 2s is around 350 bar and finally around 1140 bar due to the crossing of A$_1$ of the 1s-triplet.
Phonon transitions to the 1s-triplet can also be observed at small stress in the case of P and As: For these donors the 1s-singlet ⟨-⟩1s-triplet ground state splitting is somewhat larger than the D$^-$-threshold and shows up as a small increase in the D$^-$-signal just beyond the threshold. Such an increase is expected if the phonon density becomes higher due to scattering into or trapping within the region sensitive for D$^-$-detection. At "zero" stress we find splittings of the 1s-triplet differing from run to run which we attribute to residual <u>uniaxial</u> stress from mounting (typically 30 bar) as mentioned in the preceding paragraph. Taking account of this initial stress we obtain from the stress dependence of this signal the ground state splitting in accordance with the optically determined values. (Phonon scattering by the ground state splitting of the Sb donor in Ge was first observed by Dynes et al (1971) by a 1 meV fixed frequency setup with Sn junctions as emitter and detector and variing the stress.)

From the shift of the threshold (with the initial stress accounted for) we obtain the stress dependence of the binding energy as shown in Fig. 6 for the case of P. The full lines are the expected energy differences taking (i) the optically determined binding energy at zero stress, (ii) the deformation potential constants (DPC) for the ground state splitting as determined from the level crossings (see below), and (iii) the same DPC for the shift of the CBs. Our values are consistently smaller. The large discrepancy for the ⟨100⟩ direction is not understood. There is a slight misorientation of the stress since the number of level crossings indicates that the ground state is fully split corresponding to a nonsymmetry direction. Since, however, this direction does not deviate much from ⟨100⟩ as checked repeatedly with X-ray orientation and also estimated from the experiment the band shifts and level splittings and consequently the change in binding energy should be small. Extrapolating back to zero stress we obtain 12.4±.05 meV and 13.4±.1 meV for P and As, respectively (high stress value for Sb: 9.9±.05 meV). These values are between the optical values and those derived from the temperature dependence of Hall measurements (Lopez and Koenig 1968).

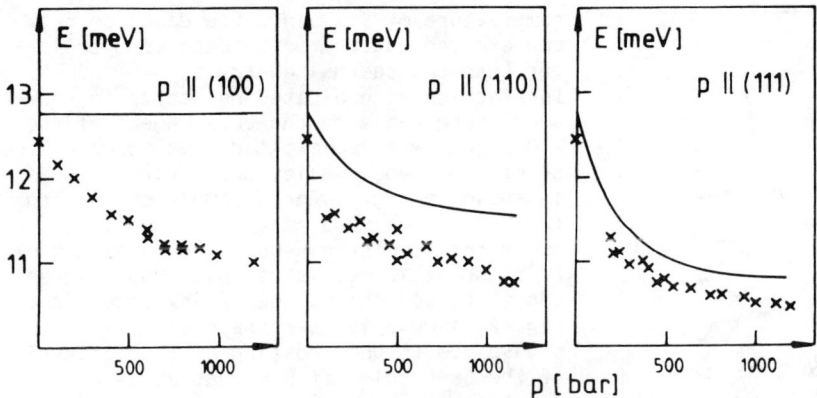

Fig. 6 Stress dependence of the PO–PIC threshold. Full lines are calculated with the optical values for the binding energies. The discrepancy for <100>-direction is discussed in the text.

By the combination of the DPC for the band shift, the value of the ground state splitting at zero stress and the stress and threshold energy for level crossing we obtain the DPCs for the ground states of Sb, P, and As as 16.5 ± .5 eV, 16.0 ± .5 eV, and 15.7 ± .5 eV, respectively, i.e. there seems to be a small chemical shift of these constants.
As regards the one-phonon ionization the donors in Ge may well be a special case because of the many valley structure of the CB supporting the k-conservation. Under similar experimental conditions we did not succeed to observe a PIC threshold for the acceptor Ga in Ge even under stresses high enough to shift the ionization energy down to about 8 meV. Also for donors in GaAs with a binding energy of nearly 6 meV we got only negative results so far. In both cases the neighbouring band extremum is at k = 0. Nevertheless, it seems promising to extend the technique to phonon-photon or phonon-phonon combination experiments to gain a direct access to the problem of phonon interaction with shallow states in semiconductors.

Financial support by the Deutsche Forschungsgemeinschaft is gratefully acknowledged. We are indebted to E.E. Haller for the supply of a P- and a Ga-doped sample and to M. Kobayashi for an As-doped sample. We are grateful to K. Thonke for his help with luminescence characterization of the donor content of the samples.

References

Burger W and Laßmann K 1986 Phys. Rev. B33 5868
Dynes R C, Narayanamurti V and Chin M 1971 Phys. Rev. Lett. 26 181
Eisenmenger W 1976 Physical Acoustics XII 80
Gummel H and Lax M 1955 Phys. Rev. 97 1469
Haller E E 1987 Physica 146B 201
Lopez A A and Koenig S H 1968 Proc. IX Int. Conf. on the Physics of
    Semiconductors, Moscow ed S M Ryvkin (Leningrad: Nauka) pp 1061-6
Nagasaka K and Narita S 1973 J. Phys. Soc. Jap. 35 788 and 797
Taniguchi M, Hirano M and Narita S 1975 Phys. Rev. Lett. 35 1095
Thonke K 1988 private communication

Nonlinear effects in impurity pumping and in impurity ionization

T. Theiler, F. Keilmann and E.E. Haller*

Max-Planck-Institut für Festkörperforschung, D-7000 Stuttgart 80, Fed. Rep. Germany
*University of California, Berkeley CA 94720, USA

ABSTRACT: The power-broadening of the lineshapes of far-infrared impurity transitions is studied experimentally. This effect allows the determination of the pertinent relaxation dynamics. The discovery of lifetime-dominated lineshapes of the stress-insensitive D(H,O) donor complex in ultrapure germanium (Navarro et al. 1988) made these experiments possible. We find that the broadening functions for both, the 1s-2p$_+$ and the 1s-3p$_-$ transitions, are in agreement with a simple two-level model, giving an energy relaxation longer than 100 ns. Furthermore we find a rather peculiar effect when pumping the D(H,O) centers with ionizing far-infrared radiation. After complete photoionization of the donors in a short pulse the generated free electrons vanish rapidly into an as yet unidentified non-conducting state. Relaxation back into the D(H,O) ground state takes a few µs.

1. INTRODUCTION

In our previous work (Navarro et al. 1988), we established that the D(H,O) donor complex in ultrapure germanium exhibits extremely narrow far-infrared resonances. To observe these transitions, the photothermal conductivity signal is monitored at some fixed far-infrared laser frequency, while the transition frequency is Zeeman-tuned by varying the magnetic field from 0 to 5 kOe.

The resonance lines exhibit nearly Lorentzian shape. The full linewidth (FWHM) in case of the 1s-2p$_-$ transition is $\Delta v = 8.6$ µeV corresponding to 0.07 cm^{-1}. If this linewidth is dominated by the dynamics of the 2p$_-$ state, we obtain $T_2 = (\pi \Delta v)^{-1} = 0.16$ nsec as the dipole relaxation time. This interpretation is strongly supported by applying the Nishikawa and Barrie (1963) theory based on weak electron-phonon coupling. We concluded that the observed linewidth is solely determined by intrinsic lifetime.

The basic absorption therefore can be considered in a model of two-level systems coupled by electric dipole µ with homogeneous broadening. Such a system is expected to exhibit linewidth broadening at high radiation intensity, an effect which can be used to determine also the energy relaxation time T_1 of the two-level systems.

2. THEORY

The absorption of homogeneously broadened two-level systems with resonance frequency v_0 is

© 1989 IOP Publishing Ltd

$$\alpha(\nu) = \frac{2\pi^2}{\hbar\varepsilon_0 c} * \frac{L}{n} * \frac{|\mu|^2}{3} * \nu * (N_1 - N_2) * g(\nu) \tag{1}$$

after Pantell and Puthoff (1969), where

- ν = light frequency;
- c = speed of light;
- L = Lorentz field factor;
- n = refractive index;
- $N_{1,2}$ = number densities of ground state and excited state in the absence of radiation;
- $g(\nu)$ = Lorentzian line factor,

$$g(\nu) = \frac{2}{\pi} * \frac{1/2\pi T_2}{(\nu-\nu_0)^2 + (1/2\pi T_2)^2 *(1+I/I_{sat})}, \tag{2}$$

where

- T_2 = dipole or phase relaxation time;
- I = light intensity;
- I_{sat} = saturation intensity of the medium

$$I_{sat} = \frac{\hbar^2 \varepsilon_0 c}{2} * \frac{n}{L} * \frac{3}{|\mu|^2} * \frac{1}{T_1 T_2}, \tag{3}$$

where T_1 is the energy relaxation time of the two-level systems.

The Lorentzian factor (2), normalized to $\int_0^\infty g(\nu) \, d\nu = 1$ at small intensities $I \ll I_{sat}$, shows two dependences of steady state absorption on intensity:

i) On resonance the absorption decreases as $g(0) \sim (1 + I/I_{sat})^{-1}$.

ii) The Lorentzian linewidth $\Delta\nu$ (FWHM) increases, i.e., is power-broadened according to

$$\Delta\nu = (1 + I/I_{sat})^{1/2} / \pi T_2. \tag{4}$$

In our experiment with low-concentration D(H,O) donors the absorption is too small to be measured with a direct light transmission experiment. However we can measure the conductivity σ of free electrons generated by thermal phonons interacting with excited impurity states. Assuming that the phonon processes do not depend on the excited state population we expect the conductivity to be proportional to both intensity and absorption coefficient. The lineshape in conductivity then is the same as in absorption, while the resonance photoconductivity saturates according to

$$\sigma(\nu_0) \sim I / (1 + I/I_{sat}). \tag{5}$$

3. EXPERIMENT

An ultrapure Ge single crystal with the dimensions 1x3x7 mm³ grown at the Lawrence Berkley Laboratory is studied. The long axis is orientated in [111] direction, the 7x3 mm² faces are normal to the [11$\bar{2}$] direction. The two 7x1 mm² faces are ion implanted with P (25 kV, 4x10¹⁴ cm⁻²) and annealed at 300C for 2h in Ar to form ohmic contacts.

The sample has a D(H,O) concentration of about 10^{11} cm^{-3} and a P concentration of 30 - 60% of that value. The concentration of residual shallow acceptors is of the order of 10^{10} cm^{-3}. The crystal is mounted in a liquid He bath cryostat at 4.2K with the electric field in [111] direction. The magnetic field is in [11$\bar{2}$] direction.

Pulsed far-infrared gas lasers capable of high power up to 1MW/cm^2 (Gross et al. 1987) are used for our experiments. The pulse length is about 50 nsec. The frequencies used are 87.3 cm^{-1}, 91.3 cm^{-1} and 111 cm^{-1}. The laser is incident parallel to the magnetic field and illuminates the crystal surface homogeneously. The laser power is varied in precision steps over several orders of magnitude by using a broadband attenuator (Keilmann 1986). The photoconductivity signal is amplified by a fast MOS transistor (Siliconix SD 211) situated near the sample in liquid helium. It is recorded on a high speed digital oscilloscope.

4. RESULTS

The photoconductivity is found to follow the pulse shape of the far-infrared laser.
Figure 1 shows the conductivity obtained at the peak of the pulse with a laser frequency of 87.3 cm^{-1}.

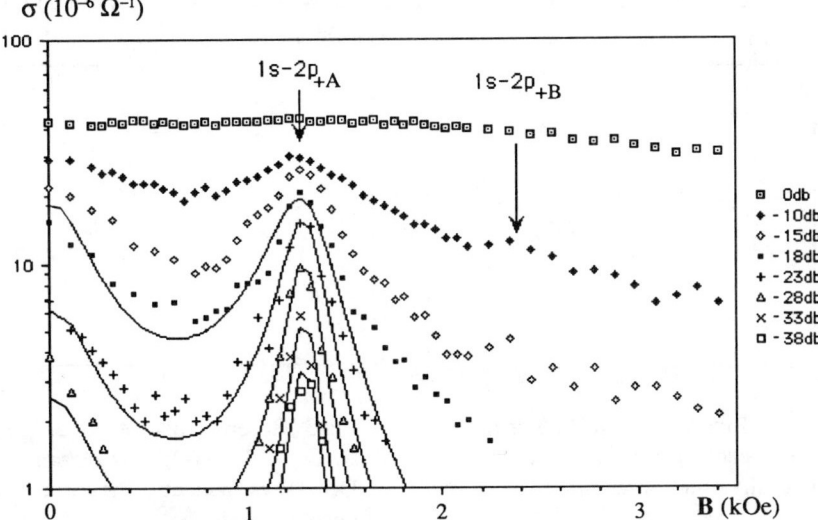

Fig.1 Saturation of the 1s-2p$_+$ resonance measured by the time-resolved photo-thermal ionization technique. The far-infrared intensity is given in db where 0 db amounts to about 100 W/cm^2 in the sample ($\nu = 87.3$ cm^{-1}).

The photoconductivity σ exhibits a well defined resonance at 1.31 kOe (equivalent to a Zeeman shift of 0.84 cm^{-1}) and a weak resonance at 2.4 kOe. Both are readily identified as shown in Fig.1 using the Zeeman data obtained earlier (Navarro et al. 1988). The 1s-2p$_+$ resonance is quite narrow at low intensity and broadens with rising intensity.

The resonance seems to ride on a broad background which falls off with increasing magnetic field. At the lower intensities, however, this background is seen to be intimately related to the resonance line itself. This relationship can be explained with the magnetoresistance effect of the

photogenerated free electrons. To elucidate this point we consider, for the lowest five intensities, the contribution of the 1s-2p$_{+A}$ resonance only. Equation (5) can be rewritten to incorporate the expression for the magnetoresistance given by Kuhrt and Lippmann (1968)

$$\sigma(B) \sim \frac{I}{(B-B_0)^2 + (\Delta B/2)^2 * (1+I/I_{sat})} * \frac{1}{(1+\beta B^2)} , \qquad (6)$$

where B is the magnetic field, B_0 is the resonance field, $\Delta B = 0.26$ kOe is the resonance width at low intensity. $\beta = (n_A/n_D) * \mu_p * \mu_n$ is the sole fitting parameter. By taking the theoretical mobilities $\mu_n(4.2K) = 3.4 \times 10^6$ cm^2/Vs and $\mu_p(4.2K) = 3.9 \times 10^6$ cm^2/Vs (Blankenship 1973, and Brown and Bray 1962) we derive the compensation ratios n_A/n_D to lie in the range 0.018 to 0.027. That these values are slightly smaller than expected for our sample may be explained by the uncertainty in the mobilities. The curves in Fig. 1 are calculated using $n_A/n_D = 0.024$.

Figures 2 and 3 show the linewidth Δv (FWHM) and the resonance photoconductivity $\sigma(v_0)$, taken from Fig.1, as a function of intensity. The observed line broadening shown in Fig.2 is compared to the theoretical model of homogeneous two-level systems by fitting the data using eq. (4). The agreement is satisfactory and we obtain the small signal linewidth to be 10.6 µeV and the saturation intensity to be 0.32 W/cm^2. When we substitute this value for I_{sat} into eq. (5) and calculate the general shape of the resonant conductivity we obtain the curve in Fig.3.

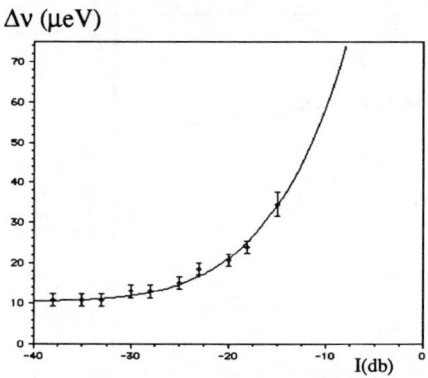 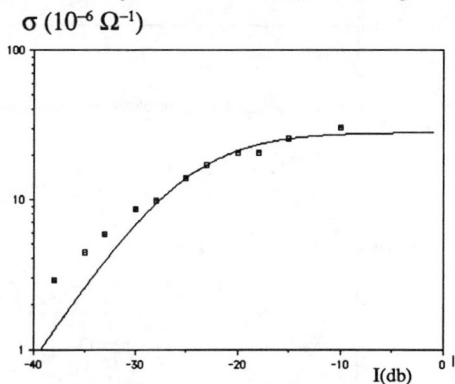

Fig.2 Linewidth (FWHM) of the 1s-2p$_{+A}$ resonance in Ge: D(H,O) vs. intensity. The curve is a theoretical fit.

Fig.3 Resonant photothermal conductivity of the 1s-2p$_{+A}$ transition in Ge: D(H,O) vs. intensity. The curve is calculated using eq. (5).

We observe that the model does not suffice to explain the photoconductivity of Fig. 1 at the highest intensities, 0 and -10 db. This is possibly due to neglecting the contribution of the 1s-2p$_{+B}$ resonance, but also other mechanisms of photogeneration of free carriers may start to play a role at the highest intensities, such as multiphoton absorption.

We can now use the derived value of $I_{sat} = 0.32$ W/cm^2 for the 1s-2p$_{+A}$ transition to extract the energy relaxation time using eq.(3). For this we have to insert the value for the transition moment µ. We use published absorption data for simple donors in Ge by Ramdas and Rodriguez (1984) and by Aggarwal et al. (1965), and calculate the absorption coefficients assuming that the oscillator strengths remain constant. Using eq. (1) we obtain for the matrix elements values in the order of 1×10^{-29} Cm. From this we obtain the energy relaxation time to be 100 nsec.

This value is of the same order of magnitude as one laser pulse duration, and therefore,we do not quite reach the steady-state saturation conditions which are however a prerequisite for our analysis. Thus we conclude that we have established a lower limit for T_1 only.

Furthermore, we measured the saturation of the 1s-3p$_{-A}$ transition with a far-infrared frequency of 91.3 cm^{-1}. These data can be analyzed in the same way as above. In this case, we derive a saturation intensity of 0.9 W/cm^2. For the dipole relaxation time T_2, we obtain 0.12 nsec (corresponding to $\Delta\nu$ (FWHM) = 10.6 μeV). When we assume that the energy relaxation time is identical for both transitions we find for the ratio between the two matrix elements ($|\mu_{2p}|^2 / |\mu_{3p}|^2$) = 3 ± 1. This is in good agreement with the theoretical value in germanium of 2.16 (Ovando 1988).

5. COMPLETE IONIZATION

In order to obtain direct information on the dynamics of the ionized donor - free electron system we conducted a photoionization experiment at 111cm^{-1}, i.e., without the involvement of a phonon-mediated step. Here we observe that at low laser power (Fig. 4c) the conductivity very nearly follows the laser pulse (Fig. 4d). At high powers, however, the conductivity is seen only during the initial part of the laser pulse, but vanishes even before the peak of the laser is reached (Fig. 4a).

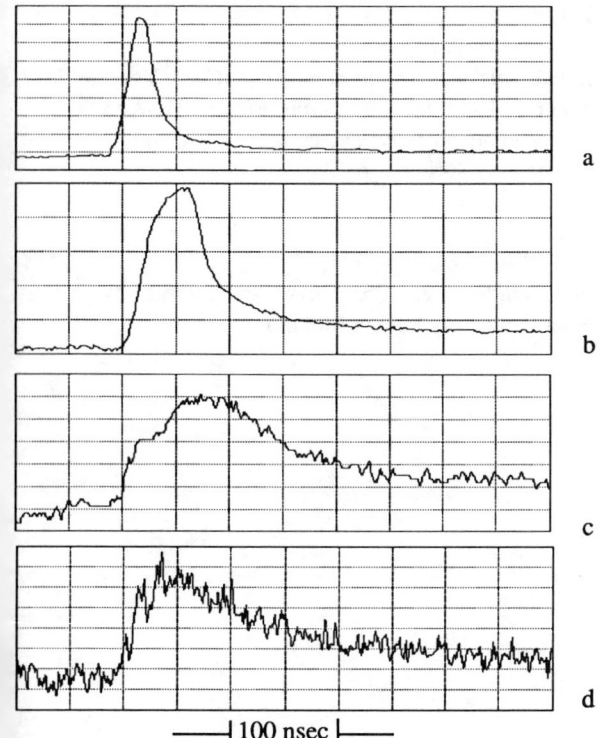

Fig. 4 Far-infrared photoconductivity σ (arbitrary units), a-c, and laser pulse shape, d. The peak laser power in the sample is 10 W/cm^2 (a), 1 W/cm^2 (b) and 0.1 W/cm^2 (c).

We explain this new phenomenon by assuming that the ground state becomes totally depleted. Further measurements at higher intensities (>50 W/cm^2) show that after rapid total depletion of the impurity ground state the conductivity vanishes in about 5 nanoseconds. Thus the electrons are transferred to a yet unidentified non-conductive state. Further observations show that the ground state population becomes fully restored after a few µs.

6. SUMMARY

We have shown experimentally that saturation of shallow impurity transitions can be understood in the framework of a homogeneously-broadened two-level model. A rather intriguing effect is observed with the D(H,O) donors studied here, viz., the complete but transient disappearance of the conductivity which immediately follows a rapid, fully ionizing far-infrared excitation. This effect calls for an as yet unidentified metastable state.

ACKNOWLEDGEMENTS

We are indebted to H.J.Queisser for numerous discussions. One of us (E.E.H.) acknowledges the partial support of the US NSF grant DMR 8502502.

REFERENCES

Navarro H, Haller E E and Keilmann F 1988 *Phys. Rev. B* **37** 10822
Nishikawa K and Barrie R 1963 *Can J. Phys.* **41** 1135
Pantell R H and Puthoff H E 1969 *Fundamentals of Quantum Electronics* (New York: Wiley)
Gross C T, Kiess J, Mayer A and Keilmann F 1987 *IEEE J. Quant. Electr.* **QE-23** 377
Keilmann F 1986 *SPIE Vol.***666** *Far-Infrared Science and Technology* p.213
Kuhrt F and Lippmann H 1968 *Hallgeneratoren* Springer p.70
Brown D M and Bray R 1962 *Phys. Rev.* **127** 1593
Blankenship J L 1973 *Phys. Rev. B* **7** 3725
Ramdas A K and Rodriguez S 1984 *Rep. Prog. Phys.* **44** 1324
Aggarwal R L, Fisher P, Mourzine V and Ramdas A K 1965 *Phys. Rev. A* **138** 882
Ovando M A P 1988 M.S.-Thesis Puebla University Puebla Mexico

Simulation of far-IR absorption and PTI-spectra of shallow impurities in compensated germanium

E Rotsaert (*), P Clauws and J Vennik

Laboratorium voor Kristallografie en Studie van de Vaste Stof, Rijksuniversiteit Gent, Krijgslaan 281, 9000 Gent, Belgium.

(*) Aspirant with the NFWO.

L Van Goethem

Metallurgie Hoboken-Overpelt, Leemanslaan 36, 2430 Olen, Belgium.

ABSTRACT : A model has been developed allowing the calculation of the photoconductive response from the shallow impurities in compensated Ge. The distribution of electrons and holes over the shallow impurity states and energy bands is obtained when the sample is illuminated with continuous bandgaplight and chopped far-IR light.
From the equilibrium distribution, it is shown that the absorption coefficient increases and saturates with increasing bandgap illumination. The model is used to study the quantitative possibilities of PTIS of residual impurities in high-purity compensated Ge.

1. INTRODUCTION

Far-IR absorption and photothermal ionization spectroscopy (PTIS) have been used extensively to identify shallow impurities in Ge. In compensated samples the compensating impurities can be observed by simultaneous illumination with far-IR and bandgaplight.
In absorption the concentration of both minority and majority impurities can be determined from the integrated absorption coefficient of the main spectral lines. This quantitative characterization is possible for the elemental group-V donors and group-III acceptors in Ge with the use of experimentally determined integrated absorption cross sections (Rotsaert et al 1987) for concentrations $\gtrsim 10^{11}$-10^{12} at/cm^3, depending on the instrument and corresponding with the absorption detection limit in Ge. For lower concentrations, PTIS can be applied. In compensated semiconductors, the photothermal ionization of the compensating impurities usually causes negative photoconductive lines. Mechanisms that may qualitatively explain this have been put forward (Bykova 1973, Skolnick 1974, Darken 1983). The idea of a reduction of the total carrier concentration by rapid recombination is most frequently proposed. A positive photoconductive contribution from the compensating centers has also been observed in high-purity Si (Haller 1978) and Ge (van de Steeg 1983). The latter authors suspect the electronic detection technique to be responsible for this effect.
In order to obtain impurity concentrations from the magnitude of the excitation lines, the role of several factors such as temperature,

compensation degree, intensity of bandgap illumination and instrumental parameters should be evaluated. As an extension of the work of van de Steeg et al (1983), we have developed a model for calculating the photoconductive response in a compensated semiconductor illuminated with far-IR and bandgaplight. By means of a set of rate equations, the distribution of electrons and holes over the shallow impurity states and energy bands is described. The model allows to explore the influence of several parameters on the magnitude and sign of the spectral lines.

2. MODEL

A schematic diagram of the model is shown in fig.1. We consider a Ge-sample that contains one type of donor-impurity with concentration N_d and one type of acceptorimpurity with concentration N_a. These impurities are indicated by their ground state E_d and E_a which are between 10 and 15 meV in Ge. From fig.1 we can distinguish the processes governing the distribution of electrons and holes over the shallow impurity states and the energy bands when the sample is illuminated with bandgaplight : the generation of free electrons and holes by absorption of the bandgaplight (a), the electron-hole recombination (b), the emission of electrons (holes) from the ground state to the conduction (valence) band by thermal ionization and by background radiation (c), the capture of electrons (holes) into the donor (acceptor) ground state (d) and the recombination of an electron (hole) with an acceptor (donor) (e).

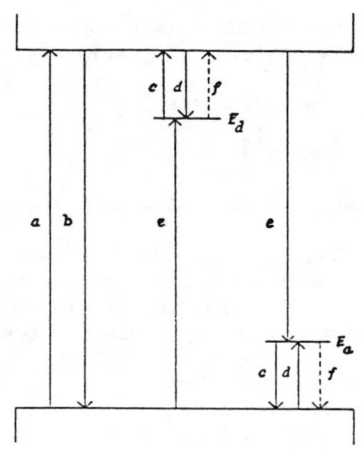

Fig. 1. Processes governing the distribution of electrons and holes over the shallow donor- and acceptorstates in a compensated semiconductor illuminated with bandgaplight.

(f) indicates the ionization by the far-IR light.
The rate equations are given by :

$$-\frac{dN_d^0}{dt} = (\alpha_n + \beta)N_d^0 + \gamma(x)N_d^0 - \delta_n n N_d^+ + R'_n p N_d^0 \quad (1)$$

$$-\frac{dN_a^0}{dt} = (\alpha_p + \beta)N_a^0 + \gamma(x)N_a^0 - \delta_p p N_a^- + R'_p n N_a^0 \quad (2)$$

$$\frac{dn}{dt} = (\alpha_n + \beta)N_d^0 + \gamma(x)N_d^0 - \delta_n n N_d^+ - R'_p n N_a^0 + G - R n p \quad (3)$$

$$\frac{dp}{dt} = (\alpha_p + \beta)N_a^0 + \gamma(x)N_a^0 - \delta_p p N_a^- - R'_n p N_d^0 + G - R n p \quad (4)$$

n and p denote the electron- and hole concentration. The neutral and ionized donor- and acceptor concentrations $N_d^0, N_d^+, N_a^0, N_a^-$ satisfy $N_d = N_d^0 + N_d^+$ and $N_a = N_a^0 + N_a^-$. The condition of charge neutrality requires $n + N_a^- = p + N_d^+$ (5).
In a first stage, the absorption coefficient of compensated n and p-type Ge as a function of the intensity of the bandgaplight has been calculated from the model. The absorption coefficient is proportional to the neutral donor- and acceptorconcentration which can be determined from the

equilibrium situation, i.e. putting the left-hand side of equations
(1)-(4) to zero. We mention that the photothermal process (f) is not
included in this calculation since it causes only a small disturbance of
N_d^0 and N_a^0.
In a second stage, the model has been used to calculate the time dependence
of the photoconductive response when the sample is exposed to continuous
bandgap illumination and chopped far-IR light. The photothermal ionization
process is simulated with the parameters γ and γ'. Since our experiments are
performed with a Fourier transform spectrometer with stepping motor, γ and γ'
are functions of the optical path x, and give for each optical path the
number of photothermally excited carriers per unit of time. Their value
depends on the incident fotonflux, the absorption cross section for transitions from the ground state to excited states and the thermal ionization
probability from the excited states into the band. The calculation of these
parameters is simplified by limiting each donor- and acceptorspectrum to
two lines with a Gaussian shape and a full width at half maximum of 0.8 cm^{-1}
for the donor- and 0.5 cm^{-1} for the acceptorlines. For each optical path,
the values of γ and γ' are substituted into the equations (1)-(4), and the
time dependence of n, p, N_d^0 and N_a^0 over one chopping period is obtained by
numerically solving the system of differential equations. The time dependence
of the conductivity σ (or n+p taking the mobilities of electrons and holes
equal) allows to simulate the interferogram and by Fourier transformation
the PTI-spectrum is obtained.

3. RESULTS AND DISCUSSION

In fig.2 we show the results of a calculation of the integrated absorption
coefficient as a function of G, the number of electrons and holes created
per unit volume and per unit of time by the bandgaplight, for a p-type
Ge sample with $1.3 \cdot 10^{13}$ cm^{-3} Ga compensated by $8.3 \cdot 10^{12}$ cm^{-3} As at 8K.
The emission and capture parameters are noted with the figure.

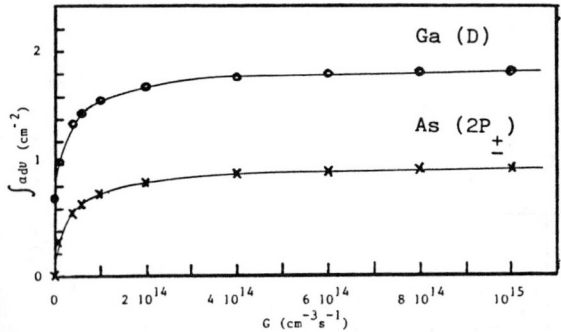

Fig.2. Integrated absorption coefficient
of p-type Ge as a function of the intensity
of the bandgaplight.
$N_{As} = 8.3 \cdot 10^{12}$ cm^{-3}, $N_{Ga} = 1.3 \cdot 10^{13}$ cm^{-3}
$T = 8K$
$\beta = 10^3$ s^{-1}
$\alpha_n = 10^2$ s^{-1} $\alpha_p = 1.3 \cdot 10^3$ s^{-1}
$\delta_n = 3.75 \cdot 10^{-6}$ cm^{3}s^{-1} $\delta_p = 2.89 \cdot 10^{-6}$ cm^{-3}s^{-1}
$R'_n = 2.89 \cdot 10^{-9}$ cm^{3}s^{-1} $R'_p = 3.75 \cdot 10^{-9}$ cm^{3}s^{-1}

The thermal ionization coefficients are calculated according to:
$$\alpha = \frac{\sigma^i \langle v_{th} \rangle N_{eff}}{g} \exp\left(-\frac{\Delta E}{kT}\right)$$
with σ^i the capture cross section for an electron (hole) by an ionized donor (acceptor), $\langle v_{th} \rangle$ the mean thermal velocity, N_{eff} the density of states in the conduction (valence) band, g and ΔE the degeneracy and binding energy of the ground state of the donors (acceptors). The capture coefficients are given by $\delta = \sigma^i \langle v_{th} \rangle$ and $R' = \sigma^o \langle v_{th} \rangle$ with σ^i, σ^o the capture cross sections for an ionized and neutral center respectively. We have taken $\sigma^i = 10^{-12}$ cm^2 and $\sigma^o = 10^{-15}$ cm^2. The ionization due to background radiation has been estimated to be of the order of 10^3 s^{-1}. The integrated absorption coefficient for the D-line of Ga is calculated as $1.45 \cdot 10^{-13} N_a^0$ and for the $2P_+$ of As as $1.2 \cdot 10^{-13} N_d^0$.

We can see on fig.2 that with increasing intensity of the bandgaplight, the integrated absorption coefficient increases and saturates at a value corresponding with N_d and N_a. This is in agreement with the experimental results.

For the calculation of the photoconductive response, the change in the electron- and hole concentration due to the on- and off-switching of the far-IR light (by chopping) is monitored. The far-IR light causes the ionization of electrons and holes from neutral donors and acceptors into the bands through a photothermal process. The conductivity change thus resulting is measured with a lock-in amplifier using phase-sensitive detection. In the model, the time dependence of n and p is calculated at each position of the interferometer over one chopping period (5 ms). The results of these calculations are shown in fig.3 for a p-type Ge sample containing Ga and As. The curves are determined at zero optical path, and correspond to different values of the bandgaplight.

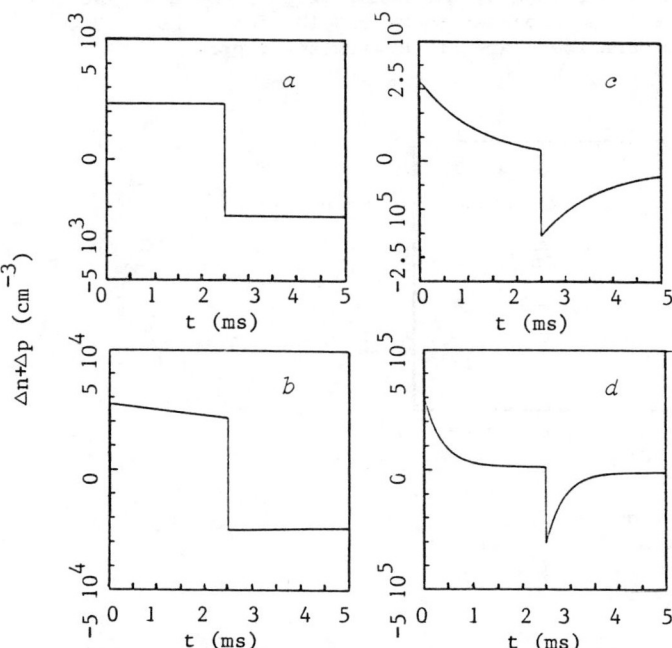

Fig.3. Time dependence of the photoconductive change $\Delta n + \Delta p$ over one chopping period, at zero optical path.
a) G=0 cm^{-3}s^{-1} b) G=10^{14} cm^{-3}s^{-1} c) G=$5 \cdot 10^{14}$ cm^{-3}s^{-1} d) G=10^{15} cm^{-3}s^{-1}

Without bandgap illumination (G=0) all minority impurities are ionized by compensation and the photoconductive signal is due totally to the ionization of majority impurities. The majority carriers respond instantaneously to the on- and off-switching of the far-IR light, as can be seen from the square wave of fig.2 (a). With bandgap illumination, part of the ionized impurities is neutralized by the capture of an electron or hole. The photoconductive response is now composed of both Δn and Δp, and as can be seen from fig.2 (b,c,d) this results in a relaxation process with a time constant which decreases for increasing G.

In fig.4, a series of simulated spectra is shown for a p-type sample with Ga and P. In the spectrum recorded without bandgaplight (a), only the lines of the majority impurities (Ga) are observed. The relative intensities of these lines corresponds with the ratio of the products of the absorption cross section with the thermal ionization probability. The latter is calculated as
$$\left(1+\frac{1}{gg'}\exp\left(\frac{\Delta E}{kT}\right)\right)^{-1}$$
(Jongbloets 1979).

In the spectra (b,c,d) recorded with bandgaplight, the lines of the compensating impurity (P) are seen to be negative. The ratio of the P-lines to the Ga-lines increases with G increasing up to 10^{16} cm^{-3}s^{-1}, above which this ratio decreases again.

Other spectra have been simulated both for n- and p-type Ge containing one of the group-V donors and one of the group-III acceptors. In some examples, negative lines are seen for the compensating impurities; in other examples, these photoconductive lines are positive.

Up to now it is not possible to specify exactly the parameters that are responsible for these results, and how quantitative information can be obtained. The results are however promising in that is possible to simulate negative lines for the compensating impurities.

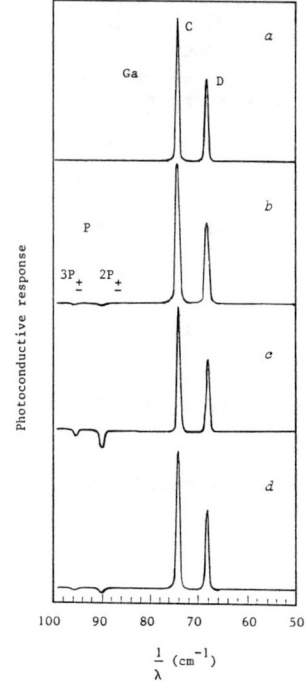

Fig.4. Simulated PTI-spectra for p-Ge containing 2.1 10^{13} cm^{-3} Ga and 4.3 10^{12} cm^{-3} P at 8.5K, for different intensities of bandgap illumination G.
a) G=0 cm^{-3}s^{-1}
b) G=10^{15} cm^{-3}s^{-1}
c) G=10^{16} cm^{-3}s^{-1}
d) G=3 10^{17} cm^{-3}s^{-1}

4. CONCLUSION

Compensated Ge with impurity concentration $\gtrsim 10^{12}$ cm^{-3} can be analyzed quantitatively by means of absorption spectroscopy. For lower concentrations, PTIS can be applied disposing quantitative information only for the majority impurities (Kogan 1977, Darken 1983). For compensated Ge, we can study the influence of several parameters such as emission and capture coefficients, compensation, temperature, bandgaplight, detection method on the photoconductive response by means of a theoretical model. The results show that a negative response for the compensating impurities can be calculated conform with experimental observations; the simulated time dependence of the photoconductive signal is qualitatively in agreement with the experimental results obtained from the analysis by means of a signal averager.

Further calculations are planned in which the influence of certain combinations of the parametervalues will be established and an extension of the simulations to the case of a rapid-scan interferometer is also considered.

5. REFERENCES

Bykova E M, Lifshits T M and Sidorov V I 1973 Sov. Phys. Semicond. 7 671
Darken L S 1982 J. Appl. Phys. 53 3754
Darken L S and Hyder S A 1983 Appl. Phys. Lett. 42 731
Haller E E 1979 Bull. Acad. Sci. USSR Phys. Ser. 42 1131
Jongbloets H W H M, Stoelinga J H M, van de Steeg M J H and Wyder P 1979 Phys. Rev. B20 3328
Kogan Sh M and Lifshits T M 1977 Phys. Stat. Sol. (a) 39 11
Rotsaert E, Clauws P, Vennik J and Van Goethem L 1987 Physica 146B 75
Skolnick M S, Eaves L, Stradling R A, Portal J C, Askenazy S 1974 Sol. State Comm. 15 1403
van de Steeg M J H, Jongbloets H W H M, Gerritsen J W and Wyder P 1983 J. Appl. Phys. 54 3464

ature.

The structure of the NL10 thermal donor in silicon

C.A.J. Ammerlaan, T. Gregorkiewicz and H.H.P.Th. Bekman

Natuurkundig Laboratorium der Universiteit van Amsterdam
Valckenierstraat 65, 1018 XE Amsterdam, The Netherlands

ABSTRACT: The Si-NL10 is a prominent thermal-donor-related EPR spectrum. Electron nuclear double resonance measurements on ^{29}Si nuclei show that the defect centre generating this spectrum has very shallow, delocalised character. At the same time it is found that the spectrum is inhomogeneously broadened as a superposition of a series of very similar components. It is concluded that the spectrum originates from a series of similar centres which are subsequently generated during the heat treatment at about 450 °C. The exhaustive study unravels the electronic and microscopic structure of these centres.

1. INTRODUCTION

Already more than 3 decades ago it was discovered that the heat treatment of oxygen-rich silicon at about 450 °C leads to the formation of shallow donor states, called thermal donors - Fuller et al. (1954), Fuller and Logan (1957), Kaiser et al. (1958). The first publication on thermal donors appeared in 1954 and since then a vast amount of data was gathered on the subject - for a recent review see e.g. Bourret (1985). It is nowadays established that upon heat treatment in the 300-500 °C temperature region a series of very similar shallow double donor centres is created - Wruck and Gaworzewski (1979). They can be observed by standard infrared absorption techniques and up to 9 different species have been reported - Pajot et al. (1983).

Among other experimental techniques also electron paramagnetic resonance has been used in the thermal donor studies. It has been noted that the formation of thermal donors (as monitored in infrared) is accompanied by the simultaneous generation of several electron paramagnetic resonance (EPR) spectra of predominantly 2mm point group symmetry - Muller et al. (1978), (1979). Further studies showed, on the basis of the production characteristics, that the Si-NL8 and Si-NL10 spectra were practically the only ones which could be related to the thermal donor centres as observed in infrared - Gregorkiewicz et al. (1987), Bekman et al. (1987). For one of these - the Si-NL8 spectrum - a direct identification with the singly ionised state of the infrared double donors was made - Lee et al. (1985). Following that a further studies of the Si-NL8 center by the electron nuclear double resonance (ENDOR) technique were undertaken - Michel et al. (1986). At the same time the origin of the other thermal-donor-related and usually the more prominent Si-NL10 centre remained unclear and required further investigation. Consequently the ENDOR study of the

© 1989 IOP Publishing Ltd

Si-NL10 centre has been performed - Gregorkiewicz et al. (1988).

2. EPR STUDIES OF THE Si-NL8 AND Si-NL10 CENTRES

2.1 Symmetry

The paramagnetic centre is characterised by the g-factor which is a basic spin-Hamiltonian parameter and therefore acts as a characteristic identification mark discriminating various EPR spectra. The symmetry of the centre is reflected directly by the symmetry of its g-tensor. The g-values for the Si-NL8 and Si-NL10 spectra in their early and late transformation stages are presented in Table 1.

Table 1. Principal g-tensor values for the Si-NL8 and Si-NL10 EPR spectra in early and late transformation stages (10 and 200 h heat-treatment time, respectively).

Spectrum	transformation stage	g_1 ‖ [011]	g_2 ‖ [0$\bar{1}$1]	g_3 ‖ [100]
Si-NL8	early	1.9926	2.0012	1.9999
	late	1.9938	2.0008	1.9999
Si-NL10	early	1.9975	1.9996	1.9996
	late	1.9980	1.9995	1.9998

For both Si-NL8 and Si-NL10 TD-related EPR spectra the symmetry was found to be orthorhombic, point group 2mm. This means that the defects generating those spectra must have two nonequivalent symmetry planes. This information is of basic importance for the modeling of the thermal donor core structure; following this result the most prominent OSB and Ylid models could be constructed. One should however note here that the conclusion concerning the symmetry of the centre is valid only within the resolution of the EPR experiment; it can never be excluded that lower symmetry interactions exist but are small for some reasons and therefore remain hidden within the resonance linewidth or lead to its inhomogeneous broadening.

2.2 Shallow character

The shallow character of the centre can already be deduced from its g-tensor value. Although the g-tensor of the defect is an important EPR parameter unfortunately its theoretical interpretation presents a very complicated problem. Even for relatively simple cases as the shallow donor and acceptor impurities, whose wave functions are known in sufficient detail no satisfactory calculations of g-values are available. For the more complicated systems of lattice defects in silicon, generally of deep level character, at most qualitative considerations can be given. On the other hand, Lee and Corbett (1973) proposed an empirical classification of g-values, which was further extended by Sieverts (1983). In this classification scheme 4 different types of defects are distinguished:

type A: vacancy-type defects with one dangling bond or two parallel dangling bonds,
type B: vacancy-type defects with two or more dangling bonds under tetrahedral angles,
type C: interstitialcies,
type D: impurities on substitutional and interstitial sites.

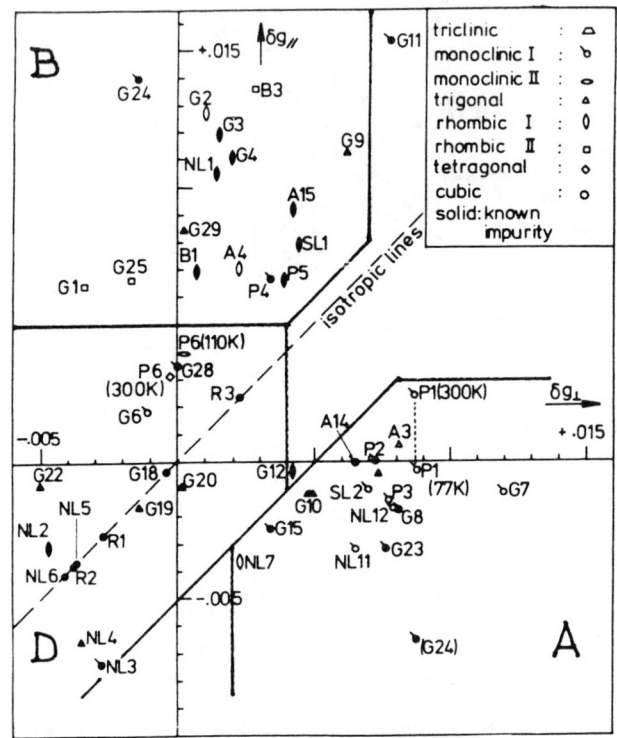

Figure 1. Plot of δg-values ($\delta g = g-2.0023$) of EPR spectra from A) defects with one or two parallel broken bonds, B) defects with two or more broken bonds under tetrahedral angles, C) interstitial-type defects, and D) substitutional and interstitial impurities and their complexes (closed symbols - known impurity) - Sieverts (1983).

In the proposed scheme g-values of defect centres are represented in a 2-dimensional plot of reduced g-values. The reduced g-values are constructed from the principal g-values by assuming them to be axially symmetric. Then the most deviating g-value is defined as the parallel value $g_\parallel$ and the average of the other two principal g-values is the perpendicular value $g_\perp$. If then $g_\parallel$ is plotted against $g_\perp$ different types of defects will group in different sections of the diagram as depicted in Fig.1. In Fig.2 the small section of the diagram grouping shallow defects and impurities of type D (the "Dutch corner") is shown. As can be seen the

heat-treatment centres Si-NL8 and Si-NL10-13-17 (presently identified to be one centre) both have their g-values in this particular region. The identification of the Si-NL8 spectrum as originating from the singly ionised TD$^+$ state of the thermal donor directly confirms the shallow character of the centre. Thus the conclusion that also the Si-NL10 spectrum originates from a loosely bound electron is supported.

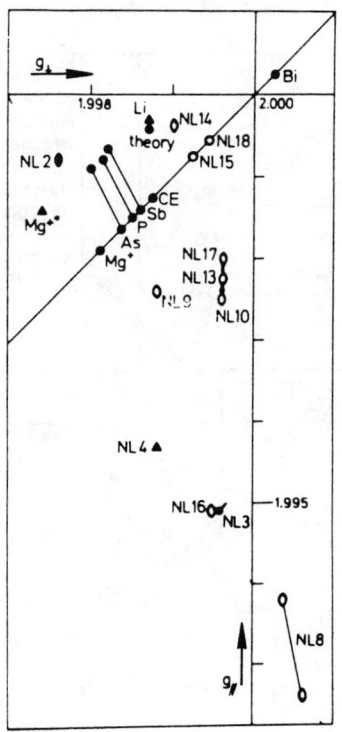

Figure 2. The "Dutch corner" - plot of shallow, effective-mass donor defects - Sieverts (1983).

2.3 g-shifting effect

The most peculiar feature of the two TD-related EPR spectra is the gradual, semicontinuous change of their g-tensor values upon duration of the heat-treatment time. Such a feature is rather unique and the fact that it is being found for both centres indicates close similarities of their structure. The shifting of the g-values can be explained either by subtle changes within the centre itself or by gradual changes of the environment - e.g. stress accumulation or release connected with oxygen aggregation in the direct vicinity of the paramagnetic centre. The EPR studies could not discriminate between those two drastically different possibilities which, at the time that the g-shifting phenomenon was first noticed, seemed equally probable. Such a distinction was only possible in the electron nuclear double resonance studies that followed.

3. ENDOR STUDY OF THE Si-NL10 CENTRE

For the Si-NL10 centre whose origin was particularly intriguing an extensive ENDOR study has been performed. This included analysis of the hyperfine interactions with ^{29}Si, ^{17}O and ^{27}Al nuclei. As a result both the microscopic - Gregorkiewicz et al. (1988) - and the electronic - Bekman et al. (1988) - structure of the Si-NL10 centre has been unraveled. It was confirmed that, as already indicated in EPR by the g-shifting phenomenon, the Si-NL10 spectrum originates from a series of gradually developing, very similar centres.

3.1 Shallow defect character

Figs. 3a and 3b show the angular dependence of the $\overleftrightarrow{A}$-tensor as determined from a ^{29}Si ENDOR experiment for two shells within one species of the Si-NL10 centre. The tensors are of orthorhombic and triclinic symmetry, respectively.

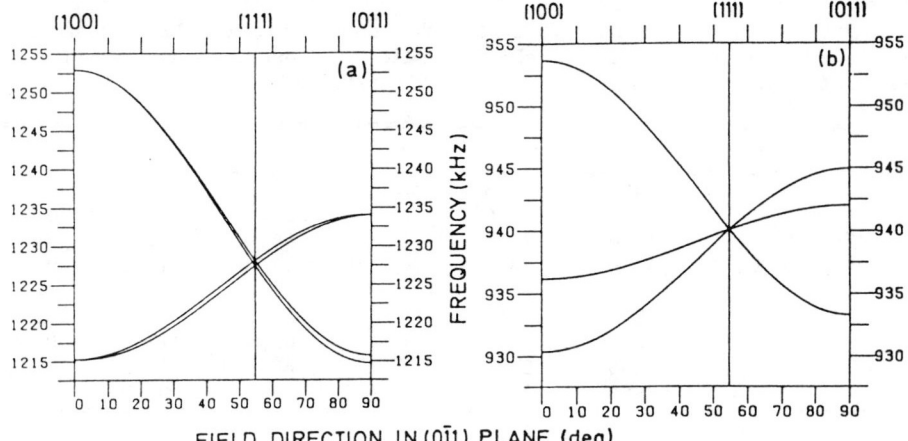

Figure 3. Computer simulations of the angular dependences of two silicon hyperfine tensors of the Si-NL10 spectrum of (a) orthorhombic (tensor T2) and (b) triclinic symmetry (tensor G2). Microwave frequency 23 GHz.

The experimental ENDOR data are usually analysed employing the linear combination of atomic orbitals (LCAO). In the LCAO analysis the wave function at the ligands is approximated by a linear combination of 3s and 3p silicon atomic orbitals, centred on that ligand. The LCAO treatment works well for deep, localised defects. A deep defect produces a ground-state wave function with big gradients over the ligand atoms. For a deep defect the wave function can quite often be approximated by almost one sp^3-hybridised orbital. This results in a hyperfine tensor which is approximately axial along a <111> bonding direction. For an electron in an orbital pointing in any of the four <111> directions a sp^3- hybridised orbital with s-character α^2 - 25%, p-character β^2 - 75% and α^2/β^2 - 0.33 is expected.

Table 2a. Hyperfine parameters in kHz for some typical <111>-axial silicon hyperfine tensors of two characteristic deep defects V⁻ and VV⁻ after Sprenger et al. (1987) and Sieverts et al. (1978).

centre	tensor	a	b	c	a/b	α^2/β^2	η^2(%)
Si:V⁻	G1	13366.0	2039.3	64.4	6.55	0.16	2.08
	G26	239.3	36.7	4.0	6.53	0.16	0.04
	Mad1	355845.6	22303.4	1262.9	15.95	0.40	27.28
	Mad17	203.7	23.1	1.5	8.80	0.22	0.02
Si:VV⁻	G1	31470	2230	460	14.11	0.35	2.97
	G20	297	59	3	5.03	0.13	0.07
	M1	195200	23300	100	8.38	0.21	27.71
	M12	701	97	2	7.23	0.18	0.11

Table 2b. Hyperfine parameters in kHz for some typical silicon hyperfine tensors of two characteristic shallow defects phosphorus and arsenic on substitutional positions - after Hale and Mieher (1969).

centre	tensor	a	b	c	a/b	α^2/β^2	η^2(%)
Si:P	B	4508	70.6	18.6	63.9	1.58	0.160
	C	3298	5.0	0.0	659.6	16.7	0.076
	I	1370	21.4	2.2	64.0	1.59	0.049
	R	758	18.4	7.6	41.2	1.02	0.033
Si:As	B	6000	105.2	22.2	57.0	1.42	0.223
	C	4074	5.8	0.0	702.4	17.5	0.094
	I	1436	23.1	1.3	62.2	1.55	0.051
	R	856	23.1	10.7	37.1	0.92	0.039

Table 2c. Hyperfine parameters in kHz for two prominent silicon interactions for the Si-NL8 - after Michel et al. (1986), and the Si-NL10 spectrum. Only one species is concerned.

centre	tensor	a	b	c	a/b	α^2/β^2	η^2(%)
Si-NL8	1	9890	70	-30	141.3	3.5	0.277
	2a	8530	60	-30	142.2	3.5	0.221
Si-NL10	T2	2455.6	25.1	1.0	97.8	2.4	0.08
	G2	1880.3	13.6	8.3	137.8	3.3	0.05

For a shallow defect all four sp³-hybridised bonds contribute. If the four sp³-bonds are equally occupied the hyperfine tensor is isotropic and in the LCAO analysis we will find only an s-contribution. For a shallow defect we will observe therefore mainly the s-part of the wave function, i.e. $\alpha^2 > 25\%$, or $\alpha^2/\beta^2 \gg 0.33$, and most of the p contribution,

expressed by β^2, cancels. This results in the underestimation of the localisation η^2, $\eta_{total}^2 \ll 100\%$. Therefore, a shallow defect is characterised by:

1) $\alpha^2/\beta^2 \gg 0.33$

2) $\eta_{total}^2 \ll 100\%$

The first criterion can be replaced by a/b >> 13 (where a and b are the reduced hyperfine parameters), which makes a comparison with literature values easier. The above criteria are shown in tables 2a,2b and 3 for a few tensors of two typical deep centres - V⁻ and VV⁻ - and two typical shallow donor centres - P and As. In tables 2c and 3 the corresponding values are shown for the heat-treatment centres Si-NL8 and Si-NL10.

Table 3. Total localisation values η_{total}^2 as obtained by LCAO analysis for several defects in silicon.

	V⁻	VV⁻	P	As	Si-NL8	Si-NL10
η_{total}^2 (%)	114.76	118.90	14.58	18.86	5.64	0.36

For deep centres with missing or dangling bonds the LCAO analysis is successful. The total localisation adds up to values close to 100%. Sometimes localisations over 100% are found, which is due to effects not accounted for in the LCAO treatment, e.g. core polarisation. However for the shallow donors, like P, As, or Sb, the LCAO analysis turns out to be less applicable. In our experiment the biggest hyperfine interaction is found to have an isotropic part a = 2.51 MHz. This corresponds to a localisation of 0.07%. In this way the LCAO analysis shows that both heat-treatment centres fall into the category of shallow defects. A comparison of the Si-NL10 data with the silicon ENDOR results for the Si-NL8 spectrum obtained by Michel et al. (1986) shows that the wave function of the Si-NL10 defect is probably more shallow. For Si-NL8 the biggest hyperfine interaction gives a localisation of 0.277%.

3.2. Symmetry of ground-state wave function

Wave functions of shallow centres can be described with the effective mass theory - Kohn and Luttinger (1955). The wave function of an electron in its ground state can be approximately represented by a linear combination of six wave functions corresponding to the six minima of the conduction band which for silicon lie along the <100> directions. Table 4 gives the combinations of one-minimum wave functions which are allowed in the case of orthorhombic symmetry. The wave functions of types A_2, B_1 and B_2 are zero on one or two of the mirror planes. The probability density of the electron on atoms which are situated in those "forbidden" mirror planes would not only be low because of the widely spread character of the shallow donor wave function, but even zero by symmetry.

The biggest hyperfine interaction found in the present ENDOR experiment has 2mm symmetry indicating that none of the two mirrorplanes is symmetry forbidden. Therefore the ground state wave function of the Si-NL10 defect

has A_1 symmetry. As can be seen from Table 4, three combinations of the six conduction band minima are then possible.

All the hyperfine interactions determined for the Si-NL10 centre in the ^{29}Si ENDOR experiment have an approximate [100] axial character, with the two-fold axis along [100]. The axial direction of the tensors coincides therefore with the two-fold axis of the defect. However this feature of the ground state wave function offers no extra information to discriminate between the three available A_1 states; all three of them allow the [100]-axiality.

Table 4. Symmetry-allowed combinations of wave functions of conduction band minima in 2mm point group symmetry. The two-fold axis is along x.

wave function symmetry	coefficients for conduction band valleys					
	x	-x	y	-y	z	-z
A_1	0	0	1	1	1	1
A_1	0	1	0	0	0	0
A_1	1	0	0	0	0	0
A_2	0	0	1	1	-1	-1
B_1	0	0	1	-1	-1	1
B_2	0	0	1	-1	1	-1

The results of the ^{29}Si-ENDOR study of the Si-NL8 centre by Michel et al. (1986) appear very similar. The biggest localisation is observed for atoms on the two-fold axis, while most of the tensors show near [100] axiality. Therefore the ground state wave function of the Si-NL8 defect also has A_1 symmetry. In case of the Si-NL8 defect additional information is available from infrared (IR) measurements under uniaxial stress. Stavola and Lee (1986) concluded from the stress response of the IR that the ground state of the thermal donor (the Si-NL8 spectrum is related to TD$^+$) is constructed from a single pair of conduction band valleys along the two-fold axis. From the wave functions consistent with the 2mm symmetry of the defect only two of the A_1 symmetry type can be constructed in this way. In case of the Si-NL10 centre, despite its large concentrations, no identification with levels has been made in IR. Therefore no additional information is available about the ground state wave function and three A_1-symmetry combinations are possible.

3.3 Microscopic structure

The microscopic atomic structure of the Si-NL10 centre has been established by detailed analysis of the hyperfine interactions with ^{17}O and ^{27}Al nuclei supplemented by the field-stepped-ENDOR experiments. It was further confirmed by the ^{29}Si ENDOR study.

From the ^{17}O ENDOR measurements up to eight oxygen shells could be distinguished. All of them were of the same mirror plane symmetry type proving that the oxygen structure of the centre is planar with all the oxygen atoms lying in one of the mirror planes. The localisation of the defect electron on oxygen nuclei was found to be very low. The quadrupole interactions were almost identical for all the shells and indicated that

the oxygen atoms were in the usual puckered bonded interstitial position.

^{27}Al ENDOR showed that although aluminium was not really necessary for the creation of the Si-NL10 centres it participated, when present, actively in the oxygen aggregation process leading to the significant enhancement in the generation rate and concentration of the Si-NL10 centres. In such case, for the early stage of the Si-NL10 formation process

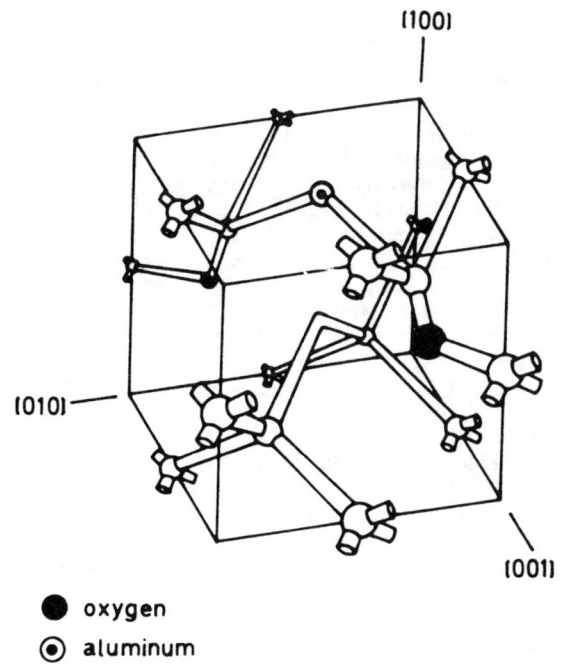

Figure 4. Structural model for the Si-NL10 centre.

the aluminium atom takes a position on the two-fold axis of the defect. The field-stepped-ENDOR technique uses the high resolution of nuclear resonance to unravel unresolved structure of the EPR signal. Applied in the current study it showed that the Si-NL10 EPR spectrum originated from a series of gradually developing similar centres, each characterised by its own spectrum with slightly different g-values. Field-stepped-ENDOR allowed also mutual correlation of the hyperfine interactions originating from different kinds of magnetic nuclei thus establishing the number of oxygen atoms participating at various stages of the transformation of the Si-NL10 centre. The smallest possible Si-NL10 species was found to incorporate 2 oxygen atoms; in the growth process a **single** oxygen atom (at a time) is added along the [0$\bar{1}$1] directed silicon chain. The structure of the centre remains planar while its overall symmetry is lowered upon growth from orthorhombic for the smallest species to monoclinic for the later ones. The structural model of the Si-NL10 centre as emerging from the ENDOR study by Gregorkiewicz et al. (1988) is depicted in Fig.4. As can be seen from the figure it is postulated that a vacancy is created in the core of the defect in order to release the stress accumulated by the

oxygen clustering. On the basis of the measurements no conclusion could be reached concerning the position of the aluminium atom on the two-fold axis nor its possible bonding to the lattice.

ACKNOWLEDGMENT

This work received support from the Netherlands Foundation for Fundamental Research on Matter (FOM).

REFERENCES

Bekman H H P Th, Gregorkiewicz T, van Wezep D A and Ammerlaan C A J 1987 J. Appl. Phys. **62** 4404
Bekman H H P Th, Gregorkiewicz T and Ammerlaan C A J 1988 to be published
Bourret A 1985 Proc. Thirteenth International Conference on Defects in Semiconductors, Coronado, California, 1984, ed. by L C Kimerling and J M Parsey, Jr. (The Metallurgical Society of AIME, Warrendale, Pennsylvania) p 129
Fuller C S, Ditzenberger J A, Hannay N B and Buehler E 1954 Phys. Rev. **96** 833
Fuller C S and Logan R A 1957 J. Appl. Phys. **28** 1427
Gregorkiewicz T, van Wezep D A, Bekman H H P Th and Ammerlaan C A J 1987 Phys. Rev. B **35** 3810
Gregorkiewicz T, Bekman H H P Th and Ammerlaan C A J 1988 Phys. Rev. B **38** 3998
Hale E B and Mieher R L 1969 Phys. Rev. **184** 739
Kaiser W, Frisch H L and Reiss H 1958 Phys. Rev. **112** 1546
Kohn W and Luttinger J M 1955 Phys. Rev. **98** 915
Lee K M, Trombetta J M and Watkins G D 1985 Microscopic Identification of Electronic Defects in Semiconductors, ed. by N M Johnson, S G Bishop and G D Watkins (Materials Research Society, Pittsburgh, Pennsylvania) p 263
Lee Y H and Corbett J W 1973 Phys. Rev. B **8** 2810
Michel J, Niklas J R, Spaeth J-M and Weinert C 1986 Phys. Rev. Lett. **57** 611
Muller S H, Sprenger M, Sieverts E G and Ammerlaan C A J 1978 Solid State Commun. **25** 987
Muller S H, Sieverts E G and Ammerlaan C A J 1979 Defects and Radiation Effects in Semiconductors 1978 (Inst. Phys., Bristol) p 297
Pajot B, Compain H, Lerouille J and Clerjaud B 1983 Physica **117B** & **118B** 110
Sieverts E G, Muller S H and Ammerlaan C A J 1978 Phys. Rev. B **18** 6834
Sieverts E G 1983 Phys. Status Solidi (b) **120** 11
Sprenger M, Muller S H, Sieverts E G and Ammerlaan C A J 1987 Phys. Rev. B **35** 1566
Stavola M and Lee K M 1986 Oxygen, Carbon, Hydrogen and Nitrogen in Crystalline Silicon, ed. by J C Mikkelsen, Jr., S J Pearton, J W Corbett and S J Pennycook (Materials Research Society, Pittsburgh, Pennsylvania) p 95
Wruck D and Gaworzewski P 1979 Phys. Status Solidi (a) **56** 557

ns*t. Phys. Conf. Ser. No 95: Chapter 3
Paper presented at Int. Conf. Shallow Impurities in Semiconductors, Linköping, Sweden, 1988

ENDOR investigations on thermal donors in silicon

J Michel, N Meilwes, J R Niklas and J-M Spaeth

University of Paderborn, Fachbereich 6, Physik, Warburger Str. 100 A
D - 4790 Paderborn, F.R.G.

ABSTRACT: Previous ENDOR investigations on ^{29}Si ligands of singly ionised thermal donors (TD$^+$'s) are briefly reviewed. New ENDOR experiments on TD$^+$'s (NL8) in ^{17}O diffused float zone Si are reported. ^{17}O is clearly involved in the core of the TD's. From the ENDOR analysis follows, that two different oxygen shells are incorporated, each containing two oxygen atoms leading to an O_4 cluster as core of the TD. First experiments on NL10 in P, B and Al doped Si show, that NL10 is not a unique defect and has different structures depending on the dopant.

1. INTRODUCTION

In Czochralski (Cz) grown Si, which contains a high concentration of interstitial oxygen ($\sim 10^{18}$ cm^{-3}), electrically active shallow defects are formed upon annealing at about 450 °C (Kaiser et al 1958). They are called thermal donors (TD's). The structure of these defects is not precisely known yet, not even, whether oxygen is an integral part of them (Newman 1985). In the last years considerable progress was made in the understanding of TD's. From IR measurements it became clear that TD's are double donors and that different species are formed with increasing annealing time, the energy levels of which become increasingly shallow. The binding energies differ only a few meV from species to species. In ESR the singly ionised TD's can be observed. They are formed in samples containing acceptors like B, Al etc. The dominating defect for short annealing times is the so-called 'NL8' defect, whereas after long annealing times (~ 100 hrs) the NL10 defect prevails. Depending on the purity of the sample and annealing conditions, a number of further paramagnetic defects after annealing were observed (Muller et al 1978). In the ESR spectrum no hyperfine (hf) interaction with ^{29}Si nor other magnetic nuclei is resolved. From the g-anisotropy it was concluded that NL8 has C_{2V} symmetry, while NL10 has C_{2V} symmetry in early annealing stages (Bekman et al 1988). With electron nuclear double resonance (ENDOR) of NL8 ^{29}Si ligand hf interactions could be resolved in Cz grown material (Michel et al 1986b) and the possible TD structures could be narrowed down considerably from the analysis of the ENDOR spectra. New experiments on float zone (FZ) silicon into which the magnetic isotope ^{17}O (I = 5/2) could be diffused under extremly clean conditions, show that oxygen is indeed incorporated into the core of the TD's (NL8). This paper first briefly reviews the major results obtained earlier from the ^{29}Si ENDOR investigation of NL8, before it describes then the new results obtained from ^{17}O diffused FZ Si (Michel et al 1988). Finally we discuss the model of the TD core derived from the ^{17}O ENDOR investigations. Comparison of B and Al doped Si, respectively, shows, that NL8 does not contain any acceptors. Finally, ENDOR experiments on NL8 and NL10 defects show that NL8 and NL10 are not related to each other (as was proposed by Bekman et al 1988) and that there are several different defects which show the same NL10 ESR spectrum.

© 1989 IOP Publishing Ltd

2. ^{29}Si ENDOR AND ^{17}O ENDOR OF NL8

Fig. 1 shows the angular dependence of the NL8 ESR spectrum for rotation of the magnetic field in a (110) plane. The ESR spectrum has the following g-values:

g_{zz} = 1.99991 g_{xx} = 1.99323 g_{yy} = 2.00091

with the main axis z being a [001] direction (Muller et al 1978). It clearly shows C_{2V} symmetry. The defect contains two (110) mirror planes which intersect along [001]. There are 6 center orientations in the crystal. When rotating the magnetic field in a (110) plane two pairs of center orientations remain equivalent as indicated in fig. 1 by their numbers. There is no hf splitting resolved in ESR. This was investigated very carefully (Michel et al 1986b). Therefore all hf interactions must be contained in the ESR line width of approximately 0.3 mT. Fig. 2 (a-c) shows a part of the Si ENDOR spectrum above 6 MHz in Cz grown Si containing 4.5×10^{15} cm^{-3} B for several annealing times at 460°C. It is not possible to determine the center concentration from the absolute ENDOR line intensities. However, the observed dependence of relative intensities on annealing time shows that a superposition of several TD$^+$'s is present in agreement with the IR data (Wagner et al 1984). In fig. 2 five TD$^+$'s are distinguishable and labelled A to E. All TD's have within Δg = 0.0001 the same g-tensors as NL8 as measured by ENDOR-induced ESR (for the method see Niklas and Spaeth 1980). All ENDOR lines are due to ^{29}Si. No B signals were detected. Fig. 2d shows the ^{29}Si ENDOR lines observed in FZ Si into which the magnetic oxygen isotope ^{17}O (I = 5/2) was diffused. For this, B doped FZ Si was used with a B concentration of 5×10^{15} cm^{-3}. The sample was diffused at 1400°C with an oxygen pressure of 3 bars for 14 days. An elliptical heat chamber (model E4-2 of Research Inc) was used. The final total oxygen concentration was 1.1×10^{18} cm^{-3}, 55% of it was ^{17}O as determined by IR spectroscopy.

The ^{29}Si ENDOR lines of the FZ diffused sample are obviously exactly the same as those in the Cz sample. In fig. 2d the ENDOR lines of the species labeled A and B (see Michel et al 1986b) appear. No new ^{29}Si ENDOR lines were found in the frequency range from 3 MHz to 30 MHz. Therefore it can be concluded, that the same TD's are produced in both Cz and FZ Silicon.

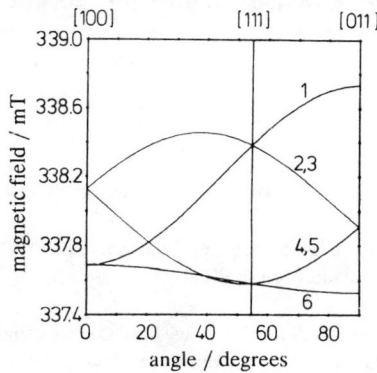

Fig. 1: Angular dependence of the NL8 ESR spectrum for rotation of B_0 in a (110) plane. The numbers refer to the different center orientations.

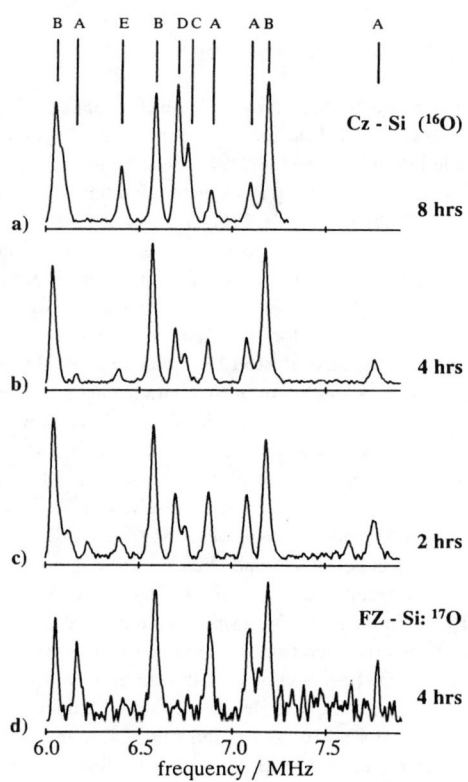

Fig. 2: Part of the ENDOR spectrum of NL8. The ENDOR lines are due to ^{29}Si lhf interactions.

The ESR spectrum in the ^{17}O diffused FZ sample is within experimental error not distinguishable from that of Cz Si, in particular, there is no hf splitting resolved due to ^{17}O nuclei. This indicates, that the ^{17}O interactions are small and contained in the ESR linewidth. Fig. 3 shows ENDOR spectra of Cz Si (lower trace) and of the ^{17}O diffused FZ Si (upper trace) in the frequency range between 1.6 and 3.9 MHz. Both Larmor frequencies ν_n of ^{29}Si and ^{17}O are marked by dashed lines. Clearly in the ^{17}O sample new lines appear around $\nu_n(^{17}O)$. They belong indeed to ^{17}O nuclei (see below) which proves, that oxygen forms an integral part of the TD (NL8) structure (Michel et al 1988).

3. ANALYSIS OF THE ENDOR SPECTRA AND DISCUSSION

The ENDOR frequencies for $S = 1/2$ in pertubation theory of first order are given by
$$\nu^\pm = h^{-1} \left| \tfrac{1}{2} W_{Shf} \mp \nu_n \pm m_q W_q \right| \qquad (1)$$
with $\nu_n = h^{-1} g_I \mu_n B_0$ the Larmor frequency and $m_q = \tfrac{1}{2}(m_I + m_{I'})$. $\qquad (2)$
The signs $\nu^\pm$ refer to the $m_S = \pm 1/2$ quantum numbers. W_q is the quadrupole interaction, W_{Shf} the ligand (super) hf interaction. Every ^{29}Si neighbor nucleus gives rise to two ENDOR transitions, since ^{29}Si ($I = 1/2$) has no quadrupole interaction whereas each ^{17}O nucleus with $I = 5/2$ gives rise to 10 ENDOR lines due to the quadrupole interaction. ENDOR lines are symmetricaly grouped around ν_n if W_{Shf}, $W_q < h\nu_n$, which is the case for both ^{29}Si and ^{17}O (see fig. 3). In Cz material some ^{29}Si ENDOR lines are labelled $a^+ - d^+$ for $m_S = -1/2$ and $a^- - d^-$ for $m_S = 1/2$. It is clearly seen that exactly the same ENDOR lines arise for the Cz and FZ material TD$^+$'s. The ENDOR lines around $\nu_n(^{17}O)$ in FZ material are due to a superposition of ^{17}O and ^{29}Si ENDOR lines. The ^{17}O ENDOR lines are stronger. The fact, that they are indeed due to ^{17}O could be shown by measuring the ENDOR lines for different values of the magnetic field. The ENDOR frequencies differ according to equation (1) by the variation of ν_n with B_0, which is proportional to $g_I(^{17}O)$. The necessary change of B_0 was achieved by varying the microwave frequency.
In order to determine the shf and quadrupole interaction tensors the angular dependence of the ENDOR spectrum must be measured. The angular dependences and their analysis of the ^{29}Si neighbors was already discussed in detail (Michel et al 1986a) and is therefore not repeated here. The major results are: there are three different symmetry types of neighbor shells. Shell 1 contains only one nucleus with the principal shf tensor axis (largest interaction) along [001]. Shell 2 nuclei are on (100) planes, shell 3 and 4 nuclei on the (110) mirror planes. There are two different shells of type 3 and 4 (3a,b and 4a,b).

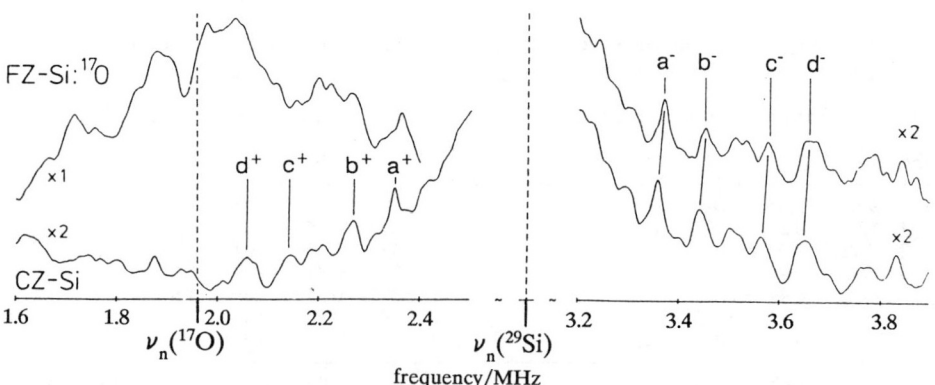

Fig. 3: ENDOR spectra of an ^{17}O diffused Fz-Si sample and a Cz-Si sample containing TD$^+$ defects A and B. The labels a to d refer to ^{29}Si ENDOR lines for $m_S = \pm 1/2$.

In fig. 5a those nuclei are indicated. The angular dependence reflects the C_{2V} symmetry of the defect. Therefore nuclei 3a and b are assigned to nuclei above and below the defect core. It is unknown which are above, which below. The higher interactions are tentatively assigned to the nuclei nearer to the core. Similarly it was done for type 2 and 4 nuclei. One of the most striking results was that the different TD species differ only in the magnitude of the interaction parameters, not in the tensor orientations. This is apparent in the ENDOR angular dependence by parallel angular patterns due to the different species and means that the core of all species is the same. Only the delocalisation of the unpaired electron increases in accordance with the decrease of TD binding energy. This shows up as a decrease of interaction constants. Table 1 contains the interaction constants for species A and B (the data for species C to E are given in Michel et al 1986b).

Whether or not the C_{2V} symmetry is precisely present, cannot be stated from the ENDOR analysis of each of the ^{29}Si ENDOR shells. If the core of the defect is a point defect, then crystal symmetry together with the ESR spectra would let no other choice then to have C_{2V} symmetry. However, if the core contains a cluster, which would not have the C_{2V} symmetry, then in principle the observations in shells 3 a,b and 4 a,b could originate from nuclei which are not related to each other by two mirror planes. If shells 3a and 3b (or 4a and 4b) contained *one* nucleus each instead of *two*, then one would still see the same angular dependence pattern, but only if two different centers are there which are related to each other by a 180° rotation about [001]. In ESR this would be seen as a fortuitous C_{2V} symmetry. If this low symmetry was there, nuclei 3a and 3b (4a and 4b) would be either above or below the core, but only either or the other. If they were due to the nearer nuclei above the core, one would have to claim, that the "lower" nuclei (below the core) would then have to have a much smaller interaction (below 3 MHz) so that they are buried in the many superimposed ENDOR lines, the angular dependence of which could not be analysed. If this was the case, then the ESR spectrum would contain a superposition of lines with g-values hardly different, so that this is not resolved. However, the ENDOR angular dependence of the shell 2 nuclei would not be consistent with such a symmetry breaking, as will be discussed below.

Fig. 4a shows the angular dependence of the ^{17}O ENDOR lines. It was measured following the ESR line of orientations 2,3 (see fig. 1), therefore the ENDOR spectra of two center orientations and species A and B (see fig. 2) are measured simultaneously. This is why so many ENDOR lines appear for each magnetic field orientation. Each center orientation gives rise to 2 ($m_s = \pm 1/2$) x 5 (I=5/2) = 10 ENDOR lines according to equation (1). The two orientations measured cause 20 lines for a *single* ^{17}O nucleus for each orientation of the magnetic field. Since two species A and B are present, for each B_0 40 ENDOR lines are measured for a single ^{17}O in the core.

Table 1: Ligand hyperfine interaction constants and quadrupole constants [in MHz] for the two TD's A and B. ϑ shows the tensor orientation [in degrees].

	shell	a	b	b'	ϑ	q	q'	a	b	b'	ϑ	q	q'
				TD$^+$A						TD$^+$B			
^{29}Si	1	9.89	0.07	-0.03	0			8.72	0.07	-0.02	0		
	2	8.53	0.06	-0.03	0			7.52	0.05	-0.03	0		
	2	8.12	0.06	-0.03	0			6.45	0.04	-0.03	0		
	3	6.74	0.45	-0.10	3.5			5.73	0.39	-0.08	3.3		
	3	6.36	0.49	-0.19	9.2			5.57	0.35	-0.08	11.6		
	4	6.03	0.17	0.08	11.5			5.02	0.14	0.07	10.8		
	4	5.87	0.15	0.08	12.5			4.59	0.12	0.07	11.0		
^{17}O	1	-0.52	0.08	-0.07	58	0.11	0.09	-0.47	0.09	-0.08	58	0.11	0.09
	2	-0.18	0.08	-0.07	62	0.12	0.05	-0.11	0.08	-0.06	62	0.11	0.04

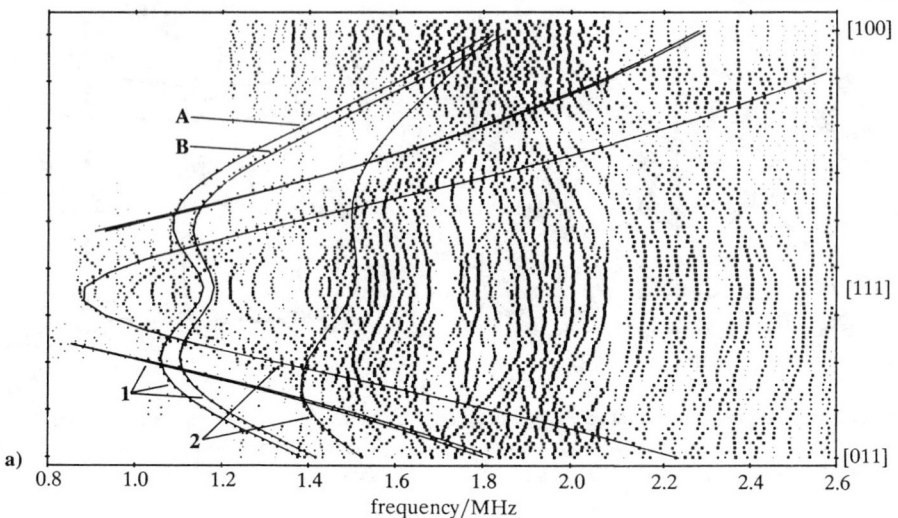

Fig. 4a: Angular dependence of part of the ENDOR spectrum for rotating the magnetic field in a (110) plane. By selecting only one branch of the ESR angular dependence only two TD center orientations contribute to the ENDOR angular dependence. The drawn lines show some of the calculated ENDOR branches of fig 4 b/c. A and B denote the two TD's, the numbers 1 and 2 the two oxygen nuclei.

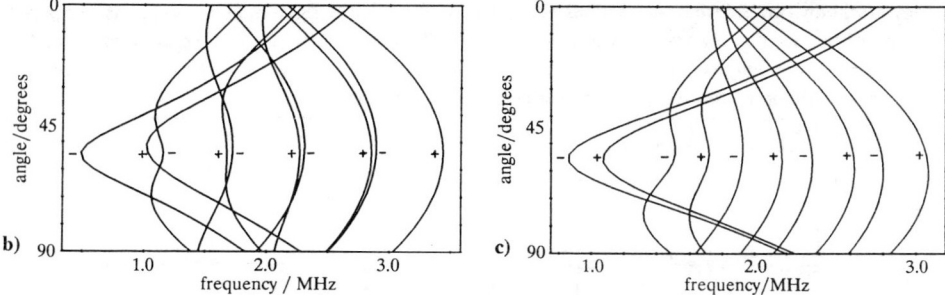

Fig 4b/c: Calculated ENDOR angular dependence for one TD center orientation for each of the different ^{17}O nuclei in approx. <111> symmetry separately. + refers to $m_s = +1/2$, - to $m_s = -1/2$.

In fig. 4a it is seen as was observed for ^{29}Si, that the (two) species differ only in the magnitude of their ^{17}O interaction constants, their angular dependence patterns are "parallel". The analysis of the spectrum yielded, that the core of each species contains two different ^{17}O shells with different interactions (see table 1), both having a tensor orientation along approximately a <111> direction. In fig. 4b and fig. 4c the calculated angular dependences for each shell for one species are shown.

We now address the question, how many oxygen atoms are in the TD core. The minimal number is two since two different ^{17}O interactions were seen. However, two inequivalent O atoms with approximately <111> hf and quadrupol tensor orientation cannot be accomodated in the core without breaking the C_{2V} symmetry. Since both interactions are very similar, their site cannot be too different which would be the case if one oxygen is in the core and one outside. One oxygen inside the core would have to have [001] symmetry and be on a Si lattice site or along a [001] direction in order not to break the C_{2V} symmetry. Such an oxygen was not seen.

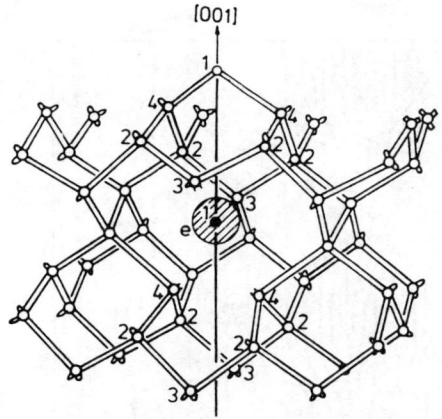
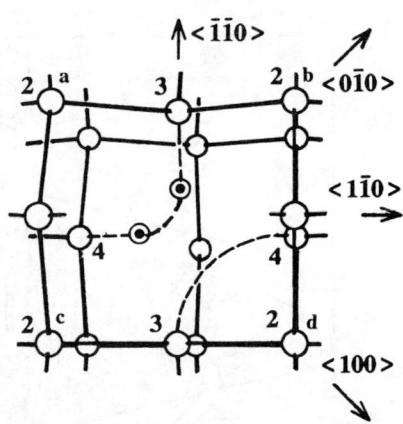

Fig. 5: a) Lattice model with tentative assignment of ^{29}Si ligand nuclei according to the symmetries deduced from the ENDOR analysis.

b) View of the defect core along a <001> axis. The numbers 2 - 4 refer to the neighbor shells given in table 1.

In fig. 5b it is assumed, that indeed there are only two oxygens in the core and it is to be dicussed, how this lower than C_{2V} symmetry affects the ^{29}Si interactions. Clearly, the ^{29}Si shell 2 would have four different interactions of comparable size in this situation, taking only the nearest shell 2 nuclei (that are those above the core) into account. According to the geometrical situation perhaps nuclei 2b and 2c (Fig 5b) have the same or nearly the same interactions. Then three different shell 2 interactions would be expected. There are, however, only two different interactions observed (shell 2 a/b). In fig. 6 the observation is compared to the expectation for the low symmetry configuration. Fig. 6b shows schematically the pattern expected for the low symmetry configuration for shell 2 nuclei. Since no more than two type 2 shells were observed, the core cannot contain only two oxygen as schematically drawn in fig. 5b.

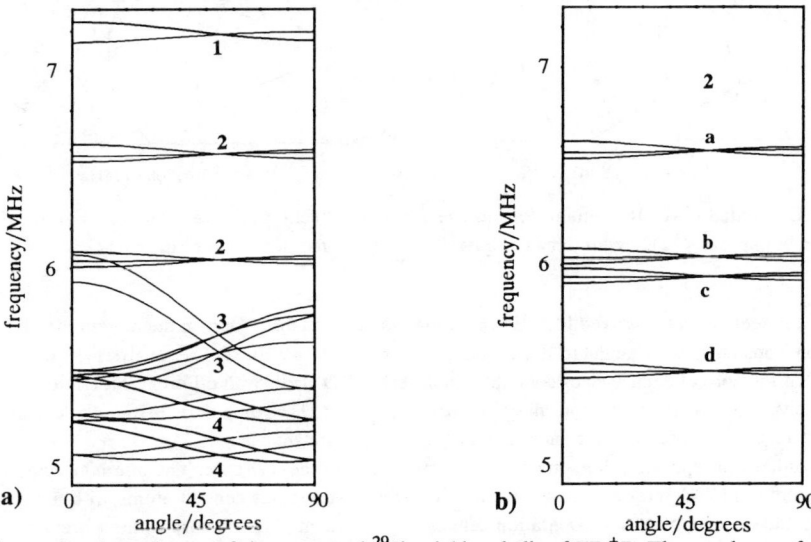

Fig. 6: a) Angular dependences of the measured ^{29}Si neighborshells of TD$^+$B. The numbers refer to the neighbor shells given in table 1.

b) Expected angular dependence (schematic) of four type 2 ^{29}Si neighbors for the assumption of lower than C_{2V} symmetry.

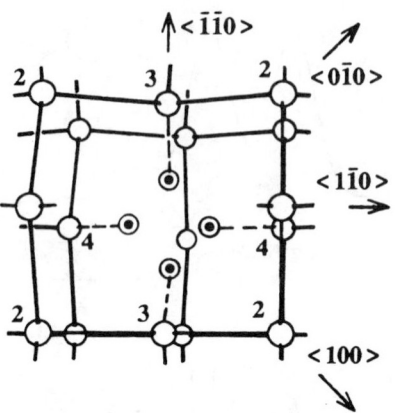
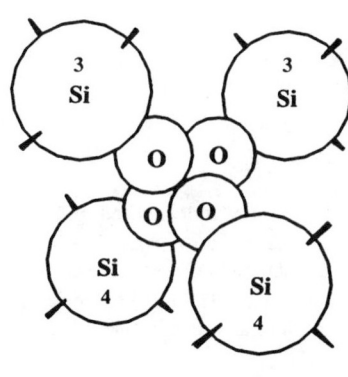

Fig. 7a): View along a <001> axis of the defect core containing four O atoms in C_{2V} symmetry.

b): Core of the thermal donors. The four oxygen atoms occupy a substitutional Si position. The Si are relaxed outward by ~ 15% assuming the atomic radii for O (0.66 Å) and Si (1.17 Å).

Two oxygen nuclei could also be arranged in the core in one of the two (110) mirror planes. This configuration is known as the dioxygen-vacancy (VO_2) in Si and it was shown, that there is no connection between VO_2 and the TD's (Svensson et al 1986).
Fig. 7a shows the four oxygen configuration in C_{2V} symmetry. All shell 2 interactions above the core are the same. Similarly those of shell 2 nuclei below the core in agreement with our observations. Therefore, it has to be concluded, that in the TD's the C_{2V} symmetry is indeed maintained in the atomic scale. This implies, that the TD core contains four oxygen atoms as shown in fig. 7b where the Si and O atoms are drawn according to their atomic radii. The four oxygen atoms can be easily accomodated. An outward distortion of only approx. 15% is necessary for the Si neighbors. The single ^{29}Si of shell 1 is probably the Si which occupied the site now shared by the four oxygens before an replacement reaction took place (Newman et al 1983). The number of four oxygens is consistent with the O_i concentration dependence of the initial formation rate (Kaiser et al 1958).
The fact, that there are two different oxygen interactions can be qualitatively understood, since the single Si along [001] will cause a slight distortion of the defect in this direction.
The <111> hf tensor orientations are suggestive for the assumption, that the O's are bonded along <111> directions to the four Si neighbors. If they form a bonding and an antibonding state between themselfes along the <111> directions, these states would be occupied by four electrons. The two electrons in the antibonding state would be the two electrons which cause the double donor nature of the TD's. The remaining four oxygen-2p electrons will form spin paired p-orbitals perpendicular to those <111> bonding directions. Although this picture seems to give a qualitative explanation of the TD electrical character, the fact, that the oxygen hf interactions observed are to small, remains as big problem.
All cluster calculations of previous TD-models (e.g. Robertson et al 1985) yielded unpaired spin densities, which were 2 - 3 orders of magnitude too high compared to the observations made by ENDOR. On the other hand, the shallow nature of TD's suggests a description of the wave function in the EMT framework. It was suggested that it can be approximately described by a two valley combination along [001] to explain stress experiments in ESR (Lee et al 1985). If one compares the order of magnitude of spin densities at shallow donors in Si (P, As, Sb) and calculates the expected spin density at a hypothetic single ^{17}O as "shallow" donor, one would expect to measure $a(^{17}O)_{EMT} \approx 50$ MHZ (the a-value for P is 117 MHz). This is a discrepancy of two orders of magnitude compared to our observations.

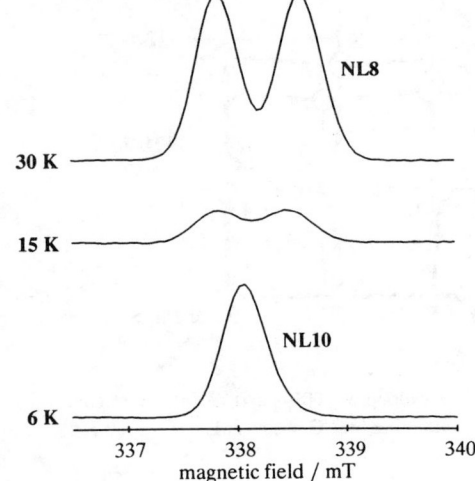

Fig. 8: ENDOR-induced ESR spectra of an Al doped Cz-Si sample at different temperatures. $B_0 \parallel [111]$. The sample was annealed 16 hrs at 460°C.

To our knowledge this is the first time that a shallow 'cluster' has been identified. There is no theory yet available for such a situation. Apparently the EMT-theory will have to be modified to account for these new results.

Our results are in conflict with the models proposed earlier for the TD core. In particular, the OBS model and the Y-lid model would have required the observation of a ^{17}O with <001> tensor symmetry in the core with the largest observed oxygen interaction, which was not seen. They can therefore not be the correct models.

Whether or not the TD growth occurs by addition of further O atoms, cannot be decided yet. Oxygen ENDOR lines of nuclei with very small interaction constants occur very near to $\nu_n(^{17}O)$ and cannot be resolved any more. Further experiments are needed on the smallest two TD species to perhaps say more about this question. It is only clear, that the C_{2v} symmetry is not changed by the TD growth, which implies that any addition of atoms must occur in pairs.

4. INFLUENCE OF ACCEPTORS ON TD-FORMATION

Recently the question came up whether the acceptor doping influences the formation and structure of the TD's. We therefore compared the NL8 spectra in both B and Al doped material, respectively (B: 4.5×10^{15} cm^{-3}; Al: 3.2×10^{15} cm^{-3}). Fig. 8 shows the ENDOR-induced ESR spectra of NL8 and NL10 in Al doped Si after an annealing time of 16 hrs at 460°C. NL8 is measured with optimal signals at 30 K, while NL10 requires 6 K because of different spin lattice relaxation times. At 6 K the ENDOR spectrum of NL8 is no more observable. NL10 is not yet saturated at 30 K and thus not observable in stationary ENDOR at this temperature. Fig. 9 a/b shows the ENDOR spectrum of NL8 in Cz grown B and Al doped Si, respectively. In both cases the spectra are completely identical. No signals of B were observed (B occurs in two magnetic isotopes: ^{10}B with 20% natural abundance and ^{11}B with 80% natural abundance), nor those of Al (^{27}Al, 100% natural abundance). Therefore it is clear, that neither B nor Al can be incorporated as a part of the core or as vital part of the TD. On the other hand, ENDOR measurements of NL10 show clearly that Al is part of the NL10 core. Fig. 9c shows the ENDOR spectrum of NL10 in Al doped Cz-Si. The ENDOR lines shown there are all due to ^{27}Al.

^{29}Si lines are too weak in comparison to Al to be seen, they are probably buried under the Al lines. The Al hf interaction is about 2 - 3 MHz. Fig. 10 shows a comparison of NL10 ENDOR in Al and B doped samples with approximately the same NL10 concentration. The NL10 ENDOR spectrum in the B doped sample is shown in fig. 10b. Only one ENDOR line could be measured, which is due to the second harmonic of the "distant" ^{29}Si ENDOR line at $\nu_n(^{29}Si)$, but no B was seen. The second harmonic of $\nu_n(^{29}Si)$ appears for technical reasons since the distant ENDOR line at $\nu_n(^{29}Si)$ is very strong. It is two orders of magnitude higher than the Al ENDOR lines. Apparently NL10 is not a unique defect. In P doped material the NL10 ESR spectrum shows again different spin lattice relaxation times compared to B or Al doped material.

Generally 'NL10' is formed after NL8 is formed. It is formed faster in Al doped Si (already after a few hours) compared to B doped Si with the consequence, that in Al doped Si the NL8 concentration is smaller through the competition of 'NL10' formation upon oxygen precipitation. Details of the 'NL10' ENDOR investigation will be published elsewhere.

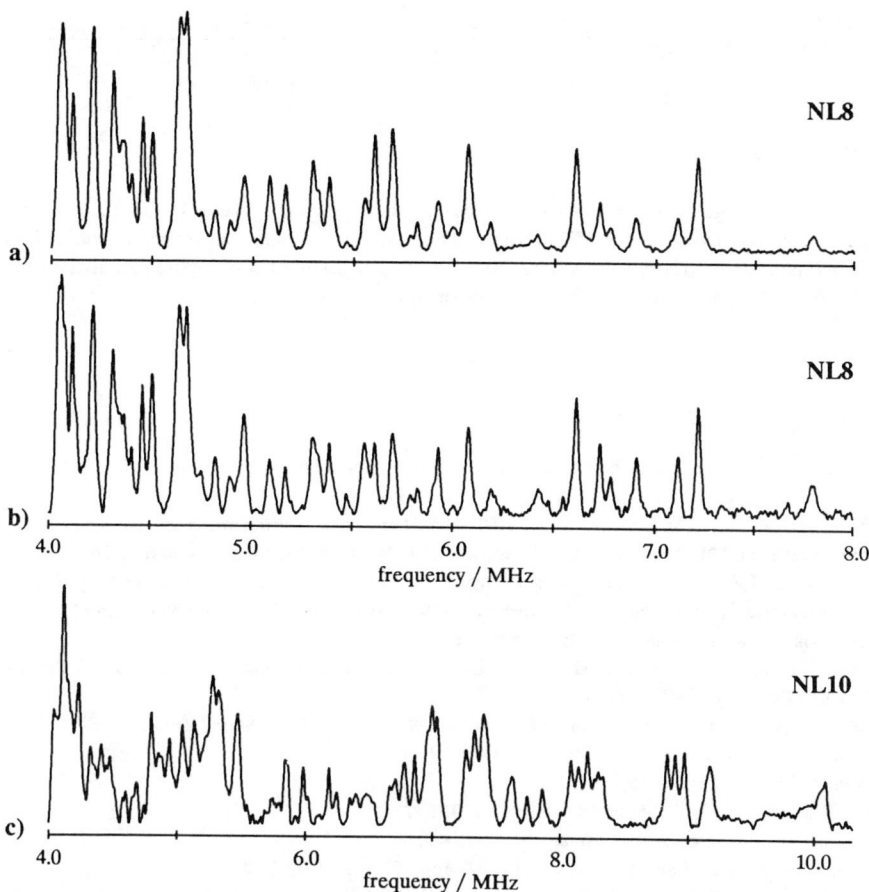

Fig. 9: ENDOR spectra of NL8 and NL10 measured in B and Al doped samples are shown up to the highest ENDOR frequencies. $B_0 \parallel [111]$.
a) shows NL8 in B doped Cz-Si (4.5×10^{15} cm^{-3}) annealed for 4 hrs at 460°C.
b) shows NL8 in Al doped Cz-Si (3.2×10^{15} cm^{-3}) annealed for 16 hrs at 460°C.
c) shows NL10 measured in the same sample as b).

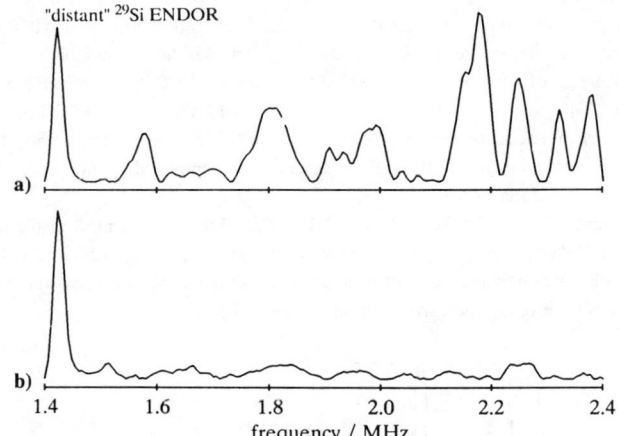

Fig. 10: Comparison of part of the ENDOR spectrum of NL10 in differently doped materials.
a) Al doped Si, annealed 100 hrs at 460°C
b) B doped Si, annealed 100 hrs at 460°C

5. CONCLUSION

Our ENDOR investigation clearly shows that oxygen is incorporated in the TD NL8, that the core and symmetry remains unchanged upon TD growth and that the core contains four oxygen atoms. TD's do not contain acceptor atoms. There is no unique 'NL10' defect. 'NL10' does contain Al in Al doped Si. 'NL10' is not related to NL8 as was proposed recently (Bekman et al 1988).

References:

Bekman H H P Th, Gregorkiewicz T and Ammerlaan C A J 1988 *Phys. Rev. Lett.* **61** 277
Kaiser W, Frisch H L and Reiss H 1958 *Phys. Rev.* **112** 1546
Lee K M, Trombetta J M and Watkins G D 1985 *Microscopic Identification of Electronic Defects in Semiconductors* ed N M Johnson, S G Bishop and G D Watkins (Pittsburgh: MRS) pp 263-8
Michel J, Niklas J R and Spaeth J-M 1986a *Oxygen, Carbon, Hydrogen and Nitrogen in Crystalline Silicon* ed J C Mikkelsen Jr, S J Pearton, J W Corbett and S J Pennycook (Pittsburgh: MRS) pp 111-24
Michel J, Niklas J R and Spaeth J-M 1986b *Phys. Rev. Lett.* **57** 611
Michel J, Niklas J R and Spaeth J-M 1988 *Defects in Electronic Materials* ed M Stavola, S J Pearton and G Davies (Pittsburgh: MRS) pp 185-8
Muller S H, Sprenger M, Sieverts E G and Ammerlaan C A J 1978 *Solid State Comm.* **25** 987
Newman R C, Oates A S and Livingston F M 1983 *J. Phys. C* **16** L667
Newman R C 1985 *J. Phys. C* **18** L967
Niklas J R and Spaeth J-M 1980 *phys. stat. sol. (b)* **101** 221
Robertson J and Ourmazd A 1985 *Appl. Phys. Lett.* **46** 559
Svensson J, Svensson B G and Lindström J L 1986 *Appl. Phys. Lett.* **49** 1435
Wagner P, Holm C, Sirtl E, Öder R and Zulehner W 1984 *Festkörperprobleme: Advances in Solid State Physics 24* ed P Grosse (Braunschweig: Vieweg) pp 191

Inst. Phys. Conf. Ser. No 95: Chapter 3
Paper presented at Int. Conf. Shallow Impurities in Semiconductors, Linköping, Sweden, 1988

The kinetics of thermal donor formation in silicon

R.C. Newman and M. Claybourn
J.J. Thomson Physical Laboratory, University of Reading, Whiteknights, PO Box 220, Reading RG6 2AF, U.K.

ABSTRACT : It is shown that the activation energies for TD-formation and the loss of interstitial oxygen from solution are 1:75 and 2.5eV in the range 350-500°C. The kinetics indicate that there are only four Oi atoms per TD-centre formed at 350°C. Samples heated in an R.F. plasma show correlated enhanced TD-formation and long range oxygen diffusion, probably due to interactions with intrinsic defects.

1. INTRODUCTION

Thermal donors (TD) are formed when Czochralski (Cz) Si containing interstitial oxygen ($[Oi] \simeq 7\text{-}10 \times 10^{17}$ atom cm^{-3}) is heated at a temperature in the range 300-500°C. Most emphasis to date has been given to anneals near 450°C for times up to 200h. As the initial rate of TD-formation is proportional to $[Oi]^4$, it was inferred originally that four Oi atoms diffused together to form a stable complex to be identified with a unique TD-centre, but that larger complexes were not electrically active (Kaiser et al 1958) (Model A). However, to explain the kinetics of the process it was necessary to assume that the normal rate of Oi diffusion was somehow enhanced by a factor of about ten (Helmreich & Sirtl 1977).

This first model appears to be oversimplified as later measurements indicated the sequential formation of a heirarchy of double donors TD(1), TD(2) -- TD(9) (Bean and Newman 1972; Wruck and Gaworzewski 1979; Wagner 1986). It was supposed that TD(N+1) centres were created from TD(N) by the capture of an additional diffusing Oi atom, but that centres with N>9 were not electrically active (Model B). A TD(9) donor would have to incorporate twelve Oi atoms with the assumption that TD(1) involves four such atoms. The kinetics for the Model B show that the Oi diffusion coefficient, D, would have to be enhanced by a factor of 10^4 (Ourmazd et al 1984).

Electron nuclear double resonance (ENDOR) measurements made on silicon containing isotopically enriched ^{17}Oi show that oxygen atoms are indeed present in the TD-centres, and that the evolution of TD(N+1) from TD(N) maintains the C$_{2v}$ symmetry (Michel et al 1986, 1988). An agglomeration model would therefore require the sequential capture of two, rather than one, migrating atoms. It is implied that D is enhanced by more than a factor of 10^4 and that TD(9)-centres would incorporate twenty Oi atoms (Model C). The latter number appears to be impossibly large as experimental results for samples heated at the higher temperature of 500°C for the longer time of 1500h, indicate that the number of oxygen atoms in each precipitate is still only twenty (Messoloras et al 1987).

© 1989 IOP Publishing Ltd

Speculative suggestions to explain the rapid kinetics at 450°C have included (a) the enhancement of D by a process that has not been identified (Ourmazd et al 1984), (b) the rapid diffusion of O_2 dimers produced in the first stage of agglomeration (Gösele and Tan 1982) and (c) the identification of TD-centres with clustered self-interstitials (I-atoms), possibly nucleated around a few oxygen atoms (Model D) (Newman 1986, 1988). The I-atoms were assumed to be released during dimer formation because there is direct evidence for their generation after long heating times from studies using electron microscopy (Reiche and Breitenstein 1988). The suggestions (a)-(c) and the models A-D now have to be critically re-assessed.

Our understanding of the basic oxygen precipitation process in the range 500-1100°C emerged by correlating the measured kinetics of the oxygen lost from solution with the measured kinetics of the growth of SiO_2 particles (Newman 1988). Near 450°C, the oxygen clusters that form are too small to be detected directly. Instead we observe TD-centres with an as-yet unknown structure. Priority should therefore be given to comparing the kinetics of oxygen loss with the rate of TD-formation over as wide a range of temperature as possible. It is necessary to find the correlations which must exist as TD-centres are not found in low-oxygen Float Zone (FZ) silicon. At the lowest temperatures, attention should be refocussed on understanding the anomalously high diffusion jump rates of single O_i atoms observed as a transient process in certain samples (Stavola et al 1983; Newman 1988).

It is crucially important that all the relevant measurements be properly quantified. The loss of O_i atoms from solution resulting from the anneals can be measured from the decrease in the strength of the 9μm infrared (IR) absorption band. The ASTM calibration of this band is unlikely to be in error by more than a few percent (ASTM 1983). The total donor concentration, $\Sigma TD(N)$, is determined from measurements of the resistivity, ρ. Although it has been recognized that the defects are double donors, not all the centres will be doubly-ionized at 300K Account must also be taken of the reduction in the electron mobility produced by the scattering from the centres that are doubly-ionized. This topic has been discussed recently (Claybourn and Newman 1988a) and we show a calculated relationship between $\rho(\Omega\text{-cm})$ and $\Sigma TD(N)$ (cm^{-3}) in Figure 1. We shall present new resistivity and optical data and relate the oxygen loss to the increase in $\Sigma TD(N)$ for a range of samples. The aim is to answer the questions, "How many oxygen atoms are there in each TD-centre and which of the models A-D, if any, comes closest to providing a satisfactory explanation?"

Further insight has been obtained from recent measurements on samples heated in a hydrogen plasma (Brown et al 1988). Correlated enhancements in D and the rate of TD-formation have also been found (Messoloras et al 1988). Thus, a direct link between the two processes has been established for the first time, together with the demonstration of long-range enhanced oxygen diffusion.

2. THE TEMPERATURE DEPENDENCE OF TD-FORMATION

The growth of each of the species TD(2), TD(3)--TD(6) as a function of heating time at 395, 415, 450, 475 and 490°C, has been derived from measurements of the strengths of the electronic IR absorption lines from $TD(N)^0$ centres in samples held at 4.2K (Claybourn and Newman 1987). TD(1)

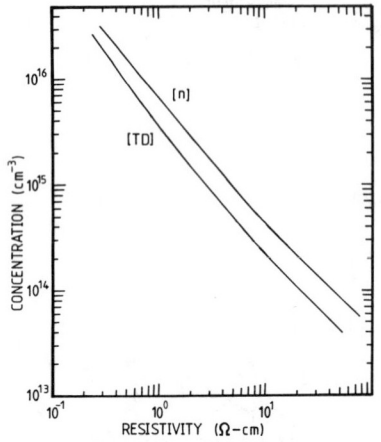

Figure 1. Concentration of TD centres versus sample resistivity. n is the number of conduction band electrons.

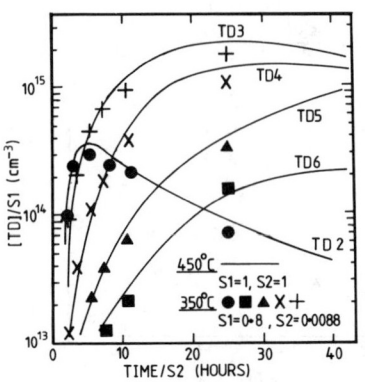

Figure 2. Concentrations of individual TD centres from IR measurements as a function of heating time at 350°C. Solid lines for data at 450°C.

was not observed in our experimental arrangement because of its bistability (Markevich et al 1986). With suitable scaling of the ordinate or strength axis (factor S_1) and the time axis (factor S_2), superimposed plots were obtained for the growth of each individual TD-centre for all five temperatures. The scaling factors which are functions of temperature are the same for each TD defect. An Arrhenius plot of the ratio S_1/S_2 yielded an activation energy $E_{TD}=1.7\pm0.1$eV which is valid over the whole period of heating. This derivation of E_{TD} did not depend on the accuracy of the calibration factors relating the strengths of the IR lines to the concentrations of TD(N)-centres as the same factors were used for all the samples : the factors were those given by Wagner (1986).

Measurements have now been made on another similar sample with $[O_i]=1.2 \times 10^{18}$ cm^{-3} and undetectable carbon after heat treatments at 350°C. The new data can be superposed on the previous results (Figure 2) with values of $S_1=0.72$ and $S_2=8.8 \times 10^{-3}$ for heating times up to 3000h (see Claybourn and Newman (1987) for further details). At this stage TD(3) and TD(4)-centres clearly dominate, while there are significant concentrations of TD(5) and TD(6). On the other hand, the concentrations of TD(1) and TD(2) have already decayed after passing through their maxima. The temperature range of our previous Arrhenius plot has been extended and a revised activation energy $E_{TD}=1.8\pm0.1$eV is obtained (Figure 3). The resistivities of all the samples discussed above, together with another heated at 315°C, were measured and converted to values of $\Sigma TD(N)$ using the calibration given in Figure 1. Linear variations with heating time were found in the early stages. A plot of $\ln[d\Sigma TD(N)/dt]$ versus reciprocal temperature (Figure 4) yielded an activation energy of 1.74 ± 0.10eV in agreement with the optical data.

These results are consistent with, but considerably extend, previous direct Hall effect measurements of the growth of TD(1) and TD(2) made on samples heated for short times (Markevich et al 1986) in the range 325-470°C which led to an activation energy $E_{TD}=1.8\pm0.2$eV. Other

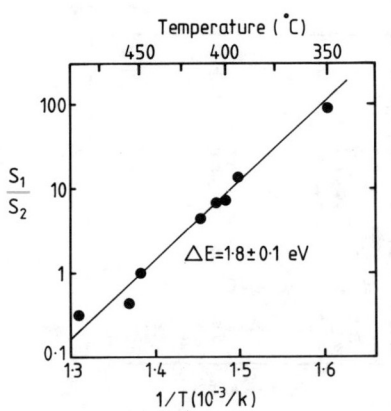
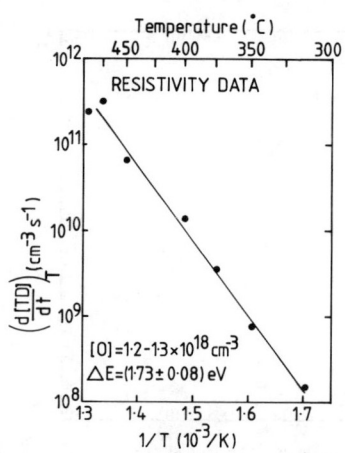

Figure 3. Arrhenius plot for growth of TD-centres (optical data)

Figure 4. Arrhenius plot for growth of TD-centres (electrical data)

electrical and optical measurements (TD(2) and TD(3)) made in the range 375-450°C also yielded values of E_{TD}=1.7±0.1eV (Stein et al 1986). It appears that a value of E_{TD}=1.75±0.1eV is now well-established for the temperature range 315 to 490°C.

3. THE KINETICS OF OXYGEN LOSS FROM SOLUTION

Precise measurements of the strength of the 9μm band can be made, preferably using a dispersive IR spectrometer with the sample held at room temperature, allowing the loss of oxygen from solution to be determined as a function of heating time. Other measurements made on samples held at 4.2K have not revealed the growth of underlying absorption from oxygen agglomerates as is found for high temperature heat treatments (Livingston et al 1984).

A "common sense" argument is that the first stage of oxygen agglomeration must be the formation of dimers (Newman 1985:1988). In that case, plots of the reciprocal oxygen concentration $[O_i(t)]^{-1}$ versus time should be linear with a slope $K=8\pi D r_c$, where r_c is the capture radius for pair formation. If r_c is chosen to be 10Å, values of $D=3.2 \times 10^{-19}$ and 6.7×10^{-20} cm^{2}s^{-1} were obtained previously for samples heated at 450 and 420°C respectively. These values agree extraordinarily well with those of 3.4×10^{-19} and 5.9×10^{-20} cm^{2}s^{-1} calculated from the expression for normal diffusion with $D=0.11\exp(-2.51eV/kT)$ cm^{2}s^{-1} (Newman 1988).

Such measurements have now been carried out for samples with $[O_i] \approx 1.25 \times 10^{18}$ cm^{-3} heated at 400°C and 350°C (Figure 5) : data for samples heated at lower temperatures are not yet available. We find values of D of 2.0×10^{-20} and 7.2×10^{-22} cm^{2}s^{-1} respectively which also agree with calculated values of 1.83×10^{-20} and 5.7×10^{-22} cm^{2}s^{-1}. All of our data are shown in Figure 6 from which it may be deduced that the activation energy for oxygen diffusion remains close to 2.5eV down to 350°C. Notwithstanding the apparent simplicity of this result, there is a complication which we shall consider in section 5.

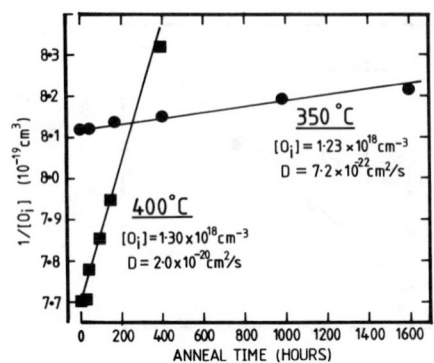

Figure 5. Oxygen loss from solution at 350°C and 400°C as a function of anneal time.

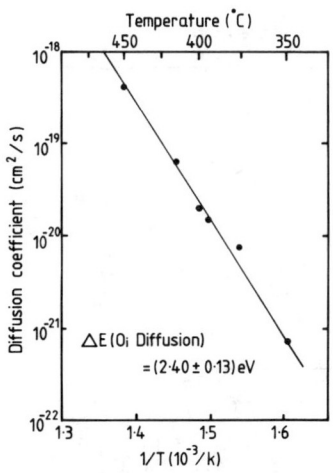

Figure 6. Arrhenius plot of oxygen diffusion coefficient (450°C-350°C).

4. THE AVERAGE NUMBER OF OXYGEN ATOMS IN EACH TD-CENTRE

Previously we indicated that at 450°C about 8-9 Oi atoms were lost per conduction band electron gained due to TD-formation at 450°C. With the assumption of completely ionized double donors and making no allowance for changes in the mobility, we obtain a revised number of some 16-18 Oi atoms lost per TD-centre formed (Newman 1988). Using the calibration given in Figure 1, this number drops to 12-13 Oi atoms, even in the early stages of the process. None of the models A-D would predict such a high number and it has to be concluded that not all of the Oi atoms lost lead to TD-formation. As the temperature is decreased the overall kinetics of TD-formation is unchanged, except for the increased time scale of the process (Claybourn and Newman 1987). However the activation energy for Oi diffusion is 2.5eV, while E_{TD} has a smaller value of 1.75eV. It follows that at the lower temperatures the ratio of the number of oxygen atoms lost per TD-centre formed must decrease. Our experimental results are shown in Figure 7. At the highest temperature of 475°C some 20-25 Oi atoms are lost for each donor formed, but at the lowest temperature of 350°C for which we have data, the ratio drops to 3.7. Extrapolation to even lower temperatures suggests that TD-centres may involve only three, two or one Oi atoms!

The results for 350°C are of considerable significance. For this reason, we show the variation of Δ[Oi] versus Δ[TD] in Figure 8, and the kinetics of oxygen loss in Figure 9. It is clear that if TD-centres are formed <u>exclusively</u> from isolated interstitial oxygen atoms present in as-grown Cz Si there cannot be more than <u>four</u> atoms per centre. This result would imply immediately that models B and C are incorrect : model D must also be incorrect since it is predicted that only one I-atom would be generated for every two oxygen atoms that combine (Newman 1988) and so large clusters of I-atoms cannot form at 350°C. We are left with Model A for which it may be necessary to invoke a modestly enhanced oxygen diffusion rate. Some evidence for this possibility exists (Sections 5 and 6).

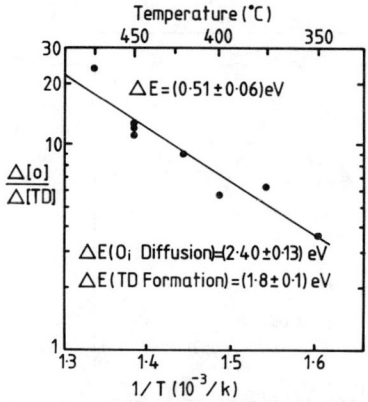

Figure 7. Arrhenius plot of Δ[Oi]/Δ[TD] for 475-350°C.

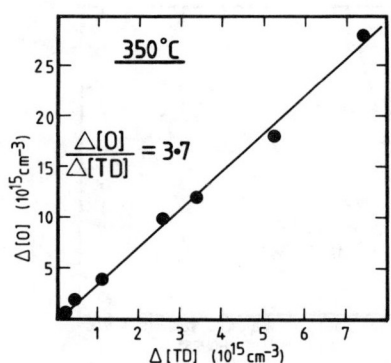

Figure 8. Loss of oxygen versus growth of TD-centres at 350°C.

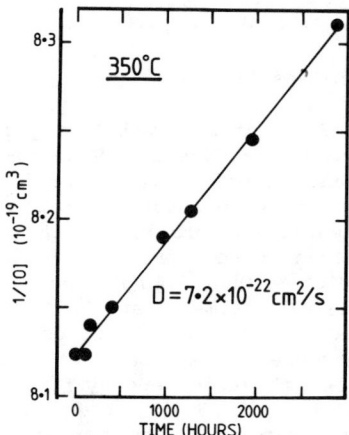

Figure 9. The loss of oxygen as a function of time at 350°C.

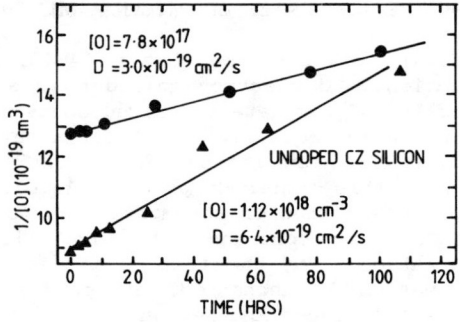

Figure 10. The loss of oxygen as a function of time at 450°C for two samples with different initial oxygen concentrations.

5. THE DEPENDENCE OF OXYGEN DIFFUSION ON CONCENTRATION

It was found by Tan et al (1986) that the rate of dimer formation increased when $[Oi^\circ]$ was increased from 7.5×10^{17} to 10^{18} cm^{-3}. The data were actually derived from a more complex computer model of the aggregation process. We have confirmed this result by determining the rate of oxygen loss in two samples with oxygen contents of 7.8×10^{17} and 1.12×10^{18} cm^{-3}, following anneals at 450°C. According to second order kinetics the two sets of measurements yielded values of $D = 3.0 \times 10^{-19}$ and 6.4×10^{-19} cm^{2}s^{-1} respectively (with rc=10Å) (Figure 10). The data of Tan et al. indicate that D might be proportional to $[Oi]^2$ (Figure 11), implying that there is a mechanism for the enhancement of D by small factors of 2-3. There is no real conflict with our previous statements that D was "normal" as we were then demonstrating that the enhancement factor was not

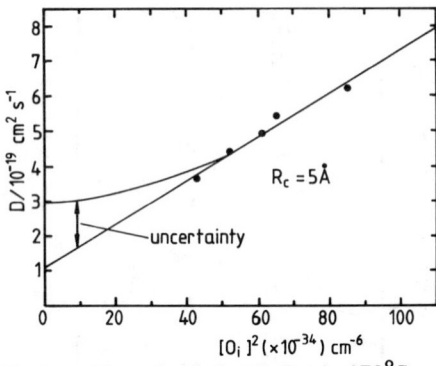
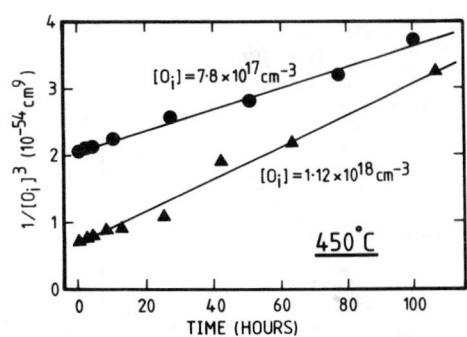

Figure 11. A plot of D at 450°C for dimer formation versus $[O_i]^2$ using the data of Tan et al (1986)

Figure 12. Data of Figure 10 replotted as $[O_i]^{-3}$ versus time (see text).

10^4 (see section 1) (Newman 1988). Thus we could write an effective diffusion coefficient $D_{EFF}=D_N+D_{ENH}$, where D_N is the "normal" value, which might be between 1×10^{-19} and 3×10^{-19} cm^{2}s^{-1} (450°C) and $D_{ENH}=\alpha[O_i]^2$, where α is a constant. The second term would become the more important at higher values of $[O_i]$. The initial rate of loss of oxygen from solution with the formation of dimers would then be "closely" proportional to $[O_i]^4$ and would be consistent with the observations for the initial rate of formation of TD-centres (Kaiser et al 1958) if this is linked to the oxygen diffusion rate. This proposal would not be in conflict with our plots of second order kinetics. Equally "good" straight lines may be obtained by plotting $[O_i]^{-3}$ versus time (Figure 12). Because the total loss of Oi from solution is small compared with $[O_i]$ for the annealing times given to samples, the functional form of the kinetics cannot be determined (see also Tan et al 1986).

Three crucial questions emerge: (a) Is the rate of TD-formation linked directly to the rate of Oi loss, (b) Can we demonstrate long range enhanced Oi migration; and (c) What mechanism would explain an $[O_i]^2$ dependence of D_{ENH}? Answers to questions (a) and (b) have been obtained from some unconventional investigations which are now outlined.

6. TD-FORMATION IN HEATED Si EXPOSED TO A HYDROGEN PLASMA

Recently we have found that heating samples in an inductively coupled 13.56 MHz plasma in hydrogen gas at 1 Torr pressure rather than in a standard furnace at 450°C leads to TD production enhanced by a factor of five (Brown et al 1988). The rate of oxygen loss is also enhanced by a similar factor and a plot of $\Delta[O_i]$ versus $\Delta[\Sigma TD]$ yields a value of twelve, the same as for samples heated in the conventional way (Figure 13) (Messoloras et al 1988). At 400°C and 350°C the enhancement factors are larger (Figure 14).
These observations show that long range diffusion of oxygen must occur, unless an in-diffusing species combines with Oi atoms to form complexes to a depth of 0.5mm from the top surface of the sample. The latter possibility is rejected as a denuded zone of TD-centres is found near a surface (Hahn 1986). It seems unlikely that hydrogen is involved

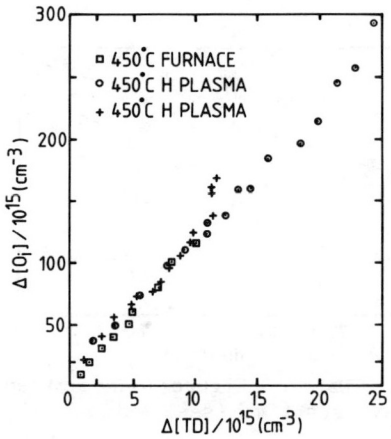 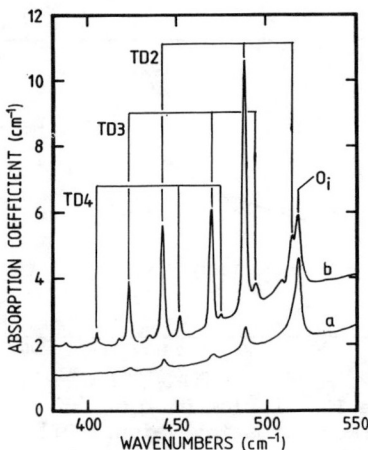

Figure 13. The correlated loss of Oi and TD formation for furnace and plasma treatments

Figure 14. IR electronic spectra after heating samples for 2h at 400°C in (a) a furnace (b) a plasma

directly, as enhancement effects, although somewhat smaller, have been found for samples heated in helium or argon plasmas and, in addition, the maximum donor concentration has the normal value. The plasma enhancement effect is thought to be due to the injection of vacancies and/or I-atoms at the surface where damage will be produced by the ion bombardment (Ponce et al 1987). The possibility that there was metallic contamination has also to be considered. Work is in progress to identify the mobile defect.

It is interesting to note that a similar enhancement for TD generation was reported in the early work of Fuller and Logan (1957) for Cz crystals grown in a hydrogen atmosphere. Since such crystals are known to contain voids and/or cracks, the involvement of point defects is again likely to provide an explanation.

7. CONCLUSIONS

It has been shown that the average number of isolated Oi atoms incorporated into TD centres by diffusion is limited to about four at 350°C. Smaller numbers are predicted at lower temperatures. However, we cannot rule out the possibility that grown-in oxygen clusters act as nuclei for TD-centres, as invoked for example by Mathiot (1988). In that case, we should also expect the as-grown material to contain some clustered I-atoms generated during cooling of the crystal. We have used as-grown silicon in our experiments but if samples are given an oxygen dispersal treatment consisting of a quench from a high temperature, care must be taken to prevent the introduction of metallic contamination and the quenched-in concentrations of intrinsic defects cannot be measured. Under the circumstances it would be unwise to make definitive statements about the validity of TD models A-D at the present time.

Exciting new results are the observation of enhanced long range diffusion of oxygen produced by exposing samples to a hydrogen plasma during heating and the correlated enhancement of TD-formation. Previously we have argued that enhanced oxygen diffusion is due only to vacancy interactions (Newman 1988). However it is known that the Oi-I complex is thermally stable up to

room temperature, and the binding energy E_B has been estimated to be 0.8eV (Brelot 1972a:1972b). The activation energy for diffusion of oxygen via the interstitial process would be $(E_F(I)-E_B+E_M)$, where $E_F(I)$ is the energy for the formation of I-atoms. The equilibrium number of I-atoms would be negligible at 450°C, but if they are produced by ion bombardment at the surface of the sample, very high values of D would occur since E_M is expected to be small (say 0.1eV). It follows that this mechanism might also explain the high rate of diffusion jumps measured from the relaxation of stress-induced dichroism at temperatures near 300°C (for a summary see Tipping et al (1986)).

We have pursued this argument because it has been suggested that I-atoms are also produced during dimer formation. D_{ENH} would then be proportional to the concentration of I-atoms present, which in turn would be proportional to $d[O_2]/dt$, and hence to $[Oi]^2$. The $[Oi]^4$ dependence of TD-formation should not then be attributed to the clustering of four Oi atoms, but only to two atoms. The operative value of E_F would correspond to the activation energy for oxygen diffusion, namely 2.5eV, leading to a value of $E_{ENH} \simeq 1.8$eV. Unfortunately, this value is not in accord with the results shown in Figure 6, and it is recognized that our argument is not really self-consistent. However, the ideas may be helpful in understanding the results of Tan et al (1986)(Figure 11). We do not consider further speculation to be justified until more information has been obtained about the nature of the defects which produce enhanced diffusion in the plasma or in pre-heated samples used for stress-dichroism experiments. Neither do we have any evidence for rapid diffusion of oxygen dimers.

8. APPENDIX

We have made no comments here on TD-formation in p-type Si, as used for ENDOR measurements. An analysis of new results for silicon doped with boron (7×10^{16}cm^{-3}) indicates that TD-formation is the same as in undoped silicon but that there is simultaneous loss of substitutional boron during heating at 450°C. The behaviour of silicon doped with aluminium to the same concentration is more complicated with rapid complexing of aluminium with oxygen, and the formation of donor centres which must involve aluminium atoms. Details of this work are presented elsewhere (Claybourn and Newman 1988b).

ACKNOWLEDGEMENTS

The authors wish to thank Professor E.C. Lightowlers, Dr. G. Davies and Dr. R. Murray for helpful comments, and SERC/Alvey for financial support of this work.

REFERENCES

ASTM 1983 Annual Book of ASTM Standards, (ASTM:Philadelphia) procedure F121-83
Bean, A.R. and Newman, R.C., 1972, J. Phys. Chem. Solids **33** 255
Brelot, A. 1972a Inst.Phys.Conf.Ser.**16** 191
Brelot, A. 1972b Thèse de Doctorat d'État, Université de Paris
Brown, A.R., Claybourn, M., Murray, R., Nandhra, P.S., Newman, R.C. and Tucker J.H. 1988 Semicond. Sci. and Tech. **3** 591

Claybourn, M. and Newman, R.C. 1987 Appl. Phys. Lett. **51** 2197
Claybourn, M. and Newman, R.C. 1988a, Appl. Phys. Lett. **52** 2139
Claybourn, M. and Newman, R.C. 1988b, Proc. 15th Int.Conf. Defects in Semiconductors Budapest, to be published
Fuller, C.S. and Logan, R.A. 1957, J. Appl. Phys. **28** 1427
Gösele, U. and Tan, T.Y. 1982 Appl. Phys. A **28** 79
Hahn, S.K. 1986 Proc. Mater. Res. Soc. Symp. **59** 181
Helmreich, D. and Sirtl, E. 1977, Semiconductor Silicon Mat. Res. and Devices, ed by H.R. Huff and E. Sirtl (Electrochem Soc.:Princeton N.J.) p 626
Kaiser, W., Frisch, H.L. and Reiss, H. 1958 Phys.Rev. **112** 1546
Livingston, F.M., Messoloras, S., Newman, R.C., Pike, B.C., Stewart, R.J., Binns, M.J. and Wilkes, J.G. 1984 J. Phys. C:Solid St. Phys. **17** 6253
Markevich, V.P., Makarenko, L.F. and Murin, L.I. 1986, phys. stat. sol. (a) **97** K173
Mathiot, D. 1988 Proc. Mater. Res. Symp. **104** 189
Messoloras, S., Newman, R.C., Stewart, R.J. and Tucker, J.H. 1987 Semicond. Sci. and Tech. **2** 1414
Messoloras, S., Newman, R.C., Stewart, R.J., Brown, A.R., Claybourn, M., Murray, R. and Wilkes, J.G. 1988, UK IT 88 Conference Publication, IEE/BCS (Swansea, July 1988) 529-32
Michel, J., Niklas, J.R. and Spaeth, J.-M. 1986 Proc. Mater. Soc. Symp. **59** 111: 1988 Proc. Mater. Soc. Symp. **104** 185
Newman, R.C. 1985 J. Phys. C:Solid St. Phys. **18** L967
Newman, R.C. 1986 Mater. Res. Symp. Proc. **59** 205
Newman, R.C. 1988 Mater. Res. Symp. Proc. **104** 25
Ourmazd, A., Bourret, A. and Schröter, W. 1984 J. Appl. Phys. **56** 1670
Ponce, F.A., Johnson, N.M., Tramontana, J.C. and Walker, J. 1987 Inst. Phys. Conf. Ser. **87** 49
Reiche, M. and Breitenstein, O. 1988 Semicond. Sci. & Tech. **3** 529
Stavola, M., Patel, J.M., Kimerling, L.C. and Freeland, P.E. 1983 Appl. Phys. Lett. **42** 7373
Stein, H., Hahn, S.K. and Shatus, S.C. 1986 J. Appl. Phys. **59** 3495
Tan, T.Y., Kleinhenz, R. and Schneider, C.P. 1986 Proc. Mater. Res. Symp. **59** 195
Tipping, A.K., Newman, R.C., Newton, D.C. and Tucker, J.H. 1986 Defects in Semiconductors Ed. H.J. von Bardeleben, Mat. Sci. Forum **10-12** 887
Wagner, P. 1986 Proc. Mater. Res. Soc. Symp. **59** 125
Wruck, D. and Gaworzewski, P. 1979 phys. stat. sol. (a) **56** 557

Molecular cluster and cyclic cluster calculations on models for the core of the 450 °C oxygen thermal donor in silicon

L C Snyder, P Deak*, R Wu and J W Corbett**

Chemistry and Physics** Departments, State University of New York at Albany, 1400 Washington Avenue, Albany, New York 12222
*Physical Institute of the Technical University of Budapest, Budafoki ut 8, H-1111 Budapest, Hungary

ABSTRACT: We have employed MINDO/3 molecular cluster calculations with a tetrahedral $Si_{35}H_{36}$ basic cluster to explore several models for the core of the 450°C oxygen thermal donor in silicon. These include the Spaeth model in which four oxygen atoms are bonded to silicon atoms of a vacancy, with and without an additional interstitial silicon atom at a tetrahedral position on the [001] axis above the vacancy. We have also explored the early SiO_4 model in which four interstitial oxygen atoms are bound to the central silicon. Our computations indicate that the Spaeth model with an interstitial silicon is 21.6 ev higher in energy than the early SiO_4 model.

1. INTRODUCTION

We have been employing semi-empirical quantum chemical methods to explore various models for the core of the 450°C oxygen thermal donor in silicon (Kaiser et al 1958). This initial paper by Kaiser proposed the first model for the thermal donor core, an SiO_4 cluster in which four interstitial oxygen atoms are bonded to a common silicon atom of the silicon crystal lattice.

Recently Michel, Niklas and Spaeth (1988) proposed a model for the thermal donor core based on ^{17}O and ^{29}Si electron nuclear double resonance (ENDOR) experiments. They proposed a model for the core in which a silicon atom is shifted along a [001] axis from its initial lattice site to the nearest tetrahedral interstitial site, and four oxygen atoms bond to the silicon atom dangling bonds of the resulting vacancy. We refer to this as the Spaeth model (VO_4Si). The Spaeth model satisfied major features of the ENDOR spectra. The first is that two sets of equivalent pairs of oxygen nuclei with substantial hyperfine couplings are found on (011) planes. Second, no oxygen is observed on the [001] rotation axis through the center of the defect.

The objective of our theoretical calculations has been to estimate the thermochemical stability of the Spaeth model relative to the Kaiser (1958) SiO_4 model. We also have undertaken to estimate whether either defect would show a donor state in the gap of silicon.

© 1989 IOP Publishing Ltd

2. THEORETICAL METHODS

To estimate the geometry of the thermal donor core models considered in this work, we have employed the Molecular Cluster Method (MCM) using the tetrahedral $Si_{35}H_{36}$ cluster depicted in Figure 1a and denoted by (Si). We have made molecular orbital computations employing the Modified Intermediate Neglect of Differential Overlap method (MINDO/3) developed by Dewar and coworkers (Bingham Dewar and Lo 1975). We have employed the MINDO/3 parameters coupling silicon and oxygen atoms developed by Edwards and Fowler (1985). The use of the MCM in calculations of properties of defects in silicon has been discussed by Deak et al (1987).

For the geometry of the $Si_{35}H_{36}$ cluster we have taken all bond angles to be tetrahedral. The Si-Si bond length is taken to be 2.372 Angstroms and the Si-H bond length is 1.474 Angstroms. These lengths minimize the computed energy of Si_5H_{12}, tetrasilylsilane.

We have incorporated the defect geometries determined from the MCM-MINDO/3 calculation into a cyclic-cluster based on 32 silicon atoms. We have made cyclic cluster model (CCM) calculations on these systems using the semiempirical electronic structure method CNDO/S with d orbitals, which gives an improved description of the conduction band of silicon. We have previously described our cyclic cluster programs (Deak and Snyder 1987), and reported the evaluation of several semiempirical quantum-chemical methods in solid-state applications. We have previously reported our application of CCM in MINDO/3 calculations to describe the state and motion of hydrogen in silicon (Deak, Snyder and Corbett 1988).

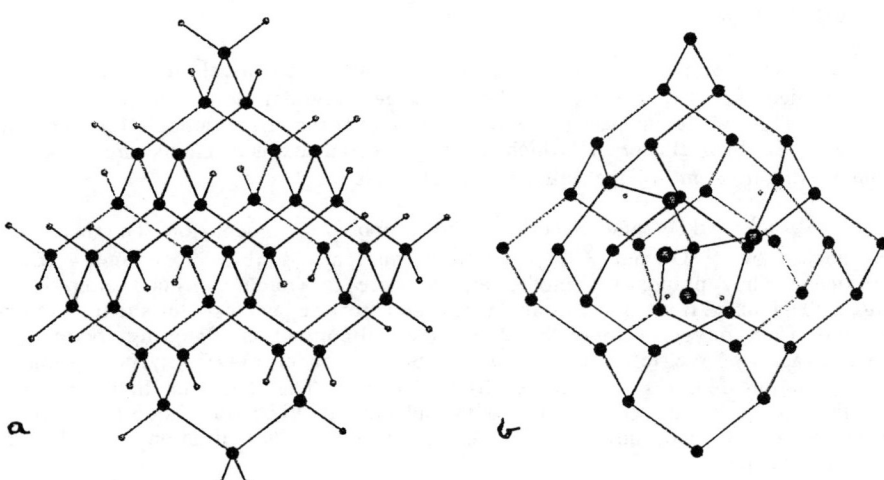

Figure 1a. Geometry of the basic $Si_{35}H_{36}$ cluster(Si).

Figure 1b. The computed structure of the cluster (SiO_4).

3. COMPUTED STRUCTURES

Defects were created in the basic cluster (Si). Computed defect geometries minimize the computed total energy subject to the condition that all hydrogen atoms and the silicon atoms to which they are bonded are fixed at their positions in the basic cluster. In Figure 1a, the large dark atoms are silicon, the small light atoms are omitted for clarity in drawings of defect structures where we have also denoted the lattice positions of displaced silicon atoms by small dots.

Our computed structure for the cluster (SiO_4) corresponding to the Kaiser (1958) SiO_4 model for the thermal donor core is depicted in Figure 1b. We have bonded four interstitial oxygen atoms to the central silicon atom and have minimized the total energy with respect to all variations of the position of the central silicon atom, the four interstitial oxygen atoms and the silicon atoms to which they are bonded.

Our computed structure for the cluster (VO_4Si) corresponding to the Spaeth model was obtained by computing a structure for the system with the interstitial silicon atom absent, the cluster (VO_4). This VO_4 structure has the oxygen atoms attached to dangling bonds of silicon atoms of the vacancy. The positions of the four oxygen and four silicon atoms were varied subject to tetrahedral symmetry. This structure with a silicon atom placed at the neighboring tetrahedral interstitial site on the [001] axis through the center of the vacancy is depicted in Figure 2a. When an attempt was made to minimize the energy of this VO_4Si system, the interstitial silicon atom moved into the vacancy and pushed two oxygens into neighboring interstitial positions to form a metastable structure (SiO_4m) having assumed C_{2v} symmetry, and depicted in Figure 2b.

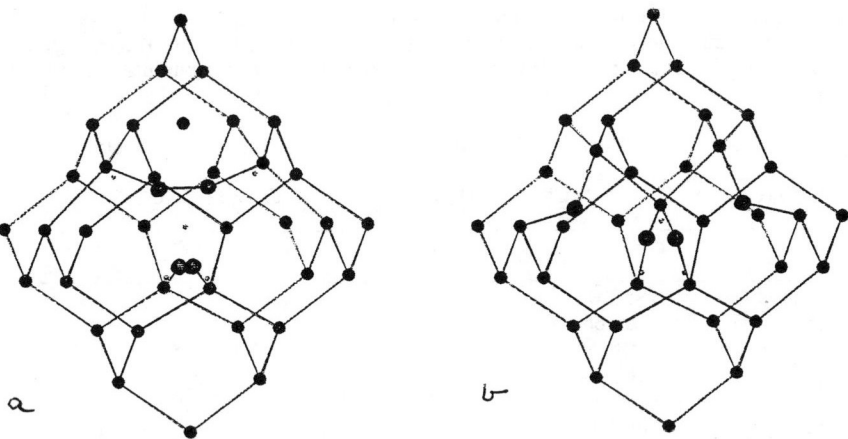

Figure 2. (a) Computed structure for the Spaeth model (VO_4Si) and (b) The metastable structure (SiO_4m).

We have also computed a structure for the cluster (V) having a vacancy at the center of our basic cluster. The positions of the four silicon atoms surrounding the vacancy have been relaxed. A structure for a cluster with a single interstitial oxygen was computed and is denoted by (SiOint).

4. THERMOCHEMISTRY

We report in Table 1 the total energy and Koopman's theorem ionization potentials computed for clusters in this study. We now comment on the thermochemistry of three cluster reactions. The reaction of four interstitial oxygen atoms to form an SiO_4 cluster is endothermic by 0.32 ev.

$$4(SiOint) \rightarrow (SiO_4) + 3(Si) \quad \Delta E = +0.32 \text{ ev}$$

The formation of the Spaeth model with an interstitial silicon from an SiO_4 is endothermic by 21.6 ev.

$$(SiO_4) \rightarrow (VO_4Si) \quad \Delta E = +21.6 \text{ ev}$$

The Spaeth model without an interstitial silicon atom would react to form a vacancy and four interstitial oxygen atoms releasing 9.4 ev of energy.

$$4(Si) + (VO_4) \rightarrow (V) + 4(SiOint) \quad \Delta E = -9.4 \text{ ev}$$

We can conclude that both with and without an interstitial silicon atom, the Spaeth model for the thermal donor core is very unstable. Silicon-oxygen bonds are much stronger than the oxygen-oxygen bond. Too many Si-O bonds have been eliminated in the Spaeth models.

We note that the metastable structure (SiO_4m) is only 2.5 ev less stable than the (SiO_4) cluster and may also be a candidate for the core of the oxygen thermal donor in silicon, since it conforms to the symmetry requirements of the ENDOR spectra.

Table 1 Computed Total Energies (E) and Ionization Potential (IP) in Electron Volts

Cluster	MCM-MINDO/3		CCM32-CNDO/S
	E	IP	IP
(Si)	-3852.07	7.9	6.3
(SiOint)	-4165.66	7.3	
(SiO_4)	-5106.15	7.3	6.8
(SiO_4m)	-5103.64	5.7	5.2
(VO_4Si)	-5084.52	5.8	4.8
(VO_4)	-4997.11	7.0	6.6
(V)	-3752.08		

5. ON STATES IN THE GAP OF SILICON

We have computed Koopman's theorem ionization potentials for our molecular clusters. The ionization potential for the silicon crystal in a full band structure using MINDO/3 is 5.3 ev. (Ricart and Illas 1985). We recall that in our molecular clusters, the hydrogen atoms terminate surface dangling bonds. These bonds are polar with a net negative charge on the hydrogen atoms. This can increase computed ionization potentials by about a volt (Deak et al 1987). Thus, for these molecular clusters only the Spaeth model (VO$_4$Si) and (SiO$_4$m) have MINDO/3 ionization potentials low enough to have states in the gap of silicon and be candidates for the thermal donor core. This is also true for the CCM-CNDO/S computed ionization potentials where only (VO$_4$Si) and (SiO$_4$m) have computed ionization potentials less than (Si). For this reason we have examined the CCM32-CNDO/S wave function of the highest occupied orbital of the (VO$_4$Si) and (SiO$_4$m) clusters. The square of atomic orbital coefficients in this orbital gives a first approximation to the electron spin distribution observed in ENDOR.

For the Spaeth model cluster (VO$_4$Si), the highest filled molecular orbital has the symmetry of the pair of conduction band valleys along the direction [001] of the cluster symmetry axis. It is clearly derived from near the conduction band X point. The largest 3s coefficient (0.25) is at the silicon on the apex of the cluster. The interstitial silicon 3s coefficient is much smaller (0.07). The next largest 3s coefficient (0.21) is located on the four equivalent (220) silicon atoms in the plane of the vacancy. Despite the fairly good agreement with the ENDOR spectra (Michel et al 1986), we believe that its thermodynamic instability rules it out as a possible thermal donor core.

For the metastable cluster (SiO$_4$m), the highest filled molecular orbital is also derived from the pair of conduction band valleys in the [001] direction near the X point. Again the wave function is widely distributed over atoms of the cluster. We think more extensive studies should be made of the structure and electronic state of (SiO$_4$m). It remains a possible core of the oxygen thermal donor.

6. CONCLUSIONS

We find the Spaeth model (VO$_4$Si) to be too unstable with respect to the silicon interstitial returning to the vacancy, for this system to be considered a likely candidate for the core of the 450°C oxygen thermal donor in silicon, although this structure probably has a state in the gap of silicon. The structure (VO$_4$) is unstable with respect to the dissociation of the oxygen atoms as interstitial oxygen with the formation of a vacancy. It has no tendency to have a filled state in the gap and function as a donor.

ACKNOWLEDGEMENT

This work was supported in part by the U.S. Army Research Office.

REFERENCES

Bingham R C, Dewar M J S and Lo H C 1975 J. Am. Chem. Soc. 97 1285

Deak P and Snyder L C 1987 Phys. Rev. B36 9619

Deak P, Snyder L C and Corbett J W 1988 Phys. Rev. B37 6887

Deak P, Snyder L C, Singh R K and Corbett J W 1987 Phys. Rev. B36 9612

Edwards A H and Fowler W B 1985 J. Phys. Chem. Solids 46 841

Kaiser W, Frisch H L and Reiss H 1958 Phys. Rev. 112 1546

Michel J, Niklas J R and Spaeth J-M 1986 in Oxygen, Carbon, Hydrogen and Nitrogen in Crystalline Silicon
edited by Mikkelson J C, Pearton S J, Corbett J W and Pennycook S J, Vol. 59 of MRS Symposium Proceedings (Materials Research Society, Pittsburgh) p. 111

Michel J, Niklas J R and Spaeth J-M 1988
in Defects in Electronic Materials,
edited by Stavola M, Pearton S J and Davies G, Vol. 104 of MRS Symposium Proceedings (Materials Research Society, Pittsburgh) p. 185.

Ricart J M and Illas F 1985 J. Mol. Struct. 120 309

Oxygen related photoluminescence lines in 450 °C annealed silicon

A. Henry, H. Weman and B. Monemar.
Department of Physics and Measurement Technology
Linköping University
S-581 83 Linköping, SWEDEN.

ABSTRACT: Photoluminescence (PL) and resistivity measurements are made on oxygen-rich boron-doped silicon annealed at 450 °C, to investigate the origin of the previously reported PL "O-line" spectrum of bound exciton lines with no-phonon energy around 1.143 eV at 2 K. The comparison of the data from the both techniques rules out the previous association with the thermal donors (TD:s). The effect of various parameters in the PL experiments is reported, and some line-structure within the envelope of this "O-line" spectrum is observed. It is tentatively suggested that the lines are due to bound exciton recombination at boron-TD (B-O) complexes.

1. INTRODUCTION.

It is now well established that in 450 °C annealed Czochralski Silicon (Cz-Si), the so-called thermal donors (TD:s) are created (Fuller et al. 1954). In spite of numerous experimental and theoretical studies, the detailed microscopic structure of the TD:s is not yet known. It is generally accepted that the TD generation involves the formation of small oxygen clusters, since the TD:s are clearly related to the early stage of oxygen precipitation (Bourret 1985).
When Cz-Si is annealed at about 450 °C, photoluminescence (PL) studies have shown that besides the TD:s, other oxygen related defects are generated, like the so called H and P lines, due to neutral complex defects (Minaev et al. 1981) and the previously reported PL "O-line" spectrum of bound exciton lines with no-phonon energy around 1.143 eV at 2 K (Tajima et al. 1979, Nakayama et al. 1981).

This paper is a contribution to a better understanding of the "O-line" spectrum. It will be shown that the "O-line" spectrum is not directly related to the TD:s, and that some structure can be observed within the envelope of this "O-line" spectrum if the experimental conditions are carefully chosen. There are also some other PL lines in Si at ~ 1.117 eV labelled as the "O-lines" (Weber et al., 1986), therefore the PL lines described in this work will be labelled as the "BO" lines, since the BO-line spectrum appears on initially oxygen-rich boron-doped silicon annealed at 450 °C.

2. SAMPLES AND EXPERIMENTAL PROCEDURE.

The samples used in this work were cut from a p-type boron doped ([B] ~ 6×10^{14} cm^{-3}) Cz-Si slice. The initial concentration of oxygen (1.4×10^{18} cm^{-3}) and carbon (0.4×10^{16} cm^{-3}) were

deduced from room-temperature infrared absorption using the 9 and 16 μm lines respectively, and the ASTM calibration factor. The annealings at 450 °C were performed in an ambient air quartz tube for different times. After each treatment, the samples were etched in a 1:3 mixture of $HF:HNO_3$ and thereafter cleaned with ethanol and deionized water.

The sample resistivity was determined with a four-point probe and the TD concentration created after annealing was obtained using Irvin's curve (Irvin 1962).

The PL measurements were done at temperatures rangeing from 1.8 to 10 K, with the 514.5 nm line of an Ar^+ ion laser for optical excitation. The luminescence was dispersed with a SPEX 1404 0.85 m double grating monochromator fitted with two 600 grooves/mm gratings blazed at 1.6 μm. A liquid nitrogen cooled North Coast EO 817 Ge detector was used with a mechanical chopper and conventional lock-in technique to recover the signal.

3. EXPERIMENTAL RESULTS.

A typical near band gap PL spectrum at 2 K for boron-doped Cz-Si annealed at 450 °C during 120 hours is presented in Figure 1, which shows the no-phonon (NP), the tranverse acoustic (TA) phonon and transverse optical (TO) phonon replicas of the boron bound exciton, $B_{NP}(BE)$, $B_{TA}(BE)$ and $B_{TO}(BE)$ respectively and also those of the BO lines, BO_{NP}, BO_{TA} and BO_{TO} respectively. Additional PL-lines were also observed at lower photon energies than those in Figure 1; these lines (P: 767 meV, H: 926 meV) are due to neutral complex defects involving oxygen and carbon (Minaev et al. 1981). As shown in Figure 1, the BO-line spectrum is dominated by the TO phonon replica at ~ 1.085 eV. The NP recombination and the TA phonon replica are about 20 times lower in intensity than the TO phonon replica. For the various samples studied in this work, we have found that the TA to NP BO-line intensity ratio is generally about 1.3. The BO-line spectrum appears easily on initially boron doped silicon after long term annealing (72 - 240 hours) at 450 °C, but it is very rarely observed in phosphorus doped samples, in this work, although a similar TD concentration is created by the treatment. This means that there is no direct correlation between the BO-lines and the TD centers, and this behaviour is confirmed by the resistivity data.

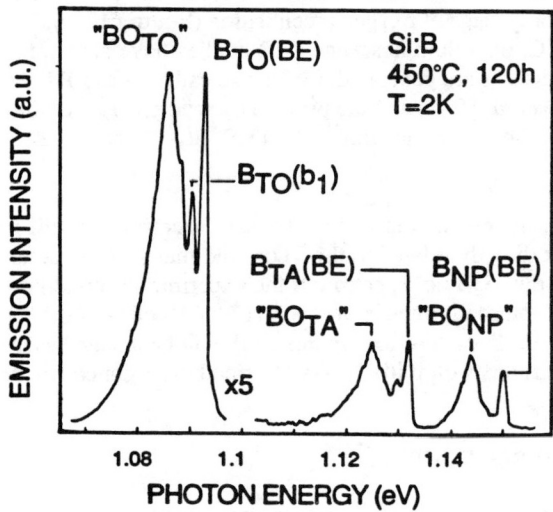

Fig 1 : Photoluminescence spectrum at 2 K of a B-doped Cz-Si sample annealed at 450 °C for 120 hours producing thermal donors with a concentration ~ 8.7×10^{15} cm^{-3}. The broad BO-lines spectrum is clearly seen as described in the text, together with the well known bound exciton line associated with boron (B_{TO}) and bound multiexciton complexes associated with boron ($B_{TO}(b_1)$). The laser beam was defocussed to a spot size of approximately 1.5 mm in diameter and the laser power was ~ 300 mW on the sample.

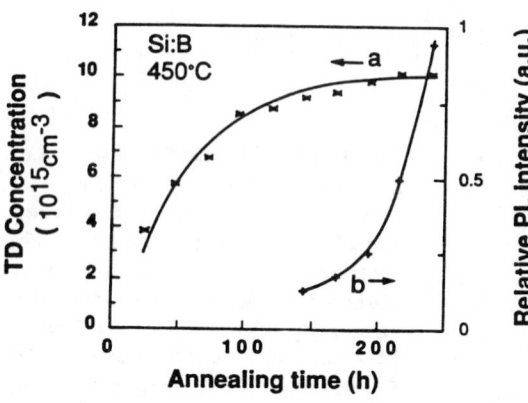

Fig. 2 : Evolution as a function of the annealing time at 450 °C for a B doped silicon sample of the:
a) TD concentration deduced from electrical measurements.
b) BO-line PL intensity at 2 K in the TO replica region. The experimental conditions were the same in all cases: the laser beam was defocussed to a spot size of approximately 1.5 mm in diameter and the laser power was ~ 100 mW on the sample.

In Figure 2, the TD concentration and the BO-line PL intensity are shown as a function of the annealing time. After 24 hours of annealing the conversion from p- to n-type is observed and the TD concentration increases when the annealing time increases, until a steady state concentration is reached after ~ 144 hours of annealing. On the other hand, the BO-lines are detected weakly only after 72 hours of annealing, for this sample and with the excitation power used for this experiment, and its intensity increases with increasing annealing time without any tendency of saturation.

Figure 3 shows, for a sample annealed for 240 h, the PL intensity of the BO-spectrum and of the boron bound exciton ($B_{TO}(BE)$) spectrum versus the laser excitation intensity without the formation of the electron-hole droplet (EHD). As expected for a rather light doping, the $B_{TO}(BE)$ intensity reaches a saturated value for high laser excitation powers, since all the substitutional boron acceptors get occupied with BE:s. On the other hand, the BO-lines spectrum intensity is found to vary linearly as a function of the laser intensity.

In Figure 4, we present several spectra obtained in the TO phonon replica region at different temperatures. Some line-structure can be observed within the envelope of the BO lines, with

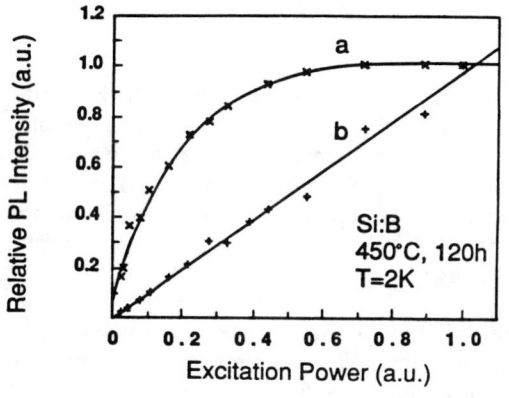

Fig. 3 : Luminescence intensities of the :
a) $B_{TO}(BE)$ line and
b) BO_{TO}-lines
as a function of the laser excitation power for the boron doped Si sample annealed at 450 °C for 120 hours.

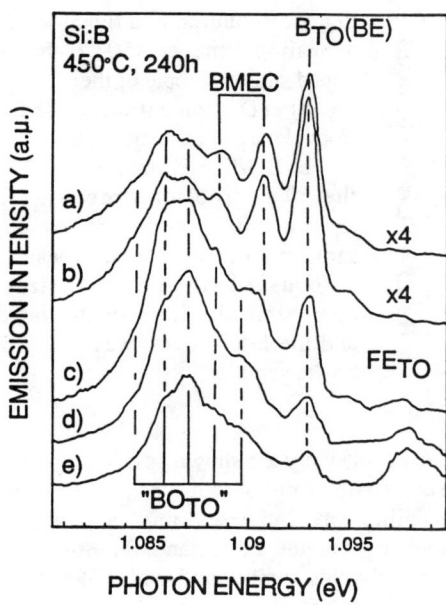

Fig. 4 : Photoluminescence spectra at a) 1.5 K, b) 4.2 K, c) 6 K, d) 8 K and e) 10 K of B-doped Cz silicon annealed at 450 °C for 240 hours. The structure within the BO-band is observed as shown in the figure with an energy separation of ~1.3 meV between each line. FE_{TO} denotes the free exciton luminescence line and BMEC the bound multiexciton complexes associated with boron.

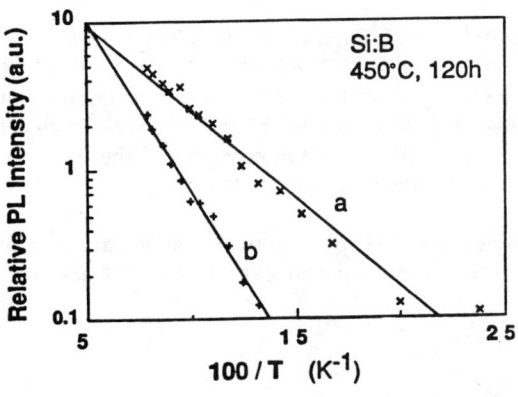

Fig. 5: Photoluminescence intensity relative to that of the FE line for
  a) the boron BE line and
  b) the BO-line (at the peak value at 1.087 eV)
versus the reciprocal sample temperature for B-doped Cz silicon annealed at 450 °C for 240 hours.

five lines at about 1084.3, 1085.8, 1087.0, 1088.4 and 1089.7 ± 0.3 meV. With a better resolution than that used to obtain the spectra shown in Figure 4 (1 meV), it was not possible to resolve these BO-lines more accurately. As the observed broadening of the BO-lines can be caused by phonon coupling, we have carefully studied the BO-lines also in the NP region, in spite of the relatively weak intensity of this spectrum. The spectra obtained in the NP region for various samples are very similar to these in the TO phonon replica region, which means that the natural widths of the BO-lines are rather large.

The relative PL intensity of the boron BE line and of the BO-line (at the peak value at 1.087 eV) are plotted versus the reciprocal sample temperature in Figure 5. From a least square fit to the data points, the thermal activation energies are found to be 5.0 ± 0.5 meV and 10.2 ± 0.5 meV for the boron BE line and the BO-line (1.087 eV) respectively.

4. DISCUSSION AND ALTERNATIVE MODEL FOR THE BO-LINE SPECTRUM.

The BO-lines (Figure 1 and 5) have previously been ascribed to bound exciton recombination at different TD:s (Tajima et al. 1980, Nakayama et al. 1981, Drakeford et al. 1987). The last authors found that the broad band which is on the same energy position range than the BO-lines in this work, contain detailed structure which can be resolved. In the weak NP spectrum, they show at least 14 sharp lines separated by ~ 0.5 meV. Also, since the BO-lines are observed, in this work, only weakly after 72 hours of annealing when the TD concentration is ~ 6×10^{15} cm^{-3}, and with a high intensity for long annealing times, we conclude that the BO-lines described in this work are not related to the same complexes as studied by Drakeford et al. 1987 and Thewalt 1988. Therefore the BO-lines cannot be ascribed to the BE recombination at differents TD:s.

This behaviour is confirmed by the comparison of the TD kinetics deduced via the resistivity data and the PL kinetics of the BO-lines (Figure 2) where no direct correlation can be found. Moreover in initially phosphorus doped samples, we have very rarely observed the BO-line spectrum, although a similar TD concentration (~ 10^{16} cm^{-3}) has been created by the annealing at 450 °C.

It should be noticed that Nakayama et al. (1981) found a saturation of the BO-lines kinetics at ~ 100 hours of annealing and van Kooten et al. (1987) a decrease of the BO-lines intensity after only 20 hours. However, in the latter reference, the "5" and "6" lines (often attributed to the decrease of TD centers) are observed in the same sample as the BO-lines. In our study these lines ("5" and "6") were never observed even after 240 h of annealing. We interpret these disagreements as due to the different "thermal history" of the samples and the initial characteristics of the material.

Further, the observation of infrared absorption lines at low temperature reveals up to nine specific double TD:s (Pajot et al. 1983), while from Figure 4 five lines can be observed in the envelope of the BO-lines.

During the annealing at 450 °C, the substitutional boron as well as the interstitital oxygen concentration can decrease as a function of the annealing time. That the interstitial oxygen concentration continues to decrease at long annealing times, even when the TD concentration has reached a saturated value, has been observed by Henry et al (1988).

We suggest that the boron atoms can be slightly mobile at 450 °C to interact with oxygen complexes. This behaviour can explain the fact that Weber et al (1986) observed a decreasing boron BE line intensity after annealing at 450 °C. It means then that the acceptor dopant in the starting material can play an important role in the formation of some complexes involving oxygen clusters, as was mentionned by van Kooten et al. (1987). We therefore believe that the BO-lines is due to bound exciton recombination at boron-TD donor complexes. This complex B-TD donor is expected to be in a concentration at least two orders of magnitude less (< 10^{14} cm^{-3}) than the TD complex itself (~ 10^{16} cm^{-3}), since it has not been observed in absorption measurements, as far as is known to the authors.

A similar excitation power dependence as observed in Figure 3, has been reported previously by Tajima et al (1979). To explain the excitation power dependence the capture cross section of excitons to the neutral BO-line defects must be larger than that to the neutral boron impurity, e.g. close to that obtained for excitons bound to deep impurities. In this case the recombination lifetime of the BO-lines should be close to that of deep impurity BE:s in silicon (Schmid 1977) which gives a much less pronounced saturation of the BO(BE) intensity via the excitation power, than of the B(BE) intensity.

The phonon coupling observed for the BO-lines (Figure 1) is characteristic for BE:s associated with shallow donors. This phonon coupling indicates that the BO-related defect is not an isoelectronic neutral defect, but rather a donor or an acceptor. The ratio of the intensity of the envelope of the BO_{TO}-lines to that of the BO_{NP}-lines is found ~ 20 in this work. Considering that the localization energy of the BO-lines is ~ a factor 2 larger than those of group V donors, the ratio of BO_{TO} to BO_{NP} intensity should be close to ~ 0.3, which is very far from the observed value. However, this is not expected to be true for complex donors where the localization of the wavefunction in k-space will be weaker than for a regular substitutional deep donor in silicon. It seems like the BO-line spectrum is due to the BE recombination at complex donors.

The energy difference between the FE PL line and the five BO-line positions is found to be ~ 12.3, 10.8, 9.6, 8.2 and 6.9 ± 0.5 meV (Figure 4). The BO-line at the peak value at 1.087 eV, was found to have a thermal ionization energy of 10.2 ± 0.5 meV (Figure 5), which corresponds within experimental errors to the spectroscopically determined binding energy of the BE. This behaviour is typical for excitons bound to neutral donors or acceptors in silicon, whereas for excitons bound to isoelectronic centers, quite different energies are found compared to the relative spectroscopic position to the FE.

5. SUMMARY

We have reported new PL data from a study of boron-doped oxygen-rich silicon annealed at 450 °C. We show that the BO-lines with no-phonon energy at ~ 1.143 eV consists of at least five lines. We suggest that these lines are due to bound exciton recombination at boron-thermal donor complexes.

ACKNOWLEDGMENTS

We wish to thank M.L.W. Thewalt for useful discussions and a preprint of a related manuscript.

REFERENCES

Bourret A 1985 *Thirteenth International Conference on Defects in Silicon* ed L C Kimerling and J H Parsey Jr (The Metallurgical Society of AIME, Warrendale, PA) pp 129-146
Drakeford A C T and Lightowlers E C 1987 in *Defects in Electronic Materials, Mat. Res. Soc.* **104** pp 209-213
Fuller C S, Ditzenberger J A, Hannay N B and Buehler E 1954 *Phys. Rev.* **96** 833
Henry A, Saminadayar K, Pautrat J L and Magnea N 1988 *phys. stat. sol. (a)* **106** 644
Irvin J C 1962 *Bell System Technical Journal* **41** 387
Minaev N S, Mudryi A V 1981 *phys. stat .sol .(a)* **68** 561
Nakayama H, Katsura J, Nishino T and Hamakawa Y 1980 *Jpn. J. Appl. Phys.* **19** L547
Pajot B, Compain H, Lerouille J and Clerjaud B 1983 *Physica* **117B** 110
Schmid W 1977 *phys. stat. sol. (b)* **84** 529
Tajima M, Kanamori A and Iizuka T 1979 *Jpn. J. Appl. Phys.* **18** 1401
Tajima M, Kishino S, Kanamori A and Iizuka T 1980 *J. Appl. Phys.* **51** 2247
Thewalt M L W 1988 *private communication*
van Kooten J J, Gregorkiewicz T, Blaakmer A J and Ammerlaan C A J 1987 *J. Phys. C : Solid State Phys.* **20** 2183

Optically detected magnetic resonance study of Si donors in hetero-epitaxial layers of $Al_xGa_{1-x}As$ on GaAs

E. Glaser, T.A. Kennedy, and B. Molnar

Naval Research Laboratory
Washington, D.C. 20375 U.S.A.

Optically detected magnetic resonance experiments have been performed at 1.6K on Si-doped epitaxial layers(1-1.5 μm) of $Al_xGa_{1-x}As$ grown on (001) GaAs substrates. The Si donor g-values exhibit an axial anisotropy about the [001] growth direction. The anisotropy increases as x increases from 0.4 to 0.8. This anisotropy is attributed to a valley repopulation effect due to stress in the heterostructure from the mismatch of the lattice constants. Analysis of these results yields the single-valley $g_\perp$-value (1.984±0.002) associated with the X-point conduction band minima.

I. INTRODUCTION

The role of shallow and deep impurity levels in the electrical and optical properties of semiconductors continues to be an important basic research issue. In particular, the $GaAs/Al_xGa_{1-x}As$ material system is currently under much experimental and theoretical scrutiny. This is due to 1) the excellent growth capabilities for layered structures from this alloy system 2) the promise for state-of-the-art opto-electronic devices with fast switching times and 3) its ability to provide a model system for the study of many basic physics problems (e.g. shallow/deep instabilities, one and two dimensional carrier localization, interface scattering, and many-body Coulombic interactions).

In the present study, we examine the optical and magnetic properties of shallow donors in epitaxial layers(1-1.5 μm) of $Al_xGa_{1-x}As$ on (001) GaAs substrates via optically detected magnetic resonance (ODMR). The motivation for this work is to study the nature of the DX center in n-doped $Al_xGa_{1-x}As$ crystals. Concurrent ODMR investigations are being performed by Montie and Henning (1988). Previous studies (Wartewig et al. 1975) of shallow levels in $Al_xGa_{1-x}As$ epitaxial layers have been carried out by electron paramagnetic resonance experiments (EPR) on much thicker layers (50-100 μm) and for a narrower range of alloy composition (0.56 ≤ x ≤ 0.80).

The technique of optically detected magnetic resonance is valuable to a study of shallow impurity levels in epitaxial layers due to its high sensitivity while probing a depth of approximately 0.3 micrometers. The present results from our ODMR studies on Si-doped layers of $Al_xGa_{1-x}As$ reveal clear evidence for the hetero-epitaxial stress present in the nearly lattice-matched $GaAs/Al_xGa_{1-x}As$ alloy system. In addition, the orbital triplet nature of the donor ground state in $Al_xGa_{1-x}As$ with high AlAs mole

© 1989 US Government

fraction (x≥0.35) has been confirmed. Also, analysis of these results yields the perpendicular g-value ($g_\perp$) associated with the X conduction band valleys. Furthermore, these results suggest strongly the importance of random strain due to alloy disorder in $Al_xGa_{1-x}As$ and of possible band-mixing effects near the direct-indirect crossover region.

II. EXPERIMENTAL DETAILS

The samples investigated in this work were thin(1-1.5 μm)epitaxial layers of $Al_xGa_{1-x}As$ (0 < x < 0.8) grown by molecular beam epitaxy (MBE) on standard semi-insulating, 500 μm thick (001) GaAs substrates. The samples were Si-doped to nominal concentrations of 10^{16}-10^{18} cm^{-3}. At these doping levels the impurity atoms can be treated as isolated. Also, a single, undoped liquid phase epitaxy (LPE) sample was studied before and after a Si ion-implantation procedure. A summary of the pertinent parameters of the samples investigated in this work is presented in Table 1. The high optical quality of these samples was confirmed by the photoluminescence spectra (Kennedy et al. 1988a).

Double crystal x-ray measurements were performed on these samples at room temperature with Cu K_α x-rays in order to determine the aluminum mole fraction (x). This technique (Bartels and Nijman 1978) is quite valuable in that it provides a <u>direct</u> measure of the strain (Δa/a) between the $Al_xGa_{1-x}As$ layer and the GaAs substrate (Δa/a-1.5x10^{-3} at room temperature for AlAs grown on GaAs). The x-values of these layers were determined from the shift of the $Al_xGa_{1-x}As$-associated (400) Bragg diffraction peak relative to the GaAs-associated (400) Bragg diffraction peak, comparison with recent double crystal x-ray results of epitaxial layers of AlAs on (001) GaAs (Tapfer et al. 1986), and Vegard's law for the entire range of aluminum mole fraction (0 < x < 1). The relatively sharp $Al_xGa_{1-x}As$ Bragg diffraction peaks (FWHM-15 arc seconds) of these samples indicate a uniform aluminum concentration in the growth direction across the entire epilayer.

As noted above, one of the samples was investigated after a Si ion-implantation procedure. This was carried out via a ^{29}Si multiple energy implantation followed by a 10 second, 850 °C close-contact anneal in an inert atmosphere (Molnar 1980). This method was employed to produce a spatially uniform dopant profile extending approximately 4000Å from the $Al_xGa_{1-x}As$ surface.

Table I. Summary of the pertinent parameters of the $Al_xGa_{1-x}As$ epitaxial layers investigated in this work.

Sample Designation	Layer Thick.(μm)	Si Conc.(cm^{-3})	AlAs Mole Frac.
M213	1.5	5x10^{16}	0.41
M214	1.5	5x10^{16}	0.50
LPE2 (Si-implanted)	1.0	1x10^{17}	0.59
M334	1.5	1x10^{18}	0.72

Shallow Impurities in III–V Compounds 235

ODMR measurements were obtained also on two of the $Al_xGa_{1-x}As$ layers after removal from the parent GaAs substrates. The GaAs substrates were removed by lapping with 5 μm grain size aluminum oxide powder, followed by a fast non-selective $GaAs/Al_xGa_{1-x}As$ etch, and, finally, by a slow GaAs selective etch.

The magnetic resonance was detected synchronously as changes in the total intensity of donor-acceptor pair photoluminescence coherent with chopping (77-570 Hz) of 50 mW of microwave power in a K-band (24 GHz) spectrometer. The photoluminescence of the $Al_xGa_{1-x}As$ layers was excited with above bandgap radiation provided by a Kr^+ laser at 476 nm with power densities between 0.1 and 1 W/cm^2. The detectors employed were Si and LN_2-cooled Ge photodiodes in the visible and near infrared spectral regions, respectively. The samples were studied under pumped helium conditions (T~1.6K) in an optical cryostat.

The data presented in this work were obtained with a 9" pole-face electromagnet with a maximum field of 1.1T. The measurements were obtained in the Voight geometry with the applied magnetic field (B) perpendicular to the wavevector of the excitation radiation. A calibration of the g-values for our system was carried out via ODMR measurements of the donor lines in phosphorous-doped Si crystals and by EPR measurements of DPPH. The stability of the magnetic field calibration from run to run was checked by nuclear magnetic resonance (NMR) measurements of ^7Li.

III. BACKGROUND

It will help to review the current understanding of the nature of the Si donor ground state in $Al_xGa_{1-x}As$ with high (x≥0.35) AlAs mole fraction (Morgan 1986). Briefly, in the effective-mass approximation (Kohn and Luttinger 1955), the wavefunction of the donor ground state is derived from 1) Bloch functions for the three equivalent x, y, and z valleys associated with the X-point conduction band minima and 2) 1s envelope functions that satisfy the effective-mass equation. In addition, due to the symmetry of the group III site relative to the X minimum (Morgan 1968), the ground state of an electron bound to the Si donor is an orbital triplet (T_2 in the T_d group). Therefore, neglecting spin-orbit interactions and random strains due to alloy disorder in these crystals, the group IV donor is three-fold degenerate. However, the degeneracy can be lifted by a large uniaxial stress applied in a direction which removes the equivalence of the three X-point conduction band minima. The compressive stress in the x-y plane of $Al_xGa_{1-x}As$ is equivalent to a tensile stress along the growth direction in these heterostructures and lowers the x and y valleys and raises the z valley. This leads to a re-distribution of the Si-donor electron spins among the three valleys (referred to as a "valley repopulation effect;" Wilson and Feher 1961, Wartewig et al 1975). As a result, the g-values found in the present ODMR studies should be a highly sensitive function of the angle of the applied magnetic field with respect to the [001] stress direction.

Detailed EPR studies of orbitally degenerate donors have been performed for interstitial Li in Si (Watkins and Ham 1970) and Sn on the Ga site in GaP (Mehran et al. 1972). For Si in $Al_xGa_{1-x}As$, we follow specifically the model employed by Mehran et al. (1972) due to the similarity in the orbital degeneracy (i.e. triplet) of the ground states for Sn donors in GaP and Si donors in $Al_xGa_{1-x}As$ with high AlAs mole fraction (x≥0.35) and

similar spin-orbit strength. In this model, the wavefunctions and g factors were derived for an orbital triplet split by a weak spin-orbit coupling and externally applied uniaxial stresses. In addition, the Zeeman term is treated by degenerate perturbation theory since the energies associated with the spin-orbit and applied stress interactions (~1-10 meV) are much larger than the Zeeman energy (~0.1 meV) for typical microwave frequencies (9-35 GHz).

Simple expressions were derived for $g'_\parallel$ and $g'_\perp$ as a function of the applied uniaxial stress (T) in the <u>limit</u> of large stresses such that the ratio of the stress splitting (D) to the effective spin-orbit splitting parameter (λ') of the donor state is much larger than unity. For our case in which the x and y valleys associated with the X-point conduction band minima are depressed and the z valley is raised, the expression for the measured value of $g'_\parallel$ as a function of the applied uniaxial stress (T) is

$$g'_\parallel = -g_\perp + 2g_L + (g_a - g_L)/(3KT)^2, \qquad (1)$$

where $g_a = (1/2)(g_\parallel + g_\perp)$, $g_\parallel$ and $g_\perp$ are the single-valley g-values associated with the long and short axes, respectively, of the X-point conduction band constant energy ellipsoids, g_L is the inter-valley orbital g-factor, and the parameter K is given by

$$K = C(\theta) / \lambda', \qquad (2)$$

where $C(\theta)$ depends on the shear deformation potential for the donor state and the cubic compliance constants (s_{11} and s_{12}) of the host lattice. As can be seen in Eq. (1), a value for $g_\perp$ can be deduced in the limit of infinite stress (i.e. $T^{-2} \approx 0$) with a knowledge of g_L.

IV. RESULTS AND DISCUSSION

The ODMR measurements on the Si-doped samples reveal a relatively sharp, single line. This feature was observed <u>only</u> for samples with aluminum mole fraction greater than ~0.35. Representative spectra for a Si-implanted $Al_xGa_{1-x}As$ epi-layer are shown in Fig. 1. The g-value associated with this feature was determined from the magnetic resonance condition

$$h\nu = g\mu_B B, \qquad (3)$$

where μ_B is the Bohr magneton and $h\nu(\nu=24 \text{ GHz}) \sim 0.1$ meV. The g-value and linewidth for the top spectrum in Fig. 1 with the external magnetic field oriented perpendicular to the [001] growth direction are 1.942 and 10 mT, respectively. This feature has been reported in prior works by our group (Kennedy et al. 1988a,1988b) and has been observed also by Montie and Henning (1988) in their recent ODMR experiments on Si-doped $Al_xGa_{1-x}As$ epitaxial layers grown by OMVPE. The assignment of this feature to a shallow Si-donor bound state associated with the X conduction band minimum in these $Al_xGa_{1-x}As$ layers is based upon the following evidence. The signal is observed for values of x (x≥0.35) for which the donor level tied to X is lowest in energy. The small, negative g-shift of this feature with respect to the g-value for free electrons (2.0023) is characteristic of shallow donors studied in other indirect bandgap semiconductors such as interstitial Li in Si (Watkins and Ham 1970) and Sn in GaP (Mehran et al.

Shallow Impurities in III–V Compounds

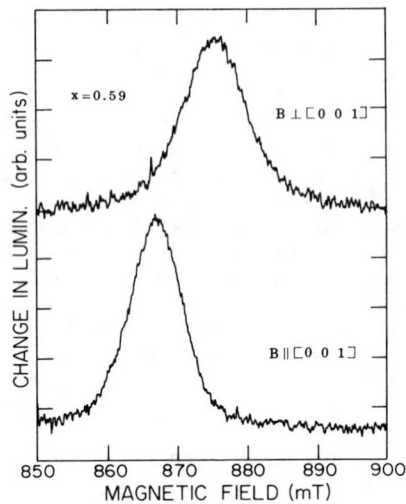

Fig. 1. ODMR spectra at 24 GHz of a Si-doped $Al_xGa_{1-x}As$ layer on a (001) GaAs substrate for two field directions. T=1.6K.

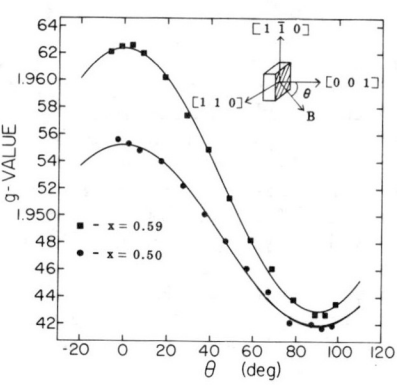

Fig. 2. Donor g-values as a function of the angle (θ) between the applied magnetic field (B) and the [001] growth direction for two layers. The solid lines are fits to the data (see Eq. 4) as described in the text.

1972) which have conduction band minima at or near the X-point. In addition, earlier electron paramagnetic resonance studies (Bottcher et al. 1973) on Si-doped solution-grown $Al_xGa_{1-x}As$ crystals with $0.68 \leq x \leq 0.84$ revealed a single isotropic line with a similar g-value (1.963) as observed in our ODMR experiments. This line was assigned by Bottcher et al. (1973) to Si donors through correlation of its intensity with the electron concentration. The identification of this ODMR feature in our work with Si donors has been verified also by the Si-implantation studies. The line was observed only very weakly (most likely due to residual shallow donor defects) for the $Al_xGa_{1-x}As$ layer (x=0.59) discussed above before it was implanted with Si impurities.

The Si donor ODMR feature was observed most strongly on the deep photoluminescence emission. This infrared band has been assigned in the Si-doped $Al_xGa_{1-x}As$ layers studied by Montie and Henning (1988) to a recombination process between the relaxed level associated with the DX center and a shallow acceptor. A similar Si-related deep broad luminescence band in $Al_{0.3}Ga_{0.7}As$ epitaxial layers has been reported recently by Souza et al (1988). These authors suggest that Si donors recombine with deep acceptors formed by a complex involving Si_{As} and an unknown defect. It is difficult to choose between these models based on present evidence.

Angular rotation studies were performed on these samples to acquire information about the symmetry properties of the shallow donor state. A representative spectrum obtained for the Si-implanted $Al_{0.59}Ga_{0.41}As$ sample with the applied magnetic field parallel to the [001] direction is shown in the bottom trace of Fig. 1. It is found that the ODMR line shifts to

lower fields by almost a full linewidth as the magnetic field is rotated to the [001] direction. This behavior corresponds to an increase in the donor g-value (see Eq. 3). The observed anisotropic behavior of the donor g-values was reported previously by Wartewig et al. (1975) from their EPR measurements on Si-doped $Al_xGa_{1-x}As$ epitaxial layers and, more recently, from the ODMR measurements on similar samples by Montie and Henning (1988). The g-value anisotropy is attributed to a valley repopulation effect (Wilson and Feher 1961, Wartewig et al. 1975) due to stress in the heterostructure from the mismatch of the $GaAs/Al_xGa_{1-x}As$ lattice constants at 1.6K. The stress leads to a biaxial compression in the plane of the layer and an elongation along the [001] direction due to the Poisson effect. In addition, there is evidence that the lattice mismatch is accomodated elastically (Rowland and Smith 1977) and that the stress (a tensile stress as opposed to a compressive stress) is nearly uniform across the entire epi-layer along the growth direction for $0 \leq x \leq 1$ (Rozgonyi et al. 1974).

The g-values deduced from the angular rotation studies on two $Al_xGa_{1-x}As$ epitaxial layers with different aluminum alloy concentrations are shown in Fig. 2. The experimental geometry employed for these studies is shown in the inset. The studies reveal a maximum g-value (g_{max}.) with the external magnetic field (B) oriented along the [001] direction, a smooth variation to lower values as the field is rotated away from [001] towards the [110] direction, and a minimum g-value (g_{min}.) for B along [110]. Due to the hetero-epitaxial stress in these layers, the symmetry of the $Al_xGa_{1-x}As$ lattice is lowered from cubic to tetrahedral about the growth direction (i.e. an axial symmetry). Fits have been made to these data with the usual expression for the g-values in the case of axial symmetry

$$g' = (g'^2_{\parallel} \cos^2\theta + g'^2_{\perp} \sin^2\theta)^{1/2} \quad (4)$$

where θ refers to the angle between the applied magnetic field and the axis of symmetry ([001]) and $g'_{\parallel}$ and $g'_{\perp}$ are the g-values associated with the magnetic field oriented parallel and perpendicular to the [001] crystallographic direction, respectively. As shown by the solid lines, the fits are quite reasonable with $g'_{\parallel}$ and $g'_{\perp}$ as the only adjustable parameters. Note, also, that $g_{max}.-g_{min}.$ increases with increasing aluminum mole fraction.

A compilation of the g-value data as a function of aluminum mole fraction (x) determined from the present ODMR experiments on several $Al_xGa_{1-x}As$ epitaxial layers is shown in Fig. 3. Several features of these data are important. First, the dominant trend is the monotonic increase of $g'_{\parallel}$ (solid circles) with increasing aluminum mole fraction. The value of $g'_{\perp}$ (solid squares) is observed to increase also with increasing x, but at a much lower rate compared to the behavior of $g'_{\parallel}$ with x. Second, the total g-value anisotropy as measured by $g'_{\parallel}-g'_{\perp}$ is found to increase strongly with increasing Al mole fraction. This is consistent with the increase in hetero-epitaxial stress in the $Al_xGa_{1-x}As$ layers with increasing x. The observation that $g'_{\perp}$ increases only weakly with x may be due to the fact that the in-plane $Al_xGa_{1-x}As$ lattice constant for the entire range of Al mole fraction is held to the lattice constant of the much thicker GaAs substrate.

The results of previous EPR studies on Si-doped $Al_xGa_{1-x}As$ layers (50-100 μm thick) by Wartewig et al. (1975) are shown by the open symbols in Fig.

Shallow Impurities in III–V Compounds

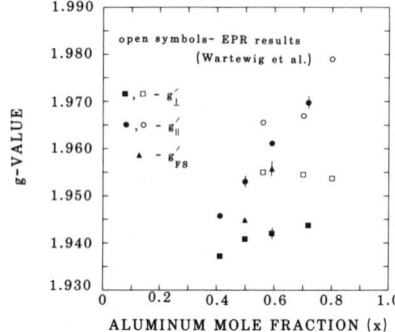

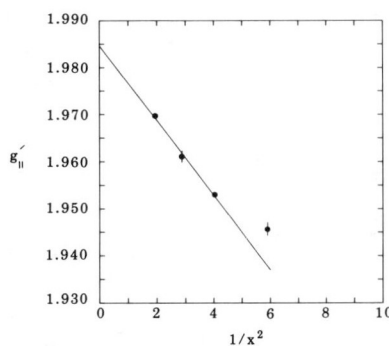

Fig. 3. Compilation of the ODMR g-values (solid symbols) as a function of the aluminim mole fraction (x) with B oriented parallel ($g'_{\parallel}$) and perpendicular ($g'_{\perp}$) to the [001] growth axis. The open symbols are the results of EPR studies (Wartewig et al. 1975) on similar samples. The triangles are from free-standing samples.

Fig. 4. Plot of the donor g-values with B along the [001] growth direction vs. $1/x^2$. The solid line is a least-squares fit to the data with $1/x^2 \leq 4$ (i.e. $x \geq 0.5$). The y-intercept gives the single-valley $g_{\perp}$-value associated with the X minima.

3. Similar trends for the behavior of $g'_{\parallel}$ and $g'_{\perp}$ with Al mole fraction are observed from these studies as found in the present ODMR results. However, semi-quantitative agreement between the EPR and ODMR results is obtained only for the $g'_{\parallel}$ data. The larger discrepancy found upon comparison of the two sets of data for $g'_{\perp}$ is not fully understood at the present time. There may be Al concentration gradients in the 50-100 μm layers leading to a more complicated strain distribution.

We have carried out a stress analysis of our results similar to that employed by Mehran et al. (1972) due to the similarity in the nature of the donor ground states of Sn in GaP and Si in $Al_xGa_{1-x}As$. In our analysis, the aluminum mole fraction (x) is treated as a uniaxial stress parameter based on the linear dependence of the strain ($\Delta a/a$) with x for the full range of aluminum mole fraction ($0 \leq x \leq 1$). Since the ODMR linewidths are smaller than the anisotropy (see Fig. 2), the random strain in the $Al_xGa_{1-x}As$ layers due to alloy disorder is small compared to the hetero-structure stress (particularly at very high x). In accordance with this model, the values for $g'_{\parallel}$ determined from the donor ODMR feature in the present Si-doped $Al_xGa_{1-x}As$ layers are plotted as a function of x^{-2} in Fig. 4. As seen by the solid line in Fig. 4, a reasonable least-squares fit can be made for this data with $1/x^2 \leq 4$ (i.e. $x \geq 0.5$). If we assume that g_L is quite small ($g_L \sim 0.005$ for GaP), we find directly that the single-valley (X) value for $g_{\perp}$ is 1.984 ± 0.002. In addition, the large deviation from the x^{-2} dependence of the $g'_{\parallel}$ data for x near 0.4 seen in this plot suggests some band-mixing effects between the orbital triplet states derived from the L- and the X-conduction band minima.

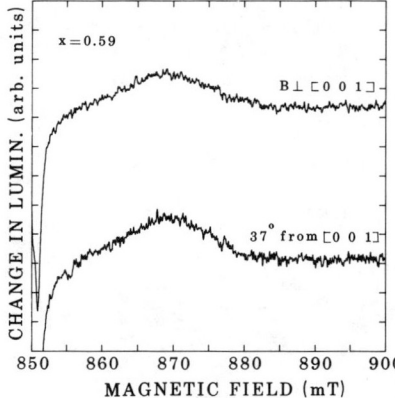

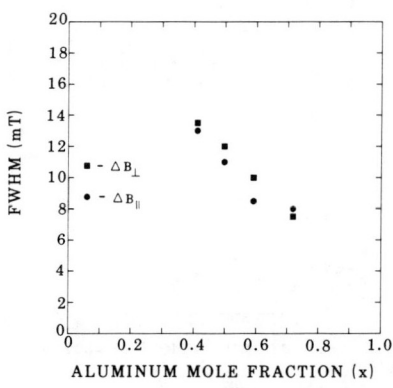

Fig. 5. ODMR spectra at 24 GHz of an Al$_x$Ga$_{1-x}$As layer removed from the parent GaAs substrate for two field directions.

Fig. 6. Plot of the ODMR linewidths with B oriented parallel and perpendicular to the [001] growth direction for the Al$_x$Ga$_{1-x}$As layers on the GaAs substrates.

As mentioned above, ODMR studies were performed also on the Si-doped Al$_x$Ga$_{1-x}$As samples <u>after</u> removal from their parent GaAs substrates. The 1-1.5 μm thick Al$_x$Ga$_{1-x}$As layers were held down in one corner with rubber cement thinned with toluene on phosphorous-doped Si substrates in order to facilitate the measurement and to have an in-situ g-value marker. No effects due to additional interface strain were apparent with this procedure. Representative spectra observed for the Si-implanted Al$_x$Ga$_{1-x}$As sample with x=0.59 discussed above (see Fig. 1) removed from the GaAs substrate are shown in Fig. 5. The strong, negative ODMR feature observed at the start of these magnetic field sweeps corresponds to the higher-lying magnetic field component of the hyperfine-split donor resonances in the phosphorous-doped Si substrate (Feher et al. 1961). The g-values associated with the Si shallow donor states in these free-standing layers of Al$_x$Ga$_{1-x}$As (solid triangles in Fig. 3) are found to be isotropic within experimental error over a full rotation of the applied magnetic field from the [110] to the [001] diretions. The g-values for the free-standing Al$_x$Ga$_{1-x}$As layers are found to lie between g'$_\parallel$ and g'$_\perp$ (see Fig. 3). This result is consistent with an appropriate weighted average of g'$_\parallel$ and g'$_\perp$ for the isotropic donor g-values observed in these layers. This result confirms that the origin of the g-value anisotropy in these heterostructures is the nearly uniform stress along the growth direction in the Al$_x$Ga$_{1-x}$As layers due to the lattice mismatch.

Additional measurements on a free-standing Al$_x$Ga$_{1-x}$As layer with a lower aluminum mole fraction, x=0.50, also reveal an isotropic g-value. However, the donor g-values determined for the free-standing Al$_x$Ga$_{1-x}$As layers are found to decrease slightly with decreasing aluminum mole fraction. Even though the hetero-epitaxial stress has been removed, the random strain due to alloy disorder still remains in these layers. Hence, the random strain may very well play a role in the observed behav-

ior of the g-values with x in the free-standing $Al_xGa_{1-x}As$ samples. However, the present results support the interpretation that the <u>dominant</u> contribution to the observed g-value anisotropy in these layers studied on the parent GaAs substrates is the hetero-epitaxial stress along the growth direction.

An analysis of the linewidths of these ODMR spectra helps us to confirm the symmetry properties of the donor ground state in $Al_xGa_{1-x}As$ with high AlAs mole fraction ($x \geq 0.35$). A comparison of the linewidths of the ODMR spectra from the same Si-doped $Al_{0.59}Ga_{0.41}As$ layer with (see Fig. 1) and without (see Fig. 5) hetero-epitaxial stress reveals that the linewidth of the donor line increases by approximately 50% after the layer is removed from the parent GaAs substrate. This leads us to conclude that the donor ground state associated with the X-point conduction band minimum is an orbital triplet, as opposed to an orbital singlet, due to the much larger strain sensitivity predicted for such states. This identification is in agreement with theoretical predictions by Morgan (1986) based on symmetry arguments for a group IV donor on a group III site in III-V semiconductors.

An analysis of the linewidths of the ODMR spectra obtained for the Si-doped $Al_xGa_{1-x}As$ epitaxial layers on the parent GaAs substrates was also carried out. The ODMR linewidths observed with the applied magnetic field oriented parallel and perpendicular to the growth direction, [001], are plotted as a function of x in Fig. 6. The linewidths for both geometries decrease with increasing aluminum mole composition. This observation is again consistent with the much larger strain sensitivity of the orbital-triplet donor ground state. However, the increase in linewidths observed for samples with x near the cross-over region ($0.37 \leq x \leq 0.42$) may arise in part from band-mixing effects since the donor ground state wavefunction near this range of x may be an admixture of X- and L-derived conduction band states with the same symmetry. The donor g-values associated with the L conduction band minimum have been determined previously from EPR experiments of phosphorous- and arsenic-doped Ge crystals (Feher et al. 1959). However, these donor g-values (~1.57) differ significantly with the donor g-values (~1.95) associated with the X-point conduction band minima found in the present Si-doped $Al_xGa_{1-x}As$ layers with high aluminum mole fraction.

It will be of much interest to re-examine the nature of the DX center in n-doped $Al_xGa_{1-x}As$ crystals with $x \geq 0.2$ grown on GaAs substrates in light of the present evidence for hetero-epitaxial stress along the growth direction due to lattice mismatch and, also, random strains in these alloys. Recent models (Morgan 1986) for the DX center suggest that the deep level is associated with a simple substitutional donor driven by a Jahn-Teller-like displacement of the atom at the central core from its centered lattice position. In addition, it is proposed that the deep level (binding energy~160 meV) is derived from an orbital triplet state of symmetry T_2 associated with the L-point conduction band minima. The present ODMR results suggest that the energies of deep (or shallow) levels that are <u>orbitally degenerate</u> will be a particularly sensitive function of uniaxial and inhomogeneous strains. Thus, the above and similar models that associate the DX center with an orbital degenerate state should take into account the effects of heterostructure stress and random strain due to alloy disorder present in the $Al_xGa_{1-x}As$ crystals.

V. SUMMARY

Optically detected magnetic resonance experiments have been performed on Si-doped epitaxial layers (1-1.5 µm) of $Al_xGa_{1-x}As$ grown on (001) GaAs substrates. Angular rotation studies reveal an anisotropy of the Si donor g-values about the [001] growth direction which increases as x increases from 0.4 to 0.8. This anisotropy is attributed to valley repopulation due to the hetero-epitaxial stress from the lattice mismatch between the $Al_xGa_{1-x}As$ layer and the GaAs substrate. Analysis of these results yields the single-valley $g_\perp$-value (1.984±0.002) associated with the X-point conduction point minima. The origin of the g-value anisotropy in these heterostructures is verified by ODMR measurements on the layers removed from their parent GaAs substrates. In addition, a comparison of the ODMR linewidths observed with the $Al_xGa_{1-x}As$ layers on and removed from the GaAs substrates confirms the predicted orbital triplet nature of the Si donor ground state in $Al_xGa_{1-x}As$ with high AlAs mole fraction ($x \geq 0.35$). Furthermore, the random strain due to alloy disorder in these $Al_xGa_{1-x}As$ layers also seems to play a role which is not fully understood at present. These results may have important implications for an understanding of the DX center in n-doped $Al_xGa_{1-x}As$ crystals.

ACKNOWLEDGEMENTS

We would like to thank M.G. Spencer of Howard University for the growth of the MBE and LPE samples, A.G. Crockett for help in data taking, M. Fatemi for the double crystal x-ray measurements, S. Kirchoefer for technical assistance in the removal of the $Al_xGa_{1-x}As$ epitaxial layers from the GaAs substrates, and B.V. Shanabrook for helpful discussions. This work was supported in part by the Office of Naval Research.

REFERENCES

Bartels W J and Nijman W 1978 J. Crystal Growth 44 518
Bottcher R, Wartewig S, Bindemann R, Kuhn G and Fischer P 1973 Phys. Stat. Sol.(b) 58 K23
Feher G, Wilson D K and Gere E A 1959 Phys. Rev. Lett. 3 25
Kennedy T A, Magno R, Glaser E and Spencer M G 1988a Defects in Electronic Materials ed. M. Stavola, S.J. Pearton and G. Davies (Pittsburg: Materials Research Society) pp. 555-560
Kennedy T A, Magno R and Spencer M G 1988b Phys. Rev. B37 6325
Kohn W and Luttinger J M 1955 Phys. Rev. 98 915
Mehran F, Morgan T N, Title R S and Blum S E 1972 Phys. Rev. B6 3917
Molnar B 1980 Appl. Phys. Lett. 36 927
Montie E A and Henning J C M 1988 J. Phys. C: Solid State Phys. 21 L311
Morgan T N 1968 Phys. Rev. Lett. 21 819
Morgan T N 1986 Phys. Rev. B34 2664
Rowland M C and Smith D A 1977 J. Crystal Growth 38 143
Rozgonyi G A, Petroff P M and Panish M B 1974 J. Crystal Growth 27 106
Souza P, Rao E V K, Alexandre F and Gauneau M 1988 J. Appl. Phys. 64 444
Tapfer L, Stolz T, Fischer A and Ploog K 1986 Surf. Sci. 174 88
Wartewig S, Bottcher R and Kuhn G 1975 Phys. Stat. Sol.(b) 70 K23
Watkins G D and Ham F S 1970 Phys. Rev. B 1 4071
Wilson D K and Feher G 1961 Phys. Rev. 124 1068

Inst. Phys. Conf. Ser. No 95: Chapter 4
Paper presented at Int. Conf. Shallow Impurities in Semiconductors, Linköping, Sweden, 1988

Shallow donor time-resolved magnetospectroscopy using a nanosecond pulse-slice laser: measurements and theory for high purity n-InP

G. L. J. A. Rikken[+][o], P. Wyder[o], J. M. Chamberlain[*] and G. A. Toombs[*]

[o]Max Planck Institut fur Festkorperforschung, BP 166X, F38042, Grenoble Cedex, France.
[*]Physics Department, Nottingham University, Nottingham NG7 2RD, United Kingdom.

ABSTRACT: The time-resolved photoconductive response of high mobility n-InP ($\mu > 10^5$ cm^2 V^{-1} s^{-1}) is measured at 4 K and in magnetic fields up to 8 T, using short (10-50 ns) far infra red pulses. The decay of the photosignal is interpreted in terms of effective lifetimes assuming phonon-assisted recombination processes. A calculation of transition rates in a magnetic field is outlined which is used to explain the experimental observations.

1. INTRODUCTION

Despite the considerable efforts which have been expended over the past two decades to determine in detail the shallow impurity levels of III-V semiconductors (Armistead et al 1984), comparatively little use has been made of pulsed far infra-red (FIR) spectroscopic techniques to investigate the recombination dynamics of these systems. Such techniques (Salzmann et al 1983, Rikken et al 1988a) offer a relatively direct means of probing decay mechanisms: in addition, the dynamics of individual transitions may be monitored by appropriate choice of laser frequency and magnetic field. The experimental conditions are chosen to ensure that the electron population is in quasi-equilibrium with the lattice (Bimberg et al 1985): this condition cannot always be assured in an electrical pulse measurement (Asche et al 1979), or when using a shorter wavelength laser excitation (Norton and Levinstein 1972).

In this communication we report the use of a novel nanosecond pulse-slice laser technique (Salzmann et al 1983, Rikken et al 1988a) to investigate recombination processes and effective lifetimes in a series of well-characterised high purity n-InP samples. We study the photoconductivity (PC) of the samples at magnetic fields of up to 8 T for both the resonant 1s-2p$_+$ and cyclotron resonance (CR) conditions, and monitor the time resolved (TR) decay of the PC signal. The samples exhibit differing degrees of compensation, and the systematics of the characteristic decay times (τ_{eff}) for the CR and 1s-2p$_+$ processes are investigated. We report a calculation of the transition rates for acoustic phonon assisted decay in a magnetic field which is used to explain qualitatively our data.

© 1989 IOP Publishing Ltd

2. EXPERIMENTAL

Samples of high purity n-InP grown by chloride VPE and MOCVD (Taylor and Anderson 1983, Bass and Young 1984) were provided with ohmic contacts: the sample characteristics are given in Table One. Pulses of FIR radiation with controllable width (10-50 ns) and sharp cut-off (~1 ns) were generated using a pulse-slice laser arrangement (Salzmann et al 1983). The PC response of the samples at low temperatures (2 K-20 K) and in field of up to 8 T was measured in the constant voltage mode. A Transient Analyser was used to monitor the decay of the photosignal after the termination of the exciting laser pulse. On occasions up to 2000 pulses were averaged to obtain a typical decay as shown in Figure 1. The conditions of sample bias and laser intensity were adjusted to ensure that the decay time was of a simple exponential form with one characteristic decay time, τ_{eff} (Rikken et al 1988b).

TABLE I

Sample	μ(77 K) $(cm^2 V^{-1} s^{-1})$	$N_d - N_a$ (77 K) $(\times 10^{14} cm^{-3})$	N_a/N_d
#1	123 000	1.2	0.61
#2	133 000	3.7	0.05
#3	76 000	3.7	0.62

Electrical properties of the samples used. μ, N_d and N_a are respectively the mobility and donor and acceptor concentrations.

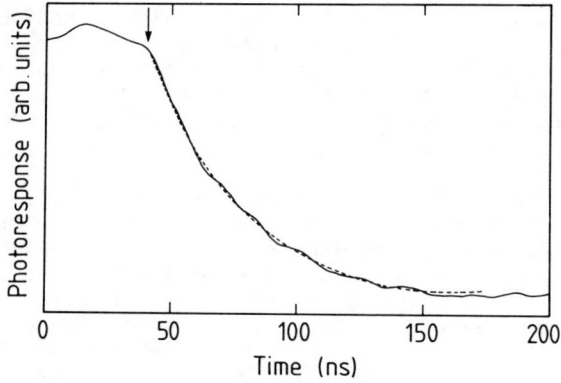

Figure 1
Decay of the resonant $1s-2p_+$ photosignal after termination of a λ = 148.5 μm wavelength pulse for sample #2 at 4.2 K under low bias conditions. The magnetic field is 5.90 T. The dashed line is an exponential fit to the data.

3. RESULTS

Figure 2 shows the decay time as a function of magnetic field for three samples of n-InP at 4.2 K under resonant 1s-2p$_+$ and CR conditions. For the lowest compensation material (sample #2) it is evident that the two values of τ_{eff} decrease together monotonically. For all samples, and for both PC processes, the curves indicate a common high magnetic field limit for τ_{eff}. However, for samples #1 and #3 at lower fields there is an initial rise in τ_{eff}, before a similar smooth decrease occurs. The zero field values of τ_{eff}, arising from the 1s-2p process, depend only on N_A (Rikken et al 1988c).

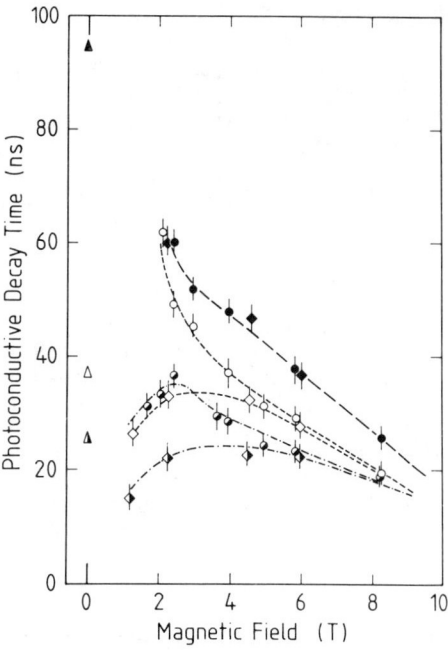

Figure 2
Decay time as a function of magnetic field for three samples at 4.2 K under resonant 1s-2p$_+$ and cyclotron resonance (CR) conditions. ◯: sample #1, CR; ◇: sample #1, 1s-2p$_+$; ●: sample #2, CR; ◆: sample #2, 1s-2p$_+$; ⊘: sample #3, CR; ◈: sample #3, 1s-2p$_+$.

4. DISCUSSION

In the subsequent analysis it is assumed that the decay of the photosignal may be ascribed entirely to the change in free carrier concentration. This implies that there are no significant changes in the carrier mobility within the measured decay time, which in turn implies that the electron population is in near-equilibrium with the lattice prior to recombination. The justification for this assumption is also given more fully elsewhere (Rikken et al 1988b). The interpretation given to the measured quantity τ_{eff} is of an effective lifetime in the $N = 0$ Landau level (LL) under conditions of carrier density equal to 10^{13} cm^{-3} and bias voltage sufficiently low to prevent hot electron effects. De-excitation from higher LLs will occur on a much shorter timescale than τ_{eff}. In our interpretation, the determining factor for τ_{eff} is the rate for capture into bound states from the $N = 0$ LL (Γ_B). This is briefly discussed below, and more fully elsewhere (Rikken et al 1988b).

The rates for electronic transitions into and between hydrogen-like bound impurity states, accompanied by phonon absorption and emission, have been previously calculated in zero magnetic field (Brown and Rodriguez 1967, Beleznay and Pataki 1966). We have calculated the transition rates in a magnetic field using a Yafet et al (1956) choice of Gaussian wavefunction for the bound states and free electron Landau-like wavefunctions for the continuum states. The results are only qualitative since the wavefunctions are always less accurately known than the energy levels in a variation calculation. We are not aware of any other report of this calculation.

All of the transition rates we calculated have a factor of the form:

$$q_0^n \exp[-2a^2 q_0^2] \tag{1}$$

where n is typically 2, $q_0 = \Delta E/\hbar v_s$ and v_s is the sound velocity in InP. ΔE is the separation of the two participating levels: the optimum level for efficient removal of electrons is of the order 0.5-1 meV below the N = 0 LL. The parameter a relates to a_λ, the length scale of the cylindrical orbit (Yafet et al 1956, Wallis and Bowlden 1958) for the bound state trial function (Wallis and Bowlden 1958) through the relation:

$$a = a_\lambda/[1 + a_\lambda^2/\ell_B^2] \tag{2}$$

where ℓ_B is the magnetic length $(\hbar/eB)^{\frac{1}{2}}$. In zero field, $a = a_\lambda = 3.0 \times 10^{-8}$ m, typically, and $q_0 = 1.9 \times 10^8$ m^{-1} for $\Delta E = 0.6$ meV. At a field of 10 T, a_λ will have decreased by 2/3 (Yafet et al 1956), and $\ell_B = 8 \times 10^{-9}$ m. Therefore the transition rate will have increased approximately 70-fold. The other factors will reduce this somewhat, so that the observed order of magnitude change in τ_{eff} over this field range is reasonable. It may be seen that the decrease arises directly from the shrinkage of the bound state perpendicular to the magnetic field. This shrinkage increases the matrix element since there are then less phonon wavelengths within the spatial extent of the bound state.

The smooth decrease in τ_{eff} seen in Figure 2 for sample #2 is adequately accounted for by a variation of the form given in (1). This discussion is appropriate when $\ell_B \ll d$, where d is the inter-ionised donor spacing.

Figure 3 illustrates diagramatically the transition rates we have calculated for phonon assisted recombination following FIR excitation to a 2p$_+$ state in n-InP. The form of the transition rate, Γ_c, indicates that electrons predominantly scatter into N = 0 LL states $\psi(\ell,m,k_z)$ with m values from m = 0 down to approximately m = -4. The magnitude of this transition rate is, from the analysis, approximately the same as the rate for the transitions from N = 0 states into <u>one</u> bound state. However, the measured PC rate $(1/\tau_{eff})$ represents an average over transitions for all impurities within range. If we average these rates over the range of k_z wavevectors, k_m, which are occupied by electrons participating in the PC process and over all the initial values of m, then we obtain an expression for Γ_B $(= 1/\tau_{eff})$ of the form:

$$\Gamma_B \sim \pi\Gamma_C/(1 + M)k_m d \tag{3}$$

where d is typically 10^{-7} m, $M = 4$ (the maximum $|m|$ value for an occupied LL) and $k_m = 5 \times 10^7$ m^{-1}. From our calculations, $\Gamma_C \approx 10^8$-10^9 s^{-1}, leading to $1/\Gamma_B \approx$ 10-100 ns, comparable with the experimental values. Furthermore, this discussion justifies our assumption made above that the determining factor for the effective conduction band lifetime is the slower rate Γ_B.

For samples #1 and #3 an initial rise in τ_{eff} is noted. As discussed above, electrons are deposited into N = 0 LL states with a range of angular momenta (typically from m = 0 to m = -4). The spatial extent of the relevant wavefunction is of order (Garstang 1977) $2\ell_B(1-2\ m)^{\frac{1}{2}} \approx 10\ \ell_B = 100$ nm when m = -4 (for a representative $\ell_B = 10$ nm). This is comparable with d in samples #1 and #3. Furthermore, the spatial extent of the N = 0 LL wavefunction increases at lower field, so the electron has a choice of going into the bound states of several impurities. The decay times are therefore shorter for these samples, as observed, and also increase with field as the number of available impurity states reduces. At the higher field (> 3 T) the situation reverts to the case considered for sample #2, and τ_{eff} falls with field.

At the highest fields, the lifetimes for CR and 1s-2p$_+$ processes converge. This is expected, since the 2p$_+$ level pins to the N = 1 LL at high fields (Boyle and Howard 1963). It is also noteworthy that the present work indicates the independence of τ_{eff} on N_A at high field. For typical experimental parameters, the cyclotron orbit centre drifts rather less than the inter-ionised donor distance before recombination. The electron therefore recombines with the original ionised donor, irrespective of the presence of other ionised donors.

ACKNOWLEDGEMENTS: JMC and GAT acknowledge the hospitality of the Max Planck Institut fur Festkorperforschung, where this work was performed, and travel funds provided by the Science and Engineering Council (U.K.). We are obliged to L. L. Taylor, D. Anderson and S. J. Bass (Royal Signals and Radar Establishment Procurement Executive, U.K. Ministry of Defence) for loans of samples. H. Krath is thanked for technical assistance.

[†]Present address: Philips Research Laboratories, 5600 JA Eindhoven, The Netherlands.

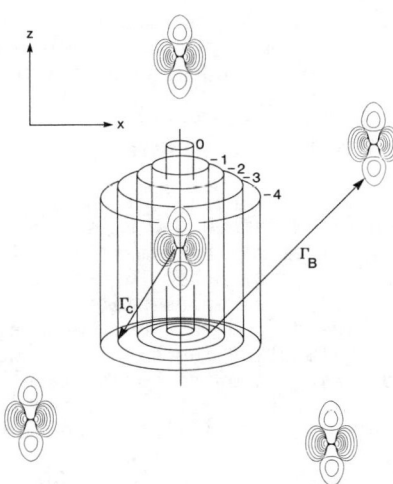

Figure 3
Schematic in the x-z plane of the calculated phonon-assisted recombination rates. The magnetic field lies along the z axis, and the gauge is chosen to coincide with an impurity centre. Transitions from $2p_+$ states at a rate Γ_C are shown, together with recombination to a representative 2s state at a rate Γ_B. Recombination to the original impurity, and to those lying along the same z axis are omitted for clarity. The m values are indicated.

REFERENCES

Armistead C J, Knowles P, Najda S P and Stradling R A 1984 J. Phys. C. **17** 6415
Asche M, Kostial H and Sarbey O G 1979 Phys. Stat. Solidi (b) **91** 521
Bass S J and Young M L 1983 J. Cryst. Growth **68** 311
Beleznay F and Pataki G 1966 Phys. Stat. Solidi **13** 499
Bimberg D, Munzel H, Steckenborn A and Christen J 1985 Phys. Rev. **B31** 7788
Boyle W S and Howard R E 1961 J. Phys. Chem. Solids **19** 181
Brown R A and Rodriguez S 1967 Phys. Rev. **153** 890
Garstang R H 1977 Rep. Prog. Phys. **40** 105
Norton P and Levinstein H 1972 Phys. Rev. **B6** 489
Rikken G L J A, Wyder P, Chamberlain J M and Taylor L L 1988a Europhysics Letters **5** 61
Rikken G L J A, Wyder P, Chamberlain J M, Grimes R T and Taylor L L 1988b Phys. Rev. **B** - to be published
Rikken G L J A, Wyder P, Chamberlain J M, Grimes R T and Halliday D P 1988c Semiconductor Science & Techology **3** 302
Salzmann H, Vogel T and Dodel G 1983 Optics Comm. **47** 340
Taylor L L and Anderson D 1983 J. Cryst. Growth **64** 55
Wallis R F and Bowlden H J 1958 J. Phys. Chem. Solids **7** 78
Yafet Y, Keyes R W and Adams E N 1956 J. Phys. Chem. Solids **1** 137

Optically detected magnetic resonance investigations of shallow donors in GaP

J J Lappe, B K Meyer and J-M Spaeth

University of Paderborn, Fachbereich Physik, Warburger Str.100A, D-4790 Paderborn, F.R.G.

ABSTRACT: Magneto-optical investigations on shallow intrinsic and extrinsic donors in GaP will be presented. In an optically detected ENDOR study in GaP:Zn,S an isolated $P_{Ga}P_4$ antisite defect was identified.

1. INTRODUCTION

The optical properties of shallow extrinsic donors in GaP (S,Te,Si) have been extensively studied using optical absorption, luminescence and Zeeman spectroscopy (see e.g. Dean 1966, 1974), whereas the ground state properties of the neutral donors were determined using Electron Spin Resonance (ESR) and Electron Nuclear Double Resonance (ENDOR) (see e.g. Landolt Börnstein 1982). For the intrinsic defects far less experimental information is available. More than 10 years ago the ESR spectrum of the intrinsic P antisite defect PP_4 has been detected, apart from the Ga vacancy it is the most intensively studied intrinsic defect in GaP. During the last years ODMR studies of the Linköping group focussed on the role of intrinsic P-antisite defects complexed with impurities such as Cu and Li diffused in the crystal. In the early ESR studies a P-antisite defect was also detected in Zn doped GaP overcompensated with S. In the ESR spectrum the interactions with the four nearest P neighbours were not resolved in contrast to the PP_4 spectrum. It was identified as a P_{Ga} antisite defect on the basis of its large central P hyperfine interaction with $I=1/2$, which is, however, reduced compared to the PP_4 defect. Pairing with a shallow background impurity i.e. Zn or S could be a possible explanation for this reduction. Optically detected ESR and ENDOR investigations using the MCD technique (see below) were performed to identify the defect structure of this particular antisite defect. In connection with this also the isolated chalcogen impurities in GaP were studied.

2. EXPERIMENTAL

The magnetic circular dichroism of the absorption is proportional to the population difference of the ground state zeeman levels of a paramagnetic defect. ESR transitions diminish the population difference and hence decrease the MCD. At fixed microwave frequency the magnetic field is swept and the decrease in the MCD is monitored as a function of the magnetic field to obtain the ODESR spectrum. In triple resonance experiments (ODENDOR) radio frequency induced nuclear magnetic transitions are detected as a further decrease of the ODESR signal and hence the MCD. The ODESR/ODENDOR spectrometer is a custom built computer controlled spectrometer working at 24 GHz (K-Band) and at 1.5 K.

3. EXPERIMENTAL RESULTS

3.1 P_{Ga} ANTISITE DEFECT

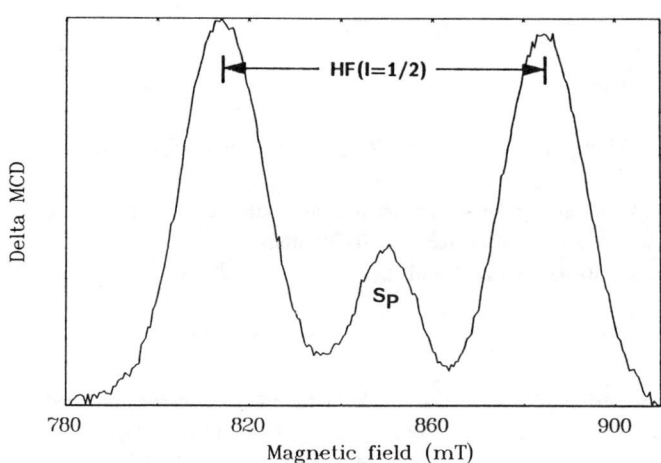

Fig.1.: ODESR spectrum of the P_{Ga} antisite and the S donor in GaP:Zn,S

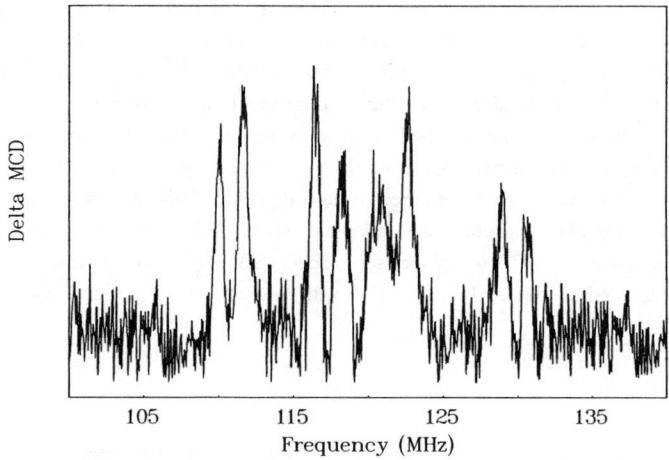

Fig.2: ODENDOR spectrum of the 4 nearest ^{31}P neighbours

The ODESR spectrum of the P-antisite defect in GaP:Zn,S is shown in fig.1. Compared to the well known PP_4 defect (Kaufmann et al 1976) the central hf interaction is reduced by 30 %. The ligand hf interactions are not resolved, therefore no further structure details can be derived from the ESR spectrum. A section of the ODENDOR spectrum detected in the high field hf line of the ODESR spectrum is shown in fig.2. The 8 ENDOR lines correspond to the interactions of the four nearest P neighbours. To determine their symmetry and their ligand hf (lhf) interaction constants the angular dependence of the ODENDOR lines was measured as shown in fig.3.b. Since P has $I=1/2$ the angular depencence has a very simple structure. A calculated angular dependence taking into account isotropic and anisotropic lhf interactions is shown in fig.3.a. for one m_s state ($m_s = -1/2$) and 4 P neighbours in tetrahedral symmetry for a rotation in a (110) plane from (100) to (110) in perturbation theory of first order. Experimentally only few angles could be measured as indicated in fig.3.a. Due to the pseudo dipolar coupling of the 4 P ligands (fig.3.c) each of the lines is split. The central line is due to two nuclei (2,3 in fig.3.a) which remain equivalent when rotating the crystals. The pseudo dipolar coupling for these two nuclei is 4.4 MHz as indicated in fig.3.c.

Shallow Impurities in III–V Compounds

Due to the pseudo dipolar coupling of all four P nuclei there is an additional splitting of 1.4 MHz which is observed on the ENDOR lines of all four nuclei. These effects were also found in the ENDOR investigations of the EL2 defect in GaAs (Meyer et al 1987), the Gallium vacancy in GaP (Hage et al 1987) and the P_{In} antisite defect in InP (Jeon et al 1987). Although these splittings complicate the ENDOR analysis, their presence have the advantage to prove the number and the symmetry of the nuclei in the first shell.

The rotation pattern is due to 4 nearest P neighbours in tetrahedral symmetry. No symmetry lowering distortion can be found. The same pattern is found in the frequency range between 80-90 MHz for the $m_s = 1/2$ state. The only other ENDOR lines found are around 24 MHz and 32 MHz and belong to the interactions with the next nearest Ga shell (12 nuclei, two isotopes ^{69}Ga and ^{71}Ga). The experimentally determined interaction constants of the first shell P and 2nd shell Ga are given in table 1 in comparision with the data of the PP_4 defect as obtained from the ESR experiments labelled from now on $PP_4(ESR)$.

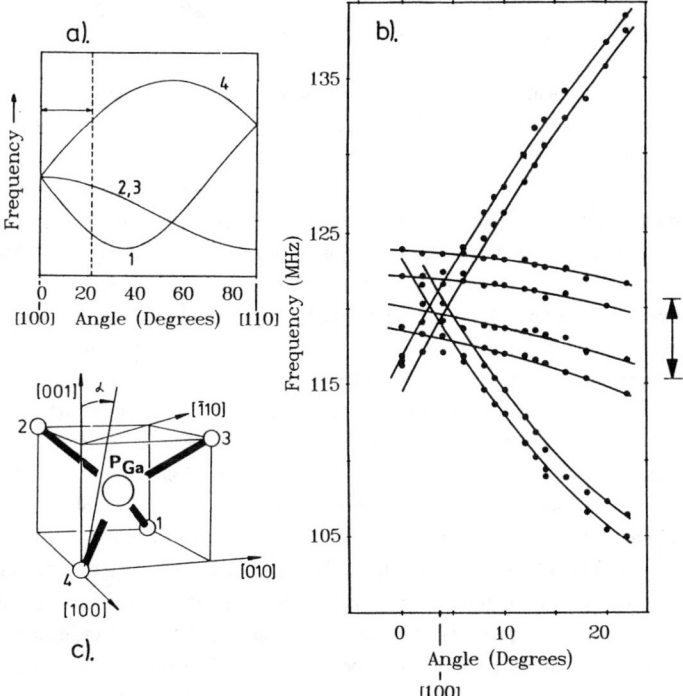

Fig.3: a) Simulation of the angular dependence of the 4 n.n. ^{31}P of the P_{Ga} antisite defect (c) for a rotation in a (110) plane; b) measured rotation pattern for $m_s = -1/2$ (full dots), the drawn lines are calculated with the values given in table 1.

In a LCAO analysis these values are compared with the values of the free P and Ga atom respectively to obtain the spin density distribution. For the $PP_4(ESR)$ defect only the values for the central P atom and next nearest P ligands are available, all the other interactions are included in the ESR linewidth of 6.3 mT. The 12 Ga nuclei with $I=3/2$ due to the presence of two isotopes with the interaction constants given in table 1 cause an individual ^{31}P lhf ESR linewidth of 12.8 mT compared to 6.3 mT found for the $PP_4(ESR)$. This is the reason why the interactions with the four nearest P neighbours are no longer

resolved in the ESR/ODESR spectrum. A simulation of the ODESR linewidth (26 mT) taking into account the measured values for the first and second shell gives excellent agreement, the deviation is

	PP_4(ESR)			PP_4(ENDOR)		
	a/MHz	b/MHz	spin local.	a/MHz	b/MHz	spin loc.
central P atom	2896	-	26%	1904	-	17%
1. shell 4 P nuclei	224	45	66%	209	34	51%
2. shell 12 Ga nuclei + ^{69}Ga, *^{71}Ga	not resolved			40* 31+	3.7* 2.9+	24%

smaller than 1 mT. This indicates that all interactions were resolved with ENDOR. In any case small interactions with higher shells or with a single nucleus bound to the defect would not contribute significantly to the linewidth.

This particular antisite defect PP_4(ENDOR) was observed in any undoped or doped (V,Ti,S) GaP material whenever the sample was slightly n-type conducting. Its level 0/+ must be very close to the level of the S donor in GaP, i.e. ca. 100 meV below the conduction band. Its ESR signal is absent when all S donors are occupied and the Fermi level is above E_c - 0.1 eV. Both defects can be observed, however, simultaneously (see fig.1). Its lower level corresponding to +/++ is close to the 0/+ level of the PP_4 at E_c-0.6 eV (Kreissl et al 1987) (the corresponding +/++ level of PP_4(ESR) is at E_c-1.1 eV) as determined from ODESR investigations of different s.i. and n-type GaP samples.

It has been remarked in the review of Kennedy 1986 that two aspects of the PP_4(ESR) seemed to be anomalous, the splitting of the ESR spectrum under uniaxial stress not to be expected for a S=1/2 defect and the very strong temperature dependence of the central hf interaction. One major difference between both defects is the distribution of the electron over the ligands as seen in table 1. The considerably smaller individual linewidth of 6.3 mT for the PP_4(ESR) defect indicates, if it is due to the interactions with the n.n. Ga neighbours, a localisation of only 8% in this shell. For the PP_4(ENDOR) defect the spin density is not so concentrated at the central part of the cluster (the central atom and the 4 P ligands). The reason for this delocalisation is at present not understood. A pairing with an extrinsic impurity at a second nearest or third nearest neighbour position could be a possible explanation, since impurities with a low abundance e.g. Zn_{Ga} or S_P would not contribute significantly to the ODESR line width. PP_4(ENDOR) cannot be due to a donor-acceptor pair. The concentration of the defect was independent of the acceptor impurity content in the samples. Also the chemical nature of the acceptors was different (Zn, Ti, V). A donor-donor pair (e.g. S_P) is even less likely, the coulomb interaction between both species could not account for the difference in the binding energy of 0.5 eV. Thus we identify the PP_4(ENDOR) defect with an isolated P antisite defect. The two anomalous aspects of the PP_4(ESR) defect are a strong suspicion that a pair defect is present, probably involving an intrinsic point defect. The equivalence of the 4 nearest P neighbours of the PP_4(ESR) is infered from the ESR spectrum alone and T_d symmetry is assumed. A small distortion of the nearest neighbours, as in the case of EL2 defect in GaAs (Meyer at al 1987), could not be resolved in the ESR spectrum. Future ENDOR experiments will concentrate on the structure identification of the PP_4(ESR) defect.

3.2 CHALCOGEN IMPURITIES IN GaP

When codoping GaP with the chalcogens like S, Se or Te the excitons bound to the neutral donors can be studied. It is of interest to compare the P site donors with those on the Ga site i.e. Si or Ge. Whereas the exciton spectra are very similar, this is not the case for the ir absorption spectra resolved with the MCD technique. Fig.4 shows the MCD spectra of the neutral bound exciton for S and Ge in GaP.

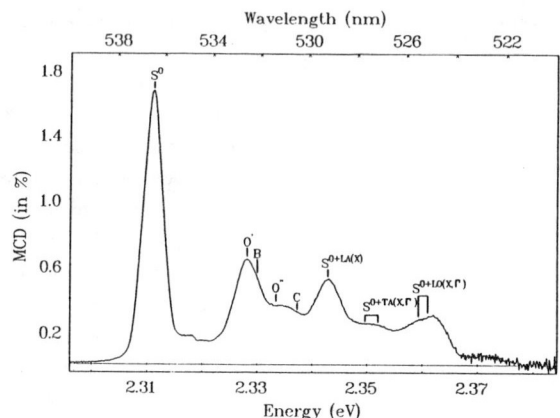

The spectra are dominated by the no-phonon line followed by a series of phonon replicas (Dean 1966, 1974). On each of these lines, i.e. the total MCD spectrum, the ODESR spectrum of the respective donor is observed, which definitely proves that all these lines belong to only one defect which is not obvious from a luminescence or transmission experiment alone. Whereas all these lines originate from the paramagnetic donors, which can easily be verified by the temperature and magnetic field dependence of the MCD in some samples we also detected MCD signals which were only dependent on the strength of the magnetic field and not on the temperature. They come from defects with a spin paired ground state, the MCD is diamagnetic. As an example not shown here, the isoelectronic trap GaP:N has been observed which opens new possibilities in the study of defects with a diamagnetic ground state using the MCD technique.

For the Ga site donors Ge and Si only the bound exciton spectra were observed in contrast to the chalcogens S, Se and Te. Fig.5 shows

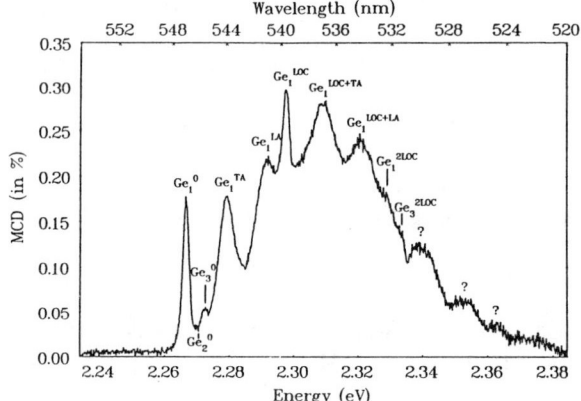

Fig.4: MCD spectra of the neutral S (a) and Ge (b) donor bound exciton in GaP

as an example the near ir MCD spectrum of GaP:S^0. It has an onset at 0.53 eV with a maximum at 0.78 eV and a continous decrease to higher photon energies. The shape of the MCD band is very similar to the photoionisation cross sections $\sigma(h\nu)$ of hydrogenic donors predicted by the theory to be

$\sigma(h\nu) = (h\nu - E_d)^{3/2} (h\nu)^{-5}$ (Blakemore 1984). This equation purports to show a maximum at $h\nu = 1.4\, E_d$ where E_d is the binding energy of the donor. A calculated dependence is shown in fig.5. with full dots. It differs somewhat from the experimental curve at higher energies indicating a steeper decrease i.e. a

higher power law dependence. However from the maximum of the MCD at 0.78 eV the onset is calculated to be at 0.55 eV in agreement with the experiment. 0.55 eV would then correspond to the binding energy. S_P has a binding energy of 107 meV with respect to the lowest conduction band minimum X_1. One has to have in mind that also transitions to the higher conduction band minimum X_3 are possible, both, X_1 and X_3, are separated by 0.36 eV (Landolt Börnstein 1982). Including this energy difference and the separation of the S level from X_1 (107 meV) this transition should occur at 0.467 meV.

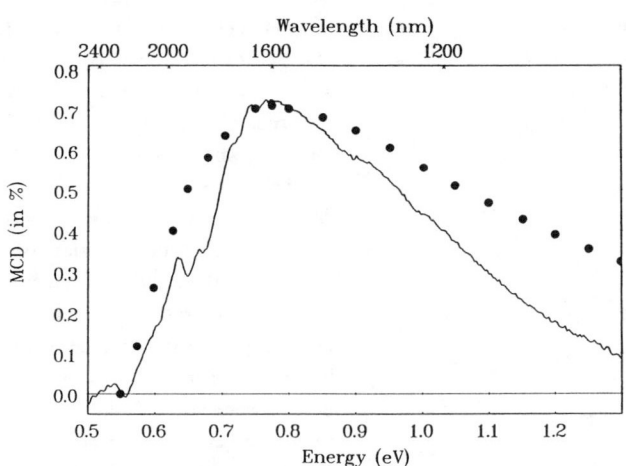

Fig.5: MCD spectrum of S^0 at H = 1.4 T and T = 1.45 K
The MCD line shape is compared with the theoretical spectral dependence (full circles) of a photoionisation transition to the conduction band

In a configuration coordinate diagram this corresponds to the transition of the zero phonon line, the electron phonon interaction can shift the optical transition to higher energies. In the theoretical description of the photoionisation transition electron phonon coupling is neglected. In the excitonic spectra of the S donor coupling to TA, LA and LO phonons is resolved. Assuming the electron phonon coupling as the reason for the discrepancy of the onset (0.54 -0.467 eV) its amount would be of the order of 80 to 90 meV and thus for a Huang Rhys factor of 3 to 4 would result in a phonon frequency of 20 meV close to the LA(X_3) phonon.

A small lattice relaxation is what one expects for the substitutional P-site donors in GaP. Very similar results were obtained for Te^0 in GaP. They are, however, absent for the Ga site donors Ge^0 and Si^0.

REFERENCES

Blakemore J S and Rahimi S 1984 Semiconductors and Semimetals, Vol.20 (Academic Press, Inc) p 250
Dean P J, Schairer W, Lorenz M and Morgan T N 1974 J.Lumin. 9 343
Dean P J 1966 Phys. Rev. 157 655
Hage J 1987 Thesis Paderborn
Jeon D Y, Gislason H P, Donegan J F and Watkins G D 1987 Phys. Rev. B 36 1324
Kaufmann U 1988 private communication
Kaufmann U, Schneider J and Räuber A 1976 Appl. Phys. Lett. 29 312
Kennedy T A 1986 Materials Science Forum Volumes 10-12 283
Kreissl J, Gehlhoff W and Ulrici W 1987 phys. stat. sol.(b) 143 207
Landolt-Börnstein 1982 Numerical Data and Functional Relationships in : Science and Technology
    Vol.17 a, b ed. (Springer: Berlin)
Meyer B K, Hofmann D M, Niklas J R and Spaeth J M 1987 Phys. Rev. B 36 1332

Inst. Phys. Conf. Ser. No 95: Chapter 4
Paper presented at Int. Conf. Shallow Impurities in Semiconductors, Linköping, Sweden, 1988

Search for the zero-field ODMR transition in the J=2 state of the bound exciton in GaP:N

M.C.J.M.Donckers and J.Schmidt

Huygens Laboratory, University of Leiden,
P.O.Box 9504, 2300 RA Leiden, The Netherlands.

ABSTRACT: A systematic search for the zero-field transitions between the E and T_1 components of the J=2 state of the excitons bound to N impurities in GaP has been undertaken using ODMR techniques.

1. INTRODUCTION

The optical spectra produced by the excitons bound to iso- electronic nitrogen traps in GaP have been studied in detail following pioneering work by Thomas and Hopfield (1966) and Cuthbert and Thomas (1967). In weakly doped materials the emission spectrum is dominated by two sharp zero-phonon lines A and B near 535 nm associated respectively with transitions from excited J=1 and J=2 states to the J=0 ground state of the bound exciton. These excited states arise from the coupling of an electron with spin angular momentum S=1/2 with a hole with total angular momentum J=3/2. The A line is electric-dipole allowed and the excited J=1 state has a lifetime of 38 ns. The B line however, which is 0.88 meV (7 cm^{-1}) lower in energy, is forbidden and the excited J=2 state has a lifetime of about 4 μs (Cuthbert and Thomas 1967). The assignment of the A and B line as corresponding with transitions between respectively J=1 and J=0 and J=2 and J=0 states is based on the Zeeman experiments of Thomas, Gershenzon and Hopfield (1963), Yafet and Thomas (1963) and Merz, Faulkner and Dean (1969).

The observation of ODMR transitions in the J=2 state would be extremely interesting to further our understanding of the properties of the exciton bound to the N impurity. To our knowledge experiments in the presence of a magnetic field have so far been unsuccessful (Cavenett, Brunwin and Nicholls 1977). As an alternative we have tried to find the zero-field transitions between the E and T_1 components (in the symmetry group T_d) of the J=2 state using ODMR techniques. In this contribution we report the results of this search which unfortunately has not yielded a definitive conclusion whether the observed signals have to be attributed to transitions between the two components of the J=2 state.

2. EXPERIMENTAL

The experiments were performed with a conventional zero-field ODMR spectrometer already described by Schmidt (1971). The resonator is a re-entrant cavity the resonance

© 1989 IOP Publishing Ltd

frequency of which can be varied over one octave (1-2 GHz, 2-4 GHz or 4-8 GHz). The sample is placed at a position where the microwave magnetic field is maximal and the microwave electric field is minimal. Light access is provided by two holes in the walls of the cavity allowing optical excitation and detection at right angles. The cavity and the sample are immersed in liquid helium pumped down to 1.2 K. The microwave source consists of a Hewlett Packard 8690 B sweep oscillator in combination with one of the HP 8690 B series of backward wave oscillators. With these oscillators microwaves can be generated between 1 GHz and 8 GHz. The output can be amplified to 10 Watt with Varian traveling wave tube amplifiers.The microwave power is amplitude modulated with the help of a PIN diode. The light source is a Spectra Physics 2016 Ar-ion laser pumping a SP 375 B dye laser. The excitation bandwidth is about 1 cm^{-1}. The GaP:N samples were obtained from Dr.A.Vink from the Philips Research Laboratories and were reported to contain a N concentration of 4×10^{16} cm^{-3}.

3. RESULTS AND DISCUSSION

When irradiating the GaP:N sample with the dye laser directly into the $J=1 \leftarrow J=0$ transition a weak and very broad ODMR signal is observed extending from 1 GHz to 2 GHz. This signal is obtained by amplitude modulation of the microwave power at frequencies varying between 120 Hz and 10 kHz and synchronous detection on the zero-phonon B line as well as on the strong phonon side bands. When irradiating over the band gap with the 488 nm line of the Ar-ion laser a similar behaviour is observed albeit that the fractional change in the light intensity then is almost one order of magnitude larger (10^{-2} versus 10^{-3}). An important observation is that in either case the signals are only observable at microwave powers as high as 1-5 Watt corresponding with microwave magnetic fields of the order of 0.1 mT (1 G).

Although it is very tempting to attribute the observed signals to transitions between the E and T_1 components of the $J=2$ state extreme care has to be exercised because as yet it cannot be excluded that heating of the sample by the high microwave power is responsible for the effect. However if the observed signals do correspond with this zero-field transition several problems remain. First the unusual broad resonance which points to a very large inhomogeneous broadening which may be the result of random stress in combination with Jahn-Teller type distortions of the N- site. Second the saturation behaviour which indicates a very rapid spin-spin or/and spin-lattice relaxation. Such rapid relaxation could be the result of thermal excitation to the nearby $J=1$ state as well as fast energy transfer as suggested by the hole-burning experiments of Brockelsby, Harley and Plaut (1987). A mystery remains why the signals are so much stronger when using optical excitation over the band gap.

Concluding we can say that we have no certainty yet as to the origin of the observed signals. Further investigations, including ODMR experiments at 60 GHz in the presence of a magnetic field are in preparation to elucidate this problem.

4. ACKNOWLEDGEMENT

This work is part of the research program of "de Stichting voor Fundamenteel Onderzoek der Materie (F.O.M.)" financially supported by "de Nederlandse Organisatie voor Wetenschappelijk Onderzoek (N.W.O.)".

5. REFERENCES

Brockelsby W S, Harley R T and Plaut A S 1987 Phys. Rev. B 36 7941.

Cavenett B C, Brunwin R F and Nicholls J E 1977 Solid State Comm. 23 71.

Cuthbert J D and Thomas D G 1967 Phys. Rev. 154 763

Merz J L, Faulkner R A and Dean P J 1969 Phys. Rev. 188 1228

Schmidt J 1971 Thesis University of Leiden The Netherlands

Thomas D G, Gershenzon M and Hopfield J J 1963 Phys. Rev. 131 2397

Thomas D G and Hopfield J J 1966 Phys. Rev. 150 680

Yafet Y and Thomas D G 1963 Phys. Rev. 1312 2405

Electronic structure and optical property of a D^- centre in a strong magnetic field

A Natori and H Kamimura

University of Electro-Communications, Chofu, Tokyo 182
University of Tokyo, Hongo, Tokyo 113

ABSTRACT: The new variational wave function of a D^- ion applicable in strong magnetic fields is presented. It takes account of the correlation effect in the direction of the magnetic field explicitly, and the calculated binding energy is improved appreciably. The optical absorption spectrum is calculated in both Farady and Voigt configurations, and the characteristic features are clarified.

1. INTRODUCTION

Recently the existence of a D^- ion in Ⅲ-V semiconductors has been confirmed by the measurement of the photoconductivity (Armistead, Najda and Stradling 1985) in strong magnetic fields. In Ⅲ-V semiconductors the condition of strong magnetic field can be realized easily, due to the light effective mass and the large dielectric constant. With the increase of magnetic field the orbital radii of a D^- ion decrease, and then the correlation effect between two electrons becomes important. Therefore it is interesting to investigate the electronic structure of a D^- ion and the accompanying optical absorption spectra in strong magnetic fields.

2. GROUND STATE OF A D^- ION

The effective Hamiltonian of a D^- ion in a magnetic field B applied parallel to z-axis is written as follows:

$$H(r_1, r_2) = H_1(r_1) + H_1(r_2) + \frac{e^2}{\kappa |r_1 - r_2|} , \quad (1)$$

$$H_1(r) = \frac{p^2}{2m} + \frac{1}{2} \hbar \omega_c L_z + \frac{1}{8} m \omega_c^2 (x^2 + y^2) - \frac{e^2}{\kappa r} ,$$

where m is an isotropic effective mass, κ a static dielectric constant, ω_c a cyclotron frequency defined by $\omega_c = eB/mc$, and L_z the z component of the orbital angular momentum around a donor ion. Here $-e$ is the charge of an electron, and c the light velocity. The Chandrasekhar type variational orbital function is adopted for a spin singlet ground state of a D^- ion:

$$\Psi^S(r_1, r_2) = \frac{1}{\sqrt{2}} \{\psi(r_1)\psi'(r_2) + \psi'(r_1)\psi(r_2)\}\{1 + \frac{1}{c^2}(z_1 - z_2)^2\} , \quad (2)$$

where we use the following trial functions for an inner orbital $\psi(r)$ and an outer orbital $\psi'(r)$ according to the YKA approximation:

$$\psi(r) = \{(2\pi)^{1.5} a^2 b\}^{-0.5} \exp\{-(\frac{x^2+y^2}{4a^2} + \frac{z^2}{4b^2})\} , \quad (3)$$

$$\psi'(r) = \{(2\pi)^{1.5} a'^2 b'\}^{-0.5} \exp\{-(\frac{x^2+y^2}{4a'^2} + \frac{z^2}{4b'^2})\} .$$

The last factor in eq.(2) describes the correlation effect between two electrons in the direction of z-axis. It was not contained in our previous calculation (Natori and Kamimura 1978), in which the correlation effect was considered only through the difference between the radii of two orbitals. In the strong magnetic field the longitudinal radius b' of the outer orbital becomes much larger than b of the inner orbital, while both transverse radii, a and a', are reduced to the cyclotron radius and are much smaller than b'. Therefore the correlation effect becomes enhanced in the direction of z-axis in the strong magnetic field, and this is taken into account explicitly in the present calculation. The energy expectation value $E^S(B) = \langle\Psi^S|H|\Psi^S\rangle/\langle\Psi^S|\Psi^S\rangle$ can be calculated analytically. The condition which minimize $E^S(B)$ with respect to five variational parameters a, b, a', b' and c yields the ground state of a D^- ion. The binding energy of a D^- ion in a magnetic field B is given as

$$\varepsilon^S(B) = \frac{\hbar\omega_c}{2} + E^0(B) - E^S(B) , \qquad (4)$$

where $E^0(B)$ is the ground state energy of a neutral donor left behind. $E^0(B)$ is obtained by the variational method with a wave function $\psi^0(r)$ of the similar form to eq.(3). The first term of eq.(4) is the energy of the ground Landau level.

The results of numerical calculation are shown in Figures 1 and 2. Here, $Ry = me^4/2\hbar^2\kappa^2$ and $a_B = \hbar^2\kappa/me^2$ are used as units of the energy and length, respectively. The strength of the magnetic field B is expressed with use of the dimensionless parameter $\gamma = \hbar\omega_c/2Ry$. In Figure 1 $\varepsilon^S(B)$ is shown as a function of γ in the range from 0.1 to 10.0. The prominent magnetic freeze-out effect is observed already in weak field regime. This is caused by the strong shrinkage effect of the outer orbital in the transverse direction in magnetic fields. Our variational wave function is

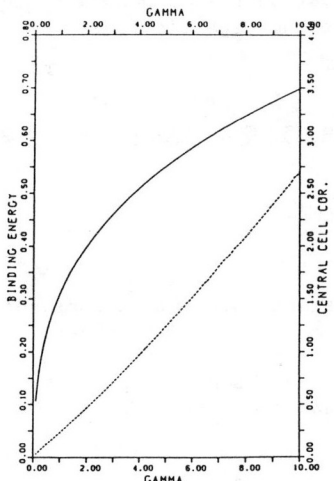 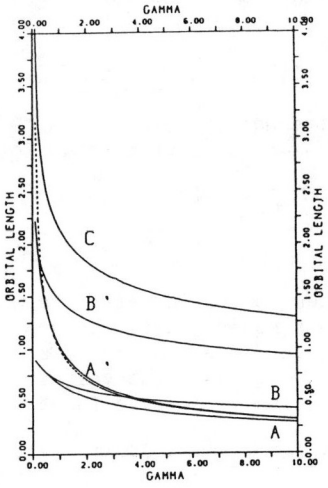

Fig.1 Binding energy (solid line) of a D^- ion in a unit of Ry, and the central cell correction (broken line) in a unit of Δ_0 as a function of γ.
Fig.2 Orbital radii a, b, a' and b', a correlation parameter c and the cyclotron radius (broken line) in a unit of a_B as a function of γ.

not appropriate in weak magnetic fields, especially for an inner orbital. However it is expected that the induced errors for $E^0(B)$ and $E^S(B)$ are partly cancelled for the binding energy $\varepsilon^S(B)$. In comparison with our earlier calculation, the binding energy is improved by 0.048Ry at $\gamma = 10$ by the introduction of the correlation effect c. In Figure 2, the obtained optimized values of the five parameters are shown together with the cyclotron radius $a_B/\sqrt{\gamma}$. Both transverse radii a and a' decrease monotonically with increase of γ and approach the cyclotron radius in high fields, while both longitudinal radii b and b' decrease with slower rate. The caused anisotropy becomes much larger for the outer orbital than inner one, because of the weak effect of Coulomb potential of a donor ion. The z-axis correlation parameter c also decreases with the increase of γ, and the effect becomes larger in higher fields. For the sake of the correlation effect c, the longitudinal radius b' of the outer orbital becomes shrinked by about 76% of that obtained without c at $\gamma = 10$, while the other three radii a, b and a' are little affected. This shrinkage of b' is the origin of the increase of the binding energy.

Finally the effect of the central cell correction is remarked. For the ground state of a neutral donor in III-V semiconductors, the central cell correction is small because of the large extension of the orbital. Therefore the effect of the central cell correction can be estimated by the perturbation method. For simplicity, the following form is assumed for the correction Hamiltonian of the central cell.

$$H_c(r) = V_0 a_B^3 \delta(r) \qquad (5)$$

The correction energy $\Delta\varepsilon^S$ for the binding energy of a D^- ion can be calculated from eq.(4) in the first order perturbation theory as follows.

$$\Delta\varepsilon^S = <\psi^0|H_c(r)|\psi^0> - \frac{<\psi^S|H_c(r_1) + H_c(r_2)|\psi^S>}{<\psi^S|\psi^S>} \qquad (6)$$

For the exact wave function of the ground state of a neutral donor in zero field, the central cell correction Δ_0 becomes V_0/π. In Figure 1, $\Delta\varepsilon^S/|\Delta_0|$ is also shown as a function of γ. The sign of V_0 is considered to be negative and then $\Delta\varepsilon^S$ has a positive sign. The value of $\Delta\varepsilon^S$ increases monotonically with increasing γ, since the inner orbital of a D^- ion is shrinked more largely than that of a neutral donor by the repulsion effect of the electron in the outer orbital. The central cell correction is enhanced for a D^- ion in high fields.

3. OPTICAL ABSORPTION SPECTRUM

The optical absorption spectrum of a D^- ion is proportional to the following spectral function $I(\omega)$.

$$I(\omega) = \sum_f \hbar\omega \, |M_{if}|^2 \, \delta(E_f - E_i - \hbar\omega), \qquad (7)$$

where M_{if} is the dipole transition matrix element between the initial ground state $\psi^S(r_1, r_2)$ and the final one electron excited state $\Psi_f(r_1, r_2)$.

$$M_{if} = \frac{<\psi^S(r_1,r_2)|e \cdot (r_1 + r_2)|\Psi_f(r_1,r_2)>}{\sqrt{<\psi^S|\psi^S>}}, \qquad (8)$$

where e is the polarization vector of the light. In the final state one electron of a D^- ion is excited to a conduction band state, while another electron is left behind in the ground state of a neutral donor. The conduction band state is specified by the Landau quantum number N, the

azimuthal angular momentum M and the wave vector k along the magnetic field. The wave function of the conduction band state with a set of quantum numbers N, M and k is written with the cylindrical coordinate (ρ,ϕ,z) as

$$\phi_{1Mk}(r) = \frac{1}{\sqrt{L_z}} e^{ikz} \frac{1}{\sqrt{2\pi}} e^{iM\phi} \sqrt{\frac{\gamma l!}{(1+|M|)!}} \frac{1}{a_B} \sigma^{\frac{1}{2}|M|} e^{-\frac{1}{2}\sigma} L_1^{|M|}(\sigma) , \quad (9)$$

where $\sigma = \gamma\rho^2/2a_B^2$, and L_1^M is the Laguerre polynomial. The Landau quantum number N is related to l and M by the relation,

$$N = 1 + \frac{(M + |M|)}{2} . \quad (10)$$

For the dipole transition the total spin is conserved, and only the spin singlet excited states contribute to the absorption spectrum. The final state of the spin singlet state specfied by three quantum numbers (l,M,k) is expressed as

$$\Psi_{1Mk} = \frac{1}{\sqrt{2}} \{\psi^0(r_1)\phi_{1Mk}(r_2) + \psi^0(r_2)\phi_{1Mk}(r_1)\} . \quad (11)$$

The corresponding energy E_{1Mk} can be written as

$$E_{1Mk} = \frac{\hbar^2 k^2}{2m} + \hbar\omega_c(\frac{1}{2} + N) + E^0(B) . \quad (12)$$

With use of these wave functions, we can calculate the transition matrix element analytically similarly to the previous calculation (Natori and Kamimura 1979). The selection rule for the transition is obtained with respect to the azimuthal angular momentum M. That is, the value of M is conserved for e_z polarization parallel to the magnetic field, and is changed by ± 1 for $e_\pm$ polarizations. Here e_+ and e_- mean the clockwise and counterclockwise circular polarizations, respectively. They are related to the linear polarizations e_x and e_y, by $e_\pm=(e_x\pm ie_y)/\sqrt{2}$. Secondly the k-selection rule is remarked. For e_z polarization the transition matrix element is proportional to k near the edge of each Landau band, due to the inversion symmetry about z-axis. On the other hand for e_x and e_y polarizations, the matrix elements remain finite at k = 0. In the neighbourhood of the edge of each Landau band, the density of states of the final states diverges as 1/k. This divergence remains in the

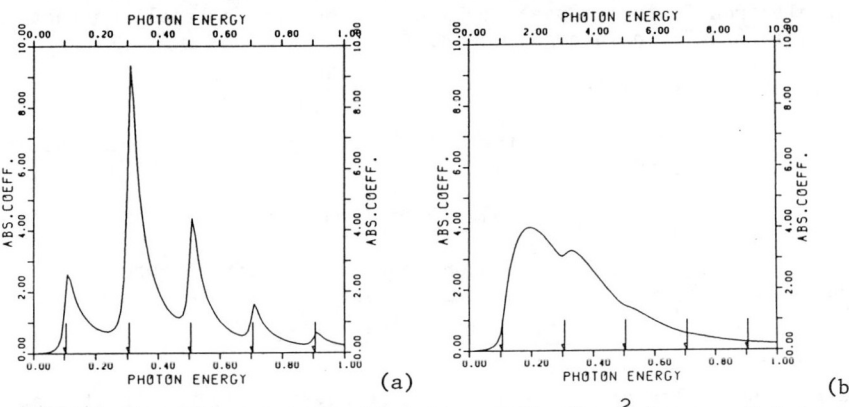

Fig.3 Optical absorption spectra $I(\omega)$ in a unit of a_B^2 for γ = 0.1 in two polarizations e_x(a) and e_z(b). The threshold energies of the transitions to N = 0, 1, 2, 3 and 4 Landau levels are shown by the arrows.

Shallow Impurities in III–V Compounds

absorption spectrum for e_x and e_y polarizations, while vanishes for e_z polarization due to k-selection rule. Really the weak divergence can be suppressed easily by the broadening effect. In this calculation the broadening effect is treated only phenomenologically, and the following conventional form is adopted for the spectral function $I(\omega)$.

$$I(\omega) = \sum_f \hbar\omega \, |M_{if}|^2 \, \frac{1}{\pi} \, \frac{\Delta}{(E_f - E_i - \hbar\omega)^2 + \Delta^2} \qquad (13)$$

This form corresponds to the life time broadening of relevant energy levels, and it agrees with eq.(7) in the zero limit of Δ.

In the first place, the optical absorption spectrum $I(\omega)$ is shown as a function of photon energy ω for a fixed value of γ. In Figure 3, γ is taken to be 0.1 and the broadening parameter Δ is assumed to be 0.01Ry. The absorption spectra are presented for two linear polarizations e_x and e_z. Each peak in e_x polarization shows a transition to each Landau level of $N = 0, 1, 2, 3$ and 4, and the shape reflects the divergent form of the density of states at each Landau level shown by the arrows. On the other hand for e_z polarization, the peaked structure becomes more fatigued, and the position of each peak is shifted considerably to higher energy side compared to each Landau level. This is because of the k-selection rule mentioned above. As for the intensity of each peak, it is sensitive to the strength γ of a magnetic field. The dipole transition matrix element decreases monotonically as l increases, and the rate of decrease accelerates with the increase of γ. In the high field limit, only the transition to the lowest l state l = 0 survives.

In the next place, we calculate the optical absorption spectrum $I(\gamma)$ for a fixed value of photon energy ω. This case just corresponds to the experimental situation of the measurement of photoconductivity (Armistead, Najda and Stradling 1985, Najda, Natori, Kamimura, Maan and Stradling 1988). In Figure 4, $I(\gamma)$ is shown for $\omega = 0.6$Ry with Δ of 0.02Ry in two polarizations e_x and e_z. $I(\gamma)$ is not shown in the region of γ smaller than 0.05, because our variational function can not be applicable in weak field regime. The characteristic saw-tooth shape is seen for e_x

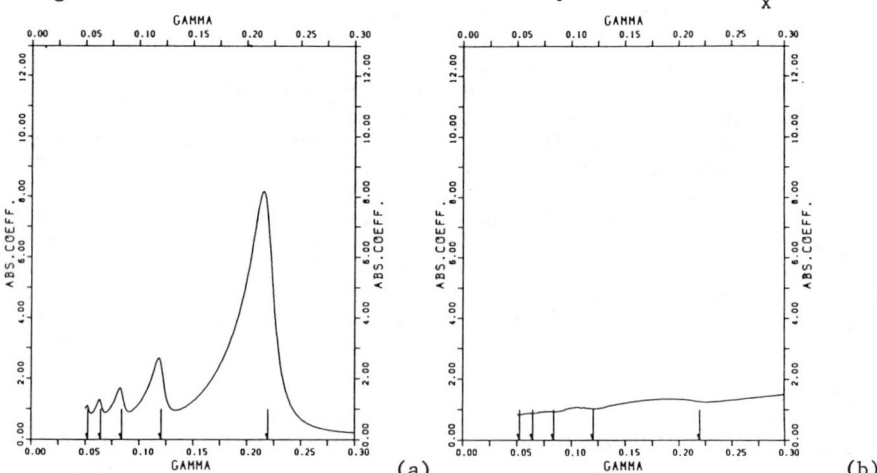

Fig.4 Optical absorption spectra $I(\gamma)$ in a unit of a_B^2 for $\omega = 0.6$Ry in two polarizations, e_x(a) and e_z(b). The threshold values γ of the transitions to N = 1, 2, 3, 4 and 5 Landau levels are shown by the arrows.

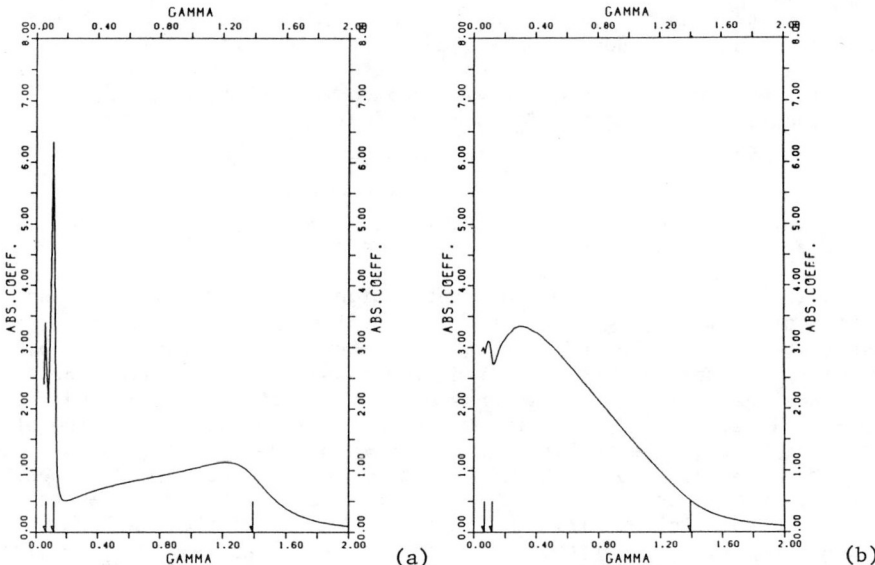

Fig.5 Optical absorption spectra $I(\gamma)$ in a unit of a_B^2 for $\omega = 0.35$Ry in two polarizations, e_x(a) and e_z(b). The threshold values γ of the transitions to $N = 0$, 1 and 2 Landau levels are shown by the arrows.

polarization and each peak corresponds to the transition to the Landau levels of $N = 1, 2, 3, 4$ and 5, respectively, from high value of γ. Each peak position is shifted a little to a lower value of γ, compared to the position of the transition to the edge of each Landau band shown by the arrow. This is due to the broadening effect, and it becomes more prominent for larger value of Δ. For e_z polarization, the saw-tooth character disappears and only the broad spectrum is seen. In Figure 5 ω is chosen to be 0.35Ry with the same value of Δ, and now the transitions to $N = 0$, 1 and 2 are seen in the range of γ greater than 0.05. For either polarizations, the transition to $N = 0$ Landau level shows broad peaked structure corresponding to the wide transition range of k. The intensity in e_x polarization becomes smaller than that in e_z polarization, because only the e_- component of e_x polarization can contribute to the transition to $N = \bar{0}$ Landau level. On the other hand, both components e_- and e_+ can contribute to the transitions to higher Landau levels. Finally it should be remarked that the slight hump structure is seen in lower γ tail of $N = 0$ transition in e_x polarization, unlike those of transitions to higher Landau levels in Figure 4. This is thought due to the nonlinear dependence of the threshold energy of the $N = 0$ transition in the present range of γ as seen in Figure 1. On the other hand, the threshold energy of the transition to the higher Landau level N changes approximately linearly with γ due to the extra term $N\hbar\omega_c$ of the excitation energy.

REFERENCES

Armistead C J, Najda S P and Stradling R A 1985 Solid State Comm. 53 1109
Najda S P, Natori A, Kamimura H, Maan J C and Stradling R A 1988 the following paper on this proceedings
Natori A and Kamimura H 1978 J.Phys.Soc.Jpn. 44 1216
Natori A and Kamimura H 1979 J.Phys.Soc.Jpn. 47 1550

Spectroscopic study of D⁻ state transitions in GaAs and InSb

S.P.Najda [1], A.Natori [2], H.Kamimura [3], J.C.Maan [1] and R.A.Stradling [4].

1 SNCI-CNRS/MPI-FKP, 25 ave des Martyrs, BP166X Grenoble, France.
2 Department of Physics, University of Electro-communications, Chofu, Tokyo, Japan.
3 Department of Physics, University of Tokyo, Bunkyo-ku, Tokyo 113, Japan.
4 Blackett Laboratory, Imperial College of Science, Prince Consort Road, London, U.K.

D⁻ state transitions in GaAs and InSb are interpreted by a lineshape calculation and compared with theory of the D⁻ ion binding energy up to the high field limit. Central-cell shifts and non-parabolic effects significantly influence the D⁻ ion binding energy for GaAs and InSb respectively. Impurity clusters or complexes are observed when the temperature is reduced from 4K to 2K for GaAs. D⁻ triplet state transitions are tentatively identified in InSb.

1. INTRODUCTION

A D⁻ state is a shallow donor impurity that captures an extra electron and can be regarded as the solid state analogy of the hydrogen ion, i.e. H⁻ ion. D⁻ states have been of considerable interest for many years. However, the experimental evidence is principally restricted to the elemental semiconductors, Si and Ge, where the multivalley, indirect band complicates the analysis of the data. With these materials a magnetic field has relatively little effect on the D⁻ ion wavefunction, therefore experiments are limited to the low field range, i.e. $\gamma \ll 1$, where $\gamma = \hbar\omega_c/2R_y^*$. Thus it is difficult to distinguish transitions involving D⁻ states from a multitude of interexcited D⁰ state transitions, i.e. n→m where the principal quantum numbers n and m are greater than 1, since the experimental conditions required to populate either state are very similar.

D⁻ state transitions have recently been observed in high purity, low compensation GaAs and InP using high sensitivity far-infrared photoconductivity (FIRPC) techniques up to 24T

(Armistead et al (1985), Najda et al (1988) hereafter referred to as ref.1). A magnetic field has a much greater influence on the D⁻ wavefunction in these direct gap materials, and the intermediate magnetic field regime can be reached, i.e $\gamma \sim 1$. The D⁻ state to lowest Landau level transition is observed as a broad, very distinctive lineshape, rendering an accurate determination of the D⁻ ion binding energy difficult. The photoionisation transition is interpreted by means of a lineshape calculation. A significant central-cell shift of the D⁻ peak transition is observed with crystals grown by different growth techniques. As the temperature is reduced from 4K to 2K impurity cluster effects are observed.

D⁻ state transitions are observed in InSb, thus enabling experiment to be compared against a high field calculation of the D⁻ binding energy. Correlation and non-parabolic effects are important. Transitions from D⁻ triplet states are tentatively observed.

2. RESULTS AND DISCUSSION FOR GaAs

High sensitivity FIRPC techniques are employed to study D⁻ state transitions at magnetic fields up to 24T. D⁻ states are created by a dynamic process that involves the photo or thermal generation of carriers into the conduction band followed by recombination with neutral donor impurities. In this respect the electric field applied to the sample plays an important role in the recombination process since it effectively governs the probability of electron capture by a D⁰ state. A large D⁻ state population can be achieved if the experimental conditions are chosen carefully, i.e. low temperature (<4K), band-gap radiation, moderate electric field bias (Ref.1).

Figure 1 shows the FIRPC spectrum of a VPE GaAs sample orientated in the Faraday geometry, irradiated with 570µm and band gap radiation at 4K (top trace). The B peaks are interexcited D⁰ state transitions observed by the 'photothermal ionisation' mechanism (Lifshits and Kogan, 1977), the cyclotron resonance (CR) is observed by the 'cross-modulation' process.

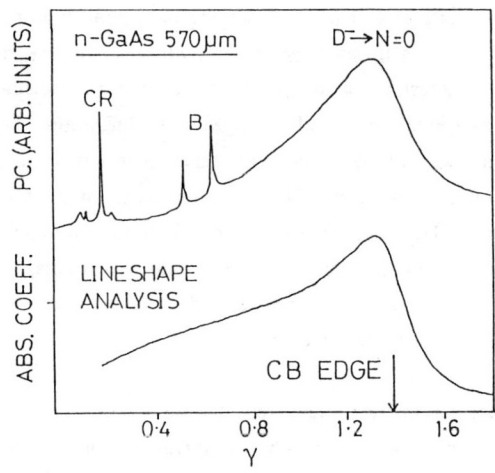

The photoionisation D⁻ state to conduction band transition is observed as a very broad peak at ~8T (γ~1.2) with a very long 'tail' to low field. The 'tail' is a consequence of the D⁻ binding energy changing slowly with field. It is difficult to accurately determine the D⁻ ion binding energy from the lineshape, since the peak value would give a better estimate if a transition from a 'D⁻ band' is mainly responsible for the broadening, compared to a half intensity point or 'edge' if the broadening is due to a photoionisation transition into the conduction band. A lineshape calculation of the photoionisation transition is performed by Natori and Kamimura (1988) using a Lorentzian broadening parameter for both the initial and final states. The lineshape calculation (bottom of fig.1) is compared to the experimentally observed D⁻ state to Landau level transition. A striking similarity between experiment and calculated lineshape is achieved. In the lineshape calculation, the conduction band edge is offset from the peak position to higher field suggesting that the conduction band may introduce significant broadening.

Figure 2 is a display for the same sample subject to similar experimental conditions but at much lower fields and irradiated with 393μm radiation. The characteristic 'sawtooth' lineshape of transitions from the D⁻ state to higher Landau levels is well reproduced by the lineshape analysis. The A_2 line is an interexcited state transition.

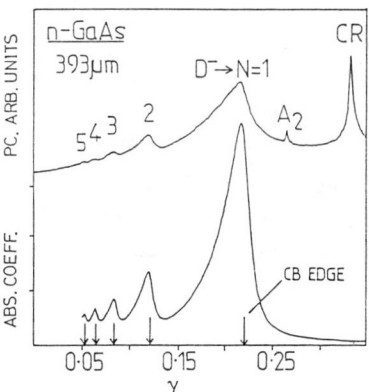

Figure 3 is a diagram of the D⁻ ion binding energy calculated for two different magnetic field regimes versus γ, the dimensionless field parameter. The low field calculation (dashed line) is performed by Larsen (1979), the high field (solid line) by Natori and Kamimura (1988).

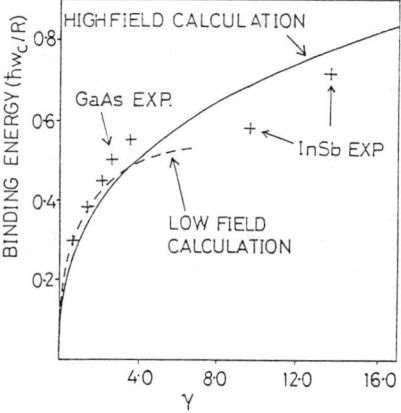

The D⁻ ion binding energy is compared to a low field calculation by Larsen (1979) in figure 3. The GaAs results coincide rather well with the low field calculation if the high field edge of the D⁻ state transition is taken rather than the peak for values of $\gamma \sim 1$. However, for higher values of γ there is an increasing deviation between theory and experiment.

A central-cell shift of the D⁻ transition is observed (fig.4) between samples grown by different techniques, where each growth system introduces a characteristic 'signature' of residual impurities (Armistead et al, 1984). The top trace is the FIRPC response of a VPE grown crystal with the deepest shallow donor (X_3) most abundant. In comparison a LPE crystal (bottom) introduces more shallow residual impurities during growth.

Figure 4 compares the FIRPC spectra of two GaAs samples grown by different methods. The D⁻ state transition for the VPE grown crystals is at lower field compared to the LPE material, illustrating the central-cell shift. The CR and B lines (effectively independent of central-cell shifts) are used as reference.

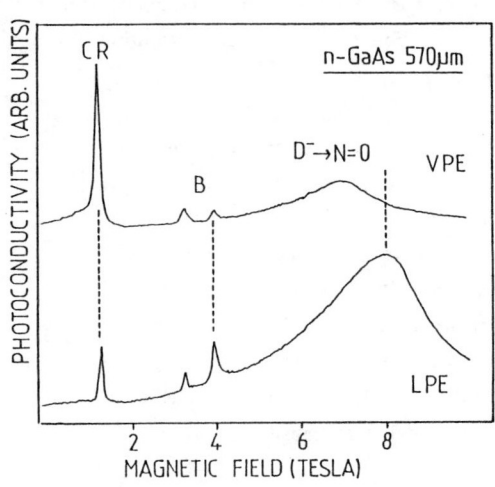

The observed lineshape represents a convolution of transitions originating from each individual residual donor, thus the D⁻ to conduction band transition is shifted to lower field if deep donors are more predominant. An estimate of the central-cell shift has been made by Natori and Kamimura (1988). The D⁻ ion binding energy at zero field is small ($0.055R^*_y$) and central-cell shifts are negligible, since the extension of the outer orbital is approximately four times that of a neutral donor (chemical shifts comprise about 1% of the ground state energy of a neutral donor in GaAs). However, as the field increases central-cell effects become more important because the inner orbital is compressed more than for a neutral donor due to the repulsion of the outer electron. The magnitude of the D⁻ central-cell splitting becomes comparable to a neutral donor at values of $\gamma \sim 4$. Central-cell effects may explain some of the discrepancy between low field theory and experiment at values of $\gamma > 1$,

since the calculation does not include central-cell effects. However, the divergence is mainly due an inappropriate choice of wavefunction at higher field. The observed lineshape appears to be due to photoionisation transitions from a central-cell split D⁻ level into the conduction band. The impurity 'signature' of residual impurities appears to have a significant influence on the lineshape.

As the temperature is lowered from 4K to 2K an additional broad peak is observed at low field (fig.5). This feature is believed to be due to a complex of the D⁻ state with neighbouring impurities. The geometry of these complexes or impurity clusters is unclear.

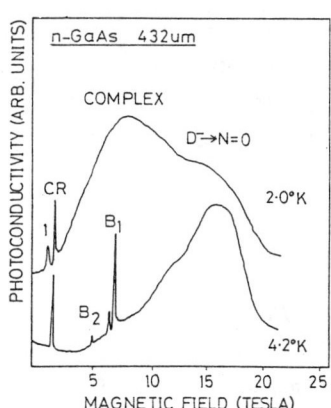

Figure 5 A large broad transition is observable at ~4T at low temperature 2K. The broad peak is due to a negatively ionised complex or cluster.

3. RESULTS AND DISCUSSION FOR InSb

D⁻ state transitions are studied in a low compensation, high purity InSb sample using millimeter wave IMPATT diodes. The advantage of studying D⁻ states in InSb is that the high magnetic field regime can be achieved, i.e.$\gamma \gg 1$, since the effective mass is small ($m^* = 0.0139m$). However the conduction band is strongly non-parabolic and therefore the lineshape obtained from a photoionisation transition is expected to be significantly different from that of GaAs. Figure 5 shows the photoconductivity spectra of the InSb sample irradiated with 110GHz (top) and 88GHz (bottom). The lower field peak is a transition from a D⁻ state into the conduction band. The broad lineshape is different compared to GaAs reflecting the InSb non-parabolic band. The peak position is plotted in fig.2 against a high field calculation performed by Natori and Kamimura 1988 since it is difficult to determine a high field edge. This calculation is expected to give an more accurate estimation of the D⁻ ion binding energy in the high field limit since a more suitable variational wavefunction is chosen and correlation effects are included in the calculation. However, the wavefunction has

the disadvantage that it does not become 'hydrogen-ion like' in the limit B->0 and is therefore not expected to be accurate in the low field range. The calculation underestimates the D⁻ ion binding energy in InSb by ~10% since non-parabolic effects are neglected. Central-cell effects are also expected to significantly influence the lineshape and hence the binding energy. At higher fields an additional peak is observed. This peak is tentatively assigned to a transition from a D⁻ triplet state to the conduction band.

Figure 6 is a display of the photoconductivity response of InSb for two different frequencies.

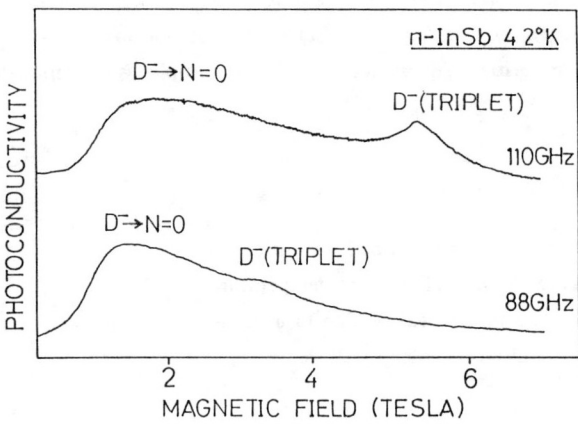

4. CONCLUSION

A lineshape analysis of the D⁻ state to conduction band photoionisation transition reveals that the final state may contribute significant broadening. In addition, the lineshape is sensitive to the residual impurity contaminants that are inadvertently introduced during growth and the chemical shifts give rise to a significant perturbation of the D⁻ ion binding energy at high fields. For the case of InSb, nonparabolic effects increase the D⁻ ion binding energy and influence the lineshape.

Armistead C.J, Knowles P, Najda S.P and Stradling R.A 1984 J.Phys.C: Solid State Phys., 17, p6415.
Armistead C.J, Najda S.P, Stradling R.A and Maan J.C. 1985 Solid State Commun., Vol.53, No.12, p1109.
Larsen D.M 1979 Phys.Rev.B, Vol2, No.12, p3904.
Lifshits T.M and Kogan M 1977 Phys. Stat. Sol(a), 39, p11.
Najda S.P, Armistead C.J, Trager C and Stradling R.A 1988 Semicon.Science.Tech, to be published.
Natori A and Kamimura H 1988 this conference.

Line shape dependence on temperature for photothermal ionization of donors in GaAs

S D Baranovskii, B L Gel'mont, V G Golubev, V I Ivanov-Omskii and A V Osutin

A F Ioffe Physical-Technical Institute, USSR Academy of Sciences, 194021 Leningrad, USSR

ABSTRACT: It is shown that the donor and acceptor concentrations in GaAs can be derived separately from the temperature dependence of the shallow donor $1s-2p_0$ photoexcitation line shape.

1. INTRODUCTION

During the two recent decades, photoelectric spectroscopy (PES) has been applied widely to study the chemical composition of shallow impurities in semiconductors (Kogan and Lifshits 1977). The method consists in investigating the spectrum of photoconductivity (PC) corresponding to the resonance photoexcitation (PE) of shallow impurities. An important asset of the method is its high sensitivity permitting one to obtain PE spectra down to impurity concentrations $\sim 10^5 cm^{-3}$. PES in a magnetic field is best suited to studies of A_3B_5 compounds (Armistead et al 1984). Application of magnetic field results in a considerable narrowing of the spectral lines, thus permitting resolution in ultrapure materials of donor PE lines whose binding energies differ by less than 10 μeV (Golubev et al 1987b). The studies of donor PE spectra were focused, in particular, on the line shape, it being connected with the concentration and degree of compensation of impurities. The dependence of the line shape on the impurity compensation and concentration was investigated both experimentally (Korn and Larsen 1973; Berman and Kogan 1987) and theoretically (Kalfa and Kogan 1973; Larsen 1973, 1976; Kogan and Lien 1981). However no satisfactory agreement of numerical results with theory has yet been reported.

Baranovskii et al (1987) reported on the possibility of determinning the concentration and compensation of impurities in GaAs from studies of the temperature dependence of the $1s-2p_0$ donor PE line shape and a comparison of the experimental data with results of computer simulations. The present publication analyzes in a comprehensive way the experimental data on the donor PE line shape and a method of separate determination of the donor and acceptor concentrations.

© 1989 IOP Publishing Ltd

2. EXPERIMENTAL DETAILS

The experimental line shape measurements of the donor PE were performed on a high-resolutions ($\sim 0.1 \mu$eV) submillimeter - range laser magnetic spectrometer by the technique described by Golubev et al (1986). CH_3OH and CH_3OD lasers were used as light sources. During PE spectrum recording, the samples were screened from interband illumination. The temperature was varied within $2K \leqslant T \leqslant 13K$. We studied VPE-grown high purity GaAs layers whose $1s-2p_0$ and $1s-2p_{-1}$ photoexcitation spectra revealed clearly resolved lines of chemically dissimilar donors.

3. THEORETICAL BACKGROUND

The donor PE line broadening in GaAs originates from the presence of electric fields and their gradients arising because of the effect of charged donors and acceptors on the energy states of neutral donors (Larsen 1973). The energy position and shape of a spectral line originating from an intraimpurity transition from state j to state k is described by a statistical distribution of transition energies

$$\mathcal{E}_{kj} = \mathcal{E}_k - \mathcal{E}_j - \frac{1}{4}(Q_k - Q_j)\frac{\partial E_z}{\partial z} - e^2[(C_\perp^{(k)} - C_\perp^{(j)})E_\perp^2 + (C_z^{(k)} - C_z^{(j)})E_z^2] \quad (1)$$

where $\mathcal{E}_{kj}$ are the states of an isolated donor, E_z, $E_\perp$ are electric field projections at the neutral donor site on the magnetic field z and perpendicular to it, respectively, $C_{z(\perp)}^{k(j)}$ are the corresponding polarizabilities, $Q_k(j)$ are the quadrupole moments, $Q_k(j) = e\langle \varphi_{k(j)}|(3z^2-r^2)|\varphi_{k(j)}\rangle$, where $\varphi_{k(j)}$ are the wave functions of the states, e is the electronic charge. The quantities $\mathcal{E}_{k(j)}$, $Q_k(j)$, $C_{z(\perp)}^{k(j)}$ are functions of the applied magnetic field.

Thus the line shape depends on the distribution of charged relative to neutral impurities in a crystal. The neutral and charged impurities were usually considered to be randomly distributed one with respect to another (Kal'fa and Kogan 1971; Larsen 1973, 1976). However at temperatures below the characteristic level spread in the impurity band (which at not too large compensations is on the order of the Coulomb interaction energy of the charges for a mean distance between the impurities $e^2 N_D^{1/3}/\mathcal{x}$, N_D is the donor concentration, $\mathcal{x}$ is the dielectric constant) there should be a correlation between the distributions of the neutral and charged donors. Such temperatures ($T_0 = e^2 N_D^{1/3}/\mathcal{x} k_B \simeq 6K$ for $N_D = 10^{14} cm^{-3}$) were reached in the experiment. The effect of electron correlation on line broadening was studied by Kal'fa and Kogan (1973) in the limiting case of very strong compensation at H=0 and by Kogan and Lien (1981) in the case of arbitrary compensation for $\gamma = \hbar\omega_c/2R\dot{y}$ ($\hbar\omega_c$-cyclotron energy, $R\dot{y}$-effective Rydberg). A considerable temperature broadening of the transitions between excited donor states in GaAs and CdTe abserved by Golka et al (1977)was accounted for by a transition from a correlated to random electron distribution over the impurities.

4. RESULTS AND DISCUSSION

We will deal here primarily with the $1s-2p_0$ donor PE line shape. The $1s-2p_0$ line shape at low temperatures is close to symmetrical (see Figure 1a). This suggests (Kogan and Lien 1981) that the major reason for the broadening lies in the interaction of the donor quadrupole moments with electric field gradients. For a more correct calculation one should include also the quadratic Stark effect (the last term in (1)) resulting in a line asymmetry. The degree of asymmetry of the experimentally observed $1s-2p_0$ donor PE lines, i.e. the difference between the linewidth at half maximum on the left and right of the maximum at low temperature (T ≤ 4K) was not more than 20% which is much less than that for the $1s-2p_{-1}$ lines (see Figure 1b) whose broadening comes primarily from the quadratic Stark effect. Thus with only the quadrupole broadening included the shape of the $1s-2p_0$ line is described by the statistical distribution of the quantities

$$\Delta_{k,j} = \frac{1}{4}(Q_k - Q_j) \frac{\partial E_z}{\partial z} \qquad (2)$$

over the neutral donors, i.e. by the statistical distribution of the component $V_{zz} = \partial E_z / \partial z$ of a tensor constructed of the second derivatives of the potential $W(V_{zz})$ produced by charged impurities at the neutral donors. The dependence of the $1s-2p_0$ linewidth on magnetic field is described by a similar relation $Q_{1s}-Q_{2p_0}$. Figure 2 compares the experimentally measured relative variation of the halfwidth Δ_0 for sample 2 (see Figure 3) and the $Q_{1s}-Q_{2p_0}$ calculated by Larsen (1973). The satisfactory agreement between experiment and calculation leads additional support to the quadrupole broadening dominating in the case of the $1s-2p_0$ transition. In comparing

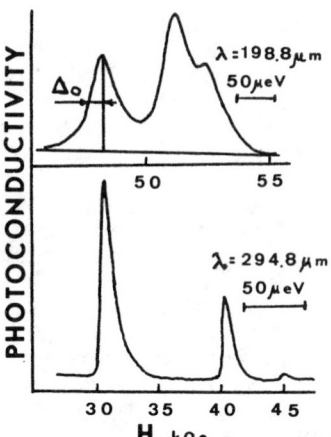

Figure 1

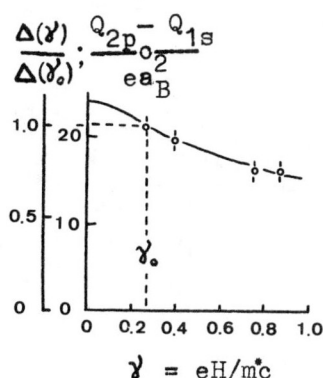

Figure 2

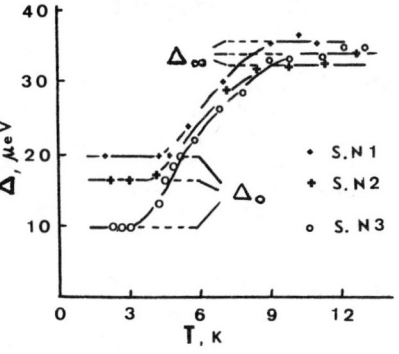

Figure 3

experimental and theoretical data we will use here the width (Δ) at half-maximum of the left-hand wing of the 1s-2p$_0$ line (see Figure 1a) since the contribution of the broadening due to the quadratic Stark effect is the smallest for it. We point out that in superpure GaAs ($N_D+N_A \lesssim 10^{13}$cm^{-3}) in a magnetic field $\gamma \lesssim 1$ linewidth of 1s-2p$_0$ transitions is very sharp ($\lesssim 10\,\mu$eV). Thus for line shape analysis the splitting of the lines associated with the difference between g factors of ground and excited impurity levels should be taken into account. The splitting is about 3 μeV (Golubev et al 1987a).

Figure 3 displays a Δ(T) dependence for three samples. For T $\lesssim$ 4K the halfwidth (Δ_0)- is minimal and constant. It grows with increasing T reaching again a constant value (Δ_∞). The ratio Δ_∞/Δ_0 is different for the samples studied. The two shelves in the temperature dependence correspond to the two types of electron distribution over the impurities, a correlated and an uncorrelated one.

For a correlated distribution over the impurities (in the case where the degree of compensation $K=N_A/N_D \ll 1$) only the nearest-neighbor donors at the acceptors are ionized. Nearly all the charges are combined in dipoles with an arm on the order of the mean donor separation. The electric field E created by them at the neutral donors is lower by a factor $K^{1/3}$ (Kogan and Lien 1981) than that which would be produced by a random charge distribution. Correlation was shown to reduce the characteristic fields at the neutral impurities by 4-5 times also at K=0.5-0.7 (Kogan et al 1980). Thus a temperature-induced breakdown of correlation results in an increase of the gradients V$_{zz}$ (approximately proportional to E) and an increase of the quadratic Stark effect (proportional to E^2) which affect the donor PE linewidth. The relative contribution of the quadratic Stark effect increases with T compared with the quadrupole term. This leads to an increased asymmetry of the 1s-2p$_0$ lines observed experimentally. For low temperatures T $\ll$ T$_0$ where the distribution function W(V$_{zz}$) is dominated by correlation the 1s-2p$_0$ line shape was studied theoretically by Baranovskii et al (1987). For the case of small K an analytical calculation was done within the dipole model (Kogan et al 1980). The linewidth turned out to be proportional to $N_D K^{4/3}$.

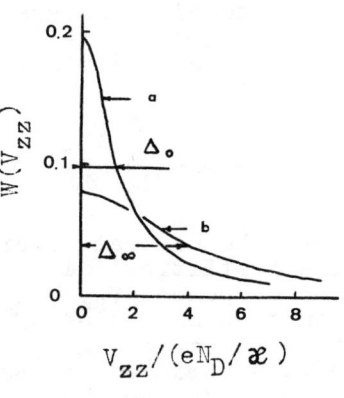

Figure 4

Figure 4 presents W(V$_{zz}$) obtained by computer simulation (curve a). A comparison of the analytical calculation with the computer simulation data suggests that the dipole model describes fairly well the W(V$_{zz}$) distribution for K$\lesssim$0.5. Curve b in Figure 4 corresponds to an analytical calculation for the model of a random charge distribution. This model was

analyzed by Kal'fa and Kogan (1971) and Larsen (1973) who obtained the following expression for the $W(V_{zz})$

$$W_{zz} = \frac{1}{\pi} \frac{\Gamma}{V_{zz}^2 + \Gamma^2}, \qquad \Gamma = 10.13 \frac{eN_D}{\mathscr{æ}} K \qquad (3)$$

Theoretical calculations valid for the two limiting cases suggest a possibility of deriving the degree of compensation from the ratio Δ_∞/Δ_0. The results of this analysis are displayed in Figure 5 presenting the relation $K(\Delta_\infty/\Delta_0)$. In the limiting case of small compensations ($K \ll 1$) these results are easy to explain. For a random distribution of electrons over the impurities, the $1s$–$2p_0$ linewidth depends only on the charged impurity concentration, $\Delta_0 \sim N_D \cdot K = N_A$ (see (3)). For a correlated distribution $\Delta_0 \sim N_D K^{4/3} = N_A K^{1/3}$. Hence $\Delta_\infty/\Delta_0 \sim K^{-1/3}$. For $K \leq 0.4$ this relationship coincides with the one shown in Figure 5. Δ_∞/Δ_0 reaches a maximal value for $K \to 0$. For $K \sim 0.9$, $\Delta_\infty/\Delta_0 \sim 1$. The effect of the correlation decreases with increasing compensation.

The $\Delta(T)$ relationships presented in Figure 3 were used to determine the degrees of compensation for the three samples studied (for sample 1, $K=0.8$; for sample 2, $K=0.77$, and for sample 3, $K=0.23$).

Having derived the degree of compensation, one can find the donor concentration N_D (and, hence, N_A) by comparing the measured halfwidth Δ_0 for $T \ll T_0$ with the results of a computer simulation run for the given degree of compensation, i.e. with the halfwidth of the theoretical line shape $W(V_{zz})$. Figure 6 presents the dependence of the halfwidth (Δ_0 theor) of the $W(V_{zz})$ distribution in units of $V_{zz}/(eN_D/\mathscr{æ})$ on the degree of compensation K.

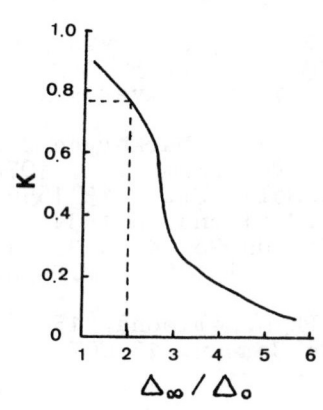

Figure 5

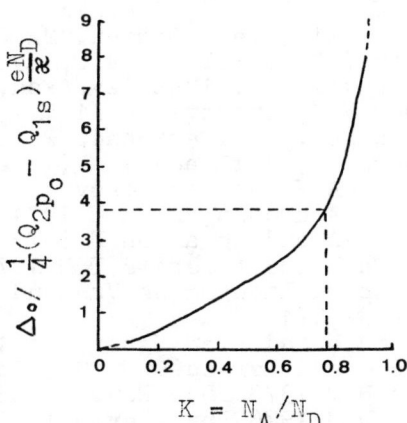

Figure 6

Thus in order to find N_D (as seen from (2)) one should substitute Δ_o theor derived from graph 6 into the expression

$$\Delta_o \text{ exp} = \Delta_o \text{ theor} \frac{eN_D}{æ} \frac{1}{4} (Q_{2p_o} - Q_{1s}) ea_B^2 \qquad (4)$$

(a_B is the Bohr radius which is $\sim 100\text{Å}$ for GaAs).

We use sample 2 to explain the procedure of determination of K, N_D and N_A. From $\Delta(T)$ we find $\Delta_\infty/\Delta_o = 2$, and from the graph $K(\Delta_\infty/\Delta_o)$ we obtain K=0.77. For the given K, the $\Delta_o(K)$ graph yields $\Delta_o = 3.6$ (in units of $eN_D (Q_{2p_o} - Q_{1s})/4æ$). For the magnetic field H=50 kOe where the $1s-2p_o$ line is measured ($\lambda = 198.8 \mu m$), $Q_{2p_o} - Q_{1s} = 16.5$ (in units of ea_B^2). Next from (4) we find $N_D = 8.9 \times 10^{13} \text{cm}^{-3}$, $N_A = N_D \cdot K = 6.9 \times 10^{13} \text{cm}^{-3}$. The data obtained from the temperature dependence of the Hall effect by the method of Van der Paw are $N_D = 1.64 \times 10^{14} \text{cm}^{-3}$, $N_A = 1.47 \times 10^{14} \text{cm}^{-3}$, K=0.9. The agreement is satisfactory.

Thus the PES possessing a unique sensitivity unreachable in the other methods can permit one, in cases where the impurity PE line shape depends on electric fields and their gradients, not only to perform a chemical analysis of impurities but to determine separately donor and acceptor concentrations. Knowing the total concentration, one can use the spectral line intensity ratio to find the relative content of chemically dissimilar impurities in the samples studied.

References

Armistead C J, Knoles P, Najda S P and Stradling R A 1984 J.Phys.C 17 6415
Baranovskii S D, Gel'mont B L, Golubev V G, Ivanov-Omskii V I and Osutin A V 1987 Pisma ZhETF 46 405
Berman L V and Kogan Sh M 1987 Sov.Phys.Semicond. 21 933
Golka J, Trylski J, Skolnick M S, Stradling R A and Couder Y 1977 Solid State Commun. 22 623
Golubev V G, Gorelenok A T, Ivanov-Omskii V I, Minervin I G and Osutin A V 1986 Izv.Akad.Nauk SSSR Ser.Fiz. 50 282
Golubev V G, Ivanov-Omskii V I, Osutin A V and Polyakov D G 1987a Sov.Phys.Semicond. 21 18
Golubev V G, Zhilyaev Yu V, Ivanov-Omskii V I, Markaryan G R, Osutin A V and Chelnokov V E 1987b Sov.Phys.Semic. 21 1074
Kal'fa A A and Kogan Sh M 1971 Sov.Phys.Solid State 13 1656
Kal'fa A A and Kogan Sh M 1973 Sov.Phys.Semicond. 6 1839
Kogan Sh M and Lifshits T M 1977 Phys.Status Solidi A 39 11
Kogan Sh M, Lien Nguen Van and Shklovskii B I 1980 Sov.Phys. JETP 51 971
Kogan Sh M and Lien Nguen Van 1981 Sov.Phys.Semicond. 15 26
Korn P M and Larsen D M 1973 Solid State Commun. 13 807
Larsen D M 1973 Phys.Rev.B 8 535
Larsen D M 1976 Phys.Rev.B 13 1681

Ionized impurities and the FIR-photoconductivity spectrum of n-GaAs

A van Klarenbosch, J Burghoorn, T O Klaassen, W Th Wenckebach

Kamerlingh Onnes and Huygens Laboratories
P.O.Box 9506, 2300 RA Leiden, The Netherlands

C T Foxon

Philips Research Laboratories, Redhill, Surrey, UK

> The far infrared spectrum of hydrogenlike donors in partially compensated n-GaAs has been studied at T=4.2 K for magnetic fields up to 6 T. The complicated spectrum at high energy is found to originate from transitions to metastable states.
> The influence of bandexcitation on position and shape of the photoconductivity peaks, and the violation of electric dipole selection rules can be understood from mixing of states due to ionized impurities.

As is well known from many experiments absorption of far infrared radiation by shallow donors in semiconductors at low temperatures changes the electrical conductivity. This so called photothermal effect is still not well understood. Recently we have started an extensive investigation into the mechanism of extrinsic photoconductivity in n-GaAs. First experiments on the 1s → $2p_{+1}$ transition in GaAs:Si have been performed using the high intensity FIR radiation of the Free Electron Laser at UCSB. Analysis of the saturation behaviour of the resonant photoconductivity shows for instance that electron tunneling between excited donor states contributes significantly to the FIR induced conductivity (Kaminski et al. 1988).
Apart from saturation- also time resolved experiments will be performed on a variety of donor transitions. Of extreme importance for the understanding of the experimental results will be the detailed knowledge about the nature of the states between which FIR transitions occur. Here we report the preliminary results of experiments that will yield the necessary spectroscopic information.
For the experiments a conventional molecular gas FIR laser has been used together with a superconducting magnet in the Faraday configuration at T=4.2 K. Although most of the experiments have been performed for fields up to 6 T, some spectra up to 15 T have been recorded at the High Magnetic Field Laboratory of the University of Nijmegen. A 2 mW HeNe-laser was used for excitation over the bandgap.
The n-GaAs sample (G42) is a 10 μm epitaxial layer on a LEC substrate, intentionally doped with Si: $N_d = 5.10^{14}$ cm^{-3}, $N_a = 2,2.10^{14}$ cm^{-3}.
Much detailed spectroscopic work on shallow donors in GaAs has been published. In most cases however with the emphasis on the low energy part of the spectrum (e.g. Armistead et al. 1986). Work involving higher energy states has recently been published by Labrujere (1988) and Wagner (1988).
Theoretical work on the hydrogen states in a magnetic field have been largely restricted to the description of the bound states (see for instance Makado, 1986). Only Simola

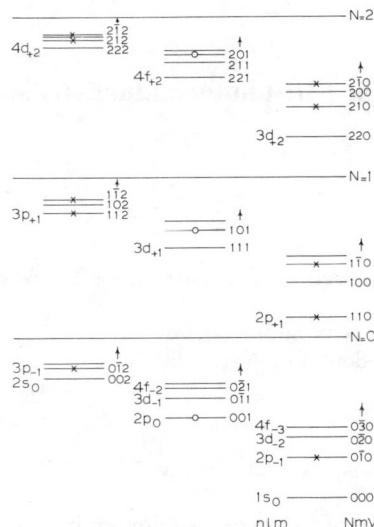

Figure 1: Schematic diagram of the hydrogen states in a magnetic field according to Simola (1978). Landau levels are indicated by N. Allowed dipole transitions out of the $1s_0$ groundstate are indicated by x/o for the Faraday/Voight configuration.

and Virtamo (1978) give a more elaborate discussion of all hydrogen states in magnetic fields. They show that for each value of the magnetic quantum number $m \leq 0$ there is an infinite number of bound states below the N=0 Landau level. For $m > 0$ only those states (N,m,ν) for which m=N are truly bound, and extrapolate to hydrogenic states at B=0. States for which $m < N$ are coupled through the Coulomb potential with free states with equal m but lower N and are therefore metastable in finite fields. No correspondence with hydrogenic states at B=0 exists for these states.

In fig.1 a schematic diagram of the bound and metastable states is given. They are grouped according to the quantum number ν in order to focus on the trends of the levelscheme in high fields. The bound states are labelled by both their high- (N,m,ν) and zero field (n,l,m) quantum numbers.

Electric dipole transitions between the ground state $1s_0 = (0,0,0)$ and excited states are allowed only for odd parity final states for which m=0 or ± 1 (Voight or Faraday configuration). The state parity is given by $\pi(N,m,\nu) = (-1)^{\nu+m}$.

In fig.2 the experimentally observed transitions (without bandexcitation) between the $1s_0$ and excited states are given. The final states have been indicated. The solid lines are fits based on the calculations of Makado and McGill (1986) using $R^* = 46.1$ cm^{-1}, $m^* = 0.0665$ m_e and a central cell correction of 0.6 cm^{-1} with a field dependence similar to that described by Fetterman et al.(1971). Also shown is the cyclotron resonance (C.R.) and the theoretical field-dependence based on the m* value.

The theoretical curves are found to fit the experimental points below the N=0 level very well. Strictly spoken however transitions to the $3d_{-2}$, $4f_{-3}$, $3d_{-1}$, $2s_0$ and (in the Faraday configuration) the $2p_0$ states are forbidden. The explanation for this apparent violation of the selection rules has been given already by Stillman et al.(1971) and Larsen (1973). Electric fields at the neutral donor site, due to ionized impurities cause a mixing of donor states with different m and ν quantum numbers. As a result, initially forbidden transitions become (weakly) allowed. Because of the approximate

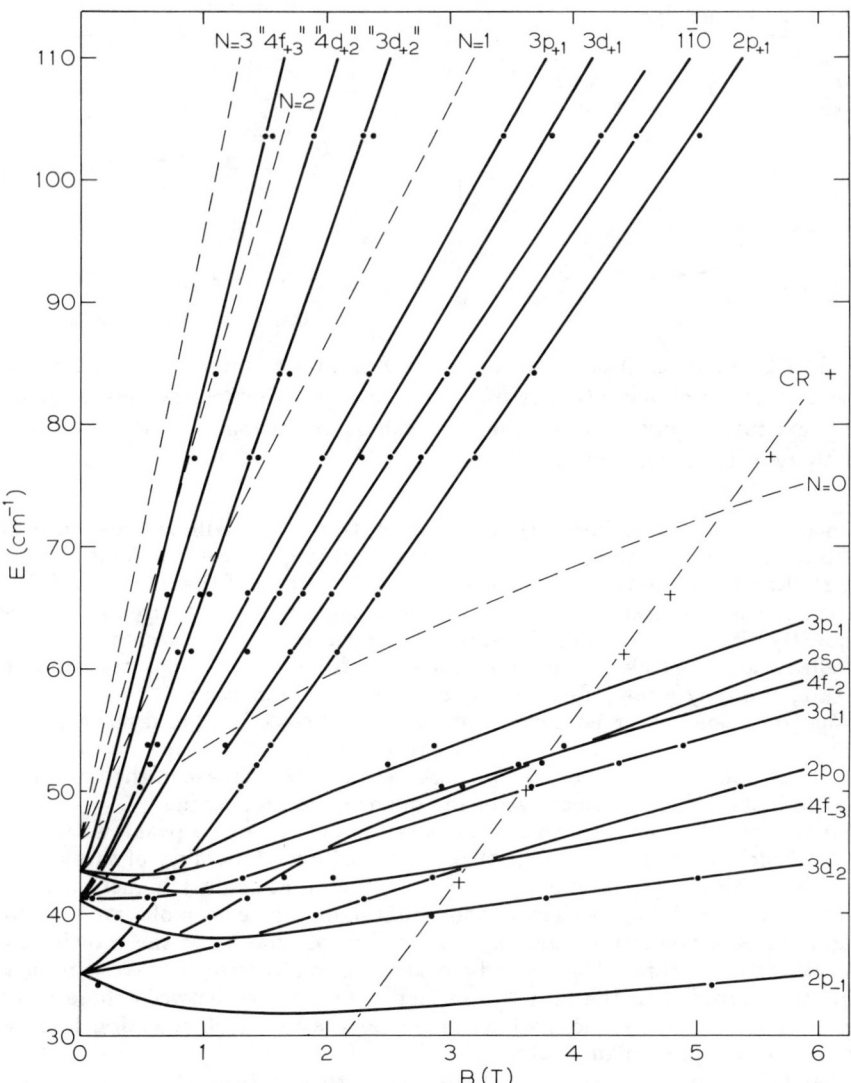

Figure 2: Magnetic field dependence of all observed lines. Final states of the transitions are indicated. Broken lines show the Landau level transition energies. Solid lines are theoretical fits. CR indicates the N=0→N=1 cyclotron resonance data.

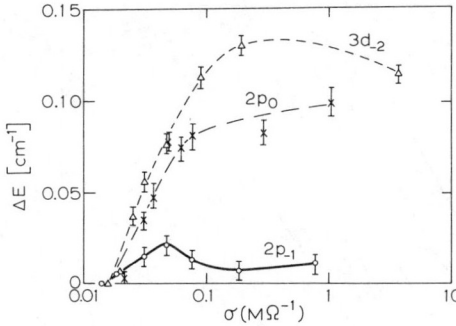
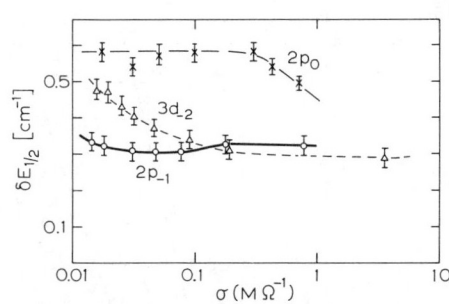

Figure 3: The energy shift of 3 transitions as a function of the intensity of the bandexcitation (indicated by the conductivity of the sample at B=0)

Figure 4: Linewidth of 3 transitions as a function of the intensity of the bandexcitation

randomness of the ionized impurity distribution, transitions will be broadened too.
The effects of mixing on the spectrum can be observed quite well. In fig.3 and fig.4 the energy shift and the linewidth of 3 transitions as a function of the intensity of the band excitation are shown. Band excitation decreases the number of ionized impurities and consequently diminishes mixing of states and the energy shift connected to the mixing. These effects are most clearly observed for the "strongly forbidden" transition to the $3d_{-2}$ state, and only weakly for the fully allowed $2p_{-1}$ transition.
From the transitions observed between the N=0 and N=1 Landau levels the $2p_{+1}$ and $3p_{+1}$ peaks are easily identified. According to Larsen (1973) the $3p_{+1}$ state has a large electric quadrupole moment. Indeed we observe a strong decrease of the linewidth with HeNe illumination due to a decrease of the quadrupole broadening.
The parity forbidden $3d_{+1}$ transition appears to consist of two peaks when observed under bandexcitation. This suggests that the transition probability of this state arises from mixing with the (1,0,1) metastable state situated at slightly higher energy.
Between the $2p_{+1}$ and $3d_{+1}$ peaks two additional signals have been observed. According to Wagner (1988) the intense signal nearest to the $2p_{+1}$ has to be the transition to the $(1,\bar{1},0)$ metastable state. The second weaker one might then be a transition to the $(1,\bar{2},0)$ state mixed with the $(1,\bar{1},0)$. In fact the difference between these two peaks decreases under bandexcitation, which might be the result of demixing leading to a smaller repulsion between the states.
At higher fields the observed transition-energies to the $2p_{+1}$, $3p_{+1}$ and $3d_{+1}$ states are slightly, but consistently lower than the calculated values. This results from the influence of band nonparabolicity and polaron effects on the separation between the N=0 and N=1 Landau levels. The same effect can be seen in the discrepancy between experimental C.R. data and the theory (see fig.2). Using the C.R. data on the field dependent value of m^* the theoretical curves for the donor transitions do fit the experimental points much better.
In fig.5 the experimental spectrum above the N=1 level is shown at a FIR energy of 142 cm^{-1}. As can be seen in fig.6 the transition indicated by "$3d_{+2}$" in fig.2 in fact shows a complicated structure. Originally we indentified these higher energy peaks as arising from transitions to the $3d_{+2}$, $4f_{+3}$ etc. states. However, as noted correctly by Wagner (1988) the FIR absorption in that energy region must be due to the presence of the (2,1,0) and $(2,\bar{1},0)$ metastable states. Analogously the absorption near "$4d_{+2}$" stems from the (2,1,2) and $(2,\bar{1},2)$ state, and that near "$4f_{+2}$" from the (2,0,1) state.

Shallow Impurities in III–V Compounds

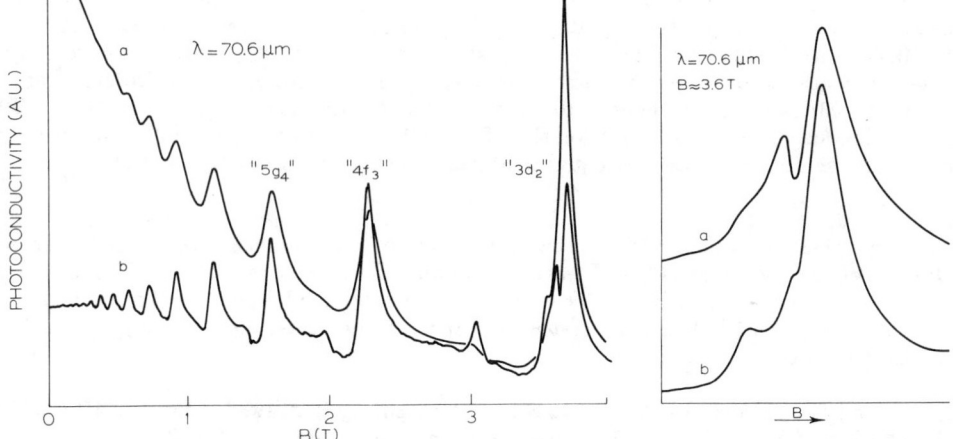

Figure 5: Photoconductivity spectrum at $\lambda = 70.6$ μm with (b) and without (a) bandexcitation

Figure 6: The shape of the "$3d_{+2}$" transition at $\lambda = 70.6$ μm with (b) and without (a) bandexcitation

Similar considerations hold for the peaks towards field i.e. higher Landau levels: a large peak at "$4f_{+3}$" ($(3,1,0)$ and $(3,\bar{1},0)$), a small peak at "$5f_{+3}$" ($(3,1,2)$ and $(3,\bar{1},2)$), etc. Although no quantitative information concerning the energies of the metastable states exists, the shape of the spectrum between N=1 and N=2 (repeating itself towards higher N) can be understood qualitatively. According to Canuto and Kelly (1972) for high fields the binding energy of the $(N,m=N,\nu)$ state (i.e. the energy difference between Landau level N and the bound state m=N) is roughly proportional to $(1+\nu/2)^{-2}$. Moreover it decreases towards higher N. Consequently the energy region in which the set of states (N,m,ν) with m=N,N-1,...,-∞ falls, decreases strongly towards high ν and N. Energy separations between states will scale accordingly.

From the energy difference between $2p_{+1}=(1,1,0)$ and $(1,\bar{1},0)$ states the $(2,2,0)$ and $(2,\bar{1},0)$ separation can be estimated to be about 4 cm^{-1}. Because of these small energy differences between the states and the large radial extension strong mixing due to the presence of ionized impurities will occur. The result can be seen in the photoconductivity spectrum as a broad peak with structure with overall width in the order of 4 cm^{-1}. Because of the strong mixing it is not well possible to give a clear assignment to the various peaks. Application of bandexcitation has a considerable influence (see fig.6). Part of the structure disappears and the two remaining peaks shift nearer to each other. Probably in the limit of no ionized impurities (high bandexcitation level or no compensation) two distinct peaks belonging to the $(2,1,0)$ and $(2,\bar{1},0)$ states will be observed.

For the $(N,m,\nu)=(2,m,2)$ set of states the energy separation between successive states will be about 1/4 of that for the $(2,m,0)$ set. Therefore one expects the "$4d_{+2}$" transition to have a linewidth of about 1 cm^{-1}, without noticeable structure, in complete agreement with experiment.

The intensity of this signal is much smaller than that for the $(2,m,0)$ transition. This is in accordance with expectations. Forster et al.(1984) calculate that for not too small magnetic fields the $\Delta\nu = 0$ transitions (in our case transitions towards ν=0 states) have by far the largest transition probabilities.

Based on the given discussion and the levelscheme (fig.1) one expects the separation between the transitions to the $3d_{-2}$ state and the "$3d_{+2}$" state to be at least equal to $2\hbar\omega_{CR}$, the Zeeman splitting between $3d_{-2}$ and the true $3d_{+2}$ state. However, even when corrections for nonparabolicity and polaron effects are applied (Lindeman et al. 1983), the $3d_{-2} -$ "$3d_{+2}$" energy difference is too small (about 2 cm^{-1} at B=3.5 T). An analogous problem is found for the $4f_{-3} -$ "$4f_{+3}$" energy separation. Surprisingly however, the $4d_{-2} -$ "$4d_{+2}$" at equal field has the correct value.

Acknowledgements:
This investigation is part of the research program of the "Stichting voor Fundamenteel Onderzoek der Materie (FOM)" which is financially supported by the "Nederlandse Organisatie voor Wetenschappelijk Onderzoek (NWO)". The authors like to acknowledge discussions with W.Prettl concerning metastable states and making available a preprint before publication.

Armistead C J, Makado P C, Najda S P and Stradling R A 1986 *J.Phys.C* **19** p 6023
Canuto V and Kelly D C 1972 *Astrophysics and Space Science* **17** p 277
Fetterman H R, Larsen D M, Stillman G E and Tannenwald P E 1971 *Phys.Rev.Lett.* **26** p 975
Forster H, Strupat W, Rösner W, Wunner G, Ruder H and Herold H 1984 *J.Phys.B* **17** p 1301
Kaminski J, Spector J, Labrujere A C, Klaassen T O, Wenckebach W Th and Foxon C T 1988, to be published
Labrujere A C, Klaassen T O, Wenckebach W Th and Foxon C T 12th Int.Conf.on Infrared and Millimeterwaves, Orlando, 1987, accepted for publication in Int.J.Infrared and Millimeterwaves
Larsen D M 1973 *Phys.Rev.B* **8** p 535
Lindeman G, Lassnig R, Seidenbusch W and Gornik E 1983 *Phys.Rev.B* **28** p 4693
Makado P C and McGill N C 1986 *J.Phys.C* **19** p 873
Simola J and Virtamo J 1978 *J.Phys.B* **11** p 3309
Stillmann G E, Larsen D M and Wolfe G M 1971 *Phys.Rev.Lett.* **27** p 989
Wagner H P and Prettl W 1988, to be published

Shallow positron traps in gallium arsenide

S. Dannefaer, P. Mascher and D. Kerr

Department of Physics, University of Winnipeg, Winnipeg, MB R3B 2E9 Canada

ABSTRACT: The presence of shallow **positron** traps has been found in GaAs and likely also in Si. These traps are shallow in the same sense as shallow impurity levels in that the positron binding energy is small ($\sim$20meV) and that the positron wavefunction is rather delocalized. For the latter reason a shallowly trapped positron and a free positron (Bloch- state) will both annihilate with nearly the same rate. On the other hand trapping by shallow traps modifies significantly trapping into deep traps such as vacancies, as demonstrated by low temperature measurements.

1. INTRODUCTION

Defects in GaAs have been investigated intensively by means of positron annihilation. Vacancy related (deep traps) positron lifetimes have been reported in the 250 to 260 ps range and in the 290 and 300 ps range by Dannefaer and Kerr (1986) and by Corbel et al. (1988). The interpretation of these two lifetime ranges in terms of their physical origin is, however, under current debate. A further, and even more fundamental problem, deals with the so-called bulk lifetime. This is the lifetime one would directly observe in perfect GaAs and corresponds to the positron being in a Bloch state. Stucky et al. (1986) claim a value of 233$\pm$2 ps while Dannefaer et al. (1984) claim a lower value of 220ps.

In this contribution we will show that shallow positron traps **commonly** play a significant role in GaAs even at room temperature. A positron binding energy close to 23 meV has been found for Cr-doped GaAs. In electron irradiated Si we find that A centers (both A^0 and A^-) are also shallow positron traps albeit only observed below room temperature. Failure to include shallow traps in the analyses can easily result in contradictory conclusions.

© 1989 IOP Publishing Ltd

2. RESULTS

Positron lifetime measurements were performed on a wide range of GaAs samples. All samples contained vacancy-like defects for which reason the bulk lifetime, τ_B, is not **directly** observed but it can be calculated from the trapping model (West 1977) according to:

$$1/\tau_B = I_1/\tau_1 + I_2/\tau_2 + I_3/\tau_3 \tag{1}$$

where $\tau_{1,2,3}$ and $I_{1,2,3}$ are determined by least squares fitting to the experimental lifetime spectrum. These values varied significantly for different samples but τ_B should nevertheless stay constant *if* the analysis is correct.

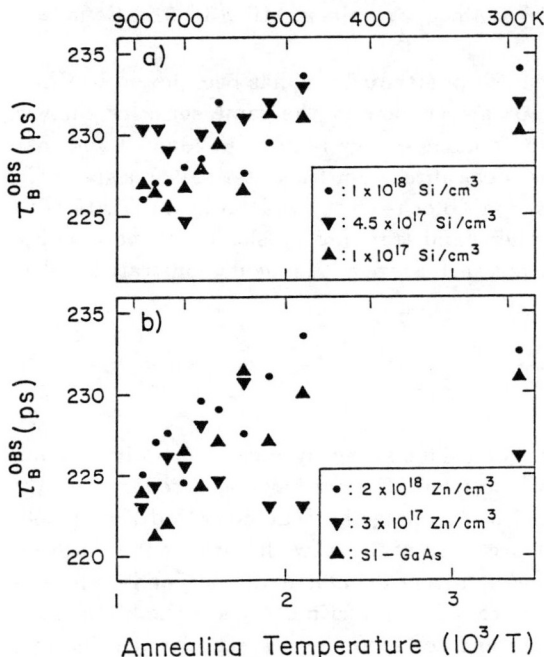

FIG. 1. Bulk lifetimes as calculated from Eq.(1) based on data by Dannefaer and Kerr (1986). All GaAs samples were LEC grown. Samples in upper panel are all n-type while those in the lower panel are p-type or semi-insulating (SI). Positron measurements were all performed at room temperature.

The data in Fig. 1 show that the calculated τ_B values (τ_B^{OBS}) apparently vary between 221 and 233 ps which is a significant variation. The lower value (220±1 ps) was also found in Liquid Phase Electro Epitaxially grown GaAs (LPEE GaAs) which proved exceptionally defect free (Dannefaer et al. 1988). The systematic variation in τ_B upon isochronal annealing for the wide range of samples shown here indicates that something is missing in the analysis of the data, i.e., the application of eq. (1) is not quite correct due to the omission of another lifetime. Low temperature measurements also indicate the existence of another lifetime component which modifies the observed bulk lifetime (see Fig. 2) towards higher values at low temperatures.

In the case of 2MeV electron irradiated Cz-Si the influence of shallow traps are

shown in Fig. 3 for a P-doped ($10^{17} cm^{-3}$) sample. In such a sample A centers are formed predominantly but also divacancies, E centers and V_2O are formed.

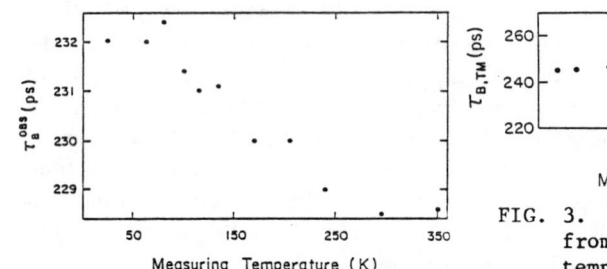

FIG. 2. Bulk lifetime as calculated from Eq.(1) for Cr-doped semi-insulating GaAs.

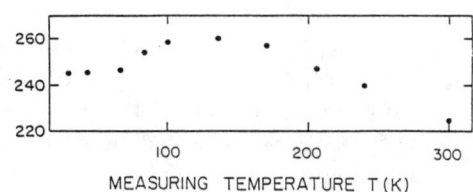

FIG. 3. Bulk lifetime as calculated from Eq.(1) as a function of temperature for electron irradiated Si. This sample was P-doped ($10^{17} cm^{-3}$) and irradiated to a dose of $10^{18} e^-/cm^2$.

The latter three defects give rise to deep positron traps and account combined for a 300 to 325 ps lifetime. Using eq. (1) one can calculate τ_B^{OBS} as shown in Fig. 3 and we find values significantly larger than the established bulk lifetime of 218 ps for Si.

3. DISCUSSION

Let us first consider the inclusion of shallow traps into the simple trapping model. In Fig. 4a is shown the most simple case, ie. no shallow traps.

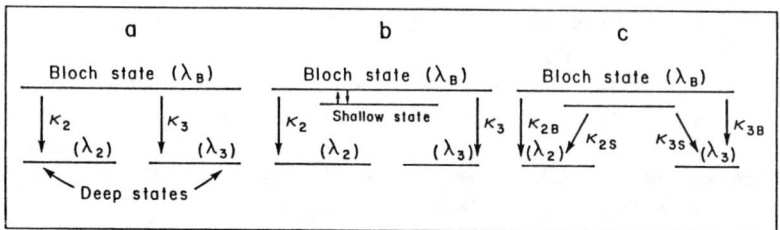

FIG. 4. Different types of positron trapping models.
(a): Trapping from Bloch state into two different deep traps with rates K_2, K_3. (b): Addition of a shallow trap with trapping and detrapping rates (short arrows).
(c): Shallow traps considered effectively as a part of the Bloch state with trapping taking place into deep traps also from the shallow state.

The upper horizontal line designates the energy level for the positron Bloch state from which an ensemble of positrons can either annihilate with rate $\lambda_B (\equiv 1/\tau_B)$ or be trapped with rates κ_2 or κ_3 into (here) two different deep levels where annihilation takes place with rates λ_2 or λ_3. Note that the experimentally determined disap-

pearance rate from the Bloch state is the sum $\lambda_B + \kappa_2 + \kappa_3$ not just λ_B, which is why τ_B is calculated from eq. (1) whenever defects are present.

In Fig. 4b, a shallow trap has been added with specific trapping and detrapping rates (short arrows), but trapping into deep traps still only takes place from the Bloch state. Finally, in Fig. 4c a slightly different situation is envisaged namely that the shallow trap is effectively just a part of the bulk state (trapping and detrapping rates very fast compared to annihilation rates) and trapping into deep traps can take place from both states albeit with different rates due to different initial positron states.

To distinguish between the two shallow trap models we note that at low temperatures (detrapping is zero) the situation depicted in Fig. 4b will predict that the lifetime arising from the shallow state should be experimentally observable. The Bloch state disappearance rate should be determined by λ_B plus all trapping rates. Positrons trapped in the shallow state should yield disappearance rates close to λ_B, and positrons in the deep trap(s) will yield λ_2 and λ_3 (both smaller than λ_B). In the model shown in Fig. 4c, however, the shallow traps do not yield a separable lifetime, but they do modify the bulk lifetime according to:

$$\tau_B^{OBS} = \alpha \tau_{ST} + (1-\alpha)\tau_B, \qquad (2)$$

where τ_{ST} is the lifetime for a positron trapped in the shallow trap and α is the fraction of positrons occupying the shallow traps as determined by the Boltzmann distribution:

$$\alpha = [1 + (N_B/N_{ST})\exp(-\Delta E/kT)]^{-1} \qquad (3)$$

Here N_B and N_{ST} are, respectively, the density of positron states in the bulk and the shallow traps, and ΔE is the positron binding energy. It is further noted that by populating the shallow traps the trapping rate into deep traps is affected, since κ_{2S} need not be the same as κ_{2B} (see Fig. 4c).

Careful analysis of the lifetime spectra showed that a distinct shallow lifetime close to the bulk lifetime in GaAs (220 ps) could not be found, so we suggest that the model in Fig. 4c applies in this case. As shown in Fig. 2, τ_B^{OBS} is constant at 232 ps at less than 90 K which according to eq. (2) yields $\alpha = 1$ and $\tau_{ST} = 232$ ps, which is 12 ps longer than the bulk lifetime for GaAs.

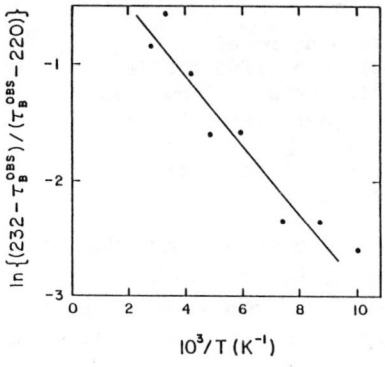

FIG. 5. Arrhenius plot of data from Fig. 2. The 232ps value is the maximum value of τ_B^{OBS} from Fig. 2 and the 220ps value is the bulk lifetime entering into Eq. 2 which is the bulk lifetime in GaAs.

Using eqs. (2) and (3) and an appropriate Arrhenius plot (Fig. 5) one finds a positron binding energy, $\triangle E$, of 23 meV. Further analysis indicates that the trapping cross-section into deep traps is about 5 times larger from the bulk state than from the shallow trap.

These values are based on detailed measurements on Cr-doped semi-insulating GaAs and we do not yet know if similar numerical values apply for other types of GaAs. However, the presence of shallow traps appears to be general according to Fig. 1, and judging from the slight increase in the apparent bulk lifetimes, slightly undersized impurities such as boron and nitrogen may well be the cause for the traps. The concentration of these unintentionally added impurities may well be in the range of $10^{17} cm^{-3}$. According to Fig. 1 there is a tendency for a reduction in this concentration after annealing above 450°C apparently more pronounced in p-type and semi-insulating type materials than in n-type (Si-doped) materials.

Evidence for shallow traps has also been found in electron irradiated Cz-Si, where A centers yield a lifetime only 7 ps longer than the bulk lifetime (Mascher et al. 1988). Both A^0 and A^- are shallow traps, with A^- being much more effective than A^0. Interestingly, it was found that the model 'c' in Fig. 3 which explained the data for GaAs cannot explain the data for the A centers. Instead model 'b' must be applied which might mean that in GaAs there is a spatial correlation between shallow and deep traps. Finally, we note that for charged shallow traps (A^-) the trapping cross-section at low temperatures ($\sim$100K or less) becomes so large that concentrations down to $10^{15} cm^{-3}$ become easily detectable.

4. CONCLUSION

The two examples reported in this paper show the importance of shallow traps in positron annihilation. Their very detection is somewhat difficult since the shallow traps only show up via a deviation from the expected behaviour when monitoring deep traps. Shallow traps are, however, at least as efficient traps as deep traps and are likely to be associated with undersized substitutional impurity atoms like B and N in GaAs and O in Si. Positrons trapped at shallow traps have lifetimes which are very close to the bulk (ie. untrapped) state to within 10 ps.

5. ACKNOWLEDGEMENTS

This work was supported by the Natural Sciences and Engineering Research Council of Canada. We are indebted to L. Lindström, Försvarets Forskningsanstalt Linköping for supplying the electron irradiated silicon samples.

6. REFERENCES

Corbel C, Stucky M, Hautojärvi P, Saarinen K and Moser P 1988
  Mat. Res. Soc. Symp. Proc. **104** 475
Dannefaer S, Hogg B, Kerr D 1984 *Phys. Rev. B* **30** 3355

Dannefaer S and Kerr D 1986 *J. Appl. Phys.* **60** 591
Dannefaer S, Mascher P and Kerr D 1988 *Mat. Res. Soc. Symp. Proc.* **104** 471
Mascher P, Dannefaer S and Kerr D 1988 Presented at the
15th International Conference on Defects in Semiconductors, Hungary
Stucky M, Corbel C, Geffroy B, Moser P and Hautojärvi P 1986 In:Defects in Semiconductors edited HJ von Bardeleben
Materials Science Forum **Vol 10-12** (Trans Tech Publ, Switzerland) p 265
West RN 1977 *Adv in Physics* **22** 263

Quantitative determination of the carbon content in MOMBE p-GaAs by low temperature photoluminescence

S. Ambros[1], M. Kamp, K. Wolter, M. Weyers, H. Heinecke[2], H. Kurz and P. Balk

Institute of Semiconductor Electronics, Technical University Aachen, Sommerfeldstr. D-5100 Aachen, FRG
[1]Institut für Schichten- und Ionentechnik, KFA Jülich, Postfach 1913, D-5170 Jülich
[2]Siemens Research Laboratories, Otto-Hahn-Ring 6, D-8000 München, FRG

ABSTRACT: Several features of low temperature (2 K) photoluminescence (PL) spectra are used for a quantitative determination of the carbon concentration below 10^{18}cm^{-3} in MOMBE p-GaAs samples. We present a close correlation between specific PL features (like peak energy, integrated intensity and FWHM of the luminescence line) and hole concentration together with a calibration of the carbon content with respect to the hole concentration. The method permits a fast and nondestructive determination of the carbon content from PL measurements.

1. INTRODUCTION

Carbon is a convenient acceptor dopant in metalorganic molecular beam epitaxy (MOMBE) of GaAs, which utilizes a Ga alkyl and AsH$_3$ as starting materials. In contrast to triethylgallium (TEG) the use of trimethylgallium (TMG) in the growth process leads to heavily doped p-type films (Pütz et al (1985)). Weyers et al covered the entire range of hole concentrations from 10^{15} to 10^{20}cm^{-3} by employing mixtures of TMG and TEG.

In order to evaluate MOMBE grown samples it is important to be able to determine the carbon concentration in the deposited epilayer directly. We will show that low temperature PL is a convenient nondestructive tool for the quantitative determination of the carbon content in GaAs from 10^{18} down to 10^{15}cm^{-3}. Specific PL characteristics were correlated with the hole concentrations evaluated from Hall measurements. To achieve a calibration of the carbon content with respect to the hole concentration we used local vibrational mode absorption (LVM) and secondary ion mass spectrometry (SIMS) as well as carbon implanted standards.

2. EXPERIMENTAL

The 17 GaAs epilayers used in our study were grown from TEG, TMG and precracked AsH$_3$ in a modified RIBER 1000 MBE system. The used TMG pressures are listed in table 1 as well as hole concentrations and mobilities evaluated from room temperature Hall measurements on van der Pauw structures.

The samples were kept at 2 K in a pumped liquid helium cryostat for PL measurements. With a krypton ion laser as source the excitation densities used range from 0.06 to 120 W/cm^2. The luminescence signal was collected by a Cassegrain type

Table 1: Characteristics of the samples investigated

Sample	p_{TMG} ($\times 10^{-5}$ Pa)	p ($\times 10^{16}$ cm^{-3})	μ ($^{Vs}/_{Am}$)
623	-	0,15	273
178	0,4	0,87	270
606	0,4	1,08	269
625	1,7	2,57	271
205	1,7	2,65	337
607	1,7	5,68	224
180	3,5	5,99	274
624	5,1	16,3	235
568	3,5	19,0	188
176	9,1	22,8	245
602	3,5	24,1	171
570	9,1	36,1	79
569	9,1	36,8	208
567	9,1	57,1	64
603	9,1	95,1	159
622	9,1	127,0	144
604	0,2	179,0	137

optic onto an 0.5 m Czerny-Turner monochromator and detected by a SIT (Silicon Intensified Target) camera for simultaneous collection of a complete near band edge spectrum.

The LVM absorption measurements were carried out at room temperature with a Bomem DA3 Fourier Spectrometer. SIMS measurements were performed with an Atomika 3000 instrument using a cesium source for the primary ion beam.

3. RESULTS

Figure 1 shows typical PL spectra of five samples with hole concentrations between 9×10^{15} cm^{-3} and 1×10^{18} cm^{-3}. Certain features can be readily assigned to known transitions.

For hole concentrations of 3×10^{16} cm^{-3} and below, the finger print of free excitons (FE), as well as of excitons bound to neutral carbon acceptors (C^o,X) at an energy of 1.5153eV and 1.5127eV, respectively, are clearly observed (Bogardus and Bebb (1968)). In addition, weak features due to defect induced bound excitons (g) appear at an energy of 1.5110eV (Künzel and Ploog (1981)). The free electron to carbon acceptor transition (e,C) is visible in the entire spectral range investigated. It shifts from 1.4937 eV to lower energies for $p > 6 \times 10^{17}$ cm^{-3} due to an overlap of the acceptor band levels with the valence band (Lucovsky and Varga (1964)). Identification of carbon as the acceptor involved in this transition is possible from the peak energy at low concentration, and from calibration measurements on highly doped samples described later on. An additional transition appears at 1.487 eV in the spectra of the lightly doped samples (spectra a,b). Its temperature dependence indicates it as the donor to carbon acceptor transition (D,C) (Dean (1982)).

At hole concentrations between 3×10^{16} and 6×10^{17} cm^{-3} the spectra are dominated by the (e,C) line and an acceptor-acceptor pair transition designated as (g-g) (Makita et

al (1986)), which shows a strong shift to lower energies with increasing hole concentrations. The transition energies versus hole concentration are summarized in figure 2.

To determine the hole concentration by luminescence intensity a careful check is required whether the PL signal saturates at a higher excitation level. As demonstrated in figure 3 the luminescence intensity increases linearly with the excitation level at higher densities. At low excitation densities the PL lines are not well resolved and separated.

The integrated intensity I(e,C) (integration over FWHM) of the (e,C) line is shown in figure 4 as a function of hole concentration. An empirical relation:

$$\log p = 0.7 \times \log I(e,C) + 15.7$$

between the integrated intensity and hole concentration p can be derived from this plot for $p < 10^{17}$ cm^{-3}.

For higher doping levels the intensity levels off and decreases above 5×10^{17} cm^{-3}, pre-

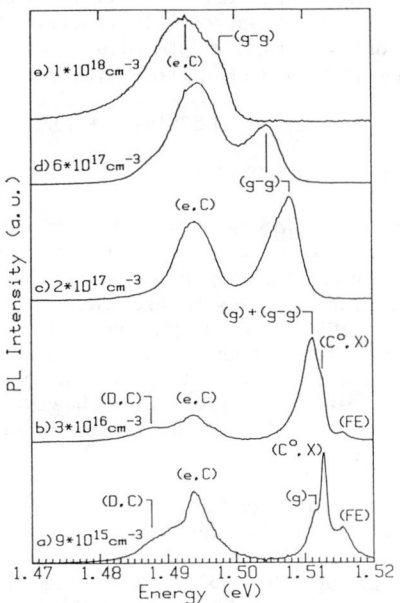

Fig. 1. Typical PL spectra for several hole concentrations

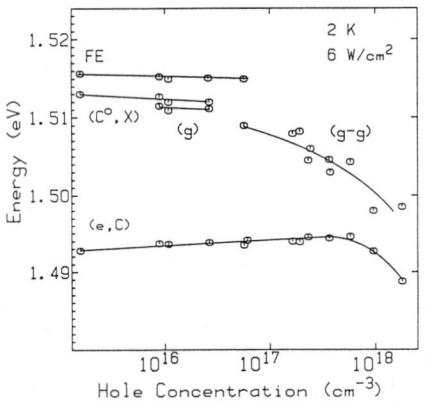

Fig. 2. Variation of emission energy as a function of hole concentration

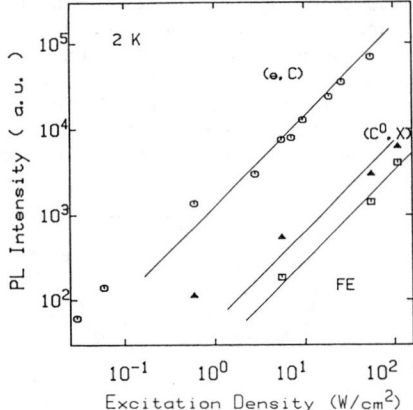

Fig. 3. Dependence of PL intensity on excitation density

sumably due to nonradiative processes. As illustrated in figure 5, a significant broadening of the (e,C) line occurs at these doping levels, which may be used for an estimate of the concentration in this range.

In addition to the (e,C) line the (C°,X) transition may serve for quantitative analysis at low doping concentrations. To correct for possible variations of intrinsic emission properties the intensity of this signal is normalized to the strength of the free

exciton transition. From figure 6, where the normalized luminescence intensity is plotted against the measured hole concentration, a second empirical relationship for the hole concentration can be derived:

$$\log p = 1.57 \times \log I + 15.9,$$

with $I = I(C^0,X)/I(FE)$.

LVM and SIMS measurements have been performed to achieve a calibration of the carbon content with respect to the hole concentration. The results are listed in table 2. Above 10^{18}cm^{-3} good agreement between hole and carbon concentrations is achieved. At lower concentrations LVM and SIMS permit only estimates. The values however, are still compatible with the measured hole concentrations.

To calibrate the PL-Signals at lower concentration carbon implanted standards have been prepared. To achieve a carbon concentration of $1 \times 10^{16} \text{cm}^{-3}$ with the required flat depth profile within the 1 μm thick epilayer we used a multiple implantation technique: $5 \times 10^{11} \text{cm}^{-2}$ at 350 keV, $2 \times 10^{11} \text{cm}^{-2}$ at 180 keV and $2 \times 10^{11} \text{cm}^{-2}$ at 80 keV. After flash lamp annealing, Hall measurements reveal a hole concentration of $1.0 \times 10^{16} \text{cm}^{-3}$. The integrated (e,C) lines of this samples indicate a carbon concentration of $1 \times 10^{16} \text{cm}^{-3}$ (table 3). These results prove that the carbon and the hole concentration are iden-

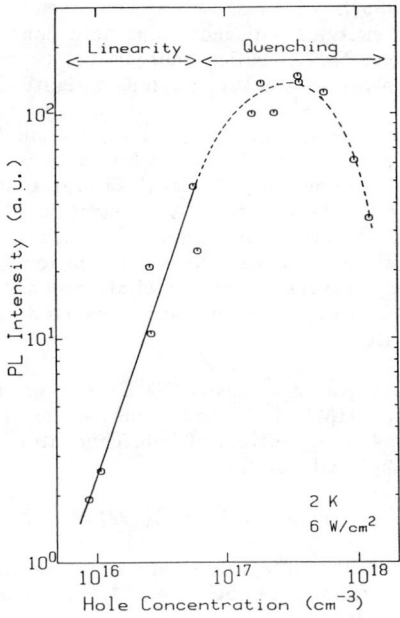

Fig. 4. Integrated intensity of (e,C) transition vs. carrier concentration

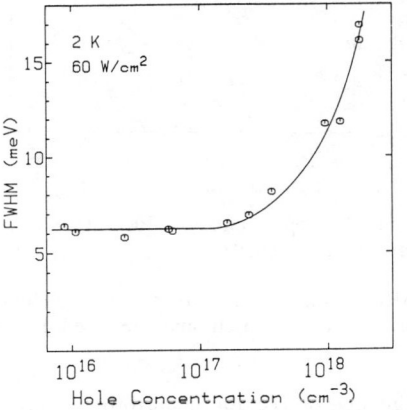

Fig. 5. Change of FWHM of (e,C) transition with increasing carrier concentration

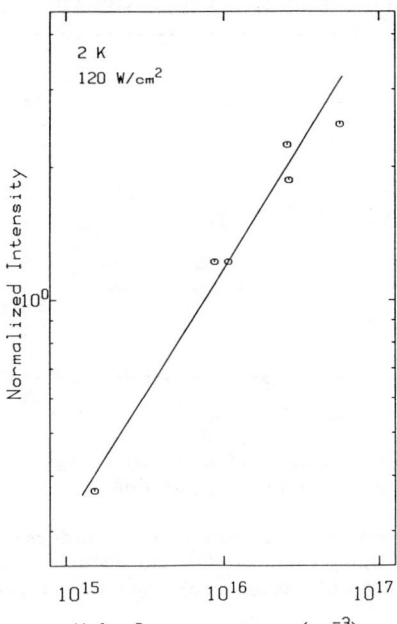

Fig. 6. Increase of integrated intensity of (C^0,X) transition (normalized to FE) with carrier concentration

tical within the accuracy of the experimental methods; they demonstrate the suitability of PL as nondestructive method for determining the carbon concentration in epitaxial layers. It should be pointed out that the experimental PL set-up should be calibrated with samples of known carbon content. The carbon content of these samples may be obtained by one of the three methods discussed in this section. It may also be obtained from the measured hole concentration in samples where carbon is known to be the dominant impurity species.

Table 2: Carbon concentration in highly doped samples determined by LVM and SIMS.

Hole concentration (cm^{-3})	Carbon concentration (cm^{-3})	Method
1.8×10^{18}	$2.4^{+}/_{-}0.7 \times 10^{18}$	LVM
1.3×10^{18}	$1-2 \times 10^{18}$	SIMS
5.7×10^{17}	$7^{+}/_{-}0.3 \times 10^{17}$	LVM
1.6×10^{17}	$1-6 \times 10^{17}$	SIMS

Table 3: Carbon concentration evaluated from integrated (e,C) lines and compared to implanted doses.

Sample Nr.	Implanted doses (cm^{-3})	Carbon concentration (cm^{-3})
1	1.0×10^{16}	0.8×10^{16}
2	1.0×10^{16}	1.1×10^{16}

4. DISCUSSION

Among the features summarized in figure 2 the most valuable lines for the carbon calibration are those corresponding to bound excitons (C^{0},X) and free electrons (e,C). Most favourable for the nondestructive analysis of the carbon content is the intensity of (e,C) transition, while the integrated intensity of (C^{0},X) transition (normalized to FE) can only be used to measure low carbon concentrations. The FE, g, and g-g signatures may serve to confirm these data.

However, this straight forward analysis of PL spectra to measure carbon concentrations assumes the absence of interference by other acceptors. For MOMBE-samples we can exclude other acceptors, as demonstrated in figure 1. Even the (D,C) transition had vanished in samples grown several months later from the same source material. This suggests that an unidentified donor impurity was removed by distillation from the TEG, which is the major Ga compound in the growth process. An additional confirmation for the low level of residual impurities is supplied by the fact, that similar weak traces of acceptors are found in high mobility LPMOCVD-samples, which were grown with the same source materials.

5. CONCLUSION

Low temperature PL is a convenient nondestructive tool for the quantitative determination of the carbon content in GaAs epilayers in doping regions where other techniques like LVM and conventional SIMS fail. Quantitative evaluation of features like peak energy, integrated intensity and FWHM is possible if a calibration for the individual experimental setup is made with samples of known carbon concentration. However, at doping levels above $10^{17} cm^{-3}$ interference with other acceptor impurities by peak broadening has to be carefully ruled out (for example by SIMS analysis). Above $10^{18} cm^{-3}$ other techniques have to be employed.

6. ACKNOWLEDGEMENT

We are indebted to M. Maier, J. Wagner and J. Seelewind from the Fraunhofer IAF, Freiburg, for the SIMS measurements and for LVM evaluations.

7. REFERENCES

Bogardus E H and Bebb H B 1968 Phys. Rev. 176 993
Dean P J 1982 Prog. Cryst. Growth Charact. 5 89
Künzel H and Ploog K 1981 Inst. Phys. Conf. Ser. No. 56 519
Lucovsky G and Varga D J 1964 J. Appl. Phys. 35 3419
Makita Y, Takeuchi Y, Ohnishi N, Nomura T, Kudo K, Tanaka H, Lee H C, Mori M and Mitsuhashi Y 1986 Appl. Phys. Lett. 49 1184
Pütz N, Heinecke H, Heyen M, Balk P, Weyers M and Lüth H 1986 J. Cryst. Growth 74 292
Weyers M, Pütz N, Heinecke H, Heyen M, Lüth H and Balk P 1986 J. Electron. Mat. 15 57

Inst. Phys. Conf. Ser. No 95: Chapter 4
Paper presented at Int. Conf. Shallow Impurities in Semiconductors, Linköping, Sweden, 1988

Residual impurities in autodoped n-GaAs grown by MBE

M.B. Stanaway, R.T. Grimes, D.P. Halliday, J.M. Chamberlain, M. Henini, O.H. Hughes, M. Davies, and G. Hill*

Physics Department, Nottingham University, Nottingham NG7 2RD, United Kingdom
*Department of Electronic Engineering, Sheffield University, Sheffield S1 3JD, United Kingdom

ABSTRACT: A modification to the source geometry of a molecular beam epitaxy machine is shown to produce GaAs with exceptionally low residual acceptor concentration (8.7×10^{12} cm^{-3}) and with peak mobilities up to 255,000 cm^2 V^{-1} s^{-1}. Hall effect, photoluminescence and far infra red photoconductivity are used to assess the layers and to identify residual contaminants. The residual donors are silicon, sulphur and selenium. The principal residual acceptor is carbon.

1. INTRODUCTION

The characterisation and control of the residual shallow impurities found in GaAs grown by molecular beam epitaxy (MBE) is essential to the production of high mobility, low compensation material (Skromme et al 1985). Larkins et al (1986 and 1987) recently described the growth of 10 μm thick n-GaAs layers by MBE having residual acceptor and donor concentrations of 2.4×10^{13} cm^{-3} and 1.5×10^{14} cm^{-3} respectively. The complex growth procedure employed resulted in extremely high mobility material (peak value 216,000 cm^2 V^{-1} s^{-1}). Very recently, Chand et al (1988) have reported that the use of a solid arsenic source improves the mobility of silicon doped GaAs (peak value 295,000 cm^2 V^{-1} s^{-1}) with low residual acceptor concentration. In comparison, the best material produced by the organo-metallic vapour phase epitaxial (OMVPE) technique (Shastry et al 1988) has residual acceptor and donor concentrations of 4×10^{12} cm^{-3} and 1.0×10^{14} cm^{-3} respectively, with a peak mobility of 310,000 cm^2 V^{-1} s^{-1}.

In this communication we report the growth of high purity n-GaAs by MBE using a rather simpler procedure than that of Larkins et al (1987), but which nevertheless delivers samples of at least equal quality. The best specimens have peak mobilities of 255,000 cm^2 V^{-1} s^{-1}, and in one case a lowering of the residual acceptor concentration to 8.7×10^{12} cm^{-3} is achieved. In unintentionally doped ('autodoped') samples the background donor concentration is as low as 6×10^{13} cm^{-3}. The combination of the present technique with the use of solid source arsenic (Chand et al 1988) might produce MBE samples whose quality approaches that of the best OMVPE specimens.

© 1989 IOP Publishing Ltd

2. EXPERIMENTAL

The samples were grown in a commercial Varian GEN-II MBE system. The source-furnace/cryopanel arrangement, however, departed from the standard design: the radial gap between a heated furnace and its supporting tube was increased from 3 to 16 mm and the optical path from the tube walls into the growth chamber was restricted by cryopanelling. The tube walls were thus subject to less radiant heating (and so outgassed less carbon acceptors): additionally, any such residual acceptors were more efficiently trapped on the cryopanels. In addition to the unintentionally doped layers #1, #3, #4 and #5 listed in Table I, a further layer (#2) was grown with intentional silicon doping. In all cases growth took place at 585°C at a rate of 1 μm per hour. The As_4 flux was just above the minimum necessary to give 2-fold reconstruction in the (100) RHEED pattern using a [100] azimuth. Van der Pauw (vdp) samples were prepared from several parts of the wafers. Au-Ge-Ni contacts were formed on the samples and, in some cases, thinned samples were made with thicknesses of 3 μm and 6 μm.

Hall effect measurements were performed in a homogeneous magnetic field up to 0.75 T using a gas-flow cryostat with careful monitoring of sample temperature. The controlling software also performed a self-consistent correction to allow for both substrate and surface depletion effects (Chandra et al 1979). Low sample currents were used to minimise carrier heating effects.

Photoluminescence (PL) measurements were performed using the 514.5 nm line of an Ar ion laser at 1.8 K and 4.2 K using standard techniques. Studies of the two electron sattelite spectra (TES) were made in a 10 T field with samples in the Faraday configuration. Finally, far infra red photoconductivity (FIRPC) measurements were made at 4.2 K on vdp samples of thickness 3, 6 and 15 μm both with and without bandgap light. FIR laser magnetospectroscopy was performed on samples in the Faraday orientation up to 10 T.

TABLE I

Sample	N_A (cm^{-3})	N_D (cm^{-3})	μ(77 K) cm^{2}V^{-1}s^{-1}	μ(peak)
#1	1.7x10^{13}	6.3x10^{13}	180,000	255,000
#2	8.7x10^{12}	2.9x10^{14}	140,000	175,000
#3	2.2x10^{14}	4.3x10^{14}	130,000	155,000
#4	2.9x10^{14}	3.8x10^{14}	115,000	135,000
#5	8.0x10^{13}	1.5x10^{14}	125,000	145,000

Table I Electrical properties of the MBE samples. All wafer thicknesses are 15 μm. N_A and N_D are best fits to the corrected data shown in Figure 1.

3. RESULTS

Table I lists the electrical properties of the samples. The solid line in Figure 1 is a theoretical fit to the experimental values for electron density (Anderson and Apsley 1986), duly corrected for surface and substrate depletion effects (Chandra et al 1979). As a check on this procedure C-V profiling (Ambridge and Faktor 1975) was used in some

instances to confirm that the self-consistently corrected Hall effect densities agree with those obtained from C-V plots to within 10%. The mobility values are displayed in Figure 2: depletion effects provide a lower temperature limit in several cases. A sixth layer was also grown which was fully depleted: from the curves given by Chandra et al (1979), $N_D - N_A$ is estimated at 2×10^{13} cm^{-3}. Uniformity checks of the electron sheet density at various points in the central area of wafers #1 – #4 indicate variations of < 15%. The mobility value for #5 appears anomalously low based on the value of $N_D + N_A$ although this wafer exhibited a greater non-uniformity and inhomogeneity effects cannot be excluded. Table 1 indicates significant variations in mobility amongst the unintentionally doped wafers. This is a direct result of recent machine history, optimisation of growth conditions having been obtained for layer #1.

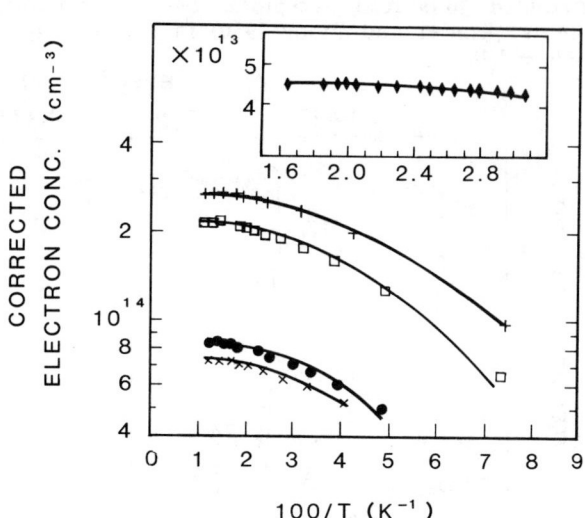

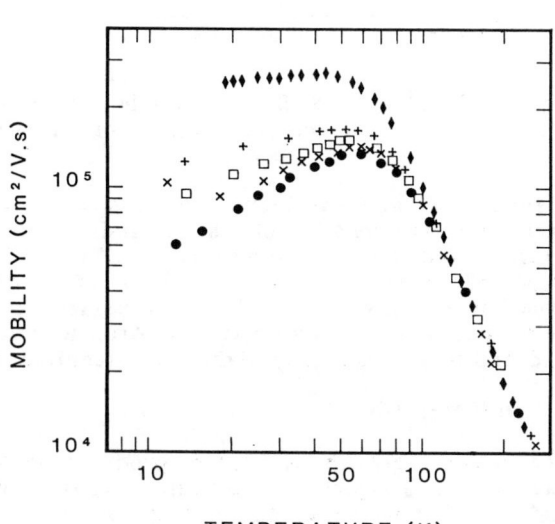

Figure 1, above: Corrected electron density as a function of temperature, and theoretical fits to data points.

Figure 2, lower: Electron mobility as a function of temperature.

For both figures,
♦: #1, +: #2, □: #3, ●: #4, x: #5.

Figure 3(a) displays PL measurements at 1.8 K in the exciton region: the full width at half maximum (FWHM) of the donor-bound exciton transition (D_0,X) is 0.22 meV, and the ratio of the (D_0,X) to acceptor-

bound excition (A_0,X) signals is 18:1. The free exciton (FE) feature appears as a slight dip. The different spin states of the (A_0,X) transition can just be resolved, as can the excited states of the $(D_0,X)_{n=1}$ transition. The two electron satelite (TES) (Dean et al 1967) structure $(D_0,X)_{n=2}$ is also clearly visible. Magnetic field studies of this structure were undertaken for samples #1 and #2; these confirmed the presence of Si and S as residual donors. Figure 3(b) displays the band-acceptor region PL spectra: in addition to the prominent carbon (D_0,A_0) peak at 1.489 eV, very weak structure is observed at approximately 1.481 eV which is attributed to silicon acceptors. There is an absence of "defect"-related PL signals (Kunzel and Ploog 1980) in the present spectra (1.504 - 1.512 eV and 1.466 - 1.482 eV).

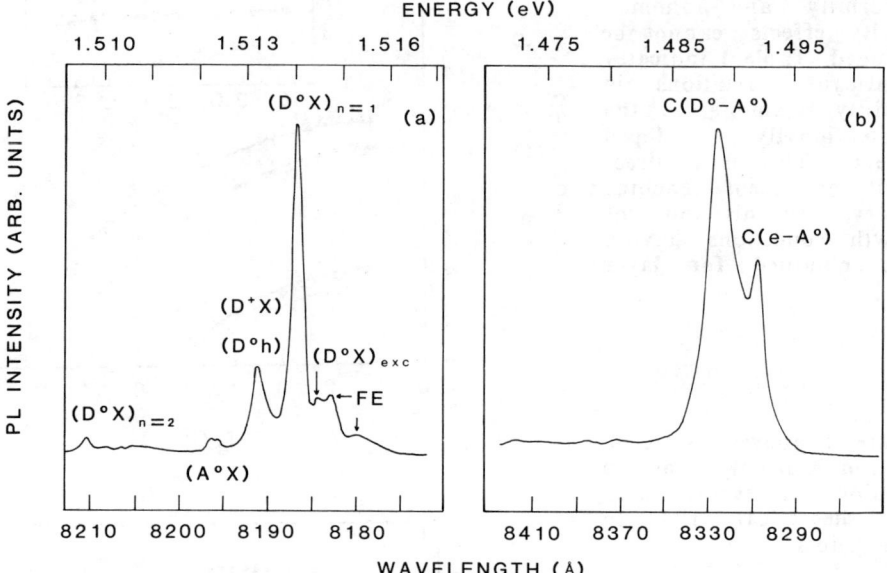

Figure 3: PL at 1.8 K (a) Sample #2 in exciton region: 10 mW cm^{-2} intensity. (b) Band-acceptor region for sample #1: 1 mW cm^{-2} intensity.

Figures 4(a) and (b) show FIRPC spectra of several samples using laser excitation wavelengths of 118 µm and 302 µm which excite the 1s-2p+ and 1s-2p- transitions respectively. Other laser wavelengths were also employed to ensure proper tracking of spectral features with field. The experimental caveats which are necessary to interpret FIRPC spectra have been thoroughly summarised by Armistead et al (1984). Thinned samples and additional band gap light were required to obtain the present results.

4. DISCUSSION

The present growth technique produces material with a consistently low background acceptor concentration: in the most favourable case (8.7x10^{12} cm^{-3}) this approaches the quality of OMVPE samples. This reduction of ionised acceptor background is borne out by the shape of the FIRPC lines (see below) and the $(A_0,X)/(D_0,X)$ ratio in the PL measurements. The trends in sample electrical properties shown in Table I and Figures 1 and 2 (e.g. peak mobility shifts with temperature and variations with N_D+N_A), are consistent with previous reports (Skromme et al 1985 and

Larkins et al 1987) and with the very recent study of Chand et al (1988). From Figure 4(a) it is evident that sample #1 has Si as a major contaminant. The chemical shift for Si (X_1) is confirmed from the spectrum of the Si back-doped sample #2. The information obtained from the magnetic field study of TES PL spectra also confirms the presence of Si as the major contaminant, with S present in small quantities. Additional inadvertant contaminants are identified in Figure 4 using the reliable catalogue of chemical shifts established by Armistead et al (1984) in studies of LPE and VPE material. The presence of Se/Sn (Afsar et al 1981, Cooke et al 1978), is noted in autodoped samples #1 and #4, but not in sample #3, which was grown with a different As source. It is not always straightforward to draw quantitative conclusions about relative amounts of impurities in view of the sensitivity of the measured peak heights to laser intensity, sample bias, additional illumination, layer thickness and precise sample temperature. The experimental lineshapes for 1s–2p_ spectra are better comparators with theoretical lineshapes (Larsen 1973) than 1s–2p+ spectra. The FIRPC linewidths quoted by Skromme et al (1985) for MBE material are found to be similar to present values. However, the narrowest linewidth presently observed is approximately 50% greater than that noted by Armistead et al (1984) under comparable circumstances for a VPE sample of similar mobility.

The PL features shown in Figure 3(a) are similar to those reported elsewhere (Larkins et al 1986) for high purity MBE samples. The FE dip observed in the present measurements, however, contrasts with its manifestation as a sharp peak (Chand et al 1988) in slightly higher mobility samples. It has been argued that a low donor concentration may lead to a reduction of FE polariton scattering and thus to a peak at this energy (Lee et al 1985, Steiner et al 1986). However, the precise lineshape of the FE feaure is known to be very dependent on sample history, excitation intensity, surface preparation and residual strain. These factors

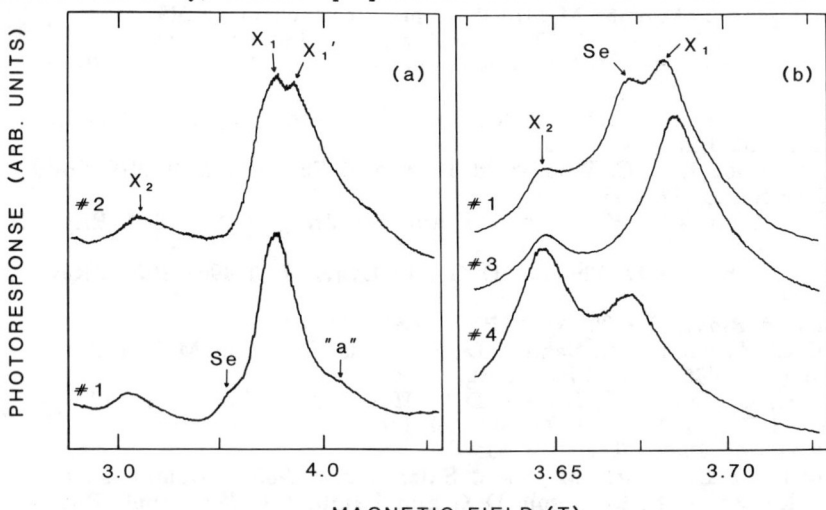

Figure 4: FIRPC at 4.2 K for 3 μm samples. (a) 302 μm laser wavelength (b) 118 μm. In addition to the major contaminants X_1 (Si), X_2 (S) and Sn/Se, the high field shoulder on X_1 (X_1') and the feature "a" noted by Armistead et al (1984) in VPE samples are consistently observed in MBE material. Their assignments are unknown.

were found to be important in the present study and it is suggested that an increase in the scattering of the exciton–polariton may reduce the PL emitted from the surface of the present samples. The predominant residual acceptor in these samples is carbon, which may originate from the As source in view of the reduction of "background" C achieved with the new source geometry. A weak $Si(D^o-A^o)$ peak is noted at 8370 Å but the concentration of Si_{As} is negligible in comparison with C_{As} in these samples.

Although the present growth technique produces MBE samples with electrical properties approaching that of VPE and LPE material, the quality of the PL and FIRPC spectra (as judged by (D_o,X) FWHM, $1s-2p_-$ linewidths etc.) appears somewhat degraded. Possible causes for this may be lower surface velocity recombination effects for exciton–polaritons which reduce PL efficiencies and random strains increasing field broadening effects for FIRPC spectra. Such effects would appear to be inherent to MBE material.

ACKNOWLEDGEMENTS

The general support of the Science and Engineering Research Council, U.K., is gratefully acknowledged. M.B. Stanaway acknowledges financial assistance from Marconi Electronic Devices, Lincoln. We are grateful to M.S. Skolnick, RSRE, Malvern, for performing TES measurements on these samples. We are obliged to A.J. Page and J.R. Middleton for expert technical assistance.

REFERENCES

Afsar M N, Button K J, Cho A Y and Morkoc H 1981 Int. J. Infrared and Millimeter Waves 21 113
Ambridge T and Faktor M M 1975 J. Appl. Electrochem. 5 319
Anderson D A and Apsley N 1986 Semicond. Sci. Technol. 1 187
Armistead C J, Knowles P, Najda S P and Stradling R A 1984 J. Phys. C. Solid State Phys. 17 6415
Chand N, Miller R C, Sergent A M, Sputz S K and Lang D V 1988 Appl. Phys. Lett. 52 1721
Chandra A, Wood C E C, Woodward D W and Eastman L F 1979 Solid State Electronics 22 645
Cooke R A, Hoult R A, Kirkman R F and Stradling R A 1978 J. Phys. D 11 945
Dean P J, Cuthbert J D, Thomas D G and Lynch R T 1967 Phys. Rev. Lett. 18 122
Kunzel H and Ploog K 1980 Appl. Phys. Lett. 37 416
Larkins E C, Hellman E S, Schlom D G, Harris J S, Kim M H and Stillman G E 1986 Appl. Phys. Lett. 49 391
Larkins E C, Hellman E S, Schlom D G, Harris J S, Kim M H and Stillman G E 1987 J. Cryst. Growth 81 344
Larsen D M 1973 Phys. Rev. B 8 535
Lee J, Koteles E S, Vassell M O and Salerno J P 1985 J. Lumin. 34 6
Shastry S K, Zemon S, Kenneson D G and Lambert G 1988 Appl. Phys. Lett. 52 150
Skromme B J, Bose S S, Lee B, Low T S Lepkowski T R, De–Jule R Y, Stillman G E and Hwang J C M 1985 J Appl. Phys. 58 4685
Steiner T, Thewalt M L W, Koteles E S and Salerno J P 1986 Phys. Rev. B 34 1006

Electron beam doping of Zn into GaAs (impurity depth profiles in GaAs and Zn)

Takao WADA, Akihiro TAKEDA, Masaya ICHIMURA, Michihiko TAKEDA[*], and Hisashi MORIKAWA[*]

Department of Electrical and Computer Engineering
Nagoya Institute of Technology, Nagoya 466, Japan

[*]Government Industrial Research Institute, Nagoya 462, Japan

ABSTRACT

Zn atoms were introduced into GaAs by using the electron-beam doping method. A Zn sheet was sandwiched between two wafers of GaAs, and the surface of the GaAs was irradiated with 7MeV electrons. The distribution of Zn in the GaAs was observed by secondary-ion-mass-spectroscopy (SIMS). The Zn concentration is very high near both the front and back surfaces, while it is relatively low and almost constant except near the surfaces. C-V profiling measurements were also carried out, and the results agree qualitatively with the results of SIMS.

INTRODUCTION

By far the most important compound semiconductor for device applications is GaAs. The greatest number of implantation experiments have therefore been made using this material. As p-type doping elements, beryllium, cadmium, magnesium and especially zinc have been studied (for example, Hunsperger 1972). During implantation, ions come to rest at random positions in the crystal lattice; many different type of radiation damage and dislocations are produced by nuclear collisions and collision cascades, sometimes going as far as the formation of amorphous areas. Therefore it is necessary, by a suitable heat treatment (annealing), to restore the crystal lattice and place the implanted ions on electrically-active lattice sites (Gibbons 1980).
Electron irradiation avoids the complication attendant upon the generation of complex damage regions. Electron beam doping (EBD) (Wada 1981a, 1981b, 1984, 1988a), oxidation (Wada 1988b) and epitaxy (Wada 1987, 1988c), which use high energy electron irradiation, were recently reported by the author and others. For the EBD samples, U-shaped diffusion profiles of impurities in substrates obtained experimentally have been explained by considering the "kick-out" mechanism and the surface diffusion process (Wada 1984).
In the present paper, Zn impurity concentration profiles in GaAs are investigated by using the EBD technique and the Zn impurity sheet.

© 1989 IOP Publishing Ltd

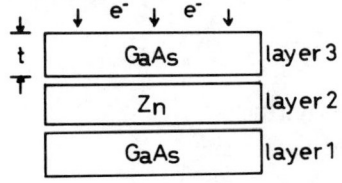

Fig. 1 Schematic diagram for the electron-beam irradiation experiments of the sandwich array of GaAs/Zn/GaAs (array III).

EXPERIMENTAL PROCEDURES

Wafers used in the experiments were (100)-oriented undoped semi-insulating GaAs grown by liquid encapsulated Czchoralski (LEC) with an area of 6x6mm^2. The thickness t of the wafer was about 0.6mm. As shown in Fig. 1, the Zn sheet (99.9%) with t=0.5mm was sandwiched between the two wafers of GaAs, constituting a GaAs(layer 3)/Zn(layer 2)/GaAs(layer 1) structure (array III). Each layer was in contact with the other layer surface. It was shown that Zn doping into GaAs is more effective in this sandwiched structure than in a simple two-layer structure of Zn/GaAs (array II) (Wada 1988a). In this paper we have concentrated on the sandwich structure. The surface of layer 3 was irradiated with a fluence of 5×10^{17} electrons cm^{-2} at 7MeV from an electron linear accelerator with a pulse width of 3.5 μs, a 200 Hz duty cycle and an average electron-beam current of 40 μAcm^{-2}. During the irradiation, the samples were put in a circulating water bath or in a vacuum ($10^{-5} \sim 10^{-6}$ Torr). In the case of irradiation in water, the temperature of the bath was kept at a constant value of 50°C by a thermoregulator, but the temperature of the sample cannot be precisely controlled in the case of irradiation in a vacuum; it will rise to above 140°C. Since the purpose of this paper is to show that enhanced diffusion occurs at low temperatures, we have mainly described the results of the irradiation in the water bath.

The concentration profiles of impurity atoms in each layer were measured with a secondary-ion-mass spectrometer (SIMS) (CAMECA IMS-3f). A focused oxygen ion beam (diameter 40μm) with an ion energy of 10 keV was used in a 5×10^{-9} Torr vacuum.

Measurements of electrochemical C-V profiling were also carried out on GaAs wafer for array III after conventional furnace annealing with a SiO_2 cap layer.

EXPERIMENTAL RESULTS

The results of SIMS measurements were all plotted in arbitrary units as functions of depth from the front and back surfaces. The concentration profiles of Zn atoms in GaAs wafers are shown for layer 1 and layer 3 in Figs.2 (a) and (b), respectively. They are quite similar to each other. The profiles seem symmetric with respect to the middle of the wafers. The concentration is very high near the surfaces, while it is

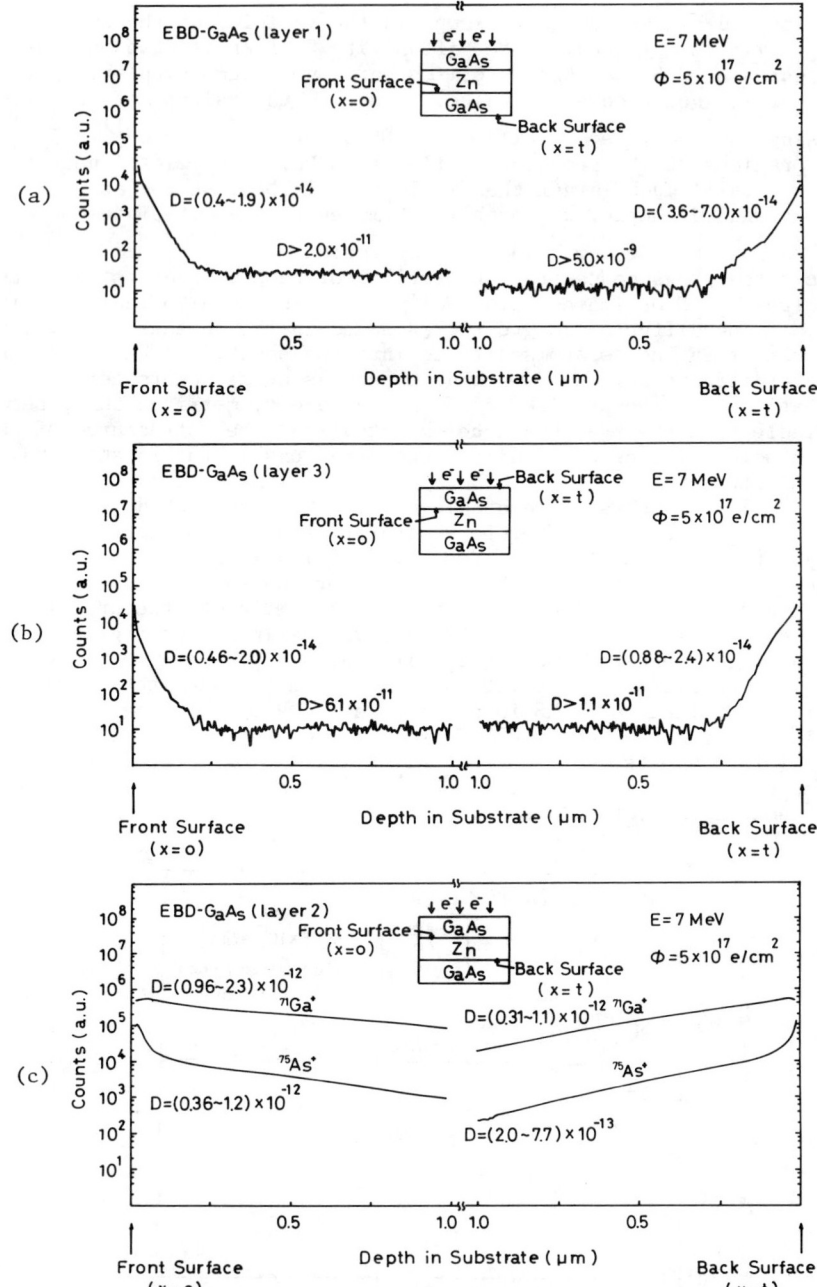

Fig. 2 Concentration profiles of impurity atoms (in arbitrary units) measured by SIMS as functions of depth from both the front and back surfaces. (a) and (b): profiles of Zn in GaAs wafers. (c): profiles of Ga and As in a Zn sheet. Values of diffusivity D are also indicated in $cm^2 sec^{-1}$.

relatively low and almost constant except in the vicinity of the surfaces, i.e., the profile is U-shaped. The diffusivity D of Zn is also indicated in the figures. Since the profiles are not complementary error functions, D is not a constant value. Here, the values of D were obtained using the analysis of Wei(1961). D is very large ($>10^{-11}$ $cm^2 s^{-1}$) in the flat regions of the profiles, while it is relatively small near the surfaces but still much larger than would usually be expected at 50 °C. Comparing a standard sample, we roughly estimated that an absolute concentration of Zn is about $10^{19} cm^{-3}$ in the flat region.

Figure 2 (c) shows concentration profiles of Ga and As atoms in a Zn sheet. Values of D estimated using Wei's analysis are also indicated in the figure. The diffusion is greatly enhanced in the Zn sheet, too, and a large number of Ga and As atoms diffuse into the sheet. This suggests that the interdiffusion of Zn atoms and, Ga and As atoms occurs across the Zn/GaAs interface. The profiles of Ga and As are symmetric with respect to the middle of a sheet as would be expected from the structure of a sample. Absolute values of impurity concentrations in the Zn sheet have not been estimated, yet.

Similar SIMS results were obtained for the sandwiched structure irradiated in a vacuum. Therefore, interdiffusion is not strongly affected by water or oxide layers formed by reaction with water.

Figure 3 shows a profile of the net acceptor concentrations of layer 1 after 20 min annealing at 750°C. It is very similar to the profile of Zn atoms shown in Fig. 2(a). Its absolute values in a flat region seem smaller than SIMS results, but it is difficult at present to evaluate rigorously the electrical activation ratio of Zn because the absolute concentration obtained by SIMS is not accurate enough. The acceptor concentration after annealing at 800°C is little different from that shown in Fig.3. This suggests that the thermal activation of a Zn acceptor is nearly completed by annealing at 750°C. The results for layer 3 are almost the same as those for layer 1.

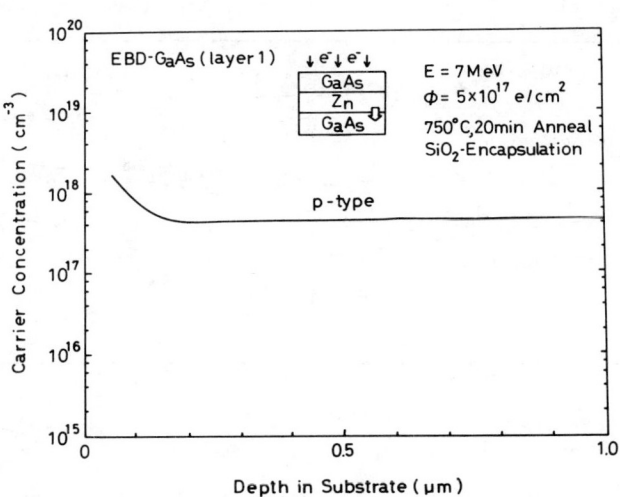

Fig. 3 Net acceptor concentration profile in the GaAs wafer (layer 1) measured with an electrochemical C-V profiler as a function of depth from the front (upper) surface.

DISCUSSION

The extrapolated range of electrons at 7MeV is so long that the incident electrons penetrate through the sample (Tabata 1972). Since the migration of atoms occurs even in the direction opposite to the electron flux, diffusion enhancement due to direct collision may be excluded from consideration. We believe that the diffusion observed is mainly due to the concentration-enhanced diffusion.

In a GaAs crystal, a Zn atom may become a substitutional atom occupying a group III lattice site or an interstitial atom. A Zn interstitial atom is rather free to move. In the thermal equilibrium state, the ratio of Zn interstitials to Zn substitutionals is very small (Weisberg 1963), because the substitutional state is energetically favorable. Thus, though Zn interstitials will have a large diffusivity D, the effective diffusivity of Zn atoms is not very large.

In the case of electron beam oxidation, it has been reported that the chemical shift for Ga_{3d} and Ga_A (Auger electron) measured by X-ray photoelectron spectroscopy was nearly equal to that of a case of the conventional plasma grown oxide GaAs structure and thus its mechanism was similar to that of plasma oxidation on semiconductors (Wada 1988b). In the present experiments, during irradiation, plasma may also be produced by a full or partial ionization that includes electrons, holes and some ions at the interfaces of water, the Zn sheet and GaAs wafer. Thus, recoil Zn atoms in a Zn sheet may be introduced into GaAs wafers and move as Zn interstitials. The interstitials may change position with Ga atoms or recombine with Ga vacancies and become Zn substitutionals, but some portion of Zn substitutionals will be displaced again. Therefore, the ratio of the interstitials will be much larger during the irradiation than in the equilibrium state, and thus the effective diffusivity will also become much larger. In addition, the diffusion of the Zn interstitials can be enhanced via the energy release mechanism and the Bourgoin mechanism, since a great number of electron-hole pairs are created during irradiation (Bourgoin 1978). However, at present it has not been certain whether these mechanisms are really operative: it is not confirmed that saddle point change, which is necessary for the Bourgoin mechanism, or nonradiative carrier capture, necessary for the energy release mechanism, actually occurs for a Zn interstitial in GaAs.

Near the surfaces of substrates, Zn concentration is very high and D is relatively small. Since Zn substitutionals have larger solubility and smaller D than Zn interstitials, it is expected that most Zn atoms are in the substitutional state there. This may be explained by considering the kick-out mechanism, originally proposed for Au in Si (Frank 1980). According to this mechanism, the following reaction occurs

$$Zn_s + Ga_i \rightleftarrows Zn_i \qquad (1)$$

where subscripts i and s mean interstitial and substitutional states, respectively. Considering the law of mass action, we can see that a concentration of Zn_s is not high where a large number of Ga_i exist. During irradiation, many Frenkel pairs and thus Ga_i's are created, and therefore the Zn_s concentration will not be high in the bulk region. However, Ga_i concentration is reduced near the surfaces since a surface acts as a sink of interstitials. Hence, Zn_s concentration can be high near the surfaces. Fast surface diffusion will make the Zn profiles

symmetric with respect to the middle of the wafer, as discussed in the previous paper (Wada 1984).

The enhanced diffusion of Ga and As atoms in a Zn sheet will be due to the increase of vacancies; both Ga and As atoms have larger atomic radii than Zn, and thus they will migrate via the vacancy mechanism rather than the interstitial mechanism (Tuck 1974). Since the effectiveness of Zn doping is greatly improved by adopting array III (GaAs/Zn/GaAs structure) rather than array II (simple Zn/GaAs structure), atom migration through the Zn sheet will play an important role in the doping process. A study on atom migration in the Zn sheet is now in progress.

SUMMARY

Zn atoms were introduced in undoped GaAs by using the electron-beam doping method. Sandwiched arrays of GaAs/Zn/GaAs were irradiated with 7MeV electrons. Concentration profiles of Zn in GaAs were measured by SIMS and found to be U-shaped. Concentration profiles of acceptors obtained by C-V measurement agree qualitatively with the SIMS results.

ACKNOWLEDGEMENTS

This work was partly supported by the Scientific Research Grant-in-Aid for Special Project Research on "Alloy Semiconductor Physics and Electronics" from the Ministry of Education, Science and Culture.

REFERENCES

Bourgoin J C and Corbett J W 1978 Radiation Effects 36 157
Frank W, Seeger A and Gosele U 1980 Defects in Semiconductors (Amsterdam: North-Holland) pp 31-54
Gibbons J F 1980 Handbook on Semiconductors (Amsterdam: North-Holland) Vol.3, pp 601-40
Hunsperger R G, Wilson R G, and Jamba D M 1972 J. Appl. Phys. 43 1395
Tabata T, Ito R and Okabe S 1972 Nucl. Instr. and Meth. 103 85
Tuck B 1974 Introduction to Diffusion in semiconductors (Southgate House: Peter Peregrinus) Chap.5
Wada T 1981a Nucl. Instr. and Meth. 182/183 131
Wada T 1981b Proc. 3rd Int. Conf. on Neutoron-Trasmutation Doped Si (New York: Plenum Press) pp 447-71
Wada T and Hada H 1984 Phys. Rev. B 30 3384
Wada T and Maeda Y 1987 Appl. Phys. Lett. 51 2130
Wada T and Takeda A 1988a presented at 6th Int. Conf. Ion Beam Modification of Materials 1988
Wada T 1988b Appl. Phys. Lett. 52 1056
Wada T and Maeda Y 1988c Appl. Phys. Lett. 52 60
Wei L Y 1961 J. Phys. Chem. Solids 18 162
Weisberg L R and Blanc J 1963 Phys. Rev. 131 1548

The DX center in GaAs and $Al_xGa_{1-x}As$

T.N. Theis

IBM Research Division, T. J. Watson Research Center, Yorktown Heights, NY, USA

ABSTRACT: The simple donor in GaAs and $Al_xGa_{1-x}As$ exhibits metastability, binding electrons in either shallow hydrogenic states, or a deep highly localized state. Recent proposals of "negative U" behavior of the donor serve to focus the debate over the origin of the metastability. Here I review some experimental results which support these proposals. A negative U model naturally explains why electron capture to the deep level should proceed via an intermediate excited state, and might also resolve an apparent conflict between photoluminescence and far-infrared absorption studies of the hydrogenic levels.

1. INTRODUCTION

Over the last twenty years, increasingly detailed studies of donors in III-V ternary alloys have revealed increasingly complex electronic properties. The key to this complexity is the coexistence in these materials of both shallow effective mass donor levels, and a deep or highly spatially localized level. Lang et al. (1979) found the optical ionization energy for the deep level to be much larger than the thermal ionization energy, indicating a large energetic relaxation of the lattice upon charge capture. Because such behavior was not expected for the simple donor, the deep level was attributed to a complex defect consisting of the donor (D) and an unknown defect (X), hence the name DX center. Greatly improved optical ionization studies (Legros et al. 1987, Mooney et al. 1988a) continue to support the idea of large lattice relaxation. Indeed, Mooney et al. found no detectable photoionization cross section for the deep level below 0.8 eV, even though a tunable infrared laser allowed the cross section to be measured eight orders of magnitude below its peak value. [Apparently contradictory results reported by Henning and Ansems (1987a, 1987b) have since been reevaluated (Henning et al. 1988a, 1988b, Montie and Henning 1988).] Other characteristics of the deep level that are readily explained only by a large lattice relaxation include the very small and strongly thermally activated electron capture cross section (Lang et al. 1979), the large and weakly thermally activated hole capture cross section (Watanabe et al. 1987) and the large vibrational entropy term in the thermal activation energy (Theis et al. 1988c). The hydrostatic pressure dependence of the capture and emission rates also indicates a large lattice relaxation (Li et al. 1987). On the other hand, occupation of the donor-related deep level in GaAs under hydrostatic pressure, at concentrations approaching the total dopant concentration (Mizuta et al. 1985, Maude et al. 1987), indicate that the center is just the simple donor. It is well known that the majority of donors normally occupy substitutional sites in GaAs, [See, for instance, Maguire et al. (1987).] but the location of the donor when the deep level is occupied is uncertain. X-ray absorption fine structure experiments suggest that Se, a group VI donor, remains on the substitutional site when the deep level is occupied (Mizuta and Kitano 1987), while Mossbauer spectroscopy suggests a significant local distortion of the atoms around Sn, a group IV donor (Gibart et al. 1988).

Along with this experimental activity there has been a growing debate concerning the role of lattice relaxation in explaining the metastability. Many have argued that the simple donor can bind an electron in a highly localized state, but that the lattice relaxation will be small (Hjalmarson and Drummond 1986, Yamaguchi 1987, Oshiyama 1988). A related approach from the point of view of effective mass theory emphasizes the role of intervalley scattering in giving rise to the deep level (Bourgoin and Mauger 1988). That the simple donor might undergo a large lattice relaxation has also been suggested, although not convincingly demonstrated

© 1989 IOP Publishing Ltd

(Morgan 1986). Very recently, Chadi and Chang (1988) and Khachaturyan et al. (1988) have proposed that a large lattice relaxation may be stabilized when *two* electrons are localized by the donor. Khacahturyan et al. give a general argument for such "negative U" behavior. (U is the energy difference between the two-electron and one-electron states.) Chadi and Chang present a microscopic model where the lattice relaxation involves movement of either a group IV donor away from a nearest neighbor As into a threefold coordinated interstitial site, or movement of nearest-neighbor Ga (or Al) atom toward a group VI donor and into an interstitial site. In part 2 of this review, I briefly discuss some important experimental results concerning the deep level which are consistent with this model.

Whatever the microscopic nature of the center, so long as large lattice relaxation occurs, it is expected that the long range Coulombic component of the defect potential must give rise to shallow or hydrogenic bound states, separated from the deep state by an energetic barrier (Toyozawa 1983). Although the coexistence of shallow and deep levels in III-V semiconductor alloys has long been known, the assumption has often been that the two types of levels arise from chemically distinct species. That the *same* donor species gives rise to both shallow and deep levels was first demonstrated by Theis et al.(1985) in direct gap $Al_xGa_{1-x}As$. The existence of shallow effective mass levels, separated from the deep level by an energetic barrier, has since been confirmed in indirect gap $Al_xGa_{1-x}As$ by Hall effect measurements (Mizuta and Mori 1988) and by far-infrared (FIR) spectroscopy (Dmochowski et al. 1988). In part 3 of this review, I discuss these experiments in some detail and compare them with photoluminescence (PL) studies of the shallow levels (Dingle et al. 1977, Henning et al. 1988a). A possible resolution of conflicting interpretations of PL and FIR data, consistent with a negative U model of the deep level, is tentatively suggested.

2. THE DEEP LEVEL

Variable-temperature Hall measurements from many different laboratories, for both group IV and group VI donors, reveal carrier freeze-out to a deep level which *approximately* tracks the L conduction band edge as a function of alloy composition. [See Theis (1986b, 1988b) for primary references.] With decreasing Al mole fraction, x, the level moves above the direct band edge and becomes a resonant state in GaAs (Theis et al. 1988a, 1988c) and $Al_xGa_{1-x}As$ of low Al content (Theis et al. 1986a, Kirtley et al. 1988, Mooney et al. 1988b). The resonant state can be brought into the fundamental gap under hydrostatic pressure (Mizuta et al. 1985, Tachikawa et al. 1985, Li et al. 1987, Maude et al. 1987). The deep (or DX) level is readily identified by its characteristic thermally activated electron capture and emission. Activation energies for thermal capture, E_c, and thermal emission, E_e, determined by DLTS, are usually larger than the thermal ionization energy E_{DX}. Thus capture can be viewed as occurring over an effective repulsive barrier as discussed by Lang et al. (1979). In Si doped $Al_xGa_{1-x}As$, over a wide range of alloy compositions, E_e is found to be constant within experimental error while E_c exhibits a pronounced minimum at the direct to indirect crossing (Mooney et al. 1987). Extrapolation of this behavior to GaAs suggests $E_c > E_e$, giving the effective barrier of a metastable state (Theis 1986b). Experiments confirm this phenomenological picture; both capture and emission by the resonant state in GaAs are thermally activated (Theis et al.1988a, 1988c). The hydrostatic pressure dependence of the emission and capture barriers is qualitatively similar to the alloy composition dependence (Mooney et al.1986, Calleja et al. 1988a, Li et al. 1988). E_e is relatively insensitive to pressure, while E_c exhibits a minimum at the direct-to-indirect crossing.

A negative U model of the deep level may account for this seemingly complex phenomenology. A configuration coordinate diagram appropriate for the deep level in GaAs is shown in figure 1. The lattice relaxation of the deep one-electron state is assumed negligible. E_{DX}^- is defined as the *average* energy per electron of the negatively charged two electron state, following the notation used by Theis et al. (1988c). Simultaneous multiphonon capture of two electrons from the lowest lying conduction valley to the deep level ground state ($2U_\Gamma \rightarrow 2U_{DX}^-$) is improbable. The dominant capture process ($2U_\Gamma \rightarrow U_\Gamma + U_{DX}^0 \rightarrow 2U_{DX}^-$) is via the one-electron state. If this

state equilibrates rapidly with the conduction band electrons, and slowly with the deep level, then the effective capture rate will depend on the the quasi-equilibrium population of this excited state. The model thus predicts an exponential dependence of capture rate on the quasi-Fermi level. Mooney et al. (1987) not only observed such a dependence, but modeled the effect in terms of capture via an excited state. Note that the experimentally measured capture barrier will be $E_c = (E_{DX}^0 - E_\Gamma) + E_b$. To explain the pressure and alloy composition dependence of the emission and capture barriers, it is only required that the energy difference between one and two electron states be weakly dependent on band structure. *No special symmetry of the deep level is required to explain the phenomenology*, in contrast to previous explanations in terms of multiphonon emission and capture through intermediate states associated with the L valley (Theis 1988b, Calleja et al. 1988a).

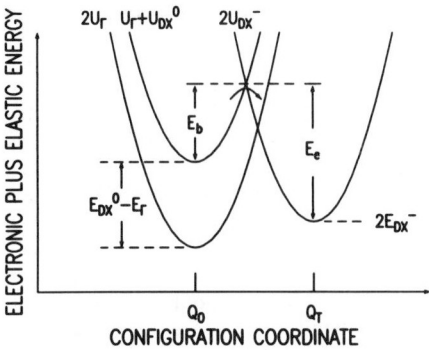

Fig. 1. Simple configuration coordinate diagram appropriate for a negative U model of the DX level.

Other recent results also appear consistent with the model of Chadi and Chang. Mooney et al. (1988b) have obtained DLTS emission spectra in Si-doped GaAs and $Al_xGa_{1-x}As$ of very low Al content ($x \lesssim 0.08$), and have, for the first time, observed discrete peaks in the spectra, attributed to discrete changes in the atomic configuration around the donor. The first substitution of an Al for a Ga atom results in a large change in E_e (0.33 eV → 0.43 eV) as well as a large change in the pre-exponential factor. The second substitution changes *only* the pre-exponential factor, and further substitutions are not discretely resolved. In agreement with the results of Calleja et al. (1988b), it is found that inhomogeneous broadening of the emission transients in the middle of the alloy composition range must be accounted for by a distribution of pre-exponential factors, *not* a distribution of activation energies as is commonly assumed. Perhaps more important, Morgan (1988) argues that the large perturbation caused by a single Al for Ga substitution, and the resolution of only a few discrete DLTS peaks, is consistent with a threefold (Ga or Al) coordinated interstitial site for the Si donor rather than a substitutional site with twelve Ga or Al second nearest neighbors. Thus these results support the proposal of an interstitial site for the donor in its deep state (Chadi and Chang 1988, Morgan 1988), but do not indicate the charge of this state. In principle, a negatively charged two-electron ground state can be distinguished from a neutral one-electron ground state by the differing statistics of occupation. Theis et al. (1988c) have therefore carefully measured the thermal occupation of the deep level as a function of temperature. They find that the vibrational entropy term in the activation energy, an unknown parameter, can be adjusted to fit either model to the data. Both models give large, but physically reasonable values for the entropy. While this study fails to resolve the question of the charge state, it does indicate, as do the other results discussed above, that a negative U model is consistent with a good body of experimental evidence.

3. SHALLOW LEVELS

In reviewing experimental observations of hydrogenic states related to the DX center, it is helpful to first discuss the relationships between the band edges and the alloy composition of $Al_xGa_{1-x}As$. Of greatest importance is the direct band gap, $E_\Gamma(x)$, and the Al mole fraction where the direct and indirect band edges cross, denoted by x_c. Optical techniques can accurately determine the band gap of a sample, but the sample composition must be independently determined by some reliable analytic procedure. Recently Kuech et al. (1987) used nuclear reaction profiling as a primary reference standard to calibrate the sample composition. After correcting their PL data for the bound exciton binding energies, one finds for direct gap $Al_xGa_{1-x}As$,

$$E_\Gamma(x) = 1.517 + 1.456x, \quad x_c = 0.37 \pm 0.015. \tag{1}$$

$E_\Gamma(x)$ is very close to linearized versions of relationships reported by Aspnes et al. (1986) and Bosio et al. (1988). Both $E_\Gamma(x)$ and x_c are extremely close to the relationship, based on photoluminescence and high precision electron microprobe measurements, presented by Oelgart et al. (1987). $E_\Gamma(x_c)$ from equation 1 also agrees to within a few meV with the value found by Dingle et al.(1977), yet x_c is much lower than the commonly accepted value of $x_c=0.43$, based on that work. Oelgart et al. (1987) and Kuech et al. (1987) therefore conclude that the discrepancy must arise from the procedure used by Dingle et al. to determine sample composition.

The commonly accepted values for the indirect band gaps in the $Al_xGa_{1-x}As$ alloy system are based primarily on extrapolation between band gaps determined or estimated for GaAs and AlAs, and band gaps determined near x_c by photoluminescence for the X valley and by Hall measurements for the L valley. In the same spirit, the band gaps can now be reestimated based on the more accurate value of x_c. Taking $E_X(x_c)$ from equation 1 and interpolating linearly to E_X in GaAs (Aspnes 1976), one obtains

$$E_X = 1.981 + 0.20x, \tag{2}$$

very close to the relationship suggested by Oelgart et al.(1987). Taking the Γ-L band edge seperation, $\Delta E_{\Gamma L}$, near x_c from Hall measurements by Lee et al. (1980) or Saxena (1981) and correcting to T=0 K with the temperature coefficients recommended by Lee et al. (1980) or Aspnes (1976) produces a range of values, $\Delta E_{\Gamma L}(x_c) = 0.04 \pm 0.01$ eV. Caluclating $E_L(x_c)$ with the help of equation 1 and linearly interpolating to E_L in GaAs (Aspnes 1976) yields

$$E_L = 1.815 + 0.757x. \tag{3}$$

These linear interpolations should be fairly accurate as long as they are not extrapolated too far into the indirect gap alloy composition range. However, I make no claim to extraordinary accuracy for the value of x_c upon which these equations are based. In comparing results from different laboratories, it may suffice to know the "local" value of x_c, determined in a particular laboratory by a particular analytical technique.

Probably the first systematic study of hydrogenic levels in the $Al_xGa_{1-x}As$ alloy system was that of Dingle et al. (1977) in Te doped material. Photoluminescent donor-acceptor (DA) transitions were identified, from which it was concluded that the responsible donor level was Γ-like ($x<0.35$), L-like ($0.35<x<0.6$), and X-like ($x>0.6$). Ionization energies of these donor levels could be determined from the temperature dependence of the PL intensity. The maximum ionization energy was found to be ~60 meV near x_c. This value was noted to be much smaller that the value ~130 meV determined from Hall measurements (SpringThorpe et al. 1975). In retrospect, it is clear that the Hall measurements were dominated by freeze-out to the deep level. On the other hand, because of its small, thermally activated electron capture rate, the population of the deep level would have been insignificant at low temperatures under the intense illumination of a PL experiment. Only the population of the hydrogenic states would have been temperature dependent, as electrons were thermally excited to the band edges. Clearly Dingle and co-workers were observing these hydrogenic levels.

The first study to clearly distinguish between hydrogenic and deep levels, and to show that they arise from the same attractive center, was that of Theis et al. (1985). The experiments were done on lightly Si doped $Al_xGa_{1-x}As$ ($0.0 \leq x < 0.35$). Capacitance-voltage techniques were used to show that in samples with $x \gtrsim 0.25$ cooled slowly in the absence of illumination, nearly all electrons were frozen into the deep level. In other words, all of the uncompensated donors acted as DX centers. The FIR absorption spectrum at 4.2 K was then used to monitor the population of the lowest lying hydrogenic state. As expected, the samples at first exhibited little or no FIR absorption. Upon illumination with a GaAs LED, an absorption line developed monotonically with the accumulated light exposure, signifying extremely persistent transfer of electrons from

the deep level to shallow states. (Electrons returned to the deep level only if the sample was warmed.) After low temperature illumination, the energy integrated absorption intensity indicated that all of the uncompensated Si acted as shallow donors. Thus the same attractive center gives rise to both a hydrogenic and a deep level. The energetic barrier separating the two states is a natural consequence of a large lattice relaxation associated with the deep state.

Figure 2 plots the energy of the shallow level absorption peak as a function of alloy composition. The dashed line is a fit to the data assuming an anticrossing between effective mass levels of A_1 symmetry associated with the Γ and L valleys. The fit assumes the alloy dependence of the Γ effective mass and dielectric constant suggested by Casey and Panish (1978), and the band edges given by equations 1-3. An equally good fit was previously obtained (Morgan 1986) using the band edges of Casey and Panish (1978). The depth of $A_1(L)$ (50 meV below the L edge) and the strength of the interaction matrix element (12 meV) are similar to the previously reported values (Morgan 1986), so the identification of the anticrossing is insensitive to assumptions about the band structure. A similar anticrossing inferred by Dingle et al. (1976) suggests $A_1(L)$ is slightly deeper for Te than for Si.

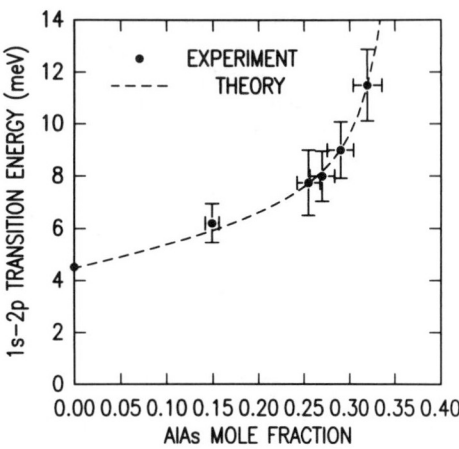

Fig. 2. Hydrogenic level transition energies in Si-doped $Al_xGa_{1-x}As$ as a function of x. The theoretical fit assumes an anticrossing of $A_1(\Gamma)$ and $A_1(L)$.

The existence of hydrogenic levels, separated from the deep level by an energetic barrier, has also been demonstrated by variable temperature Hall measurements (Mizuta and Mori 1988). This work was motivated by the report of Chand et al. (1984) of very little persistent photoconductivity for Si-doped indirect-gap $Al_xGa_{1-x}As$ samples at 4.2 K. Mizuta and Mori clearly showed that this phenomenon does *not* indicate any reduction in the concentration of deep levels in indirect gap material. Just as in direct gap material, illumination of a sample at low temperatures persistently transfers electrons from deep to hydrogenic states. However, the hydrogenic states are sufficiently deep in indirect gap material that at low temperatures these states are not fully thermally ionized. However, as the sample is warmed, electrons are activated to the conduction band and large persistent photoconductivity effects *are* observed.

Very similar phenomena are illustrated by the previously unpublished Hall data of Figure 3, obtained from a lightly Si-doped $Al_{0.29}Ga_{0.71}As$ sample. The sample was cooled in the dark, and the Hall free carrier concentration was determined as a function of temperature during a slow warm-up. Distinctly different behavior is observed at high and low temperatures. At high temperatures, the population of the deep level equilibrates rapidly on the time scale of the measurements. No persistent photoconductivity is observed, and carrier freeze-out is dominated by the deep level. At lower temperatures, where the deep level population equilibrates

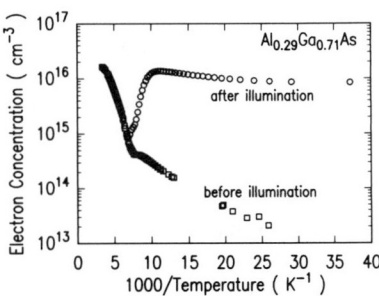

Fig. 3. Hall free carrier concentration in lightly Si-doped $Al_{0.29}Ga_{0.71}As$.

exceedingly slowly, the slope of the Arrhenius plot changes drastically. Carrier freeze-out is to

the much shallower hydrogenic level with a thermal ionization energy ~14 meV. This energy agrees well with the value inferred from the FIR data (figure 2) obtained on the same sample. Furthermore, illumination of the sample persistently transfers additional carriers to the hydrogenic state. As the temperature is then raised the electrons eventually surmount the capture barrier to the deep level and the free electron concentration drops rapidly to its equilibrium value.

Note that after low temperature illumination of this sample, a meaningful thermal ionization energy can no longer be determined from the Arrhenius plot. This may indicate that the sample has passed to the metallic side of the metal-insulator transition and that delocalized states therefore dominate the very low temperature conduction. Indeed, this light-induced control of the carrier concentration has been exploited by Katsumoto et al. (1987) in an elegant study of critical exponents of the metal-insulator transition in n-$Al_xGa_{1-x}As$. It is interesting to note that FIR spectra from the same sample, under the same post-illumination condition, (Theis et al. 1985) exhibit a broad, but well defined shallow level absorption peak. Theis et al. (1985) also compared GaAs samples above and below the metal-insulator transition and found negligible change in the peak energy corresponding to the 1s-2p transition. These results are fully consistent with the findings of Lee et al (1988) that the impurity band in a small effective mass system such as GaAs remains distinct from the main conduction band until well above the metal-insulator transition. Ionization energies inferred from FIR data should therefore be accurate, even when samples are close to or slightly above the metal-insulator transition. On the other hand, ionization energies found from transport measurements in heavily doped samples, or even lightly doped samples after illumination, must be treated with caution. Finally, it should be noted that observation of the metal-insulator transition is additional evidence of a hydrogenic state. The deep (DX) level shows no perturbation of its emission and capture kinetics (no tendency to form an impurity band) at free carrier concentrations $>1 \times 10^{19}$ cm^{-3} (Theis et al. 1988a).

Dmochowski et al. (1988) have studied the FIR absorption spectra of the lowest lying hydrogenic level in the the indirect gap alloy. These experiments were similar to those of Theis et al. (1985) but had the advantage of probing the X-like effective mass level, rather than the very shallow Γ-like level. The samples were far from the metal-insulator transition, and quantitative line shape fits to the FIR spectra could be made. The two FIR studies appear to nicely compliment each other, as can be seen from figure 4. In comparing the results, it should be remembered that Dmochowski and co-workers studied Te doped material, while Theis and co-workers studied Si doped material. Nevertheless, one may conclude that the lowest lying donor level is first Γ-like and then X-like as the Al mole fraction increases.

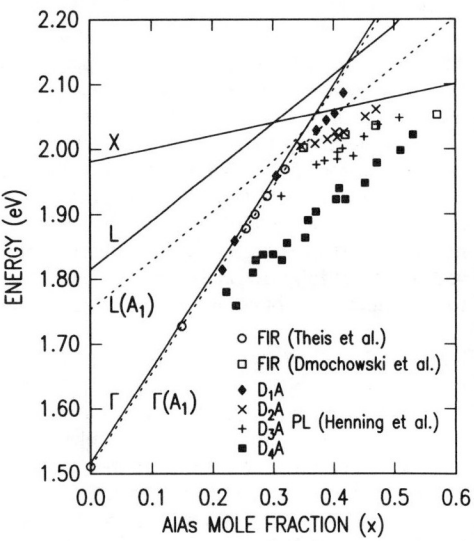

Fig. 4. Comparison of hydrogenic level ionization energies derived from far-infrared transitions (Theis et al. 1985, Dmochowski et al. 1988) and photoluminescence (Henning et al. 1988).

Henning et al. (1988a) have performed an extensive study of donor-acceptor photoluminescence in Si doped $Al_xGa_{1-x}As$, and have come to quite different conclusions. In figure 4 this PL data is compared with the ionization energies inferred by Dmochowski et al. (1988), and the ionization energies inferred from the data analysis of figure 2 above. The alloy composition in the FIR experiments

was determined by electron microprobe, so the band edges versus alloy composition can be taken to good accuracy from equations 1-3. In plotting the PL results versus the same band edges, I have scaled their alloy compositions by a multiplicative factor, $x_c/x_c^H = 0.37/0.40$, where $x_c^H = 0.40$ is the "local" value for the direct to indirect gap crossing given by Henning (1988c). I have also added the estimated acceptor depth, 30 meV, to the PL transition energies.

That this coordinate transformation is reasonable is indicated by the excellent match between the D_1A PL, identified by Henning et al. as $A_1(\Gamma)$, and the corresponding FIR points (Theis et al. 1985). I believe, considering the unavoidable scatter in the PL data, that the PL is also consistent with the $A_1(\Gamma - L)$ anticrossing inferred from the data of figure 2 and indicated by the dashed lines in figure 4. There is also reasonable agreement between the D_3A PL, identified by Henning et al. as $T_2(X,L)$ and the lowest lying hydrogenic donor level found by Dmochowski et al. (1988). (Hall measurements in our laboratory indicate a binding energy of ~75 meV for the lowest lying hydrogenic donor level in Si-doped $Al_{0.6}Ga_{0.4}As$. Thus the Si level may be deeper than the Te level studied by Dmochowski and coworkers.)

The identification of D_4A as a hydrogenic level, $A_1(L)$ lying ~200 meV below the L valley (Henning et al. 1988a) directly contradicts all of the FIR and Hall studies reviewed above. Those experiments measure the depth of the lowest lying hydrogenic level in quasi-equilibrium after the DX level is photoionized. Either this level lies at least 100 meV closer to the L edge than is suggested by the interpretation of the PL data, or it is separated from the hydrogenic states by a large energetic barrier. I speculate that D_4A may be related to a highly localized state with small lattice relaxation. Such a state is predicted by various theoretical approaches (Hjalmarson and Drummond 1986, Yamaguchi 1987, Oshiyama 1988), including the calculation of Chadi and Chang (1988). Since good arguments have been presented that D_4A is a zero-phonon transition (Henning and Ansems 1987) further experiments to determine the thermal depth and possible lattice relaxation of this state are needed. Hydrostatic pressure studies might resolve the question of whether the level is hydrogenic or localized. Clearly the hydrogenic level spectrum, as well as the possible spectrum of excited deep levels must be understood before we can say that we understand the DX center.

4. CONCLUDING REMARKS

I have reviewed a number of experiments which conclusively demonstrate the shallow/deep bistability of the DX center in the $Al_xGa_{1-x}As$ alloy system. There is no doubt that the deep level is highly localized, and there is, at the time of this writing, no explanation, other than a large lattice relaxation, which accounts for the properties of this deep level. The evidence that the DX center is the simple donor is also quite persuasive. The calculation of Chadi and Chang (1988) indicates that a large lattice relaxation can occur for the simple donor, and I have reviewed various experimental results which might be explained by a negative U model. I wish to emphasize, however, that none of these experiments confirms the model. Neither the charge state nor the nature of the lattice distortion has been conclusively established. The most profitable experiments in the immediate future are likely to be those which address these questions.

REFERENCES

Aspnes D E 1976 Phys. Rev. B **14** 5331
Aspnes D E, Kelso S M, Logan R A and Bhat R 1986 J. Appl. Phys. **60** 754
Bosio C, Staehli J L, Guzzi M, Burri G and Logan R A 1988 Phys. Rev. B **38** 3263
Bourgoin J C and Mauger A 1988 Appl. Phys. Lett. **53** 749
Calleja E, Gomez A and Munoz E 1988a Appl. Phys. Lett. **52** 383
Calleja E, Gomez A, Munoz E and Camara 1988b Appl. Phys. Lett. **52** 1877
Casey H C Jr. and Panish M B 1978 Heterostructure Lasers (New York: Academic Press) pp 188-194
Chadi D J and Chang K J 1988 Phys. Rev. Lett. **61** 873

Chand N, Henderson T, Klem J, Masselink W T, Fischer R, Chang Y C and Morkoç H 1984 *Phys. Rev. B* **30** 4481
Dingle R, Logan R A and Arthur J R Jr. 1977 *Inst Phys. Conf. Ser.* **33a** 210
Dmochowski J E, Langer J, Raczynska J, and Jantsch W 1988 *Phys. Rev. B* **38** 3276
Gibart P, Williamson D L, El Jani B, and Basmaji P 1988 *Inst. Phys. Conf. Ser.* **91** 379
Henning J C M and Ansems J P M 1987a *Semicond. Sci. Technol.* **2** 1
Henning J C M and Ansems J P M 1987b *Appl. Phys. A* **44** 245
Henning J C M, Ansems J P M and Roksnoer P J 1988a *Semicond. Sci. Technol.* **3** 361
Henning J C M and Ansems J P M 1988b *Phys. Rev. B* to be published
Henning J C M 1988c private communication.
Hjalmarson H P and Drummond T J 1986 *Appl. Phys. Lett.* **48** 656
Katsumoto S, Komori F, Naokatsu S and Kobayashi S 1987 *J. Phys. Soc. Japan* **56** 2259
Khachaturyan K A and Weber E R 1988 unpublished
Kirtley J R, Theis T N and Wright S L 1988 *J. Appl. Phys.* **63** 1541
Kuech T F, Wolford D J, Potemski R, Bradley J A, Kelleher K H, Yan D, Farrell J P, Lesser P M S and Pollack F H 1987 *Appl. Phys. Lett.* **51** 505
Lang D V, Logan R A and Jaros M 1979 *Phys. Rev. B* **19** 1015
Lee H J, Juravel L Y, Wooley J C and SpringThorpe A J 1980 *Phys. Rev. B* **21** 659
Lee M W, Romero D, Drew H D, Shayegan M and Elman B S 1988 *Solid State Commun.* **66** 23
Legros R, Mooney P M and Wright S L 1987 *Phys. Rev. B* **35** 7505
Li M F, Yu P Y, Weber E R and Hansen W 1987 *Phys. Rev. B* **36** 4531
Li M F, Shan W, Yu P Y, Hansen W and Bauser E 1988 *Proc. 15th Int. Conf. on Defects in Semiconductors*, August 22-26, Budapest
Maguire J, Murray R and Newman R C 1987 *Appl. Phys. Lett.* **50** 516
Maude D K, Portal J C, Dmowski L, Foster T, Eaves L, Nathan M, Heiblum M, Harris J J and Beall R B 1987 *Phys. Rev. Lett.* **59** 815
Mizuta M, Tachikawa M, Kukimoto H and Minomura S 1985 *Jpn. J. Appl. Phys.* **24** L143
Mizuta M and Kitano T 1987 *Appl. Phys. Lett.* **52** 126
Mizuta M and Mori K 1988 *Phys. Rev. B* **37** 1043
Montie E A and Henning J C M 1988 *J. Phys. C: Solid State Phys.* **21** L311
Mooney P M, Calleja E, Wright S L and Heiblum M 1986 *Defects in Semiconductors* ed H J von Bardeleben (Switzerland: Trans Tech) pp 417-22
Mooney P M, Caswell N S and Wright S L 1987 *J. Appl. Phys.* **62** 4786
Mooney P M, Northrup G A, Morgan T N and Grimmeiss H G 1988a *Phys. Rev. B.* **37** 8298
Mooney P M, Theis T N and Wright S L 1988b *Proc. 15th Int. Conf. on Defects in Semiconductors*, August 22-26, Budapest
Morgan T N 1986 *Phys. Rev. B.* **34** 2664
Morgan T N 1988 *Proc. 15th Int. Conf. on Defects in Semiconductors*, August 22-26, Budapest
Oelgart G, Schwabe R, Heider M and Jacobs B 1987 *Semicond. Sci. Technol.* **2** 468
Oshiyama A 1988, *Proc. 15th Int. Conf. on Defects in Semiconductors*, August 22-26, Budapest
Saxena A K 1981 *Phys. Stat. Sol. (b)* **105** 777
Springthorpe A J, King F. D. and Becke A 1975 *J. Elec. Mat.* **4** 101
Tachikawa M, Fujisawa T, Kukimoto H, Shibata A, Oomi G and Minomura S 1985b *Jpn. J. Appl. Phys.* **24** L893
Theis T N, Kuech T F, Palmateer L F and Mooney P M 1985 *Inst. Phys. Conf. Ser.* **74** 241
Theis T N, Parker B D, Solomon P M and Wright S L 1986a *Appl. Phys. Lett.* **49** 1542
Theis T N 1986b *Defects in Semiconductors* ed H J von Bardeleben (Switzerland: Trans Tech) pp 393-8
Theis T N, Mooney P M and Wright S L 1988a *Phys. Rev. Lett.* **60** 361
Theis T N 1988b *Inst. Phys. Conf. Ser.* **91** 1
Theis T N, Morgan T N, Parker B D and Wright S L 1988c *Proc. 15th Int. Conf. on Defects in Semiconductors*, August 22-26, Budapest
Toyozawa Y 1983 *Physica* **116B** 7
Watanabe M O, Ahizawa Y, Sugiyama N and Nakanisi T 1987 *Inst. Phys. Conf. Ser.* **83** 105
Yamaguchi E 1987 *Jour. Phys. Soc. Japan* **56** 2835

Studies of the DX centre in heavily doped n⁺GaAs

L. Eaves*, T. J. Foster*, D. K. Maude*, J. C. Portal°, R. Murray⁺,
R. C. Newman⁺, L. Dmowski°, R. B. Beall#, J. J. Harris#, M. I. Nathan'
and M. Heiblum'

*Department of Physics, University of Nottingham, NG7 2RD, U.K.
°Dept. de Genie Physique, INSA, 31077 Toulouse and SNCI-CNRS, Avenue des Martyrs, 38042 Grenoble, France.
⁺J. J. Thomson Physical Laboratory, University of Reading, RG6 2AF, U.K.
#Philips Research Laboratories, Redhill, Surrey RH1 5HA, U.K.
'IBM, T. J. Watson Research Center, Yorktown Heights, NY 10598, U.S.A.

ABSTRACT: Shubnikov-de Haas measurements on n⁺GaAs heavily doped with either Si or Sn show that with increasing pressure, electrons are trapped out of the Γ-conduction band into localised DX states ($D^+ + e_\Gamma^- \to DX^\circ$), with an increase in mobility. These trapped electrons are returned to the band under optical illumination for $T \leq 100$ K. In the most heavily doped layers ($1-2 \times 10^{19}$ cm⁻³) the DX level is partially filled at atmospheric pressure, thus limiting the free carrier concentration. The energy of DX relative to the Γ-minimum is deduced and is found to be doping-dependent. Comparison with local vibration mode measurements for n⁺GaAs(Si) indicates that the DX level is due to simple substitutional Si_{Ga}.

1. INTRODUCTION

The properties of the DX level in the n-(AlGa)As alloy system have been a subject of great interest for more than a decade. Recent work (Mizuta et al 1985, Maude et al 1987, Eaves et al 1987, Theis et al 1986, 1987, 1988 and references therein) has shown that a donor-like level of DX character (Lang 1986 and refs. therein) strongly influences the electrical properties of moderately doped n-GaAs under the application of hydrostatic pressure and acts to limit the free carrier concentration of heavily doped material at around $1-2 \times 10^{19}$ cm⁻³ at ambient pressure. This and other work has also indicated that the DX level is associated with the isolated simple substitutional donor (Si or Sn) rather than a complex (e.g. one involving a simple donor and a native defect). The degree of lattice relaxation of the donor atom in the DX state remains a matter of controversy (Henning 1987, Mizuta and Kitanu 1988, Li et al 1988). The experimental work on DX has been complemented by theoretical investigations of its electronic structure (e.g. Morgan 1986, Hjalmarson 1986, Yamaguchi 1987).

This paper is arranged as follows. Sections 2 and 3 describe a series of Shubnikov-de Haas measurements under hydrostatic pressure of up to 15 kbar on a range of n⁺GaAs layers doped with either Si or Sn. The

© 1989 IOP Publishing Ltd

measurements demonstrate pressure-induced trapping of free carriers, the pinning of the Fermi energy at the DX level and the persistent photoconductivity (PPC) effect with which the DX level is closely associated. From these measurements we estimate (section 4) the energy of the DX level relative to the conduction band edge of GaAs. We find that this energy is doping-dependent.

In section 5, we describe how secondary ion mass spectroscopy (SIMS) and far infrared local vibration mode (LVM) measurements can be used to calculate the concentration and the site distribution of the shallow donors in Si-doped material. From these measurements we conclude that the DX level which gives rise to the Fermi level pinning, the pressure-induced trapping and the PPC in our Si-doped layers must be associated with the isolated simple substitutional donor Si_{Ga}.

2. EXPERIMENT

A series of six n-GaAs layers heavily doped (0.2-1.8 x 10^{19} cm^{-3}) with either Si or Sn were investigated. The layers (labelled a to f in Figure 1) were prepared by molecular beam epitaxy. Samples a, d, e and f were grown at Philips and samples b and c at IBM. Enhanced doping can be achieved by lowering the substrate temperature during growth (Neave et al 1983). This technique was used for samples a and d, for which the substrate temperature was 400°C. The Si- or Sn-doped n$^+$ layers varied in thickness between 0.15 and 12 µm. The carrier concentration was measured at low temperature and with magnetic fields up to 18 T using the Shubnikov-de Haas effect (SdH). This measures the radius of the Fermi sphere in k-space and yields directly the volume density (n) of conduction electrons in the Γ-minimum. Pressure was applied by means of a liquid clamp cell manufactured by Unipress (Warsaw). An important feature of these cells is that the pressure can be applied at room temperature only. When measurements are subsequently made at low temperatures the applied pressure is "frozen in". Cooling to low temperatures is accompanied by a small decrease of pressure (typically < 10%). The pressure on the sample is continuously monitored in situ with a calibrated n-InSb resistance manometer. The cooling procedure to the helium temperatures required for the SdH measurements involved a precool to 77 K of ~20 minutes' duration. The subsequent cooling to 4 K took place over ~5 minutes. The occupation of the DX centre is fixed at some intermediate temperature (the "freeze-in" temperature ~120 K) at which the capture and emission rates become so slow that the electron population of the trap is no longer in equilibrium with the free electron density. Above this temperature the well-known PPC effect is quenched (e.g. Theis, 1987). In order to study PPC, samples could be illuminated under pressure at low temperatures by means of a visible light-emitting diode mounted inside the cell.

As P is increased, the period, $\Delta(1/B) = 2e/\hbar (3\pi^2 n)^{-2/3}$, of the SdH oscillations increased, corresponding to a decrease in carrier concentration. The amplitude of the oscillations also increased with pressure and the oscillatory structure became observable down to lower magnetic fields (Maude et al 1987). This is a clear indication that the mean collision time τ and mobility µ increase with increasing pressure since the amplitude of SdH structure is controlled by the factor $\omega_c\tau$, where $\omega_c = eB/m^*$ is the cyclotron frequency. Since the period $\Delta(1/B)$ of the SdH structure and the zero field resistance give, respectively, the

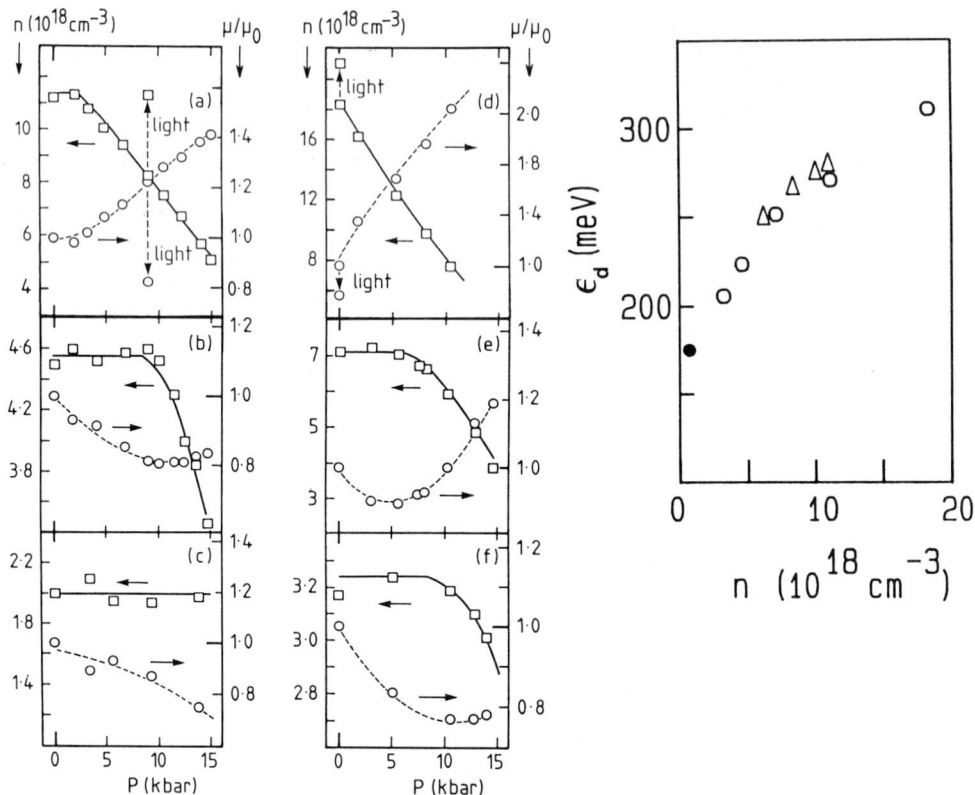

Figure 1 (left) Carrier concentration n (□) and mobility μ (o) versus pressure P for a range of Si (a,b,c) and Sn (d,e,f) doped samples. The mobility is normalised to its atmospheric pressure value. The solid lines show the expected variation of n with P calculated using Fermi-Dirac statistics as described in the text. The effect of illumination on n and μ is indicated in Figure 1(a) for the most heavily Si-doped sample at 9 kbar and in (d) for the most heavily Sn-doped sample at atmospheric pressure.

Figure 2 (right) Energy of the DX level relative to Γ minimum, ε_d (o) versus carrier concentration calculated using the procedure described in the text. Values reported by Theis et al for ε_d (△) are shown for comparison. The energy of the DX level deduced by Mizuta and coworkers is also indicated (●).

electron carrier concentration n and the conductivity σ, the variation of μ with P can be determined. Typical results are shown in Figure 1. For the most heavily doped sample, n falls immediately on increasing the pressure above atmospheric (i.e. Figure 1(d)). This indicates that for this doping the DX level is very close to the Fermi energy and is partially occupied even at atmospheric pressure. In the more lightly doped samples, e.g. Figure 1(b), (e) and (f), the carrier concentration remains roughly constant up to a critical value P_c and then falls

sharply. The sample doped at 2×10^{18} cm^{-3} (Figure 1(c)) shows no significant change of n up to 15 kbar. It can be seen that in all samples a decrease in n with P is accompanied by an increase in μ. In lighter-doped samples, over the range of pressure for which n remains constant, μ is found to decrease slowly with pressure. This effect is associated with the increase of the electron effective mass of the Γ-conduction band. These results indicate that the loss of free carriers with increasing pressure is associated with capture at donor-like localised resonance levels located between the Γ- and L-conduction band minima. This capture process is described by the relation $D^+ + e^- \rightarrow D^\circ$. The neutralisation of the ionised donors by electron capture and the corresponding decrease in ionised impurity scattering explains the increases in mobility and in amplitude of the SdH structure.

3. PERSISTENT PHOTOCONDUCTIVITY

To test if the donor-like level in our n$^+$GaAs showed the PPC effects which characterise the DX centre in (AlGa)As, we investigated the effect of pulses of light on the carrier concentration and mobility with the sample maintained under pressure at low temperature. Here it is important to note that the pressure in these experiments was initially applied at room temperature and the sample was cooled slowly. Therefore, if the pressure is above the threshold value for trapping out onto the donor-like resonance levels, a fraction of the DX donor states corresponding to thermal equilibrium (T $\simeq$ 120 K) should be filled with electrons after cooling in the dark to low temperatures. By illuminating the sample with light from an LED under pressure at low temperatures, the free electron concentration, measured by the SdH effect, was restored to its zero pressure value. The changes in n and μ following illumination at 9 kbar are indicated by broken vertical arrows in Figure 1(a) for the most heavily Si-doped sample. Similar results are shown in Figure 1(d) for the most heavily Sn-doped sample, cooled under ambient pressure and then illuminated. This increase in n persisted long after switching off the light. It was accompanied by a persistent decrease of μ and of the amplitude of the SdH structure. Such behaviour is fully consistent with photo-ionisation of a donor level ($D^\circ \rightarrow D^+ + e^-$). The PPC indicates that the centre involved has the same character as the DX centre in n-(AlGa)As. Note that on illuminating under pressure, the mobility does not fall to its zero pressure value but to a somewhat lower value due to the increase of effective mass with pressure.

Of particular note is the most heavily doped sample shown in Figure 1(d). The pressure dependence of n indicates that the DX level is at or very close to the Fermi energy at atmospheric pressure since even low pressure ($\lesssim$ 2 kbar) causes a fall in n. This is confirmed by the observation of a persistent increase in n and decrease in μ on illuminating the sample at 4 K and atmospheric pressure following a cool-down in the dark. The change in n and μ are indicated by broken vertical arrows in Figure 1(d). On illuminating n changes from 1.83×10^{19} to 2.10×10^{19} cm^{-3} and μ falls by ~20%. As discussed below, the increase of n, $\Delta n = 2.7 \times 10^{18}$ cm^{-3} is consistent with the number of electrons that we estimate to be trapped on DX at the "freeze-in" temperature of ~120 K.

In considering the origin of PPC, it is important to note that the effect can arise from the macroscopic separation of photo-excited charge carriers. Queisser and Theodorou have demonstrated convincingly that this is the mechanism of PPC in lightly doped layers. However, this model cannot explain the large absolute changes of carrier concentration in our heavily-doped samples. This can be seen by applying Gauss's law for the PPC shown in Figure 1(a) in which the carrier concentration changes by $\Delta n = 3 \times 10^{18}$ cm^{-3}. The n$^+$ layer thickness $t = 12$ μm. The associated electric field and built-in potential if there were a macroscopic separation of charge would be $E = e\Delta nt/2\varepsilon_r\varepsilon_0 = 5 \times 10^{10}$ V/m and $V = e\Delta nt^2/2\varepsilon_r\varepsilon_0 = 3 \times 10^5$ V, respectively. These values are unphysical and clearly show that macroscopic charge separation makes a negligible contribution to the observed change in n. Similar remarks apply to the PPC in Figure 1(d).

4. DISCUSSION AND ANALYSIS

It is clear from our measurements that the level responsible for the trapping is present at a concentration comparable to that of the Si or Sn doping level. The role of the pressure in producing the observed trapping-out of electrons from the Γ-minimum is due to its effect on the energy of the deep donor level and on the conduction band structure. The analysis in this section determines the energy, ε_d, of the DX level relative to the Γ-minimum at atmospheric pressure for samples (b)-(f). As shown in Figure 2, ε_d is doping-density dependent.

We have previously shown (Maude et al 1987, Portal et al 1988) that, for samples doped at $n < 5 \times 10^{18}$ cm^{-3}, the variation of n with P can be fitted by assuming that electrons capture at a donor level whose energy relative to the Γ-minimum decreases at a rate of 4.8 meV/kbar. This value is close to the pressure coefficient of the energy difference between the Γ and L minima (Blakemore 1982), which suggests that the level may have some of the character of the L-minima. As the pressure is increased above a critical value the DX level moves to a lower energy than the Fermi energy ε_F of the electrons in the Γ-conduction band, which then remains pinned at the energy of the deep level. The decrease of n with P is governed by the above pressure coefficient and by the density of states of the Γ minimum at ε_F. It is important to note that although the carrier concentration is measured at 4.2 K the actual occupation of the DX centre and hence the carrier concentration is fixed at the intermediate "freeze-in" temperature (~120 K) at which the capture and emission rates become so slow that the electron population in the DX level is no longer in equilibrium with the free electron density. Using Fermi-Dirac statistics it is possible to calculate the position of the DX centre relative to the Γ conduction band at 120 K. We assume that the number of DX centres N_{DX} is equal to the donor doping N_D. Using the usual Fermi-Dirac expression together with the charge neutrality equation, it is easy to show that, in the absence of compensation, and assuming a degeneracy g = 2, the energy of DX above the Γ minimum is given by

$$\varepsilon_d = \varepsilon_F - kT \ln\left[\frac{1}{2}\left(\frac{N_{DX}}{n} - 1\right)\right]. \qquad (1)$$

As will be discussed later in the paper, we have evidence (LVM measurements) that the compensation of the Philips-grown samples is

small, ~ 0.1. (A compensation ratio $N_A/N_D \lesssim 1/3$ has very little effect on the calculated value of ε_d). The Fermi energy ε_F in the Γ conduction minimum is calculated taking into account non-parabolicity and the pressure dependence of the effective mass (Shantharama et al 1984, Raymond et al 1979). A value of N_D slightly greater than the zero pressure carrier concentration is used for the two most heavily doped samples ($N_D = 2.1 \times 10^{19}$ cm^{-3} for sample d - see below). The calculated values of the energy of the DX level together with the pressure dependence of the Γ, X and L minima (Blakemore 1982) and of ε_F are shown in Figure 3 for two of the samples. The positions of the X and L minima illustrated in this diagram are the positions for undoped GaAs. Band gap renormalisation (Abram et al 1978) at the high doping levels used here probably shifts the X and L minima to higher energies relative to the Γ conduction minimum and may also affect the pressure coefficients, even at constant carrier concentration.

As can be seen from Figure 3, for a particular sample, the variation of the position of DX as a function of hydrostatic pressure is approximately linear. From the slope, we can estimate the pressure coefficient for the energy of the DX level relative to the Γ-minimum. With increasing doping the DX level is shifted to higher energies. For a particular dopant the pressure coefficient of DX also increases with increasing doping. These results are tabulated in Table 1. At these high doping levels it is possible that the observed reduction of the binding energy of the electron on DX arises from Coulomb and screening effects. These effects are known to occur for the much more extended shallow donor states of the Γ-minimum at low doping levels ($\sim 10^{16}$ cm^{-3}) (Leloupe et al 1978). Above the critical pressure, the electrons become trapped and neutralise ionised donors, thereby reducing the strength of the coulomb interactions. Hence the level would become deeper relative to the L-minima, giving rise to an increased rate of shift of the donor level relative to the Γ-minimum. The zero pressure values of ε_d as a function of doping shown in Figure 2 are calculated by assuming that

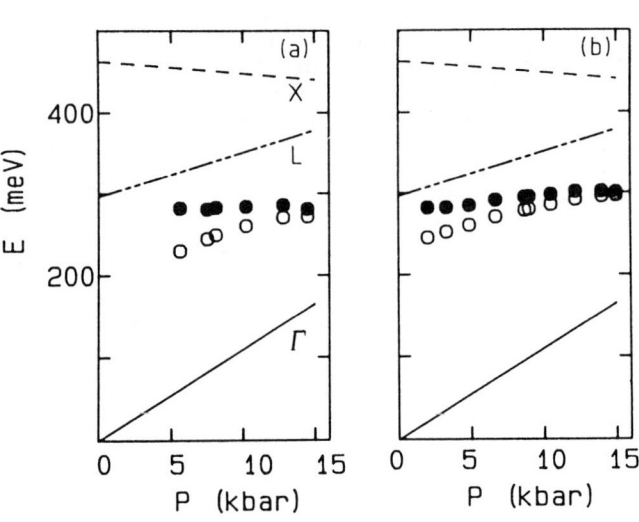

Figure 3 Energy of the DX level (●) and the Fermi energy (o) versus pressure for (a) 7.1×10^{18} cm^{-3} Sn-doped sample and (b) a 1.1×10^{19} cm^{-3} Si-doped sample. Note that the positions of the X and L minima illustrated are for undoped GaAs. Band gap renormalisation at these high doping levels probably shifts the X and L minima to higher energies relative to the Γ conduction band.

DX Centre in III-V Compounds and Related Problems

for pressures below the critical pressure the energy difference between DX and L remains fixed. Values of ε_d reported by Theis et al and Mizuta et al (1985) using DLTS are also plotted for comparison. The degree of consistency between the three sets of data is notable.

The solid lines in Figure 1 show the expected variation of n with P, calculated using an iterative method since equation (1) cannot be solved analytically for n. The variation of ε_d with pressure was calculated using the pressure coefficient in Table 1. Non-parabolicity and pressure dependence of effective mass were taken into account when calculating the Fermi energy ε_F.

Zero Pressure Carrier Conc. ($\times 10^{18}$ cm^{-3})	$E_F(P=0)$ (meV)	P_c (kbar)	N_D ($\times 10^{18}$ cm^{-3})	ε_d (meV)	$d\varepsilon_d/dP$ (meV/kbar)
(a) [Si] 11.3	226	2	12.00	270	-9.4
(b) [Si] 4.6	126	10	4.60	223	-9.1
(c) [Si] 2.0	73	pressure has no effect on n.			
(d) [Sn] 18.0	287	0	21.00	311	-13.3
(e) [Sn] 7.1	175	7	7.15	251	-10.7
(f) [Sn] 3.2	109	12	3.25	206	-6.0

Table 1 The measured carrier concentration n, at P = 0, for the samples studied are indicated in column 1. The Fermi energy ε_F is calculated taking into account non-parabolicity. P_c is the critical pressure at which n starts to fall. N_D is the donor doping level used in the Fermi-Dirac calculation. The energy of the DX level above the Γ minimum(ε_d) is calculated using the Fermi-Dirac statistics as described in the text. The final column is the pressure coefficient of the energy of the DX level relative to the Γ minimum used to obtain the fitting curves in Figure 1.

It can be seen that this procedure, taking full account of the Fermi statitics, provides a good fit to the experimental data shown in Figure 1. Note that the fit is a significant improvement over that given in our earlier paper (Maude et al 1987) where, for simplicity, it was assumed that the Fermi level was precisely pinned at the DX level for pressures above the critical pressure at which the level starts to fill.

Our analysis of the occupancy of the Γ-conduction band and the DX level in the most heavily Sn-doped sample (d) indicates that at the "freeze-in" temperature (~120 K) and atmospheric pressure the DX level has a significant occupancy (~3 $\times 10^{18}$ cm^{-3}). Thus the DX level is

acting to limit the free carrier concentration. The observed persistent increase in n ($\Delta n = 3 \times 10^{18}$ cm^{-3}) on illuminating this sample (Figure 1(d)) is in excellent agreement with the occupancy of DX expected from our analysis. This interpretation also explains why the carrier concentration deduced from Hall measurements at room temperature for the most heavily doped samples is slightly lower than that obtained from the low temperature SdH measurements at atmospheric pressure: as the sample is cooled from room temperature, the occupancy of the DX level decreases as electrons thermalise into unoccupied states of the Γ-conduction band (see also Theis 1987).

In the above analysis, we have followed other authors by neglecting the so-called band gap renormalisation. At high doping levels ($\gtrsim 5 \times 10^{18}$ cm^{-3}) this effect has been estimated to decrease the direct band gap by up to $\Delta E_G \simeq 40$ meV (Abram et al 1978). However, over the range of dopings of interest here, the change in direct bandgap is small, ~10 meV. The uncertainties about the size of the renormalisation do not affect our estimates of the energy ε_d of the donor level relative to the Γ-minimum plotted in Figure 3. The pressure dependence of the carrier concentration in the lighter-doped samples indicates that the level has the same pressure coefficient as the L-minima.

5. THE MICROSCOPIC NATURE OF THE DX CENTRE IN GaAs

In order to investigate the microscopic nature of DX, we have performed Fourier Transform Infrared (FTIR) absorption measurements on the most heavily Si-doped sample (a) to determine the local vibrational mode (LVM) energies of the Si impurities. Similar measurements on the heavier Sn atom are prevented by lattice absorption. The samples are irradiated with 2 MeV electrons to eliminate free carrier absorption before taking the spectra. These measurements give detailed information about the atomic location of the impurities, since the vibrational frequencies depend on the local force constant and the masses of neighbouring host lattice atoms or other impurities, as well as the mass of the impurity itself. This is illustrated in Figure 4(a) for heavily Si-doped bulk GaAs, where seven Si-related absorption lines are present, including those due to complexes Si-X, Si-Y (Chen et al 1980, Maguire et al 1987 and Woodhead et al 1985, Murray et al, to be published).

Figure 4(b) shows the same spectral region for sample (a) (see Figure 1(a)). Despite the high doping level there is no detectable absorption from Si-X or Si-Y defects, although Si-X has been observed in layers grown at higher temperatures (Maguire et al 1987). The concentrations of impurities giving rise to the remaining LVM lines of sample (a) are:

Line	Position (cm^{-1})	Concentration (x 10^{19} cm^{-3})
Si$_{Ga}$	384	1.25 $\pm$ 0.21
Si$_{Ga}$-Si$_{As}$	393	0.06 $\pm$ 0.01
Si$_{As}$	399	0.13 $\pm$ 0.02

The calibration of Woodhead et al (1985) for Si has been used, and it has been assumed that the same calibration holds for the other lines. These LVM results indicate a total Si content of 1.44×10^{19} cm^{-3}, in good agreement with SIMS and the intended Si doping level. The carrier

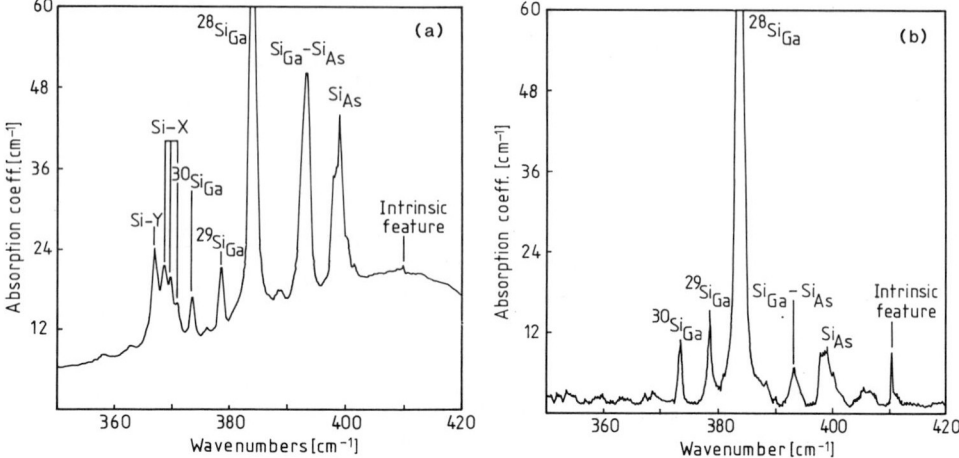

Figure 4 (a) Far infrared vibration mode absorption spectrum of Si doped bulk GaAs - see text. (b) Corresponding spectrum for the Si-doped MBE layer.

concentration, assuming the Si impurities to be the only electrically active species, is n = [Si$_{Ga}$]-[Si$_{As}$] = 1.12 x 10^{19} cm^{-3}, in excellent agreement with the SdH results (1.13 x 10^{19} cm^{-3} at P = 0).

This is very strong evidence that the DX centres observed in this sample through the pressure-induced trapping and PPC are not due to a Si complex, since we have shown that ~90% of Si is present as the substitutional donor Si$_{Ga}$; the remaining 10% is much smaller than the reduction in carrier density observed under hydrostatic pressure.

To conclude this section, we note the recent work on hydrogen passivation (Nabity et al 1987, Pajot et al 1987) supports our conclusion that DX arises from simple Si$_{Ga}$ (see Eaves et al 1988).

6. CONCLUSION - THE DX LEVEL IN NARROW GaAs QUANTUM WELLS

The measurements reported here show clearly that when the Fermi energy of the free carriers in heavily-doped n$^+$GaAs is high enough ($\gtrsim$ 250 meV), the DX level of the simple substitutional donor, Si$_{Ga}$, traps out electrons from the Γ minimum.

In degenerate systems, electrons are forced into high kinetic energy states as a consequence of the Pauli Principle. In quantum wells the kinetic energy of an electron is increased as a result of the quantum confinement. It is therefore interesting to consider if, in a sufficiently narrow well, the lowest energy Γ-like conduction electron state is forced above the energy of the DX level. In this case electrons should be trapped out onto DX. With this in mind we have examined a multi-quantum well structure consisting of the following layers grown on a semi-insulating substrate: 2 µm undoped GaAs buffer layer; 50 nm undoped (AlGa)As,[Al] = 0.33; 20 periods of [1.7 nm GaAs, Si = 4 x 10^{18} cm^{-3}; 20 nm undoped (AlGa)As]; 20 nm undoped GaAs cap. The

total sheet dopant density is high, 1.36×10^{13} cm^{-2}. At the growth temperature of 600 K, some of the Si donors introduced into the well diffuse into the neighbouring (AlGa)As barrier. Although the structure conducts at room temperature with a total sheet density of 1.5×10^{12} cm^{-2}, the carriers completely freeze out on cooling. In comparison, wider quantum well structures with similar doping remain conducting down to 4 K. Thus we attribute the observed freeze-out to trapping of carriers into DX states of Si$_{Ga}$ donors in the quantum wells.

ACKNOWLEDGEMENTS - We are grateful to Professor J. Dow for an informative discussion concerning DX levels in narrow quantum wells. This work is supported in part by SERC (U.K.) and CNRS (France).

REFERENCES

Abram R A, Rees G J and Wilson B L H 1978 Advances in Physics **27** 799.
Blakemore J S 1982 J. Appl. Phys. **53(10)** R123.
Chen R T, Rana V and Spitzer W G 1980 J. Appl. Phys. **51** 1532.
Eaves L et al 1987 Proc. of Int. Symp. GaAs and Related Compounds, Heraklion, Greece; in Inst. Phys. Conf. Ser. **91** 4 355 (1988).
Henning J C M and Ansems J P M 1987 Semicond. Sci. and Technol. **2** 1.
Hjalmarson H P and Drummond T J 1986 Appl. Phys. Lett. **48** 656.
Lang D V and Logan R A 1977 Phys. Rev. Lett. **39** 635. See also Lang D V, Logan R A and Jaros M 1987 Phys. Rev. B **19** 349.
Leloupe J, Djerassi H, Albany J H and Mullin J B 1978 J. Appl. Phys. **49(6)** 3359.
Li M F, Yu P Y, Weber E R and Hansen W 1987 Appl. Phys. Lett. **51** 349.
Maguire J, Murray R, Newman R C, Beal R B and Harris J J 1987 Appl. Phys. Lett. **50(9)** 516.
Maude D K et al 1987 Phys. Rev. Lett. **59** 815.
Mizuta M, Kitano T 1988 Appl. Phys. Lett. **52** 126.
Mizuta M, Tachikawa M, Kukimoto H and Minomura S 1985 Japan. J. Appl. Phys. **24** L143. See also Tachikawa M et al, ibid. **24** L821.
Morgan T N 1986 Phys. Rev. B **34** 2664.
Murray R, Newman R C, Sangster M J L, Beall R B, Harris J J, Wright P J, Wagner J and Ramsteiner M; to be published in Phys. Rev.
Nabity J C, Stavola M, Lopata J, Dautremont-Smith W C, Tu C W and Pearton S J 1987 Appl. Phys. Lett. **50** 921.
Neave J H, Dobson P J, Harris J J, Dawson P and Joyce B A 1983 Appl. Phys. A **32** 195.
Pajot B, Newman R C, Murray R, Jalil A, Chevallier J and Azoulay R 1987 Phys. Rev. B **37** 4188.
Portal J C et al 1988 Superlattices and Microstructures **4** 33.
Queisser H J and Theodorou D E 1986 Phys. Rev. B **33** 4027.
Raymond A, Robert J L and Bernard C 1979 J. Phys. C **12** 2289.
Shantharama L G, Adams A R, Ahmad C N and Nicholas R J, 1984 J. Phys. C: Sol. State Phys. **17** 4429.
Theis T N 1986 Proc. Int. Conference on Defects in Semiconductors, ed H J von Bardeleben, Material Science Forum **10-12** 393-398.
Theis T N, Mooney P M and Wright S L 1988 Phys. Rev. Lett. **60** 361.
Theis T N 1987 Proc. of Int. Symp. GaAs and Related Compounds,, Heraklion, Greece; in Inst. Phys. Conf. Ser. **91** 4 1 (1988).
Woodhead J, Newman R C, Tipping A K, Clegg J B, Roberts J A and Gale I 1985 J. Phys. D **18** 1575.
Yamaguchi E 1986 Japan J. Appl. Phys. **25** L643.

Inst. Phys. Conf. Ser. No 95: Chapter 5
Paper presented at Int. Conf. Shallow Impurities in Semiconductors, Linköping, Sweden, 1988

Metastable effective mass states of centers exhibiting large lattice relaxation

Janusz E. Dmochowski[a], Jerzy M. Langer[a] and Wolfgang Jantsch[b]

[a] Institute of Physics, Polish Academy of Sciences, 02-668 Warsaw, Al. Lotników 32/46, Poland, [b] Institut für Experimentalphysik, Johannes Kepler Universität, A-4040 Linz, Austria

ABSTRACT: This paper summarizes the results of a successful search for bistability for two classes of defects exhibiting large lattice relaxation phenomena : trivalent dopants in highly ionic CdF_2 and DX centers in III-V compounds. At high temperatures, under thermal equilibrium conditions, the properties of defects are governed mainly by the deep localized ground state, while at low temperatures, the vibronic barrier leads to metastability effects: in this case, the shallow effective-mass state plays the major role. The role and properties of these states are discussed.

1. INTRODUCTION

All defect states in semiconductors can be fairly safely divided into two groups. The first is formed by effective-mass states, while the second by the so-called *deep* states. The most significant difference between them is the electron wavefunction localization. The Bohr-radius a of the first class is of the order of simple hydrogenic estimations, $a=0.5 Å \varepsilon (m_o/m_*)$. For the second group of states, good knowledge of the wavefunction is scarce, but there is quite substantial experimental evidence that the deep states are localized (the Bohr radius is of the order of the interatomic distance in a semiconductor). This dychotomy has immediate consequences for lattice-defect vibronic coupling. The delocalized states couple predominantly to the optical modes, while those which are localized, to large $\mathbf{k}$-vector (predominantly acoustic) modes. This is a consequence of the much stronger dependence of the coupling parameter on the inverse of the bound carrier radius a for the acoustic phonons ($1/a^2$) than for the optical phonons ($1/a$) (Stoneham 1979). A dramatic increase of the strength of the electron-phonon coupling for localized defects may lead to quite unusual phenomena known as the Large-Lattice-Relaxation (LLR) effects (Langer 1980a,b, 1983, Lang 1986, Dobaczewski et al 1988). Among the many manifestations of LLR are: metastable occupancy of some defect states and the resulting persistent photoconductivity, the so-called negative-U behaviour (Watkins 1984), and recombination-enhanced defect reactions (Kimerling 1978, Itoh 1982, Sheinkman 1988). The most general framework for the theoretical description of all these phenomena, and especially the by now well-established dychotomy in the electron wavefunction localization at defects, has been developed in a series of papers by Toyozawa (1978, 1980, 1983a,b). The discontinuity in localization results from the competition between the long-range (Coulombic) and short-range (local defect potential

© 1989 IOP Publishing Ltd

and acoustic phonon coupling) forces. In the defect energy functional $E(\lambda)$, in which λ is the localization parameter ($\lambda \sim 1/a$), the former leads to an attractive term proportional to λ, while the latter, being proportional to local charge density, gives an attractive term proportional to λ^3. Such a functional produces either delocalized effective-mass states pinned to the relevant band minimum, and corrected for polaron and central-cell effects, or highly localized states much more strongly coupled to host lattice vibrations. If the thermal ionization energies for both states are of the same order, they should be separated by a barrier. This barrier facilitates metastable population of the higher-lying state and the observation of various persistency effects. Quite an important consequence of this most general approach to defect states in crystals is that, in principle, the deep defects should possess delocalized effective-mass states. For defects exhibiting LLR phenomena these *hydrogen-like* states should, in most cases, be metastable states.

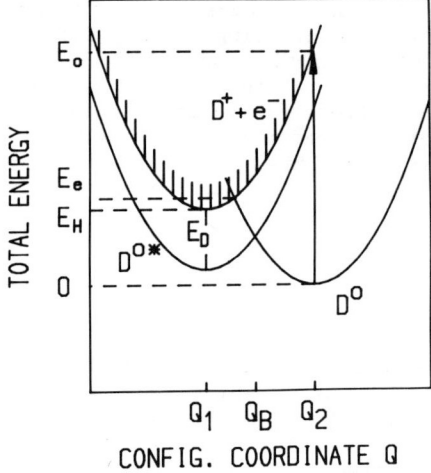

Fig.1 Configuration coordinate diagram for donor exhibiting LLR.

Strong defect-lattice coupling is best visualized by the configuration-coordinate diagram (CCD), describing the total energy of a defect-lattice system as a function of its electronic state and a single (often fictitious) lattice parameter Q (Fig.1). In CCD, a barrier separating two electronic states appears when the difference in Q for the two minima of the two states is large (LLR).

The first system in which both types of states, localized and delocalized, have been seen are F-centers in alkali-halides (Wood and Öpik 1969). In semiconductors, only recently has existence of effective-mass states of deep defects been finally experimentally proven (Grimmeiss et al 1981). In all these cases, the effective-mass states are normal excited states, i.e., they do not exhibit any metastability effects. For defects exhibiting LLR phenomena, the first, and for a long time the only, case has been In in CdF_2 (Piekara et al 1977, Langer et al 1979, Dmochowski et al 1986). Quite recently a similar observation has been made for DX-centers in GaAlAs, in which hydrogenic Γ-like states have been seen in the direct-gap region (Theis et al 1984) and *metastable* effective-mass states pinned to the X-conduction band minimum in the indirect-gap region of GaAlAs (Dmochowski et al 1987b, 1988b)

The key difficulty in observation of these states lies in finding the proper dopant concentration, which must be large enough to facilitate observation of the localized states and, simultaneously, low enough to avoid banding effects for the effective-mass excited states.

In the following part of this review we shall summarize the most essential facts on the two abovementioned defect systems, namely, donors in CdF_2 and DX-centers in GaAlAs. In the summary we shall comment on microscopic models of these, and other defects exhibiting LLR phenomena.

2. CdF2: Me^{3+}

CdF2 is a highly ionic broad-band ($E_g \cong 8eV$) crystal which, when doped with many trivalent metals (Y, Sc, many RE^{3+}), reveals semiconducting properties i.e. these impurities produce stable effective-mass shallow donor states. The ionization energy is fairly large for these states, 110 meV, (Langer et al 1974) and thus their Bohr radius a is small ($a \cong 7 \text{Å}$), resulting in critical Mott concentration well in excess of 10^{18} cm^{-3} (Potemski 1985). These stable donors produce broad IR absorption bands of photoionization origin, with constant oscillator strength in the 300-4 K temperature range clearly showing their usual stable shallow donor character (Langer et al 1974). Among the trivalent dopants, two: In (Piekara et al 1977, Langer et al 1979) and the more recently investigated Ga (Dmochowski et al 1988a), have been found to produce instead of shallow states highly localized deep ground states exhibiting LLR.

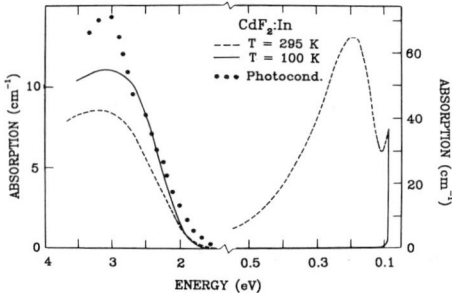

Fig. 2 Absorption bands of CdF2: In.

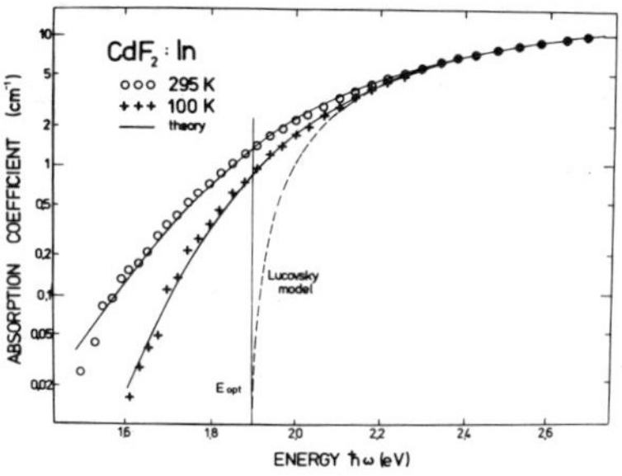

Fig. 3 Temperature broadening of the 3-eV photoionization band.

In room-temperature absorption spectra of CdF2: In, two strongly asymmetric bands are seen (Fig.2, Piekara et al 1977): one of these is peaked at 3 eV, and the second one at 0.2 eV. The high-energy photoconductivity cross section has identical shape as the 3-eV absorption band, thus proving its photoionization origin. Below 50 K , after cooling the sample in darkness, only the 3-eV band is seen, while, after illumination with light of an energy contained in this band, the 0.2-eV band grows at the expense of the 3-eV band until its complete bleaching. This indicates a possibility of a total inversion of occupancy of states responsible for both bands. The photoinduced population of the higher state is metastable below 50 K. A detailed shape analysis of the UV-VIS band (Fig.3) (Piekara et al 1977, Langer et al 1979) yields an optical ionization energy of its initial state of 1.9 eV at 10 K. The thermal ionization energy of this state measured at temperatures above the metastability threshold, is only 0.2 eV. The temperature dependence of the ratio of the oscillator strengths of the 0.2-eV and the 3-eV bands measured close to thermal equilibrium implies that the state responsible for the 0.2-eV band lies only 0.1 eV above the ground In^{2+} state. Therefore the thermal ionization energy of the shallower state is 0.1 eV, equal to its optical ionization energy. The energy barrier separating the two states, and leading to the metastable effects, is slightly temperature dependent. At 100 K it is about 0.17 eV.

From these results it is clear that the ground In^{2+} state must be very localized, since no delocalized state can produce such an enormous Stokes shift ($E_{opt} - E_{th} = 1.7$ eV). From the temperature dependence of vibronic

broadening of the 3-eV photoionization band, a local phonon energy of 22 meV is estimated for the ground state of In^{2+} (Langer et al 1979), while for the final state (the ionized In^{3+} state plus one electron in the conduction band), it should be close to 30 meV, as observed in resonant Raman scattering (O'Horo and White 1973, Ciepielewski and Langer 1985). Therefore, the Huang-Rhys factor S is between 50 and 80. Strong localization of the ground In^{2+} state is further evidenced by its strong pressure dependence (Langer et al 1979), which does not occur for delocalized effective-mass states (Jantsch et al 1982), and an almost 1eV "chemical shift" of the UV-VIS absorption band between In and Ga (Dmochowski et al 1988a) showing the high value of the electron wavefunction in the vicinity of the impurity. Lattice relaxation must be even larger for Ga, since its ground state is thermally only 0.1 eV deeper than that of In. All the aforementioned facts show that the extension of the electron wavefunction in the initial state must be of the order of the inter-atomic distance, as it is in the ground (1s) state of the F centers in alkali halides (Wood and Öpik 1969).

The effective-mass large-radius character of the metastable exited state and its weak coupling to the lattice can be deduced from spectroscopic investigations of high-purity crystals (Dmochowski et al 1986,1987a,1988a). Photoinduced metastable IR absorption of Ga and In reveals many well resolved discrete transitions (Fig.4). All of these features are common for metastable Ga and In, and similar to those found earlier for the stable shallow donor state represented, e.g. by Y (Lee and Moser 1971, Dmochowski et al 1983). The well resolved zero-phonon lines lying below the thermal ionization energy (series A in Fig.4) are characteristic of transitions to the higher excited states within a large radius "hydrogenic" polaron. A quantitative interpretation of the spectra is thwarted by the large polaron

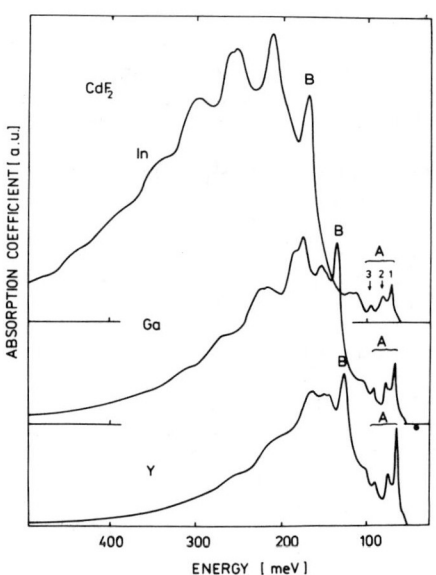
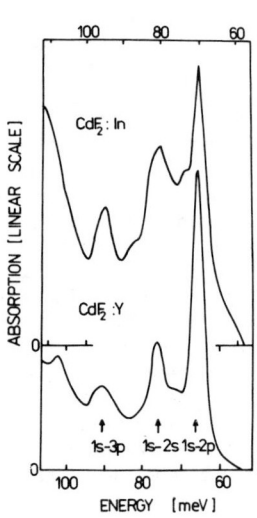

Fig.4 Metastably photoinduced infrared absorption of In and Ga and stable Y absorption in lowly doped CdF2.

Fig.5 Low energy zerophonon intradonor transitions.

coupling constant (Fröhlich coupling constant α≅3) of CdF2. The assignment of the low-energy series proposed in Fig.5 is based upon the following facts: all the three peaks lie below the photoionization continuum: the ionization energy estimated by Potemski (1985) from the Hall-effect data extrapolated to zero impurity concentration is 110∓5 meV for Y. A crude application of the hydrogenic model yields, for Y, E_D=110 meV (112.5 meV and 115 meV for Ga and In, respectively) and a positive "chemical" shift of 73 meV (71.5 meV and 68 meV for Ga and In). For the 2s state it is 1/7, close to the 1/8 predicted by the hydrogen-like model. The small difference in the ground state energy, 2.5 meV, between Y and Ga and, next, 2.5 meV between Ga and In, are evidence of the very small electronic wavefunction amplitude at the impurity, characteristic of a large-radius state. This amplitude is even smaller than that expected for a purely hydrogenic 1s ground state, as shown by absence of hyperfine splitting of the ESR line (Eisenberger et al 1968). The high energy tail of the IR absorption has, for all three spectra, a $h\nu^{-3.5}$ dependence, characteristic of photoionization of 1s hydrogenic states (Langer et al 1974). The small coupling to the lattice for the transitions responsible for the IR absorption is evidenced by a small Huang-Rhys factor, estimated from the ratio of the oscillator strengths of zero-phonon lines and their phonon replicas. For the A series, it is clearly below unity for all the dopants. The transitions B, to an unidentified excited state resonant with the conduction band, are coupled more strongly to the lattice but the H-R factor is of the order of 2 for In, and smaller for Ga and Y, respectively. The two phonon modes resolved in the phonon replica of this series, 42 and 50 meV, are most likely optical modes from the Γ point, in agreement with large dimensions of the states involved. A very small Stokes shift for ionization of metastable states is evidenced by observation of IR photoconductivity (Fig.6), photoinduced by white light for bistable In donor (Dmochowski and Langer 1987a). This photoconductivity, with metastability properties analogous to IR absorption, has a low energy edge practically equal to the thermal ionization energy of the shallow state. Again, it is similar to that observed for stable Y donor.

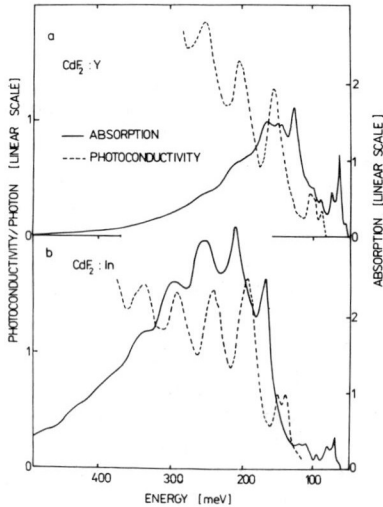

Fig.6. Metastably photoinduced oscillatory IR photoconductivity of CdF2:In and its stable analogue for shallow Y donor.

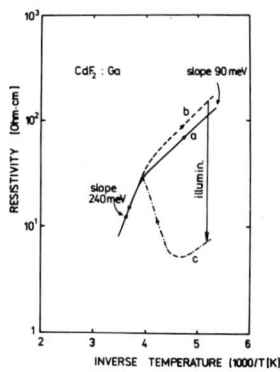

Fig.7 Temperature dependence of resistivity of CdF2:Ga in darkness and under PPC condition.

For a metastable effective-mass state the small lattice relaxation upon its ionization plays a crucial role in understanding the properties of bistable donors with LLR at low temperatures. Photoinduced carriers are captured by these states in a picosecond time scale, as evidenced by oscillatory (Fig.6), with period LO(Γ), structure (Stradling 1980). Thus, when the ground state is switched off from thermal equilibrium by a vibronic barrier, the upper, effective-mass state remains in thermal equilibrium with the conduction band down to the lowest temperatures. As a consequence, it starts to control the carrier concentration in the crystal. The evidence for this is given by the exponentially activated character of the photoinduced persistent conductivity (PPC), as well as dark conductivity under metastability condition (Fig.7).

3. DX-CENTERS IN AlGaAs.

Some 10 years ago, Lang and Logan (1977) and Nelson (1977) discovered that all the group IV and VI donors in AlGaAs exhibit unusual behavior, e.g., a large Stokes shift, in their ionization and metastable photogeneration of the free carriers optically released from these defects, called persistent photoconductivity (PPC). It should be mentioned, however, that their observation, as well as the proposed interpretation, were preceded by a series of papers on background donors in InSb by Porowski et al.(1974), who showed that under strong hydrostatic pressure new-type donor states emerge from the conduction band. These were found to exhibit metastability phenomena similar to those of DX's and their unusual properties were ascribed to the LLR phenomenon.

The key property of the DX centers indicating a departure from the effective mass description is their large Stokes shift. For example, for Te-DX in AlGaAs the following characteristic energies has been extracted from different experiments: E(Hall)=0.1eV, E(photoionization)=0.85eV, E(electron emission)=0.28eV, and E(electron capture)= 0.18eV. There have recently been some reports questioning these, but as shown by the very careful measurements of photoionization cross-section by Mooney et.al.(1988), the early findings of Lang and Logan are correct.

Sometimes, metastable phenomena have been ascribed to the existence of macroscopic potentials resulting from composition fluctuations in the AlGaAs alloy. Although, in principle, possible, this effect is not the source of the metastability there. Mizuta et al. have shown that application of hydrostatic pressure to GaAs causes emergence of defect states otherwise resonant with the conduction band (a result similar to the much earlier observation of the Porowski group in InSb (Porowski et al 1974) and CdTe:Cl (Dmowski et al 1976). Their properties are very much the same as those of the DX-centers seen without pressure for AlGaAs with the Al content above 20%. Moreover, their energetic position coincides well with a linear extrapolation of the DX ground energy level to GaAs. These states control the Fermi-energy position, limiting the carrier concentration in GaAs, and are efficient traps for hot carriers (Theis et al 1986). A final proof of the validity of the LLR model of a DX-center is obtained by looking into its basic properties expected from the general discussion of the Toyozawa model of defects, namely, the electron localization in the ground state of a DX center leading to metastability, and existence of the regular hydrogen-like effective mass states of these defects, separated from the ground localized states by a barrier.

The first has been unequivocally shown by Dobaczewski and Langer (1986) from a study of the alloying effect on the level broadening, as seen in

DLTS measurements. The analysis of the nonexponential kinetics led to the estimation of the number of the cation sites encompassed by the DX center wavefunction. For a DX center, at x=0.3, this equals ≅15, which indicates a truly localized character of its wavefunction.

The missing component of the LLR-case proof has been direct observation of metastable shallow effective mass states of DX centers. Such states have finally been observed in the indirect-gap range of AlGaAs by Dmochowski et al (1987b,1988a) (Fig.8). Photoinduced IR optical absorption of thick (20-50μm) GaAlAs:Te/SIGaAs(Cr) (N_{Te}≅5 10^{16} cm^{-3}) LPE layers has been measured with a Fourier spectrometer after cooling the sample in darkness down to 10K, and then illuminating it with tungsten lamp white light. The photoinduced spectra remain metastable for a one day measurement run (until heating of up to about 100 K).

The most remarkable feature of the photoinduced absorption is the independence of its peak position and of the low energy cutoff of composition. The shape of the absorption band can be well reproduced by the general form of impurity photoionization cross-section:

$$\alpha = C(\hbar\omega - E_D)^A / \hbar\omega^{A+B}$$

with A≅1.5 and B≅2.5. Values of A and B similar to observed for shallow donors in GaP (Kopylov and Pikhtin 1977), the effective X-valley pinning of the state ($E_D(x)$≅const., Fig.9), as well as the value of the ionization energy itself (E_D≅45meV) (Morgan 1986), strongly indicate an X-like effective-mass origin of the metastable donor state responsible for this absorption.

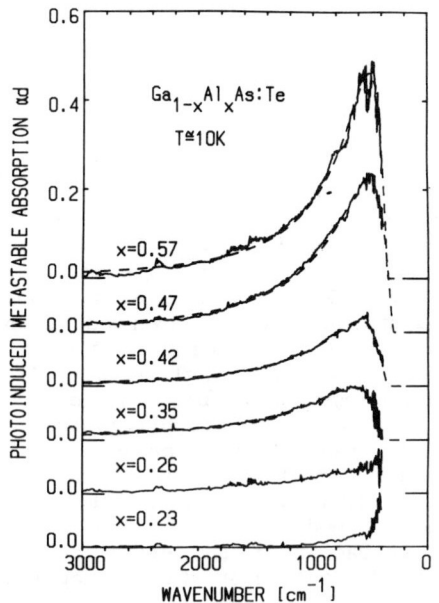

Fig.8 Photoinduced metastable IR absorption in AlGaAs:Te.
----- photoionization theory.

Our observations suggests a simple and natural explanation of an apparent disappearance of persistent photoconductivity in indirect gap AlGaAs, observed by Chand et al (1984). For x<0.4, the only nonresonant excited state of a DX center is a Γ-like state of vanishing thermal ionization energy for donor concentrations larger than $10^{16}cm^{-3}$. For x>0.4, an X-like state, reported above, emerges from the conduction band. This state, as one exhibiting no LLR, is in thermal equilibrium with the conduction band at any temperature. Thus, all photoinduced carriers must be retrapped by DX centers, but now by the effective mass X-like states. Strong evidence for such behaviour comes from Hall-effect measurements carried out under no modulation doping condition (Fig 10). At high temperatures, a freeze-out on the deep localized DX level is observed, as reported previously (Chand et al 1984). Below 100 K, when the ground state is switched off from recombination due to the immeasurably long capture time

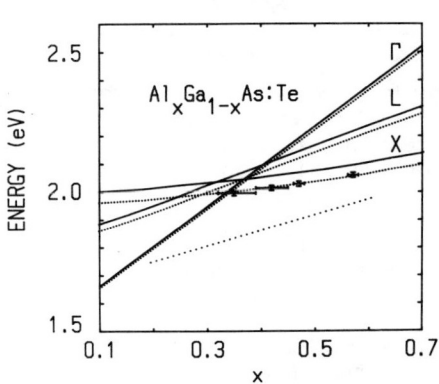

Fig.9 Conduction band minima and DX donor levels in AlGaAs.

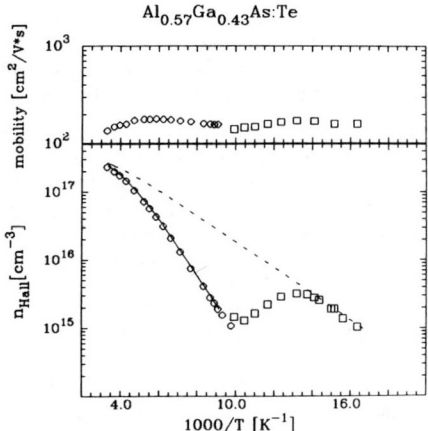

Fig.10 Electron concentration and mobility in indirect-gap AlGaAs:Te in darkness (circles) and under PPC condition (squares).

illumination of the sample leads to a PPC. In contrast to the direct-gap AlGaAs, PPC now becomes exponentially temperature activated, clearly indicating that the metastable excited shallow state is in thermal equilibrium with the conduction band down to the lowest temperatures. Thus, there is no barrier for electron capture to this state, i.e., small lattice relaxation during this process. An analysis of the Hall data within a simple two-level donor model shows that the number of shallow states must be the same as the number of deep states.

The effective-mass excited states of DX centers must control, at low temperatures, not only the transport properties, but also all optical recombination. In particular, the dominant recombination channel in the III-V compounds involving donors is the donor-acceptor pair emission. The ground, localized state of DX's can participate in this process, but due to the large lattice displacement its probability must be very low and the spectrum must be broadened in the same manner as the ground state photoionization. In contrast, the hydrogen-like excited states should produce standard D-A recombination with a phonon coupling governed by the more localized particle, i.e., by the acceptor. Such recombination via donors pinned to the X-minimum has been seen in AlGaAs (Dingle et al 1977). Henning et al (1987) have observed low-temperature photoluminescence involving donors related to Si in AlGaAs. As the phonon coupling observed was rather weak, the authors considered this observation as proof of small lattice relaxation of the Si-DX centers. Direct observation of the effective-mass X-like states and the above discussion explains that this observation is not contradictory with the LLR model of DX centers. It is also worth pointing out that metastable character of the effective-mass excited state of DX centers should also be reflected in the optical transitions. In particular, this should lead to slow transient effects in $D_{hydrog}(DX)$-A recombination at low temperatures, due to the metastable optical repopulation of the DX states. Such an effect has, in fact, been observed recently (Jantsch et al 1988), but as all observations have been made on highly Si doped AlGaAs, separation of the band-tails effects due to

the high doping level from the DX-center specific properties is necessary.

4. MICROSCOPIC MODELS

The observations summarized above validate a general description of defects in solids by the Toyozawa model. The model cannot, however, help in building a realistic microscopic model of a defect and the LLR involved in it. Its validity does not depend explicitly on the ionicity of the host solid, which, as is clear from the wide range of solids in which LLR phenomena have been observed, can play only a secondary role. The ionicity could be much more important in a microscopic model of the lattice relaxation involved. Again, In or Ga in highly ionic CdF2 can be well contrasted with the DX centers in the covalent III-V semiconductors. As suggested by one of us (Langer 1980a,b,1983), there should be two general types of LLR. One of them, found

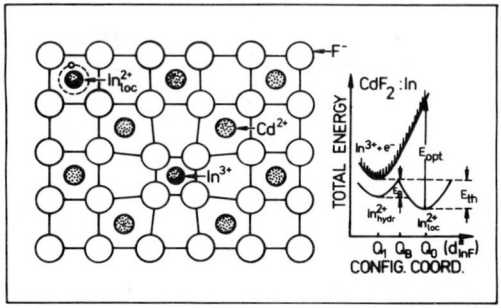

Fig.11 Microscopic model for CdF2: In.

only in crystals with predominantly ionic bonding, is a symmetrical local lattice collapse due to change of the core screening after ionization of a defect (Fig.11). The corresponding local distortion is of the order of 10% of the n.n. distance. The examples are In and Ga in CdF2, F-centers in alkali halides, self-trapped exciton states in alkali halides (here, the lattice distortion need not to be symmetrical) and, maybe, transition metals in II-VI compounds. The second group consists of defects in which LLR is a local rearrangement, in which the atoms move even from one lattice site to another (e.g. substitutional to interstitial). We believe that all the bizarre defect states found in the more covalent crystals are of this nature. Thus, the LLR occurring for DX's can be classified, as suggested by Langer (1983),"*a local defect reaction, quite similar to the valence alternation at negative-U defects in chalcogenide glasses*". It is by now established that there is no strong change of mobility accompanying DX-center ionization (Fig.10). Such a change is expected during ionization of a normal donor. There is also absence of any EPR signal due to DX's, expected for a single-electron defect state. Both of these facts are well understood if DX-centers are, in fact, two electron donors in their ground state, exhibiting negative-U behavior. This model requires, however,a much more detailed study, similar to that performed for the vacancy and interstitial boron in silicon (Watkins 1984).

5. REFERENCES

Chand N, Henderson T, Klem J, Masselink W T, Fischer R, Chang Y -C and
 Morkoç H 1984 *Phys. Rev.* **B30** 4481
Ciepielewski P and Langer J M 1985 *Acta Phys. Polon.* **A67** 89
Dingle R, Logan R A and Arthur J R 1977 *Gallium Arsenide and Related
 Compounds*, IOP Conference Proceedings **33a** (Bristol: Hilger) p 210
Dmochowski J E, Kosacki I, Langer J M 1983 *Radiat. Eff.* **72** 139
Dmochowski J E, Langer J M, Kaliński Z and Jantsch W 1986 *Phys. Rev. Lett.*
 56 1735

Dmochowski J E and Langer J M 1987a *The Physics of Semiconductors* ed O Engström (Singapore: World Scientific) p 867
Dmochowski J E, Langer J M, Raczyńska J and Jantsch W 1987b *6th Int. Conf. on Deep Impurity Levels, Lund 87, Sardinia, Italy* (unpublished)
Dmochowski J E, Jantsch W, Dobosz D and Langer J M 1988a *Acta Phys. Polon.* **A73** 247
Dmochowski J E, Langer J M, Raczyńska J and Jantsch W 1988b *Phys. Rev.* **B**
Dmowski L, Kończykowski M, Piotrzkowski R and Porowski S 1976 *Phys. Stat. Sol. (b)* **73** K131
Dobaczewski L and Langer J M 1986 *Materials Science Forum* **10-12** 399
Dobaczewski L, Dmochowski J E and Langer J M 1988 *8th Int. School on Defects, Szczyrk, Poland* (Singapore: World Scientific) (in print)
Eisenberger P, Pershan P S 1968 *Phys. Rev.* **167** 292
Grimmeiss H G, Janzén E, Ennen H, Schirmer O, Schneider J, Wörner R, Holm C, Sirtl E, Wagner P 1981 *Phys. Rev.* **B24** 4571
Henning J C and Ansems J P M 1987 *Semicond. Sci. Technol.* **2** 1
Itoh N 1982 *Radiat. Eff.* **64** 161
Jantsch W, Wünstel K, Kumagai O and Vogl P 1982 *Phys. Rev.* **B25** 5515
Jantsch W 1988 (unpublished)
Kimerling L C 1978 *Solid State Electron.* **21** 1391
Kopylov A A and Pikhtin A N 1977 *Fiz. Tekh. Poluprovod.* **11** 510 [*Sov. Phys. Semicond.* **11** 510]
Lang D V and Logan R A 1977 *Phys. Rev. Lett.* **39** 635
Lang D V 1986 *Deep Centers in Semiconductors* ed S T Pantelides (New York: Gordon and Breach) p 489
Langer J M, Langer T, Pearson G L, Krukowska-Fulde B and Piekara U 1974 *Phys. Stat. Sol. (b)* **66** 537
Langer J M, Ogonowska U and Iller A 1979 *Physics of Semiconductors* ed B L H Wilson, *IOP Conference Proceedings* **43** (Bristol: Hilger) p 277
Langer J M 1980a *Lecture Notes in Physics* **122** (New York: Springer) p 123
Langer J M 1980b *J. Phys. Soc. Jpn.* **49** Suppl.A 207
Langer J M 1983 *Radiat. Eff.* **72** 55
Lee T M, Moser F 1971 *Phys. Rev.* **B3** 347
Mizuta M, Tachikawa M, Kukimoto H, and Minomura S 1985 *Jap. J. Appl. Phys.* **24** L143
Mooney P M, Northrop G A, Morgan T N, and Grimmeiss H G 1988 *Phys. Rev.* **B**
Morgan T N 1986 *Phys. Rev.* **B34** 2664
Nelson R J 1977 *Appl. Phys. Lett.* **31** 351.
O'Horo M P and White W B 1973 *Phys. Rev.* **B7** 3748
Piekara U, Langer J M and Krukowska-Fulde B 1977 *Solid State Commun.* **23** 583
Porowski S, Kończykowski M and Chroboczek J 1974 *Phys. Stat. Sol.(b)* **63** 291
Potemski M 1985 Ph.D. thesis (unpublished)
Sheinkman M K 1988 *Acta Phys. Polon.* **A73** 925
Stoneham A M 1979 *J. Phys. C* **12** 891
Stradling R A 1980 *Lecture Notes in Physics* **122** (New York: Springer) p 1
Theis T N, Kuech T F, Palmateer L F, and Mooney P M 1984 *Gallium Arsenide and Related Compounds, IOP Conference Proceedings* **74** (Bristol: Hilger) p 241
Theis T N, Parker B D, Solomon P M and Wright S L 1986 *Appl. Phys. Lett.* **49** 1542
Toyozawa Y 1978 *Solid State Electron.* **21** 1313
Toyozawa Y 1980 *Relaxation of Elementary Excitations* ed R Kubo and E Hanamura (New York: Springer-Verlag) p 3
Toyozawa Y 1983 *Physica (Amsterdam)* **116B** 7
Toyozawa Y 1983 *Semicond. Insul.* **5** 175
Watkins G D 1984 *Festkörperprobleme* **24** (Braunschweig: Vieweg) p 163
Wood R F and Öpik U 1969 *Phys. Rev.* **175** 78

Paper presented at Int. Conf. Shallow Impurities in Semiconductors, Linköping, Sweden, 1988

Crossover between shallow and deep levels in pressurized gallium arsenide

Atsushi Oshiyama

Fundamental Research Laboratories, NEC Corporation
Miyazaki, Miyamae-ku, Kawasaki 213 Japan

ABSTRACT: Occurrence of deep levels associated with substitutional impurity in GaAs under hydrostatic pressure is explored by a state-of-the-art Green's function calculation within the local-density-functional formalism. It is found that a singlet A_1 level induced by Si impurity and located in the conduction bands in GaAs appears in the energy gap by applying hydrostatic pressure, and that Al impurity, on the contrary, induces no deep level in pressurized GaAs.

1. INTRODUCTION

In III-V compound semiconductors, both group IV and group VI elements behave as dopants and induce shallow levels just below the conduction-band bottom. In GaAs, for example, Si and Sn substitute cation sites, whereas S, Se and Te do anion sites. Those impurities induce typical effective mass states at several meV below the conduction-band bottom. In alloy semiconductors, e.g. $Al_xGa_{1-x}As$, however, the shallow donor level becomes deep at more than hundred meV below the conduction-band bottom: DX center. Energy barriers for carrier capture and emission were measured, and persistent photoconductivity related to the deep level has been also observed (Lang and Logan 1977, Nelson 1977). On the other hand, a group V element, nitrogen, is supposed to substitute the anion site, and also induces electronic levels in the energy gap (isoelectronic trap) of the alloy semiconductors (Wolford et al 1976). Another isoelectronic impurity, a group III atom Al, behaves in totally different way in GaAs. It has high solubility in GaAs, and constitutes the alloy semiconductor itself: $Al_xGa_{1-x}As$.

Microscopic reason for the occurrence of the deep levels above described is still obscure. Yet, recent extensive work on DX center (for a review, see Theis 1987) has provided us with suggestive information. In particular, variation in conduction band structure with alloying is expected to play an important role in the appearance of the deep level. Hydrostatic pressure also produces variation, similar to that upon alloying, in the conduction-band structure (Lifshitz et al 1980): with increasing pressure, the conduction-band edges at Γ and L symmetry points shift upwards, while the edge at X point slightly shifts downwards. Under this circumstances, occurrence of the deep level under more than 24 kbar hydrostatic pressure has been observed for Si impurity in GaAs (Mizuta et al 1985).

To identify these deep levels, I have done a state-of-the-art Green's-function calculation for Si impurity under both hydrostatic and atmospheric pressure.

© 1989 IOP Publishing Ltd

Calculation for Al impurity is also performed for comparison. Among the new results from the present calculation are (i) Si substitutional impurity induces a sharp resonant state of singlet A_1 symmetry near the edge of the conduction bands in GaAs, (ii) that the A_1 resonant state shifts downwards into the energy gap by applying hydrostatic pressure, (iii) that the A_1 state has character of an antibonding state between s orbitals of Si and As, (iv) and that breathing distortion around the Si impurity plays only minor roles in determining level structure.

2. CALCULATION

In the calculation, at first, norm-conserving pseudopotentials (Hamann et al 1979, Bachelet et al 1982) are constructed by solving all-electron atomic (ionic) Dirac equation. They are placed at each atomic site and then valence electron charge density is determined selfconsistently within the local-density-functional formalism (Hohenberg and Kohn 1964, Kohn and Sham 1965) with the Gaussian-orbitals basis set. The exponents of Gaussian orbitals are determined by fitting the atomic orbitals numerically obtained. Improvements of the exponents have also been done so as to reproduce the results from the well-converged plane-wave-basis-set calculation for GaAs and Si crystal (Oshiyama and Saito 1987a 1987b). I have used three exponents for the s orbital, two exponents for the p orbital and one exponent for the d orbital of each atom: Ga s (0.14, 0.34, 0.82), Ga p (0.14, 0.24), Ga d (0.18), As s (0.14, 0.53, 0.89), As p (0.14, 0.31), As d (0.18), Si s (0.14, 0.30, 1.0), Si p (0.14, 0.3) and, Si d (0.18) in atomic unit.

At first, the total energy of host GaAs is calculated as a function of lattice constant. The total-energy minimization leads to the theoretical values for the lattice constant and bulk modulus: 5.61 A and 0.7 Mbar in the present calculation. The agreement between the theoretical and experimental values is satisfactory so that we can mimic pressurized circumstances by varying the lattice constant as an external parameter. The overall features of the conduction band structure under hydrostatic pressure (Lifshitz et al 1980) are well reproduced in the calculation. In particular, the behavior of the conduction-band edges at Γ, L and X points are well described in this calculation, although the absolute value of the energy gap is considerably underestimated in the local-density approximation.

Next, a Ga atom is replaced by an impurity atom (Si or Al). Re-distribution of the valence electrons and then the induced change in density of states (DOS) and in the total energy are calculated by the Green's-function method (Car et al 1984) within the local-density-functional framework. The Gaussian orbitals are placed at each atomic sites around the impurity to express the change in distribution of electron density. I have retained the Gaussian orbitals located at first three shells surrounding the impurity (total of 29 atoms). The extended basis set (the gaussian orbitals at first four or five shells) have been also examined. The results on the level structure from the extended basis set are found to be identical to the results presented here. Amount of breathing distortion (symmetric mode) of the surrounding four As atoms is also calculated from total-energy-minimization.

3. RESULTS AND DISCUSSION

The calculated change in DOS induced by the Si substitutional impurity is shown in Figure 1. Without pressure there is no deep level in the energy gap of GaAs: i.e. only shallow effective mass state exists just below the conduction-band

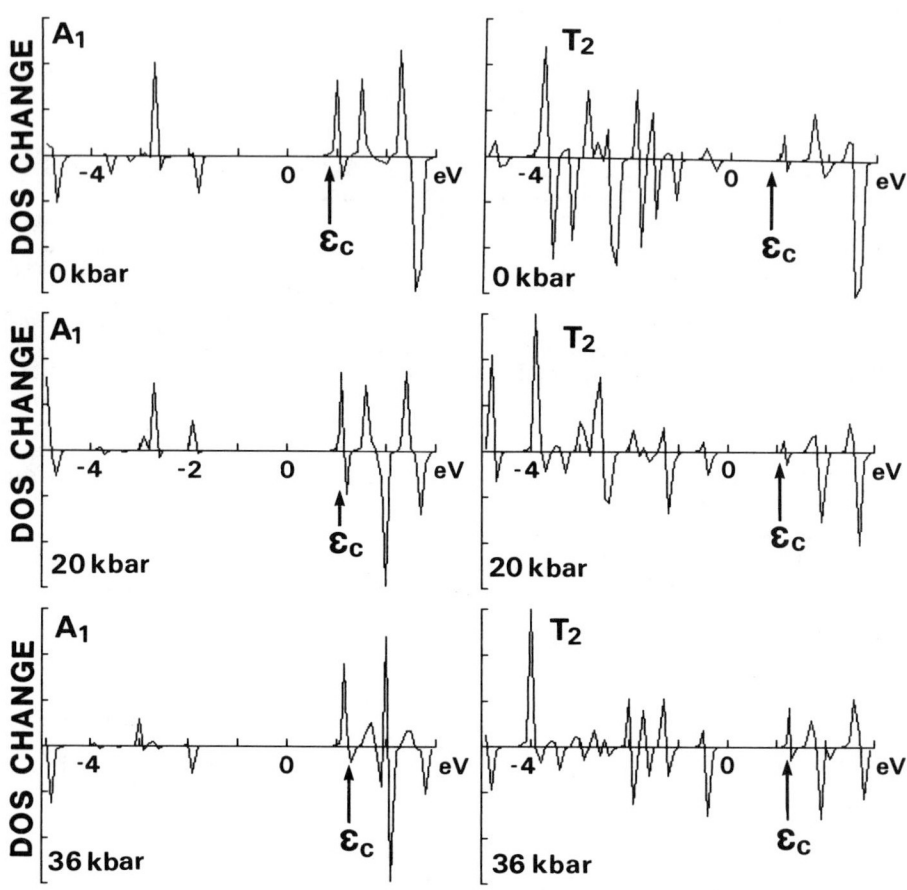

Figure 1: Calculated DOS change for A_1 and T_2 symmetry states induced by Si substitutional impurity in pressurized GaAs. Valence band top is the energy reference, and ε_c denotes conduction band bottom.

bottom. Yet, a sharp resonant state of A_1 symmetry is found to be located at 0.15 eV above the conduction-band bottom. By applying pressure the energy-gap becomes wide and the resonant state appears deep in the energy gap: 0.1 eV below the conduction-band bottom at 36 kbar. A triplet resonant state of T_2 symmetry also exists near the conduction-band bottom, although the resonance peak in Figure 1 is not so prominent as the peak of the A_1 symmetry. As energy gap increases with pressure, the triplet resonant state becomes close to the conduction-band edge. The locations of the resonant or deep levels are plotted as a function of the applied pressure in Figure 2, along with the calculated conduction-band edges at Γ, L and X points of host GaAs. It is clearly seen that the A_1 level induced by the substitutional Si impurity appears in the energy gap

under hydrostatic pressure. Moreover, The lowest A_1 and T_2 states (denoted by a_1 and t_2 hereafter) apparently track the L conduction-band edge. However, the a_1 and t_2 states in Figure 2 are not simple effective mass states: the energy-difference between the L edge and the a_1 state is much larger than the value expected from effective mass theroy. Further, the Gaussian orbitals placed at first three shells in this calculation cover rather small region with radius of about 5 A, while the Bohr radii of the typical effective mass states ranges about 100 A for the Γ point or about 10 A for the L or X points in GaAs.

To clarify the nature of these resonant or deep states, I have investigated the characteristic features of the impurity wavefunctions $\Psi_i = \Sigma\ C_{ij}\Phi_j$ which are expanded in terms of the atomic Gaussian orbitals. It is found that the a_1 state is rather localized: the atomic orbital of s character at the central Si site and the orbital of, again, s character at the first neighbor As sites constitute 25% of the total norm of the wavefunction. Examination of the sign of the coefficients C_{ij} leads to the conclusion that the a_1 state is primarily an antibonding state of the Si s orbital and the first neighbor As s orbitals. I have also calculated the contribution from the Bloch state of n'th band at k point: $P_i^{nk} = |<\Psi_{nk}|\Psi_i>|^2$. The largest P_i^{nk} comes from the four valleys of the L point of the lowest conduction band, and the next comes from the Γ point of the lowest valence band. Yet, P_i^{nk} of the lower symmetry points and of another bands has comparable value. It should be noted that the largest contribution (13% of sum of P_i^{nk} over n and k, ΣP_i^{nk}) comes from the lowest conduction band, and the next (6%) comes from the lowest valence band. This finding supports the notion that the a_1 state is the antibonding state of the impurity and the first neighbor orbitals.

The t_2 state has somewhat different character from the a_1 state. This is also an antibonding state. But, the atomic orbitals at the central Si site contribute little to the wavefunction of the t_2 state. Instead, the s orbitals at the first neighbor As atoms and at the second neighbor Ga atoms constitute the main contribution. Projection analysis on the Bloch states shows that P_i^{nk} for the L point of the lowest conduction band has largest value. The lowest conduction band contributes 9 % to the summation ΣP_i^{nk}, and the four valence bands do also 9 %.

Since the wavefunction of the t_2 state is mainly localized at the first neighbor As and the second neighbor Ga sites, this state is expected to be quite insensitive to the potential disturbance from the

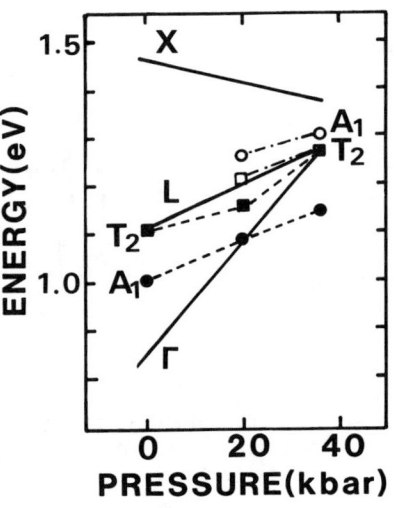

Figure 2: Calculated Γ, L and X conduction band edges (solid lines) and the lowest A_1 and T_2 resonant (deep) levels induced by Si (solid circles and squares with dashed lines) and by Al (open circles and squares with dash-dotted lines) impurities in pressurized GaAs. The energy is measured from the valence band top.

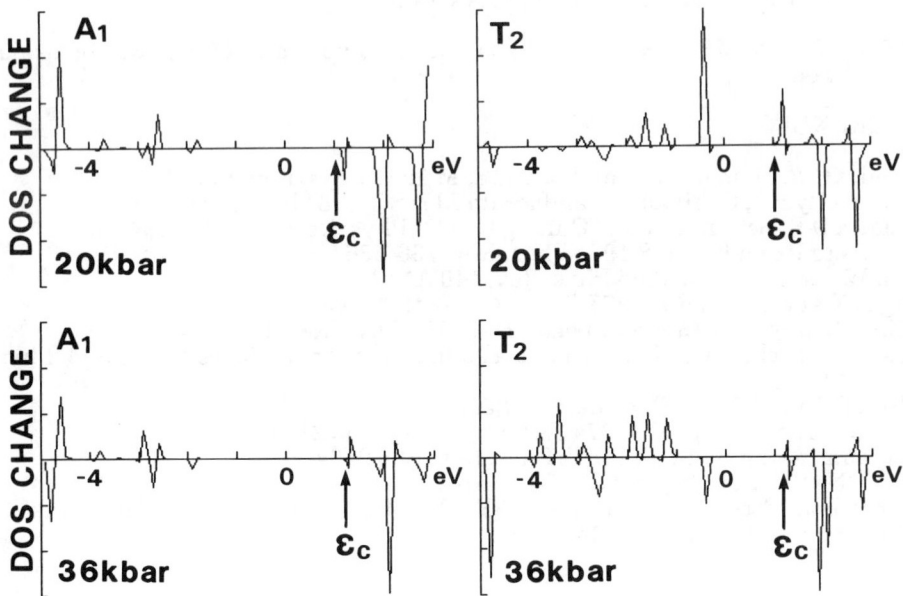

Figure 3: Calculated DOS change for A_1 and T_2 symmetry states induced by Al substitutional impurity in pressurized GaAs. Energy is measured from valence band top, and ε_c denotes conduction band bottom.

impurity substituting the cation site. In Figure 3, Calculated change in DOS induced by the Al impurity in GaAs is shown. The locations of the corresponding a_1 and t_2 states are plotted in Figure 2. It is clear that the t_2 states for the two types of impurities are similar to each other in their energy-level locations. This is the consequence from the nature of the wavefunction of the t_2 state described above.

Comparison between Figure 1 and Figure 3 suggests general difference between Si and Al substitutional impurity. It is noticed that the change in DOS near the energy gap is not so prominent for Al impurity as for Si impurity. This implies that the resonant states induced by the Al impurity is rather extended compared with the Si impurity. Then the energy level of the a_1 state for Al is considerably higher than the Si-induced a_1 level so that no state appears in the energy gap even under pressure.

Lattice relaxation around the Si impurity has been also examined by the present Green's function scheme. The total-energy minimization favors the outward breathing distortion (symmetric mode) of surrounding four As atoms. The amount of distortion, however, is found to be less than 0.1 Å. This outward breathing distortion makes the a_1 state slightly deeper, but the essential features of the level structure are unchanged.

In conclusion, the present state-of-the-art Green's function calculation reveals that the A_1 resonant state induced by Si substitutional impurity and located at

0.15 eV above the conduction-band bottom of GaAs appears in the energy gap by applying hydrostatic pressure. This A_1 state has the character of the antibonding state of the Si s and the first neighbor As s orbitals.

ACKNOWLEDGMENTS - I wish to thank M. Mizuta and M. Ogawa for useful conversation.

REFERENCES

Bachelet G B, Hamann D R and Schlüter M 1982 Phys. Rev. B26 4199.
Car R, Kelly P J, Oshiyama A and Pantelides S T 1984 Phys. Rev. Lett. 52 1814.
Hamann D R, Schlüter M and Chiang C 1979 Phys. Rev. Lett. 43 1494.
Hohenberg P and Kohn W 1964 Phys. Rev. 136 B864.
Kohn W and Sham L J 1965 Phys. Rev. 140 A1133.
Lang D V and Logan R A 1977 Phys. Rev. Lett. 39 635.
Lifshitz N, Jayaraman A and Logan R A 1980 Phys. Rev. B21 670.
Mizuta M, Tachikawa H, Kukimoto H and Minomura S 1985 Jpn. J. Appl. Phys. 24 L143.
Nelson R J 1977 Appl. Phys. Lett. 31 351.
Oshiyama A and Saito M 1987a J. Phys. Soc. Jpn. 56 2104.
Oshiyama, A and Saito, M 1987b Phys. Rev. B36 2104.
Theis T N 1987 *Inst. Physics Conf. ser.* No. 91 p1.
Wolford D J, Streetman B G, Hsu W Y, Dow J D, Nelson R J and Holonyak Jr N 1976 Phys. Rev. Lett. 36 1400.

Inst. Phys. Conf. Ser. No 95: Chapter 6
Paper presented at Int. Conf. Shallow Impurities in Semiconductors, Linköping, Sweden, 1988

Magnetospectroscopy in HgCdTe: shallow donors and localization

V.J. Goldman[a] and M. Shayegan
Princeton University, Princeton, NJ 08544, U.S.A.
H.D. Drew and J.B. Choi
University of Maryland, College Park, MD 20742, U.S.A.

We report the observations of impurity cyclotron resonance as well as intra-Landau-level impurity transitions in n-type $Hg_{1-x}Cd_x Te$ ($x \simeq 0.2$). These data provide the first conclusive evidence that at low temperatures and high magnetic fields the conduction-band electrons are bound on shallow donors in this semiconductor.

Introduction

The low temperature transport properties of n-type HgCdTe have long been considered anomalous in comparison with other semiconductors, such as InSb, with similar electronic structure. Until recently it was believed that the carriers in HgCdTe did not freeze out at low temperatures and high magnetic fields. This circumstance led to the suggestion of a more exotic ground state in HgCdTe in high magnetic fields. In particular, Wigner crystallization of the free carriers has been conjectured and several authors have reported evidence for such a state based on transport data [1,2]. This claim is suspect, however, since there is much inconsistency in the transport data that has been reported in the literature which is a consequence of several severe material problems in HgCdTe [3]. In addition to the usual problems of donor density inhomogeneity in doped semiconductors, in HgCdTe there is the possibility of Hg and Cd concentration inhomogeneity and a tendency for the formation of conducting surface layers which tend to short out the bulk conductance, especially for low carrier densities at high magnetic fields and low temperatures. Only recently have samples of sufficiently high quality become available that it has been possible to convincingly observe the magnetic-field-induced localization [2,4]. However, the clearest evidence for a donor-bound electron ground state comes from the observation of impurity cyclotron resonance [5].

Magnetotransport

We present here magnetotransport data for a representative HgCdTe sample (n = $2.7 \times 10^{14} cm^{-3}$, $\mu = 2.7 \times 10^5 cm^2/Vs$). The sample was mounted in the tail of the mixing chamber of a dilution refrigerator which was placed in a Bitter magnet. The Hall (ρ_{xy}) and transverse (ρ_{xx}) resistivities were measured as a function of magnetic field B at several temperatures (Fig. 1). The sample is in the extreme magnetic

---

[a] Present and permanent address: Dept. of Physics, SUNY, Stony Brook, NY 11794.

© 1989 IOP Publishing Ltd

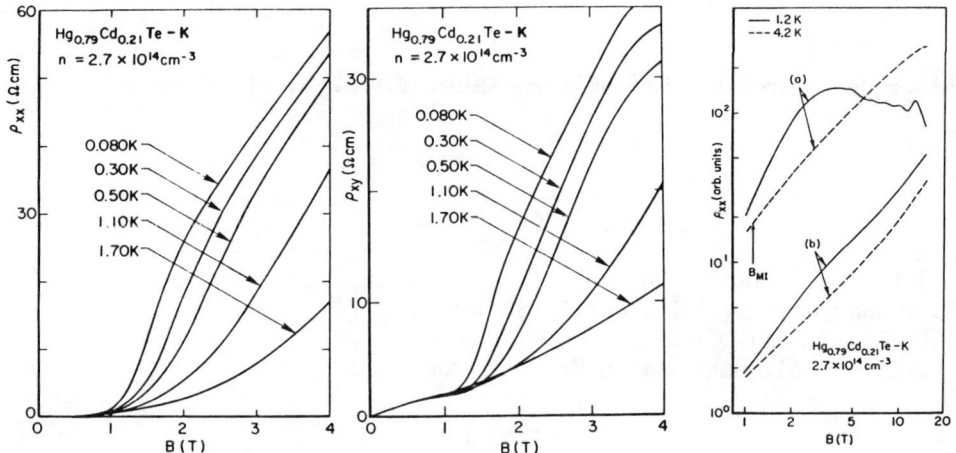

FIG. 1 The ρ_{xx} and ρ_{xy} versus magnetic field data are shown for a $Hg_{0.79}Cd_{0.21}Te$ sample at low temperatures down to 80 mK.

FIG. 2 The ρ_{xx} data are shown for a $Hg_{0.79}Cd_{0.21}Te$ sample (K). In (a) oscillations in ρ_{xx} are observed at low temperature and at magnetic fields well above B_{MI}. These are attributed to a high-mobility surface layer. In (b) the remeasured data for the same sample are shown after the surface was roughened (see text).

quantum limit (where only the lowest, spin polarized Landau level is occupied) at B = 0.3T. The metal-insulator (MI) transition occurs at B_{MI}=1.1T and is accompanied by a sharp rise in ρ_{xy}. The observed B_{MI} is in a good agreement with the scaling criterion which relates B_{MI} to the electron concentration and the material properties (effective mass and the dielectric constant) [6].

The measured magnetotransport coefficients in Fig. 1 display anomalously weak dependencies on B and T at high magnetic field (B $>$ 2T). As pointed out previously, this behavior is caused by the shorting of the bulk by a conducting surface layer. This effect is illustrated further in Fig. 2. In the solid trace labeled (a) clearly visible is a series of Shubnikov-de Haas oscillations which would imply a bulk electron density of $\sim 10^{18}$ cm^{-3}, four orders of magnitude higher than that for the sample. After a sand-blasting damage of the surface of the sample, these oscillations disappeared (lower traces).

Impurity Cyclotron Resonance in $Hg_{1-x}Cd_xTe$

Impurity cyclotron resonance (ICR) is an optically induced transition of a donor-bound electron from the ground state, related to the lowest (N=0, spin up) Landau level states, to an excited bound state, related to the N=1, spin up Landau level. The energy separation of the donor-bound states is slightly greater than that of the Landau levels and, therefore, ICR is shifted from the free-carrier resonance. The first observation of impurity cyclotron resonance was reported for InSb in 1957 [7] and this phenomenon has been studied extensively since then [8-10].

Here we report on the observation of ICR in $Hg_{1-x}Cd_xTe$ -- the first conclusive evidence, to our knowledge, that at low enough temperatures and sufficiently high

Shallow Impurities in II–VI Compounds

magnetic fields the conduction band-electrons are bound on donors [5]. The energy splitting between the ICR and the conduction band cyclotron resonance (CCR) is consistent with calculations performed within the hydrogenic donor model [11,12] and can be used to determine the binding energy of the electrons. The saturation of the ICR absorption with the incident radiation intensity, previously studied in InSb [9,10], was used here to measure the lifetime of the electrons accumulating in the lowest Landau level; we found $T_1 \sim 10^{-6}$ s - nearly two orders of magnitude longer than in InSb.

Samples were cut from 15 mm-diam. wafers, thinned, polished and briefly etched in a 3% bromine-methanol solution minutes before mounting in the cryostat [13]. Final thicknesses were d = 305, 290, and 260 μm for samples Ha, Hb, and Y, respectively. Measurements were made in Faraday configuration using far-infrared (FIR) radiation from an optically pumped cw laser. The radiation was circularly polarized using a FIR linear polarizer and a crystal quartz $\lambda/4$ plate and was detected by a Ge:Ga composite bolometer. The samples were mounted on a wedged germanium substrate and were immersed in pumped liquid ^{4}He to achieve temperatures below 4K. A FIR absorber with a small, 3mm-diam. hole was placed immediately in front of the samples to ensure the uniformity of the incident radiation.

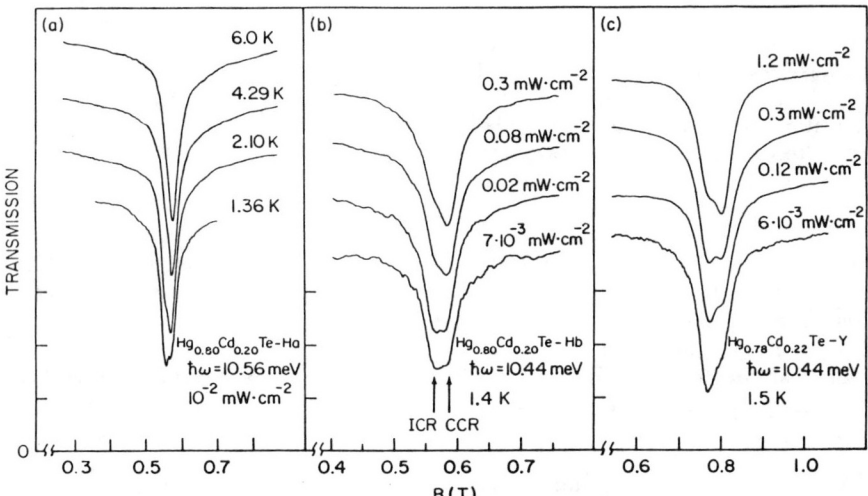

FIG. 3 Magnetotransmission spectra of n-$Hg_{1-x}Cd_x$Te samples at different temperatures and FIR radiation intensities. The arrows show the absorption peak assignment. Zero transmission levels, corrected to 100% cyclotron resonance-active circular polarization, are shown by horizontal tick marks on the left axis for each respective trace. Wafer H has x=0.204 ($E_g \simeq 75$meV), n = 3×10^{13}cm^{-3} and $\mu = 2.7 \times 10^5$ cm^2/Vs; wafer Y has x = 0.224 ($E_g \simeq 105$ meV), n = 6×10^{13}cm^{-3} and $\mu = 1.2 \times 10^5$cm^2/Vs.

Typical magnetotransmission spectra are shown in Fig. 3. The ICR absorption decreases and the CCR absorption increases as the temperature (Fig. 3a) or the radiation intensity (Figs. 3b and 3c) is raised. There is no ICR absorption in the cyclotron resonance-inactive circular polarization within the experimental uncertainty of $\sim 2\%$, in agreement with the theoretical prediction [11] that the selection rules for

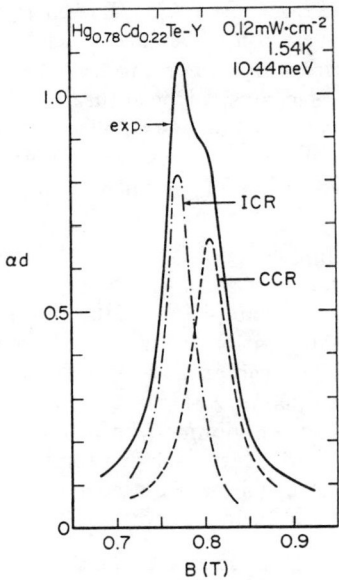
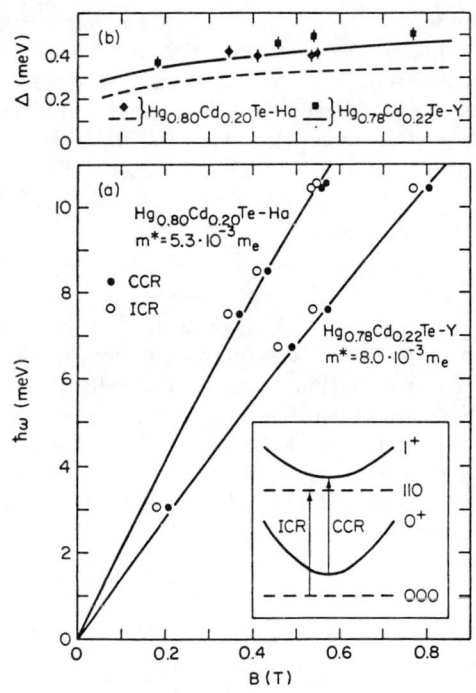

FIG. 4 An example of FIR absorption analysis (described in the text). About 10 magnetotransmission spectra at different intensities were used at every photon energy to obtain a pair of data points in Fig. 5 and the experimental uncertainties given in the text.

FIG. 5 a) Resonance magnetic fields at several FIR photon energies $\hbar\omega$. The lines give nonparabolic fits for $0^+ \rightarrow 1^+$ transition, with m^* used as fitting parameter. The inset shows the relevant energy-level scheme.
b) The energy splitting $\Delta = \hbar\omega_{ICR} - \hbar\omega_{CCR}$ vs. magnetic field. The lines show scaled theoretical prediction of Ref. 12. Effective Rydbergs for samples Ha and Y are, respectively, 0.25 meV and 0.38 meV.

the two processes are the same.

From the transmission data we calculated αd, where $\alpha(B)$ is the absorption coefficient and d is the sample thickness. We neglected multiple reflections because of the wedged substrate. At high temperatures ($\sim$ 10K) where only CCR is present, the data can be fitted reasonably well with a Lorentzian absorption line, which indicates high compositional uniformity of the samples. The line-width is consistent with the 77K Hall mobility. From the low-temperature experimental absorption we subtract a Lorentzian line with the resonance field, B_{CCR}, and the half-width, Γ, obtained from the high-temperature fit and use the peak CCR absorption, α_C, as a fitting parameter to get a smooth, approximately Lorentzian ICR absorption (cf. Fig. 4) [14]. We estimate the resulting uncertainty in the resonance field splitting, $B_{CCR} - B_{ICR}$, as 2.5mT for x = 0.224 and 1.5mT for x = 0.204. The peak absorption $\alpha_I d$ is accurate to within 5-10%.

Shallow Impurities in II–VI Compounds

The resonance fields, B_{CCR} and B_{ICR}, are plotted vs. photon energy, $\hbar\omega$, in Fig. 5a; the lines show a nonparabolic, Bowers-Yafet model [15] calculation with the band bottom effective mass, $m^*(B=0)$, used as a fitting parameter. The energy splitting between the ICR and CCR [16], $\Delta \equiv (E_{110} - E_{000}) - (E_{1+} - E_{0+})$, was calculated from

$$\Delta(B) \simeq [d(E_{1+} - E_{0+})/\,dB]_{B=B_{CCR}}(B_{CCR} - B_{ICR}),$$

and is shown in Fig. 5b. The lines in Fig. 5b give results of Larsen's nonparabolic model calculation [12], adapted for our samples by using m^* given in Fig. 5a and the static dielectric constant $\kappa = 17$. In the experimental range of magnetic fields, the data is in qualitative agreement with the theory [12] developed for an isolated donor. The quantitative discrepancy ($\sim 20\%$) is similar to that observed in InSb samples [8,17].

Figure 6 gives the dependence of the peak ICR absorption, $\alpha_1 d$, on the FIR radiation intensity I. A three-level model [9] predicts

$$\frac{\alpha_1(I)}{\alpha_1(0)} = [1 + \frac{\alpha_1(0)\cdot I}{\hbar\omega n}\,T_1]^{-1}, \qquad (1)$$

where T_1 is related to the electronic lifetime of the 0^+ Landau level. In this model the electrons are excited by the FIR radiation from (000) state to (110) state, from which they relax to the 0^+ level and directly to the ground state (000) with time constants τ_{32} and τ_{31}, respectively. Denoting the lifetime of the 0^+ Landau level by τ_{21} and making the same assumptions as Ref. 9 ($\tau_{32} \ll \tau_{31}$, $\tau_{32} \ll \tau_{21}$) at low enough temperatures $T_1 \simeq \tau_{21}$.

The reported lifetimes [9,10] for InSb fit the empirical relation $T_1 \simeq 2.5 \times 10^{-9}\gamma$ s, where $\gamma \equiv (a^*/l)^2$ and the magnetic length $l = (\hbar/eB)^{1/2}$. For samples Y and Hb we obtained $T_1 \simeq 1.0 \times 10^{-7}\gamma$ s and $1.4 \times 10^{-7}\gamma$ s, respectively. These are substantially longer lifetimes than for InSb. We believe that the relative uncertainty in values of I is $\sim 10\%$, however, the absolute calibration of FIR intensity is accurate only to within a factor of 2. A more extensive study in this alloy semiconductor can determine the band gap and/or effective mass dependence of the inelastic lifetime of the lowest Landau level and the chemical shifts [17] in the donor binding energies.

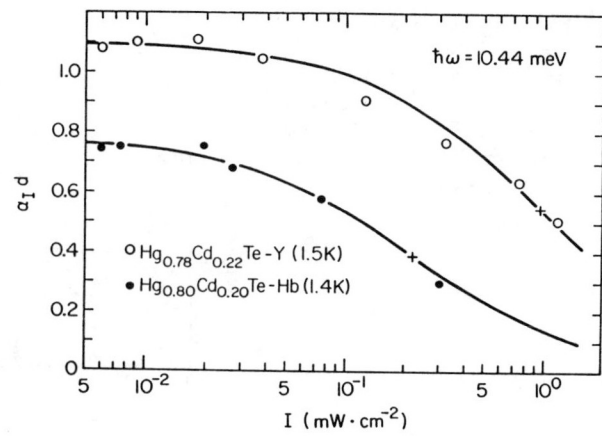

FIG. 6 Measured peak ICR absorption of the samples as a function of FIR radiation intensity. The curves are fits of Eq. (1) with the sample parameters given in the text. Crosses denote intensities $I_{1/2}$ such that $2\alpha_1(I_{1/2})=\alpha_1(0)$. The fits give times $T_1=3.4\times10^{-6}$ s and 1.6×10^{-6} s for samples Hb and Y, respectively.

Several factors combine to make the experimental observation of ICR and data analysis more difficult for $Hg_{0.8}Cd_{0.2}Te$ than for InSb. Lower electron effective mass and low $\hbar\omega_{LO} = 17$ meV limit the measurements to lower magnetic fields, which, in turn, combined with smaller effective Rydberg, leads to smaller relative splitting between ICR and CCR and requires lower experimental temperatures and impurity concentrations. The alloy potential fluctuations may broaden both ICR and CCR. In addition, the surface of $Hg_{1-x}Cd_xTe$ attracts electrons from the bulk. We have performed a study of this phenomenon using CR absorption to determine the fraction of the electrons left in the bulk. It has been found that with our sample preparation procedure the bulk electron per square concentration decreased linearly with the sample's thickness as $N = nd - N_s$, where $N_s \simeq 6 \times 10^{11} cm^{-2}$ was concluded to be the surface electron density. Since the width of CCR did not change appreciably as a sample was thinned, down to $N \simeq 3 \times 10^{10} cm^{-2}$, we also concluded that the bulk concentration, n, does not change, but rather, that the surface electrons leave behind depletion layers, at each side of the sample.

We note here that the cyclotron resonance of an accumulation layer would be shifted to higher fields due to nonparabolicity and, more important, would be much broader than the bulk CCR due to the high scattering rate of the surface electrons. We do observe a weak and broad background absorption in the magnetotransmission data which can be due to the surface-bound electrons.

Shorting of the bulk by the surface conduction prevented us from obtaining useful transport data in these low n samples. However, by extrapolation from the data on higher n samples [6], we can estimate that the magnetic-field-induced metal-insulator transition field, B_{MI}, for sample Y is approximately 0.35T. The FIR data at $\hbar\omega = 2.99$ meV show the ICR absorption present at a lower field (cf. Fig. 7) with the integrated absorption strength equal to that at higher fields within the experimental uncertainty. This experimental observation does not fit into any of the MI transition pictures considered in the literature and suggests that even on the metallic side of the MI transition, the delocalized electrons are in donor band states which are distinct from the conduction band states.

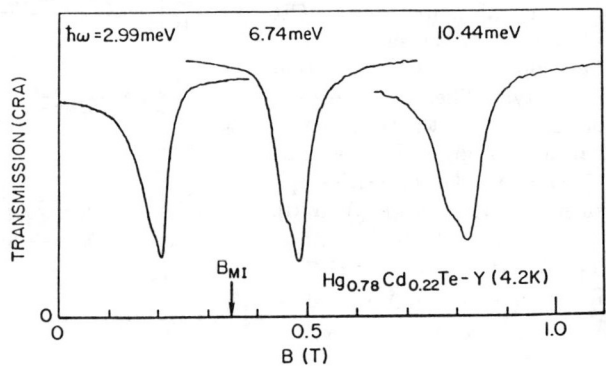

FIG. 7 Magnetotransmission spectra of a $Hg_{0.78}Cd_{0.22}Te$ sample (Y) at different FIR wavelengths [cyclotron-resonance-active (CRA) circular polarization] showing the presence of ICR at magnetic fields below B_{MI}.

Intra-Landau-level Transitions

In this paper we also present our results on the (000)→(001) transition which occurs at energies far from cyclotron resonance and has different selection rules [18]. Magnetooptical transmission was studied using far-infrared radiation from an optically pumped cw laser in the spectral regions $\lambda = 118.8 \mu m$ to $888.9 \mu m$. The Voigt configuration cryostat was equipped with a linear polarizer which can be rotated from the outside.

Typical traces of the Voigt geometry magnetotransmission in the E ∥ B polarization for the sample with x=0.237 are shown in Fig. 8 for different photon energies. When the angle of polarizer is changed so that E is perpendicular to B, the impurity transition disappears and the conduction band cyclotron resonance appears at much lower magnetic fields. This selection rule confirms the identification of the (000)→(001) impurity transition. Fig. 9 shows a plot of transition energy vs. magnetic field and the observed transition energies are compared to the predictions of non-parabolic theory for the (000+)→(001+) transition energy. The unusual

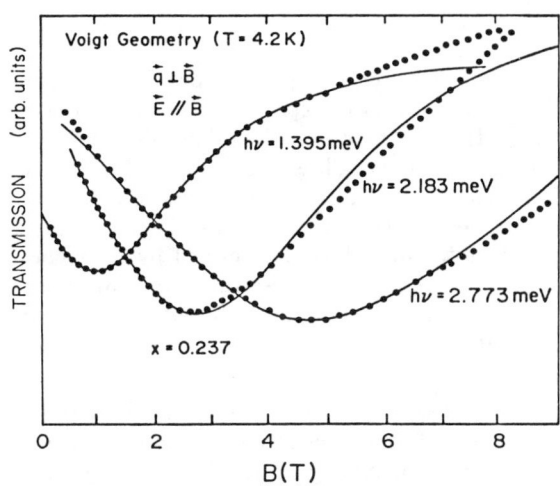

FIG. 8 Typical magnetotransmission spectra for the x=0.237 HgCdTe sample in the Voigt geometry. The data are the dotted curves; the solid lines give the fits to the Lorentzian oscillator model dielectric function.

broadening of the absorption lines, seen in Fig. 8, can be understood in terms of the effects of the nonlinear dependence of the transition energy on magnetic field in these fixed-frequency, field-sweep measurements. Because of the sublinear dependence of transition energy on magnetic field, as seen in Fig. 9, the width, in magnetic field, of the absorption lines increases for larger photon energies. This also leads to asymmetric lineshapes for these impurity transitions. To confirm this qualitative argument we have fitted the data to calculations of the transmission based on a general oscillator model dielectric function for a bound electron given by

$$\epsilon = \epsilon_0 - \omega_0^2/(\omega^2 - \omega_B^2 + i\omega\Gamma)$$

where ϵ_0 is the static dielectric constant, $\omega_0^2 = 4\pi n e^2/m^*$ and Γ is the linewidth arising from lifetime broadening or finite bandwidth. The oscillator strength is assumed

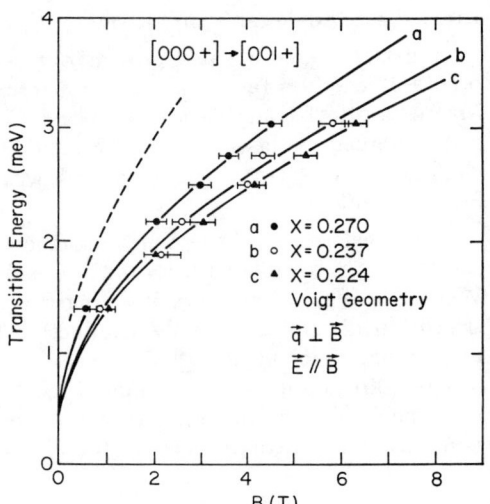

FIG. 9 The magnetic field dependence of the impurity transition energy for samples with different values of x. The solid lines correspond to the best fits including the central cell effect. The fitting parameters are: $\epsilon_0 = 16.3$ and $c_i = 2.4 \times 10^{-17}$ meVcm^3 for $x = 0.237$; $\epsilon_0 = 16.1$ and $c_i = 2.3 \times 10^{-17}$ meVcm^3 for $x = 0.27$; and $\epsilon_0 = 16.2$ and $c_i = 2.7 \times 10^{-17}$ meVcm^3 for $x = 0.224$, where c_i give the magnitude of the central cell effect (cf. Ref. 18). The dashed curve represents the dispersion of the $(0\bar{1}0) \rightarrow (0\bar{1}k_z)$ photoionization transition observed in the $x = 0.224$ sample at 4.2K.

to be unity which is a good approximation since the next strongest impurity transition $(0\bar{1}0) \rightarrow (0\bar{1}1)$ is expected to be very weak compared to the $(000) \rightarrow (001)$ transition [11]. The magnetic-field-dependent transition frequency, ω_B, used in the fit was determined from a least square fit of the experimental data in Fig. 9 to a simple functional form. The resulting transmission curve fits, shown by solid lines in Fig. 8, correspond to values of $n = 6 \times 10^{13}$cm^{-3} and of a.c. mobility $= 10^5$cm^2/Vsec which are close to the values determined by transport measurements. The good agreement of the fits, therefore, successfully accounts for the unusual broadening of the absorption lines. It follows that in broad-band spectroscopy, in which the magnetic field is fixed and the photon energy scanned, narrower and more symmetric lines should be observed.

We have also observed the photoionization transition $(0\bar{1}0) \rightarrow (0\bar{1}k_z)$. The dispersion of the observed transition for the $x = 0.224$ sample is indicated by the dashed curve in Fig. 9. This absorption peak, which is observed only in the Voigt (E ∥ B) geometry, vanishes at 1.5K. Its identification comes from the position, temperature dependence and selection rules: 1) The position is close to that predicted from YKA [15] theory. 2) Its disappearance at 1.5K corresponds to the freezeout of the electrons from the $(0\bar{1}0)$ level to the (000) ground state. 3) Calculations of the absorption strength of this transition indicate that its strength in the Faraday geometry (E ⊥ B) should be weaker by a factor of 10 compared with the Voigt geometry (E ∥ B) as observed in the experiment.

Conclusions

We believe that the selection rules, the magnitude of the splitting and its magnetic field dependence together with the absorption saturation effect allow us to identify the low-field peak in the magnetotransmission spectra of $Hg_{1-x}Cd_xTe$ as the $(000) \rightarrow (110)$ ICR transition which is direct and conclusive evidence for hydrogenic donors in this material. This is further confirmed by the Voigt geometry measurements of the $(000) \rightarrow (001)$ transitions. These observations allow us to conclude that the Wigner condensation of the conduction band electrons does not occur in $Hg_{1-x}Cd_xTe$.

Acknowledgements - We thank D.A. Nelson for the $Hg_{1-x}Cd_xTe$ samples, and acknowledge support through an NSF PYI Award (Grant No. ECS-8553110) and an NSF Grant No. DMR-8705002 for M.S., an NSF Grant No. DMR-8704670 and an ONR Grant No. N00014-86-K-0273 for H.D.D.

References

1. G. Nimtz, B. Schlicht, E. Tyssen, R. Dornhaus, and L.D. Haas, Solid State Commun. **32**, 669 (1979); J.P. Stadler and G. Nimtz, Phys. Rev. Lett. **56**, 382 (1986); G. Nimtz and J. Gebhardt, in *Proc. 18th Intl. Conf. Phys. Semicond.*, edited by O. Engström, (World Scientific, Singapore, 1987), p. 1197.
2. T.F. Rosenbaum, S.B. Field, D.A. Nelson, and P.B. Littlewood, Phys. Rev. Lett. **54**, 241 (1985); S.B. Field, D.H. Reich, B.S. Shivram, T.F. Rosenbaum, D.A. Nelson, and P.B. Littlewood, Phys. Rev. B **33**, 5082 (1986).
3. J.B. Mullin and A. Royle, J. Phys. D **17**, L69 (1984); W. Zhao, C. Mazuré, F. Koch, J. Ziegler, and H. Maier, Surf. Sci. **142**, 400 (1984); O.G. Balev, P.I.Baranskii, G.V. Beketov, R.M. Vinetskii, and O.P. Gorodnichii, Sov. Phys. Semicond. **21**, 625 (1987).
4. M. Shayegan, V.J. Goldman, H.D. Drew, D.A. Nelson, and P.M. Tedrow, Phys. Rev. B **32**, 6952 (1985); M. Shayegan, V.J. Goldman, H.D. Drew, N.A. Fortune, and J.S. Brooks, Solid State Commun. **60**, 817 (1986).
5. V.J. Goldman, H.D. Drew, M. Shayegan, and D.A. Nelson, Phys. Rev. Lett. **56**, 968 (1986).
6. M. Shayegan, V.J. Goldman, and H.D. Drew, Phys. Rev. **B38** (1988).
7. W.S. Boyle and A.D. Brailsford, Phys. Rev. **107**, 903 (1957).
8. R. Kaplan, Phys. Rev. **181**, 1154 (1969); E.J. Johnson and D.H. Dickey, Phys. Rev. B **1**, 2676 (1970); F. Kuchar, E. Fantner, and G. Bauer, J. Phys. C**10**, 3577 (1977).
9. T. Murotani and Y. Nisida, J. Phys. Soc. Japan **32**, 986 (1972); K.L.I. Kobayashi and E. Otsuka, J. Phys. Chem. Solids **35**, 839 (1974).
10. E. Gornik, T.Y. Chang, T.J. Bridges, V.T. Nguyen, and J.D. McGee, Phys. Rev. Lett. **40**, 1151 (1978); W. Muller, E. Gornik, T.J. Bridges, and T.Y. Chang, Solid State Electronics **21**, 1455 (1978).
11. R.F. Wallis and H.J. Bowlden, J. Phys. Chem. Solids **7**, 78 (1958).

12. D.M. Larsen, J. Phys. Chem. Solids **29**, 271 (1968).
13. Samples Ha and Hb were cut from the same wafer; the difference in the positions of CCR yields composition variation $\delta x = [dx/dm^*]_{x=0.2} \cdot \delta m^* \simeq 1 \times 10^{-3}$.
14. There is additional absorption in the low field tail, probably (see second paper in Ref. 9) due to the photoionization transition $(000) \rightarrow (1^+)$.
15. R. Bowers and Y. Yafet, Phys. Rev. **115**, 1165 (1959).
16. We use donor level notation of H. Hasegawa and R.E. Howard, J. Phys. Chem. Solids **21**, 179 (1961).
17. R. Kaplan, R.A. Cooke, R.A. Stradling, and F. Kuchar, in *The Application of High Magnetic Fields in Semiconductor Physics*, ed. J.F. Ryan (Clarendon Laboratory, University of Oxford, 1978), p. 397.
18. J.B. Choi, L.S. Kim, H.D. Drew, and D.A. Nelson, Solid State Commun. **65**, 547 (1988); J.M. Perez, J.E. Furneaux, and R.J. Wagner, J. Vac. Sci. Technol. **A6**, 2681 (1988).

Inst. Phys. Conf. Ser. No 95: Chapter 6
Paper presented at Int. Conf. Shallow Impurities in Semiconductors, Linköping, Sweden, 1988

The intrinsic and extrinsic doping of mercury cadmium telluride grown by molecular beam epitaxy

M. Boukerche and J.P. Faurie
University of Illinois, Dept of Physics, Chicago, Illinois 60680

ABSTRACT: The electrical properties of the mercury cadmium telluride semiconducting alloy grown by the molecular beam epitaxy technique are reviewed. The intrinsic doping produced by stoichiometry deviation during growth, as well as the extrinsic doping resulting from incorporation of indium, lithium, silicon, arsenic and antimony are described. The element from column I (Li) is a p-type dopant whereas the others dope the material n-type. We will show that these impurities mostly substitute in the metal site. We will conclude that the growth occurs under tellurium rich conditions.

1. INTRODUCTION

The mercury cadmium telluride material (MCT) is recognized as the most important infrared material for detector applications in the 8-12um and 3-5um wavelength atmospheric transparency windows. Since its direct bandgap can be tuned by adjusting its cadmium composition, it was also suggested as a candidate for photonic applications in a much wider energy range. The molecular-beam epitaxy (MBE) technique is now established as a possible technique to grow high quality MCT epitaxial layers.

Very important improvements have been made in the growth control and understanding (Sivananthan et al 1988) since it was first reported seven years ago (Faurie and Millon 1981). The best as-grown MBE material compares with the best post-annealed material produced by the other techniques. Highly uniform hetero-epitaxial MCT-MBE layers have been demonstrated on two-inch GaAs substrates (Lange et al 1988). Characterizations of heterojunctions and homojunctions made in-situ have already been reported (Boukerche et al 1988a). High performance photovoltaic operation has also been shown recently on homojunction diodes made ex-situ by ion implantation on unannealed MBE material (Faurie et al 1988, Arias et al 1988).

The highly flexible MBE growth technique is an ideally versatile tool to study the correlations between the growth conditions and the electrical properties of the layers. It is also being considered for industrial applications. We will review the few studies published on the doping of the MBE-MCT material, and will present the latest results obtained in our laboratory.

2. INTRINSIC DOPING

Like for the other techniques, the MBE technique is facing all the difficulties associated with the ionicity of this II-VI pseudo-random alloy. The large

© 1989 IOP Publishing Ltd

stoichiometry deviation range ($\delta y \leq 10^{-5}$) can generate intrinsic dopings in the 10^{17} cm^{-3} range. We can see that the growth control is even more critical than for the III-V materials. A few degrees change in substrate or Hg cell temperature can change substantially the doping concentration of the material being grown. The precision of the growth controls will determine the minimum extrinsic doping level achievable with reproducibility.

The best compromise between the growth requirements of CdTe and HgTe sets the substrate temperature (Ts) around 190C (Sivananthan et al 1988) for Cd compositions (X) close to 0.2-0.3. Monitoring the growth by high energy electron diffraction (HEED) and establishing correlations with Hall measurements showed that high quality epitaxy can only be achieved within narrow ranges of Hg pressures and substrate temperatures for fixed X and growth rates in the (111)B orientation. If the Hg flux is too low or Ts too high the p-type character of the layer increases. When rings become detectable on the HEED screen, which is attributed to tellurium precipitation, the Hall mobility decreases to 100 cm^2/V/s or less and the doping level is p-type in the low 10^{17}cm^{-3} range for X=0.2-0.3 and growth along the (111)B orientation. In the same crystallographic orientation, if the mercury pressure is too high or Ts is too low, twin dots appear along the diffraction lines of the HEED screen, and the electrical properties of the layers degrade rapidly (Faurie et al 1986a). They usually show n-type conduction. We think that the related defects are twins. Abnormal behaviors were reported for such layers after isothermal anneals under mercury pressure (Arias et al 1987). The fact that excess Hg produces n-type material even before twins are visible is consistent with the well recognized properties of Hg in MCT (Vydyanath 1981a).

The growth requirements vary with the growth orientation. These differences will also be reflected in their respective electrical properties. The minimum Hg pressure found necessary to maintain a monocrystalline growth was found to be an order of magnitude higher for the (111)A cadmium face than for the (111)B tellurium face (Sivananthan et al 1986). The same results were also reported for the (112)Cd and Te faces (Koestner and Schaake 1988). The case of the (100) face lies somewhere in between.This is one of the reasons why the (111)B orientation was studied first since high vacuum conditions were difficult to sustain in non specifically designed MBE systems. The price of high purity mercury is also to be taken into consideration for industrial applications.

In addition to the effects of the Hg flux and Ts previously mentioned, an increase in Cd composition tends to produce p-type material (Boukerche et al 1986). In the (111)B orientation good quality layers can only be grown p-type for Cd compositions more than .2, and n-type for X less than 0.2 (Sivananthan et al 1988). The p-type intrinsic doping level can reach the mid-10^{17}cm^{-3} range (Boukerche et al 1986). Fittings of the Hall data with a model based on two band Kane theory and a single shallow doping level singly ionized were performed. Very good agreements could be obtained between the carrier concentrations measured and calculated in the hole temperature range, as well as for the Cd compositions measured by Fourier transform infrared spectroscopy (FTIR) at room temperature and the ones fitted. The donor compensation was estimated to be in the low 10^{17}cm^{-3} range for this particular set of samples. Automatic fittings made with the same model on crystals grown more recently gave compensation levels in the mid-10^{16}cm^{-3} range (Fig. 1). Interestingly, a two level model with a doubly ionized state was not required to reach such lower compensation magnitudes. The non-uniform or suspicious layers were obviously not considered in this analysis. We should point out that surface inversion in as-grown p-type

MBE layers is very unusual, even for doping levels of 1×10^{16}cm^{-3}. The best p-type Hall mobilities reached consistently 1100cm^2/V.s or more (Table I) without two dimensional confinement effects due to HgTe interfaces (Faurie et al 1986b). They compare favorably with published data for the other growth techniques (Gold and Nelson 1986, Meyer et al 1987). Doping levels in the 10^{15}cm^{-3} range were also observed. Excess carriers lifetimes were measured on these materials being illuminated by a solid state laser. The photosignal was always checked for exponentiality. The electron lifetime in some of these p-type layers could exceed 700ns at 220K (Fig. 2) and was very well fitted with a Shockley-Read model including a recombination level near mid-gap (DeSouza et al 1988). The analysis of such material versus magnetic field with a three band model confirmed the expected band structure with one light electron, one light hole and one heavy hole bands (Wijewarnasuriya et al 1988).

High mobility n-type layers can also be achieved in the (111) direction by stoichiometry adjustment. However, the carrier concentration determined by the Hall measurement monotically decreases with temperature and never exceeds the low 10^{16}cm^{-3}

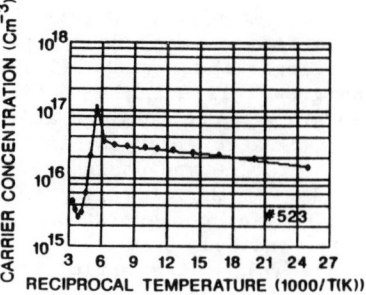

Fig. 1. Hall curve calculation for a p-type intrinsic layer parameters used: x = .259, N_d = 2.9×10^{16}cm^{-3}, N_A = 6×10^{16}cm^{-3}, E_A = 3 meV, the donors are assumed fully ionized.

SAMPLE	SUBSTRATE	X	THICKNESS	T	CARRIER CONC. N_D-N_A(cm^{-3})	MOBILITY μ_H(cm^{2}V^{-1}s^{-1})
(111)B ORIENTATION							
196 481	CdTe	0.20	4.8µm	30K	6.0x10^{15}	1.1x10^3
198 484	CdTe	0.21	4.3µm	30K	5.1x10^{15}	1.0x10^3
242 602	CdTe	0.22	11.9µm	23K	2.0x10^{15}	1.1x10^3
215 527	CdTe	0.25	11.9µm	30K	6.2x10^{15}	8.7x10^2
205 516	CdTe	0.29	15.6µm	30K	7.5x10^{15}	8.7x10^2
216 528	CdTe	0.34	12.1µm	77K	3.6x10^{15}	8.0x10^2
667 532	GaAs	0.20	2.1µm	23K	4.1x10^{15}	1.2x10^3
681 540	GaAs	0.21	3.7µm	23K	3.6x10^{15}	1.1x10^3
583 453	GaAs	0.22	5.4µm	30K	1.4x10^{15}	8.7x10^2
654 508	GaAs	0.25	3.8µm	30K	1.1x10^{15}	7.3x10^2
393 244	GaAs	0.28	1.3µm	40K	2.8x10^{15}	5.2x10^2
500 308	GaAs	0.31	2.3µm	30K	1.1x10^{15}	4.5x10^2
4 319	CdTeSe	0.31	7.6µm	30K	2.4x10^{15}	8.4x10^2
2 310	CdTeSe	0.32	9.4µm	30K	1.2x10^{15}	6.7x10^2
33 601	CdZnTe	0.21	12.1µm	23K	1.5x10^{15}	1.6x10^3
34 605	CdZnTe	0.24	9.2µm	23K	1.4x10^{15}	8.0x10^2
(100) ORIENTATION							
673 534	GaAs	0.21	4.2µm	23K	3.4x10^{15}	5.1x10^2
125 304	CdTe	0.24	3.0µm	77K	2.0x10^{15}	2.5x10^2
127 306	CdTe	0.29	2.7µm	77K	1.5x10^{16}	1.2x10^3

TABLE 1: Electrical characteristics of p-type HgCdTe grown by MBE between 190-200°C (No HgTe layer at the interface).

range at low temperatures (Boukerche et al 1986a). When the Cd composition is less than .2, the carrier concentration at high temperatures can be higher than the expected intrinsic carrier concentration. Such layers were confirmed to be n-type by capacitance measurements on MIS structures (Yoo et al 1988) and thermoelectric probing at cryogenic temperatures. Such a behavior looks similar to bandgap narrowing due to the formation of an impurity band. We suggest that it could also be related to a self-compensated native defects together with a low electron to hole mobility ratio. This subject is still being investigated.

MBE growth along the (100) orientation can also produce high quality n or p-type MCT material. However, the layers are n-type for Cd compositions less than .35. P-type material can be achieved only when the substrate temperature exceeds 200C (Arias et al 1987, Sivananthan et al 1988). No twinning is observed on the HEED screen during the growth. As previously

mentioned the magnitude of the Hg flux necessary to maintain monocrystallinity is larger than in the (111)B case. The growth is actually easier to control since the suitable relative range of Hg flux is wider. As in the (111)B case, the n-type doping level is limited to the low $10^{16} cm^{-3}$ range. The carrier concentration and resistivity deduced from the Hall measurements follow the anticipated behaviors of the narrow bandgap n-type MCT material versus temperature (Hansen and Schmit 1983). The donor level is merged in the conduction band and is always fully ionized. The resistivity dips at the onset of extrinsic conduction. Faceting during growth along the (100) direction resulting in pyramidal voids present in the layers was reported recently

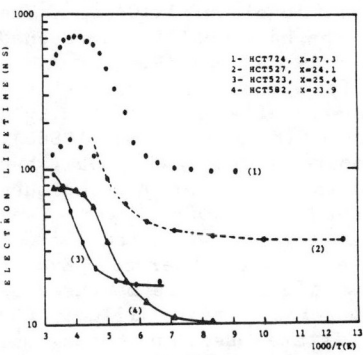

Fig. 2. Example of photodecay lifetime measurements on p-type intrinsic material. ---: calculation with the Shockley-Read theory, $E_T/E_g=.49, \tau_{no}=35ns$.

(Million et al 1988). They were shown to originate at the CdTe interface. The density of these defects was function of the substrate used and the surface preparation. We did observe the same defects. Whether they will remain a serious limitation for the growth in the (100) orientation, or are linked to the quality of the growth control, is still to be determined.

We see that the physical and electrical properties of the layers are totally dependent on the growth orientation. It was previously suggested that the Hg pressure requirements for the (111)Te face were smaller because the Te growing front was acting as a protective layer limiting the Hg (and Cd) reevaporation (Sivananthan et al 1986). We think that Te is responsible for the different intrinsic properties of the MCT-MBE material in the two directions described above. The (111)B growth seems to be intrinsically p-type. The negative sign of the Hall constant at low X is more linked to the narrow bandgap effects (electron to hole mobility and density of states ratios) than to the actual doping density. An opposite situation prevails in the (100) orientation since the layers are definitely doped n-type. This difference cannot be explained by a model based on the concentration of mercury vacancies since more Hg is required to grow the (100) direction. Furthermore the (100) growth orientation leads to p-type material only when the substrate temperature exceeds 200C, which is known to induce substantial tellurium reevaporation (Sivananthan et al 1986). We conclude that Te is indeed responsible for the different doping properties of these two growth orientations. Substitutional Te atoms on the metal site might explain the (100) n-type (Faurie 1987). We will now see that it also controls its extrinsic doping properties.

3. DOPING WITH INDIUM

Controlled In evaporation during MBE growth was found to produce n-type layers on CdTe(111)//GaAs(100) combination substrates (Boukerche et al 1986b). The maximum electron concentration measured by Hall could reach $2 \times 10^{18} cm^{-3}$, which is two orders of magnitude more than what can be achieved by stoichiometry deviation. The carrier concentration at low temperatures was compared to the atomic In concentration determined by secondary ion mass spectrometry (SIMS) calibrated with implanted reference samples. Even

though the Cd composition of the samples varied from .2 to .3 and some uncertainty was unavoidable, a clear trend was detected. The data measured could be very well fitted with a model assuming that part of the In concentration precipitated in the form of In_2Te_3, the rest of it being singly ionized. The formation of this compound should actually not be seen as a singular defect since it has been reported in solid solution in HgTe for compositions between 0 and 0.2 (Spencer 1964). The model also assumed a fixed concentration of compensating acceptors in the low $10^{17} cm^{-3}$ range. This value was not surprising since the layers were intentionally grown with conditions adjusted to produce intrinsic p-type material without In. More recent results analyzed with the same model show that the In population is fully ionized for concentrations of $10^{17} cm^{-3}$ or below (Sou et al 1988) as can be seen in Fig. 3. The compensating acceptor concentration will be precisely determined when more measurements will be available. At the present time it could only be estimated to be at least a decade lower than previously reported. The electrical efficiency of In in MCT made by MBE is larger than for other growth techniques (Vydyanath 1981b). This is probably related to the lower substrate temperature used during growth decreasing the rate of In_2Te_3 precipitation. SIMS profiling of In junctions grown during two hours or more showed diffusion profiles around 2000A wide probably limited by the instrument itself (Fig. 4). Uncorrected spreading resistance data measured at cryogenic temperature gave a transition width of 3000A for a junction 2.5um deep. A layer with a Cd composition 0.55 was doped with In and was measured having $10^{15} cm^{-3}$ electrons. This could never be achieved by stoichiometry adjustment. In general the Hall mobilities of the In doped layers look promising in view of their doping levels and Cd compositions (Fig. 5). We discovered that the layers grown after In doping was used tend to be n-type. SIMS measurements proved that In was still present in the subsequently grown material. This limitation is obviously critical if junctions are to be realized. However we think that it is a technological rather than physical limitation linked to the high vapor pressure of In versus temperature. This problem has now nearly been solved since intrinsic p-type layers could be grown after In doped material. Thanks to its exceptional electrical properties, this element will most certainly

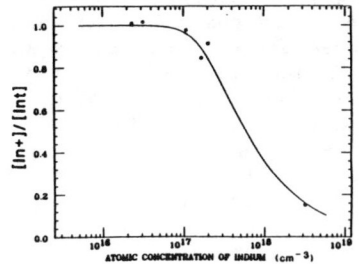

Fig. 3. Fraction of ionized indium atoms versus the atomic concentration of indium.

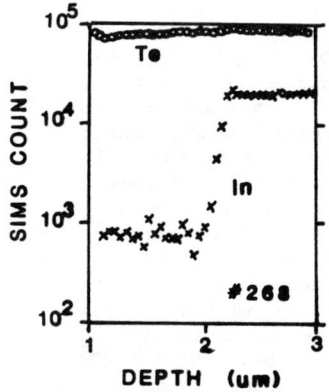

Fig. 4. SIMS profile of sharp In doping transition made during growth.

remain as a one of the major n-type dopants of MBE-MCT. We should also mention that it was demonstrated as an efficient n-type dopant in the 10^{17}cm^{-3} range for CdTe grown by laser assisted MBE (Bicknell et al 1987).

4. DOPING WITH LITHIUM

The need for a p-type dopant in MBE-MCT rises first for the growth in the (100) orientation since it usually produces intrinsic n-type material as mentioned above. Since Li is a column I element, it was expected to behave as a p-type dopant. Such was indeed the case (Wijewarnasuriya et al 1987). Doping levels as high as 8×10^{18}cm^{-3} could be measured. The amount of Lithium incorporated in the MBE layers grown was estimated from the Li effusion cell temperature, the geometrical configuration of the system, and the growth rate, assuming a unity condensation coefficient. A very good agreement was found between these calculated values and the carrier concentration deduced from the Hall measurements above 5×10^{17}cm^3 (Fig. 6). We conclude that most of the Li atoms are indeed incorporated and singly ionized in the crystal. Hole freeze-out was only detected once for a doping level of 10^{16}cm^{-3}. The carrier concentration remained independent of temperature in agreement with a shallow

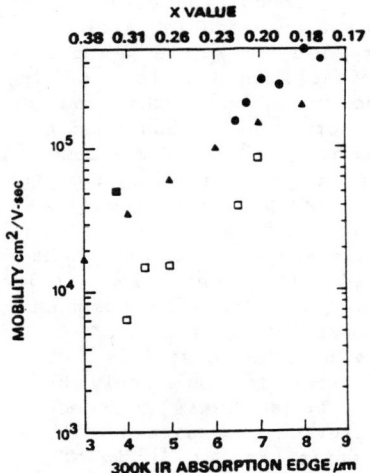

Fig. 5. Comparison of mobilities for MCT layers from different origins.

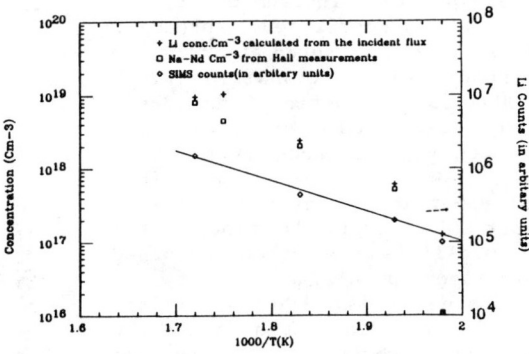

Fig. 6. Comparison of calculated Li incident flux with SIMS and electrical data.

doping level merged into the valence band for doping levels of 4×10^{16}cm^{-3} or more. Above 2×10^{18}cm^{-3} the carrier concentration started to decrease slightly along a curved shape with reciprocal temperature showing the onset of semi metallic impurity band conduction. Impurity scattering calculations similar to the ones made by Vydyanath (1981a) were attempted. When the mobility without In is assumed to be around 500cm^2/V.s, the calculated and measured values, assuming that the donors are singly ionized at low temperatures, are in reasonable agreement (Fig. 7). Lithium was also found to dope the (111)B MBE grown material p-type, even for Cd compositions as low as 0.16. The Hall constant measured on such low X MBE material is always negative without lithium incorporation. It thus seems

that Li is a very efficient p-type dopant probably substituting in the metal site. Unfortunately we discovered that the MCT material grown above or below Li doped regions had basically the same Li content as detected by SIMS. Lithium is highly mobile in the MBE layers. It can also diffuse out of the material since some Li doped crystals were measured n-type after annealing. This result is not surprising since the Li atomic radius is small, and it obviously compromises any electronic device application.

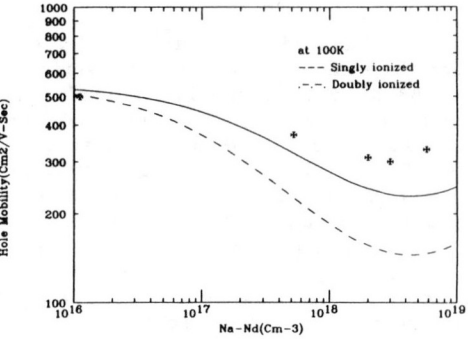

Fig. 7. Measured and calculated hole mobilities versus Li doping level.

5. DOPING WITH OTHER ELEMENTS

Silicon, a column IV element, was found to behave as an n-type dopant when evaporated during MBE-MCT growth (Boukerche et al 1988b). The highest level achieved reached $1.7 \times 10^{17} \text{cm}^{-3}$. Unlike for the case of indium, no memory effect was observed. The reproducibility of the doping level was more difficult to achieve, probably because of the more intimate interactions between this element and the native growth conditions. The amphoterous character of this element is recognised in the growth of III-V materials, and it was previously shown to produce n-type MCT material before (Johnson and Schmit 1977). A fully ionized behavior independent of temperature was seen even for doping levels in the 10^{15}cm^{-3} range. A strong acceptor compensation was suspected since strong impurity scattering limited the mobility at low temperatures. SIMS profiles could resolve less than 1000A transitions on 3um thick junctions. Homojunctions made on intrinsic p-material with this dopant were reported recently (Boukerche et al 1988a). The doping concentration was suspected to be non-uniform close by the junction. It was also concluded that the device operation was limited by generation-recombination.

The column V elements arsenic and antimony were also found to act as n-type dopants in MBE-MCT. The doping levels achieved never exceeded the 10^{16}cm^{-3} range. The carrier concentration was nearly independent of temperature in the extrinsic range. Unlike intrinsic doping it was also the case for low X material in the (111)B orientation. Strong impurity scattering was limiting the mobility at low temperature. Since the doping magnitude remained low, we think that these elements were highly self-compensated in the MBE material. Arsenic was previously reported to be amphoteric when incorporated during tellurium rich MCT growth by the liquid phase epitaxy technique (Vydyanath et al 1986). Illumination of the crystal during the growth by an Nd-Yag green laser pulsed at 10Khz only slightly increased the Cd composition and the electron concentration in the material, but did not change its electrical type. Isothermal annealing at temperatures between 200 and 250C also kept the samples n-type, showing that the doping mechanism was stable.

This is a very important result giving insights on the MBE growth mechanism since As was found to dope MCT p-type by diffusion (Johnson and Schmit 1977).

CONCLUSION

We showed that lithium, a column I element, substitutes in the metal site to dope the material p-type. Silver, another element from column I, was also reported as doping the MBE material p-type (Wroge et al 1988). Indium, a column III element, was shown to dope efficiently the layers n-type, mostly substituting in the metal site, the rest precipitating isoelectrically in the form of In_2Te_3. When going to column IV (Si) and column V (As,Sb) the elements investigated still dope the material n-type, but the maximum doping levels achieved decrease sharply and the self-compensation increases. We conclude that none of the elements studied could substitute in the chalcogenide site. This proves that the MBE-MCT growth is occurring under very rich Te conditions so that no Te site is available for substitution. This analysis is consistent with the case of intrinsic doping discussed above since we already suggested that Te is responsible for the electrical properties of the layers.

ACKNOWLEDGEMENTS

The authors would like to thank I.K.Sou, Y.J.Kim and S.Sivananthan for the growth of the samples, P.S.Wijewarnasuriya and S.Yoo for their electrical characterizations, K.K.Mahavadi for the SIMS measurements, Z.Ali and S.Farook for their technical support.

This work was funded by the Defense Advance Research Project Agency under contract No.F49620-87-C-0021.

Arias J M, Shin S H, Cheung J T, Chen J S, Sivananthan S, Reno J and Faurie J P 1987 J.Vac.Sci.Technol. A5 3133
Arias J M, Shin S H, Pasko J G and Gertner E R 1988 Appl.Phys.Lett. 52 39
Bicknell R N, Giles N C and Schetzina J F 1987 Appl.Phys.Lett. 50 691
Boukerche M, Wijewarnasuriya P S, Reno J, Sou I K and Faurie J P 1986a J.Vac.Sci. Technol. A4 2072
Boukerche M, Reno J, Sou I K, Hsu C and Faurie J P 1986b Appl.Phys.Lett. 48 1733
Boukerche M, Yoo S, Sou I K, DeSouza M and Faurie J P 1988a J.Vac.Sci. Technol. A6 2623
Boukerche M, Wijewarnasuriya P S, Sivananthan S, Sou I K, Kim Y J, Mahavadi K K and Faurie J P 1988b J.Vac.Sci.Technol. A6 2830
DeSouza M, Boukerche M and Faurie J P 1988 (to be published)
Faurie J P and Millon A 1981 J.Cryst.Growth 54 582
Faurie J P, Reno J, Sivananthan S, Sou I K, Chu X, Boukerche M and Wijewarnasuriya P S 1986a J.Vac.Sci.Technol. B4 585
Faurie J P, Sou I K, Wijewarnasuriya P S, Rafol S and Woo K C 1986b Phys. Rev. 34 6000
Faurie J P 1987 HgCdTe MBE Workshop, DARPA-CNVEO, Washington (unpublished results)
Faurie J P, Sivananthan S, Lange M, DeWames R E, Vandewyck A M B, Williams G M, Yamini D, and Yao E 1988 Appl.Phys.Lett. 52 2151
Gold M C and Nelson D A 1986 J.Vac.Sci.Technol. A4 2040
Hansen G L and Schmit J L 1983 J.Appl.Phys. 54 1639
Johnson E S and Schmit J C 1977 J.Electron.Mater. 6 25
Koestner R J and Schaake H F 1988 J.Vac.Sci.Technol. A6 2834
Lange M D et al 1988 Appl.Phys.Lett. 52 978
Meyer J R, Bartoli F J and Hoffman C A 1987 J.Vac.Sci.Technol. A5 3035
Million A, DiCioccio L, Gaillard J P and Piaguet J 1988 J.Vac.Sci.Technol. A6 2813

Sivananthan S, Chu X, Reno J and Faurie J P 1986 J.Appl.Phys. 60 1359
Sivananthan S, Lange M D, Monfroy G and Faurie J P 1988 J.Vac.Sci.Technol. B6 788
Sou I K, Boukerche M and Faurie J P 1988 (unpublished results)
Spencer P M 1964 Br.J.Appl.Phys. 15 625
Vydyanath H R 1981a J.Electrochem.Soc. 128 2609
Vydyanath H R 1981b J.Electrochem.Soc. 128 2619
Vydyanath H R, Hampton S R, Ward P B, Fishman L, Slawinski J and Krueger T 1986 paper presented at the IRS Detector Specialty Group Meeting, NASA AMES Research Center, Moffet Field, California
Wijewarnasuriya P S, Sou I K, Kim Y J, Mahavadi K K, Sivananthan S, Boukerche M and Faurie J P 1987 Appl.Phys.Lett. 51 2025
Wijewarnasuriya P S, Boukerche M and Faurie J P 1988 (to be published)
Worge M L, Peterman D J, Morris B J, Leopold D J, Broerman J G and Feldman B J 1988 J.Vac.Sci.Technol. A6 2826
Yoo S, Boukerche M and Faurie J P 1988 (unpublished results)

Shallow impurities in diluted magnetic semiconductors

A K Ramdas

Department of Physics, Purdue University, West Lafayette, Indiana 47907 USA

Spin-flip Raman scattering from electrons bound to donors in diluted magnetic semiconductors (DMS) is a powerful probe for discovering and delineating the magnetic properties which uniquely single out DMS's.

1. INTRODUCTION

Diluted magnetic semiconductors (DMS) are semiconductors in which one of the constituent group of atoms is partially replaced with magnetic ions. In recent years the tetrahedrally coordinated, zinc blende or wurtzite II-VI compound semiconductors like CdTe, CdS, CdSe, ZnSe or HgTe, in which the group II atoms are replaced with transition metal ions, resulting in $Cd_{1-x}Mn_xTe$, $Cd_{1-x}Co_xSe$, $Hg_{1-x}Fe_xTe$,..., have attracted considerable attention. They exhibit striking magnetic phenomena not usually encountered in semiconductor physics (Furdyna and Kossut, 1988).

In the II-VI semiconductors, the transition metal ions Mn^{2+} and Co^{2+}, with their partially filled 3d shell along with crystal field effects, are characterized by a magnetic ground state with a large magnetic moment (Ramdas and Rodriguez, 1988; Bartholomew et al, 1988). Thus the Mn-based and Co-based II-VI DMS's exhibit ordinary paramagnetism. Fe^{2+} in $Cd_{1-x}Fe_xSe$ and $Hg_{1-x}Fe_xSe$, on the other hand, has a nonmagnetic ground state but with magnetic excited states energetically close to it; thus the Fe-based DMS's exhibit Van Vleck paramagnetism (Serre et al, 1982).

The II-VI DMS's exhibit spectacular magnetic phenomena such as giant Faraday rotation, and the underlying large excitonic Zeeman splitting, huge Raman shifts associated with donor bound electrons, and magnetic ordering at sufficiently low temperatures. These have their origin in (a) the spin-spin exchange interaction between the magnetic moments of the transition metal ion and those of the s-electrons of the conduction band or the p-electrons of the valence band (the so-called 'sp-d' exchange interaction) and (b) the antiferromagnetic coupling between these magnetic ions.

In the present review the focus is on the experimental observations on the spin-flip Raman scattering of electrons bound to donors in large, effective mass, orbits. The phenomenon exposes in a particularly transparent fashion the microscopic mechanisms which underly the unique combination of magnetic and semiconducting properties of DMS's.

© 1989 IOP Publishing Ltd

2. SPIN-FLIP RAMAN SCATTERING: DONORS IN Mn-BASED DMS

Spin-flip Raman scattering from electrons in the wide gap semiconductor was first observed by Thomas and Hopfield (1968); they have rather small spin splittings characterized by $|g^*| \leq 2$. The large Raman shift associated with spin-flip scattering from electrons in DMS's was first observed by Geyer and Fan (1980) in the narrow gap $Hg_{1-x}Mn_xTe$. Nawrocki et al (1981) reported the large spin-flip Raman shift of donor bound electrons in $Cd_{1-x}Mn_xSe$ as well as the finite Raman shift at zero magnetic field (H); the latter is attributed to the formation of a 'bound magnetic polaron.' A comprehensive investigation of spin-flip Raman scattering in wide gap Mn-based II-VI DMS has been carried out by Peterson et al (1985). More recently Suh et al (1987a) have observed spin-flip Raman scattering from $Cd_{1-x}Mn_xTe$:In epilayers and modulation doped $Cd_{1-x}Mn_xTe$:In/CdTe superlattices grown by photoassisted molecular beam epitaxy (PAMBE). Li et al (1988) have succeeded in observing spin-flip Raman scattering from In-doped epilayers and platelets of $Cd_{1-x}Mn_xTe$ grown by liquid phase epitaxy (LPE). Incorporation of donors and acceptors in II-VI semiconductors, notoriously difficult by conventional techniques, but now feasible with PAMBE and LPE, represents an important step.

Figure 1 shows the Raman spectrum of a gallium doped $Cd_{1-x}Mn_xTe$ (x = 0.03) in the presence of an external magnetic field. The features labeled 'SF' and 'PM' are observed in the presence of an external magnetic field.

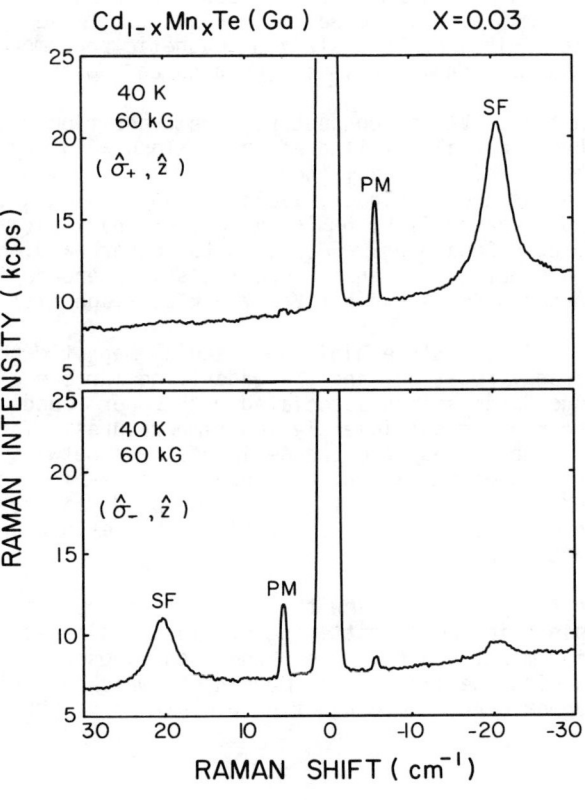

Fig. 1. Raman spectra of $Cd_{1-x}Mn_xTe(Ga)$, x = 0.03, showing the $\Delta m_S = \pm 1$ transitions within the Zeeman multiplet of Mn^{2+} (PM) and the spin flip of electrons bound to Ga donors (SF). kcps $\equiv 10^3$ counts/sec (Peterson et al, 1985)

The experiment was performed with $\vec{H} \parallel \hat{z}$, the incident laser light along $\hat{z}$, scattered at right angles and analyzed along $\hat{z}$. The observation of the Stokes Raman lines in $(\hat{\sigma}_+,\hat{z})$ and that of the anti-Stokes in $(\hat{\sigma}_-,\hat{z})$, i.e., the mutual exclusion of Stokes and anti-Stokes in this scattering geometry is characteristic of magnetic excitations. Here $\hat{\sigma}_+$ and $\hat{\sigma}_-$ represent circularly polarized radiation of positive and negative helicities, respectively ($\hat{\sigma}_\pm = (\hat{x}\pm i\hat{y})/\sqrt{2}$). The feature labeled 'PM' has been shown to be due to Raman-EPR of Mn^{2+} viz. the Raman shifts are associated with the Δm_S = +1 (Stokes) and -1 (anti-Stokes) spin-flip transitions between the adjacent sublevels of the Zeeman multiplets of Mn^{2+} yielding a Raman shift of $\hbar\omega_{PM} = g\mu_B H$ and g = 2. For further discussion of this phenomenon we refer the reader to Ramdas and Rodriguez (1988).

The Raman shift labeled 'SF' as a function of magnetic field (H) and temperature (T) is displayed in Fig. 2. It is attributed to spin-flip Raman scattering from electrons bound to gallium donors. If simply interpreted in terms of an effective g-factor (g_{eff}) representing its shift in Fig. 1, one would conclude that $g_{eff} \sim 10$. However it is clear from Fig. 2, ω_{SF} has a strong dependence on both H and T. The primary source of the spin splitting of the electronic level is the exchange coupling with the Mn^{2+} ions. Hence the Raman shift should be approximately proportional to the magnetization of the Mn^{2+}-ion system, which amplifies the effect of the magnetic field in the ion. As can be seen in Fig. 2, a finite Raman shift is observed for zero magnetic field. This effect is attributed by Dietl and Spa*l*ek (1982,1983) to the 'bound magnetic polaron (BMP).' The

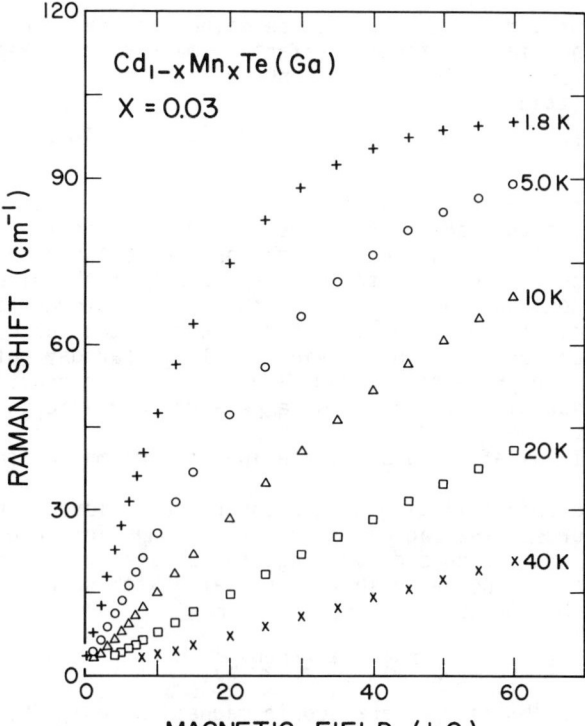

Fig. 2. Magnetic field and temperature dependence of the Raman shift associated with the spin-flip of electrons bound to donors in $Cd_{1-x}Mn_xTe(Ga)$, x = 0.03 (Peterson et al 1985)

electron, localized on a donor in a diluted magnetic semiconductor, polarizes the magnetic ions within its orbit, creating a spin cloud that exhibits a net magnetic moment. An additional effect in the binding of the electron bound to the donor originates from thermodynamic fluctuations of the magnetization and the resulting spin alignment of the magnetic ions around the donor. In Fig. 3 the zero field spin-flip Raman spectra for $Cd_{1-x}Mn_xSe$ (x = 0.05, 0.10, 0.20 and 0.30) clearly show the BMP.

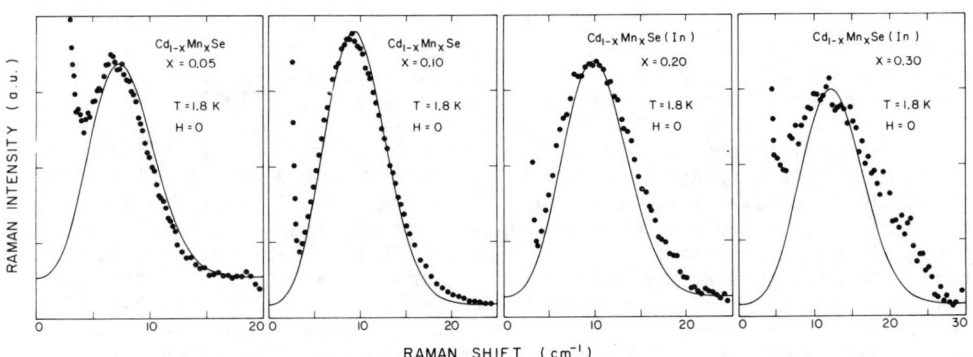

Fig. 3. Zero-field spin-flip Raman spectra for the $Cd_{1-x}Mn_xSe$, x = 0.05, 0.10, 0.20 and 0.30 samples at T = 1.8 K. The scans were recorded in the $(\hat{\sigma}_+, \hat{z})$ polarization. a.u. ≡ arbitrary units (Peterson et al, 1985)

Consider the electron to be either in the conduction band or bound to a donor in an extended, effective mass state. Representing the exchange interactions by the Heisenberg form, the Hamiltonian of this system can be written as

$$H_M(\{\vec{S}_i\}, \vec{s}) = -\alpha N_0 \sum_i \vec{S}_i \cdot \vec{s} + g^* \mu_B \vec{H} \cdot \vec{s} + g\mu_B \vec{H} \cdot \sum_i \vec{S}_i + \sum_{i<j} J_{ij} \vec{S}_i \cdot \vec{S}_j . \quad (2.1)$$

Here αN_0 is the exchange constant of the interaction between localized spins $\{S_i\}$ and that of the electron, μ_B is the Bohr magneton, J_{ij} is the anti-ferromagnetic exchange constant between Mn^{2+} ions, g and g^* are the Landé g-factor of the Mn^{2+} spins and of the electron, respectively. The second and third terms are the Zeeman interactions with the applied magnetic field. The first term leads to the large spin splittings of the electron levels underlying the large Faraday rotations in DMS. The last term represents the antiferromagnetic interaction of the magnetic ions responsible for the low temperature magnetic phases observed in DMS.

2.1 Spin-flip Raman Scattering for Low Mn Concentration

The spin splitting of the donor energy levels in the Mn-based DMS has two sources, the magnetization of the Mn^{2+} ions and the Zeeman effect. Due to the strong s-d coupling, the effects due to the magnetization will dominate. The Raman shift associated with spin-flip scattering from these donor states will have the form

$$\hbar\omega_0 = \Delta_0 = \frac{\alpha}{g\mu_B} M_0(H) + g^*\mu_B H , \quad (2.2)$$

where M_0 is the macroscopic magnetization. The magnetization will be proportional to the thermal average of the Mn^{2+} spin projection along H multiplied by the density of Mn^{2+} ions that contribute to the magnetiza-

tion, yielding

$$\Delta_0 = \bar{x}\alpha N_0 \langle S_z^{Mn} \rangle + g^* \mu_B H \ . \tag{2.3}$$

Here N_0 is the density of cations and $\bar{x}$ is the concentration of Mn^{2+} ions that contribute to the magnetization. For small x, the crystal is paramagnetic and the thermal average of the Mn^{2+} spins is

$$\langle S_z^{Mn} \rangle = \frac{5}{2} B_{5/2} \left(\frac{g\mu_B H}{k_B T} \right) \ , \tag{2.4}$$

where $B_{5/2}$ is the Brillouin function B_J for $J = \frac{5}{2}$.

2.2 Compositional Dependence of Spin-flip Raman Scattering: Mean Field Approximation

The spin-flip Raman shifts for $Cd_{1-x}Mn_xTe$ at T = 1.8 K are shown in Fig. 4 as a function of magnetic field and composition. The results for x = 0.01 show the saturation behavior characteristic of the paramagnetic phase. As the Mn concentration is increased to x = 0.03 and x = 0.05, the Raman shifts increase and the effects of saturation are still clearly evident, but less pronounced. For x = 0.10, the deviation from the paramagnetic behavior is quite evident. For H = 60 kG, the Raman shift for x = 0.10 is only four times that for x = 0.01. As x exceeds 0.10, the Raman shifts for a given field actually decrease; note that the shifts for the x = 0.20 sample lie below those for the x = 0.10 sample. And the Raman shifts for the $Cd_{1-x}Mn_xTe(Ga)$, x = 0.30, samples are significantly smaller than those for the x = 0.10 and x = 0.20 samples.

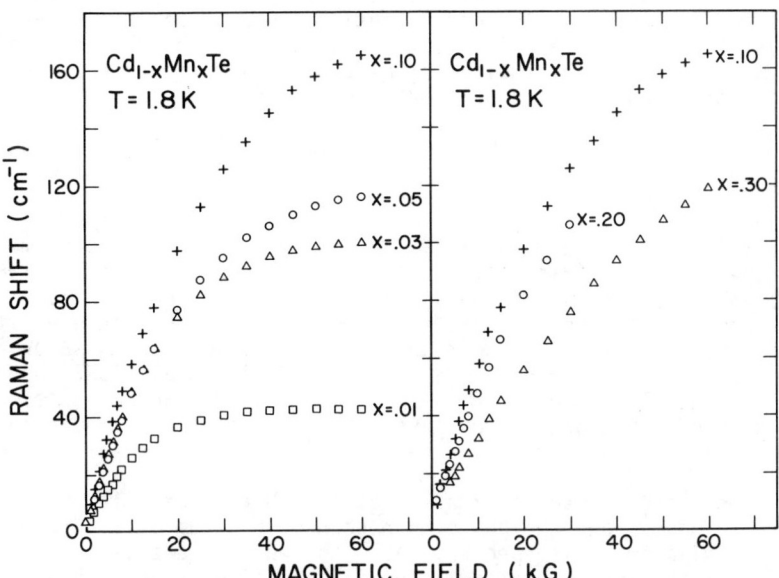

Fig. 4. Magnetic field and composition dependence of the peak spin-flip Raman shift in the $Cd_{1-x}Mn_xTe$ samples at T = 1.8 K (Peterson et al, 1985)

3. SPIN-FLIP RAMAN SCATTERING: DONORS IN Co- AND Fe-BASED DMS

Spin-flip Raman scattering from donors in $Cd_{1-x}Co_xSe$ and $Cd_{1-x}Fe_xSe$ has also been observed (Suh et al, 1987b, Bartholomew et al, 1988). The contrast between Mn^{2+}, Co^{2+}, and Fe^{2+} with respect to the nature of their ground state is worth noting. The atomic ground state of Mn^{2+}, $^6S_{5/2}$, in a tetrahedral environment characteristic of, say, a zinc blende structure undergoes a crystal field splitting into a Γ_8 quadruplet and a Γ_7 doublet. Experimentally this crystal field splitting is very small and can be ignored; as expected, the $g_{Mn^{2+}} = 2$ is consistent with atomic $^6S_{5/2}$ like ground state. Co^{2+} has an unfilled 3d electron configuration. The atomic ground state of Co^{2+}, ^4F, splits into an orbital singlet (Γ_2) and two higher lying triplets (Γ_5, Γ_4). The crystal field splitting separating the lowest Γ_2 ground state from the Γ_5 and Γ_4 states is so large that all the magnetic phenomena are controlled by Γ_2 along with its total spin of 3/2. Thus, the magnetic behavior of Co^{2+} is qualitatively analogous to that of Mn^{2+}. It is very different from that of Fe^{2+} which has complex level structure and a non-magnetic ground state; the Fe based DMS's thus exhibit Van Vleck paramagnetism (Serre et al, 1982; Heiman et al, 1988).

Raman scattering associated with the spin flip of electrons bound to donors in $Cd_{1-x}Co_xSe$ and $Cd_{1-x}Fe_xSe$ has been observed by Bartholomew et al (1988) and Suh et al (1987b). The magnetic field dependence of ω_{SF} in $Cd_{1-x}Co_xSe$ at various temperatures shown in Fig. 5(a) indicates a large s-d exchange interaction ($\alpha N_0 = 320$ meV) and a clear evidence of a bound magnetic polaron. The data for the spin flip of donor electrons in $Cd_{1-x}Fe_xSe$ are displayed in Fig. 5(b). The magnetic field and temperature dependence of ω_{SF} in $Cd_{1-x}Fe_xSe$ are qualitatively different from that in $Cd_{1-x}Mn_xSe$ and $Cd_{1-x}Co_xSe$. The non-magnetic nature of Fe^{2+} is clearly indicated by the zero Raman shift at zero field, i.e., by the absence of the bound magnetic polaron. Heiman et al (1988) have reported and discussed the lack of bound magnetic polaron in $Cd_{1-x}Fe_xSe$.

4. SPIN-FLIP RAMAN SCATTERING FROM $Cd_{1-x}Mn_xTe$:In EPILAYERS AND MODULATION-DOPED $Cd_{1-x}Mn_xTe$:In/CdTe SUPERLATTICES GROWN BY PHOTOASSISTED MOLECULAR-BEAM EPITAXY

The successful growth of high-quality heterostructures of DMS's by molecular-beam epitaxy (MBE) (Bicknell et al, 1984; Kolodziejski et al, 1984) has led to the discovery of new physical phenomena not encountered in the bulk. It should be emphasized that the large s-d and p-d exchange interactions observed in bulk DMS's continued to prevail in the DMS superlattices (Venugopalan et al, 1984). Recently, Bicknell et al (1987) have shown that a controlled doping of II-VI semiconductors can be achieved in the photoassisted MBE growth of epilayers and superlattices, where the substrate is illuminated during the deposition process.

Figure 6 shows the Raman spectrum associated with the spin flip of electrons bound to donors and the photoluminescence spectrum of a $Cd_{1-x}Mn_xTe$:In ($x = 0.15$) epilayer; this spectrum was obtained in the $z(yx)\bar{z}$ configuration with an external magnetic field of 50 kG along $\hat{x}$ where $\hat{x}, \hat{y}$ are in the layer plane, and $\hat{z}$ is along [001]. We see two luminescence peaks L1 and L2 associated with excitonic transitions, and a strong spin-flip Raman line. Because of the close match between the photon energies of the scattered light and that of excitonic features, there is a resonant enhancement of the spin-flip Raman intensity. As the external magnetic field is lowered, the exciton peak L1 moves even closer to the spin-flip

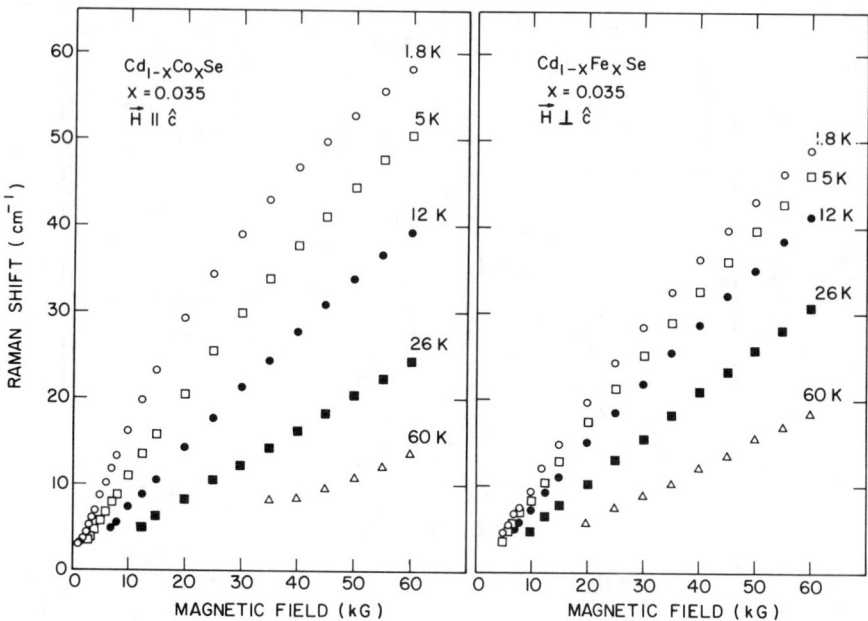

Fig. 5. Spin-flip Raman shift (ω_{SF}) from donor-bound electrons in (a) $Cd_{1-x}Co_xSe$ and (b) $Cd_{1-x}Fe_xSe$ as a function of applied magnetic field (Bartholomew et al, 1988 and Suh et al, 1987b)

line resulting in a larger resonance enhancement. In the inset (a) of Fig. 6, we show the Raman spectrum in the frequency range where the spin-flip and LO phonon features occur, the spectrum being recorded in the parallel polarization, $z(xx)\bar{z}$. The spin-flip line is forbidden in this scattering configuration, while the CdTe-like LO phonon and its combination with Raman EPR are clearly observed. In inset (b) of Fig. 6, we also show the Raman line associated with the spin-flip transition within the Zeeman split $3d^5$ multiplet of Mn^{2+}, i.e., the Raman-EPR (PM), whose intensity is also resonantly enhanced when the scattered photon energy is close to that of the excitonic transitions. These resonance enhancements of spin-flip lines demonstrate that the Raman mechanism for both donor spin-flip and Raman-EPR involves interband transitions of band electrons.

As in bulk DMS, the Raman shift of the donor spin-flip line exhibits a Brillouin function-like behavior depending on both the temperature and the magnetic field as can be seen in Fig. 7. From the slope of the linear portion of the spin-flip data shown in Fig. 6, we obtain $g_{eff} = 74$ at $T = 5$ K, which is comparable to that of bulk $Cd_{1-x}Mn_xTe$ with $x = 0.16$. We note that a zero-field spin-flip peak, the bound magnetic polaron (BMP), has been clearly observed at 9 cm^{-1}. As the temperature is lowered to 1.8 K, the BMP energy increases slightly due to the gradual transition from the high-temperature fluctuation-dominated to the low-temperature spin-aligned polaron regime (Suh et al, 1987a).

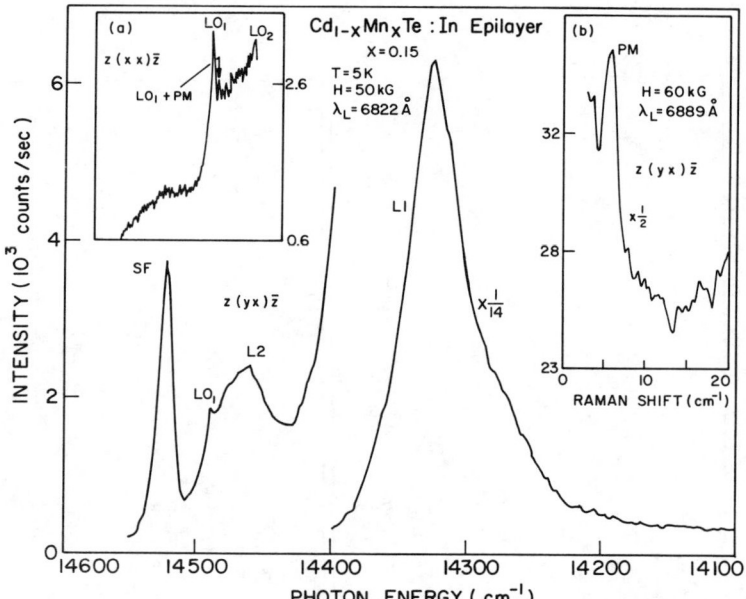

Fig. 6. The Raman spectrum of $Cd_{1-x}Mn_xTe$:In ($x = 0.15$) epilayer showing a spin-flip Raman scattering from electrons bound to donors, the LO phonons, as well as the exciton photoluminescence at T = 5 K with an external magnetic field of 50 kG in the layer plane. The spectrum is obtained in the crossed polarization $z(yx)\bar{z}$ with incident laser wavelength λ_L = 6822 Å and laser power P_L = 3.2 mW. Here z is the growth direction and (xy) is the plane of the epilayer. Inset (a) shows the $z(xx)\bar{z}$ Raman spectrum in the region of the spin-flip line and the LO phonons. Note the CdTe-like LO phonon and its combination with the Raman-EPR (PM). Inset (b) shows a resonantly enhanced Raman-EPR peak with H = 60 kG, λ_L = 6889 Å and P_L = 20 mW (Suh et al, 1987a).

In the modulation-doped $Cd_{1-x}Mn_xTe$:In/CdTe superlattices, electrons are expected to be spatially separated from their parent donors and confined in the CdTe well layers, thus resulting in high electron mobilities. In the superlattice with L_z = 155 Å, the carrier concentration and mobility (Bicknell et al, 1987) suggest that a significant number of carriers are in the CdTe well even at low temperatures. In this superlattice, no spin-flip Raman peak was observed. Had the electrons been bound to donors in the barrier or at the interface, a large s-d interaction enhanced spin-flip Raman shift would have been easily observed, whereas electrons confined to the CdTe well with wave functions having negligible penetration into the barrier and characterized with g^* = -0.75, would have spin-flip Raman shift which would escape observation. Thus the absence of an observable spin-flip Raman shift indicates that the In donors located during growth in the $Cd_{1-x}Mn_xTe$ barriers are ionized and the donor electrons are in the CdTe wells.

As the well width is decreased, e.g., in the superlattice with L_z = 74 Å, a spin-flip feature is indeed observed when the incident photon energy is resonant with the electronic transition of the well; the magnetic field and temperature dependence of the Raman shift are shown in Fig. 8. The observation of the s-d enhanced spin-flip Raman shift in this

Shallow Impurities in II–VI Compounds

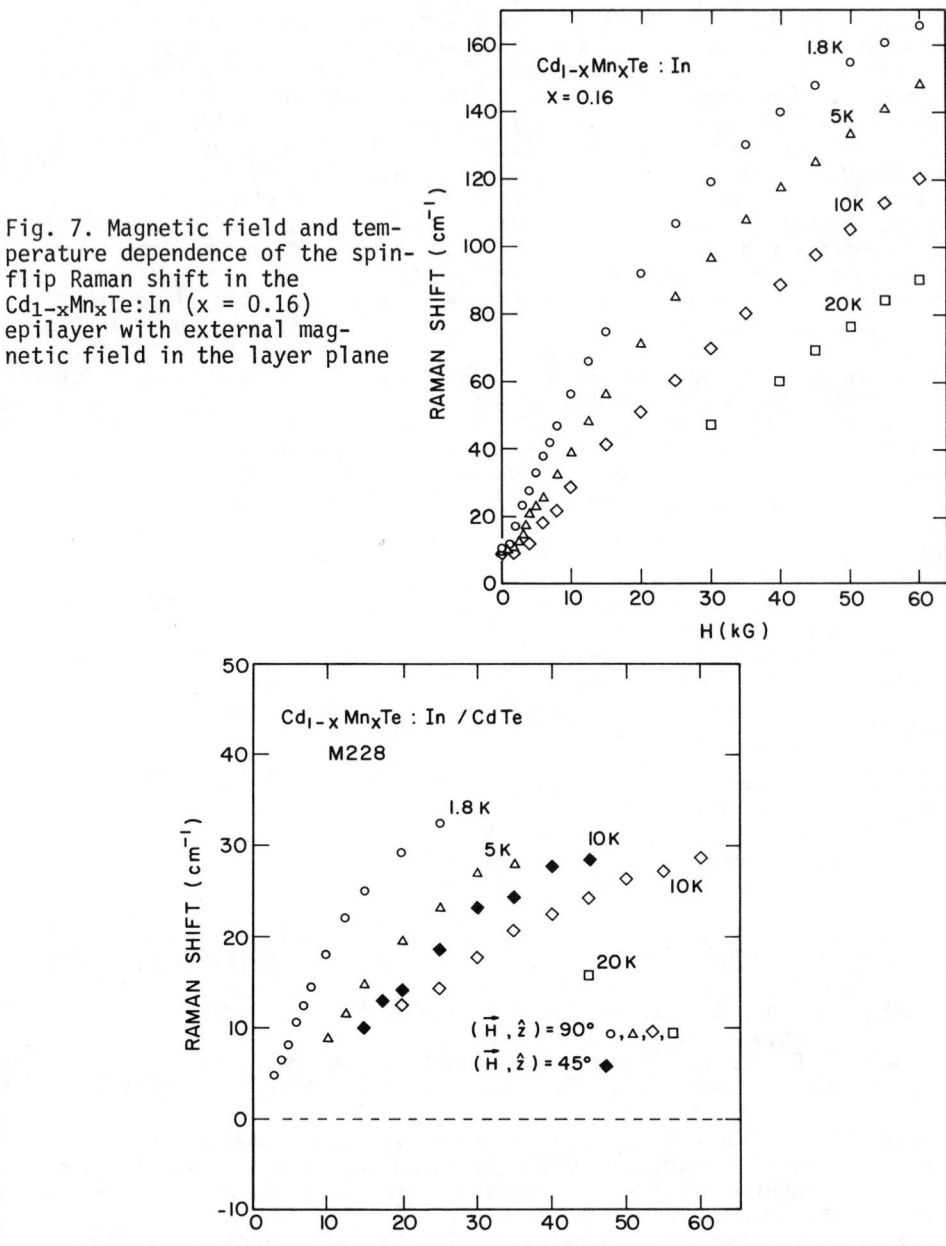

Fig. 7. Magnetic field and temperature dependence of the spin-flip Raman shift in the $Cd_{1-x}Mn_xTe$:In ($x = 0.16$) epilayer with external magnetic field in the layer plane

Fig. 8. Magnetic field and temperature dependence of the spin-flip Raman shift in the superlattice, $Cd_{1-x}Mn_xTe$:In/CdTe with $L_z = 74$ Å. Data are obtained with $\lambda_L = 7525$ Å and $P_L = 15$ mW. Open symbols are for data obtained when the angle between H and $\hat{z}$ is 90°, i.e., H is in the layer plane, while solid symbols are for those when the angle is 45°, at T = 10 K (Suh et al, 1987a)

superlattice can be attributed to the penetration of the electronic wave function into the $Cd_{1-x}Mn_xTe$ barrier layer. The g_{eff} estimated from the slope of the low field spin-flip Raman shifts in Fig. 8 is approximately 24 at T = 5 K in Fig. 7, for example. We note that the carrier concentration measured by the Hall effect decreases significantly at low temperatures suggesting that a considerable number of donors reside near the interface or perhaps in the well itself. At low temperatures, electrons are presumably bound to donors, say, located at the interface; their orbits will then penetrate the barrier to a finite distance resulting in the observable spin-flip Raman shift, the g_{eff} being then determined by the penetration distance. Such electrons bound to donors at interfaces could lead to an anisotropic g_{eff} arising from differing penetration depths depending on the direction of the external magnetic field with respect to the superlattice axis $\hat{z}$. Finally, a simple extrapolation of spin-flip Raman shift to zero field yields a negative intercept. This could indicate either a highly nonlinear behavior at lower magnetic fields or a finite-field onset for the spin flip.

Acknowledgements--The work reported in this review received support from the U S National Science Foundation (Grants No. DMR 86-16787 and DMR 85-20866) as well as from DARPA/URI (N00014-86K-0760).

References

Bartholomew D U, Suh E-K, Ramdas A K, Rodriguez S, Debska U and Furdyna J K 1988 unpublished.
Bicknell R N, Yanka R W, Giles-Taylor N C, Blanks D K, Buckland E L and Schetzina J F 1984 Appl. Phys. Lett. 45 92
Bicknell R N, Giles N C and Schetzina J F 1986 Appl. Phys. Lett. 49 1735
Bicknell R N, Giles N C and Schetzina J F 1987 Appl. Phys. Lett. 50 691
Dietl T and Spa/ek J 1982 Phys. Rev. Lett. 48 355
Dietl T and Spa/ek J 1983 Phys. Rev. B 28 1548
Furdyna J K and Kossut J 1988 eds Diluted Magnetic Semiconductors, volume 25 of Semiconductors and Semimetals eds R K Willardson and A C Beer (New York: Academic)
Geyer F F and Fan H Y 1980 IEEE J. Quant. Electron. QE-16 1365
Heiman D, Petrou A, Bloom S H, Shapira Y, Isaacs E D and Giriat W 1988 Phys. Rev. Lett. 60 1876
Kolodziejski L A, Bonsett T C, Gunshor R L, Datta S, Bylsma R B, Becker W M and Otsuka N 1984 Appl. Phys. Lett. 45 440
Li L-X, Furdyna J K, Suh E-K, Lee Y R, Alonso R G and Ramdas A K 1988 Bull. Am. Phys. Soc. 33 520
Nawrocki M, Planel R, Fishman G and Galazka R 1981 Phys. Rev. Lett. 46 735
Peterson D L, Bartholomew D U, Debska U, Ramdas A K and Rodriguez S 1985 Phys. Rev. B 32 323
Ramdas A K and Rodriguez S 1988 in J K Furdyna and J Kossut 1988
Serre H, Bastard G, Rigaux C, Mycielski J and Furdyna J K 1982 4th Int. Conf. on Physics of Narrow Gap Semiconductors eds E Gornik, H. Heinrich and L Palmetshofer (New York: Springer)
Suh E-K, Bartholomew D U, Ramdas A K, Bicknell R N, Harper R L, Giles N C and Schetzina J F 1987a Phys. Rev. B 36 9358
Suh E-K, Bartholomew D U, Furdyna J K, Debska U, Ramdas A K and Rodriguez S 1987b Bull. Am. Phys. Soc. 32 802
Thomas D G and Hopfield J J 1968 Phys. Rev. 175 1021
Venugopalan S, Kolodziejski L A, Gunshor R L and Ramdas A K 1984 Appl. Phys. Lett 45 974

Inst. Phys. Conf. Ser. No 95: Chapter 6
Paper presented at Int. Conf. Shallow Impurities in Semiconductors, Linköping, Sweden, 1988

The shallow resonant acceptor in semimagnetic, zerogap $Hg_{1-x}Mn_xTe$

R. G. Mani and J. R. Anderson

Joint Program for Advanced Electronic Materials, Dept. of Physics, University of Maryland, and Laboratory of Physical Sciences, College Park, MD 20742 U. S. A.

ABSTRACT: A magnetotransport study in a series of annealed and unannealed zero-gap (Mn,Hg)Te samples reveals the influence of a resonant acceptor state. We show that low temperature negative-magnetoresistance originates from an increase in the hole density within the resonant acceptor state rather than an increase in the hole mobility as proposed by others. Also, we have observed an unusual increase in the low temperature electron mobility with decreasing temperature which we attribute to spatial ordering of the ionized acceptors within the resonant acceptor state.

The narrow-gap semiconductor $Hg_{1-x}Cd_xTe$ (HCT) becomes a semimetal for x-values x < 0.165 at T=4.2K (Long and Schmit (1970)). Low temperature magnetotransport measurements in this zerogap system indicate multiband conduction with n-type behavior at low magnetic fields and p-type behavior at high fields. As the overlap between the conduction and heavy-hole (h.h) valence bands is negligble at zero field, and the application of a magnetic field induces a gap in the HCT system, the observation of hole transport even at the lowest temperatures has been attributed to conduction within a resonant acceptor state (Elliot et al (1972)). These results agree with theory which suggests that the lack of a gap in zerogap systems forces shallow acceptor levels, usually found in the bandgap of a normal semiconductor, to be resonant with the conduction band (Gelmont et al (1972)).

We have studied magnetotransport in zerogap $Hg_{1-x}Mn_xTe$ (HMT), i.e. x < 0.07 (Mycielski A et al (1984)), to determine if resonant acceptor states occur in this semimagnetic system and also investigate the nature of transport in zerogap systems with an exchange interaction between the carriers and localized Mn^{2+} spins. We point out that the non-magnetic properties of HMT and HCT are the same for similar gap but the bandgap increases approximately twice as fast with the x-value in HMT compared to HCT (Furdyna (1982)).

The HMT crystals used in our study were grown by the Bridgman technique and then characterized by microprobe, x-ray fluorescence, magnetization, and density measurements. Here, we shall present the transport results on two typical samples, labelled A (unannealed) and B (annealed in Hg vapor), which originated from the same wafer and thus had the same residual donor density and x-value. The x-value, x~0.06, corresponds to a gap $E_o = E(\Gamma_6) - E(\Gamma_8) \sim -50meV$ at T=4.2K. By comparing samples with different annealing histories, we hope to separate the

© 1989 IOP Publishing Ltd

effects of a resonant acceptor state originating from Hg vacancies.

Shown in Fig. 1, and 2 are V_R, V_H, σ_{xx} and σ_{xy} for samples B and A respectively. Here, V_R measures the resistance, V_H is the Hall voltage, and σ_{xx}, σ_{xy} are components of the conductivity tensor. The measured quantities, V_R and V_H, have been converted to σ_{xx} and σ_{xy} because the contribution of different bands are additive in these quantities. We point out the following features in the data for sample B (Fig.1): First, Shubnikov-de Haas (SdH) oscillations are superimposed upon positive magnetoresistance at low fields (H<5kOe). The SdH oscillations were observed down to 500 Oe (i.e. 7 cycles) and the SdH frequency (F = 3.4kOe --> n_e = 1.1×10^{15} cm^{-3}) was temperature-independent over the range 1.5K < T < 4.2K within the resolution of our measurement (~2%). Second, at fields above about 5kOe, the oscillations disappear before the onset of negative magnetoresistance and, at T=1.6K, the resistance decreases by a factor of 2.5 over the field range 5kOe < H < 40kOe. Third, the Hall voltage indicates n-type conduction over the range of fields where SdH oscillations are observed. Fourth, for H > 5kOe, the sample is p-type and V_H saturates at high fields.

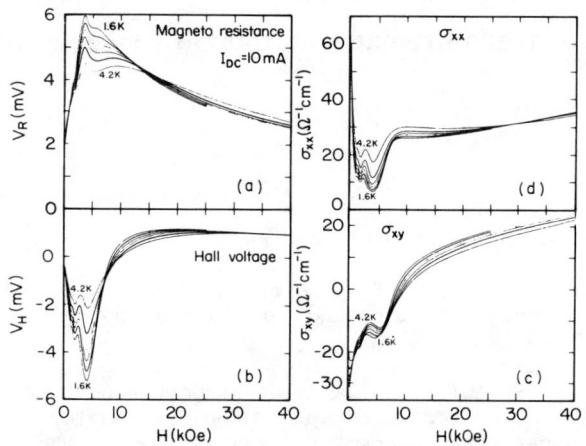

Fig. 1) From the top left counter-clockwise are the resistance (A), V_R, the Hall voltage (B), V_H, the conductivity tensor components σ_{xy} (C), and σ_{xx} (D) for the annealed sample B. The curves correspond to measurements made at the following temperatures: T = 4.2K, 3.0K, 2.4K, 2.1K, 1.8K, 1.6K.

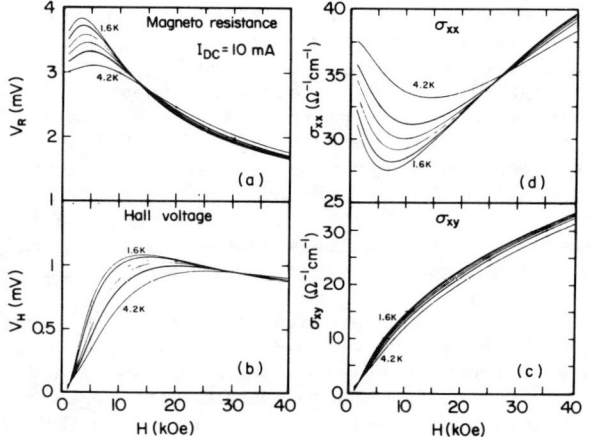

Fig. 2) From the top left counter-clockwise are V_R (A), V_H (B), σ_{xy} (C), and σ_{xx} (D) for the unannealed sample A. Refer to Fig. 1 for temperatures.

Finally, the crossover field for the n-type to p-type transition decreases with decreasing temperature and the crossover also becomes more abrupt at lower temperatures.

In comparison, the SdH oscillations and positive magnetoresistance seen in the annealed sample B are missing in the unannealed sample A (see Fig. 2). Also, sample A is p-type over the entire range of magnetic fields

and the resistance decreases by a factor of 2.5 over the range 0-40kOe just as in sample B. We suggest that differences between the two samples originate from the presence of high mobility, conduction band electrons in the annealed sample B which are negligble in the unannealed sample A due to defects originating from Hg vacancies. Thus, hole conduction dominates transport even at low fields in the unannealed sample.

The origin of hole conduction is the issue of interest in these zerogap samples. For zerogap HCT, the theory of Gelmont et al (1972) indicates that donor levels are ionized while acceptor levels are superimposed upon the conduction band with an ionization energy E_A with respect to the heavy hole (h.h) band (see Fig. 3). Thus, for large enough temperatures, (E_F/k_BT < 1), and a conduction band electron concentration, n_e, such that the Fermi energy lies in the acceptor band, it is possible to have hole conduction take place simultaneously within the acceptor band and the h.h valence band.

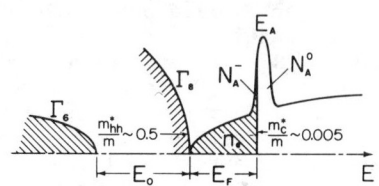

Fig. 3) Figure illustrates possible band structure for zerogap $Hg_{1-x}Mn_xTe$. The Fermi energy is pinned by the resonant acceptor band at low temperatures

The magneto-optical results of Bastard et al (1978 and 1979) would suggest that exchange effects cause the heavy hole band to cross over the conduction band at a certain value of the magnetic field and the hole conductivity observed at the lowest temperatures occurs when the Fermi energy is pinned by the high density of states of the h.h. valence band. However, this model fails to explain similar behavior in non-magnetic HCT (Elliot et al (1972)) where exchange effects do not occur. Thus, we assume that p-type conduction occurs in a resonant acceptor state and show that it explains the following features observed in our data.

First, studies of the SdH effect in zerogap HCT by Brandt et al (1975) have indicated a temperature dependent SdH frequency even for T < 4.2K due to the $T^{3/2}$ temperature dependence of the carrier density in zerogap systems. In contrast, we have found a temperature independent carrier density in sample B which is consistent with the Fermi energy, E_F, being pinned by a resonant acceptor band.

Second, positive magnetoresistance observed in the annealed sample B but missing in the unannealed sample A originates from a shift in the conduction from high mobility electrons responsible for SdH oscillations to low mobility holes in the acceptor band as the magnetic field decreases the contribution of the high mobility carriers to σ_{xx}.

Third, the non-monotonic behavior of σ_{xx} and the correlation of this effect with the last SdH oscillation suggests that electrons which condense onto the N=0, spin-up Landau level after the last SdH oscillation are emptied into the acceptor band when the Landau level energy becomes larger than the acceptor ionization energy. Here, we point out that classical multiband conduction with field independent parameters would predict that σ_{xx} is a monotonically decreasing function of the magnetic field. The abrupt increase of the acceptor-band carrier-density would be manifested as a sharp rise in σ_{xx} since the lower mobility of the carriers within this band make it less likely to satisfy the freezeout condition, $\mu_a H \gg 1$, compared to the conduction electrons at the same magnetic field. These features also imply it is incorrect to assume field independent electron and hole densities as is usually done when modelling this system (Brandt et al(1983)).

We have pointed out that both samples show similar negative magnetoresistance for H > 5kOe when p-type conduction dominates (see Fig. 1, 2). We determine the origin of negative magnetoresistance by assuming that high field transport is only due to carriers in the acceptor band, the high mobility electrons having been emptied out of the conduction band by the magnetic field. Then, assuming the holes satisfy $\mu_a H \ll 1$, we obtain $n_a = \sigma_{xx}^2 H/(\sigma_{xy} q)$ and $\mu_a = \sigma_{xy}/\sigma_{xx} H$. In Fig. 4, we plot $\log(n_a)$ and μ_a as a function of the magnetic field. The low field part of these curves should be neglected because the single band approximation is invalid in this region. The figure shows that the hole density in sample B at fields immediately after the sharp rise in σ_{xx} is approximately 1×10^{15} cm^{-3} and equals the low field electron density obtained from the S dH effect thus suggesting that holes result from transfer of conduction band carriers into the resonant acceptor band. It also supports our interpretation that the sharp rise in σ_{xx} is a signature of charge transfer from the conduction band to the resonant acceptor state which pins the Fermi energy.

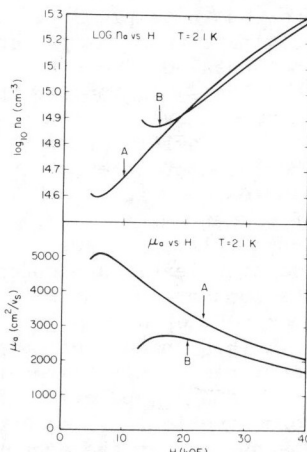

Fig. 4) The field dependence of the hole density (top) and the hole mobility (bottom) for the samples B and are shown for T=2.1K.

The model of Davydov et al(1980) attributes negative magnetoresistance in semimetallic HMT to an increase in the hole mobility due to reduced scattering from the Mn^{2+} spins as they are ordered by the magnetic field. Our data contradicts this picture and shows that negative magnetoresistance originates from an increase in the hole density in the acceptor band (see Fig. 4). This behavior possibly originates from exchange effects due to the presence of Mn^{2+} ions. We suggest the observed increase in the hole density with field may be due to activation of carriers from the h.h valence band into the acceptor states as exchange effects cause the h.h. band to approach the acceptor state as in open-gap HMT (Mycielski J (1981)). This picture is consistent with the temperature dependence of the negative magnetoresistance since the exchange interaction, which is proportional to the magnetization, is also enhanced with decreasing temperature (Furdyna (1982)).

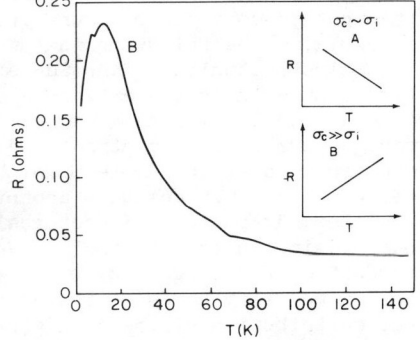

Fig. 5) The temperature dependence of the resistance for the annealed sample B. The insets illustrate the different temperature dependence observed for the annealed (bottom) and unannealed (top) sample.

Finally, at low temperatures, T < 10K, the zero field resistance of sample B shows an unusual temperature dependence (see Fig. 5) which can also be understood in a picture which includes a resonant acceptor state. The zero- field resistance shows non-monotonic behavior and increases with increasing temperature for T<10K while the unannealed sample shows opposite behavior (see inset Fig. 5). By correlating these zero-field

results with SdH measurements which imply a temperature independent electron density for T<10K, we find a dramatic decrease in the electron mobility with increasing temperature for T<10K. Similar behavior has been observed in nonmagnetic, zerogap HCT by Arapov et al(1983) and in zerogap HMT by Brandt et al(1983) but not well understood. We point out that for charged center scattering, we would expect the mobility to increase with increasing temperature at low temperatures in contradiction to these experimental results.

Our picture, which consists of a partially filled resonant acceptor state pinning the Fermi energy, allows for virtual transitions between the conduction band and the resonant acceptor state; these virtual transitions limit the lifetime of a particular ionized acceptor. Although the acceptors (vacancies) are randomly distributed at fixed sites throughout the lattice, only a fraction are charged, i.e. $N_A \gg N_A^-$. Thus, there is a large degeneracy of possible occupational configurations for the ionized acceptors on the lattice and the thermal energy tends to continuously change their spatial configuration. As the temperature is lowered and the thermal fluctuations are reduced, the system of ionized acceptors tends to minimize the total Coulomb energy of the system by condensing into a partially ordered

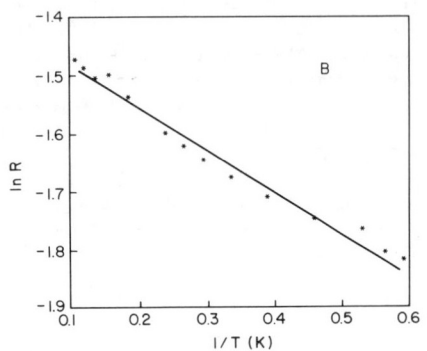

Fig. 6) A plot of the low temperature (T<10K) zero-field resistance for the annealed sample shows activated behavior indicating possible ordering of charged acceptors.

superlattice of ionized acceptors. The increased ordering of the ionized acceptors with decreasing temperature would enhance the electron mobility since ordering supresses scattering (Mycielski J (1986)). This model would also suggest an activated temperature dependence of the mobility over the range of temperatures (T<10K) where this unusual behavior is observed. A plot of ln(R) vs. 1/T indicates that the zero-field resistance follows activated behavior with an activation energy of ~0.1meV. We believe this value probably represents the mean Coulomb energy among ionized acceptors.

In conclusion, negative magnetoresistance and the associated p-type conduction in semimetallic $Hg_{1-x}Mn_xTe$ occur due to an increase in the carrier concentration, rather than an increase in the mobility, within a resonant acceptor band. Also, the increase of the electron mobility with decreasing temperatures occurs due to the spatial ordering of the charged centers in the resonant acceptor state and the resulting decrease in the scattering.

This research was partly supported by the U. S. Defense Advanced Research Projects Agency (DARPA) under grant DAAG29-85-K-0052.

REFERENCES

Arapov Yu G, Davydov A B and Tsidilkovskii I M 1983 *Fiz. Tekh. Poluprovodn.* **17** 24 ((1983) *Sov. Phys. Semicond.* **17** 14)
Bastard G, Rigaux C, Guldner Y, Mycielski J and Mycielski A 1978 *Jour. de Physique* **39** 87
Bastard G, Rigaux C, Couder Y, Thome H and Mycielski J 1979 *Phys. Stat. Sol.* (b) **94** 205
Brandt N B, Belousova O N and Ponomarev Ya G 1975 *Fiz. Tverd. Tela* **17**

3585 ((1976) *Sov. Phys. Sol. St.* <u>17</u> 2334)
Brandt N B, Moshchalkov V V, Orlov A O, Skrbet L, Tsidilkovskii I M and Chudinov S M 1983 *Zh. Eksp. Teor. Fiz.* <u>84</u> 1050 ((1983) *Sov. Phys. -JETP* <u>57</u> 614)
Davydov A B, Ponikarov B B and Tsidilkovskii I M 1980 *Phys. Stat. Sol.* (b) <u>101</u> 129
Elliot C T, Melngailis J, Harman T C, Kafalas J A and Kernan W C 1972 *Phys. Rev. B* <u>5</u> 2985
Furdyna J K 1982 *J. Vac. Sci. Tech.* <u>21</u> 220
Gelmont B L, Dyakonov M I, Ivanov-Omskii V I, Kolomiets B I, Ogorodnikov V K and Smekalova K P 1972 *Proc. 11th Intl. Conf. on Phys. Semicond.* ed Polish Academy Sciences (Warsaw: PWN) p.938
Long D and Schmit J 1970 *Semiconductors and Semimetals* ed. R K Willardson and A C Beer (New York: Academy) Vol. 16 p.175
Mycielski A, Rigaux C, Menant M, Dietl T and Otto M 1984 *Sol. St. Comm.* <u>50</u>, 257
Mycielski J 1981 *Recent Developments in Semiconductor Physics* ed. J T Devreese (New York: Plenum)
Mycielski J 1986 *Sol. St. Comm.* <u>60</u> 165

Excited states and doublet structure of the shallow-acceptor Na bound excitons in ZnSe bulk and MOCVD film

Yoichi YAMADA, Tsunemasa TAGUCHI and Akio HIRAKI

Department of Electrical Engineering, Faculty of Engineering, Osaka University, Suita, Osaka 565, Japan.

ABSTRACT: A doublet structure observed in the neutral-Na-acceptor bound-exciton (I_1^{Na}) emission line has been investigated by the temperature-dependence of photoluminescence spectra. Two-hole transitions appeared under the below-bandgap excitation; the ionization energy of the 1s Na acceptor is estimated to be 102 meV using acceptor central-cell corrections. Dependence of spectral features of the I_1^{Na} line and of related donor-acceptor pair bands on thermal diffusion conditions was found. It is confirmed from the presence of a strong I_1^{Na} line that Na was incorporated into a ZnSe layer grown on GaAs substrate at 300 °C by low-pressure metalorganic chemical-vapour-deposition (MOCVD).

1. INTRODUCTION

The I_A-group alkali impurities such as Li and Na can give rise to the shallowest acceptor levels in ZnSe when replaced substitutionally on the Zn lattice sites (Chacham et al 1987). In fact, these two acceptors have been investigated so far in the low-temperature photoluminescence (PL) spectra (Merz et al 1973, Bhargava et al 1979) as the radiative recombinations either of excitons bound to neutral acceptors or of donor-acceptor pair (DAP) bands. From the analysis of the DAP bands, the ionization energies can be estimated to be about 114 and 128 meV for Li_{Zn} and Na_{Zn}, respectively (Tews et al 1979). However, these values can not be obtained from the standard Haynes' rule or from a simple effective mass argument (Sharma and Rodriguez 1967) which is deduced from the relationship between the binding energy of the bound exciton (BE) and the localized center which binds excitons.

Bhargava et al (1979) have argued that it is sufficient to incorporate Na in ZnSe and consequently the Na-doped ZnSe is p-type. It has been, however, claimed that production of p-type ZnSe, having an appropriate electrical active hole concentration, is very difficult by the thermal diffusion of Na because of the compensation due to interstitial atoms and/or ions and unstablity of substitutional atoms (Neumark and Catlow 1984).

This paper is concerned with the fundamental radiative recombination properties of Na shallow acceptor BE emission in the thermally-Na-doped high-purity bulk ZnSe crystals and in metalorganic chemical-vapour-deposition-grown ZnSe layers on GaAs substrate. In particular, we shall concentrate our discussion on a doublet structure in the excited states of the principal BE line and the two-hole transitions of Na acceptor. Optical measurements and crystal growth procedures have already been described in our previous paper (Asao et al 1988).

2. GENERAL FEATURES OF Na ACCEPTOR BOUND-EXCITON EMISSION

The Na BE line, at an energy of 1 meV higher than I_1^X line, has been observed at about 2.791 eV at 4.2 K by Merz at al (1973) in ZnSe crystals doped with Na under excess Zn, so called I_1^{Na} ($=I_1^y$) line, where the I_1^X results from the recombination of the neutral Li acceptor BE. This line exhibited three components. On the other hand, Neumark et al (1984) have observed recently the two I_1 lines which are located at 2.7937 and 2.7929 eV at 5 K, respectively, in LPE-grown ZnSe layers using Na_2Se with excess Se. But, they did not interpret explicitly whether or not the doublet originates from Na BE emission.

We have also found the doublet (occasionally, triplet) (Asao et al 1988) at 4.2 K in the intentionally-Na-doped ZnSe crystals as shown in Fig. 1. Hereafter, we designate as I_{11}^{Na} the line at lower energy (2.7929 eV) and as I_{12}^{Na} the line at higher energy (2.7938 eV). Therefore, the exciton binding energy for the I_{11}^{Na} is 6.4 meV, provided that the position of free-exciton line is 2.7993 eV (Asao et al 1988). This value can yield that an acceptor binding energy for Na is about 106 meV using a simple effective mass argument, which is less than that of Li acceptor. In particular, PL of our crystal is dominated by very weak Al donor (I_2) BE line and strong I_1^d line compared to the free-exciton (E_x) line, but no DAP band can be observed. Moreover, following characteristics are obtained: when the crystal was excited by the 325 nm He-Cd laser (the above-bandgap excitation) at 2 K, the I_1^{Na} line completely disappears and also when the 441.6 nm excitation was used at 4.2 K (corresponding to free excitons resonantly excited), the I_{12}^{Na} line becomes very weak in intensity.

The insert shows the two-hole (2h) transitions associated with 2s and 2p transitions of the Na acceptor at the 441.6 nm excitation (below-bandgap) at 4.2 K. Transitions to the 2s-acceptor excited state are enhanced under the 441.6 nm excitation relative to the 325 nm excitation. Each energy difference for $1s_{1/2}-2p_{3/2}$, $1s_{1/2}-2s_{3/2}$ and $1s_{1/2}-2p_{5/2}$ is calculated to be 68, 78 and 80 meV, respectively. Using acceptor central-cell correction, the Na acceptor binding energy is estimated to be 102 meV. This value is certainly different from the ionization energy (128 meV) obtained previously for the Na acceptor, but is in fairly good agreement with that obtained by the effective mass argument. At present, we could not discuss the reason for such difference in detail because of an ambiguous Coulomb energy used in the DAP analysis (Neumark et al 1984, Stutius 1982).

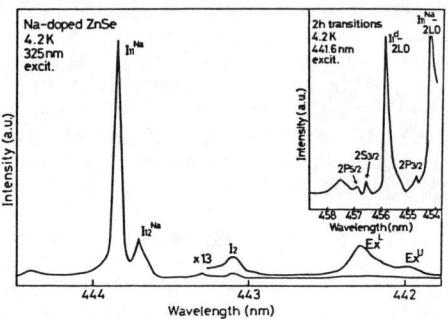

Fig. 1. Doublet structure at 4.2 K of Na acceptor BE emission. The insert shows 2h transitions obtained at the 441.6 nm excitation. The I_1^{Na} line consists of two components: I_{11}^{Na} and I_{12}^{Na} lines at 4.2 K.

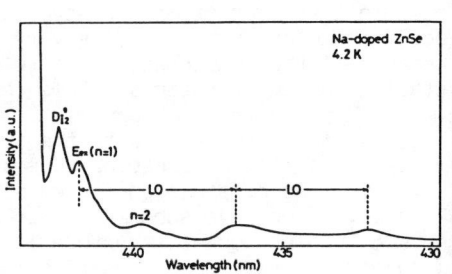

Fig. 2. PL excitation (PLE) spectrum of the I_{11}^{Na} line at 4.2 K.

Fig. 2 shows the PL excitation (PLE) spectrum of the I_{11}^{Na} line obtained at 4.2 K in a Na-doped ZnSe. Judging from the results observed in N-doped ZnSe (Dean et al 1983), the sharp lines, $D_{I_2}^*$ and E_x (n=1 and 2) are, respectively, close to donor BE excited state, and free exciton and its second excited state. The PLE sepctrum of the I_{11}^{Na} line contains broad structure with threshold energies close to $E_x + n\hbar\omega_{LO}$ (n=1 and 2) and as a result the excitons are to relax rapidly by correlated LO phonon scattering. It is therefore reasonable to consider that the formation of indirect exciton is important for the formation of the Na BE line.

3. SPECTRAL CHANGES OF Na BE LINE UNDER THERMAL DIFFUSION CONDITION

In our diffusion experiments, Na was intentionally doped into the undoped ZnSe crystals with (110) cleaved face using Na_2Se and excess Se in an evacuated quartz ampoule for several hours. The crystal was kept at a constant temperature (T_1) of 500 or 600 °C, and the temperature differences ($\Delta T = T_2 - T_1$) between the T_1 and the temperature (T_2) of Na_2Se source were varied from 120 to 340 °C.

Fig. 3 shows the dependence of 4.2 K PL spectra on ΔT at a constant temperature of 500 °C. In the case of $\Delta T=120$ °C, the Na BE line and its LO-phonon replicas were certainly seen in addition to the I_1^d line. With increasing ΔT, both the I_1^d and I_1^{Na} BE lines tend to become strong in intensity, but usual DAP bands are very weak in intensity. Moreover, we could not find out other BE lines located in the vicinity of the I_1^{Na} line at all. On the other hand, when T_1 was altered to 600 °C, the PL spectrum as a function of ΔT was dramatically changed as shown in Figs. 4(a) and (b). In Fig. 4(a), at $\Delta T=120$ °C, the I_{11}^{Na} line is relatively strong in intensity in comparison with the result of Fig. 3, as judged from the intensity ratio between the I_{11}^{Na} and I_1^d line. When ΔT was 260 or 340 °C, both the I_1^x (=I_1^{Li}) and I_2 lines as shown in Fig. 4(b) were observed and as a result the two types of DAP bands having the zero-phonon line, located at 2.693 and 2.708 eV, respectively, appeared. The Li BE lines with the doublet structure are located at an energy 1 meV below the I_{11}^{Na} line and the intensity of the Li BE becomes more intense at $\Delta T=340$ °C. We tentatively interpret that the 2.693 eV DAP band is due to the radiative recombination between Li_{Zn} acceptor and Al_{Zn} donor, while the 2.708 eV DAP band is due to the radiative recombination between Na_{Zn} acceptor and Al_{Zn} donor, because of the presence of Al I_2 BE line.

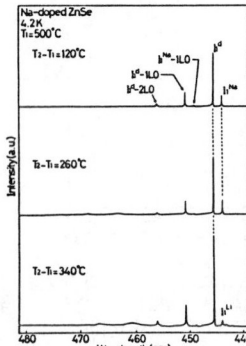
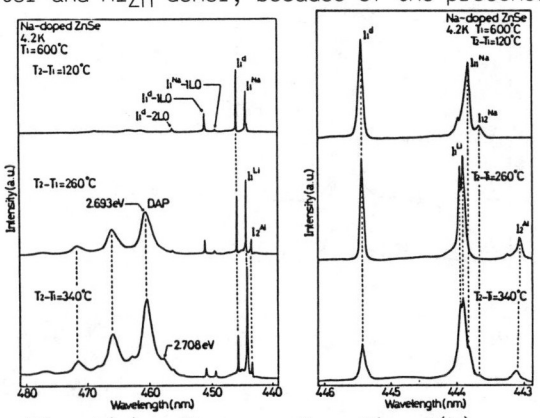

Fig. 3. Changes of the I_1^d and I_1^{Na} lines as a function of ΔT at $T_1=500$ °C.

Fig. 4(a). Changes of the BE and DAP bands at 600 °C.

Fig. 4(b) Enlarged BE spectra of Fig. 4(a).

This proposition is supported by the fact that the Li spectrum is often observed in the intentionally-Na-doped species, since the contamination with the Li impurity is more likely. It is therefore natural to consider that Li contamination is much pronounced at temperatures higher than 500 °C during Na diffusion. Our present results thus propose that in order to exclude both the Li and donor impurity contaminations, the introduction of Na acceptor must be carried out at temperatures as low as possible.

4. TEMPERATURE DEPENDENCE OF THE DOUBLET STRUCTURE

Fig. 5 shows the temperature dependence of the emission intensities in the I_1^{Na} BE doublet structure (I_{11}^{Na} and I_{12}^{Na}) where the temperature was increased from 2 to 10 K. The energy separation (ΔE) between the I_{11}^{Na} and I_{12}^{Na} lines is about 0.9 meV, and the ΔE is found to be constant over the temperatures measured. At 2 K, the I_{11}^{Na} line only appears. With increasing temerature, the relative intensities of the I_{12}^{Na} line (the higher energy component) increases. This thermalization indicates that the doublet is related to the excited states of the Na BE, and that the intensity ratio between the two lines follows the Boltzman distribution.

Fig. 6 shows the intensity ratio between two components as a function of reciprocal absolute temeprature. The activation energy obtained from the straight portion of the slope is 1.06 meV which corresponds to the ΔE. At the same time, the linewidth of the I_{11}^{Na} line (fitted by a Gaussian curve) follows the relation described by $\Delta H = A[\coth(\hbar\omega/2kT)]^{\frac{1}{2}}$, where A is a constant (=0.53 meV) and $\hbar\omega$ is the vibrational energy (=0.5 meV) in the excited state (Yamada et al 1988). It is therefore reasonable to suggest that the two components are associated with the same species of impurity, namely Na acceptor which binds excitons.

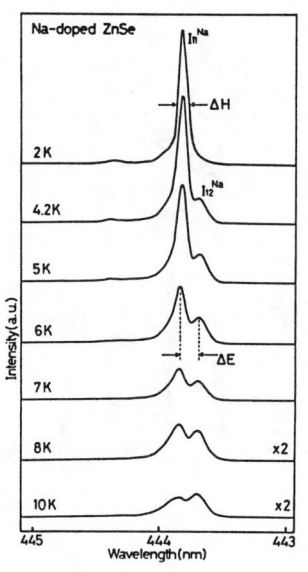

Fig. 5. Temperature dependence of the doublet (I_{11}^{Na} and I_{12}^{Na}) components in Na-doped ZnSe.

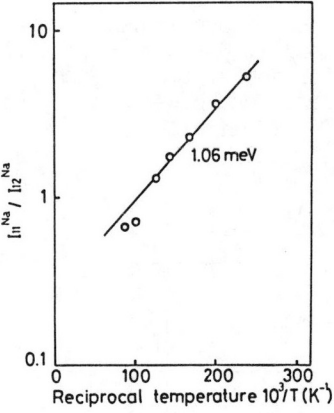

Fig. 6. Relative emission intensity ratio between the I_{11}^{Na} and I_{12}^{Na} as a function of reciprocal temperature.

It is suggested that the doublet structure originates from J-J coupling which two identical holes (Γ_8) of angular momentum j=3/2 are combined antisymmetry to form J=0 and 2 states. Due to the large effective mass of hole

(m_h^*) with respect to that of electron (m_e^*), we neglect the effect of electron in the neutral acceptor BE emission in the strong spin-orbit coupling limit, since the electron results in a small effect on the orbits of the two holes (Pan 1981). Using the variational calculation, we can estimate the theoretical ΔE to be about 2 meV when ϵ_o=8.1, γ_1=3.77 and σ=0.1 in ZnSe. Until now, the theoretical value obtained for the neutral acceptor BE line in GaAs, InP and GaP, is twice larger than the experimental value.
We thus have to consider further splitting or spectral variations in the doublet structure due to the breakdown of a sperical symmetry of the localized acceptor center in the T_d crystal. It is therefore necessary to perform the Zeeman measurement in order to confirm this model.

5. Na BE EMISSION IN MOCVD-GROWN ZnSe FILM

Na impurity was incorporated by adding Na_2Se which was placed in a quartz cell capable of heating at temperatures ranged from 300 to 450 °C in the low pressure MOCVD chamber. In our MOCVD system, Na vapour can be transported by H_2 carrier gas to the substrate. Table 1 shows the growth condition of Na-doped ZnSe film on a (100) GaAs substrate by a low-pressure MOCVD method. Principally, the growth parameters are the same as those described for the growth of the not-intentionally-doped ZnSe films (Asao et al 1988).

Fig. 7 represents the PL spectrum obtained at 4.2 K of Na-doped ZnSe/(100) GaAs (a thickness of 1.4 µm) where the Na cell temperature was about 400 °C. The I_1^{Na} line appears at about 2.7927 eV which indicates almost the same position as observed for the I_{11}^{Na} line in bulk ZnSe. The line is strong in intensity, but, several donor BE I_x and I_2 lines are seen and also the free-exciton (E_x) line, so that the crystalline quality is thought to be excellent. Due to the appearance of the I_x and I_2 lines, the DAP band with LO-phonon replicas, where the zero-phonon line appears at 2.708 eV, is observed. The present DAP band is ascribed to an electronic transition between Na_{Zn} acceptor and Al_{Zn} donor described in the section 3.

At present, we have already found that the I_1^{Na} line intensity depends upon the Na cell temperatures and that the Na can successfully incorporated in homoepitaxial ZnSe films on (110) ZnSe substrate.

Table 1. MOCVD growth condition for Na doping

Growth Conditions	
SUBSTRATE	GaAs(100)
SUBSTRATE TEMP.	300 °C
FLOW RATE OF DMZ	5.85×10^{-6} mol/min.
FLOW RATE OF H_2Se	1.44×10^{-4} mol/min.
(VI)/(II) RATIO	24.6
PRESSURE	3.0 Torr
Na CELL TEMP.	300~450 °C

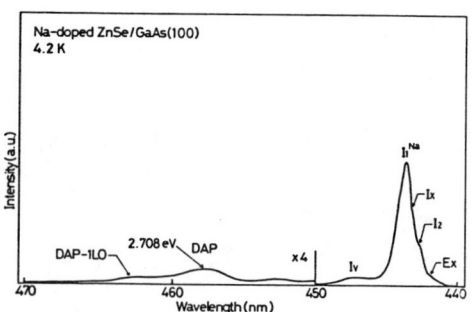

Fig. 7. PL spectrum obtained at 4.2 K of a Na-doped MOCVD ZnSe film grown on (100) GaAs substrate.

6. CONCLUSIONS

(1) There appears the doublet structure of the Na BE line which is explained in terms of J-J coupling in the excited state and originates from the same impurity center.
(2) From 2h transitions, the 1s Na acceptor binding energy is estimated to be 102 meV.
(3) Above 500 °C, the I_1^x (Li) BE line and related donor-acceptor pair band strongly appear in Na-doped species.
(4) Using the low-pressure MOCVD method, the Na acceptor can be successufully incorporated into ZnSe layers at 300 °C using Na_2Se under excess Se condition.

ACKNOWLEDGEMENTS

One of the authors (T.T) thanks H. Namba of Sumitomo Electric Corp. for supplying ZnSe CVD ingot and K. Matsumoto of Nippon Sanso Corp. for supplying DMZn gas. Part of this work was supported by a Grant-in-Aid for Scientific Research on Priority Areas, New Functionality Material-Design, Preparation and Control No. 62604583, from the Ministry of Education, Science and Culture of Japan.

REFERENCES

Asao Y, Hayamizu S, Yamada Y, Taguchi T and Hiraki A 1988 Proc. of MRS 127 68.
Bhargava R N, Seymour R J, Fitzpatrick B J and Herko S P 1979 Phys. Rev. B 20 2407.
Chacham H, Alves J L A and De Siqueria M L 1987 Solid St. Commun. 64 863.
Dean P J, Stutius W, Neumark G F, Fitzpatrick B J and Bhargava R N 1983 Phys. Rev. B 27 2419.
Merz J L, Nassau K and Shiever J W, 1973 Phys. Rev. B 8 1444.
Neumark G F, Herko S P, McGee III T F and Fitzpatrick B J 1984 Phys. Rev. Lett. 53 604.
Neumark G F and Catlow C R A 1984 J. Phys. C: Solid St. Phys. 17 6087.
Pan D S 1981 Solid St. Commun. 37 375.
Sharma R R and Rodriguez S 1967 Phys. Rev. 159 449.
Stutius W 1982 J. Crystal Growth 59 1.
Tews H, Venghaus H and Dean P J 1979 Phys. Rev. B 19 5178.
Yamada Y, Kidoguchi I and Taguchi T 1988 Jpn.J.Appl. Phys. to be published.

Excited states of acceptors in cubic semiconductors: central-cell effect in ZnTe

M A Kanehisa and M Said †

Laboratoire de Physique des Solides‡, Université Pierre et Marie Curie, Tour 13, 4 Place Jussieu, 75252 Paris Cédex 05, France

ABSTRACT: Higher excited states of acceptors in ZnTe are obtained by using the Balderschi and Lipari spherical model including the cubic correction and the central-cell effect. This is done by solving the coupled radial equations by the finite-element method with Arnoldi's algorithm, which gives several ($\sim$10–20) low-lying states simultaneously. Our procedure permits to determine very accurately the host band-structure parameters. For ZnTe, we obtain the Luttinger parameters $\gamma_1 = 3.80$, $\gamma_2 = 0.86$, $\gamma_3 = 1.30$ and the static dielectric constant $\epsilon_0 = 9.4$.

1. INTRODUCTION

The problem of shallow acceptor states has been formally solved since more than 30 years by the effective-mass approximation (for reviews, see Kohn 1957, Bassani *et al* 1974, Pantelides 1978). However, in contrast to donors, numerical calculation of acceptor states is difficult. This is because, due to the degeneracy of the valence band, one has to solve a system of coupled Schrödinger equations, which is usually done by variational methods, and only the ground state and, at best, a few low-lying excited states can be obtained. Baldereschi and Lipari (1973, 1974)'s spherical model reduces the complicated eigenvalue problem to a simple radial Hamiltonian. However, the resultant set of coupled differential equations are still to be solved variationally, and only a few low-lying states have been obtained.

Recently we have developed a new method for solving nonvariationally the Baldereschi-Lipari radial equation (Said *et al* 1986, 1987a). Our approach is based on the finite-element method and Arnoldi's algorithm for diagonalising efficiently large sparse matrices. In this way, one obtains several eigenstates simultaneouly. Here, we apply this method to ZnTe and examine the central-cell effect on higher excited states, theoretically little studied so far but experimentally known to be important. Then by analyzing hitherto unexploited experimental data on higher excited states, we determine accurately the mass parameters for the host energy band.

2. THE ACCEPTOR STATES

† Present adress: Fritz-Haber-Institut der Max-Planck-Gesellschaft, Faradayweg 4-6, D-1000 Berlin 33, FRG.
‡ Associé au Centre National de la Recherche Scientifique.

© 1989 IOP Publishing Ltd

The structure of higher excited acceptor levels deviates considerabley from the hydrogen scheme due to the level crossing (Said et al 1987a). Fig. 1 shows the acceptor S states in the effective-mass approximation as a function of the spherical "spin-orbit" coupling parameter μ without the cubic term ($\delta = 0$). For $n \geq 3$, each hydrogen-like nS state splits into two levels, nS$_a$ and nS$_b$. For small μ, these states are bonding and antibonding combinations of hydrogen-like nS and nD states. But for the μ values of interest ($\mu \approx 0.6$ for ZnTe), 3S$_b$ has S-like square amplitude of 0.12 and D-like square amplitude of 0.88. The binding energy of the nS$_a$ states increases with μ while that of nS$_b$ decreases with μ. This makes crossing and repulsion of levels for the μ value of interest and thus distribution of oscillator strength depends crucially on the μ value.

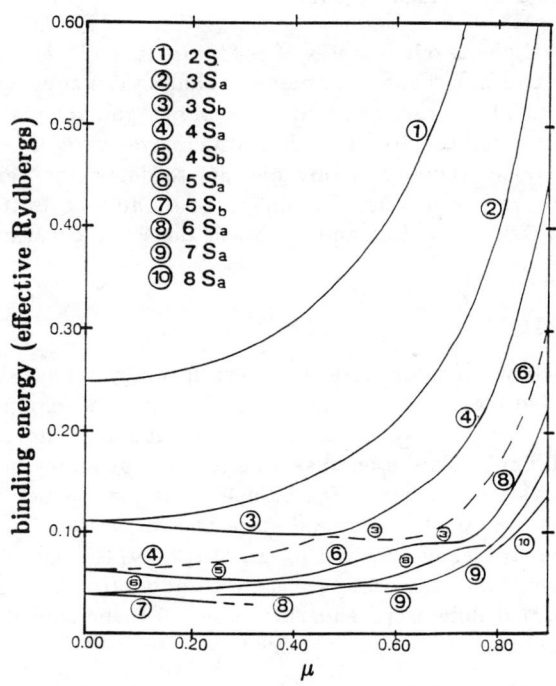

Fig. 1. General structure of the acceptor S states in the effective-mass approximation as a function of the spherical "spin-orbit" parameter μ without cubic correction (taken from Said et al (1987a)).

3. IMPURITY POTENTIAL

The Coulomb-type potential with the static dielectric constant ϵ_0 for the host material

$$V(r) = -\frac{e^2}{\epsilon_0 r} \tag{1}$$

gives the same acceptor energy independent of the acceptor species, although experimentally it differs from one impurity to another (chemical shift). This difference is

due to the deviation of the real impurity potential from the simple law (1) near the impurity center (central-cell potential). There have been many proposals for the form of the real impurity potential (Bassani *et al* 1974, Pantelides 1978), and it is by no means the purpose of the present paper to propose yet another potential.

Here we adopt a simple analytic expression containing as few parameters as possible and try to reproduce the whole set of experimetal data on higher excited states. For this purpose we have chosen the potential proposed by Lipari, Baldereschi and Thewalt (1980):

$$V(r) = -\frac{e^2}{\epsilon_0 r}\left[1 + (\epsilon_0 - 1)e^{-\alpha' r}\right]. \tag{2}$$

This potential contains only one parameter α'. Its inverse α'^{-1} is the impurity-dependent screening length and measures the effective radius of the central cell. In fact, for large r, $V(r)$ reduces to $-e^2/\epsilon_0 r$, while for small r, it reduces to the bare Coulomb potential (1). For substitutional impurity having the same core electron as the host atom (isocoric impurity), α' is expected to reduce to the screening constant α of the host dielectric function. For ZnTe, α^{-1} is estimated to be 0.64 Å by using the numerical calculation of $\epsilon(\mathbf{q})$ by Vindsome and Richardson (1970).

4. FINITE-ELEMENT METHOD

The finite-element method consists in transforming differential equation into difference equation and then to solve this as an algebraic eigenvalue problem. The eigenfunction obtained is nothing but the acceptor wavefunction in the direct r-space. The matrix to be diagonalised is large (typically 2000 × 2000) and asymmetric but its elements are concentrated near the diagonal ("sparse" matrix). We have employed a newly-developed variant of Arnoldi's method with Chebyshev acceleration (Saad 1984, Saad and Schultz 1985, Parlett and Saad 1987). Essentially, this method consists in projecting the large matrix onto a much smaller ($\sim$ 20 × 20) one so that the low-lying eigenvalues of the original matrix are contained in the projected matrix, and the latter then can be easily diagonalised.

5. RESULTS

Calculation of the acceptor states requires the knowledge of the host material constants: Luttinger parameters γ_1, γ_2, γ_3, and the static dielectric constant ϵ_0. In addition to this, for each impurity, the central-cell parameter α' is to be chosen. We have done this in the following way (Kanehisa and Said 1988).

To determine the dielectric constant, we use the donor transition energies. Using the conduction-band mass $m_e = 0.116 m_0$ determined by magneto-optical measurements (Dean, Venghaus and Simmonds 1978), we calculate the 1S-2S and 1S-3S transition energies and, by comparing them with experiment (Venghaus and Dean 1980), ϵ_0 is determined to be 9.4. As for the central-cell effect for donors, since the donor species is unknown, we simply used the host screening length $\alpha'^{-1} = 0.64$ Å.

In order to deduce the valence-band mass parameters, we use exciton data since it is intrinsic to the host. We use for the electron-hole pair the screened potential (2)

with the host screening length $\alpha'^{-1} = 0.64$ Å. By comparing with experimental data (Venghaus and Dean 1979) and using the above electron mass $m_e = 0.116 m_0$, we deduce $\gamma_1 = 3.80$, $\gamma_2 = 0.86$ and $\gamma_3 = 1.30$. We also checked that the parameters give correct acceptor P states since they are not affected by the central cell. Our values are to be compared with those determined from effective-mass theory without r-dependent screening ($\gamma_1 = 3.8$, $\gamma_2 = 0.83$, $\gamma_3 = 1.28$) (Herbert et al 1978) and from magneto-optical measurements ($\gamma_1 = 3.9$, $\gamma_2 = 0.83$, $\gamma_3 = 1.30$) (Venghaus and Jusserand 1980).

Finally, for each impurity, the central-cell parameter α' is chosen so that it gives the observed ground-state binding energy. For ZnTe, the central-cell radii α'^{-1} determined for acceptors Li (substituting Zn) and P and As (substituting Te) are 0.11 Å, 0.56 Å, and 0.61 Å, respectively. α'^{-1} is smallest for Li as is expected. For group V impurities, our value of α'^{-1} increases in going from P to As, and is expected to become close to the host screening length 0.64 Å for isocoric impurity Sb.

Fig. 2 compares our results with acceptor transitions observed by electronic Raman scattering (Said et al 1987b). The agreement is satisfactory. For As-doped ZnTe, our calculation gives central-cell corrections of 16.4 meV for the 1S state and 1.9 and 0.6 meV for the 2S and 3S states, respectively. Quantitatively, by using only one adjustable parameter α', we reproduce the whole set (10–15 states) of available experimental data with an accuracy of a few meV in the worst case.

6. DISCUSSION

It is to be noted that in case of acceptors in Si and Ge, higher excited states have been considered by Lipari et al (1980) by a somewhat tour-de-force variational method with hundreds of basis functions. This is due to the small spin-orbit coupling (comparable to typical acceptor energy) in these materials, which requires the inclusion of the split-off bands and, especially in Si, due to the exceptionally large cubic term preventing the perturbational (spherical) approach. On the contrary, in II-VI compounds in which we are interested, the spin-orbit split-off band may be neglected and the cubic term can be treated as perturbation. Our investigation in the framework of the spherical model (Baldereschi and Lipari 1973, 1974) thus permits to get physical insight into the role played by the central-cell potential.

Let us discuss some effects on acceptor states not taken into account in our theory. In polar crystals like ZnTe, acceptors are coupled with the longitudinal optical phonon. The lowest-order effect of this coupling is the replacement of the bare mass by the renormalised mass. Thus the valence band parameters we obtained are considered to be the polaron masses. There is also a contribution to screening from logitudinal optical lattice vibrations. This is partially allowed for in our treatment by the use of ϵ_0 rather than ϵ_∞. More accurate theory would use the dynamical screening $\epsilon(\omega)$ due to phonon exchange, which would require solving an integro-differential equation. Finally, in the spherical framework, the cubic term is treated perturbatively. This may cause some inaccuracy for higher states where adjacent levels are closely spaced. To remedy this and also to allow for the split-off valence subband we have neglected, we would have to expand the acceptor wavefunction in terms of a large number of L states.

ACKNOWLEDGMENT

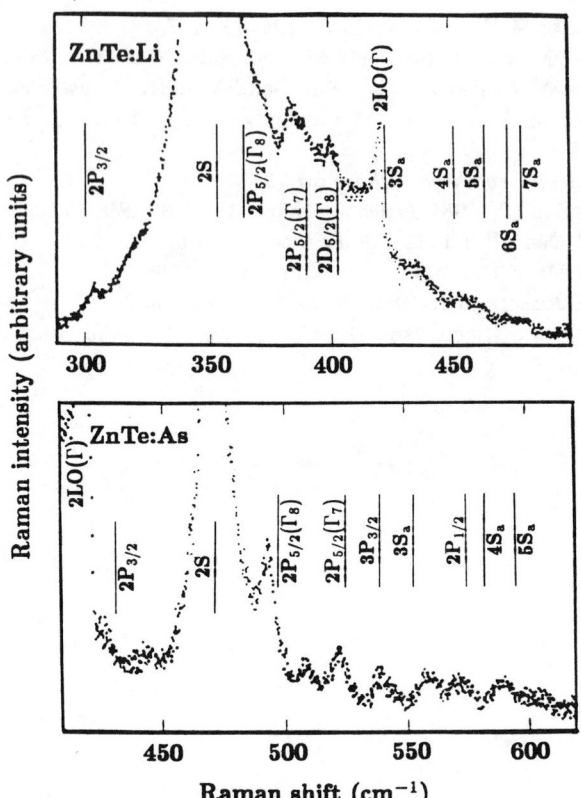

Fig. 2. Electronic Raman spectra for Li- and As-doped ZnTe at liquid helium temperature taken from Said et al (1987b) compared with calculated acceptor transition energies between the 1S ground state and indicated excited states (vertical lines).

The computer resources were provided in part by the Scientific Committee of the Centre de Calcul Vectoriel pour la Recherche (Palaiseau, France).

REFERENCES

Baldereschi A and Lipari N O 1973 *Phys. Rev.* B **8** 2697; 1974 *Phys. Rev.* B **9** 1525
Bassani F, Iadonisi G and Preziosi B 1974 *Rep. Prog. Phys.* **37** 1099
Dean P J, Venghaus H and Simmonds P E 1978 *Phys. Rev.* B **18** 6813.
Herbert D C, Dean P J, Venghaus H and Pfister J C 1978, *J. Phys. C: Solid State Phys.* **11** 3641
Kanehisa M A and Said M 1988 *J. Phys. C: Solid State Phys.* **21** 4637
Kohn W 1957 *Solid State Physics* **5** 257
Lipari N O, Baldereschi A and Thewalt M L W 1980 *Solid State Commun.* **33** 277
Saad Y 1984 *Math. Comp.* **42**, 567

Saad Y and Schultz M H 1985 *Math. Comp.* **44**, 417
Said M, Kanehisa M A and Balkanski M 1986 *Solid State Commun.* **57**, 417
Said M, Kanehisa M A, Balkanski M and Saad Y 1987a *Phys. Rev.* B **35** 687
Said M, Kanehisa M A, Jouanne M and Balkanski M 1987b *J. Phys. C: Solid State Phys.* **20** 2917
Pantelides S T 1978 *Rev. Mod. Phys.* **50** 797
Parlett B N and Saad Y 1987 *Linear Algebra Appl.* **88/89**, 575
Venghaus H and Dean P J 1979 *Solid State Commun.* **31** 897
Venghaus H and Dean P J 1980 *Phys. Rev.* B **21** 1596
Venghaus H and Jusserand B 1980 *Phys. Rev.* B **22** 932
Vindsome P K W and Richardson D 1970 *J. Phys. C: Solid State Phys.* **4** 2650

Inst. Phys. Conf. Ser. No 95: Chapter 6
Paper presented at Int. Conf. Shallow Impurities in Semiconductors, Linköping, Sweden, 1988

Energy levels of phosphorus and arsenic acceptors in CdTe: interaction with deuterium and lattice defects

L. SVOB and Y. MARFAING
Laboratoire de Physique des Solides de Bellevue, CNRS, 1, Place A. Briand
F - 92195 MEUDON-CEDEX, France.

ABSTRACT. CdTe samples coimplanted with phosphorus or arsenic and deuterium have been analysed by SIMS and photoluminescence techniques. After short annealing times (1-6 hours at 500°C) high energy donor-acceptor bands appear in addition to those associated with the simple acceptors P_{Te} and As_{Te}. For longer annealing times deuterium diffuses out of the samples and the new bands disappear. This behaviour is interpreted as the formation of complex acceptor centers P-D and As-D where the group V impurities are in the trivalent state. For very long annealing times deep donor-acceptor pair bands become dominant and are thought to originate from coupling of in-diffused Te-vacancies with the group V acceptors.

1. INTRODUCTION.

Hydrogen is a commonly used transport gas in semiconductor technology. In many technological processes it may be unintentionaly incorporated into semiconductor lattices and can cause changes in the properties of these materials. This fact indicates a requirement for the study of hydrogen as an impurity in semiconductors. Most papers published about hydrogen behaviour in semiconductors relate to silicon and gallium arsenide. The present authors have undertaken a study of hydrogen and deuterium in II-VI compounds, particularly in CdTe. Interaction with some donor and acceptor impurities was reported in previous papers (Boudoukha et al 1985, Svob et Marfaing 1986a, b). In this communication we focus our attention on group V acceptor impurities (P, As), which are shown to have specific type of interaction with deuterium.

2. EXPERIMENTAL RESULTS.

An important difference between CdTe and other semiconductors like Si and GaAs is that an RF plasma, mostly used for hydrogenation cannot be used in the case of CdTe (Svob et al 1986c). This limits the methods applicable to CdTe to high temperature annealing in gaseous hydrogen or deuterium (Svob et Marfaing 1988) and implantation (Svob et Marfaing 1986a).

For this study both deuterium and the doping impurity were implanted. The implantation energy of each ion, D (12 keV), P (75 eV), As (150keV) was adjusted to lead to identical projected ranges for the impurity and the deuterium in the coimplanted samples ($\sim 0.1 \mu m$). Doses were around $10^{15} cm^{-2}$. The samples successively implanted with P + D or As + D were then annealed at 500°C (with excess Cd at the same temperature) for diffe-

© 1989 IOP Publishing Ltd

rent times from 6 hours to 178 hours (∼7 days) for the phosphorus implanted samples, and from 1 hour to 75 hours for the arsenic implanted ones. Annealed samples have been studied using luminescence spectroscopy and SIMS. For the luminescence studies excitation was provided by an Ar-laser. The spectra were measured at 1.8 K. The effects due to impurity-deuterium interaction have been deduced from a comparison with spectra from reference samples, in which only phosphorus or arsenic was implanted. The deuterium diffusion profiles were measured by SIMS using cesium sputtering.

PHOTOLUMINESCENCE RESULTS.

The luminescence spectra of phosphorus and arsenic doped CdTe have been studied by E. Molva et al (1983). The authors reported the presence of donor acceptor pair bands (DAP) situated at 1.537 eV and 1.510 eV in the case of P and As respectively. Using also excitation spectroscopy they set the ground state energy level of the corresponding acceptor at 68.2 meV and 92 meV above the valence band.

The spectra of P - D implanted samples annealed for 6 hours show besides the usual phosphorus $(DAP)_1$ band another band at higher energy (1.555eV) (Figure 1). This new band is dominant, whereas the intensity of the $(DAP)_1$ band is strongly reduced compared to that of the reference sample.

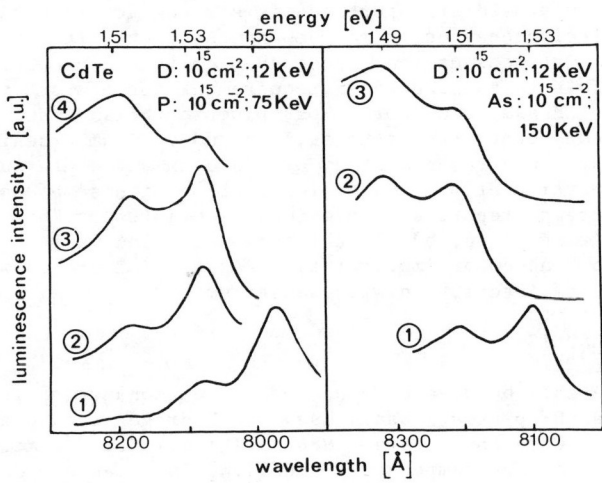

Figure 1. Photoluminescence DAP-bands in coimplanted CdTe samples annealed at 500°C for different times. Deuterium-phosphorus coimplanted CdTe : 1. 6 hours ; 2. 16 hours ; 3. 56 hours ; 4. 178 hours. Deuterium-arsenic coimplanted CdTe : 1. 1 hour ; 2. 7 hours ; 3. 75 hours.

The spectra corresponding to the longer annealing times again show the $(DAP)_1$ band to be dominant whereas the new deuterium associated band disappears completely. Similar observations have been made on the luminescence spectra of As-D implanted samples. The annealing time leading to the appearance of the high energy DAP band is only 1 hour in this case and the new band is at 1.531 eV (Figure 1).

If the annealing time of both reference and deuterium implanted samples exceeds 30 hours another DAP band appears almost superposed on the $(DAP)_1$ - LO replica corresponding to each impurity. After long time annealing at 500°C this band becomes dominant. The appearance of this band is to be related to the presence of phosphorus and arsenic, but not to that of deuterium, according to the results of deuterium profiling reported below.

SIMS PROFILING.

The profiles of deuterium concentration in the annealed samples have been measured by SIMS. The change in deuterium concentration profiles as a function of annealing time is seen in Figure 2. After 6 hours annealing at 500°C the P-D implanted sample continues to show a significant deuterium concentration compared with the unannealed one. Even after 16 hours annealing deuterium is still detectable. However in As-D implanted samples the decrease in the deuterium concentration after similar anneals is much greater (Figure 2). From these data it can be inferred that deuterium has completely out diffused for annealing times greater than about 30 hours.

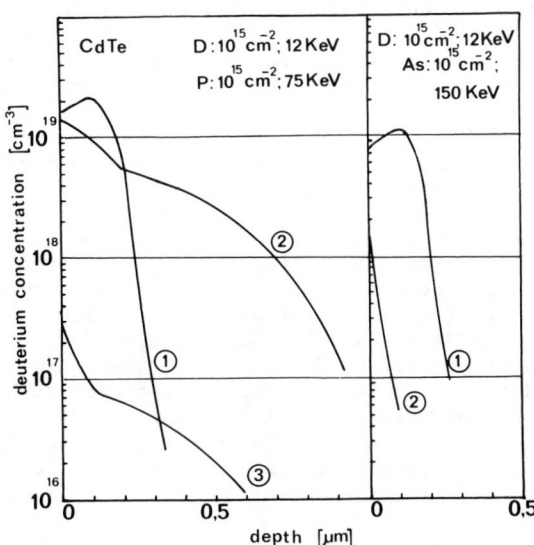

Figure 2. Deuterium depth profiles in coimplanted CdTe obtained from SIMS measurements on samples annealed at 500°C for different times : Deuterium phosphorus implanted CdTe : 1. as implanted ; 2. 6 hours ; 3. 16 hours. Deuterium-arsenic coimplanted CdTe : 1. as implanted ; 2. 1 hour.

3. DISCUSSION.

The first aspect to be pointed out is the rather strong stability of deuterium in those doped samples due to interaction with the impurities. In particular the interaction P-D appears to be stronger than As-D in agreement with the difference in the formation energy of compounds like PH_3 and AsH_3.

The second aspect is related to the emergence of new high energy DAP bands at 1.555 eV and 1.531 eV in P and As implanted samples respectively. These bands appear only in the presence of deuterium and are observed after annealing conditions which are usually sufficient to remove implantation defects : on the one hand, implanted phosphorus can be fully activated with annealing at 500°C for 1 hour (Uzan 1983), while on the other we have shown in previous work that defects associated with implanted protons are annealed at temperatures as low as 200°C (Svob et Marfaing 1986a). We are thus led to think of the formation of new acceptors resulting from interaction between deuterium and P or As. This behaviour is novel in that the incorporation of hydrogen (deuterium) in acceptor-doped materials gives commonly rise to the neutralization of the impurities as shown in the case of Si : B (Pankove 1984), GaAs : Zn (Pearton 1987), CdTe : Cu (Svob et Marfaing 1986b). The particular effect observed can be explained by considering that the bonding between P (As) and its Cd-neighbours is modified. It is supposed that deuterium preferentially binds with the group V impurity, say P, so that one P-Cd bond is broken and P relaxes in the opposite direction taking up a trivalent configuration. Because the number of bonding electrons brought about by the complex P-D is now only four, the new centre can be described as a double acceptor. This simplified picture accounts for the observed absence of acceptor neutralization and the introduction of new acceptor levels never reported before. The energy of those levels can be roughly deduced from the shift of DAP bands : we obtain $E_A - E_v$ = 45 meV and 71 meV for P-D and As-D respectively.

The third aspect of this work refers to the deep DAP bands observed at 1.514 eV and 1.482 eV in P and As implanted samples following long annealing times (> 30 hours) at 500°C under excess cadmium pressure. We relate this phenomenon to the diffusion of some species from or to the external gas phase. While contamination from an impurity (O ?) cannot be completely excluded we think rather that outdiffusion of tellurium occurs in order to establish the required Te_2 partial pressure. The diffusion coefficient of interstitial Te or the Te vacancy at 500°C is around $5.10^{-16} cm^2 sec^{-1}$ (Chern 1975), sufficiently small to fit the long time annealings. At the microscopic level V_{Te} which is supposed to be a deep double donor will interact with the phosphorus (arsenic) acceptor to form some kind of $P_{Te}V_{Te}$ donor complex. The later will be involved in the DAP bands. It is clear that a detailed spectroscopic analysis of these new bands is needed in order to determine the actual donor and acceptor energy levels.

4. CONCLUSIONS.

Deuterium has been shown to interact with P_{Te} and As_{Te} acceptors to form new acceptor centres shifted to lower energy by about 22 meV. A direct coupling of D with group V impurities and a trivalent state of the latter

are assumed to account for the non-neutralization of acceptor activity in this case. This is a distinct behaviour from that observed in hydrogenated p-type Si and GaAs.

REFERENCES.

Boudoukha A, Legros R, Svob L, Marfaing Y 1985, J. Cryst. Growth 72 226
Chern S S, Kröger A 1975, Sol. Stat. Chem. 14 44
Molva E, Saminadayar K, Pautrat J L, Ligeon E 1983 Sol. Stat. Comm. 48 955
Pankove Y I, Wance R O, Berleyheiser 1984 Appl. Phys. Lett. 45 1100
Pearton S Y, Dautremont-Smith W C, Tu C W, Nabity Y C, Swaminathan V, Stavola M, Chevallier J 1987, Inst. Phys. Conf. Ser. 83 (Inst. Phys., Bristol U.K.) p. 289
Svob L, Marfaing Y 1986a Defects in Semiconductors Ed. H.Y. von Bardeleben (Mater. Sci. Forum Vol. 10-12) pp 857-862
Svob L, Marfaing Y 1986b Solid State Comm. 58 343
Svob L, Chevallier J, Ossart P, Mircea A 1986c J. Mater. Sci. Lett. 5 1319
Svob L, Heurtel A, Marfaing Y 1988 J. Cryst. Growth 86 815
Uzan C, Legros R, Marfaing Y 1983 J. Cryst. Growth 72 252

Inst. Phys. Conf. Ser. No 95: Chapter 7
Paper presented at Int. Conf. Shallow Impurities in Semiconductors, Linköping, Sweden, 1988

Mechanism of hydrogen passivation in silicon

T. Sasaki* and H. Katayama-Yoshida

Department of Physics, Tohoku University, Sendai 980, Japan
*Present address: National Research Institute for Metals,
Meguro-ku, Tokyo 153, Japan

ABSTRACT: The atomic configuration and the electronic structure of hydrogen and boron acceptor complexes in crystalline silicon are determined using a norm-conserving pseudopotential and supercell methods taking into account lattice relaxation within the local-density approximation. The bond center model is favorable for the hydrogen passivation mechanism because the total energy of the hydrogen and boron complex at the bond center with large lattice relaxation is 0.017 Ry more stable than that at the antibonding site of a silicon-boron covalent bond. Our hydrogen passivation model based upon the three-centered bond is quite different from the classical model based upon the two-centered bond. We also discuss the site assignment of the isolated hydrogen in pure silicon comparing with the data of muon spin rotation.

1. Introduction

   HYDROGEN INDUCED PASSIVATION of shallow acceptors in crystalline silicon is one of the most fascinating problems in the physics of semiconductors (Sah et al. 1983, Pankove et al. 1983, Johnson et al. 1986). Secondary-ion-mass-spectroscopy (SIMS) measurement on p-type silicon exposed to deuterium revealed that only one deuterium atom is needed to passivate an acceptor level (Johnson 1985, Pankove et al. 1985). Thewalt et al. (1985) observed that the acceptor-bound-exciton luminescence is substantially reduced by hydrogenation, indicating that compensation alone, without complex formation, cannot explain the passivation (Cappizi and Mittiga 1987, Pantelides et al. 1986). There are conflicting models concerning the mechanism of the hydrogen passivation of shallow acceptors. Experimental and theoretical studies on this problem have proposed two possible models for the hydrogen passivation: The bond-center (BC) site model by Pankove et al. (1983) and Deleo and Fowler (1985, 1986) and the antibonding (AB) site model by Assali and Leite (1985, 1986). In the BC-site model, the hydrogen atom is located at the bond center of a silicon-boron covalent bond and hydrogen makes a strong covalent bond with a silicon atom, terminating the dangling bond of silicon, so that boron has threefold-coordinated covalent bonds with silicon atoms (Fig. 1a). In the AB-site model, the hydrogen atom resides at the interstitial anti-bonding site of a silicon-boron covalent bond and hydrogen becomes H^+ and boron becomes B^- as a result of charge transfer (Fig. 1b).

   We determine the atomic configuration and electronic structure of hydrogen and boron acceptor complexes in crystalline silicon in order to

© 1989 IOP Publishing Ltd

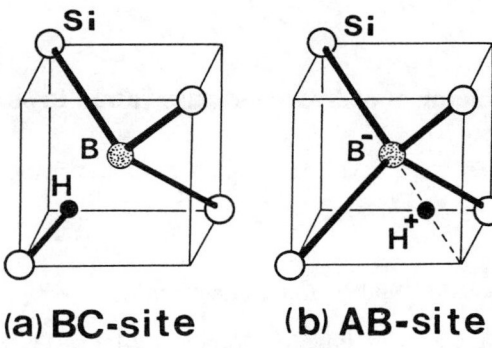

Fig. 1 Two models for the microscopic structure of B-H pairs where H atom resides at either (a) a bond center or (b) an antibonding interstitial site.

provide physical insights into the correct mechanism of the hydrogen passivation without any adjustable parameters, using a norm-conserving pseudopotential and supercell methods within the local-density-functional formalism. The supercell contains seven silicon atoms, one boron atom and one hydrogen atom. The Ceperley-Alder form (1980) parameterized by Perdew and Zunger (1981) is used for the exchange-correlation energy. While we use the norm-conserving pseudopotentials developed by Bachelet et al. (1982) for the silicon and boron atoms, the real Coulomb potential is used for the hydrogen atom.

Two possible mechanisms, the BC-site and AB-site models, for neutralizing the shallow acceptor levels in p-type silicon are considered. We calculate the Hellmann-Feynman force (Ihm et al. 1979) at each atomic site, and determine the stable atomic configuration of the hydrogen and boron complexes taking into account lattice relaxation. Since the hydrogen atom is known to be located at a site which has C_{3v}-symmetry for both BC- and AB-site models (Wichert et al. 1987), we search the stable position of the hydrogen, boron and silicon atoms along the C_{3v}-axis. The total energies are calculated by the k-space formalism (Ihm et al. 1979). We also calculate the electronic structure of an isolated hydrogen atom at the BC- and AB-sites in undoped silicon and determine the stable atomic configuration. We compare the calculated result of the isolated-hydrogen with the experimental data of muon spin rotation (Patterson et al. 1978, Estle 1985).

2. Isolated Hydrogen Atom in Silicon

2.1 Total Energy: The calculated total energies of the isolated-hydrogen are shown in Table I. It is found that the total energy at the AB-site is 0.036 Ry more stable than that of the BC-site. Comparing the electrostatic and electronic energies, we can see that the former of the BC-site is much lower than that of the AB-site owing to the strong hybridization between the hydrogen 1s orbital and sp^3-orbitals of silicon atoms as shown later. On the other hand, the electrostatic energy is responsible for the energy reduction at the AB-site. This difference of the energy reductions reflects in the lattice relaxation. In the case of BC-site, the distance between neighboring silicon atoms is 3.19 A which should be compared with the bond length of pure silicon, 2.3 A. In the case of AB-site, the hydrogen atom is located 1.92 A from the nearest neighbor silicon atom which is dis-

Table I Calculated total energies of the isolated hydrogen at the BC- and AB-sites in unit of Ry/cell.

Si:H^0	BC-site	AB-site
electronic part	3.869	4.746
electrostatic part	-68.276	-69.190
total	-64.407	-64.443

placed only 0.08 A from the position of pure silicon.

2.2 Electronic Structure of an Isolated Hydrogen at the BC-site: Figure 2 shows the contour plot of the calculated charge density and the density of states (DOS). We can see that the charge density has large value at the hydrogen site because of its real Coulomb potential. We can observe a gap state near the conduction band and plot the charge density of the gap state in Fig.2(c). The charge density is delocalized near the back-bond site of the silicon atoms neighboring with the hydrogen atom. Therefore, the gap state is regarded as the antibonding one between the dangling bond states which come from the enlarging of the Si-Si distance. The corresponding bonding state between them strongly mixes with the hydrogen 1s orbital and the mixed state is pushed down into the valence band of silicon, making a three-centered bond. Thus the charge density along the silicon, hydrogen and silicon chain has large value.

2.3 Electronic Structure of an Isolated Hydrogen at the AB-site: The charge-density contour plot and DOS of the AB-site model are shown in Fig. 3. A gap state is also made in this case. In contrast with the case of BC-site, it can be seen more clearly in the middle of the band gap. Figure 3(c) is the charge density from this state. We find that the gap state is localized at the hydrogen site and not so strongly hybridized with the wave function of host silicon such as the BC-site model.

2.4 The Assignment of Normal and Anomalous Muonium: Although no experimental data on the isolated hydrogen in silicon are available, positive muon spin rotation provides alternative experimental data (the only difference between the hydrogen and muon is in their atomic masses, m(H)=9m(Muon)). The muon spin rotation studies have shown that two kinds of muonium centers exist in silicon, normal muonium (Mu) and anomalous muonium (Mu*) (Patterson et al. 1978). The latter has an anisotropic hyperfine coupling constant while the former has an isotropic one. Moreover, the absolute value of the hyperfine coupling constant of Mu* is a few percent of that of Mu or free muonium. The atomic configuration and electronic states of Mu*, however, are still controversial problem.

In the above calculation on the isolated hydrogen atom in silicon, the stable position of the muon at the AB-site deviates from the tetrahedral (T_d) interstitial site by only 0.35 A. Although Mu was believed as a muon located at the T_d-site up to date, we found from the precise calculation that the muon at the ideal T_d-site is unstable. Using the second derivative of the adiabatic potential at the stable AB-site, we estimated the amplitude of the zero point oscillation of the muon, 0.42 A. Comparing this value with the distance between the neighboring AB-sites, 0.57 A, we conclude that Mu is a muon hopping between four equivalent AB-sites around the T_d-site. Moreover, we assign that the muon at the BC-site makes Mu* and its small hyperfine coupling constant is due to the strong hybridization mentioned above. This assignment is consistent with the recent ENDOR experimental data (Kiefl et al. 1988). Because our calculation is a spin restricted one and based on the plane wave expansion, we cannot calculate the spin density at the muon site. If the contribution to the spin density from the valence band is small, we can estimate the spin density using that of the gap state. The ratios of the charge density of the gap state at the muon site to that of free muonium are 0.0039 and 0.058 for the BC- and AB-site models, respectively. The difference between these values qualitatively explains the experimental result that the ratio of the hyperfine coupling constants of Mu* and free muonium, 0.015, is much smaller than that for Mu and free muonium, 0.450.

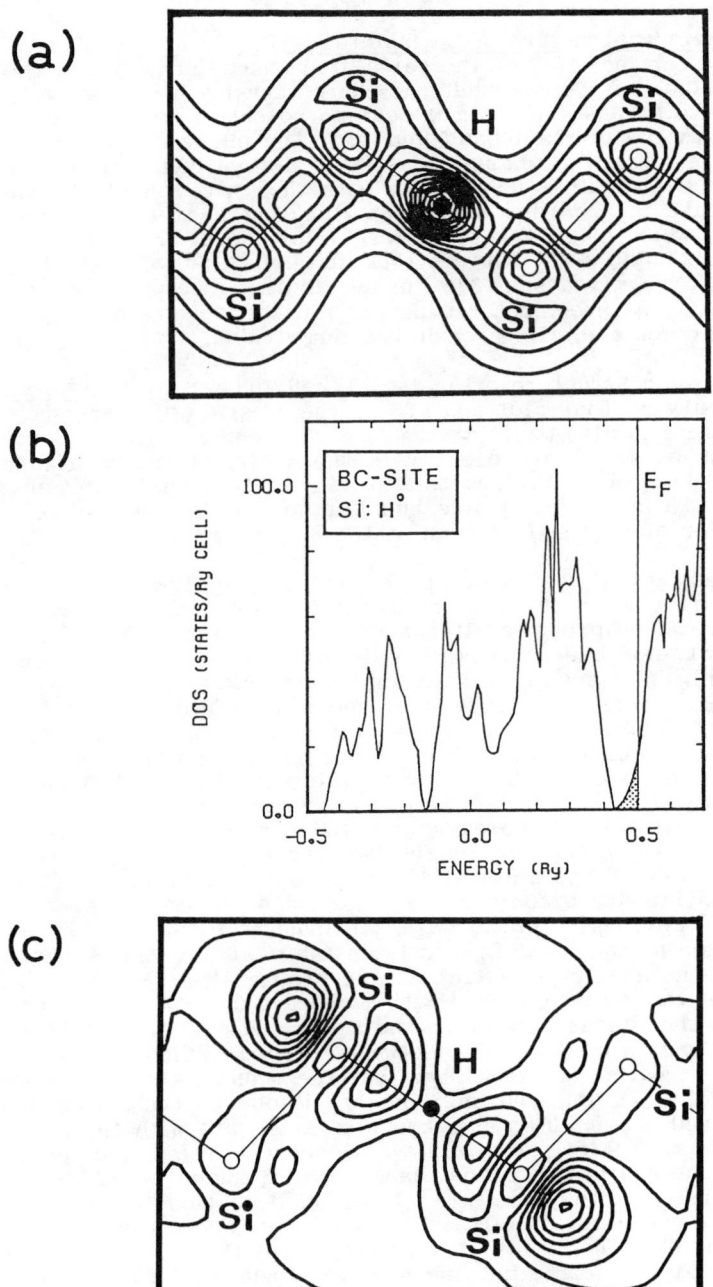

Fig. 2 (a)Contour plot of the calculated charge density in the (110) plane, (b)DOS and (c)the charge density of the gap state (shaded area in (b)) for the BC-site model. The width between lines in (a) and (c) are 0.1 and 0.005 electrons/A^3, respectively.

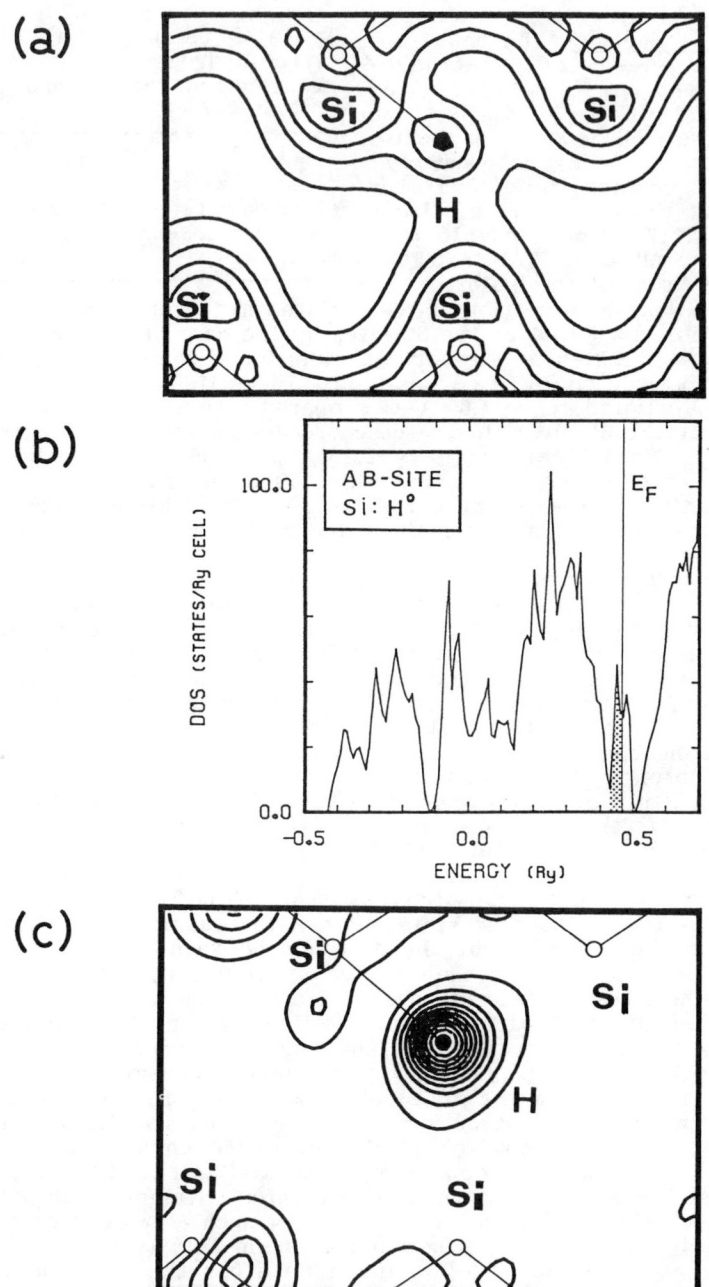

Fig. 3 Calculated results for the isolated hydrogen at the AB-site. Each figure corresponds to Fig. 2.

3. Hydrogen and Boron Complexes in Silicon.

3.1 Total Energy: The result of the total energy calculation of hydrogen and boron complexes for the BC- and AB-site models is shown in Table II. It is found that the total energy of the hydrogen atom at the BC-site between the substituted boron and the neighboring silicon atoms is 0.017 Ry more stable than that at the AB-site. This result is contrast to that of the isolated-

Table II Calculated total energies of the hydrogen and boron complexes in silicon at the BC- and AB-sites in unit of Ry/cell.

Si:(H-B)^0	BC-site	AB-site
electronic part	2.893	3.377
electrostatic part	-64.476	-64.943
total	-61.583	-61.566

hydrogen as shown in section 2; the isolated hydrogen atom at the AB-site is more stable than that at the BC-site. In the case of hydrogen and boron complexes at the BC-site, the boron atom is found to be relaxed 0.50 A from its ideal fourfold-coordinated position. The silicon-hydrogen bond distance is calculated to be 1.60 A as compared with 1.59 A for the isolated-hydrogen at the BC-site. In the case of AB-site, the bond length between the boron and silicon atoms is 2.15 A and the distance between the hydrogen and boron sites is 1.40 A which is much smaller than the value corresponding to the isolated hydrogen system. This configuration indicates the existence of a strong three-centered bond.

3.2 Electronic Structure of Hydrogen and Boron Complex at the BC-site: A contour plot of the calculated charge density and DOS are shown in Fig.4. We can see that a strong three-centered bond of the linear silicon-hydrogen-boron chain is formed and its bonding states lies deep in the valence band. There is an unoccupied gap state near the edge of the conduction band. Fermi level (E_F) for the neutral hydrogen and boron complex at the BC-site is located just in the band gap between the top of the valence band and the unoccupied gap state. The hydrogen-silicon and hydrogen-boron hybridizations are found to remove the electronic states associated with the boron acceptor to the valence band, and thus hydrogen atom neutralize the shallow acceptor of boron. We plot the charge density of the unoccupied gap states in Fig. 4(c). The charge density of the gap states is delocalized near the back-bond sites of silicon and boron atoms neighboring around the hydrogen. Therefore, the gap state is also regarded as the antibonding state between the dangling bond states which comes from the increases of the distance between the silicon and boron atoms, as was discussed for the isolated hydrogen at the BC-site in pure silicon. The bonding state strongly mixed with the hydrogen 1s orbital is pushed down into the valence band of silicon, making three-centered bond in silicon-hydrogen-boron chain, and neutralizing the shallow acceptor. This neutralization mechanism caused by the three-centered bond is completely different from the classical BC-model speculated from the experiment by Pankove et al. (1983) for the hydrogen passivation of shallow acceptors, because in Pankove's model the hydrogen atom makes a strong covalent bond with a silicon atom, terminating the dangling bond of silicon, and boron has threefold-coordinated covalent bonds with silicon atoms. Our calculated result does not show such a two-centered bond arrangement in the silicon-hydrogen-boron chain. This means that the classical two-centered bond model is inadequate for the explanation of the hydrogen passivation mechanism.

3.3 Electronic Structure of Hydrogen and Boron Complex at the AB-site: Figure 5 shows a contour plot of the calculated charge density and DOS for the AB-site model. We also can see that a strong three-centered bond of the linear hydrogen-boron-silicon chain is formed and the bonding

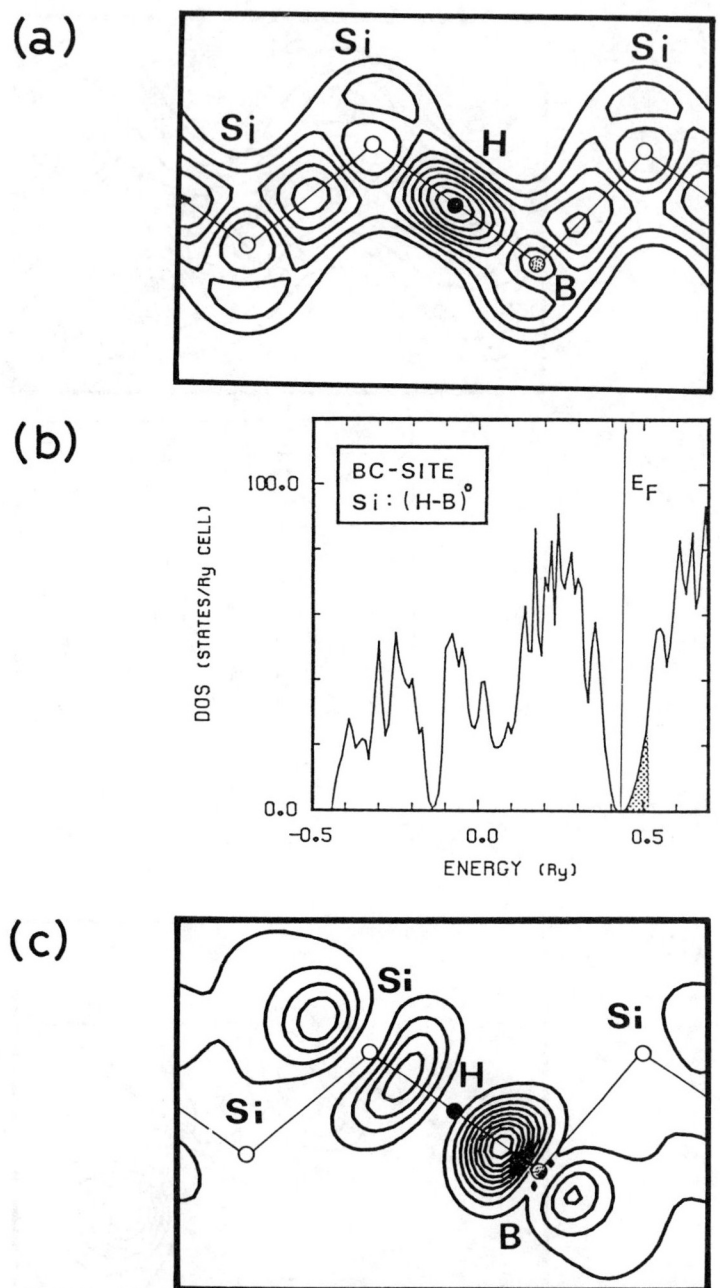

Fig. 4 Calculated results for the hydrogen and boron complex at the BC-site. Each figure corresponds to Fig. 2. The charge density map shown in (c) is the charge density of the shaded area in (b) and its width between lines are 0.01 electrons/A^3.

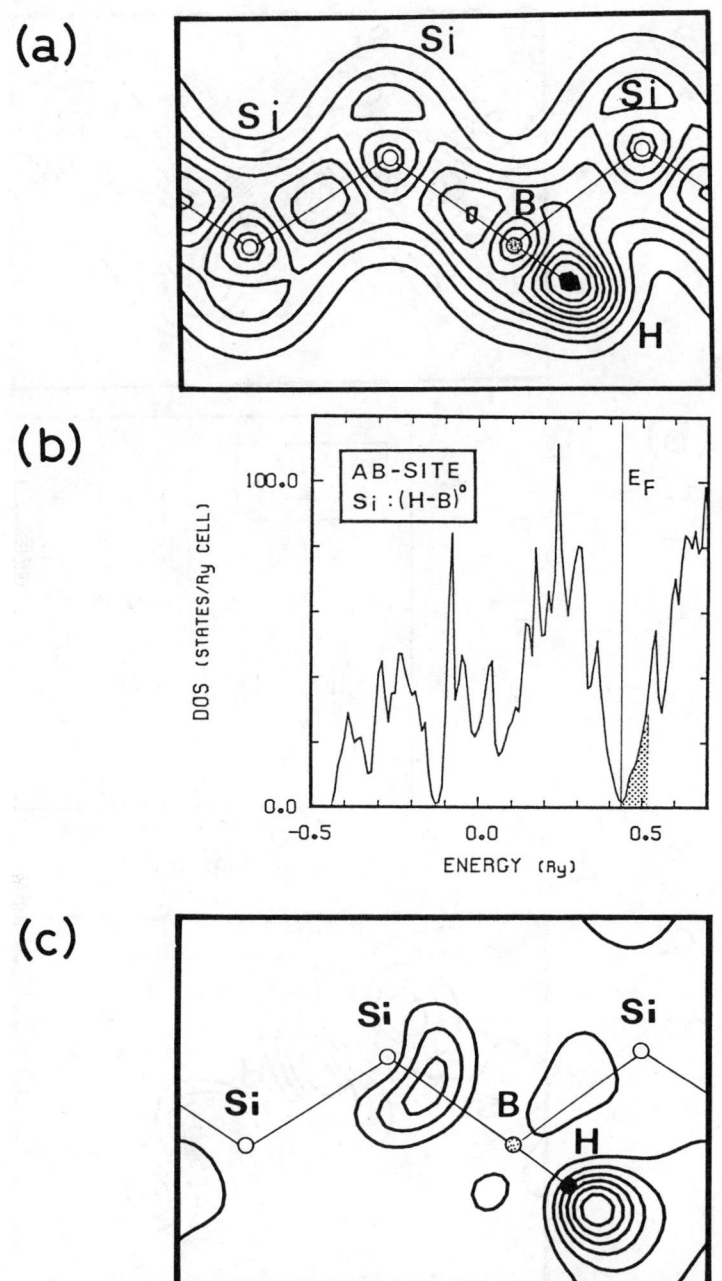

Fig. 5 Calculated results for the hydrogen and boron complexes at the AB-site. Each figure corresponds to Fig. 2 and (c) is similar to Fig. 4(c).

Hydrogen Passivation of Shallow Impurities

states lies deep in the valence band. In contrast with the case of BC-site, the unoccupied gap state can be seen more clearly in the upper part of the band gap. Figure 5(c) is the charge density from this gap state. We can see that the gap state is localized at the back-bond site of hydrogen. The hydrogen and boron complex at the AB-site is also possible mechanism for the hydrogen passivation of the neutralization, because three-centered bond in linear hydrogen-boron-silicon chain is formed and remove the electronic states associated with the boron acceptor to bonding states lying deep in the valence band. As was discussed in section 3.1, however, the total energy of hydrogen and boron complex at the BC-site is 0.017 Ry more stable than that of the AB-site. Therefore the BC-site model is favorable for the mechanism of the hydrogen passivation of neutralizing the shallow acceptors in silicon.

4. Schematic Explanation of the Hydrogen Passivation

We discuss the relation between the result of the total energy calculation and the calculated electronic structure. Figure 6 shows a schematic diagram of the electronic structures of the isolated neutral hydrogen and hydrogen-boron complexes in silicon. In this figure the energy levels in the valence band correspond to the three-centered bond states resonating in the valence band as mentioned in the section 2 and 3. In the case of the isolated hydrogen in pure silicon, a gap state appears for both the BC- and AB-site model and is occupied by one electron. Since the occupied gap state of the BC-site model exists near the conduction band edge the energy reduction with the three-centered bond is compensated. Thus the reduction of the electrostatic energy determines the stable site of the hydrogen atom. On the other hand, the gap states of the hydrogen-boron complexes are unoccupied for both the AB- and BC-site models. Hence the important part in the reduction of the total energy becomes the electronic part. That is, the creation of the three-centered bond is responsible for the reduction of the electronic energy and this bond for the BC-site model is much stronger than that for the AB-site model. Therefore the BC-site is more stable than the AB-site for the hydrogen and boron complex.

Recently, Chang and Chadi (1988) obtained the result of the calculation on the hydrogen and boron complexes which is similar to the present result. They reported that the stable site of the hydrogen and phosphorus complexes is an AB-site. Using our schematic diagram, their result

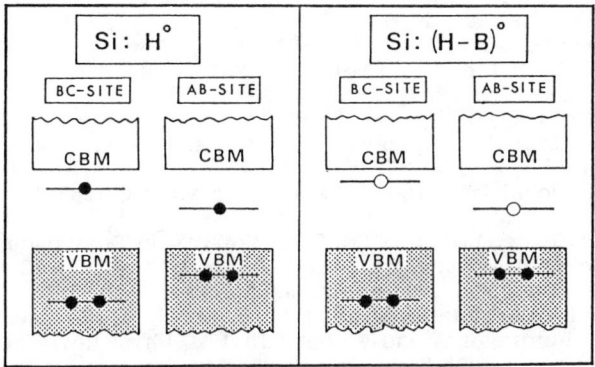

Fig. 6. Schematic diagram of the electronic structures of the isolated hydrogens and the hydrogen and boron complexes for the BC- and AB-site. VBM and CBM denote the valence band maximum and the conduction band minimum, respectively. The occupied states are shown by solid circles.

for phosphorus can be understood as follows: In the case of the hydrogen-phosphorus complex, the gap states are fully occupied and the electronic part is not largely reduced by the three-centered bond. Hence the AB-site model is more favorable than that for the isolated hydrogen system. Experimentally, the hydrogen passivation has not be seen in the case of the n-type silicon (Mikkelsen 1985). We speculate that the hydrogen and phosphorus complexes may have high total energy because of the large electronic energy contribution from the occupied gap states and therefore this complex may not be realized in the n-type silicon.

5. Conclusion

We have theoretically determined the atomic configuration and the electronic structure of hydrogen and boron acceptor complexes in silicon. We find that the BC-site model is more favorable than the AB-site model. The calculated electronic structure of the BC-site model shows that the hydrogen atom makes a strong three-centered bond with a silicon and boron atoms and thus passivates the shallow acceptor level. This mechanism is quite different from the classical BC-model based upon the two-centered bond model. The electronic structure of the AB-site model is also different from the ionic one which has been proposed and the three-centered bond picture is adequate. We have also given the site assignment of normal and anomalous muonium in silicon based upon the calculation of the isolated hydrogen.

This work is supported by a Grant-in-Aid for Scientific Research from the Ministry of Education, Science and Culture of Japan.

References

Assali L V C and Leite J R 1985 Phys. Rev. Lett. **55** 980
Assali L V C and Leite J R 1986 Phys. Rev. Lett. **56** 403
Bachelet B B Hamann D R and Schluter M 1982 Phys. Rev. B **26** 4199
Capizzi M and Mittiga A 1987 Appl. Phys. Lett **50** 918
Ceperley D M and B Alder 1980 Phys. Rev. Lett. **45** 566
Chang K J and Chadi D J 1988 Phys. Rev. Lett. **60** 1422
DeLeo G G and Fowler W B 1985 Phys. Rev. B **31** 6861
DeLeo G G and Fowler W B 1986 Phys. Rev. Lett. **56** 402
Estle T L 1985 Proc. 17th Int. Conf. Phys. Semiconductors ed by J D Chadi and W A Harrison (Springer-Verlag : New York) pp 685
Ihm J, Zunger A and Cohen M L 1979 J Phys C **12** 4401
Johnson N M 1985 Phys. Rev. B **31** 5525
Johnson N M, Herring C and Chadi D J 1986 Phys. Rev. Lett. **56** 769
Kiefl R F, Celio M, Estle T L, Kreitzman S R, Luke G M, Riseman T M and Ansaldo E J 1988 Phys. Rev. Lett. **60** 224
Mikkelsen J C (1985) Appl. Phys. Lett. **46** 882
Pankove J I, Carlson D E, Berkeyheiser J E and Wance R O 1983 Phys. Rev. Lett. **51** 2224
Pankove J I, Zanzucchi P J, Magee C W and Lucovsky G 1985 Appl. Phys. Lett. **46** 421
Pantelides S 1986 Proc. 14th Int. Conf. on Defects in Semiconductors et by H J von Bardeleben, Materials Science Forum Vols. 10-12 (Trans. Tech. Publications, Clausthal-Zellerfeld, West Germany) pp 573
Patterson B D, Hintermann A, Kundig W, Meier P F, Waldner F, Graf H, Recknagel E, Weidinger A and Wichert Th 1978 Phys. Rev. Lett. **40** 1347
Perdew J and Zunger A 1981 Phys. Rev. B **23** 5048
Sah C T, Sun J Y -C and Tzou J J -T 1983 Appl. Phys. Lett. **43** 204
Thewalt M L W, Lightowlers E C and Pankove J I 1985 Appl. Phys. Lett. **46** 689
Wichert Th, Skudlk H, Deicher M, Grubel G, Keller R, Recknagel E and Song L 1987 Phys. Rev. Lett. **59** 2087

Hydrogen diffusion and passivation of shallow impurities in crystalline silicon

C G Van de Walle, P J H Denteneer, Y Bar-Yam and S T Pantelides

IBM Research Division, Thomas J. Watson Research Center,
Yorktown Heights, NY 10598, USA

We use state-of-the-art theoretical techniques to analyze stable configurations, migration paths, charge states, etc., for H in pure and in doped Si. It is established that H acts as a donor in p-type material; the observed passivation of B acceptors is caused by compensation, followed by formation of B-H pairs. Analysis of the energy surface for the complex shows that the global energy minimum occurs for H at the center of a Si-B bond. H can move on a spherical shell around B with an energy barrier of only 0.2 eV. We also address H_2 molecule formation and H-induced damage.

1. INTRODUCTION

The interaction between hydrogen and silicon has presented a very challenging problem to both theorists and experimentalists. It has often implicitly been assumed that H would retain its atomic character in its interactions with bulk crystalline Si, i.e. no strong binding to the crystalline network would occur, and H would favor interstitial locations where the interaction with the Si charge density would be minimal. This point of view has led to the neglect of relaxation of the network in most of the earlier studies of the location of H in the Si lattice.

The role H plays in crystalline semiconductors has also emerged as an important technological problem, as discussed in a recent review by Pearton et al. (1987). Hydrogen has been known for a long time to saturate dangling bonds at defects by forming Si-H bonds. The fact that H can also passivate shallow impurities has only been appreciated more recently. Passivation of the electrical activity of p-type material was first observed in MOS capacitors by Sah et al. (1983a,b). Subsequent experiments by Pankove et al. (1983, 1984, 1985a) and by Johnson (1985a) unambiguously established the correlation between H and acceptor profiles, and the existence of H-acceptor pairs.

Passivation of shallow donors, on the other hand, was initially thought to be non-existent (Sah 1983b, Pankove 1984) or very weak (Johnson et al. 1986). Recently, however, Bergman et al. (1988) have provided conclusive evidence for passivation of samples doped with P, As, and Sb, which showed a reduction of up to 80% in carrier concentrations. This passivation, while dramatic, is still not as complete as that of p-type samples.

Understanding of the passivation mechanisms has been limited, large because reliable theoretical results were lacking. As Pantelides (1987) first pointed out, attempts to explain the observed phenomena led to a number of contradictory assumptions regarding the nature of the charge states of H along its diffusion path, and hence about the H-impurity reactions that can occur. Most of the previous theoretical studies were based on semiempirical Hamiltonians and/or small clusters, and produced contradicting results. An overview of the earlier literature has been included in a recent review by Patterson (1988).

Until recently, no experimental observations were available for isolated paramagnetic hydrogen centers. A large amount of experimental effort has been devoted, however, to the study of muonium, a pseudo-isotope of H. The muon-spin-rotation technique [see the review by Patterson (1988)] has allowed the observation of two paramagnetic forms of muonium: the so-called "normal" muonium, with an isotropic hyperfine interaction, and "anomalous" muonium with trigonal symmetry and a strong anisotropy of its hyperfine tensor. Because the muonium lifetime is only 2.2 μs, and because its mass is 1/9 that of H, caution should be taken in applying results about muonium to H. Normal muonium is usually associated with the tetrahedral interstitial site (T_d). Recently, anomalous muonium was shown to be located at the bond center (Kiefl et al. 1988). We will see that our theoretical results are consistent with these observations.

There has been one recent report of a paramagnetic hydrogen state, with indirect evidence that it would be associated with the bond center. ESR experiments by Gordeev et al. (1988) showed a paramagnetic state due to H in Si, called the AA9 center. They also showed that the characteristics of AA9 are similar to those of anomalous muonium. Since anomalous muonium is now known to be associated with the bond center (Kiefl 1988) (a fact not appreciated by Gordeev et al.) this provides indirect evidence for the presence of a paramagnetic H state located at the bond center site.

The present study, in contrast with most previous approaches, uses state-of-the-art parameter-free theoretical techniques which have proven their value in numerous studies of point defects and impurities in semiconductors. These techniques -- density-functional theory in the local-density approximation (LDA), norm-conserving pseudopotentials, and large supercells for the solution of the relevant Schrödinger equation -- will be described in Sec. 2. The results of the total-energy calculations describe the entire energy surface, where the position of the H atom is the coordinate. These surfaces are analyzed and displayed using a novel technique that provides immediate insight into the relevant processes. Section 3 will deal with the behavior of a single H atom in crystalline Si (Van de Walle et al. 1988). We explore the stable configurations and migration paths for H in its various charge states, including sites and paths that involve strong interactions with the Si network. We also determine the relative stability of the different charge states in intrinsic, n-type, and p-type material. This analysis already provides an explanation for the passivation mechanisms in doped materials. In particular, we will see that H acts as a donor in p-type material. The observed passivation of shallow acceptors can therefore be attributed to direct compensation. Pairing between H$^+$ and shallow impurities follows compensation.

The structure of the resulting H-acceptor complex will be addressed in Sec. 4, for the example of a boron (B) acceptor. The position of the H around the acceptor has been widely debated in the literature (Pankove 1983, Assali and Leite 1985, DeLeo and Fowler, 1985). Based on a complete analysis of the energy surface for the complex, we will conclusively show that the H occupies a bond-center position between B and Si. The previously proposed antibonding site corresponds to a saddle point in the energy surface and is therefore not stable. A direct connection with experiment will also be made through the calculation of H stretching vibrational frequencies.

Finally, in Sec. 5, we will explore some phenomena that arise from the interaction of two or more H atoms. This includes H_2 molecule formation, as well as cooperative reactions of several H atoms with the Si network. This allows us to examine several mechanisms which could be responsible for the recently observed hydrogen-induced damage in the near-surface region (Johnson *et al.* 1987).

2. METHODS

Our calculations are based on density-functional theory in the LDA, using *ab initio* norm-conserving pseudopotentials for Si and B, and the Coulomb potential for H (Van de Walle *et al.* 1988). The pseudopotentials were generated according to the Hamann-Schlüter-Chiang (1981) scheme. For B, the cutoff radii were adjusted so as to obtain convergence with the smallest possible number of basis functions in the plane-wave expansions, while still providing an accurate description of the Si-B-H interactions. We use supercells containing 32 Si atoms, such that the distance between neighboring defects is 9.4 Å. Relaxation of two shells of Si atoms surrounding the H impurity was included in the full calculations. Convergence as a function of supercell size was checked by performing calculations on supercells containing 8, 16, and 32 atoms. The 16-atom cells already provide adequate results in many cases; however, the 32-atom cells allow relaxation of more shells of neighbors around the impurity, and facilitate the extraction of band positions for an isolated defect. The position of this level was determined by taking a weighted average over the band position at the special points. It changed by less than 0.1 eV when the cell size was increased from 16 to 32 atoms. Two special points (or equivalent larger sets for less symmetric configurations) were used in the integrations over the first Brillouin zone.

We have performed extensive tests to establish the convergence as a function of the plane-wave basis set which is used for expanding wave functions and potentials. Inclusion of plane waves with kinetic energy up to $E_c = 12$ Ry (those above $E_c/2 = 6$ Ry being treated in second-order Löwdin perturbation theory) is sufficient for obtaining the general features of the energy surfaces. For configurations which are particularly sensitive to the energy cutoff (such as the bond-center site), we carried out calculations up to a cutoff of 18 Ry for H in Si, and 20 Ry for H-B complexes. This corresponds to basis sets of ~6000 plane waves in the 32-atom cells. The values for energy differences and barriers that will be quoted all result from these high-cutoff calculations. Test calculations at still higher cutoffs have established that our results for total-energy differences are converged to within 0.1 eV.

The main sources of uncertainty in the calculations are the well-known intrinsic deficiencies of the LDA, in particular the fact that LDA predicts conduction bands and hence conduction-band-derived energy levels to be too low. This uncertainty, combined with residual interactions between supercells, puts a substantial error bar (on the order of several tenths of an eV) on the position of any defect level. However, a qualitative distinction between deep and shallow levels can still be made. We also note that, while the absolute position of the defect level is uncertain, its relative motion induced by displacements of the impurity or by changes in the charge state is quite reliable. These observations will allow us to derive conclusions about the deep levels induced by hydrogen, as described later in the paper. Only in the section where we discuss the relative stability of various charge states as a function of the Fermi level position will we be confronted with the limitations of the LDA.

In this work, we will report local-density-functional results for total energies and defect levels; spin polarization (Katayama-Yoshida and Shindo 1983), which affects only the neutral charge state (with an unpaired electron), was investigated as follows. We carried out self-consistent spin-density-functional calculations, which are much more time-consuming, at selected sites. For the bond-center position, the inclusion of spin-polarization had very minor effects: the total energy went down by less than 0.02 eV, and the defect level was split by only 0.04 eV. The deviation from the spin-averaged results is expected to be largest for H at the tetrahedral interstitial (T_d) site, where the crystal charge density reaches its lowest value so that the impurity is most "free-atom"-like. We found that inclusion of spin polarization lowered the total energy only by 0.1 eV. The defect level was split into a spin-up and a spin-down level, which were separated by 0.37 eV. These results justify our use of spin-averaged calculations for deriving the properties of H in Si.

The total energies that are generated by our calculations describe the energy surface for a H atom in pure or B-doped Si. In order to analyze and display these results, we introduce a novel technique to generate energy surfaces. The central point is the realization that, in the case of H in pure Si, the energy surface has the full symmetry of the perfect crystal, and can therefore be expanded in a set of basis functions which exhibit this symmetry. A relatively small set of energy values at representative points is then sufficient to determine the expansion coefficients. At each of these points, the total energy of the impurity (in a particular charge state) is calculated for a configuration in which the surrounding Si atoms are completely relaxed. This database is then used to determine the expansion coefficients for the entire energy surface. We used symmetrized plane waves as basis functions and found that less than 10 stars of reciprocal lattice vectors provide sufficient accuracy. An example of the resulting energy surface for H^+ is shown in Fig. 1. Note that the Si relaxations for each position of the H atom are different but are not displayed in the figure. The energy surface for H in Si with a substitutional B atom is shown in Fig. 2. The procedure for generating this surface is similar as above, except that the basis functions now reflect the lower symmetry of the environment.

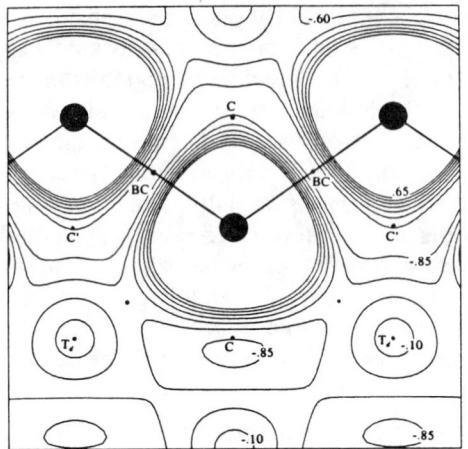 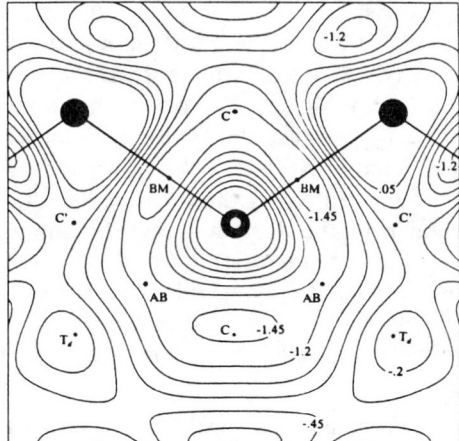

Fig. 1. Contour plot in the (110) plane of the total energy surface for H^+ in Si. The big dots indicate (unrelaxed) atomic positions and the small dots indicate high symmetry positions of special interest (see text). The unit of energy is eV and the contour step is 0.25 eV. Highest contour shown: 0.65 eV.

Fig. 2. Contour plot in the (110) plane of the total energy surface for H in Si with substitutional B (in the center of the plot). Indicators are as in Fig. 1. The BM site is the global energy minimum and the AB site is a saddle point. The unit of energy is eV and the contour step is 0.25 eV. Highest contour shown: 0.05 eV.

3. HYDROGEN IN PURE SILICON

The energy surfaces for H in the positive, neutral, and negative charge state exhibit a number of common features (Van de Walle et al. 1988). In all three charge states, there are two distinct regions in which the H atoms exhibit significantly different behavior. First, there is the region of high electron density, which includes the BC (bond-center) site, the C site (at the center of a rhombus formed by three adjacent Si and the nearest T_d), etc. In this region, the nearby Si atoms relax strongly. For example, when the H atom is placed at the BC site, the adjacent Si atoms relax out by 0.4 Å for a net gain in energy of more than 4 eV. Furthermore, in this high-density region, a H-induced defect level occurs in the upper part of the energy gap; it is identified as a state formed out of an antibonding combination of the broken bonds on the neighboring Si atoms. The second region consists of the low-electron-density "channels" and includes the high-symmetry tetrahedral (T_d) and hexagonal (H) interstitial sites. Here, the Si atoms in the vicinity of H relax very little if at all. Furthermore, a H-related level now occurs just below the top of the valence bands. The precise position of the defect levels changes only by ~0.1 eV as a function of charge state.

Figure 1 shows a contour plot of the energy surface in the (110) plane for a positively charged H (H$^+$). The global minimum is at the BC site, symmetrically located between two Si atoms. In contrast, the energy of H$^+$ in the low-density region is more than 0.5 eV higher. Of course, the state H$^+$ in the low-density region actually does not occur, because the H-related level which must be kept empty lies inside the valence bands. Note that the positive charge state does not imply that the H occurs as a bare proton; at the bond center, the missing charge is actually taken from the region near the Si atoms, corresponding to the antibonding state occurring in the band gap. A migration path in the (110) plane can be traced between the BC positions; the barrier along this path is less than 0.2 eV high. The saddle point occurs very close to the point indicated with C' in the figure; the points C are actually symmetry-related points along equivalent paths perpendicular to the plane of the figure.

It is of course impossible to pictorially represent the energy surface (a four-dimensional object) as a function of all three dimensions (although our choice of data points and our fitting procedure assure that we take the full three-dimensional character into account). The (110) plane and the indicated migration path should therefore only be considered as a representative example. In particular, we have also studied the behavior around the M-site, which is midway between two C-sites [only one of which lies in the (110) plane]. We find it to be at approximately the same energy as the bond center B, with no barrier between the two. The point M also lies on a line perpendicular to the Si-Si bond, connecting the bond center with the neigboring hexagonal interstitial site; all points between BC and M on this line have approximately the same energy. For these "buckled" configurations, the Si-H distance remains almost constant (equal to 1.6 Å), due to appropriate relaxation of the Si atoms. In all cases, the H prefers to be symmetrically located with respect to two Si atoms.

We have also generated the energy surfaces for H in the neutral and negative charge states. For neutral H, the same features and relative positions of local minima can be recognized as in the case of H$^+$, including a local minimum at the bond center. As for H$^+$, we have examined carefully whether there is any tendency for H^0 to preferentially bind to one of the Si neighbors, leading to an asymmetric configuration. We find that the symmetric situation is lowest in energy. The high-density path is again favored, but the low-density path is less than 0.2 eV higher. Thus, neutral H seems to be able to move rather freely through the network with very small energy barriers. We note that the tetrahedral interstitial site is a local maximum of the energy surface for H^0. Moving from T$_d$ towards a substitutional site, the energy first decreases and then increases in the <111> direction. However, the lowest energy in this antibonding direction (less than 0.1 eV lower than at T$_d$) does not correspond to a local minimum, but to a saddle point. For the case of negative H (H$^-$), the T$_d$ site is the lowest in energy, with the energy rising sharply outside the low-density regions. In particular, the BC site is now more than 0.5 eV higher in energy than the T$_d$ site. The barrier to migration along a path through the low-density region and going through the hexagonal interstitial site is 0.3 eV.

We now examine the *relative* energies of the different charge states of the H, in order to determine the lowest-energy state. These relative energies depend on the position

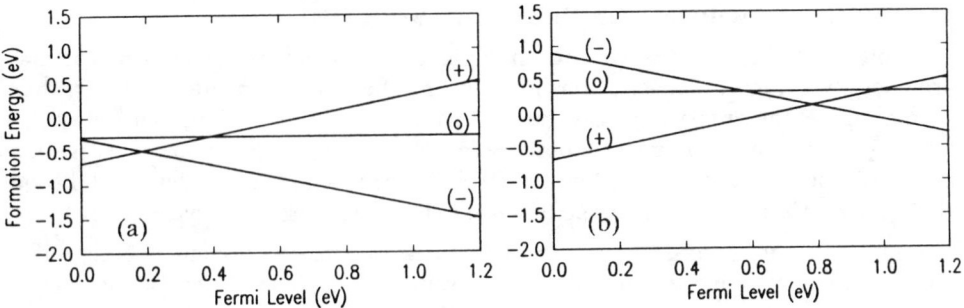

Fig.3. *Relative formation energies for different charge states of a H interstitial impurity in Si. a) shows the straight LDA values, while b) results from applying a simple correction scheme to the energy levels. The zero of energy is arbitrarily chosen as the energy of H^0 at T_d.*

of the Fermi level, with which electrons are traded in order to alter the charge state of the defect. Figure 3 shows the relative formation energies for different charge states, as a function of Fermi level position. To simplify the plot, we only show the formation energies for the impurity positions which correspond to the global minimum for a particular charge state, i.e. BC for H^+ and H^0, and T_d for H^-. Figure 3a shows the values directly obtained from the LDA calculations. As pointed out above, these suffer from an uncertainty in the position of the defect level. Rigorous calculational schemes which could eliminate these uncertainties by going beyond the LDA are presently too complex to apply to defect calculations. We have therefore applied a very simple *a posteriori* correction, amounting to a rigid shift of the defect level together with the conduction bands, to bring the band gap into agreement with experiment. For the BC position of the impurity, this should be a good approximation since the defect-induced level is clearly conduction-band derived. The result of this procedure is shown in Fig. 3b. The energies are shifted now, according to the number of electrons present in the level.

In p-type material (Fermi level at the top of the valence band), the lowest-energy state is H^+ in the high-density region; thus, H^+ diffuses via the high-density path and exhibits donor-like behavior. These conclusions are unambiguous and independent of any error bars in our LDA calculations. This result confirms the suggestion that the passivation of p-type material is a direct result of compensation, i.e. electrons from neutral H-atoms annihilate the free holes in the valence band (Pantelides 1987). Pairing between H^+ and ionized acceptors follows compensation, as discussed in the next section.

From Fig. 3 we see that our calculations predict H to be a negative-U impurity (Car *et al.* 1984). In p-type material, the stable state is H^+ in the high-density region; as the Fermi-level is raised, however, the stable state becomes H^- in the low-density region. H^0 is not the stable state for any Fermi level. However, the uncertainty in the LDA energy levels (and in our simple correction procedure) makes the error bar too large to unambiguously exclude the occurrence of H^0.

4. HYDROGEN-BORON COMPLEXES IN SILICON

From our studies of H in Si, we can unambiguously conclude that H in p-type material acts as a donor. The electrons given up by the H compensate free holes, thus accounting for the observed passivation. This mechanism provides a natural explanation for numerous experimental observations, including suppression of passivation in counter-doped material (Johnson 1985a), electric-field dependence (Johnson 1985b), and effects of a thin n-type overlayer (Pankove 1985b, Johnson 1985b).

The final result of the passivation mechanism is the formation of neutral acceptor-H pairs. We will examine the microscopic structure of such complexes for the case of B, the most widely used and studied acceptor in Si. Previous theoretical studies were controversial regarding the position of the H atom: semi-empirical cluster calculations predicted either a position near the bond center of a Si-B bond (DeLeo and Fowler 1985), or an antibonding position (AB) on the extension of such a bond (Assali and Leite 1985). Both sites are indicated in Fig. 2. We label the position near the bond center with "BM" to distinguish it from the geometrical bond center BC.

The contour plot shows the results of total-energy calculations at a large number of sites. It is clearly seen that the AB site is not a minimum, but a saddle point of the energy surface; it is 0.45 eV higher in energy than the BM site, which is the global energy minimum. At the BM site, the nearest B and Si atoms relax outward by 0.24 Å and 0.42 Å, respectively (these displacements are not shown in Fig. 2). The B atom creates an attractive well for the H; the region around it is indeed lower in energy than in Fig. 1 (the zero of both energy scales is a T_d site far away from the B atom). Studying the energy surface in other planes as well, we find a low-energy region for the H that is a spherical shell (with holes) at a radius of ~1.3 Å around the (relaxed) B atom. The H atom can move between equivalent BM sites by passing over an energy barrier of only 0.2 eV, close to the C site (see Fig. 2), if the relaxation of the host crystal adjusts accordingly.

If we assume that the complex breaks up into an ionized acceptor B^-, a free hole (which restores the conductivity), and a neutral H atom, we calculate a dissociation (or binding) energy of 1.1 eV. We have also calculated the vibrational frequencies for the H stretching mode. We find 1830 cm^{-1} for H at BM, and 1680 cm^{-1} for H at AB. Considering the error bar of ~100 cm^{-1}, the value for the BM site is in reasonable agreement with experiment (Stavola et al. 1988). Although the value for H at BM agrees very well with previously calculated values (DeLeo and Fowler 1985, Chang and Chadi 1988), the value for H at AB is close to the one at BM and disagrees with previously calculated values.

5. INTERACTIONS OF SEVERAL H ATOMS

First, we examine how two neutral H atoms may combine and form an H_2 molecule. We have found the minimum energy position for the molecule straddling the T_d site, oriented in the <100> direction, with the atoms separated by 0.86 Å (to be compared with 0.75 Å in vacuum). At the hexagonal interstitial site, which would lie on a migration path, the energy of the molecule is 1.1 eV higher. The binding energy of H_2 (as compared with isolated neutral H atoms at their lowest interstitial position, i.e. at the bond center) is 2±0.5 eV per molecule, or ~1 eV per atom.

Another phenomenon that involves the cooperative interaction of several H atoms with the Si atom is related to the recent observation (Johnson et al. 1987) that hydrogenation can induce microdefects in a region within ~1000 Å from the surface. These defects, studied with transmission electron microscopy (TEM), have the appearance of platelets along {111} crystallographic planes, range in size from 50 to 100 Å, and exhibit no net Burgers vector. Raman measurements indicated that essentially all of the H in the region was incorporated in Si-H bonds.

We have examined several possibilities for the structure of these platelets, by performing total-energy calculations in a superlattice geometry; edge effects at the platelet boundary are thus neglected. First, we explore the situation in which one H atom is inserted into each of the Si-Si bonds of a (111) plane. By inspection of the total energy for the relaxed configuration we find that the energy per H is more than 0.5 eV higher than it is for the isolated impurity, i.e the formation of this type of extended defect is unfavorable.

Another possibility for extended defect formation is the insertion of two H atoms in each Si-Si bond, i.e. the formation of two Si-H bonds out of each Si-Si bond. It is essential to place the H atoms off the Si-Si axis in order to find a favorable configuration. A representative position is for the H atoms at two M-sites associated with each Si-Si bond. The energy per H atom is now similar to that for isolated atoms. This structure would therefore be unstable to H_2 molecule formation. We conclude that these proposed configurations are energetically not favorable.

We have therefore examined a different type of mechanism, based on the removal of Si atoms from the defect region, with the resulting dangling bonds tied off by H atoms. This is based on our calculated result that H atoms can assist Frenkel-pair creation. In a perfect crystal, the creation of a Frenkel pair (vacancy-interstitial pair) normally costs about 8 eV (Bar-Yam and Joannopoulos 1985, Car et al. 1985). If, however, a sufficient number of H atoms are available in the immediate neighborhood of a particular Si atom, Frenkel-pair formation can actually be exothermic with a slight gain of energy. In the final configuration, a self-interstitial is emitted while four H atoms tie off the dangling bonds of the vacancy. The calculated energy gain for the process in which a neutral interstitial H atom passivates a dangling bond is ~2.2 eV per Si-H bond. This energy value was confirmed in a superlattice calculation modelling an extended defect in which a double row of Si atoms was removed in a {111} plane, with all dangling bonds tied off by H.

These theoretical results for the interaction of several H atoms lead us to the following conclusions. On the basis of energetic considerations, H_2 molecules are the preferred state for several neutral H atoms in pure crystalline Si. Kinetic considerations also suggest that H-assisted Frenkel-pair creation would be a rare event. However, H-assisted ejection of threefold- or twofold-coordinated Si atoms is kinetically more favorable, such that enlargement of a pre-existing defect is likely. The particular atomistic processes that lead to defect nucleation and enlargement cannot be described in more detail at this point; however, the energetic arguments given above for defect formation and extension suggest the vacancy-formation mechanism is likely to be involved in the observed hydrogen-induced damage.

ACKNOWLEDGMENT

This work was supported in part by ONR under Contract No. N00014-84-C-0396.

REFERENCES

Assali L V C and Leite J R 1985 *Phys. Rev. Lett.* **55** 980
Bar-Yam Y and Joannopoulos J D 1985 *J. Electron. Mater.* **14a** 261
Bergman K, Stavola M, Pearton S J and Lopata J 1988 *Phys. Rev. B* **37** 2770
Car R, Kelly P J, Oshiyama A, and Pantelides S T 1984 *Phys. Rev. Lett.* **52** 1814
Car R, Kelly P J, Oshiyama A and Pantelides S T 1985 *J. Electron. Mater.* **14a** 269
Chang K J and Chadi D J 1988 *Phys. Rev. Lett.* **60** 1422
DeLeo G G and Fowler W B 1985 *Phys. Rev. B* **31** 6861
Gordeev V A, Gorelkinskii Yu V, Konopleva R F, Nevinnyi N N, Obukhov Yu V and Firsov V G 1988 *to be published*
Hamann D R, Schlüter M and Chiang C 1979 *Phys. Rev. Lett.* **43** 1494
Johnson N M 1985a *Phys. Rev. B* **31** 5525
Johnson N M 1985b *Appl. Phys. Lett.* **47** 874
Johnson N M, Herring C and Chadi D J 1986 *Phys. Rev. Lett.* **56** 769
Johnson N M, Ponce F A, Street R A and Nemanich R J 1987 *Phys. Rev. B* **35** 4166
Katayama-Yoshida H and Shindo K 1983 *Phys. Rev. Lett.* **51** 207
Kiefl R F, Celio M, Estle T L, Kreitzman S R, Luke G M, Riseman T M and Ansaldo E J 1988 *Phys. Rev. Lett.* **60** 224
Pankove J I, Carlson D E, Berkeyheiser J E and Wance R O 1983 *Phys. Rev. Lett.* **51** 2224
Pankove J I, Wance R O and Berkeyheiser J E 1984 *Appl. Phys. Lett.* **45** 1100
Pankove J I, Zanzucchi P J and Magee C W 1985a *Appl. Phys. Lett.* **46** 421
Pankove J I, Magee C W and Wance R O 1985b *Appl. Phys. Lett.* **47** 748
Pantelides S T 1987 *Appl. Phys. Lett.* **50** 995
Patterson B D 1988 *Rev. Mod. Phys.* **60** 69
Pearton S J, Corbett J W and Shi T S 1987 *Appl. Phys. A* **43** 153
Sah C T, Sun J Y C and Tzou J J T 1983a *Appl. Phys. Lett.* **43** 204
Sah C T, Sun J Y C and Tzou J J T 1983b *J. Appl. Phys.* **54** 5864
Stavola M, Pearton S J, Lopata J and Dautremont-Smith W C 1988 *Phys. Rev. B* **37** 8313
Van de Walle C G, Bar-Yam Y and Pantelides S T 1988 *Phys. Rev. Lett.* **60** 2761

Hydrogen migration and complex formation in silicon

N. M. Johnson and C. Herring*

Xerox Palo Alto Research Center, Palo Alto, California 94304, USA

> ABSTRACT: New experimental results are presented which support the conclusion that hydrogen migrates in different charge states and forms stable and perhaps also metastable diatomic complexes in silicon. In addition, results are presented from *in-situ* ("real-time") electrical measurements of hydrogen migration and dopant neutralization in silicon.

1. INTRODUCTION

Two fundamental questions head the list of issues that continue to challenge our understanding of hydrogen in crystalline silicon: (1) does hydrogen diffuse as a charged species† and (2) does molecular hydrogen form at moderate temperatures (e.g., <300°C)? Previous experimental results for hydrogen diffusion and shallow-acceptor neutralization in p-type silicon have been interpreted in terms of a deep-donor model, with the hydrogen donor level well within the bandgap, such that boron neutralization proceeds by the reaction $H^+ + B^- \rightarrow (BH)^0$ (Tavendale, Alexiev and Williams 1985, Johnson 1985, Capizzi and Mittiga 1987 and Pantelides 1987). In n-type silicon, hydrogen neutralizes shallow-donor dopants (Johnson et al. 1986, Johnson, Herring and Chadi 1987, Bergman et al. 1988), and theory suggests that this arises from the formation of a complex involving the dopant and H^- (Johnson et al. 1986, Chang and Chadi 1988). It thus appears that interstitial hydrogen complexes may exist in any of three charge states, that is, as an amphoteric center whose stability regions define a donor and an acceptor level within the bandgap. Other properties, such as diffusion and chemical reactivity, should then depend on the Fermi level. However, more convincing evidence is required for these charge states.

The situation is more desperate for molecular hydrogen. Many theoretical studies have examined possible H_2 configurations (Chang and Chadi 1988, Shi et al. 1984, Mainwood and Stoneham 1984, Van de Walle et al. 1988, Deák et al. 1988); most conclude that H_2 at a tetrahedral interstitial site is more stable in the perfect lattice than the separate monatomic species. However, to date there is no direct experimental evidence for the existence and properties of H_2 in crystalline silicon; such an identification is rendered difficult by the anticipated electrical, optical, and paramagnetic inactivity of H_2. We have recently presented evidence for a non-monatomic species of hydrogen that forms during hydrogen diffusion in p-n junctions (Johnson and Herring 1988a). This new H complex is charge neutral, immobile at moderate temperatures (i.e., ≤350°C), and does not incorporate dopant impurities. The results are interpreted with the

hypothesis that after migrating largely as H^+, hydrogen can be immobilized through a chemical reaction in which H^+ is one of the reactants. The immobile species itself is likely to be molecular hydrogen.

In this paper we present some key experimental results which further support the amphoteric model for diffusing hydrogen in silicon and which suggest the formation of stable and metastable diatomic hydrogen complexes. The new data include results from hydrogen migration through p-n junctions and post-hydrogenation anneals. The hydrogenation system utilized a remote microwave plasma source so that the material was isolated from both charged-particle bombardment and illumination from the plasma. Depth profiles of the total hydrogen (i.e., deuterium) concentration were measured by secondary-ion mass spectrometry.

All experimental studies of hydrogen passivation in silicon have, to date, been based on electrical or spectroscopic measurements conducted after the hydrogen has been introduced into the material and generally after it has formed complexes that are stable at room temperature. In this paper we also present the first demonstration of electrical conductivity changes *during* hydrogenation. The principal value of such "real-time" measurements is the ability to study the chemically-reactive species as it migrates through the material rather than attempting to infer the properties of this entity from the formation kinetics of various stable hydrogen-related complexes.

2. p-TYPE SILICON

Charge-state or Fermi-level effects during hydrogen migration in Si have recently been demonstrated in a study of the variation with depth of the total hydrogen concentration in p-n junction structures (Johnson and Herring 1988a). Depth profiles of deuterium across n^+-p junctions are shown in Figure 1 for two diodes that were deuterated either with a floating potential or with an applied reverse bias V_R = 10 V. The n^+-layer was formed by P^+ implantation into p-type(1×10^{17} B/cm^3) Si followed by a furnace anneal, which located the metallurgical junction at an estimated depth of 0.5 μm. Two features in the profiles are extraordinary. First, in each case a prominent peak appears on the p-side of the junction. The peak is shifted to a greater depth in the reverse-biased diode. The shift is consistent with the peak being located near the edge of the depletion layer which increases in width upon

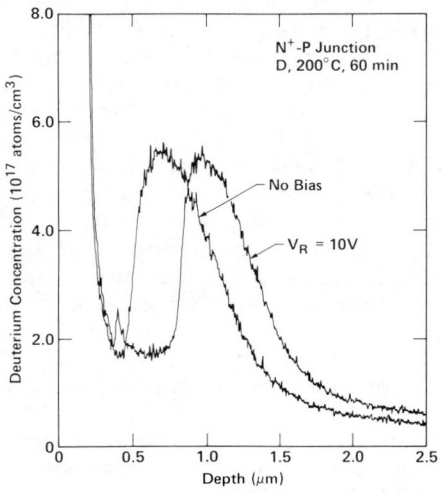

Fig. 1. Effect of a reverse bias V_R applied during deuteration (200°C, 1 h) on the depth profile of D in n^+-p junction diodes.

application of the reverse bias. Secondly, for the reverse-biased diode, a small step appears within the depletion layer near the metallurgical junction. While this step is only slightly above the noise, it is very reproducible: the step is never seen in a floating or zero-biased junction but only when a reverse bias is present. As V_R increases in magnitude, the small step shifts within the depletion layer toward the n-layer, although the large peak shifts away from the n-layer as the width of the depletion layer increases. For both the peak (or large step) and the small step in the profiles, the D concentration significantly exceeds the B concentration, so that the phenomena cannot be ascribed merely to the immobilization of D at B acceptor sites.

We have argued (Johnson and Herring 1988a, 1988b) that most of the excess hydrogen in both the large peak and the small step must be in the form of highly immobile neutral entities. In brief, neutrality is essential if overcompensation of the boron is to be avoided in the large peak, and immobility is required by the steepness of onset of this peak and has been directly demonstrated by the persistence of both features during subsequent vacuum anneals, a phenomenon we shall discuss in detail below. Essential to the formation of these marked features, however, is that they cannot represent concentrations in thermal equilibrium with a presumably rather mobile neutral species H^0, because in such case they would not be significantly affected by the electric field in the junction. Rather, they must represent a neutral species that is in the process of building up via process involving H^+: the concentration of the immobile species, which we presume to be H_2, reflects primarily the rate of a process $H^0 + H^+ \rightarrow H_2 + h^+$, where h^+ is a free hole, and therefore is responsive to the field-dependent concentration of the H^+ species.

According to the model proposed, emission and absorption of free carriers by the monatomic hydrogen species establishes a local equilibrium between the charge states of the latter in each region of an unbiased junction, in which

$$n_+ / n_0 = (1/2) \exp [(E_D - E_F) / kT], \qquad (1)$$

where n_+ and n_0 are, respectively, the concentrations of H^+ and H^0, E_D is the hydrogen donor level, E_F is the Fermi level, and the 1/2 allows for spin degeneracy. The ratio (1) rises enormously as one goes from the n-side to the p-side of the junction, so that, in spite of a gradual decrease of n_0 with depth due to the finite diffusion rate, the product $n_+ n_0$, and hence the rate of production of H_2, can increase sharply.

The seemingly mysterious appearance of the small step near the n-side of the depletion layer in the biased junctions actually accords nicely with this model(Johnson and Herring 1988a). In the presence of a reverse bias, (1) no longer holds in the depletion region, where both electrons and holes are swept out and are no longer in equilibrium. Rather, the ratio n_+ / n_0 becomes equal to the ratio of the lifetime τ_{+0} for emission of a hole by H^+ to the lifetime τ_{0+} for emission of an electron by H^0, these being the only processes by which the charge state can change in the absence of carriers to absorb. This ratio is constant across the depletion region. Thus, n_+ / n_0 starts to rise near the left (n-)

side of the junction, levels off when it reaches τ_{+0}/τ_{0+}, and resumes its rise near the right (p-) side, where holes become available to be absorbed. The first stage of this rise causes the small step. Increasing the reverse bias should shift this step to the left, and indeed, a shift of about the right magnitude is observed (Johnson and Herring 1988a). Incidentally, the sharpness of the step attests to the rapidity of the electronic equilibration (Johnson and Herring 1988a).

Most of the excess hydrogen is in the form of highly immobile neutral entities. We have proposed (Johnson and Herring 1988a) that these are hydrogen pairs, H_2. Formation of H_2 by the conventional reaction $2H^0 \rightarrow H_2$ is, by itself, incapable of accounting for the structure in the depth profiles, and it seems necessary to conclude that the competing reaction $H^+ + H^0 \rightarrow H_2 + h^+$ where h^+, denotes a free hole, becomes dominant in p-type regions, at least at 200°C. As noted above, local electronic equilibration of H^0 with H^+ seems to occur much more rapidly than the time scale (1 h) of our experiments.

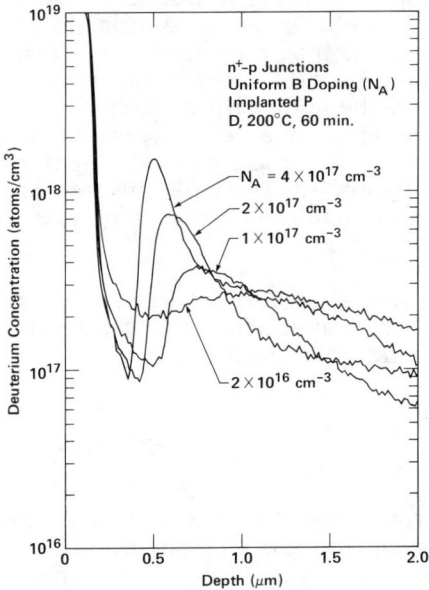
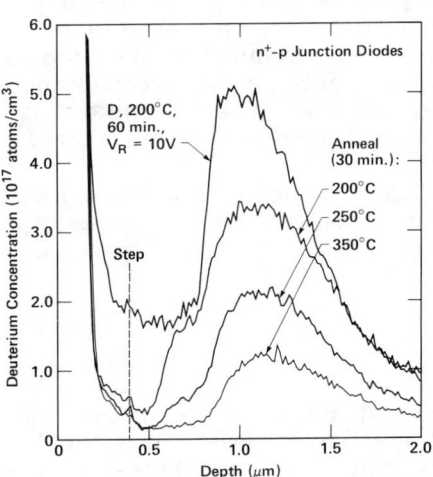

Fig. 2. Dependence of the D depth profile on the uniform B concentration in P-implanted $n+$-p junction structures.

Fig. 3. Post-deuteration anneals on $n+$-p junction diodes. The diodes were deuterated under reverse bias V_R and vacuum annealed without bias.

The interpretation just presented for Figure 1 was originally proposed (Johnson and Herring 1988a) on the basis of experiments on $n+$-p junctions constructed on substrates of a single conductivity type, namely, doped with 1×10^{17} B/cm³. According to the model described above, the height of the main peak in

deuterium concentration just inside the p-region adjacent to the depletion layer should increase as the boron concentration is increased, since the higher doping will increase n_+/n_0, hence, (if n_0 does not change greatly) n_+ and the reaction rate of $H^+ + H^0 \rightarrow H_2 + h^+$ will also increase. Figure 2 shows that this expectation is indeed fulfilled over the range of acceptor concentration N_A from 2×10^{16} to 4×10^{17} cm^{-3}. Indeed, over the higher part of this range, the peak height seems to be about proportional to N_A, as would be expected if the boron compensation is nearly complete and $n_0 \ll n_+$. However, quantitative conclusions must await a detailed analysis of these data.

Considerable insight into the species of hydrogen that is responsible for the various features in Figure 1 can be obtained from studies of the way these features change when subjected to vacuum anneals. Figure 3 shows a typical D depth profile resulting from deuterating (200°C, 1 h) an n^+-p junction *under a reverse bias* of 10V, and the distributions of D that remain after similarly deuterated specimens have been subjected to vacuum anneals at various temperatures, *with the bias removed*. Although only slightly above the noise, the most immediately revealing feature in Figure 3 is the behavior of the small step near the n-side of the depletion layer. The sharpness of the step does not change preceptibly during 200° and 250°C anneals, nor does its height above background decrease significantly, although the initial height is admittedly hard to estimate because of the noise. Thus, the sharp rise in concentration that is revealed by the step must be due to deuterium in a form so immobile that it diffuses less than a few hundred Angstroms in 0.5 h at 250°C, and so stable that most of it fails to dissociate in this time.

These conclusions are confirmed by the behavior of the large-scale features to the right of the depletion region. While a detailed quantitative model of these features remains to be developed, we shall offer here a few reasonable guesses at the contributions of different hydrogen species to these features, and show that these comprise a consistent picture. Consider first the original "main peak" near a depth of 1 μm. The contribution of H^+ to this peak, plus undissociated $(HB)^0$, is probably close to 9×10^{16} cm^{-3}, as is indicated by room-temperature capacitance-voltage profiles of the carrier concentration (Johnson and Herring 1988a), which show that ~90% of the boron is neutralized under the deuteration conditions of Figure 3. The height of the top curve within the depletion region probably represents mostly a mobile neutral species, presumably H^0, since most of this disappears with even a 200°C anneal. Thus, it would be reasonable to suppose a background H^0 contribution of roughly 1.4×10^{17} cm^{-3} in the region near 1 μm. Though these guesses give only a very crude estimate for the ratio n_0/n_+ in (1) above, we may note that the implied value of E_D, ≥ 0.2 eV below midgap, is not unreasonable. Subtracting these estimated concentrations of H^0 and H^+ [including any possible $(HB)^0$] from the observed total H at the main peak leaves $\sim 2.7 \times 10^{17}$ cm^{-3} for the contribution of deuterium in stable molecular complexes. After the 200°C anneal this may have changed by no more than a few percent, as H^+ and H^0 could still amount to a few times 10^{16} cm^{-3}. (The ratio of n_+/n_0 in the p-region will increase as loss of mobile H lessens the degree of compensation.) After the 250°C anneal, perhaps a quarter of the original molecular contribution appears to have been lost, or perhaps considerably less, if we overestimated the starting molecular contribution. More certain is the conclusion that after the 350°C anneal, half or more has been lost.

Note that the dissociation time of the molecules in the region of the main peak should be shorter than that in the small step on the left, because the process $H_2 + h^+ \rightarrow H^0 + H^+$ should be dominant over $H_2 \rightarrow 2H^0$ in the former region.

Concurrent with the above changes, the annealings produce and eventually erase a shoulder in the total H profile in Figure 3 near 0.6 μm, the edge of the depletion region for the zero-bias condition. The height of this shoulder is consistent with its being predominantly due to H^+, which after the 200°C anneal may well neutralize most of the boron in this region, with small contributions from background H^0 and newly-formed H_2. The latter is probably not very large, however: the product $n_+ n_0$ must be about an order of magnitude smaller than in the main peak of the unannealed specimen, and the time is only half as great.

The steepness of the leading edge of the remaining (molecular) deuterium after the 350°C anneal shows that this deuterium species cannot have diffused more than a few hundred Angstroms, just as we concluded above from the steepness of the small step. This yields an upper limit for the diffusion coefficient of molecular hydrogen in p-type Si of $\sim 10^{-14}$ cm²/sec at 350°C. Indeed, it appears that the removal of molecular hydrogen from the peak is dominated by dissociation of the complex and diffusion of monatomic species, rather than by diffusion of the complex itself.

3. n-TYPE SILICON

In Figure 4 is shown the effect of vacuum anneals on the depth profile of deuterium that was diffused into n-type Si (1×10^{17} P/cm³) at 150°C for 30 min; three different annealing conditions are illustrated. At first sight these curves seem very easy to interpret: beyond the first ~0.1 μm, they are in rough quantitative accord with the predictions of simple diffusion theory for diffusion of a single neutral species with a diffusion coefficient $\approx 1.8 \times 10^{-13}$ cm²/sec at 150°C, and the surface concentration equal to 2×10^{18} D/cm³ during deuteration and to zero during the vacuum anneals. [The surface peak is due to H-induced microdefects (Johnson, Ponce, Street and Nemanich 1987).] However, if nearly all the measured hydrogen is of a single species, this species must surely be neutral, since its concentration is up to an order of magnitude higher than that of the phosphorus. Actually, an appreciable fraction of negatively-charged hydrogen is expected on the

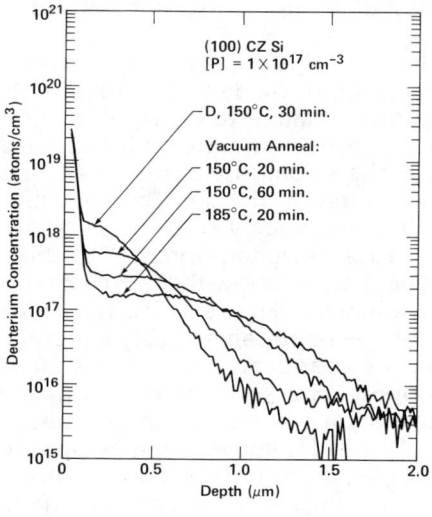

Fig. 4. Post-deuteration anneals in (100)-oriented, CZ, n-type (1×10^{17} P/cm³) Si.

basis of experiments on the variation of the hydrogen concentration with n-type doping (Johnson and Herring 1988b); by compensating the P this should cause band bending and departure from Fickian diffusion. Also, the presence of neutral atomic hydrogen in such high concentrations would have to be reconciled with the universal failure, to date, of attempts to observe the electron spin resonance of this species at or above room temperature. One could try to avoid these difficulties by postulating that most of the hydrogen in the present experiment is in the form of neutral diatomic complexes which dissociate at the temperature of 300°C that was used in the experiments of Johnson and Herring 1988b. These could not be the stable "molecules" found in the p-type material discussed above, since the latter are highly immobile; rather, one would have to postulate a different, metastable, complex H_2^*. This would probably have to be mobile as a unit, since assuming the diffusive transport to be dominated by a small fraction of monatomic species in equilibrium with the diatomic one would imply an effective diffusion coefficient strongly dependent on the concentration, hence non-Fickian profiles.

One firm conclusion can be drawn, however: the stable, immobile, molecular species that is found in p-type Si after deuteration at 200°C (see previous section) does not form appreciably in moderately n-type Si at 150°C. This implies (if the species in question is H_2) either that the reaction $2H^0 \rightarrow H_2$ has to surmount a sizable activation barrier, or that there is a similar barrier against $H^0 + H_2^* \rightarrow H_2 + H^0$, the barrier in either case being absent (or much less) if an H^0 is replaced by an H^+.

4. *IN-SITU* ELECTRICAL MEASUREMENTS ON SILICON ON SAPPHIRE (SOS)

In-situ electrical measurements require special design of the hydrogenation system and the fabrication of electrical test devices. The remote microwave plasma source offers the advantage of isolating the device from both the charged particles and the intense light of the plasma; a further serious requirement is minimizing convective heating of the device from the downstream products of the plasma. The heated substrate holder is equipped with electrical leads for contacts to the device. To monitor the electrical conductivity, a four-terminal van der Pauw structure is used, which eliminates complications from hydrogen-induced changes in the contact resistance. Sensitivity to migrating hydrogen is enhanced by restricting the current to a semiconducting surface layer. This is achieved by either isolating the surface layer with a p-n junction or using a heteroepitaxial layer such as single-crystal (SOS). In an alternative structure, the capacitance of a reverse-biased p-n junction diode can be monitored *in situ* to directly detect hydrogen-induced changes in the depletion width of the space-charge layer.

Results from *in-situ* conductivity measurements at different temperatures are shown in Figure 5 for van der Pauw devices on p-type SOS (Si layer: 0.5-μm thick, 1.5×10^{19} B/cm^3). As schematically illustrated in the inset in Figure 5, the measurement involves impressing a constant dc current through an adjacent pair of contacts and monitoring the potential drop between the other pair of contacts (with an electrometer) during hydrogenation. With the device at the specified temperature, the remote plasma is ignited at time $t = 0$; at later times the plasma is turned off and on again as indicated in Figure 5. Several hydrogen-

induced processes are observed. At all three temperatures the resistance increases when the plasma is initiated and the device is thereby exposed to free-radical hydrogen. In addition, the rate of change of the resistance is observed to increase with temperature, as expected for a thermally-activated diffusion process. At 150°C and even 200°C this rise in the resistance should reflect a combination of compensation by migrating H^+ and neutralization of the boron acceptors by the formation of $(HB)^0$ complexes. However, at 300°C the $(HB)^0$ centers should be completely dissociated (e.g., Stavola et al. 1988) so that the sharp increase in resistance is due predominantly to in-diffusion of H^+. At all three temperatures, when the plasma is turned off the resistance relaxes with a time constant that is comparable to its initial rise time; the relaxation time should depend on the dissociation rate of $(HB)^0$ complexes as well as on out-diffusion of H^+.

An additional phenomenon is evident in the time-resolved measurements of Figure 5. At both 200°C and 300°C, the initial rise in resistance is followed by a slower decay with time even though the plasma is still on. When the plasma is turned off and then back on, the resistance approaches the value reached just before the plasma was turned off, rather than the maximum resistance that was initially obtained, and then continues to decay. Insight into the mechanism responsible for this behavior is afforded by the following observations: in the 300°C data the initial sharp rise is reproducible if the device is etched in dilute HF prior to installation in the hydrogenation system; similarly, at the lower temperatures, the initial rise is also reproducible if the device is first annealed at 300°C and then re-etched before installation. If the device is not re-etched, the resistance continues along the decay curve. These observations suggest a re-structuring of the exposed silicon surface during hydrogenation which decreases the flux of mobile hydrogen from the gas phase into the silicon substrate. This surface re-structuring is most likely due to the growth of a thin oxide layer induced by free-radical oxygen during

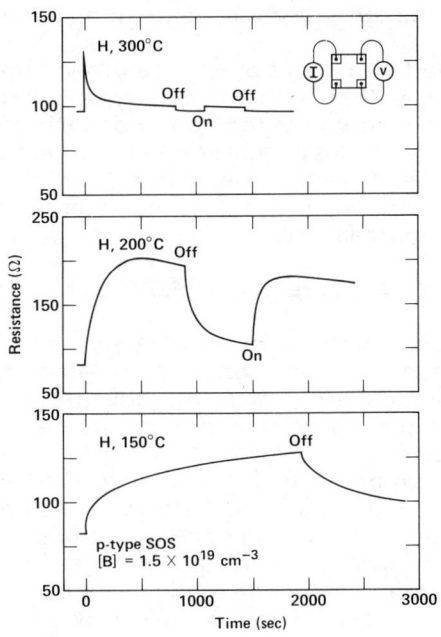

Fig. 5. Time-dependence of the resistance monitored *during* hydrogenation at different temperatures of van der Pauw devices (inset) on p-type SOS.

hydrogenation; plasma sources generally contain a small partial pressure of oxygen either intentionally introduced (Johnson and Moyer 1985) or present as a contaminant.

The general shape of the resistance-time curves in Figure 5 is of the sort one would expect from simple diffusion theory, which would predict a depth-integrated buildup of H that is proportional to $\sqrt{t}$ at small times t and saturates at large t. While accurate values for the effective diffusion coefficient D_{eff} of the migrating species cannot be deduced without further knowledge about the compensation and/or neutralization processes that are involved, $D_{eff}(T)$ appears to be in the range covered by earlier estimates for p-type Si (Capizzi and Mittiga 1987) or perhaps a little higher, that is, $\approx 10^{-11}$ cm^2/sec at 200°C.

5. SUMMARY

A large collection of experimental data on hydrogen migration and complex formation in silicon can be understood with an amphoteric impurity model for interstitial hydrogen, in which hydrogen possesses both an acceptor and a donor level within the silicon bandgap. In addition, hydrogen can form neutral entities which are describable as molecular hydrogen complexes, H_2. In p-type regions these complexes are highly immobile and their formation appears to be dominated by the reaction $H^+ + H^0 \rightarrow H_2 + h^+$. In n-type Si the hydrogen pairs may be metastable. Finally, *in-situ* electrical measurements are providing a direct approach to study the migration of charged hydrogen species and the kinetics of complex formation in semiconductors.

6. ACKNOWLEDGMENT

One of the authors (NMJ) is pleased to acknowledge support during the completion of this work from the Alexander von Humboldt Foundation, Federal Republic of Germany. The authors also thank J. Walker for technical support.

7. REFERENCES

* Permanent address: Dept. of Applied Physics, Stanford Univ., Stanford, CA, USA
† The symbols H^+, H^0 and H^- designate possible charge states of H-Si complexes
Bergman K, Stavola M, Pearton S J and Lopata J 1988 J. Phys. Rev. B **37** 2770
Capizzi M and Mittiga A 1987 Appl. Phys. Lett. **50** 918
Chang K J and Chadi D J 1988 Phys. Rev. Lett. **60** 1422
Deák P, Snyder L C and Corbett J W 1988 Phys. Rev. B **37** 6887
Johnson N M 1985 Appl. Phys. Lett. **47** 874
Johnson N M and Herring C 1988a Phys. Rev. B, in press.
Johnson N M and Herring C 1988b *Proc. 15th Int. Conf. Defects in Semicond.*, in press
Johnson N M, Herring C and Chadi D J 1986 Phys. Rev. Lett. **56** 769
Johnson N M, Herring C and Chadi D J 1987 Phys. Rev. Lett. **59** 2116
Johnson N M and Moyer M D 1985 Appl. Phys. Lett. **46**, 787
Johnson N M, Ponce F A, Street R A and Nemanich R J 1987 Phys. Rev. B **35** 4166

Mainwood A and Stoneham A M 1984 J. Phys. C: Solid State Phys. **17** 2513
Pantelides S T 1987 Appl. Phys. Lett. **50** 995
Shi T S, Sahu S N, Corbett J W and Snyder L C 1984 Scientica Sinica **27** 98
Stavola M, Pearton S J, Lopata J and Dautremont-Smith W C 1988 Phys. Rev. B **37** 8313
Tavendale A J, Alexiev D and Williams A A 1985 Appl. Phys. Lett. **47** 316
Van de Walle C G, Bar-Yam Y and Pantelides S T 1988 Phys. Rev. Lett. **60** 2761

Hydrogen-related effects in crystalline semiconductors

Eugene E. Haller

Department of Materials Science and Mineral Engineering and Lawrence Berkeley Laboratory, University of California at Berkeley, Berkeley, CA 94720 U.S.A.

ABSTRACT: Recent experimental and theoretical information regarding the states of hydrogen in crystalline semiconductors is reviewed. The abundance of results illustrates that hydrogen does not preferentially occupy a few specific lattice sites but that it binds to native defects and impurities, forming a large variety of neutral and electrically active complexes. The study of hydrogen passivated shallow acceptors and donors and of partially passivated multivalent acceptors has yielded information on the electronic and real space structure and on the chemical composition of these complexes. Infrared spectroscopy, ion channeling, hydrogen isotope substitution and electric field drift experiments have shown that both static trigonal complexes as well as centers with tunneling hydrogen exist. Total energy calculations indicate that the charge state of the hydrogen ion which leads to passivation dominates, i.e., H^+ in p-type and H^- in n-type crystals. Recent theoretical calculations indicate that it is unlikely for a large fraction of the atomic hydrogen to exist in its neutral state, a result which is consistent with the total absence of any Electron Paramagnetic Resonance (EPR) signal. An alternative explanation for this result is the formation of H_2. Despite the numerous experimental and theoretical results on hydrogen-related effects in Ge and Si there remains a wealth of interesting physics to be explored, especially in compound and alloy semiconductors.

1. INTRODUCTION

Despite the fact that semiconductor science is by now a mature field, it regularly produces unexpected discoveries which in time lead to intensive research worldwide. Electron-hole drops and the quantized Hall effect, which in turn led to the awarding of a Nobel Prize, are just two examples illustrating this point. The discovery of a multitude of effects which can be traced to hydrogen in semiconductors may not be quite as spectacular as the aforementioned examples, but it has produced a wealth of interesting and new physics. I am very pleased to see that at this year's semiconductor conferences (ICPS 19 in Warsaw, ICDS 15 in Budapest and this conference) the topic of hydrogen in semiconductors has been accorded a very prominent position.

Because the focus of this conference is on shallow levels in semiconductors, I will deliberately avoid any discussion of implantation or high energy radiation-related defects as well as passivation of deep levels.

The early hydrogen diffusion studies at high temperatures in crystalline Si (van Wieringen and Warmoltz 1956) and Ge (van Wieringen and Warmoltz 1956, and Frank and Thomas 1960) yielded important information on the diffusion coefficient and on solubility. We learned that hydrogen diffuses interstitially very rapidly, that the equilibrium solubility near the melting point is small ($<10^{16}$ cm^{-3}), and that hydrogen is dissolved atomically at high temperatures. The first serious question arose when Electron Paramagnetic Resonance (EPR) experiments could not detect the much expected spin precession signal from the hydrogen-bound electron (Jeffries 1980). The first electronic effects may be traced to unpublished work performed by Seitz and Zerbst (Winstel 1981). They found minority lifetime reductions up to 2 cm deep in a Si bulk single crystal which had been electrolytically

© 1989 US Government

etched in an aqueous KOH solution at T=150° C. They could not positively identify hydrogen as the primary cause for the lifetime reduction and did not further pursue their studies. With today's information we must conclude that it was most probably hydrogen which led to the observed effects.

The first systematic studies of electronic effects of hydrogen in a semiconductor coincided with the development of ultra-pure Ge (Haller et al 1981) which was used for gamma-ray detector applications (Haller and Goulding 1981). R. Hall discovered the "rapid quench" shallow acceptor and donor (Hall 1975) and our group showed that these were due to the hydrogen-containing complexes A(H, Si) and D(H, O) respectively (Haller et al 1980, and et al 1980). The combination of a sensitive spectroscopic technique (photothermal ionization spectroscopy [PTIS]) (Kogan and Lifshitz 1977) and active bulk crystal growth research programs in several laboratories led to the discovery and understanding of a large number of hydrogen-related impurity complexes. The exceptional purity of this material was a great advantage in obtaining high quality, definitive, spectroscopic information.

The developments in crystalline Si regarding hydrogen-related shallow level effects stagnated for many years. The major reason for this was the failure to find electrically or optically active centers which were not radiation damage related. The situation began to change when passivation of shallow levels (Pankove et al 1983) and deep levels (Benton et al 1980) by hydrogen was discovered. An enormous body of publications on H in Si has accumulated since. Much of this work has been reviewed recently (Quin et al 1986, Corbett et al 1988, Pearton et al 1987). When studying such reviews the reader should keep in mind that this field is still in a state of flux and many hypotheses and models are becoming outdated

Hydrogen-related effects were discovered a few years ago also in III-IV semiconductors (Dautremont-Smith 1988). In order to obtain clear-cut, definitive answers, many researchers have chosen to perform high resolution spectroscopic studies of well defined hydrogen-containing complexes; a wise choice indeed in these difficult materials.

In some sense the "hydrogen story" has, in retrospect and in my personal view, evolved along a very typical pattern. The simplest, best understood and most perfect (though technologically not very important) semiconductor, Ge, led the way regarding the understanding of the structure, both electronic and real space, and the composition of many hydrogen-related complexes. The investigations of shallow level effects of hydrogen in the technologically most important semiconductor, Si, followed the developments in Ge with quite a delay in time. Once the interest in H in Si was sparked by Pankove's and Sah's experiments, a very large amount of work was produced in the short time of about four years. In the early rush for dominance, much work was produced, little of which we find today to be definitive. One serious shortcoming was the absence of research-oriented experimental Si bulk crystal growth programs. Only one group in China (Shi et al 1985, Qi et al 1985, Shi et al 1986) occupied itself with H_2 and D_2 atmosphere-grown Si crystals, producing important contributions to this field.

The discovery of hydrogen diffusion at low temperatures into single-crystal Si in a H_2-plasma must be considered most fortunate. The ensuing effort was so intense that some of the important questions were answered in a relatively short time, though several fundamental, interesting puzzles remain to be solved. I will emphasize these open questions in the hope that they may stimulate work towards a deeper understanding of the physics and chemistry of H in semiconductors. The effort spent on studying the effects of H in Si appears to have reached a peak, judging from the number of invited talks, the largest number on any one topic at this conference, and from the emphasis given to this field at ICPS 19 in Warsaw in the coming week and at ICDS 15 in Budapest the week after.

It appears that in the future, the emphasis might shift from silicon to III-V, II-VI, and other semiconductor compounds and alloys. The vast variety of energy levels related to defects

and impurities, many of which have not been investigated or are not understood, makes me anticipate a wide range of new and interesting studies.

For this review I have decided to categorize the electronic effects of hydrogen in crystalline semiconductors according to the materials. This poses the danger of losing the common fundamentals. I will try to point these out wherever appropriate and to emphasize them again in the discussion section.

2. GERMANIUM

Because much of the seminal work in this material was performed some years ago I will begin this paragraph with a brief summary of the major results of the early work and will continue with a more detailed review of recent important findings.

The combination of very pure and structurally perfect crystals with the photoconductivity technique PTIS created an almost ideal setting for the discovery and the study of novel centers in ultra-pure Ge (Haller et al 1981). Many of the bi-atomic complexes form shallow levels which have unique properties. A major finding for the bound hole (electron) states of hydrogen-containing complexes which form shallow acceptors (donors) is a splitting of the 1s state into complicated multiplets (Haller et al 1980, Joós et al 1980). This unusual splitting cannot originate from the band structure. It is caused by the coupling between the electronic structure of the defect and the motion of the nearby proton. On the basis of the early experimental findings, Falicov created the tunneling hydrogen model (Falicov and Haller 1985). Recent experiments by Kahn et al (1987) using improved resolution, more uniform uniaxial stress, and better suited crystals showed that the experimental data for the acceptors A(H,Si), A(H,C), A(Be,H), and A(Zn,H) can be explained fully with static, trigonal complexes. These complexes do not align along the stress orientation up to stress values as high as one GPa and up to temperatures as high as room temperature. They remain distributed evenly among all equivalent <111> directions. Whereas only the "static" limit of the tunneling model is now required to explain A(H,Si), A(H,C), A(Be,H), and A(Zn,H) in Ge, the tunneling model has been successfully used recently to explain the bound hole spectra of the shallow acceptor complex A(Be,H) in Si (Muro and Sievers 1987). It furthermore still applies to the donor complexes D(Li,O) (Haller and Falicov 1978) and D(H,O) in Ge (Joós et al 1980). Static models have been proposed for D(H,O) (Broeckx et al 1980, Ham to be published) but these cannot explain the experimentally determined complicated structure of the 1s state (Navarro et al 1986). The stress insensitivity of the electron ground and bound excited states of the D(H,O) donor has led to the sharpest electronic transition lines in a semiconductor (Navarro et al 1988, Haller et al 1987).

An interesting discovery was made by Kahn during the study of the shallow acceptor complex $A(Cu,H_2)$. The fact that two hydrogen atoms bind to the copper triple acceptor, reducing the number of bound holes to one, was proven with isotope shifts of the ground state energy of the bound hole (Kahn et al 1986). A large difference was found between the appearance of the ground state to bound excited state hole spectra of A (Cu,H_2) and A(Cu,X,Y) where X and Y are hydrogen isotopes. The $A(Cu,H_2)$ spectrum displays at least eleven hydrogenic series of lines, indicating a very complicated set of closely spaced 1s states (Fig. 1a). These 1s states can be populated sequentially by increasing the sample temperature, proving that only one center leads to these spectra, i.e., we do not observe eleven or more similar configurations of one or more impurity complexes. All the centers containing at most one hydrogen and the complement deuterium and/or tritium, produce a single hydrogenic series of lines (Fig.1b and 1c). Kahn et al (1986) interpreted their findings using the Devonshire model (Devonshire 1936). The tunneling of the hydrogen occurring in $A(Cu,H_2)$ leads to the multiplicity in the 1s state. If one or both hydrogen isotopes have an atomic mass > 1, the hydrogen isotopes in the center cease to tunnel, leading to a single 1s groundstate level and a correspondingly simple hydrogenic spectrum.

The acceptor passivation effects which led to the extensive activity in Si exist in Ge as well, but are not as pronounced. The distribution of acceptors and donors in large, high-purity Ge single crystals grown under a hydrogen atmosphere was studied in the 1970's. Ultra-pure Ge crystal growers found that the residual chemical net-impurity concentrations were not only changing along the crystal growth axis which would have indicated simple impurity segregation, but also exhibited strong radial net-concentration variations. This unusual impurity distribution was called "coring", i.e., a n-type core was surrounded by p-type material over quite a large fraction of the length of the crystal. Prolonged annealing in a metal bath at 300 to 400° C removed the radial dependences (Hansen and Haller 1984). Today we can interpret these results with hydrogen passivation of the residual acceptors aluminum and boron. The annealing redistributes the hydrogen more evenly leading to a less pronounced radial dependence.

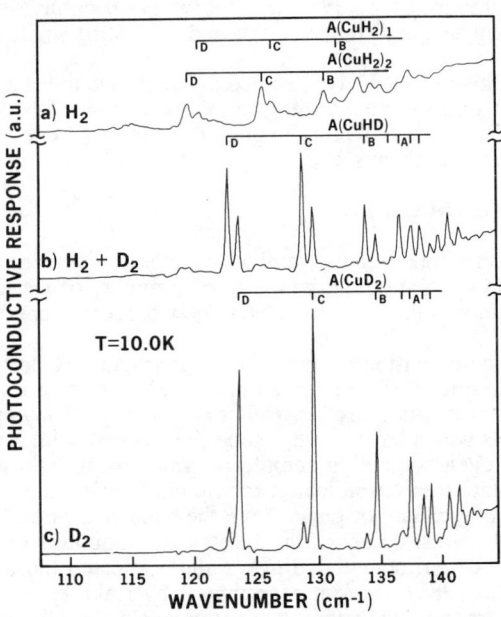

Fig. 1. PTI spectra of the copper-dihydrogen acceptors that appear in samples which were grown under atmospheres of different hydrogen isotopes. (a) Pure H_2, showing the complex spectrum of $A(CuH_2)$; (b) a 1:1 mixture of H_2 and D_2, showing $A(CuH_2)$, $A(CuHD)$ and $A(CuD_2)$ in a 1:2:1 ratio; (c) nearly pure D_2, showing $A(CuD_2)$ and a trace of $A(CuHD)$.

Though the studies of hydrogen-related effects in Ge have evolved over many years, a number of interesting questions remain:

- Why does such a large variety of hydrogen-related, electrically active shallow level complexes exist in Ge while other semiconductors exhibit a significantly smaller variety of hydrogen-related shallow levels?

- The concentrations of the electrically active hydrogen complexes are typically two to four orders of magnitude lower than the total hydrogen concentration in a bulk crystal. Is H_2, as one might expect, the dominant form of hydrogen in Ge at low temperatures?

- Is hydrogen occupying the bonding or the antibonding positions in the various shallow level complexes [A(H,Si), A(H,C), etc.]?

- Does isolated atomic hydrogen, if it ever exists at low temperatures, form a level (or levels) within the bandgap, and if so, are the levels donor- and/or acceptor-like?

- Why does hydrogen bind to Si and C--electrically neutral impurities in Ge--and what are the binding forces?

- What are the conditions which lead to static or dynamic behavior of hydrogen complexes?

3. SILICON

3.1. Shallow Acceptor Passivation

As mentioned in the introduction, the investigations regarding electronic effects of hydrogen on shallow levels in Si proceeded slowly at first. The major discovery which started an avalanche of experiments and, in time, theoretical work as well, was an increase of the resistivity of B-doped (Pankove et al 1983) and later Al-, Ga-, and In-doped (Pankove et al 1984) p-type silicon upon the introduction of hydrogen. Pankove et al introduced hydrogen at low temperatures using a radio-frequency (RF) hydrogen plasma (Pankove et al 1983, Pankove et al 1984). Sah et al (1983) interpreted their observations with hydrogen originating in the oxide of a metal oxide semiconductor (MOS) structure. The increase in resistivity was properly interpreted as passivation of acceptors by hydrogen. This was proven directly by Thewalt et al (1985) who demonstrated that the acceptor-related photoluminescence signal disappeared in hydrogen-passivated p-type Si.

Passivation is defined here as the deactivation of all electronic activity of the shallow isolated, substitutional acceptor. Passivation is a consequence of acceptor-hydrogen pair formation and is not to be confused with compensation, as has been and is still done by numerous researchers. Compensation of p-type Si means that randomly distributed donors are present which reduce (just as in passivation) the maximum available number of free holes to $N_A - N_D$ (N_A = acceptor concentration $\{cm^{-3}\}$ and N_D = donor concentration $\{cm^{-3}\}$). However, compensation increases the number of ionized scattering centers ($N_{A-} + N_{D+}$) which, in turn, reduce the hole mobility. Such a reduction has never been observed after passivation; rather an increase in mobility was detected. The experimental results are clear cut and exclude the effect of compensation as a major mechanism in the final equilibrium state of hydrogenated p-type Si. This fact does not contradict the results of recent theoretical calculations (Van de Walle et al 1988) which indicate that isolated, interstitial H may form a donor state in p-type Si. It is quite likely that this is a transitory (metastable) donor state which may affect the dynamics of the acceptor passivation process but which is not observed experimentally in the equilibrium final state. I therefore propose that, in the future, we specify clearly which state we are discussing: the equilibrium state in which acceptors are passivated by hydrogen, or the dynamic state in which the donor state of the H^+ species may play a role, partially compensating the acceptors. Experiments which could measure the relevant parameters for the metastable donor state would help very much in strengthening the theoretical results (Van de Walle 1988).

The key contributions which have led to the current state of understanding of the passivation of acceptors and of the structure of the neutral acceptor-hydrogen complexes came from IR local vibrational mode (LVM) spectroscopy (Stavola et al 1987, Stavola et al 1988, Marwick et al 1988), from electrical measurements (Pankove et al 1983, Pankove et al 1984, Sah et al 1983), from channeling experiments (Marwick et al 1987, Nielsen et al 1988), from electric field drift experiments (Tavendale et al 1985, Johnson and Herring 1988), and from theoretical calculations (Van de Walle et al 1988). The preferred model among those proposed by Pankove et al (1985) explained boron neutralization with the removal of one bond from the boron impurity, leading to a threefold coordinated boron atom retracting from the tetrahedral site while the Si dangling bond was passivated by the hydrogen atom. This model had evolved naturally from the studies of amorphous, hydrogenated Si by the RCA group. Today it is clear that this picture is, in its original form, incorrect. Boron as well as the other group III acceptors remain to within about 0.2 Å at their tetrahedral site as has been shown recently by channeling experiments performed by Marwick et al (1987) (Fig. 2). The hydrogen atom binds to both the acceptor and the Si neighbor occupying a position close to the bond center between the boron and one silicon neighbor. The position of hydrogen was also determined directly by Nielsen et al (1988) using channeling experiments. A good model for this acceptor hydrogen complex structure was proposed by DeLeo and Fowler.

(1985). They calculated LVM frequencies for a hydrogen atom next to an acceptor. Their predictions were subse-quently supported by impressive and definitive IR LVM spectroscopy performed by Stavola et al (1988, 1987, Bergman et al to be published). An especially nice discovery was low frequency excitations in acceptor-hydrogen complexes by Stavola et al (1988). These were first detected as side bands of the Al-H and Ga-H vibrations (Stavola et al 1987) (Fig. 3, Stavola et al 1988). All of today's information, both experimental as well as theoretical, supports a model which shows the hydrogen between the group III acceptor and one neighboring silicon atom, with the acceptor assuming a position very close to the T_d site (Fig. 4, Bergman et al to be published).

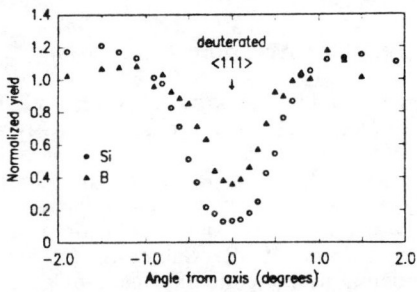

Fig. 2. Channeled dip in <111> axis showing Si and B yields. The narrowing of the B dip shows that B atoms are displaced from their sites.

This is an appropriate moment for mentioning an interesting further experimental fact. In Si, the double acceptor Be can be partially passivated by one hydrogen atom, forming the single acceptor A(Be,H) which binds one hole at low temperature. Many years ago Crouch et al (1974) had shown with absorption spectroscopy that Be binds H. Muro et al (1986) showed with IR spectroscopy of bound hole transitions that the single acceptor complex A(Be,H) has its H tunneling between four equivalent real space positions. It is difficult to

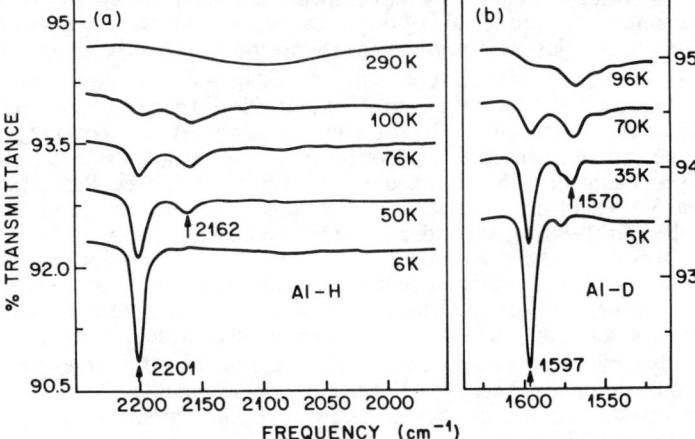

Fig. 3. Hydrogen stretching vibration as a function of temperature for (a) Al-H and (b) Al-D complexes in silicon. The small feature for Al-D at 5 K near 1575 cm^{-1} has no counterpart in the Al-H spectrum and is presumed to be unrelated to the Al-D complex.

understand how H could tunnel rapidly if it were to occupy a bond-centered position as it does in the group III acceptor-hydrogen complexes. Does hydrogen occupy an antibonding position as has been suggested by Assali and Leite (1985, 1986, da Silva et al 1988) for the acceptor-hydrogen complexes? It would be useful if the calculations which were made recently for the neutral boron-hydrogen complex were repeated (Van de Walle et al 1988) for the beryllium-hydrogen single acceptor complex. It would be most interesting to learn if the minimum energy position for hydrogen is strongly sensitive to the valency or to other properties of the particular impurity.

3.2. Shallow Donor Passivation

Johnson et al (1986) surprised the research community with their observation that shallow donors can also be passivated by hydrogen, though less effectively than shallow acceptors. Donor passivation had not previously been observed. Using resistivity and Hall effect measurements, they showed unambiguously that, as in the case of the shallow acceptors, passivation and not compensation takes place. After the incorporation of H, the free electron concentration decreased and the mobility increased, a consequence of a reduction of the total number of donors and ionized centers. If residual acceptors would have been passivated in these n-type crystals, an increase in the free electron concentration would have resulted directly opposite to the experimental results! Johnson et al (1986) also proposed a model for the donor-hydrogen complex in which they place the hydrogen atom not next to the donor ion but at a next-nearest antibonding position (Fig. 5). The recent LVM studies performed on donor (P, As, and Sb)-hydrogen complexes by Bergman et al (1988a, 1988b) fully confirm the passivation argument of Johnson et al . The observed vibrational frequencies, however, are significantly lower than the results of Johnson et al's calculations. Experiments show that the LVM frequencies are practically independent of the donor species. Improved calculations are required in order to arrive at the LVM frequency values which have been established experimentally.

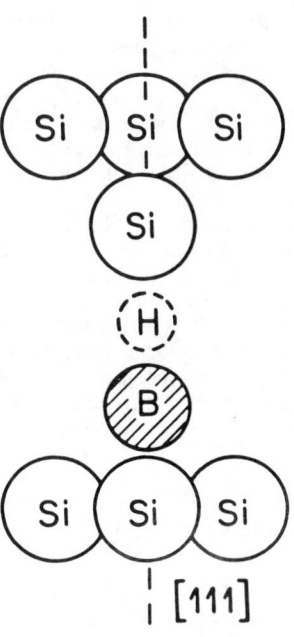

Fig. 4. Schematic model of the B-H complex.

After having accepted the experimental facts of donor passivation by hydrogen, one may ask how the well-studied double donors S, Se, and Te would interact with H. Pensl et al (1988) used deep level transient spectroscopy (DLTS) to show that the two energy levels of all chalcogen donor species disappear simultaneously and that no new levels are created. Further experiments will be required to determine if indeed only one hydrogen atom can passivate a double donor and what the microscopic structure of these chalcogen-hydrogen complexes is.

The amphoteric character of hydrogen (i.e., hydrogen can passivate acceptors as well as donors) leads us to assume that hydrogen must exist in several different charge states in semiconductors. In their recent theoretical work, Van de Walle et al (1988) suggest that H^+ is the energetically favored charge state in p-type material while H^- is the preferred charge state in n-type crystals. Tavendale et al (1985) have performed electric field drift experiments with p-n- and Schottky diodes which support the first part of the above proposal. In a recent study by Johnson and Herring (1988), a quantitative analysis of the spatial distribution of hydrogen in a p-n junction has been attempted. Both H^+ and H^0 are required to explain the data while not much can be concluded about H^-. Clearly, further experiments and theoretical work are required to arrive at a complete understanding of the hydrogen charge state question (Johnson 1988).

I want to close this section on Si by listing a few questions which require further attention before a complete understanding is reached. One question concerns the introduction of hydrogen at low temperatures. It appears that atomic hydrogen can enter a single crystal through its native oxide of at least a few tens of Ångstroms with little or no hindrance. Careful experiments by Johnson et al (1981) showed that no photons or energetic particles are required to bombard the surface in order to let atomic hydrogen enter the crystal. This finding is in agreement with hydrogenation experiments using electrolysis of phosphoric acid

(Oehrlein et al 1981). Schnegg et al (1988) have reported passivation effects which occur during the polishing of Si wafers with Syton. Schnegg et al conclude on the basis of secondary ion mass spectroscopy (SIMS) results and other measurements that hydrogen is responsible for their shallow acceptor passivation. What is unusual, is the relatively low stability of this kind of acceptor passivation at room temperature, compared to RF-plasma or electrolytical hydrogen passivation. One possible explanation for this result is to assume that the hydrogen occupies a local energy minimum, e.g., an antibonding position, and does not reach the absolute minimum at a bond center site. This would revive the Assali and Leite model (1985, 1986, da Silva et al 1988)!

Some of the more important, unanswered questions regarding H in Si are:

- What is the dependence of the charge state distribution (H^+, H^0, H^-) on the Fermi level and other parameters? A partial, theoretical answer is available (Van de Walle et al 1988) but further work is required.

- What are the conditions which lead to static or dynamic behavior of hydrogen in Si?

- Are there two stable sites for hydrogen which would explain the different stabilities of hydrogen passivation by the RF-plasma, H_2O-boiling and electrolysis process on the one hand, and Syton polishing on the other hand?

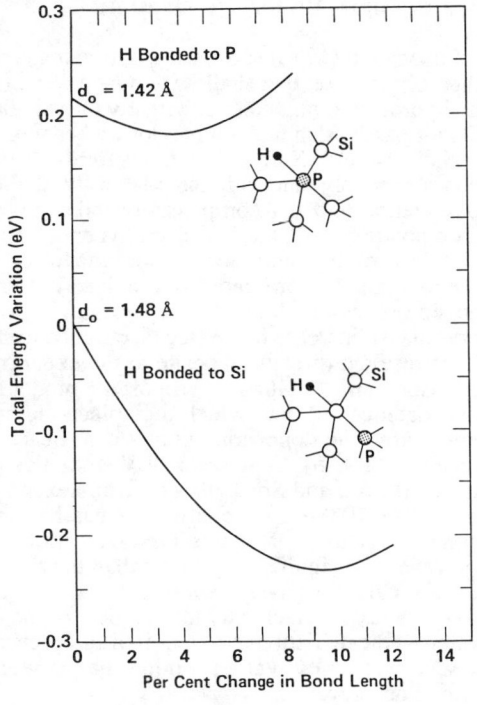

Fig. 5. Bond-length dependence of the relative binding energies for H in an antibonding position of (1) a substitutional P atom (upper curve) and (2) a Si atom that is a nearest neighbor of the substitutional P (lower curve). The distances d_0 are the P-H and Si-H bond lengths in phosphene (PH_3) and silane (SiH_4), respectively.

4. GALLIUM ARSENIDE AND OTHER III-V COMPOUNDS

Regarding shallow levels, the effects of hydrogen in GaAs are in most respects similar to its effects in Ge and Si. Chevallier et al (1985) used electrical measurements to show that Si donor activity is completely passivated by hydrogen diffusing into doped, molecular-beam-epitaxy (MBE)-grown layers. H passivation of other donors was subsequently demonstrated. In a detailed, high resolution spectroscopy study Pajot et al (1988) found isotope effects of the three stable Si isotopes with atomic weights 28, 29 and 30, proving that H binds directly to Si (Fig. 6, Pajot et al 1987). The symmetry of donor-hydrogen complexes appears to be trigonal.

Acceptors have been hydrogenated in GaAs as well. Pajot et al (1987) used LVM and H isotopes to show that H binds to an As atom with neighboring Zn acceptor. Johnson et al (1986) had demonstrated earlier acceptor passivation using the capacitance-voltage dependences of Schottky barriers on hydrogenated, Zn-doped GaAs.

Some preliminary experimental results on Sn donor passivation in InP and donor as well as acceptor passivation in AlGaAs have been reported. A good review of the still small activity in III-V semiconductors was given by Dautremont-Smith (1988). So far, the studies which have generated some conclusive results have been dominated by high-quality spectroscopy experiments (Pajot 1988). We can look forward to many interesting investigations especially in highly structured systems like superlattices and quantum wells.

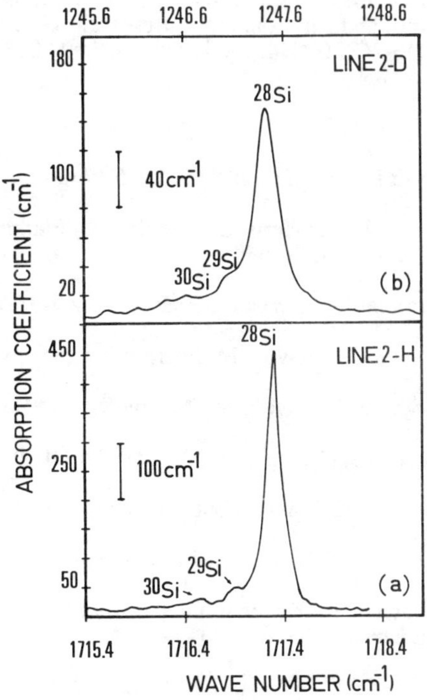

5. SUMMARY

The evolution of the understanding of hydrogen-related, shallow level effects started with the discovery of hydrogen-containing complexes in ultra-pure Ge, expanded to a large effort in Si where primarily the passivation of shallow levels is observed, and is now involving III-V compound and alloy semiconductors. Complex formation has been shown unambiguously to be the cause for passivation of the electrical activity. Compensation by deep hydrogen donors may be a metastable phenomenon which can affect hydrogen transport in electric fields. Experiments which could identify hydrogen donor levels would be very significant.

High-resolution IR spectroscopy of the electron (hole) bound to a shallow level, hydrogen-containing donor (acceptor) complex allows very fine probing of the static

Fig. 6 (a) Absorption coefficient of line 2-H(a) and 2-D(b) at 6K in passivated GaAs:Si with an apodized resolution of 0.05 cm^{-1}. Note the low-energy side asymmetry of the line. Before passivation, n=5x10^{18}cm^{-3}. (b) Same as (a) for line 1-D.

or dynamic properties of the complex. This fact makes partially passivated centers such as multivalent acceptors with a number of passivating hydrogen atoms which is smaller than the acceptor (donor) valence especially interesting. The acceptor A(Be,H) which has been studied in both Si and Ge is an example of such a case. LVM spectroscopy can provide symmetry information in those cases where the concentration of centers is sufficiently high.

Many challenging physics questions remain to be answered, and we can expect interesting experiments and results in the future. If the future is projected from the past, spectroscopic approaches promise to generate the most definitive answers.

ACKNOWLEDGMENTS

It has been a pleasure to work with many outstanding scientists over the past ten years on hydrogen-related phenomena in semiconductors. I am indebted to the following colleagues for valuable scientific inputs regarding H in Ge: F. S. Goulding, N. M. Haegel, L. M. Falicov, W. L. Hansen, J. M. Kahn, R. McMurray, A. Ramdas and P. L. Richards. I would like to thank K. Bergman, N. M. Johnson, A. Marwick, B. Pajot, and M. Stavola for letting me use material from their publications for this review. Let me apologize to all of you whose work is important and interesting but could not be included because of space limitations.

This work was supported in part by the Director's Office of Energy Research, Office of Health and Environmental Research, U.S. Department of Energy under Contract No. DE-AC03-76SF00098, and in part by the U.S. National Science Foundation under Contract No. DMR-8502502.

REFERENCES

Assali L V C and Leite J R 1985 *Phys. Rev. Lett* **55** 980; 1986 *Phys. Rev. Lett.* **56** 403

Benton J L, Doherty C J, Ferris S D, Flamm D L, Kimerling L C and Leamy H J 1980 *Appl. Phys. Lett.* **36** 670

Bergman K, Stavola M, Pearton S J and Hayes T to be published *Phys Rev B*

Bergman K, Stavola M, Pearton S J and Lopata J 1988a *Proc. Mat. Res. Soc.* **104** pp 281

Bergman K, Stavola M, Pearton S J and Lopata J 1988b *Phys. Rev. B* **37** 2770

Broeckx J, Clauws P and Vennik J 1980 *J. Phys. C: Solid State Physics* **13** L141

Chevallier J, Dautremont-Smith W C, Tu C W and Pearton S J 1985 *Appl. Phys. Lett.* **47** 108

Corbett J W, Lindstrom J L and Pearton S J 1988 *Proc. Mat. Res. Soc.* **104** pp 229

Crouch R K, Robertson J B, Morgan H T, Gilmer Jr T E and Franks R K 1974 *J. Phys. Chem. Solids* **35** 833

da Silva E C F, Assali L V C, Leite J R and Dal Pino Jr A 1988 *Phys. Rev. B* **37** 3113

For a recent review see: Dautremont-Smith W C 1988 *Proc. Mat. Res. Soc.* **104** pp 313

DeLeo G G and Fowler W B 1985 *Phys. Rev. B* **31** 6861

Devonshire A F 1936 *Proc. R. Soc. London, Series A* **153** 60

Falicov L M and Haller E E 1985 *Sol. State Comm.* **53** 1121

Frank R C and Thomas J E 1960 *J. Phys. Chem. Solids* **16**, 144

Hall R N 1974 *IEEE Trans. Nucl. Sci.* **NS-21** 260; 1975 *Inst. Phys. Conf. Series* **23** 190

Haller E E and Falicov L M 1978 *Phys. Rev. Lett.* **41** 1192

Haller E E, Joós B and Falicov L M 1980 *Phys Rev.* **B21** 4729

Haller E E and Goulding F S 1981 *Handbook on Semiconductors* Vol. 4, Ch. 6, ed. C. Hilsum (North-Holland) 799

For a review see: Haller E E , Hansen W L and Goulding F S 1981 *Adv. Phys.* **30** 93

Haller E E, Navarro H and Keilmann F 1987 *Proc. 18th Intl. Conf. Phys. Semicon.* ed. O. Engstrom (Singapore: World Scientific Publ. Co.) p. 837

Ham F, to be published in Phys. Rev. B

Hansen W L and Haller E E 1984 *Proc. Matl. Res. Soc.* **16** (New York: Elsevier) pp 1-16

Jeffries C D 1980, private communication

Johnson N M 1988 this conference

Johnson N M, Biegelson D K and Moyer M D 1981 *Appl. Phys. Lett.* **38** 995

Johnson N M, Burnham R D, Street R A and Thornton R L 1986 *Phys. Rev. B* **33** 1102

Johnson N M, Herring C and Chadi D J 1986 *Phys. Rev. Lett.* **56** 769

Johnson N M and Herring C 1988 *Phys. Rev. B* **38** 1581

Joós B, Haller E E and Falicov L M 1980 *Phys. Rev.* **B22** 832

Kahn J M, Falicov L M and Haller E E 1986 *Phys. Rev. Lett.* **57** 2077; for details see: Kahn J M 1986 *PhD Thesis*, Lawrence Berkeley Laboratory Report LBL-**22652**

Kahn J M, McMurray Jr. R E, Haller E E and Falicov L M 1987 *Phys. Rev. B* **36** 8001

For a review see: Kogan Sh M and Lifshitz T M 1977 *Phys. Stat. Solidi (a)* **39** 11

Marwick A D, Oehrlein G S and Johnson N M 1987 *Phys. Rev. B* **36** 4539; see also Marwick A D, Oehrlein G S, Barrett J H and Johnson N M 1988 *Proc. Mat. Res. Soc.* **104** pp 259

Muro K and Sievers A J 1986 *Phys. Rev. Lett.* **57** 87; 1987 *Proc. 18th Intl. Conf. Phys. Semicon.* ed. O. Engstrom (Singapore: World Scientific Publ. Co.) p. 891

Navarro H, Griffin J, Haller E E and McMurray Jr R E 1986 *Sol. State Comm.* **64** 1297

Navarro H, Haller E E and Keilmann F 1988 *Phys. Rev. B* **37** 10822

Nielsen B Bech, Andersen J U and Pearton S J 1988 *Phys. Rev. Lett.* **60** 321; see also Nielsen B Bech 1986 *Proc. Mat. Res. Soc.* **59** pp 487

Oehrlein G S, Lindström J L and Corbett J W 1981 *Physics Lett.* **81A** 246

Pajot B 1988 this conference

Pajot B, Jalil A, Chevallier J and Azoulay R 1987 *Semicon. Sci. Tech.* **2** 305

Pajot B, Newman R C, Murray R, Jalil A, Chevallier J and Azoulay R 1988 *Phys. Rev. B* **37** 4188

Pankove J I, Carlson D E, Berkeyheiser J E and Wance R O 1983 *Phys. Rev. Lett.* **51** 2224

Pankove J I, Wance R O and Berkeyheiser J E 1984 *Appl. Phys. Lett.* **45** 1100

Pankove J I, Zanzucchi P J, Magee C W and Lucovsky G 1985 *Appl. Phys. Lett.* **46** 421; note especially Fig. 4.

Pearton S J, Corbett J W and Shi T S 1987 *Appl. Phys.* **A43** 153

Pensl G, Roos G, Stolz P, Johnson N M and Holm C 1988 *Proc. Mat. Res. Soc.* **104** pp 241

Qi M W, Bai G R, Shi T S and Xie L M 1985 *Materials Letters* **3** 467

Quin Guo-Gang, Du Yon-Chang, Wu Jin and Yao Xiu-Chen 1986 *Defects in Semiconductors* ed. H.J. von Bardeleben, Mat. Sci. Forum **10-12** 563

Sah C T, Sun J V C and Tzou J J T 1983 *Appl. Phys. Lett.* **43** 204; 1983 *J. Appl. Phys.* **54** 4378

Shi T S, Xie L M, Bai G R and Qi M W 1985 *Phys. Stat. Solid (b)* **131** 511

Shi T S, Bai G R, Qi M W and Zhou J K 1986 *Defects in Semiconductors* ed. H.J. von Bardeleben, Mat. Sci. Forum **10-12** 597

Schnegg A, Prigge H, Grundner M, Hahn P O and Jacob H 1988 *Proc. Mat. Res. Soc.* **104** pp 291

Stavola M, Pearton S J, Lopata J and Dautremont-Smith W C 1987 *Appl. Phys. Lett.* **50** 1086

Stavola M, Pearton S J, Lopata J and Dautremont-Smith W C 1988 *Phys. Rev. B* **37** 8313

Tavendale A J, Alexiev D and Williams A A 1985 *Appl. Phys. Lett.* **47** 316

Thewalt M L W, Lightowlers E C and Pankove J I 1985 *Appl. Phys. Lett.* **46** 689

Van de Walle 1988 this conference

Van de Walle C G, Bar-Yam Y and Pantelides S T 1988 *Phys. Rev. Lett.* **60** 2761

van Wieringen A and Warmoltz N 1956 *Physica* **22** 849

Winstel G 1981, private communication

Inst. Phys. Conf. Ser. No 95: Chapter 7
Paper presented at Int. Conf. Shallow Impurities in Semiconductors, Linköping, Sweden, 1988

Hydrogen passivation of shallow donors and acceptors in GaAs

B Pajot

Groupe de Physique des Solides de l'Ecole Normale Supérieure, Tour 23, Université Paris 7, 2 place Jussieu, 75251 Paris Cedex 05, France

ABSTRACT: The introduction of atomic hydrogen in n- and p-type GaAs reduces the carrier concentration in the region where it penetrates. Carrier mobility measurements indicate a passivation effect rather than a simple compensation mechanism. A vibrational spectroscopy study of the H-passivated samples at low temperature reveals new lines related to the bonding of hydrogen with atoms in the crystal. I review the results obtained by this method for Si donor in GaAs and GaAlAs alloys and for different acceptors in GaAs. I show that they can be interpreted by models proposed to explain H-passivation in c-silicon and that they agree with recent *ab-initio* calculations for donors and acceptors in GaAs. Among my conclusions, the most important is that compensated dopant atoms cannot be passivated by hydrogen in GaAs.

1. INTRODUCTION

The loss of electrical activity of the shallow donors and acceptors in crystalline semiconductors under exposure to a hydrogen plasma is one of the manifestations of the interaction between this element and various defect structures in this class of material (Pearton et al 1987). The first direct evidence was reported by Pankove et al (1983) for boron in silicon. The loss of electrical activity was later explained (Pankove et al 1985) by the formation of a neutral complex involving hydrogen and the acceptor atom. The process leading to the formation of this complex was termed "neutralisation" or "passivation" by opposition to the global compensation effect where the acceptor remains isolated. One of the differences between the two states is that the neutral complex cannot trap a free hole at low temperature while a compensated acceptor can bind a hole and become electrically active under excess hole injection.

It was subsequently shown by Chevallier et al (1985) that the exposure of Si-doped n-type GaAs to a RF hydrogen plasma could also passivate the Si donors and also by Johnson et al (1986) that Zn acceptors could also be passivated in the same way. Since then, other donors and/or acceptors have been successfully passivated by hydrogen in III-V compounds and their alloys so that we can think of this process as a rather general one. We have investigated the passivation of shallow dopants by hydrogen in GaAs using vibrational spectroscopy as a tool. After recalling briefly what is needed to be known here on shallow dopants and electrical effects of passivation in III-V compounds, I will present a summary of our results and try to describe what we learned of the "static" part of passivation from them.

2. SHALLOW DOPANTS IN III-V COMPOUNDS

The usual shallow dopant elements of the III-V semiconductors with zinc blende structure are 1) Group II (acceptors) and group IV (donors) atoms on the cation c.f.c. sublattice of group III atoms and 2) Group IV (acceptors) and group VI (donors) atoms on the anion c.f.c. sublattice of group V atoms. Group IV elements are amphoteric depending on which sublattice they are located. The discussion of the physical reasons why a group IV atom goes

© 1989 IOP Publishing Ltd

preferentially on a given sublattice is interesting but outside the scope of this paper. This amphoteric behaviour is illustrated by Si which can be found on both sublattices at very high concentration (Maguire et al 1987), at a difference with C which has never been detected on the Ga sublattice. Liquid phase epitaxy at low temperature (Kachare et al 1976) can also give p-type GaAs where the dominant acceptor is isolated Si on the As sublattice. However for most technological uses the growth methods are such that Si is incorporated as a donor on the Ga sublattice to concentrations up to about 4×10^{18} at/cm^3. A group III (V) atom on the anion (cation) sublattice should act as a double acceptor (donor) but the "natural" site for these atoms seems to be their own sublattice where they are electrically inactive. However, the electrical activity of group III elements in III-V compounds seems to depend, understandably, upon the stoichiometry of the material. The presence of a double acceptor in GaAs grown from Ga-rich melts (Moore et al 1984) must very likely be ascribed to a group III atom (Ga or B) in the As sublattice (Fischer and Yu 1986) as unambiguous evidence of B in the As sublattice has been found in this kind of material (Gledhill et al 1984). Isolated lithium should behave as an interstitial donor and/or as a substitutional double acceptor in GaAs, but like hydrogen it can complex with many substitutional atoms or even form polyatomic Li complexes (Levy and Spitzer 1973) so that doping of GaAs with lithium has not been widely used except for fundamental studies.

3. HOW HYDROGEN IS INTRODUCED?

Hydrogen has been detected in doped GaAs samples exposed to a RF plasma, which contains atomic hydrogen and protons and in p-type GaAs exposed to a microwave plasma. Exposure to molecular hydrogen seems to be ineffective or much less effective. A good review of the hydrogen plasma exposure results in III-V compounds is given by Dautremont-Smith (1988). The power density of the plasma and the sample temperature are key parameters to determine the penetration depth and the concentration of hydrogen in the sample. Generally, below a perturbed surface region about one hundred nanometers thick, the passivated layer can extend to a few micrometers in the material. Secondary Ion Mass Spectroscopy (SIMS) measurements performed on deuterium plasma passivated GaAs layers show also that the concentration of deuterium introduced increases with the doping level of the material and that the D profile depend on the type of the material: In D-passivated n-type GaAs:Si, the D profile below the D-rich perturbed layer follows more or less a complementary error function below 240°C with a maximum D concentration well above the Si concentration (Jalil 1987). In D-passivated GaAs:Zn (Johnson et al 1986), the D profile at 200°C is flat and the D concentration matches closely the Zn concentration.

Another method to introduce hydrogen or deuterium in III-V compounds is proton or deuteron implantation (Newman and Woodhead 1980). One advantage of this technique is that the dose and the depth at which it is located can be controlled. A disadvantage of implantation is that high-energy bombardment with relatively heavy particles produces lattice defects in the material. Some of these defects are associated with H or D and the IR absorption of their local modes can be detected (Tatarkiewicz et al 1987). The deep levels associated with these defects can trap the free carriers and mask the passivation by mere compensation but annealing of the implanted samples can dissociate the defects and make passivation show up again.

In the IR spectra of Liquid Encapsulated Czochralski (LEC) grown III-V compounds (GaAs, InP and GaP), Clerjaud et al (1987) detected weak lines near 2000 cm^{-1}. The spectral domain was typical of the stretching mode of X-H bonds and this was an indication that hydrogen is introduced inadvertently during the LEC growth of III-V compounds. Similarly, some post-growth treatments of Metal-Organic Chemical Vapour Deposition (MOCVD) grown layers are very likely to introduce hydrogen in the epilayers.

4. ELECTRICAL PROPERTIES OF THE PASSIVATED MATERIAL

When passivation is observed, that is for H concentrations greater than or equal to the initial free carrier concentration, the H concentration and the carrier concentration are anti-correlated as a function of depth. This is an indication that hydrogen is involved in the passivation of the

dopant. The passivation efficiency depends again on the doping level and on the dopant type, but also on the chemical nature of the dopant. This last point suggests the formation of a dopant-hydrogen complex. Under annealing, donors or acceptors can be electrically re-activated with a corresponding decrease of the hydrogen concentration. This suggest a thermally activated process and electrical re-activation energies (the thermal dissociation energies of the complexes) have been obtained on this basis (Pearton et al 1986). The electrical mobility at 77 K in the passivated region is roughly the one expected for the carrier concentration measured (Jalil et al 1986). This is a point against a compensation mechanism since in that case, ionised impurity scattering would reduce the mobility value.

5. VIBRATIONAL SPECTROSCOPY OF HYDROGEN-PASSIVATED MATERIAL

Localised vibrational mode (LVM) spectroscopy is used to detect the presence of hydrogen or deuterium in semiconductors when bonded to an atom X of the material either crystalline or amorphous. The stretching mode frequency of the H atom of a X-H bond lies in the 1700-2400 cm^{-1} spectral range depending on the nature and on the mass of the X atom. In general, the lighter the X atom the higher the frequency ω_{X-H}. A bending mode of the X-H bond can also be observed in the 500-700 cm^{-1} spectral range. The attribution of a given band or line to hydrogen can be made securely only if the corresponding feature is observed in a sample treated with deuterium: In the harmonic approximation, the frequency ω_{X-H} is proportional to $1/\sqrt{\mu_{X-H}}$ where μ_{X-H} is the reduced mass of the X-H harmonic oscillator. Hence the ratio r of the frequencies of the X-H and X-D oscillators in H-mass units is:

$$r = \omega_{X-H}/\omega_{X-D} = \sqrt{(2(X+1)/(X+2))} \qquad [1]$$

This ratio indicates approximately where the D-related line should be found. This ratio depends on the mass of atom X but it cannot be used in the absolute to determine the mass of atom X because the harmonic oscillator approximation is a very drastic one in the case of an atom in a crystal lattice. The experimental r-factors always vary less than the ones calculated using [1]; however, by comparing r-values for unknown atoms with some which have been obtained experimentally for known atoms, it is possible to choose between two very different values for the mass of unknown X, as shown below.

X	Mass	r from [1]	r (measured)
Be	9	1.348	
Si	28	1.390	1.38
As	75	1.405	1.39

In some very favourable cases, when for instance atom X has more than one isotope and when the LVMs are sharp, one can even determine the identity of atom X from the isotopic structure of the X-H line.

If the mass of the X atom is smaller than the masses of the atoms of the compound, the vibration of the isolated X atom lies outside from the one-phonon spectrum of the crystal and it can be observed as a LVM (Newman 1973); When atom X is perturbed by hydrogen, the symmetry of the centre is lowered from tetrahedral (Substitutional atom in zinc blende-type cystal) to trigonal (For X-H along a <111> direction). This results in a splitting of the LVM of atom X, which is a supplementary proof of its coupling with hydrogen.

We have used low-temperature FTS spectroscopy whose sensitivity is essential as the active thickness of the samples ranges from 1 to 5 µm. A carrier concentration near 10^{18}/cm^3 in the

non-passivated region requires us to keep this region as thin as possible. For this reason, we have used doped epilayers deposited by MOCVD or by Molecular Beam Epitaxy (MBE) on semi-insulating substrates.

5.1 N-type GaAs:Si

The first IR line observed in H-passivated n-type GaAs:Si was located at 896.82 cm^{-1} at LHeT and it could be followed up to ambient where it was shifted to 889.8 cm^{-1}. The observation of a corresponding line at 641.52 cm^{-1} (LHeT) in the D-passivated sample convinced us that despite their rather low frequencies, these lines were related to element hydrogen with a ratio r_1 of 1.398 (Jalil et al 1987). These lines were consequently labelled 1-H and 1-D. The intensity of line 1-H can be appreciated from the transmission spectrum of Figure 1. The line looks intense but it is rather sharp too with a full width at half power (FWHP) of ≈ 0.07 cm^{-1} so that the cross section radius (Integrated intensity/concentration of centres) is $\approx 10^{-18}$ cm and this value is similar for instance to the one found for the stretching mode of interstitial oxygen in silicon at LHeT. Later on (Pajot et al 1988a) a second line at 1717.25 cm^{-1}, about twice less intense than line 1-H, was observed in the H-passivated sample. A corresponding line at 1247.61 cm-1 was also observed in the D-passivated sample. These two lines were labelled 2-H and 2-D and the ratio r_2 was 1.376. This latter value of r had been reported for lines observed in c-silicon samples grown in hydrogen and deuterium atmospheres (Bai et al 1985). In samples with an epilayer on each side and a slightly higher Si concentration, weak components were observed (Figure 2) on the low-energy side of 2-H which were due to the less abundant ^{29}Si and ^{30}Si isotopes of silicon (4.7 and 3.1% respectively). This prooved that line 2-H was due to the stretching mode of a Si-H bond. Was line 1 related to line 2? The fact that these two lines were found in all the samples with the same intensity ratio led us to ascribe them to the same centre.

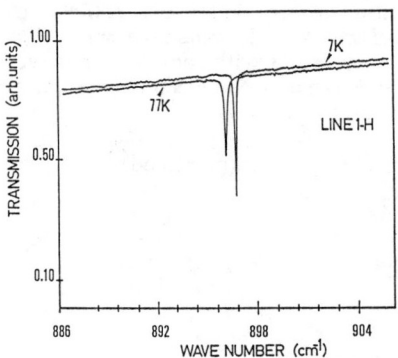

Fig. 1. Transmission of a 4 μm-thick n GaAs:Si single epilayer in the vicinity of line 1-H. Apodised resolution: 0.05 cm^{-1}

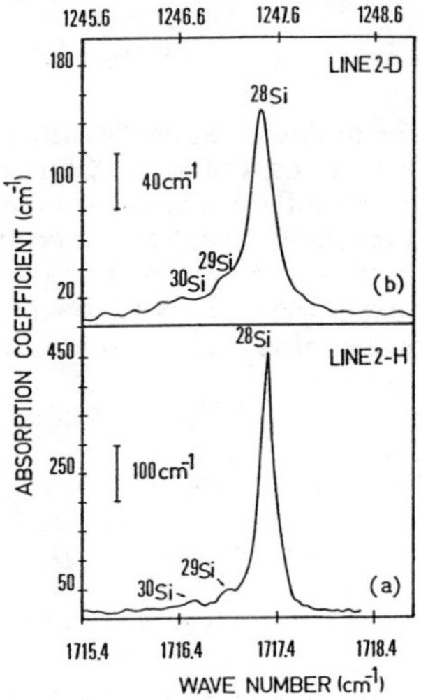

Fig. 2. Absorption of lines 2-H and 2-D in passivated n GaAs:Si. Apodised resolution: 0.1 cm^1. T $\approx$ 6 K

Samples which were passivated with equal proportions of hydrogen and deuterium showed no extra line indicating that only one H atom was associated with the Si atom. At high concentration, Si can be found on both sublattices in GaAs and we had to check the situation for the samples we investigated: At Reading, a non-passivated sample was electron irradiated and measured in the spectral domain 370-405 cm^{-1} were LVMs associated to Si on the two sublattices can be observed (Theis et al 1982). The spectrum showed that the sample contained only Si on the Ga sublattice and it could be confirmed from this that lines 1 and 2 were related to the vibrations of the same Si_{Ga}-H bond. The most plausible attribution for line 1 was the

bending mode of hydrogen with respect to the direction of the Si-H bond.

The microscopic model (Figure 3) to explain the spectroscopic data and the passivation of the Si donor (Pajot et al 1988a) bears some similarity with the one proposed by Johnson et al (1986) to explain the passivation of P donor in n-type silicon, namely that the donor electron forms a supplementary bond for hydrogen to the already fourfold co-ordinated Si atom. In this model, hydrogen is on an antibonding position along the [-1-1-1] direction and the IR results of Bergman et al (1988) on the donor-hydrogen complexes have also been explained along these lines. Recent *ab initio* calculations (Briddon and Jones 1988) actually show that for H in n-type GaAs:Si, this antibonding configuration is energetically more favourable than a bond centred one, but that the strain between the Si atom and the As atom along the [111] direction breaks the Si-As bond. The picture of a H atom bonded to the Si atom is corroborated by a change in the LVM frequency of Si on the Ga sublattice from 384 cm-1 when isolated to 410 cm^{-1} when bonded to H. A second mode is expected below 384 cm-1 but it was not observed. The absence of this mode could be explained by the relaxation of the Si atom which might lower the frequency of the motion of the Si atom perpendicular to the [111] direction.

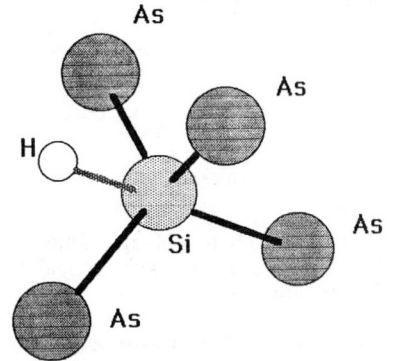

Fig. 3. Configuration of the complex of Si donor with hydrogen in GaAs

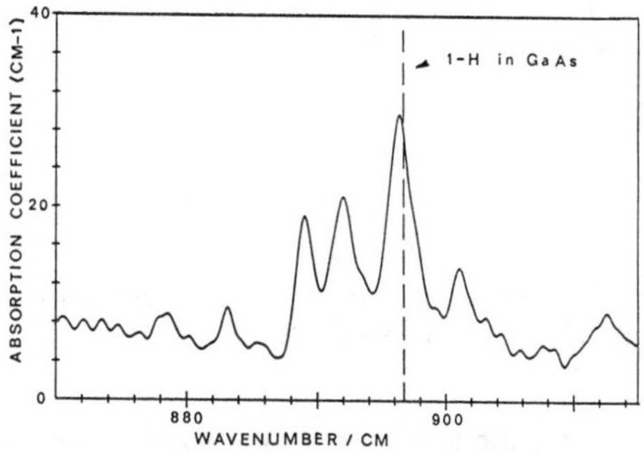

Fig. 4. Absorption of a H-passivated sample of $Ga_{0.8}Al_{0.2}As$ near line 1-H at 6 K

Hydrogen passivation of Si donors together with that of DX centres has also been obtained in $Ga_{1-x}Al_xAs$ alloys (Nabity et al 1987). In H-passivated Si-doped samples with x=0.2 we have observed a broadening and a splitting of line 1-H despite a reduction of the amplitude of the lines: The FWHP increase to about one wave number and the number of components detected is estimated to about 12. The strongest feature is at 896.4 cm^{-1} near the position of line 1-H in pure GaAs, but weak components are spread out down to 878 cm-1. For the above-described Si-H centre, the number of distinct configurations of next nearest neighbour Al atoms quickly increases with their number. For x=0.2, the probabilities of a Si atom having 0, 1, 2, and 3 nnn Al atoms are 0.069, 0.206, 0.284 and 0.236 respectively but for one nnn Al atom there are already 3 possible distinct configurations, 6 for 2 nnn Al atoms and 12 for 3 nnn Al atoms. The feature at 896.4 cm-1 should correspond to symmetric configurations with 0 and 3 nnn Al atoms. For each other configuration, the bending mode is split into two components and this can explain easily the complexity of the structure observed (Figure 4).

The LVMs can be split in a more controlled way with the use of a uniaxial stress along a well-defined crystal axis and the response of lines 1-H and 2-H to a uniaxial stress along a <110> direction of the sample was investigated. No polarisation was used as the experiment was a feasibility test on a composite sample. The spectral resolution used was 0.1 cm^{-1}. Line 1-H starts to split for a stress of 22.2 MPa. Figure 5 shows the splitting as a function of stress. At the maximum applied stress of 155 MPa, the line splits into four components with extreme separation of 1.1 cm^{-1} but the two high-energy components are not resolved and their separation is estimated to 0.2 cm^{-1}. This last figure is also approximately the FWHP of the low

energy components which remain relatively sharp. When considering a trigonal centre like the Si-H bond there are two inequivalent configurations with respect to a <110> perturbation and for each configuration, the twofold degeneracy of the bending mode is lifted. This crude analysis account for the four components observed here for line 1-H and this seems to validate the attribution of the line. Similarly, as expected, line 2-H splits into two components, but at the maximum stress, the splitting is only 0.7 cm^{-1}.

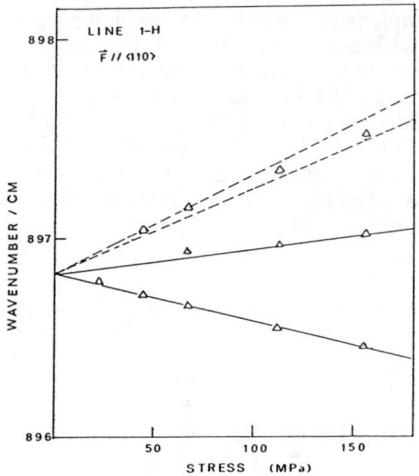

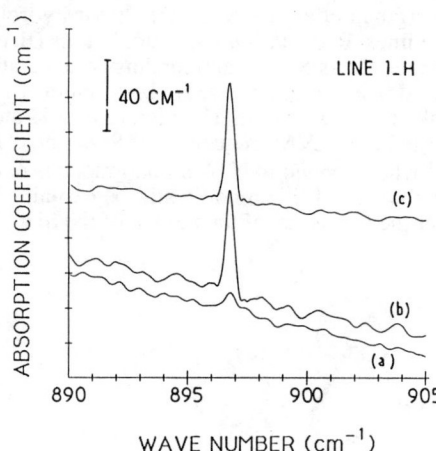

Fig. 5. Splitting of line 1-H under a uniaxial stress along a <110> axis at 6 K

Fig. 6. Line 1-H at 6 K in a H$^+$ implanted sample. (a): as-implanted (b): 200°C annealing (c): 400°C annealing

I have mentioned proton bombardment as a way to introduce hydrogen in semiconductors. This technique is used to produce high-resistivity zones in n$^+$GaAs:Si (Zavada et al 1985) and the important point is carrier removal. With the spectroscopic results in mind, we tried to investigate the possible role of H-passivation in the carrier removal of n-type GaAs:Si under proton implantation (Pajot et al 1988b). For this, 4 μm-thick epilayers doped with ≈ 3x10^{18} Si atoms/cm^3 on each side of the substrate, identical to some used in the H-passivation experiments were implanted with a dose of 5x10^{15}/cm^2 at an energy of 190 keV. This energy ensured for H$^+$ a total path length of ≈ 1.8 μm and a range straggling of ≈ 130 nm. The as-implanted samples showed features associated with H- and D-related intrinsic implantation defects (Tatarkiewicz et al 1987) but line 1-H was also observed in the protonated sample. This was an indication that passivating complexes already formed in the as-implanted material. Annealing at 200°C destroyed one of the H-related complexes (Presumably an As vacancy partially or fully decorated with hydrogen) by breaking the Ga-H bonds and annihilating the vacancy by some process. The second net result of this annealing was an increase of the intensity of line 1-H showing that hydrogen atoms and electrons released from the above defects were trapped by Si ions, forming passivating centres (Figure 6). Using line 1-H as a probe, the presence of passivating centres was still detected after annealing at 400°C but the intensity of the line indicated that the centres had already started dissociating. This study showed that the spectroscopic results on the Si-H centres could be used to observe changes in the carrier trapping during the annealing of proton-implanted n-type GaAs:Si, here, a gradual change from compensation by defects to passivation of the donors.

5.2 P-type GaAs:(Zn or Be)

We investigated first H- and D-passivated GaAs doped with zinc. A line at 2146.9 cm^{-1} in the H-treated sample was shifted to 1549.1 cm^{-1} in the D-treated one (Figure 7). The r-value was

1.386 and it could be interpreted as a Zn-H as well as a As-H binding, but the frequency of the Zn-H mode in organic compounds is near 1600 cm^{-1} compared to about 2110 cm^{-1} for As-H (Herzberg and Herzberg 1972) and the natural isotopic abundance of zinc would produce, if H was bonded to Zn, a line profile not as symmetric as the one observed. These points led us to ascribe the line at 2146.9 cm^{-1} to an As-H stretching mode. We proposed that the formation of the As-H bond resulted from the breaking of a Zn-As bond (Zn is on the Ga sublattice) followed by the saturation by hydrogen of the As dangling bond while the Zn dangling bond is annihilated by the free hole, leaving this atom tri-coordinated (Pajot et al 1987). This model is the same as the one proposed by Pankove et al (1985) to explain the passivation of the boron acceptor by hydrogen in silicon (H bonded to Si and B left tri-coordinated). DeLeo and Fowler (1986) have predicted from their calculations that for boron and aluminium in silicon, if Pankove's model holds globally, there is however some coupling between the acceptor and the H atom, small for boron, but relatively important for aluminium. We have given experimental evidence of the coupling between hydrogen and boron in silicon (Pajot et al 1988c) but it seems difficult to determine experimentally the strength of the interaction between the H and Zn atoms, in GaAs which should exist anyway.

The situation is better for beryllium, another acceptor on the Ga sublattice. There again, after passivation, a line is observed at 2037 cm^{-1} (hydrogen) and at 1471 cm^{-1} (deuterium). The r-factor is 1.385 (1.386 for Zn) and this is a confirmation of the binding of H to the As atom (Nandhra et al 1988). An interesting point is that Be is a light atom with a LVM at 482 cm^{-1} when isolated. After passivation new LVMs appeared at 556 cm^{-1} (hydrogen) and 554 cm^{-1} (deuterium). This can be understood by a change in the force constants of the Be atom after relaxation, but the hydrogen isotope dependence should imply a small interaction between H or D and the Be atom. The *ab-initio* caculations of Briddon and Jones (1988) for the (Be, H-As) complex in GaAs showed indeed a very large relaxation of the Be atom with the H atom in a bond centred position.

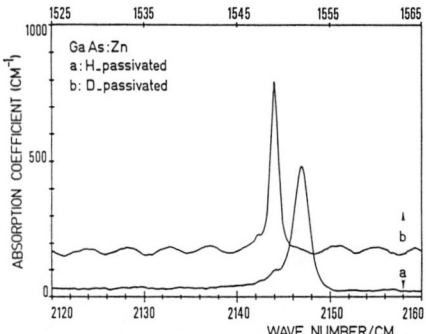

Fig. 7. Absorption of passivated GaAs:Zn The interference fringes in b are due to the substrate. T ≈ 7 K

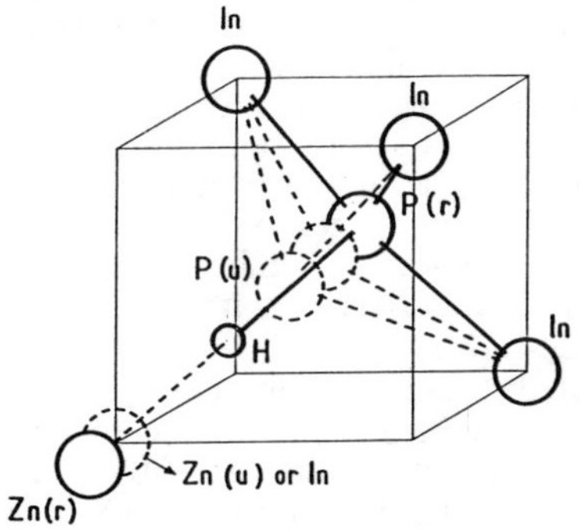

Fig. 8 Configuration of the H-passivating complex for group II acceptors in III-V compounds. Here (InP:Zn), relaxation is indicated

The FWHPs of the As-H and As-D modes in Zn- and Be-doped GaAs are about one order of magnitude larger than that of the Si_{Ga}-H mode. This can be qualitatively understood by the interstitial antibonding position of the H atom bonded to Si which limits the interactions with neighbouring atoms. Another difference is that the As-D lines are systematically sharper by a factor of 2 or more than than the As-H lines and the same phenomenon is also observed for the P-H and P-D modes in Zn-doped InP. Up to now there is no convincing explanation for this fact.

We described briefly the passivation model of acceptors in terms of H atoms and of broken bonds. Looking at the pairing and passivation efficiencies for acceptors in GaAs, it seems very likely that the driving force for complex formation is coulombic interaction. This implies that

in p-type III-V compounds like GaAs and InP, hydrogen diffuses as a proton, which means that similarly to the proposal of Pantelides (1986) for silicon, hydrogen should be regarded also as an interstitial deep donor. Then, another way to describe the passivation of acceptors by hydrogen would be to say that the proton is trapped near a Be or a Zn atom and that the electronic density between the acceptor and the group V atom is modified to produce mainly a (group V atom)-H bond while the acceptor relaxes in a tri-coordinated configuration (Figure 8). In the first H-passivated GaAs:Zn sample we investigated, we reported a weak line at 2144.2 cm^{-1} on the low-energy side of the As-H line. The corresponding line was at 1547.3 cm^{-1} in the D-passivated sample with a r-value of 1.386 (Pajot et al 1987). We argued then that it could result from the passivation of a second acceptor, Cd or Mg. Subsequently, this weak line was found in all the GaAs:Zn samples investigated and I am convinced that this shoulder is related to the zinc doping. There seems to be no physical reason to the existence of a low-energy shoulder of the stretching mode of a trigonal As-H centre: For instance, contrary to the situation for Al- and Ga-doped silicon (Stavola et al 1988), no temperature-dependent structure is observed for the As-H and As-D lines. I propose that because of the similarity between zinc and gallium a small fraction of the passivating complexes could be slightly different from the one described above: The Zn-As bond would remain unchanged, but the proton would be nested between the As atom and one of the three Ga atoms nnn of the Zn atom and this Ga atom would be tri-coordinared. A similar line is not observed in GaAs:Be. This can arise from too large an energy difference for the two minimums when the acceptor potential differs markedly from that of the host group III atom.

5.3 P-type GaAs:Si

The material used was p+ GaAs:Si with p ≈ 7 x 10^{18} cm^{-3}. The hydrogen treated material displayed the usual passivation effect. The distribution of silicon among the different possible sites (Si on the As and Ga sublattices, near neighbour Si pairs, Si complexes (Theis and Spitzer 1984)) was investigated by low-temperature LVM spectroscopy after electron irradiation to reduce the neutral acceptors concentration (Figure 9a). From the relative intensities of the features associated with Si on the two sublattices (Maguire et al 1987) an optically estimated carrier concentration of 1.4 x 10^{19} cm^{-3} was obtained due to Si acceptors alone. Lines are observed at 2095 cm-1 (H-passivated sample) and at 1515 cm-1 (D-passivated sample) with r=1.383 (Chevallier et al 1988). Figure 9b shows the LVMs associated with Si after D-passivation. The drastic reduction of the intensity of the

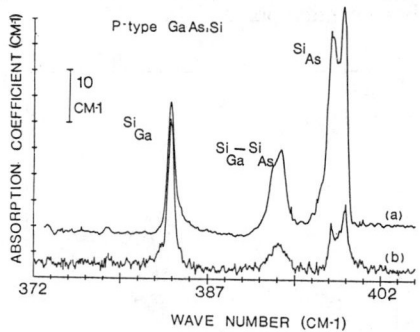

Fig. 9. Si-related LVMs in p$^+$ GaAs:Si at 6K (a): electron-irradiated (b): passivated with deuterium. Resolution: 0.1 cm^{-1}

LVM of isolated Si$_{As}$ is a good indication that the near IR lines are due to Si$_{As}$-H and -D modes. The intensity of the compensating Si$_{Ga}$ donor stays unchanged as there is no free electrons in a p-type material to build Si$_{Ga}$-H complexes. The decrease of the intensity of the feature attributed to Si pairs is unexpected at first sight as the pairs are globally neutral. This point is to be connected to the observation of lines at 972 and 704 cm^{-1} in the H- and D-passivated samples. These lines are not related to a bending mode of Si$_{As}$-H and -D as they were not observed in other p-type GaAs:Si samples; they could be due to a complex of H or D with the Si pairs whose concentration as isolated pairs has strongly decreased after deuteration. The value r=1.383 of the Si$_{As}$-H/Si$_{As}$-D complexes is relatively high. The average value for different "pure" Si-H bonds is 1.375±0.003. For the Si--H--Ga complexes in silicon, it is ≈1.378 (Stavola et al 1988). Such an anomalous value could indicate a strong relaxation of

Si_{As}-H in the plane of the three Ga atoms to which it remains bonded. This could be also the reason for the softening of the LVM of Si_{As} bonded to hydrogen which shifts this mode in the one phonon spectrum where it cannot be detected.

6. CONCLUSION

I have tried to present here what we have learned on the structure of the passivating complexes between hydrogen and shallow dopants in GaAs using high-resolution low-temperature LVM spectroscopy. These experiments were performed on samples with high doping levels for which the passivating efficiency is relatively high but the efficiency decreases with the dopant concentration. H-passivation of Si donors in high-purity GaAs has been reported by Pan et al (1987) using PTIS. Evidences for the reduction of the concentration of isolated Si donors were also derived by Weber et al (1986) from the reduction of the photoluminescence (PL) associated with recombination processes involving isolated donors. The spectroscopic results on the H-passivation of donors in GaAs is now limited to Si. For acceptors, where passivation is easier to obtain, the spectroscopic data indicate that H binds in a bond centred position to the atom on the group V sublattice as this leaves the atom of the group III sublattice quasi tri-coordinated.

Concerning the passivation of the dopants by hydrogen, we have evidence that the passivation in GaAs concerns the net carrier concentration. Compensating shallow donors in p-type material should not be passivated by hydrogen, and with them the same concentration of shallow acceptors should remain ionised. PL meaurements of H-treated GaP samples were interpreted as due to the simultaneous passivation of S donors and Zn or Cd acceptors (Weber and Singh 1988); this could be related to the increased lifetime of the e^- h^+ pairs in this indirect band gap material and to the greater stability of the H-related complexes in GaP. On the other hand, if the compensating species are deep centres and if they can be passivated by hydrogen (Pearton et al 1987) and release at the same time the carrier they trapped, the passivation can then in principle be total. The actual fraction passivated will however be limited by the passivation efficiency. One of the determining parameters is the temperature at which the complex dissociates or more correctly the dissociation energy of the complex. LVM spectroscopy has been used to determine these energies (Bergman et al 1988).

I do not know of results on the deliberate passivation by hydrogen of double donors or acceptors in III-V compounds as the control on the introduction of such dopants is poor. However interactions have been found to exist between double acceptors and hydrogen in silicon and germanium (Muro and Sievers 1986, Kahn et al 1987 and references therein). It turns out that these double acceptors are Zn and Be. We have described the passivation by hydrogen of these elements in GaAs and it could be extrapolated easily to germanium whose mass and atomic radius are not too different from those of Ga and As: In p-type Ge:(Be or Zn), a H atom should bind to a Ge atom nearest neighbour of the group II atom, leaving the latter tri-coordinated. Such complexes would be single acceptors with trigonal symmetry and their observation has indeed been reported (Kahn et al 1987) in Ge:(Be or Zn) grown in a hydrogen atmosphere; It is a nice thing that the passivation studies of acceptors in GaAs can provide a simple explanation to the structure of these complexes.

Acknowledgments: This study was made possible by a close cooperation with J Chevallier and Mrs R Azoulay. The cooperation with R Newman and R Murray was also decisive in this work. I wish to thank B Clerjaud for stimulating discussions and R Jones for sending a very interesting preprint..

REFERENCES

Bergman K, Stavola M, Pearton S J and Lopata J 1988 *Phys. Rev. B* **37** 2770-3
Briddon P and Jones R 1988, these proceedings
Chevallier J, Dautremont-Smith W C, Tu C W and Pearton S J 1985a *Appl. Phys Lett.*, **47** 108-10
Chevallier J, Pajot B, Jalil A, Mostefaoui R, Rahbi R and Boissy M C 1988b *Mat. Res. Soc. Symp. Proc.* **104** eds M Stavola, S J Pearton and G Davies (Pittsburgh: MRS) pp 337-40
Clerjaud B, Côte D and Naud C 1987 *Phys. Rev. Lett.* **58** 1755
Dautremont-Smith 1988, *Mat. Res. Soc. Symp. Proc.* **104** eds M. Stavola, S J Pearton and G. Davies (Pittsburgh: MRS). pp 313-23
DeLeo G G and Fowler W B 1985 *Phys. Rev. B* **31** 6861
Fisher D W and Yu P W 1986 *J. Appl. Phys.* **59** 1952-5
Gledhill G A, Newman R C and Woodhead J 1984 *J. Phys. C* **17** L301-4
Herzberg G and Herzberg L 1972 *AIP Handbook 3rd edn* . **7** 185
Jalil A, Chevallier J, Azoulay R and Mircea A 1986 *J. Appl. Phys.* **59** 3774-7
Jalil A, Chevallier J, Pesant J C, Mostefaoui R, Pajot B, Murawala P and Azoulay R 1987 *Appl. Phys. Lett.* **50** 439-41
Johnson N M, Burnham R D, Street R A and Thornton R L 1986 *Phys. Rev. B* **33** 1102-5
Johnson N M, Herring C and Chadi D J 1986 *Phys. Rev. Lett.* **56** 769-72
Kachare A H, Spitzer W G, Whelan J M and Narayanan G H 1976 *J. Appl. Phys.* **47** 5022-9
Kahn J M, McMurray, Jr R E, Haller E E and Falicov L M 1987 *Phys. Rev.B* **36** 8001-14
Levy M E and Spitzer W G 1973 *J. Phys. C* **6** 3223-44
Maguire J, Murray R, Newman R C, Beall R J and Harris J J 1987 *Appl. Phys. Lett.* **50** 516-9
Moore W J, Shanabrook and Kennedy T A 1984 *Semi-Insulating Materials* eds D C Look and J S Blakemore (Cheshire, England: Shiva) pp 453-8
Muro K and Sievers A J 1986 *Phys. Rev. Lett.* **57** 897-900
Nabity J C, Stavola M, Lopata J, Dautrmont-Smith W C and Tu C W 1987 *Appl. Phys. Lett.* **50** 921-3
Nandhra P S, Newman R C, Murray R, Pajot B, Chevallier J, Beall R B and Harris J J 1988 *Semicond. Sci. Technol.* **3** 356--60
Newman R C 1973 *Infrared Studies of Crystal Defects* (London: Taylor and Francis)
Newman R C and Woodhead J 1980 *Radiation Effects* **53** 41-6
Pajot B, Jalil A, Chevallier J and Azoulay R 1987 *Semicond. Sci. Technol.* **2** 305-7
Pajot B, Newman R C, Murray R, Jalil A, Chevallier J and Azoulay R 1988a *Phys. Rev. B* **37** 4188-94
Pajot B, Chevallier J, Chaumont J and Azoulay R 1988b *Mat. Res. Soc. Symp. Proc.* **104** eds M. Stavola, S J Pearton and G. Davies (Pittsburgh: MRS) pp 345-8
Pajot B, Chari A, Aucouturier M, Astier M and Chantre A 1988c *Solid State Commun.* **67** 855-8
Pan N, Lee B, Bose S S, Kim M H, Hughes S J, Stillmann G E, Arai K and Nashimoto Y 1987 *Appl. Phys. Lett.* **50** 1832-4
Pankove J I, Carlson, D E, Berkeyheiser J E and Wance R O 1983 *Phys. Rev. Lett.* **51** 2224-5
Pankove J I, Zanzucchi, P J, Magee, C W and Lucovsky, G 1985 *Appl. Phys. Lett.* **48** 421-3
Pantelides S 1986 *Mat. Sci. Forum* **10-12** ed H J von Bardeleben (Switzerland: Trans Tech Publications) pp 573-8
Pearton S J, Dautremont-Smith W C, Chevallier J, Tu C W and Cummings K D 1986 *J. Appl. Phys.* **59** 2821-7
Pearton S J Corbett J W and Shi T S 1987 *Appl. Phys A* **43** 153-95
Stavola M, Pearton S J, Lopata J and Dautremont-Smith W C 1988 *Phys. Rev. B* **37** 8313-7
Tatarkiewicz J, Krol A, Breitschwerdt A and Cardona M 1987 *Phys. Stat. Sol. (b)* **140** 369-75
Theis W M, Bajaj K K, Litton C W and Spitzer W G 1982 *Appl. Phys. Lett.* **41** 70-2
Theis W M and Spitzer W G 1984 *J. Appl. Phys.* **56** 890-8
Weber J, Pearton S J and Dautremont -Smith W C 1986 *Appl. Phys. Lett.* **49** 1181-3
Weber J and Singh M 1988 *Mat. Res. Soc. Symp. Proc.* **104** eds M Stavola, S J Pearton and G Davies (Pittsburgh: MRS) pp 325-9
Zavada J M, Jenkinson H A, Sarkis R G and Wilson R G 1985 *J. Appl. Phys.* **58** 3731-4

Inst. Phys. Conf. Ser. No 95: Chapter 7
Paper presented at Int. Conf. Shallow Impurities in Semiconductors, Linköping, Sweden, 1988

The symmetry and properties of donor-H and acceptor-H complexes in Si from uniaxial stress studies

Michael Stavola,[†] K. Bergman,[*] S. J. Pearton,[†] J. Lopata,[†], and T. Hayes[†]

[†]AT&T Bell Laboratories, Murray Hill, NJ 07974
[*]Department of Solid State Physics, University of Lund, S-221 00 Lund, Sweden

ABSTRACT: The symmetries of As-H and B-H complexes are shown to be trigonal in uniaxial stress studies. The evidence provided for the structure of donor-H complexes by the vibrational spectroscopy and the stress results is discussed. The motion of H in the B-H complex from bond centered site to bond centered site is characterized by a stress-induced dichroism technique.

1. INTRODUCTION

In silicon, the passivation of the group III acceptors, B, Al and Ga, is well known and has been studied extensively.[1] The recently discovered passivation of group V donors by hydrogen[2] is a weaker effect than the passivation of acceptors and is not as well studied or understood. The donor-H and acceptor-H complexes provide excellent model systems for the study of H-related defects. Here we examine these complexes with infrared absorption and uniaxial stress to gain insight into their structure and properties.

Of the models that have been proposed for B-H complexes, the model where the H atom is bond-centered between the boron and one of the boron's Si nearest neighbors (Fig. 1a) is gaining acceptance. This model was proposed by Pankove[3] and was later confirmed to be the lowest energy configuration in cluster calculations by DeLeo and Fowler.[4] Recent calculations by Chang and Chadi[5] and by Denteneer et al.[6] also find that this configuration has lowest energy. Ion channeling and nuclear reaction analysis experiments on B-D complexes[7,8] have determined that the D occupies the bond-centered site between boron and one of its Si neighbors.

Recent Hall effect measurements by Johnson et al. have demonstrated the weak passivation of Si:P by H_2 plasma exposure.[2] The formation of donor-H complexes was inferred to be responsible for the passivation from the increase in the Hall mobility. Donor-H complex formation was confirmed by the observation of H vibrations at passivated P, As, and Sb donors by Bergman et al.[9]

On the basis of semiempirical calculations, Johnson et al.[2] put forward a structural model for the donor-H complex in which the H atom is attached to one of the donor's Si nearest neighbors in an antibonding position along the <111> axis (Fig. 1b). Chang and Chadi[5] have recently performed ab initio pseudopotential calculations and have found that the structure shown in Fig. 1b is the most stable for P-H complexes.

Here we describe infrared absorption results for B-H and As-H complexes under uniaxial stress. Both complexes are shown to have trigonal symmetry. In addition, we have found that the H can move from bond-centered site to bond-centered site in the B-H complex as has been suggested by recent theoretical calculations.[6] We use a stress induced dichroism technique to determine the kinetics of hydrogen motion.

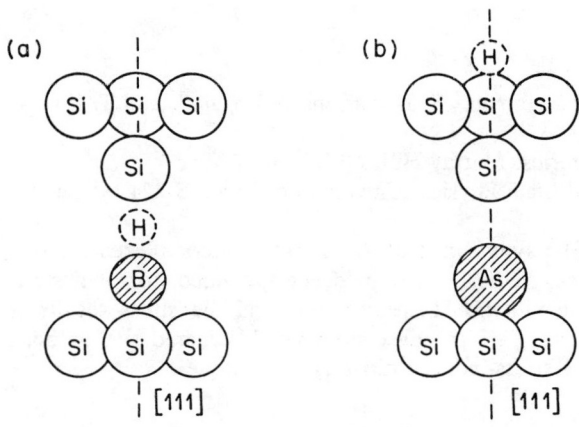

Fig. 1 Schematic models of a) the B-H complex and b) the As-H complex.

2. EXPERIMENTAL PROCEDURES

The samples for this study were made from floating zone Si that was ion implanted with B or As. For B the energies were 30,100 and 180 keV and the dose was 7×10^{14} cm^{-2}. For As, the energies were 30 and 100 keV and the dose was 1×10^{15} cm^{-2}. The implants were activated by a rapid thermal anneal at 1200°C for 60s. Stress samples were bar-shaped, typically $8 \times 2 \times 2$ mm^3, with the long axis parallel to the stress direction. Samples were hydrogenated or deuterated in a Technics Planar Etch II, plasma reactor at 120°C for 3h on each side.

The spectra were measured with a Nicolet 60SX Fourier transform spectrometer equipped with a variable temperature cryostat. The transmitted light was polarized with a wire-grid polarizer on a KRS-5 substrate. Stress was applied with a push-rod system attached to the cold finger of the cryostat.

3. DONOR-HYDROGEN COMPLEXES

The absorption bands assigned to H-stretching at donor-H complexes[9] are shown in Fig. 2 for the different donors. In spite of large differences in the donor masses and sizes, the stretching bands are all close in frequency. We also note that the frequencies of these bands are lower than the H-stretching frequencies of the acceptor-H complexes (~1560 cm^{-1} as compared to ~2200 cm^{-1}).

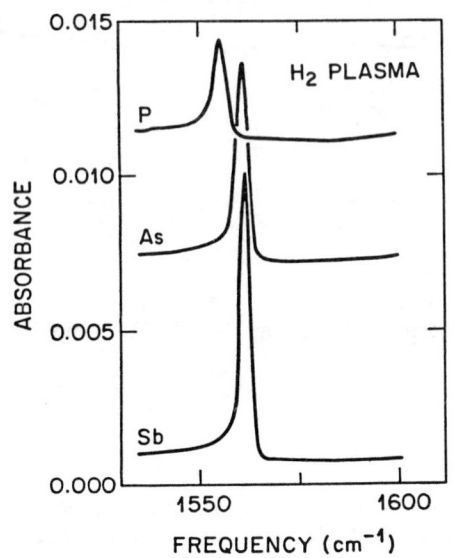

Fig. 2 H-stretching bands for donor-H complexes measured near liquid He temperature.

The uniaxial stress dependence of the H-stretching frequency and the polarization selection rules for the As-H complex are shown in Fig. 3. The stress splitting pattern is characteristic of a nondegenerate mode of a center with trigonal symmetry whose orientational degeneracy is lifted by the stress. (There is no [001] stress splitting because all the trigonal axes are at the same angle to the [001] stress axis.)

A band at 810 cm^{-1} with twice the intensity of the ~1560 cm^{-1} band has been assigned to the doubly degenerate wagging mode of the donor-H complexes.[9] The analysis of the uniaxial stress dependence of this band is more complicated than for the 1560 cm^{-1} band because the stress lifts the vibrational degeneracy as well as the orientational degeneracy. We have shown that the stress dependence of the 810 cm^{-1} band confirms the trigonal symmetry of the As-H complex and the double degeneracy of the wagging mode.[10]

The vibrational spectroscopy and uniaxial stress data support the structural model of the As-H complex shown in Fig. 1b. For such a complex we expect little or no dependence of the vibrational frequencies on the chemical identity of the donor because the H is attached to Si. The lower vibrational frequency expected for the antibonding configuration as compared to the bond-centered configuration that typifies acceptor-H complexes is in accord with experiment. The trigonal symmetry of the complex is also in accord with the uniaxial stress results. Additional support, described in detail elsewhere,[10] for the antibonding site for the H comes from the signs of the stress coupling parameters. Although we have not measured the stress dependence of the vibrations of the P-H and Sb-H complexes we expect results (and also structures) similar to As-H because of the similarity in the vibrational characteristics.

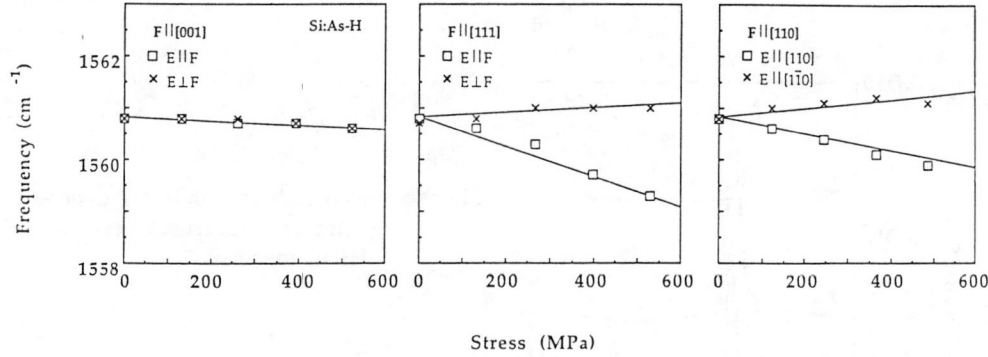

Fig. 3 Stress dependence of the H-stretching mode for the As-H complex measured near 15K.

4. B-H COMPLEX

Uniaxial stress measurements provide a further confirmation of the symmetry of the B-H complex. The frequency and polarization dependence of the 1903 cm^{-1} absorption band under stress (Fig. 4) are consistent with a trigonal complex in agreement with theoretical models[4-6] and channeling results.[7,8] The splittings, however, are much larger than is typical for local vibrational modes. (One would expect splittings of 1 or 2 cm^{-1} at these stress values like we have observed for the donor-H complexes rather than ~10 cm^{-1}). We speculate that the large stress splittings may be related to the ease of off-axis motions and distortions for these complexes. Our notion is that an off axis distortion can reduce the compression in the crowded, on-axis bonds and cause the vibrational frequencies to increase under stress.[10]

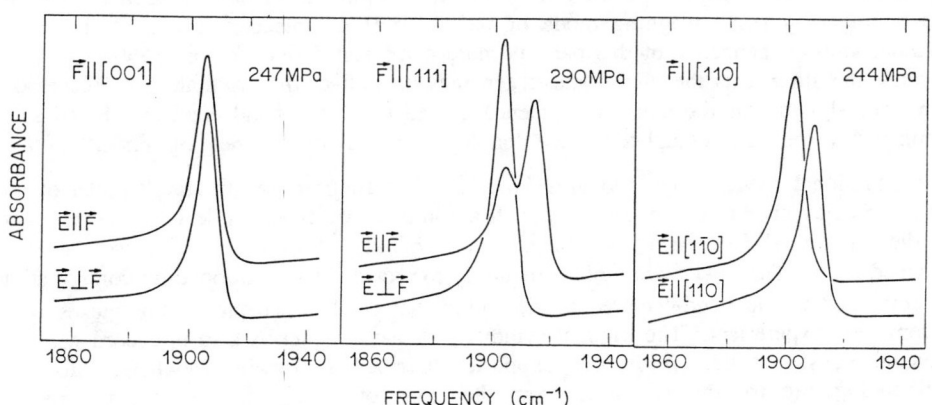

Fig. 4 The H-stretching vibration of the B-H complex under uniaxial stress measured near 15K.

In a related set of stress experiments we have found that the hydrogen can move from bond-centered site to bond-centered site.[11] A uniaxial stress applied at temperatures near 60K creates a preferential alignment of the hydrogen along those <111> axes with the least component of the applied stress along them. This alignment results in a large optical dichroism for the B-H absorption bands (i.e. a difference in the absorption strength for light polarized // and ⊥ to the applied stress.) When a sample is cooled under stress, and then the stress is removed, the hydrogen will redistribute among the bonds. The timescale for redistribution can be determined by observing the decay of the optical dichroism. In Fig. 5 we plot the time constant for the decay of dichroism vs. reciprocal temperature. The slope of this line gives an activation energy of 0.19 eV for the decay of the dichroism and hence also the H motion from one bond centered site to another.

Chang and Chadi[5] and Denteneer et al.[6] have found that a distortion perpendicular to the bond for the B-H complex requires only a small energy. In the calculation by Denteneer et al.[6] there is essentially a channel in the energy surface in which the H can move easily about the B. Our measurement confirms the existence of this low energy pathway and determines the barrier for motion (i.e. the corrugation) in this channel.

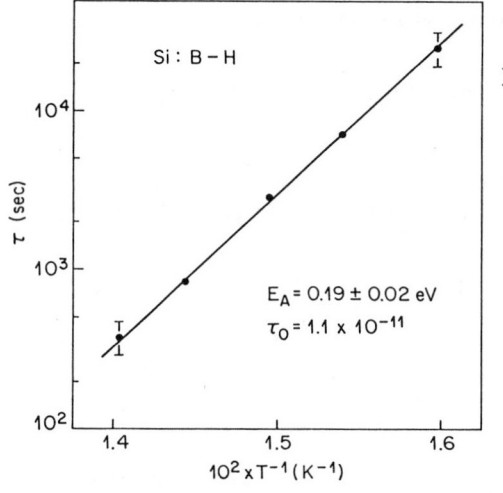

Fig. 5 Time constant for the decay of the stress-induced optical dichroism vs. T^{-1} for the B-H complex.

5. CONCLUSION

IR absorption spectroscopy and uniaxial stress techniques provide unique information on the structure and internal dynamics of H-related complexes. For donor passivation in Si, the discovery of vibrational bands has established the existence of donor-H complexes.[9] The frequencies of vibration, their dependence on the donor species, and the symmetry determined by uniaxial stress[10] help to establish the structural model for the complexes that has been proposed on the basis of calculations.[2,5] For acceptor passivation, infrared spectroscopy determines vibrational frequencies[12] that can be calculated and hence

provides a test of theoretical models. The surprising appearance of additional sideband structure on the H stretching bands points to H motions in the acceptor-H complexes that are not yet understood.[12] Furthermore, stress induced dichroism techniques have revealed the internal dynamics of H in the B-H complex,[11] in strikingly good agreement with theory.[6]

REFERENCES

1. S. J. Pearton, J. W. Corbett, and T. S. Shi, Appl. Phys. A *43*, 153 (1987).
2. N. M. Johnson, C. Herring, and D. J. Chadi, Phys. Rev. Lett. *56*, 769 (1986).
3. J. I. Pankove, P. J. Zanzucchi, C. W. Magee, and G. Lucovsky, Appl. Phys. Lett. *46*, 421 (1985). In this work the H is proposed to be bonded only to Si and the boron is suggested to be tricoordinated. The configuration found in subsequent calculations has a much smaller B relaxation and the hydrogen is bound to both B and Si.
4. G. G. DeLeo and W. B. Fowler, J. Electron. Mater. *14a*, 745 (1985).
5. K. J. Chang and D. J. Chadi, Phys. Rev. Lett. *60*, 1422 (1988).
6. P. J. H. Dentaneer, C. G. Van de Walle, Y. Bar-Yam and S. T. Pantelides, submitted, Phys. Rev. B; Proc. 15th Int. Conf. on Defects in Semiconductors, Budapest, 1988.
7. A. D. Marwick, G. S. Oehrlein, and N. M. Johnson, Phys. Rev. B 4539 (1987).
8. B. Bech Nielsen, J. U. Andersen, and S. J. Pearton, Phys. Rev. Lett. *60*, 321 (1988).
9. K. Bergman, M. Stavola, S. J. Pearton, and J. Lopata, Phys. Rev. B *37*, 2770 (1988).
10. K. Bergman, M. Stavola, S. J. Pearton and T. Hayes, submitted to Phys. Rev. B.
11. M. Stavola, K. Bergman, S. J. Pearton, and J. Lopata, unpublished.
12. M. Stavola, S. J. Pearton, J. Lopata, and W. C. Dautremont-Smith, Phys. Rev. B *37*, 8313 (1988).

Shallow acceptor action of a C-H pair in silicon and germanium

L V C Assali, V M S Gomes and J R Leite

Instituto de Física da Universidade de São Paulo, CP 20516,
01498 São Paulo, SP, Brazil

H Chacham and J L A Alves

Instituto de Ciências Exatas da Universidade Federal de Minas Gerais,
CP 702, 30161 Belo Horizonte, MG, Brazil

ABSTRACT: Electronic-state calculations have been carried out for a C_s-H_i pair complex in Ge and Si. The H 1s-derived state gives rise to an a_1 level which merges as a resonance into the valence band when the H atom moves closer to the C impurity. As a consequence, a s-like a_1 hole state is pushed into the band gap. This indirect effect induced by H explains the origin of the shallow acceptor activity observed for the C-H pair in Ge.

1. INTRODUCTION

The physical properties of hydrogen-related defects in semiconductors have deserved a great deal of attention in the last years due to their importance for the electronic device tecnology based on crystalline, polycrystalline and amorphous materials. It is well known that the addition of atomic hydrogen to doped elemental semiconductors neutralizes single shallow acceptors, converts double shallow acceptors to single shallow acceptors and induces shallow acceptor activities in an otherwise electrically inactive group-IV dopant. Hydrogenation of B- and Be-doped Si and C-doped Ge, respectively, are good examples of these phenomena. It has been currently accepted, and in some cases demonstrated, that such effects are associated to the formation of pair complexes of atomic H and the corresponding substitutional impurity. The microscopic structure and the electronic properties of such pairs have been subject of several investigations (Pearton et al 1987). Despite all these efforts, an unified view of the physical properties of these complexes is still lacking. A particularly important question to be addressed is the origin of the shallow acceptor activity observed for the C-H pair in Ge (Haller 1986, Kahn et al 1988).

In the present work rigorous self-consistent-field electronic state calculations have been carried out for the C-H pair complex in Ge and Si. The stability and electronic properties of this pair were analysed within the framework of the microscopic structure and computational technique adopted by Assali and Leite (1985) to describe the properties of the B-H pair in Si. Accordingly, the H atom is placed at an antibonding interstitial site configuration close to the C impurity and a molecular cluster model associated to the multiple-scattering (MS) $X\alpha$ technique is used to describe the pair electronic structure.

© 1989 IOP Publishing Ltd

2. THEORETICAL MODEL

Molecular clusters of 26 atoms centered at the tetrahedral interstitial site (T), were assumed to simulate the electronic structure of the perfect Ge and Si crystals. These clusters are indicated by 26Ge and 26Si, respectively. The isolated T-site interstitial H impurity was simulated by placing the H atom at the center of the clusters. This impurity in Ge and Si is then described by the 26Ge + H_i and 26Si + H_i clusters, respectively. By replacing one Ge or Si atom nearest-neighbor of H in the clusters by the C impurity, we simulate the complex C_s-H_i in Ge or Si, respectively. These clusters are denoted by 25Ge + C_sH_i and 25Si + C_sH_i. The MS-Xα cluster models discussed here have been previously used by Assali and Leite (1985) and Leite and Assali (1987) to describe the B-H pair in Si, by Assali et al (1985) to describe the Au-Fe pair in Si, by Assali and Leite (1987) to describe the Fe-B pair in Si and by Gomes and Leite (1985) to study thermal donor activities in Si. We report here the first application of the model to the study of defects in crystalline Ge (Alves and Leite 1986).

The schematic representation of the cluster model used in this investigation is shown in Figure 1. The cluster electronic-state calculations are carried out for several positions of the H impurity along the $\langle 111 \rangle$ direction indicated by the line drown in Figure 1.

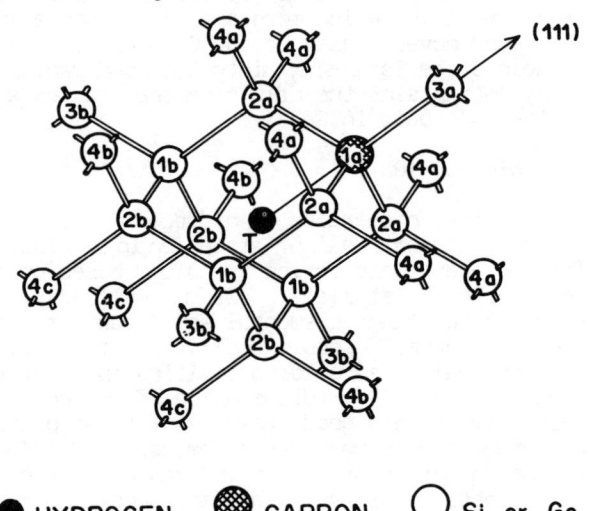

Fig. 1. Schematic representation of the cluster used to simulate the C_s-H_i pair complex in silicon and germanium. The cluster is centered at the T-site and comprises one central H atom and four shells of host atoms, numbered from 1 to 4, in T_d symmetry. The cluster symmetry is lowered to C_{3v} when ones replace a nearest-neighbor host atom by a C atom. The atoms in C_{3v} symmetry belong to classes of equivalence that are indicated by the labels a, b, and c.

● HYDROGEN   ◉ CARBON   ○ Si or Ge

3. ELECTRONIC STRUCTURE OF THE C-H PAIR IN Ge AND Si

We start by first analysing the results obtained for the 26Ge cluster which simulates the perfect Ge crystal. Although the relativistic effects have been neglected in our calculations, we conclude that the energy spectrum of the 26Ge cluster provides a fairly consistent description of the crystal band edges. The comparison between theory and experiment is made in Figure 2. The value 0.8 eV obtained for the crystal band gap is exactly the measured result reported by McLean (1960). The value 11.18 eV obtained for the valence band width is also in agreement with the experimental result, 12.6 eV, reported by Grobman et al (1975).

We discuss now the electronic structure of a T-site H impurity in Ge. The

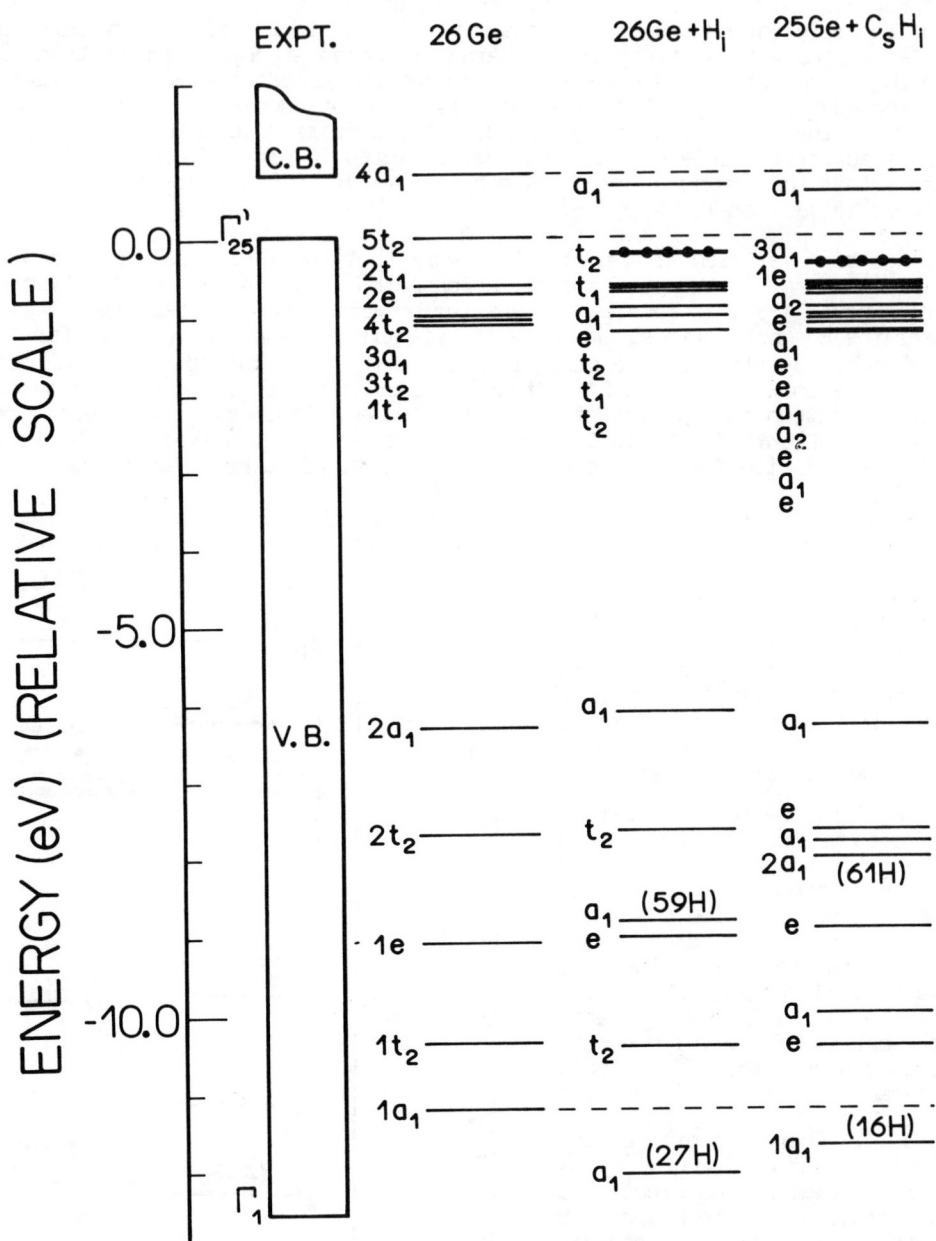

Fig. 2. Energy spectra of the 26Ge, 26Ge + H_i, and 25Ge + C_sH_i clusters simulating the electronic structure of Ge, T-site H impurity, and C-H pair complex, respectively. The experimental values for the germanium valence-band width and band gap are also shown. The filled circles indicate the occupancy of the levels close to the top of valence band. All the remaining levels below these are fully occupied. The numbers between parenthesis give the percentage of charge within the H sphere (radius = 2.32 a.u.).

energy spectrum of 26Ge+H_i cluster depicted in Figure 2 indicates that the impurity introduces an a_1 resonance in the lower half of the crystal valence band with a strong H 1s character and an a_1 hyperdeep state close to the bottom of the valence band which retains a moderate H 1s character. In the neutral charge state the impurity also introduces a delocalized hole close to the top of the valence band. The comparison between the spectrum obtained for the 26Ge cluster and that depicted in Figure 2 for the 26Ge+H_i shows that the perturbation caused by T-site H in the bulk valence band states is very small.

It has been verified from theoretical work that the T-site is a position of minimum energy in the possible diffusion paths of H in Ge and Si (Van de Walle et al 1988). Thus, we are assuming that when H reaches a T-site near the C substitutional impurity it gets trapped and a complex is formed. The energy spectrum of the cluster 25Ge+C_sH_i, simulating the C_s-H_i pair in Ge is shown in Figure 2. It is remarkable the striking similarity observed between the spectra of the 26Ge+H_i and 25Ge+C_sH_i clusters depicted in Figure 2. This interesting feature indicates that the H impurity at a T-site feels the presence of C as it was a host Ge atom.

The energy spectra of the 26Si+H_i and 25Si+C_sH_i clusters, simulating the electronic structure of a T-site H impurity and a C_s-H_i complex in Si, respectively, are shown in Figure 3. The comparison between the results shown in Figure 2 and 3 indicates clearly that the T-site H impurity in Ge and Si are similar centers. The same conclusion remains valid as far as the C_s-H_i pair in Ge and Si are concerned.

We analyse now the interesting effects that occur when H moves closer to the C impurity along the direction $\langle 111 \rangle$ indicated in Figure 1. Figure 4 shows that, as the H atom gets closer to the C site, the 1s is -derived hyperdeep level, labelled as $1a_1$, moves towards the bottom of the valence band and assumes the Γ_1 bulk character. The H 1s-derived

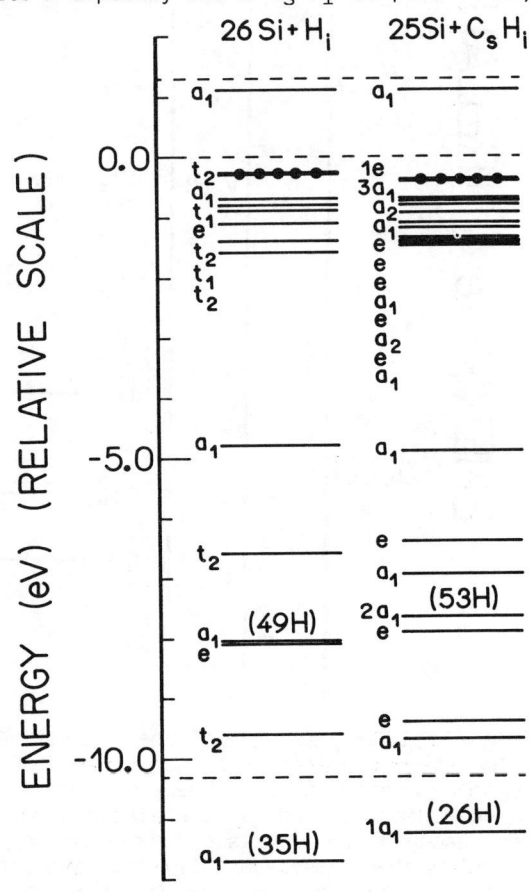

Fig. 3. Energy spectra of the 26Si+H_i and 25Si+C_sH_i clusters simulating the electronic structure of T-site H and C-H systems in silicon, respectively. The solid circles indicate the occupancy of the levels close to the top of the valence band. The numbers between parenthesis give the percentage of charge within the H sphere (radius = 2.22 a.u.).

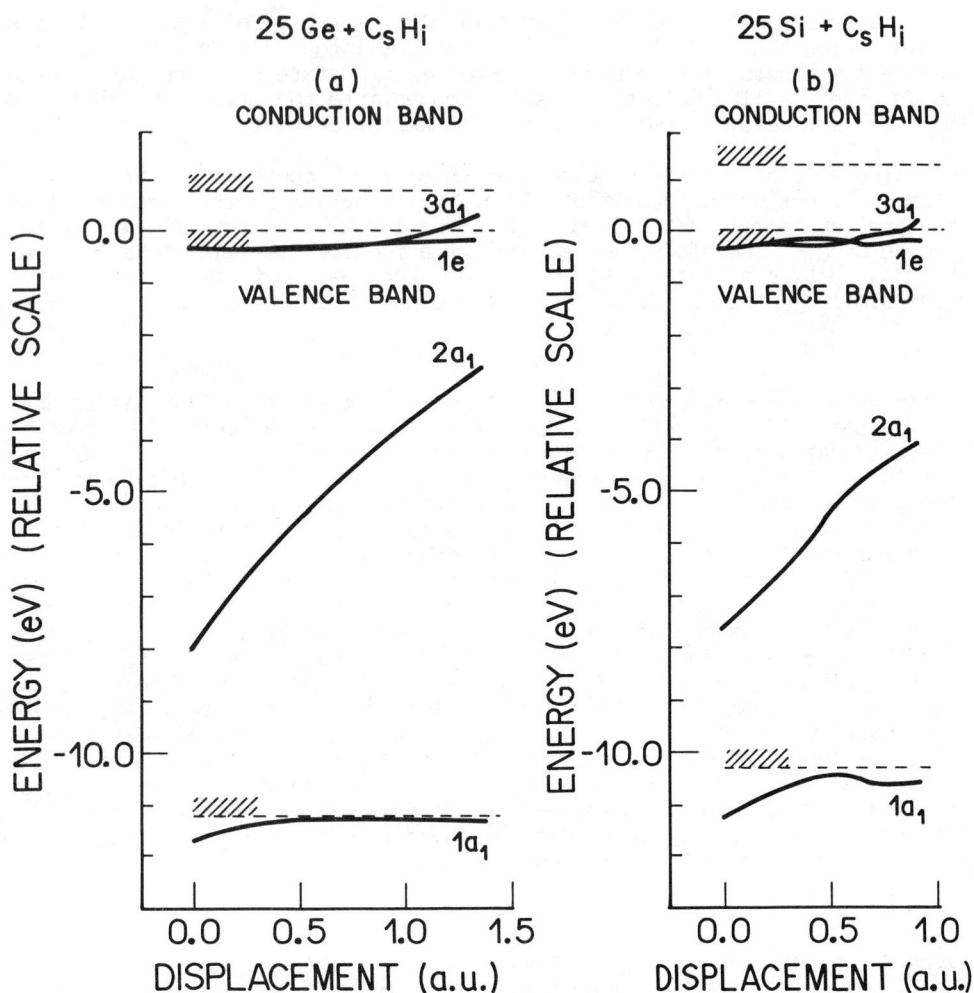

Fig. 4. Behavior of relevant energy levels of the C_s-H_i complex in (a) silicon and (b) germanium as a function of the hydrogen displacement towards the substitutional carbon impurity. The zero of displacement corresponds to the hydrogen at the T-site. The levels are those indicated in Figures 2 and 3.

resonance, labelled as $3a_1$, displaces from the lower half of the valence band upwards to the top. As the H becomes closer to the C impurity, the H 1s-derived resonance hybridizes strongly with the C 2p-derived resonant states, giving rise to a C-H bond.

It is interesting now to observe the behavior of the top of the valence band, $5t_2$ level, as the H impurity comes closer to the C impurity. When H is located at the T-site, the $5t_2$ level appears as a resonance just below the top of the valence band comprising a single hole state. The small C-H interaction is not enough to split this level as should be expected for a C_{3v} crystal field. As the C-H chemical bond starts to form, the $5t_2$ level splits into a twofold-degenerated level and a 1s-like non degenerated

level, labelled as 1e and $3a_1$, respectively. Both, the 1e and $3a_1$ levels remains below the top of the valence band until the H 1s-derived $2a_1$ resonance state gets close enough to push the $3a_1$ state into the gap. As a consequence a s-like hole state is introduced into the crystal band gap and the C-H pair becomes an active shallow acceptor center.

The calculations show that the activation of the C center in Ge or Si is an indirect effect produced by the H impurity. The hole state introduced into the band gap is originated from a perturbation of the top of the crystal valence band. Therefore, we expect an effective mass-like behavior for the $3a_1$ active acceptor state. Indeed, the $3a_1$ and $5t_2$ states have a quite similar charge distribution in the cluster regions.

4. FINAL REMARKS

As final remarks we would like to point out first the essential differences between the electronic structure of the C-H pair in Ge and Si discussed here and that obtained previously by Assali and Leite (1985) for the B-H pair in Si. The results obtained for the B-H pair show that there are strong interactions between the H 1s-derived state and the B and Si host states, even when H is located at the T-site (Leite and Assali 1987). These interactions are such that the hyperdeep $1a_1$ level induced by H at the T-site disappears and no well identified H-derived levels can be found in the spectrum. Since there is no extra states in the valence band created by the presence of the H atom, one electron is left to neutralize the B shallow acceptor activity. On the other hand, the results obtained for the C-H pair in Ge and Si show that the introduction of an extra $2a_1$ state in the valence band subtracts one electron from the $5t_2$ level, creating an active shallow hole state. Finally we would like to remark that we obtain a higher energy for the $25Ge + C_sH_i$ cluster with the H atom at the T-site than with the H located closer to the C impurity. This indicates that the C-H pair in Ge displays a total energy minimum along the $\langle 111 \rangle$ direction. This local minimum is not found for the C-H pair in Si, according to our preliminar total energy analysis.

REFERENCES

Alves J L A and Leite J R 1986 Phys. Rev. B 34 7174
Assali L V C and Leite J R 1985 Phys. Rev. Lett. 55 980
Assali L V C, Leite J R and Fazzio A 1985 Phys. Rev. B 32 8085
Assali L V C and Leite J R 1987 Phys. Rev. B 36 1296
Gomes V M S and Leite J R 1985 Appl. Phys. Lett. 47 824
Grobman W D, Eastman D E and Freeouf J L 1975 Phys. Rev. B 12 4405
Haller E E 1986 Festkörperprobleme XXVI 203
Kahn J M, McMurray, Jr. R E, Haller E E and Falikov L M 1987 Phys. Rev. B 36 8001
Leite J R and Assali L V C 1987 Proc. 18th Int. Conf. on the Physics of Semiconductors ed O Engstrom (Singapore: World Scientific) 2 pp 999-1002
McLean T P 1960 Progress in Semiconductors ed A F Gibson (New York: Wiley) 5 pp 53-102
Van de Walle C G, Bar-Yam Y and Pantelides S T 1988 Phys. Rev. Lett. 60 2761

Inst. Phys. Conf. Ser. No 95: Chapter 7
Paper presented at Int. Conf. Shallow Impurities in Semiconductors, Linköping, Sweden, 1988

Ab-initio calculations of the passivation of shallow impurities in GaAs

P Briddon and R Jones

Dept. of Physics, University of Exeter, Stocker Road, Exeter, EX4 4QL, U.K.

Abstract

The structures, vibratory modes and electronic properties of H in GaAs, GaAs:Be GaAs:Si are determined using ab-initio local density functional pseudopotential theory applied to 56 or 57 atom clusters such as $Ga_{13}As_{13}H_{30}$. Neutral H marginally favours a tetrahedral interstitial site, but H in GaAs:Be lies near a Be-As bond centre whereas it prefers an antibonding site in GaAs:Si. The vibratory frequencies of H are in reasonable agreement with those observed and the filled states are pushed towards the band edges, thus explaining the role of H in neutralising Si and Be impurities.

§ 1 Introduction

We report the results of state-of-the-art calculations on the effect of hydrogen upon shallow acceptor (Be) and donor (Si) impurities in GaAs (both replace Ga atoms). We use a self-consistent, local density functional pseudopotential method applied to a large stoichiometric cluster (~56 atoms) whose surface bonds are terminated by hydrogen atoms (Jones et. al. 1986, Jones 1987a, 1987b, Jones 1988, Briddon et. al. 1988).

Hydrogen and deuterium passivate both deep levels as well as shallow ones (Pearton 1982, Lagowski et. al. 1982, Chevallier et. al. 1985, Johnson et. al. 1986b, Pearton et. al. 1986, Jalil et. al. 1986, Weber et. al. 1986, Clerjaud et. al. 1987). The passivation depth of H is inversely proportional to the impurity concentration suggesting a pairing effect, but the main evidence for this comes from high resolution infra-red absorption experiments (Pajot et. al. 1988, Nandhra et. al. 1988). For GaAs:Si, Pajot et al (1988) conclude that H is bonded to pentavalent silicon and lies in an antibonding position along [-1-1-1]. Our work substantially confirms this, but we find that the Si-As bond along [111] is broken and Si is then tetravalent. Another possible model is that of bond centred hydrogen (DeLeo et. al. 1985, Estle et. al. 1987, Van de Walle et. al.1988, Briddon et. al. 1988). We shall show below that this defect has a higher energy than H at an antibonding site, a H stretch frequency which is far from the observed one, and a filled deep level in the

© 1989 IOP Publishing Ltd

gap which would lead to electrical activity.

For GaAs:Be, Nandra et al (1988) conclude that H is a bond centred defect lying between Be_{Ga} and a neighbouring As. We find that this gives a H stretch frequency close to the observed one and the gap completely cleared of defect levels.

§ 2 Pure Gallium Arsenide

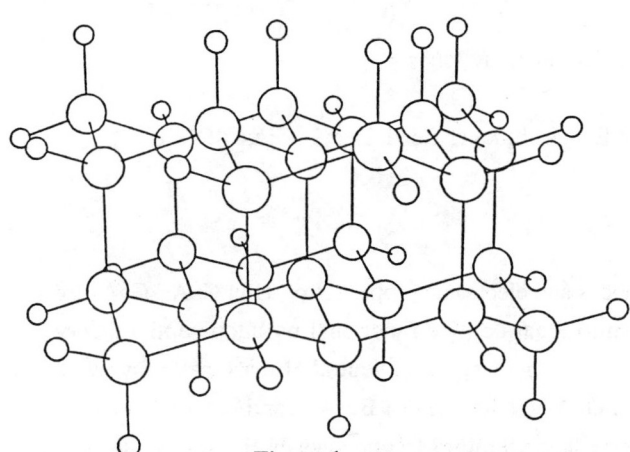

Figure 1 : 56 atom cluster

Small stoichiometric clusters of GaAs whose surface is saturated with H (bond length 1.59Å) gave Ga-As bond lengths of 2.41 Å. We therefore built a 56 atom cluster $Ga_{13}As_{13}H_{30}$ with the tetrahedral structure with all Ga-As bond lengths 2.41Å. The inner 8 atoms (4 Ga, 4 As) were then relaxed, keeping all others fixed. The average bond length between these 8 atoms was 2.40Å with a rms variation of 0.03Å or 1%. The experimental bond length is 2.45Å (2% longer). All angles were within 2% of their tetrahedral values. A Mulliken bond population analysis showed a net transfer of ~0.1e to the bond centre from each Ga atom giving each Ga atom a charge of ~0.4e while leaving the As atoms uncharged. The band gap is 3.5eV, its excessive size being typical of a cluster calculation with relaxed surface H atoms.

Fitting a valence force potential to the second derviatives of the energy yielded an optical phonon freqency (at Γ) of 8.8 THz. The experimental values for the longitudinal and transverse optic modes at Γ are 8.6 THz and 8.0 THz respectively.

§ 3 Interstitial H in GaAs

A neutral hydrogen atom was placed at the tetrahedral interstitial site with Ga nearest neighbours and relaxed, together with its 4 surrounding atoms. It remained at this site with very little distortion of the neighbouring atoms. The partially filled level lay in the lower half of the gap at $E_v + 0.9eV$.

A neutral hydrogen atom was then placed at a bond centre and relaxed together with the 8 surrounding atoms. The GaAs bond length increased by 42%. The energy of this structure is 0.2eV above the tetrahedrally placed interstitial. However as this difference is so small, futher calculations would be necessary on larger clusters to confirm it. The partially filled energy level fell in the upper half of the gap at $E_v+2.56eV$.

This suggests that in p-type GaAs the bond centred site would be more stable than the tetrahedral site, with the reverse true in n-type GaAs. We find this to be the case below.

§ 4 Passivation of the Be acceptor : $(Be)_{Ga}$

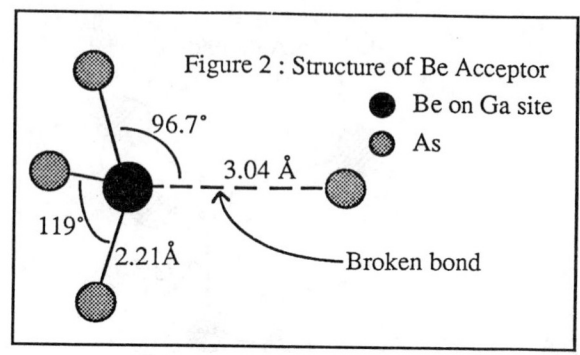

Figure 2 : Structure of Be Acceptor
● Be on Ga site
◎ As
Broken bond

Replacing the central Ga atom with Be and relaxing it together with the four surrounding atoms gave a partially filled (acceptor) level at $E_v + 0.06eV$. The structure of the defect was quite unexpected : the Be atom was displaced along [-1-1-1] creating an almost planar arrangement with three of its As neighbours (fig 2). The remaining Be-As bond along [111] was broken : its 'length' became 3.04Å. The wave function of the shallow partially filled level was substantially localised on a p-orbital on the 3-fold co-ordinated As atom and very little amplitude on the Be atom. This could imply a very large dipole moment for $(Be)^-_{Ga}$ as is indeed observed (Nandra et. al. 1988). Furthermore, it also implies a large overlap with a hydrogenic 1s wave function if the latter is located at the 'bond centre'.

We then inserted H into this 'bond centre' and relaxed it and the surrounding 8 atoms. The relaxed H atom was located 1.54Å from As and 1.77Å from Ga. The filled level occurs at $E_v + 0.04eV$ - neutralising the acceptor. The calculated vibration frequencies of the H were 2083 cm^{-1} for the stretch mode and 383, 346 cm^{-1} for the bend modes (the latter should be degenerate - the splitting is an indication of the numerical error). Our stretching frequency is in excellent agreement with the experimental value of 2037 cm^{-1}; the bending modes have not been reported. The wave function of the highest filled level is now delocalised over the cluster.

§ 5 Passivation of the Si donor : $(Si)_{Ga}$

We relaxed a Si atom placed at a Ga site together with its 4 surrounding neighbours. We found $(Si)_{Ga}$ to be a tetrahedral on-site defect with a Si-As bond length of 2.38Å. It has a partially filled level just below the conduction band.

We inserted a H atom into an antibonding site (fig. 3) along [-1-1-1] and relaxed it, together with the surrounding 8 atoms. The [111] Si-As bond became very strained, effectively breaking to leave the As 3-fold co-ordinated. The H-Si length is close to a normal one but the bond angles are heavily distorted : the three other Si-As bonds are almost planar showing sp^2 hybridisation. The highest filled level is low in the gap at $E_v + 0.3eV$ and its wave function (fig. 4) has its amplitude shared between the 3-fold co-ordinated As atom and the H. However, given the uncertainty in the energy levels implicit in LDF theory it is not possible to be certain that this level lies in the gap. The H stretch and bend frequencies are

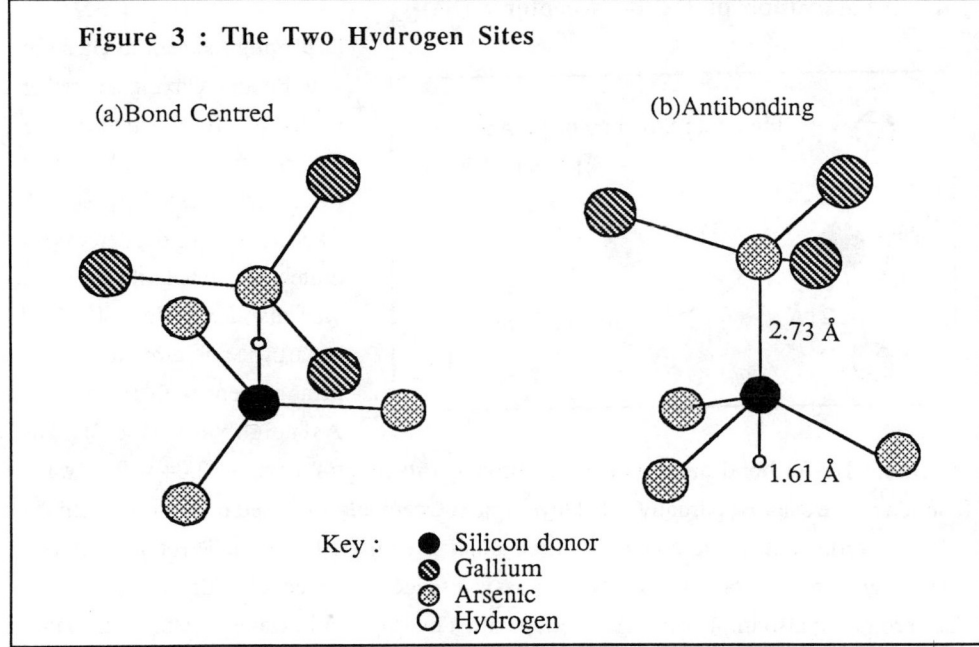

Figure 3 : The Two Hydrogen Sites

(a) Bond Centred

(b) Antibonding

2.73 Å
1.61 Å

Key : ● Silicon donor
       ◐ Gallium
       ◎ Arsenic
       ○ Hydrogen

1592 cm^{-1}, 1046 cm^{-1} and 916 cm^{-1}. Again, the last two modes should be degenerate and the splitting shows numerical errors but is almost certainly dominated by asymmetry in the relaxed structure - it should have C_{3v} symmetry. Further calculations on a larger cluster are necessary to get more accurate values. The observed LVM's are 1717 cm^{-1} and 896 cm^{-1}.

Figure 4 : Wavefunction for antibonding H

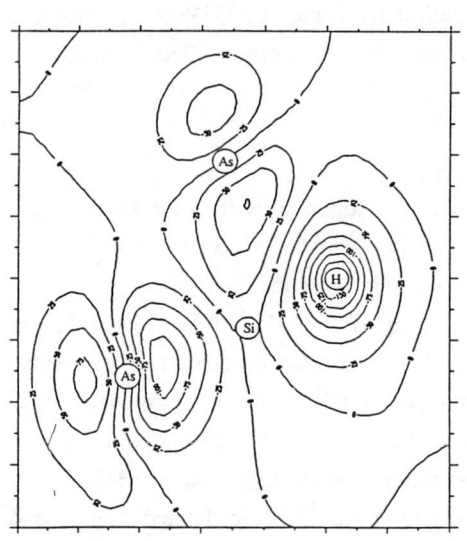

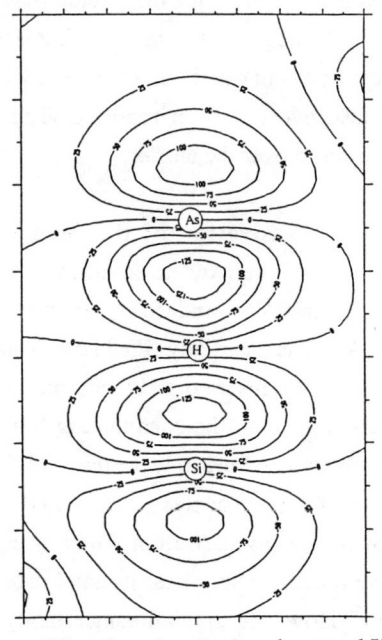

Figure 5 : Wavefunction for bond centred H

We then investigated the structure of the bond centred H. Inserting H between Si and one As neighbour and relaxing 9 atoms gave a defect with an energy higher than that of the antibonding hydrogen defect (above) by 1.2eV. There is a 42% Ga-As bond expansion typical of such bond centred defects (Briddon et. al. 1988). The highest filled level is just above mid gap at $E_V + 2.25eV$. Its wave function is made up of two antibonding p-orbitals on the adjacent Si and As atoms (fig 5). The stretch and bend H frequencies occur at 1389 cm^{-1}, 786 cm^{-1} and 792 cm^{-1} respectively. These properties make it unlikely to be the defect responsible for passivating GaAs:Si.

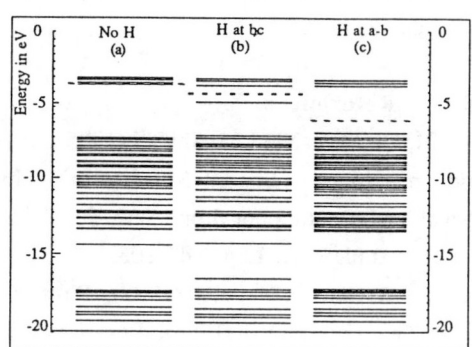

Figure 6 : The calculated enegy levels for $(Si)_{Ga}$. The dotted line separates the occupied states from those unoccupied at 0 K.

The gap at ~-15eV is due to the lack of d states in our calculation.

§ 6 Summary and Discussion

The ab-initio LDF pseudopotential method yields a GaAs bond length within 2% of the correct value and an excellent optical phonon freqency at Γ.

The stable site for neutral interstitial H appears to be the tetrahedral site, but the energy difference between that and the bond centred site is only 0.2 eV. The energy levels are however very different : bond centred H gives a level in the upper half of the gap whereas tetrahedral H gives a level in the lower half of the gap suggesting that in n-type GaAs the T-site would be preferred while in p-type GaAs the reverse would be true.

$(Be)_{Si}$ is an off-centred defect with C_{3v} symmetry. It consequently should have two localised vibratory modes but only one at 452 cm^{-1} (Nandhra et. al. 1988) has been reported. It may be that the others are broad resonances. Uniaxial stress measurements on the 452 cm^{-1} line would be invaluable. There is a very shallow level above E_V with a wavefunction having a large amplitude on the 3-fold co-ordinated As atom. Upon passivation, this overlaps strongly with the hydrogenic 1s wave function depressing the level. The As-H bond length of 1.54Å is close to that in AsH$_3$ (1.52Å) explaining the closeness of their stretch frequencies.

$(Si)_{Ga}$ is an on-centre defect with an energy level just below the conduction band. The passivated defect consists of H displaced from the tetrahedral interstitial site towards the anti-bonding site. Si is also displaced breaking the Si-As bond and leaving three Si-As bonds in a planar configuration. The calculated vibratory frequencies of H are in rough

agreement with those observed and the filled level falls very low in the gap. The alternative bond centred H defect is meta-stable having an energy 1.23 eV higher than the anti-bonding site. It also has a level just above mid-gap and a H-stretch frequency much lower than the observed one.

§7 Acknowledgements

We thank Ron Newman for sending us preprints of his work and for useful discussions. We have also benefitted from numerous discussions with our colleagues Malcolm Heggie and Grenville Lister.

§8 References

Bai GR, Qi MW, Xie LM, Shi TS (1985), Solid State Commun., **56** 277.
Briddon P, Jones R, Lister GMS, (1988), J. phys. C : Solid State Phys. Submitted.
Chevallier J, Dautremont-Smith WC, Tu CW, Pearton SJ (1985),
   Appl. Phys. Lett., **47** 108.
Clerjaud B, Côte D, Naud C (1987), Phys. Rev. Lett., **58** 1755.
DeLeo GG, Fowler WB (1985), Phys Rev, B **31** 6861.
Estle TL, Estreicher S, Marynick DS (1987), Phys Rev Lett., **58** 1547.
Jalil A, Chevallier J, Azoulay R, Micea A (1986), J. Appl. Phys., **59** 3774.
Johnson NM (1985), Phys. Rev., B **31** 5525.
Johnson NM, Herring C, Chadi DJ (1986a), Phys Rev Lett., **56** 769.
Johnson NH, Burnham RD, Street RA, Thornton RL (1986b), Phys. Rev., B **33** 1102.
Jones R, Sayyash A (1986), J. Phys. C : Solid St. Phys., **19** L653.
Jones R, (1987a), J Phys C : Solid St. Phys., **20** L271.
Jones R, (1987b), J Phys C : Solid St. Phys., **20** L713.
Jones R, (1988), J Phys C : Solid St. Phys. Submitted.
Lagowski J, Kaminska M, Parsey JM, Gatos HO, Lichtensteiger M (1982), Appl. Phys.
   Lett., **41** 1078.
Nandhra PS, Newman RC, Murray R, Pajot B, Chevallier J, Beall RB, Haris JJ (1988),
   Semicon. Sci. and Tech. - to be published.
Pajot B, Newman RC, Murray R, Jajil A, Chevallier J, Azoulay R (1988), Phys. Rev.,
   To be published.
Pearton S (1982), J. Appl. Phys., **53** 4509.
Pearton SJ, Dautremont-Smith WC, Chevallier J, Tu CW,Cummings KD (1986),
   J. Appl. Phys., **59** 2821.
Stavola M, Pearton SJ, Lopata J, Dautremont-Smith WC (1987),
   Appl. Phys. Lett., **50** 1056.
Van de Walle CG, Bar-Yam Y, Pantelides ST, (1988), Phys. Rev. Lett., **60** 2761.
Weber J, Pearton SJ, Dautremont-Smith WC (1986), Appl. Phys. Lett., **49** 1181.

Inst. Phys. Conf. Ser. No 95: Chapter 7
Paper presented at Int. Conf. Shallow Impurities in Semiconductors, Linköping, Sweden, 1988

Hydrogen passivation of deep and shallow levels in carbon-doped silicon

A E Jaworowski, J H Robison* and S R Hayden*

Physics Department, Wright State University, Dayton, OH 45435, USA
*Present address: Wright-Patterson Air Force Base, OH 45433, USA

Hydrogen passivation of carbon-related complexes and boron acceptors has been observed in p-type carbon-doped silicon. Neutralization of boron acceptors was measured by capacitance-voltage profiling of Ti/Si and Al/Si Schottky diodes. Short term reverse bias annealing was used to monitor hydrogen motion in the near-interface region of silicon. Deep level transient spectroscopy of 2-MeV electron irradiated samples revealed the effect of hydrogen motion into the silicon lattice on deep level profiles. The spectacular phenomenon of the reinjection of hydrogen into the interface region of silicon has been confirmed.

1. INTRODUCTION

The hydrogen passivation of shallow and deep levels in semiconductors is now well established (Pearton et al 1987, Corbett et al 1988). It is known that hydrogen can neutralize shallow acceptors (Pankove et al 1983, Sah et al 1983, Pankove et al 1985), and the model proposed by Pankove et al (1985) has been supported by calculations (DeLeo and Fowler 1985,1986; Bonapasta et al 1987) and experiments (Johnson 1985). In the proposed model, the hydrogen interrupts the silicon-boron bond, bonds itself to the silicon and permits the boron to relax toward the plane of its three silicon neighbors. Thus, neither the Si-H pair nor the three-fold coordinated boron is electrically active. Recently, Kazmerski (1988) has directly demonstrated the validity of the Pankove model for the hydrogen passivation of a number of shallow acceptors in silicon.
Despite the intensified research activity in the area of hydrogen/silicon effects which resulted in better understanding of the role of hydrogen in silicon technology (Seager et al 1987, Pearton et al 1987), the mechanism of deep level passivation is not clear (Corbett et al 1988). The importance of the role that hydrogen plays in modern silicon device technology has been recently recognized because its introduction into the silicon lattice at almost every stage of device processing (Seager et al 1987, Pearton et al 1987). The unintentional hydrogenation effect is now well documented (Seager et al 1987, Jaworowski et al 1985, Mu et al 1986, Tavendale et al 1985,1986; Pearton et al 1986, Tavendale et al 1988). Recently, for the first time the phenomenon of hydrogen injection into the silicon lattice has been directly observed (Jaworowski 1988). The observed formation of a subsurface hydrogen barrier appears to be a general effect and property of the near-surface

© 1989 IOP Publishing Ltd

of silicon, and probably other semiconductors.
In this paper, we present results on in-diffusion from the near-surface hydrogen layer. Short term reverse bias annealing is used to activate a rapid motion of hydrogen in the near-surface region of Schottky diodes. The effects of hydrogen on both shallow and deep levels are simultaneously observed in electron-irradiated carbon-doped silicon. Our results indicate that the effect of spontaneous injection of hydrogen into the silicon lattice (Jaworowski 1988), results in the unintentional hydrogenation of the near-interface region of Schottky diodes. The subsequent in-diffusion of hydrogen from the near-interface region has been found to be significant.

2. EXPERIMENT

The samples used in this study were Chochralski grown ($O_i=9.6\times10E17/cm^3$), carbon-doped ($C_s=11\times10E17/cm^3$), p-type ($B=4.2\times10E15/cm^3$) silicon, and were cut from adjacent positions of the wafer. Schottky contacts were made by electron-beam evaporation of titanium or aluminum. A conventional DLTS spectrometer was used for deep level characterization and capacitance-voltage (C-V) carrier profiling of Al/Si and Ti/Si Schottky diodes. The samples were mounted on a water-cooled sample holder and were irradiated at 290K with 2-MeV electrons at a flux of $5\times10E12$ e$^-$/cm^{2}s. A 2.45 GHz oscillator was employed for hydrogen plasma passivation at a beam energy of 50 eV. Alternatively, short term reverse bias annealing at 400K was used to activate hydrogen motion in the near-interface region of diodes.

3. RESULTS AND DISCUSSION

Figure 1 shows the acceptor concentration profiles measured at 300K in a Schottky diode after each of the four annealing (5 min) steps at 400K. Reverse bias of 5, 10, 15 and 20V was applied at the annealing temperature of 400K. A combination of elevated temperature (400K) and application of a reverse bias voltage to the diode results in a decrease in acceptor concentration at depth corresponding to the maximum extent of the depletion region. As seen in Figure 1, five minutes at each voltage is adequate for substantial motion of hydrogen at 400K. The maximum concentration of H is approximately $3\times10E15$ atoms/cm^3. As the reverse bias is increased, hydrogen is drawn deeper into the bulk. If an anneal is done at a voltage lower then the previous step, uniform diffusion occurs and the peak diminishes in concentration.

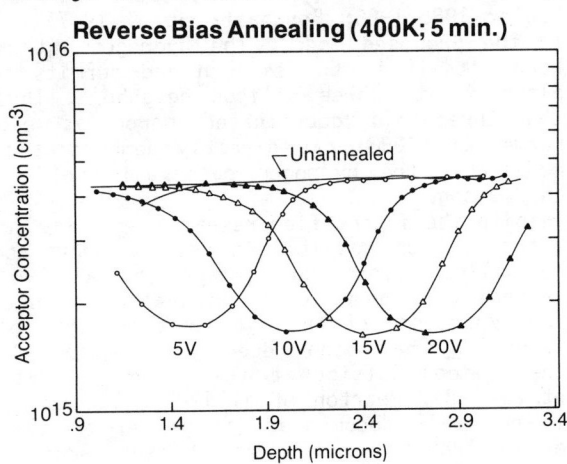

Figure 1. Acceptor profiles in a Ti/Si Schottky diode after reverse bias (5, 10, 15 and 20V) annealing at 400K (5 min).

Figure 2 shows the evolution of the electrically active boron profile as a function of the annealing time and different reverse bias voltages. The hydrogen concentration is at the beginning located at a depth corresponding to 20 V bias. Then, subsequent anneals result in: diffusion (step b: 5 min at 20V, and step c: 5 min at 2V), reverse modification (step d: 5 min at 20V), and again in diffusion (step e: 15 min at 20V). After numerous annealing steps (and reversible hydrogen redistribution) the hydrogen concentration is so low throughout the sample that reverse annealing no longer has any influence on the measured profiles.

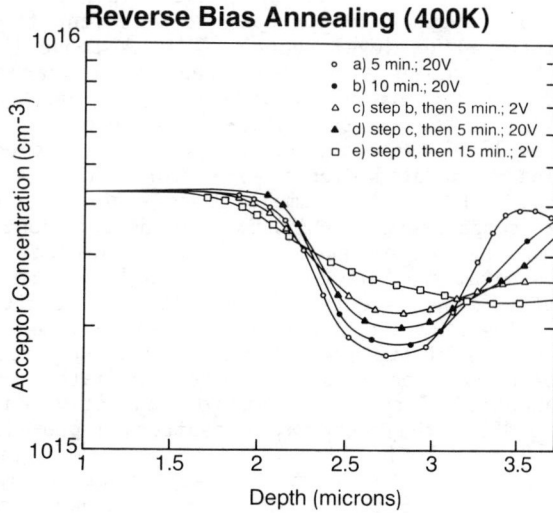

Figure 2. Acceptor profiles in a Ti-Schottky p-Si diode as a function of reverse bias (2 and 20V) and annealing time.

When the samples are left at room temperature for several hours or days, the hydrogen concentration at the near-surface <u>returns</u> and the process may be repeated, though the concentration is is not as high as present initially. The neutralization profiles were identical in both Ti/Si and Al/Si diodes.

In order to see effects of hydrogen motion on deep levels, electron irradiations were performed on the Schottky diodes. In Figure 3, a typical DLTS spectrum is shown for a sample irradiated to a dose of $2 \times 10E16$ e/cm^2. The spectrum has been measured after a 15-minutes anneal at room temperature. The dominant level H2 observed immediately after the irradiation at 0.28 eV above the valence band is attributed to the carbon interstitial [C_i] (Watkins and Brower 1976, Kimerling et al 1977, Asom et al 1987). The well known H3 level $E_v + 0.38$ eV (Mooney et al 1977) has been recently firmly indentified by Trombetta and Watkins (1987,1988) as an interstitial carbon – interstitial oxygen pair [$C_i + O_i$]. The level H1 at $E_v + 0.19$ eV corresponds to the divacancy (Corbett and Watkins 1961). During the room temperature

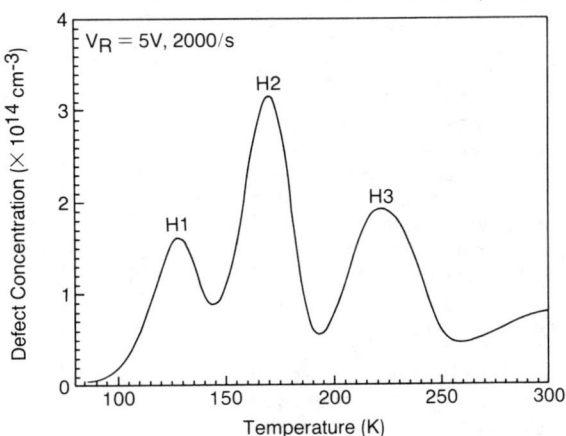

Figure 3. Typical DLTS spectrum of CZ carbon-doped p-type silicon irradiated at 290K with 2-MeV electrons to a dose of $2 \times 10E16$ e/cm^2 at a flux of $5 \times 10E12$ e/cm^2 s.

annealing of the carbon-interstitial, the interstitial carbon - interstitial oxygen pair increases almost four times in concentration. Furthermore, a new level appears 0.12 eV above the valence band. We have found this level as a dominant trap in high-carbon and low-oxygen polycrystalline silicon and therefore we tentatively correlate it with the interstitial carbon - substitutional carbon pair (Song et al 1988). The divacancy level (H1) is not affected by the room temperature annealing.

The irradiated diodes were stored at room temperature for at least one day and then the samples were exposed to hydrogen plasma while held at a temperature of 250°C. The defect concentrations and profiles were measured after each exposure to hydrogen plasma. The hydrogen plasma passivation for 20 min virtually eliminated the electrical activity of all the defects, while annealing at 250°C alone had almost no effect on the DLTS spectra of electron-irradiated samples.

For the first time, effects of hydrogen motion on the DLTS spectrum have been observed in-situ in the irradiated samples which were reverse bias annealed. The hydrogen motion was activated by the reverse bias annealing at 400K, the acceptor concentration profile was measured for the particular applied voltage, and then the DLTS profiling of deep level concentration to the depth corresponding to hydrogen in-diffusion has been performed. Figure 4 shows the concentration profile of the C_i-O_i pair. As the hydrogen is drawn into the sample, the defect concentration decreases. The defect concentration is the lowest at depth corresponding to the highest hydrogen concentration (and the lowest acceptor concentration). The observed correlation clearly demonstrates that the fast diffusing hydrogen is involved in the neutralization process of both deep and shallow levels. It has recently been recognized (Tavendale et al 1986, Pearton et al 1986) that a number of low-temperature procedures lead to significant hydrogen injection with near-surface acceptor neutralization and the electric field drift experiments (Tavendale et al 1986, Zundel et al 1987, Tavendale et al 1988) were used to monitor hydrogen motion into the bulk of silicon. In these experiments, the reverse bias annealing was performed at 80-90°C for several hours or days. We have recently reported (Robison and Jaworowski 1988) that short term (5 min) reverse bias annealling at 400K allows for a "real-time" in-situ monitoring of hydrogen motion in the near-surface of silicon. The observed rapid motion of hydrogen in the near-surface region of the Schottky diodes

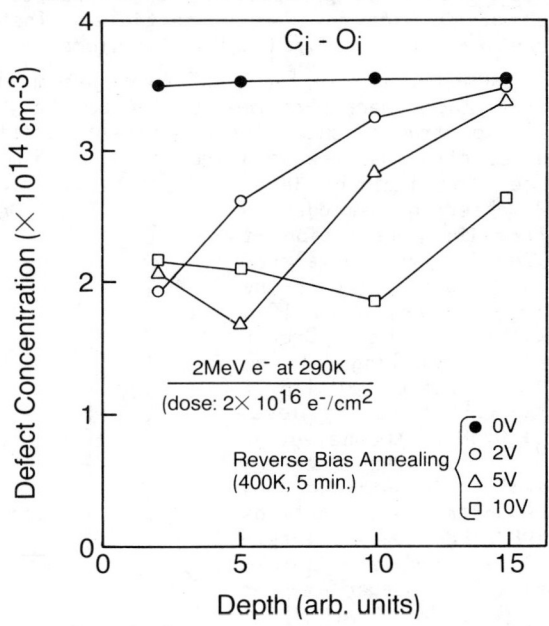

Figure 4. Profiles of the carbon interstitial - oxygen interstitial (C_i-O_i) pair in a diode subjected to reverse bias annealing, V_R= 2, 5 and 10V, at 400K (5 min).

results in hydrogen passivation of both deep and shallow levels. In this work, we have demonstarted that hydrogen which is injected into the near-interface region of silicon exposed to room air (Jaworowski 1988) can be easily driven into the bulk of silicon by short term (5 min) reverse bias annealing at relatively low (400K) temperature. We have also observed that hydrogen reappeared in the subsurface region of the samples which were annealed out and had no measurable hydrogen content. The observed <u>reinjection</u> of hydrogen into the near-surface region is rather spectacular effect which clearly demonstartes that hydrogen is easily and inadvertently introduced into the lattice of silicon, and probably other semiconductors. The exact physicochemical mechanism of of the observed hydrogen injection which results in unintentional hydrogenation of the near-interface region of Schottky diodes is not yet clear.

In conclusion, we have confirmed that hydrogen is spontaneously injected into the near-surface region of silicon exposed to room air. The hydrogen passivation of deep and shallow levels in reverse bias annealed Schottky diodes was easily achieved due to significant in-diffusion of hydrogen which was injected into the subsurface.

ACKNOWLEDGMENTS

The authors wish to thank Al Smith for fabrication of Schottky diodes used in this work. One of us (AEJ) has been supported in part by the Research Council of Wright State University.

REFERENCES

Asom M T, Benton J L, Sauer R and Kimerling L C 1987 Appl.Phys.Lett. **51** 256
Bonapasta A A, Lapiccirella A, Tomassini N and Capizzi M 1987 Phys.Rev.B **36** 6228
Corbett J W and Watkins G D 1961 Phys.Rev.Lett. **7** 314
Corbett J W, Linstrom J L and Pearton S J 1988 Mat.Res.Soc.Symp.Proc. Vol.104 229
DeLeo G G and Fowler W B 1985 Phys.Rev.B **31** 6861
DeLeo G G and Fowler W B 1986 Phys.Rev.Lett. **56** 402
Hansen W L, Pearton S J and Haller E E 1984 Appl.Phys.Lett. **44** 889
Jaworowski A E, Wielunski L S and Listerman T W 1985 Mat.Res.Soc. Symp.Proc. Vol.46 561
Jaworowski A E and Wielunski L S 1988 Mat.Res.Soc.Symp.Proc. Vol.104 304
Jaworowski A E 1988 Surf. Interface Anal. (to be published)
Johnson N M 1985 Phys.Rev.B **35** 5525
Johnson N M Herring C and Chadi D J 1986 Phys.Rev.Lett. **56** 769
Kazmerski L L 1988 (to be published)
Kimerling L C 1977 Inst.Phys.Conf.Ser. No. 31 221
Mooney P M, Cheng L J, Suli M, Gerson J D and Corbett J W 1977 Phys.Rev.B **15** 3836
Mu X C, Fonash S J and Singh K 1986 Appl.Phys.Lett. **49** 67
Pankove J I, Wance R O and Berkeyheiser J E 1983 Phys.Rev.Lett. **51** 2224
Pankove J I, Zanzucchi P J, Magee C W and Lucovsky G 1985 Appl.Phys.Lett. **46** 412
Pearton S J, Tavendale A J, Williams A A and Alexiev D 1986 Electroch. Soc.Proc. Vol.86-4 826
Pearton S J, Corbett J W and Shi T S 1987 Appl.Phys.A **43** 153
Robison J H and Jaworowski A E 1988 Bull.Am.Phys.Soc. **33** 1250

Sah C T, Sun J Y and Tzou J J T 1983 Appl.Phys.Lett. **43** 204
Seager C H, Anderson R A and Panitz J K G 1987 J.Mat.Res. **2** 96
Song L W, Zhan X D, Benson B W and Watkins G D 1988 Phys.Rev.Lett. **60** 460
Tavendale A J, Alexiev D and Williams A A 1985 Appl.Phys.Lett. **47** 316
Tavendale A J, Williams A A and Pearton S J 1986 Appl.Phys.Lett. **48** 590
Tavendale A J, Williams A A and Pearton S J 1988 Mat.Res.Soc.Symp.Proc. Vol.104 285
Trombetta J M and Watkins G D 1987 Appl.Phys.Lett. **51** 1103
Trombetta J M and Watkins G D 1988 Mat.Res.Soc.Symp.Proc. Vol.104 93
Watkins G D and Brower K 1976 Phys.Rev.Lett. **36** 1329
Zundel T, Courcelle E, Mesli A, Muller J C and Siffert P 1986 Appl.Phys. A **40** 67

Hydrogen passivation of indium acceptors in silicon

A Baurichter, S Deubler, D Forkel, M Uhrmacher[*], H Wolf, and W Witthuhn

Physikalisches Institut der Universität Erlangen-Nürnberg,
D-8520 Erlangen, Germany
[*] II. Physikalisches Institut der Universität Göttingen,
D-3400 Göttingen, Germany

Abstract: Formation and stability of molecule-like In-H pairs are studied by the perturbed angular correlation method (PAC). The p-Si samples doped with the radioactive ^{111}In probe-nuclei were hydrogenated by different conventional procedures or by high-energy proton irradiations (E_p=8.4 MeV). In the temperature range between 4.2 K and 420 K the PAC spectra reveal two distinct metastable In-H configurations along the <111> lattice directions. A third new In-H related complex involves an intrinsic defect or indicates the existence of two ionization states of one of the two previously observed In-H pairs.

1. INTRODUCTION

The electrical passivation of deep and shallow impurities in silicon by hydrogen has been a subject of numerous investigations in the last ten years (Pearton 1987). Although a large amount of results on various aspects of the system hydrogen in silicon has been reported, the experimental and theoretical efforts in determining the passivation mechanism and diffusion behaviour of hydrogen could not provide a conclusive picture so far. Since the shallow group-III-acceptor ^{111}In represents the standard nuclear probe atom of the perturbed angular correlation (PAC) spectroscopy, this nuclear technique can yield microscopic information on formation conditions, structure and stability of acceptor-hydrogen complexes via the related electric field gradient (EFG). In particular, new valuable results regarding the mechanism of shallow acceptor passivation can be expected. In the last few years this method has been successfully applied to studies of acceptor-donor-pairs in semiconductors (Forkel 1985 and 1988); first results on In-hydrogen complexes were reported by Wichert et al. (1987).

2. EXPERIMENT

The PAC measurements were performed on the isomeric 5/2$^+$ level of ^{111}Cd ($T_{1/2}$ = 84 ns), which was populated by the EC-decay of ^{111}In ($T_{1/2}$ = 2.8 d). For the time-differential detection of the γ-γ-cascade emitted from the excited Cd isotopes a standard four-detector set-up was

used. An appropriate combination of the 12 coincidence time spectra leads to an intensity ratio R(t), which is proportional to the product of the angular correlation coefficient A_{22} and the perturbation factor $G_{22}(t)$:

$$R(t) = A_{22} \sum_i f^{(i)} G_{22}^{(i)}(t), \text{ with}$$

$$G_{22}^{(i)}(t) = \sum_{n=0}^{3} s_{2n} \cos(n \omega_o^{(i)} t).$$

Here $f^{(i)}$ denotes the fraction i of probe atoms exposed to a distinct EFG, which is characterized by the basic frequency $\omega_o^{(i)}$. This frequency is related to the quadrupole coupling constant ν_Q, which in turn depends on the largest component V_{zz} of the EFG-tensor (axial symmetry is assumed, i.e. $\eta = (V_{xx}-V_{yy})/V_{zz}=0$):

$$\omega_0 = (3\pi/10) \nu_Q \text{ with } \nu_Q = eQV_{zz}/h$$

For more details concerning the PAC method applied to semiconductors see Forkel et al. (1988).

The samples were doped with the radioactive ^{111}In probe atoms and hydrogenated by different procedures, which significantly influenced the results of the PAC-measurements:

Ion Implantation

Boron doped single-crystalline samples ([B] = 10^{14}-10^{16} cm^{-3}) were implanted with ^{111}In$^+$ ions either at the IONAS implantator in Göttingen (400 keV) or at the ISOLDE facility of CERN (60 keV). The implantation doses were in the order of 10^{12} cm^{-2} resulting in local concentrations between 10^{16} and 10^{17} In cm^{-3} within the implanted regions.

The radiation damage induced by the ion implantation was annealed by thermal treatments in vacuum or under flowing inert gases. This process was monitored by PAC-spectra taken from the "as-implanted" samples and after the annealing programme. Typically 70% to 80% of the radioactive In probe atoms were finally located at substitutional lattice sites within an unperturbed environment.

In a final treatment the samples were hydrogenated by conventional procedures: Charging with atomic hydrogen was carried out in a H_2 glow-discharge tube with the sample located about 2 cm downstream off the plasma, in order to avoid plasma induced damage. The specimens were kept at 400 K during the charging process. The second way of hydrogenating was achieved by boiling the samples in water or hydrochloric acid.

High-Energy Proton Implantation

Here the ^{111}In-activity was diffused into the samples by means of a silicon direct bonding process (SDB) (Lindner 1987), which was adapted to the specific requirements of the PAC spectroscopy (Forkel 1988, Deubler 1988a). The final In-acceptor concentration was in the order of 10^{15} cm^{-3}.

The hydrogenation of the samples was then performed by irradiation with high-energy protons (8.4 MeV) at the Erlangen tandem-accelerator. The choice of this proton energy provides a maximum overlap of the In- and the proton-profiles.

3. RESULTS AND DISCUSSION

Fig. 1a-c show PAC spectra obtained from measurements after different hydrogenation procedures. The analysis of the PAC spectra acquired at ambient temperatures reveal two In-H complexes labelled $f^{(1)}$ and $f^{(2)}$, characterized by $\nu_Q = 349$ MHz, $\eta = 0$ ($f^{(1)}$), and $\nu_Q = 463$ MHz, $\eta = 0$ ($f^{(2)}$), respectively. The EFG tensors are oriented along the <111> lattice directions. These results agree with those obtained by Wichert et al. (1987). They propose that the complexes correspond to hydrogen trapped at two different interstitial sites: $f^{(1)}$ is due to hydrogen on an "antibonding site", whereas $f^{(2)}$ is assigned to hydrogen located at the "bond center site". These sites were proposed for B-H, Al-H, and Ga-H pairs (Assali 1985, Pankove 1985, De Leo 1985). Recent results of channeling measurements of deuterium in p-type silicon confirm the bond center site at low temperatures (Nielsen 1988). Since the results at ambient temperatures are still contradictory, the lattice site could not be determined by channeling so far.

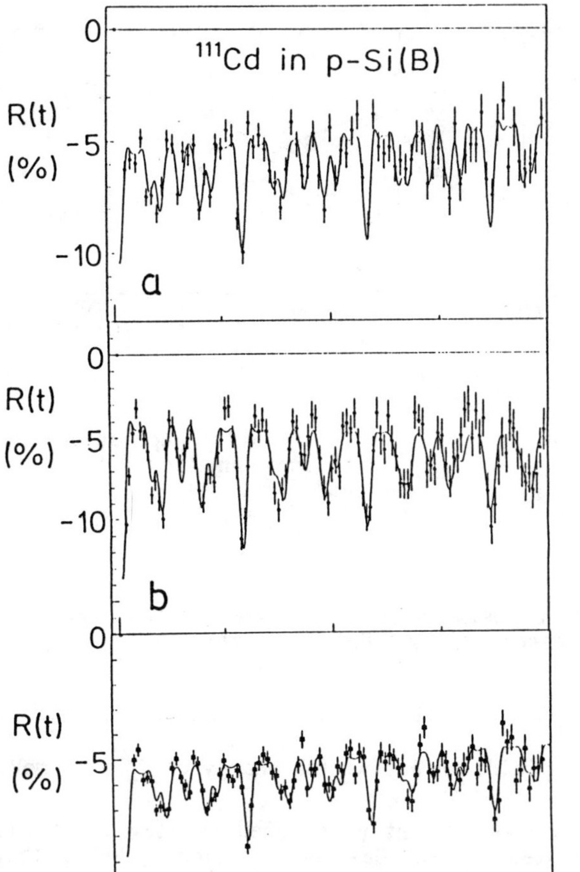

fig.1: PAC-spectra from different hydrogenated samples:
a) after plasma charging, b) after boiling in hydrochloric acid and c) after proton irradiation.
(data obtained at T = 295 K)

The electric field gradients show only a weak temperature dependence (fig. 2). However, the relative population of the two configurations changes drastically (see fig. 3): at low temperatures (T < 20 K) only fraction $f^{(2)}$ can be observed, whereas at ambient temperatures fraction $f^{(1)}$ predominates. This temperature dependent conversion of the two complexes was found to be reversible. An interpretation in terms of a thermally activated ionization of this impurity complex is very unlikely: In this case one would expect only one fraction with a strong temperature dependent EFG as has been demonstrated in the system In-As (Forkel 1988, Deubler 1988b).

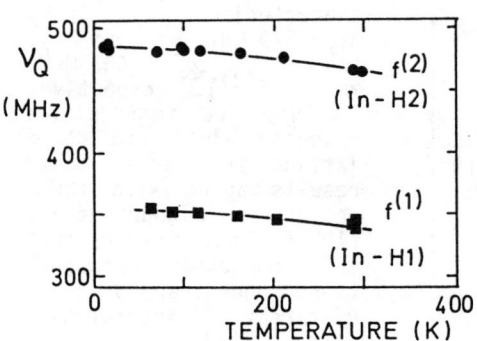
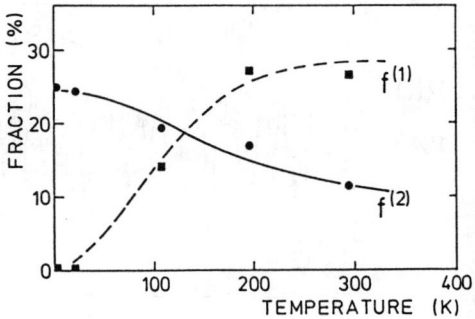

fig 2: Temperature dependence of the EFG of complex $f^{(1)}$ and $f^{(2)}$.

fig 3: The relative population of configuration $f^{(1)}$ and $f^{(2)}$.

The data indicate a metastability between two structural states. Here the temperature dependence can be described by a Boltzmann factor:

$$f^{(1)}/f^{(2)} \propto \exp(-E_a/k_BT)$$

A fit to the data (see fig. 4) yields an activation energy of E_a = 12 meV.

A similar temperature dependence was recently reported by Stavola et al. on IR measurements in hydrogenated Al- and Ga- doped silicon (1988). They observed a thermally populated vibrational sideband. The reported activation energy of 10 meV for the Al-H complex excellently agrees with our results on In-H pairs. The authors suggest that the population of the sideband is due to either a vibration of the acceptor-H-Si unit with the hydrogen off the <111> axis, or a H motion between the different <111> directions. The latter configuration can be excluded on the basis of our results since the EFG's should be identical. Also the first possibility seems to be unlikely because it hardly can produce two distinct EFG's.

In samples after hydrogenation by high energy proton implantation three defect configurations were detected during an isochronal annealing

programme. They are characterized by ν_Q= 349 MHz (295K), η = 0 ($f^{(1)}$), ν_Q= 484 MHz (77K), η = 0 ($f^{(2)}$) and ν_Q= 270 MHz (77K), η = 0 ($f^{(3)}$). The EFG tensors are oriented along the <111> lattice directions.

As a remarkable result we can state that all three defect configurations never occur simultanously: While defect complex $f^{(3)}$ only appears at 77 K $f^{(1)}$ cannot be detected at all at this temperature. After annealing at 350 K PAC measurements at room temperature show the two configurations $f^{(1)}$ and $f^{(2)}$ as observed after conventional hydrogenation techniques.

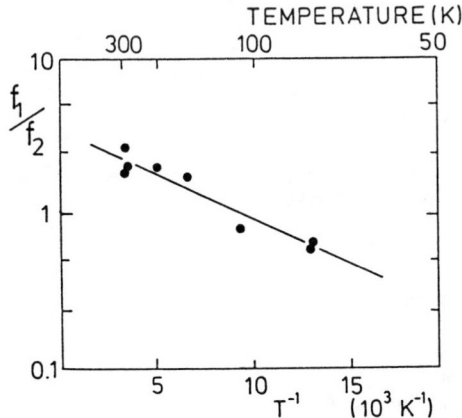

fig 4: Arrhenius plot of the ratio $f^{(1)}/f^{(2)}$ (Additional data points of a second sample not shown in fig 3 are included)

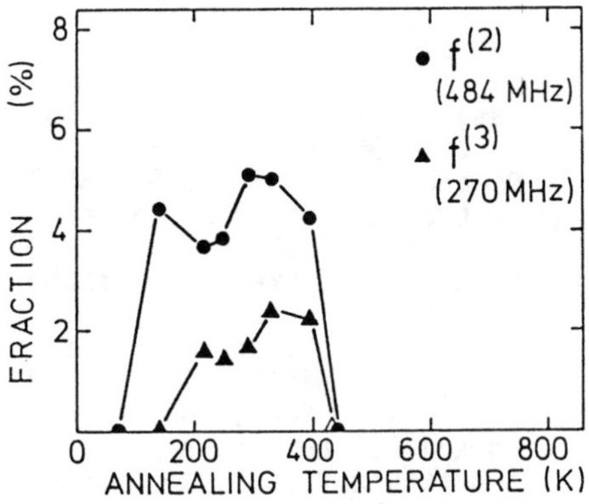

fig 5: Annealing behaviour of complex $f^{(2)}$ and $f^{(3)}$.

The defect complex $f^{(3)}$ neither appears in irradiation experiments with the protons penetrating the sample nor in other ion implantation experiments (Forkel 1988). Wichert et al. observed in highly boron doped silicon a defect complex whose quadrupole coupling constant of 270 MHz at 77 K reversibly changes to 333 MHz at 295 K (1988). The strongly damped spectra in the intermediate temperature range indicate a thermally activated ionization of the impurity complex. This can most probably be reproduced by the model given in ref. (Deubler 1988b). The similar annealing behaviour of $f^{(2)}$ and $f^{(3)}$ (fig. 5) supports the assignment of $f^{(3)}$ to an indium-hydrogen complex. However, since the temperature dependence of the quadrupole coupling constant of $f^{(3)}$ is still lacking, it is not possible to distinguish by the present PAC data between In-H and a complex, consisting of hydrogen and another irradiation induced defect.

4. CONCLUSIONS

We observed two In-H complexes after hydrogenating the samples either by boiling in water, by plasma charging or by proton implantation (at 77K) and subsequent annealing above 140 K. A third complex $f^{(3)}$ was found in proton implanted samples instead of $f^{(1)}$. We conclude that $f^{(3)}$ is either due to a different ionisation state of $f^{(1)}$ or due to a complex involving H and another intrinsic defect. The temperature dependence of $f^{(1)}$ and

$f^{(2)}$ reveals a Boltzmann type population of the antibonding and the bond center site. Our data can be explained if we assign $f^{(1)}$ to the antibonding and $f^{(2)}$ to the bond center site as it has been done by Wichert et al. (1987).

5. ACKNOWLEDGEMENT

This work was financially supported by the Bundesministerium für Forschung und Technologie.

Assali L V C, Leite J R 1985 Phys. Rev. Lett. 55 980
De Leo G G, Fowler W B 1985 Phys. Rev. B 31 6861
Deubler S, Forkel D, Lindner R, Witthuhn W 1988a, to be published
Deubler S, Forkel D, Witthuhn W, Wolf H 1988b, contribution to this conference
Forkel D, Föttinger H, Iwatschenko-Borho M, Meyer F, Witthuhn W, Wolf H 1985 Mat. Res. Soc., Vol. 46 481
Forkel D, Baurichter A, Deubler S, Wolf H, Witthuhn W 1988, to appear in Appl. Phys. A
Lindner R, Helbig R, Lehmann V, Hofmann-Tikhanen R, Glasow P A 1987 Mat. Res. Soc. (Strasbourg)
Nielsen B B, Andersen J U, Pearton S J 1988 Phys. Rev. Lett. 60 321
Pankove J I, Zanzucchi P J, Magee C W, Lucovsky G 1985 Appl. Phys. Lett. 46 421
Pearton S J, Corbett J W, Shi T S 1987 Appl. Phys. A 43 153 and references therein
Stavola M, Pearton S J, Lopata J, Dautremont-Smith W C 1988 Phys. Rev. B 37 8313
Wichert T, Skudlik H, Deicher M, Grübel G, Keller R, Recknagel E, Song L 1987 Phys. Rev. Lett. 59 2087
Wichert T, Deicher M, Grübel G, Keller R, Schulz N, Skudlik H 1988, to appear in Appl. Phys. A

Inst. Phys. Conf. Ser. No 95: Chapter 7
Paper presented at Int. Conf. Shallow Impurities in Semiconductors, Linköping, Sweden, 1988

Hydrogen passivation of shallow acceptor impurities and radiation defects in p-type silicon

Kh A Abdullin, B N Mukashev, M F Tamendarov, T B Tashenov, S Z Tokmoldin and E V Chikhrai

The Institute of High Energy Physics, Alma-Ata 480082, UdSSR

ABSTRACT. Passivation of shallow impurities and radiation defects by hydrogen has been studied with infrared (IR) spectroscopy and electrical methods (DLTS, CV-measurements). It is suggested that new and also previously observed (Stavola et al 1987) IR stretching bands are associated with two types of the passivated acceptor impurities. The introduction rate of radiation defects is considerably decreased for samples exposed to H plasma before irradiation. It is found that this process is temperature dependent: the exposure of samples at 150ºC decreases the introduction rate for all radiation defects, while exposure at $\sim$ 200ºC decreases the introduction rate of defects including interstitial atoms and slightly influences the introduction rate of divacancies. The passivation mechanisms of shallow acceptor impurities are briefly discussed.

1. INTRODUCTION

The discovery that hydrogen besides saturating dangling bonds neutralizes the electrical activity of deep levels of the introduced defects and also the one of shallow impurities, has promoted new interest in understanding its behavior. In spite of the fact that numerous studies have been made to determine the microscopic mechanism of H passivation, the specific chemical reaction and defect complex which are responsible for the passivation of impurities or radiation defects have not yet been established (Pearton et al 1987). In the present paper we report studies of hydrogen passivation of shallow acceptor impurities as well as radiation defects in p-type silicon, by IR, DLTS and CV-measurements.

2. EXPERIMENTAL

The Si samples used in this work were cut from crucible-grown and float-zoned crystals, doped with either boron ($\rho\sim 0.005$; 1; 25 Ohm cm), aluminium ($\rho\sim 0.1$; 10-20 Ohm cm) or gallium ($\rho\sim 10$-20 Ohm cm). The heavily doped samples were implanted at room temperature with different doses of 12 MeV/n hydrogen

© 1989 IOP Publishing Ltd

(Si:B, Al:H) and/or deuterium (Si:B, Al:D; Si:B, Al:H:D). The IR spectra were measured with a resolution of ~ 0.3 cm^{-1}. Schottky barriers or n$^+$ p-junctions were used for DLTS and CV-measurements. The radiation defects were introduced by protons (30 MeV) or α-particles (4.7 MeV) at room temperature. The initial and irradiated samples were exposed to monoatomic hydrogen by insertion in a high frequency gas discharge.

3. RESULTS AND DISCUSSION

Figure 1a,b shows the free carrier concentration profiles (a) as a function of plasma exposing time and isochronal annealing and the capacitance change of the n$^+$ p-junction after different plasma exposure times (b). From the free carrier concentration profiles the effective hydrogen diffusion coefficient was determined, $D = 8 \cdot 10^{12}$ cm^{2}s^{-1} at 130°C, which is in good agreement with previously published data (Pearton et al 1987). Note that a highly phosphorus doped layer slightly influences the D value. Isochronal annealing for 1 hour for the exposed samples completely reactivates the boron atoms at 200 °C (Fig.1b). Before the boron reactivation stage one can observe a reverse annealing stage at 100-150°C, which could be associated both with the phosphorus atom reactivation and with the phosphorus-vacancy annealing. We have observed an additional high temperature recovery stage for samples that were exposed 6 hours (Fig.1b). Low energy hydrogen ions (0.3 keV) created point defects in the n$^+$ region, which migrate to the volume and interact with hydrogen and impurities.

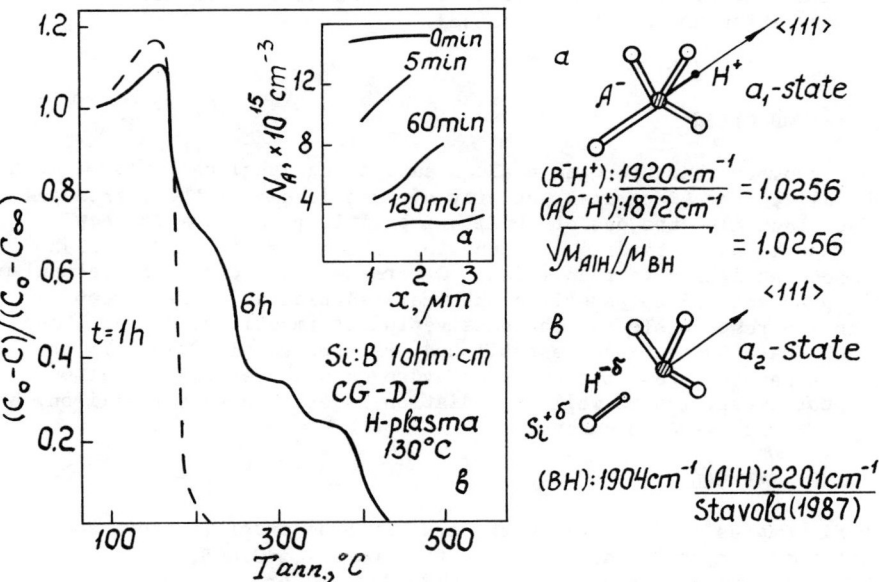

Fig.1 Explanation in the text.   Fig.2 Proposed model of the H-acceptor center.

As a consequence of this secondary passivated defects are produced. The high temperature stage is perhaps connected with the reactivation and annealing of these defects: the divacancy, the K-center, etc. (Corbett et al 1976). In spite of the fact that irradiation of Si by a plasma creates radiation defects, an increasing exposure time essentially decreases the reverse current of an n^+ p-junction. The formation of mono- and dihydrids at the Si surface increases the surface resistivity and suppresses the leakage current (Mukashev et al 1986).

We have found new absorption bands in heavily doped p-Si implanted by hydrogen or deuterium. We discuss here only the origin of the a_1 (1920 cm^{-1}⟨B⟩ ; 1872 cm^{-1}⟨Al⟩) and a_2 (1904 cm^{-1}⟨B⟩; 2201 cm^{-1} ⟨Al⟩) vibrational modes. The isotope substitution of H by D shifts the stretching frequency by a factor of 1.37 to lower energy. Note that the a_2 bands were previously observed by Stavola et al (1987). It is well known that the reactivation of acceptor atoms has two temperature stages (Sah et al 1985). The analysis of the annealing behavior and the relation of AlH/BH reduced mass allows us to suggest two models of the passivated acceptor (Fig.2a,b). The first type of these centers is $A^0 + e^- + H^+ \rightleftharpoons A^-H^+$. B^-H^+ reactivates at ~150ºC and Al^-H^+ (or Ga^-H^+) reactivates at 120-140ºC. The a_2 bands are connected with the second type of passivated acceptors $(AH)^0$. Note that the process of reactivation for the second center is dependent on impurity binding. Fig.3 displays the decrease of the divacancy annealing temperature in previously H-exposed samples, in comparison with the initial samples. These data could be explained by the annealing of the A^-H^+ centers and the formation of a passivated divacancy. Tavendale et al (1987) were the first to demonstrate that hydrogen will drift as a H^+ species under the action of an electrical field. These findings are also consistent with the discussed mechanism of acceptor passivation.

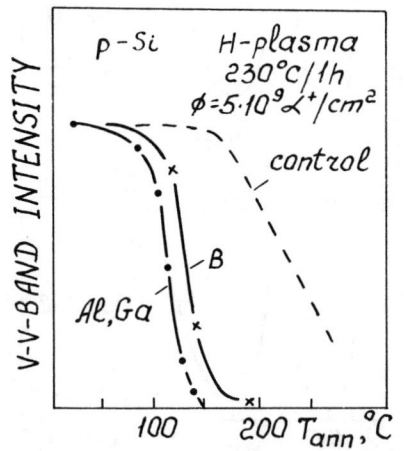
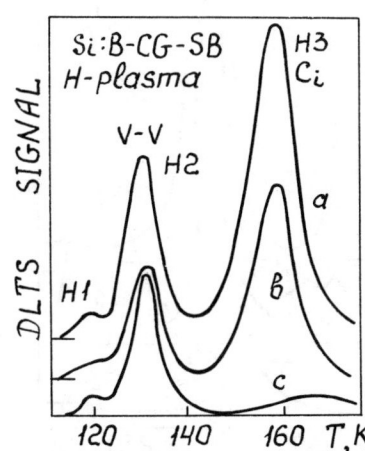

Fig.3 Thermal annealing of divacancy (DLTS band 0.2 eV) in control and H-exposed p-Si samples.

Fig.4 DLTS spectrum from control (a) and H-plasma exposed at 150ºC/2 hours (b) and 230ºC/1 hour (c) p-Si samples. Dose of α-irradiated was $5 \cdot 10^9$ α^+/cm^2 for all samples.

The exposure of the samples to H-plasma before irradiation with high energy particles decreases the secondary radiation defect production. The process is temperature dependent: the exposure of samples at $\sim$150°C decreases the introduction rate for all radiation defects (Fig.4b), at the same time the exposure of samples at $\sim$230°C considerably decreases the introduction rate of C_i atoms and slightly influences the introduction rate of divacancies (Fig.4c). The radiation enhanced hydrogen emission from passivated acceptors or from other centers is responsible for the decreasing introduction rate of radiation defects in the samples exposed at low temperature.

The exposure of samples at high temperature introduced preferentially vacancy-hydrogen related centers which are traps for self-interstitials. We have observed vacancy-related bands increasing in H implanted samples annealed at $\sim$300°C (Mukashev et al 1985). Therefore one can observe a decrease of H3 (interstitial carbon) and H4 (VV-C_iO_i) defects and a slight increase of the H2 (VV) center.

The p- and α-irradiation introduced in crucible-grown samples the following defects (Fig.5a): H1 (unknown), H2 (0.20 eV, VV), H3 (0.29 eV, C_i) and H4 (VV-C_iO_i). A 20 min exposure of irradiated samples remarkably changes DLTS spectra.

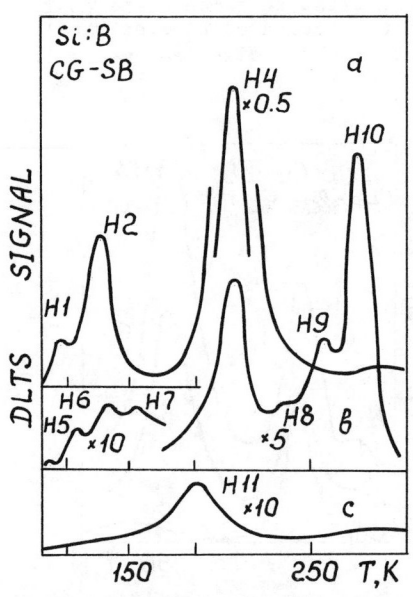
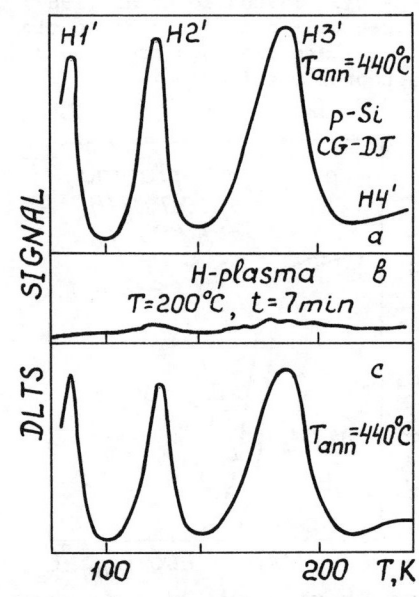

Fig.5 DLTS spectrum from α-irradiated Si (a) after H-exposured at 120°C: 20 min (b) and 120 min (c).
Fig.6 DLTS spectrum from proton-irradiated (30 Mev, $3 \cdot 10^{13}$ p/cm^2) p-Si (p=$6 \cdot 10^{14}$ cm^{-3}): after anneal at 400°C (a), H-plasma exposure (b), post hydrogenation annealing at 440°C (c).

All these defects appeared during exposure to He plasma. Hence the observed centers do not include hydrogen. After 120 min exposure in H plasma, there are no defect states in the spectrum, except for the small H11 peak. We have studied the reactivation of high temperature resistent defects because in general defect reactivation and annealing occurred in the same temperature region. After annealing of irradiated crucible-grown samples at 440°C, the following hole traps are found (Fig.6): H1⁻ (0.15 eV), H2⁻ (0.25 eV), H3⁻ (0.32 eV), H4⁻ (0.42 eV). These defect states are absent in float zone samples. It is supposed that these traps are associated with different multiple vacancy-oxygen related complexes. 5 min exposure removes these hole traps, however after annealing at 440°C these defect states appeared again. Thus one can observe reversible transformation of H1⁻-H4⁻ centers. The activation energies of reactivation (E_α) for H1⁻ and H3⁻ defects are equal ~ 2 eV. E_α for the H2⁻ defects has the value ~ 2.5 eV. Note that these values are very close to the activation energy of breaking up the Si-H bond.

REFERENCES

Corbett J W, Bourgoin J C, Cheng L J, Corelli J, Lee Y H, Mooney P M and Weigel C 1976 Def. and Rad.Eff. in Semicond. (Conf.Ser. N31, London) 1-11.
Mukashev B N, Tamendarov M F, Tokmoldin S Zh and Frolov V V 1985 Phys.Stat.Sol.A 91 509.
Mukashev B N, Kolodin L G, Smirnov V V, Tokmoldin S Zh and Tamendarov M F 1986 Fiz.Tekh.Poluprov. 20 773.
Pearton S J, Corbett J W and Shi T S 1987 Appl.Phys. A43 153.
Sah C T, Pan S C S and Hsu C C H 1985 J. Appl. Phy. 57 5148.
Stavola M, Pearton S J, Lopata J and Dautremont-Smith W C 1987 Appl. Phys. Lett. 50 1986.
Tavendale A J, Alexiev D and Williams A A 1985 Appl. Phys. Lett. 47 317.

Inst. Phys. Conf. Ser. No 95: Chapter 8
Paper presented at Int. Conf. Shallow Impurities in Semiconductors, Linköping, Sweden, 1988

Diffusion of shallow impurities in silicon

Paul Fahey

IBM Watson Research Center, Yorktown Heights, NY, 10598

ABSTRACT: The diffusion in Si of shallow impurities from columns III and V of the periodic table has been the subject of many experimental studies. However, the identification of the diffusion mechanisms responsible for these elements remains a hotly debated issue in the literature. We review the experimental evidence which shows some dopants to diffuse primarily by interstitial-assisted mechanisms while others are dominated by the vacancy mechanism and address the question of why one type of mechanism may dominate over the other for a particular dopant. We also show new data for the behavior of Ga and Ge diffusion in the presence of excess silicon interstitials or vacancies. The possible importance of Ge diffusion studies to questions concerning Si self-diffusion is discussed.

1. INTRODUCTION

In spite of many experimental investigations which have studied the diffusion of shallow impurities in silicon (namely, the dopant atoms P, As, Sb, B, and Ga) outstanding questions remain concerning the atomistic mechanisms of diffusion. Although it was assumed for many years that vacancy-assisted mechanisms were dominant, it is now clear that substitutional-interstitial(cy) interchange diffusion mechanisms can be dominant in a wide variety of experimental conditions (including equilibrium conditions). The two basic questions are: (1) How do we determine the nature of the diffusion mechanism in the first place? (2) Why would some dopants diffuse predominantly with a vacancy-assisted mechanism and others with an interstitial(cy)-assisted mechanism?

An important, related question concerns the relationship between dopant diffusion and self-diffusion. An Arrhenius plot of Si self-diffusion and the most well-studied dopants is shown in Fig. 1. Although the diffusion mechanism (or mechanisms) of Si self-diffusion has still not been determined, it is certain that in the temperature range above 1050°C, where the majority of self-diffusion experiments have been performed, the activation energy of self-diffusion Q_{Si}, is between $1.0 - 1.5$ eV greater than the activation energy of self-diffusion. Since we maintain that dopants do not all diffuse by the same mechanism, the pertinent question is: How can we explain this relatively large energy difference between self-diffusion and dopant diffusion if both take place by the same mechanism?

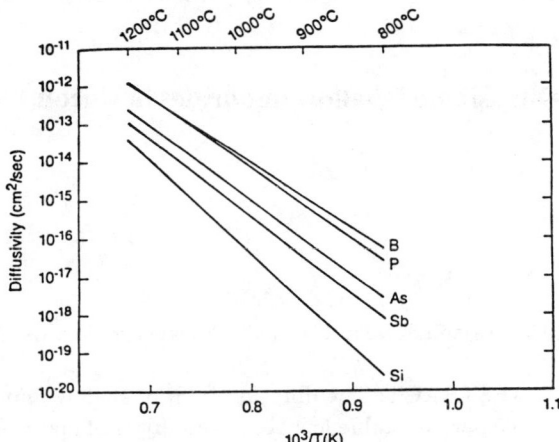

Fig. 1. Dopant diffusivities and self-diffusivity in Si. Values for dopants come from the SUPREM III computer program commonly used by engineers to simulate diffusion (Ho et al., 1984). Self-diffusion plot comes from Morehead (1987) who optimized a fit to self-diffusion data from a variety of studies.

2. DIFFUSION BY A DUAL MECHANISM

We assume that in general it is possible for a dopant A, to diffuse by either an interstitial-assisted or vacancy-assisted mechanism. The relative importance of each is determined from experiment. In the following, we denote native vacancy point defects by V. Vacancies associated with a dopant atom A, are written as AV. Silicon interstitials (Si atoms occupying interstices) and interstitialcies (consisting of two silicon atoms configured about a single substitutional site) are both denoted by I. Whether I-type defects are interstitials or interstitialcies does not matter for the experimental data discussed in this paper. Dopant atoms in a nonsubstitutional state are written AI.

The diffusivity measured by observing profile broadening of the substitutional dopants, D_A, is given by the sum of its two components:

$$\begin{aligned} D_A &= D_A^\circ \exp(-Q_A/kT) \\ &= D_{AV}^\circ \exp(-Q_{AV}/kT) + D_{AI}^\circ \exp(-Q_{AI}/kT) \, . \end{aligned} \quad (1)$$

We can write a similar equation for self-diffusion.

It is possible in principle to determine the separate contributions by making an Arrhenius plot of diffusivity over a wide enough temperature range, but this has not been accomplished as yet. We rely instead on interpretation of diffusion data from nonequilibrium studies.

3. NONEQUILIBRIUM DIFFUSION STUDIES

Experimental evidence about the mechanisms of dopant diffusion in Si comes primarily from nonequilibrium diffusion studies. In these studies, excess point defects are created by a variety of means (implantation damage, chemical reactions at the silicon surface, etc.) and the effect on dopant diffusivity is investigated. For a dopant A, which can diffuse by either vacancy or substitutional-interstitial(cy) interchange diffusion mechanisms, the diffusivity can be written

$$D_A = D_{AI} + D_{AV}$$
$$= d_{AI}\frac{C_{AI}}{C_A} + d_{AV}\frac{C_{AV}}{C_A} \qquad (2)$$

where d_{AI} and d_{AV} are the diffusivities of a labeled atom diffusing in an interstitial-type state or a vacancy-type state respectively. Equation 2 simply states that the the diffusivity is determined by how fast individual dopant atoms migrate, weighted by the fraction of atoms which are in a diffusing state at any instant of time. In nonequilibrium studies, the concentrations of AI and AV defects are changed.

If we define the fractional interstitial(cy) component of diffusion under equilibrium conditions as

$$f_{AI} \equiv \frac{D^*_{AI}}{D^*_{AI} + D^*_{AV}} = \frac{D^*_{AI}}{D^*_A}, \qquad (3)$$

where the superscript *, denotes equilibrium, then under nonequilibrium conditions the diffusivity will change as

$$\frac{D_A}{D^*_A} = f_{AI}\frac{C_{AI}}{C^*_{AI}} + (1 - f_{AI})\frac{C_{AV}}{C^*_{AV}}. \qquad (4)$$

For a given change in AI and AV, the observation of enhanced ($D_A/D^*_A > 1$) or retarded ($D_A/D^*_A < 1$) diffusion depends on the value of f_{AI}, which, it should be kept in mind, is temperature dependent.

In the most commonly performed type of nonequilibrium diffusion experiment, I are injected into the Si substrate by thermally oxidizing the sample surface; the underlying cause of the reaction generating I is still not known. Another method which has been used in the last few years utilizes thermal nitridation reactions, as pictured in Fig. 2 (Hayafuji et al., 1982; Mizuo et al., 1983; Fahey et al., 1985). As with the case of oxidizing reactions, it is not known why these reactions generate point defects. A summary of the effects of these reactions on diffusion and growth or shrinkage of extrinsic stacking faults is given in Table I. Not all studies for the

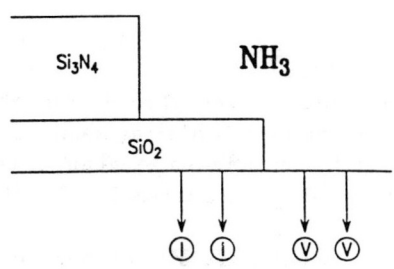

Fig. 2. Thermal nitridation. Nitridation of SiO_2 injects interstitials into the bulk while direct nitridation of the silicon surface injects vacancies.

different elements have been performed over an extensive temperature range; however, all results are valid for temperatures between 1050 − 1100°C.

Table I. Summary of surface reactions on diffusion and stacking fault growth. $I \uparrow V \downarrow$ signifies a supersaturation of I and undersaturation of V, and $I \downarrow V \uparrow$ indicates I undersaturation and V supersaturation.

	Oxidation $I \uparrow V \downarrow$	Oxynitridation $I \uparrow V \downarrow$	Nitridation $I \downarrow V \uparrow$
Stacking faults	grow	grow	shrink
Group III diffusion			
B	enhanced	enhanced	retarded
Al	enhanced		
Ga	enhanced	enhanced	retarded
In	enhanced		
Group V diffusion			
P	enhanced	enhanced	retarded
As	enhanced	enhanced	enhanced
Sb	retarded	retarded	enhanced

To interpret the results in Table I, we focus first on the oxidation and oxynitridation reactions which generate I point defects. The extrinsic stacking faults (which are extra half planes of atoms) grow by absorbing excess I. Dopant diffusion is enhanced by an increase in AI concentration, presumably through the reaction $I + A \rightarrow AI$. Diffusion can be retarded by reduction of the concentration of AV. This happens through the recombination events

$$I + V \rightarrow 0 \quad \text{and} \quad I + AV \rightarrow A . \qquad (5)$$

The interpretation of results for vacancy generation from direct nitridation of the silicon surface follow along similar lines. Stacking faults are annihilated by excess V and dopant diffusion is enhanced by increasing the concentration of AV. Diffusion is retarded by recombination of V with I or AI.

The fact that Ga, B, and P are actually retarded in the presence of excess V, shows clearly that under equilibrium conditions these elements diffuse predominantly by an I-type mechanism. Correspondingly, the retarded diffusion of Sb in the presence of excess I demonstrates the dominance of the vacancy mechanism. The observation that As diffusion is enhanced under both conditions shows that this element has significant contributions of both I-assisted and V-assisted mechanisms to its diffusivity.

4. INTERACTIONS BETWEEN POINT DEFECTS AND DOPANT ATOMS

If one accepts that diffusion of substitutional dopants is mediated by interactions with native point defects (i.e., I and V), the most intuitively obvious explanation for experimental evidence that some atoms diffuse primarily by V-type mechanisms and some by I-type is that some atoms have a greater tendency to associate themselves with V and some have a greater tendency to associate themselves with I. In Secs. 5 and 6 it is shown that the diffusivity of a dopant A which is observed experimentally is not a simple function of the strength of the interaction potential between A and a given type of point defect. In the remainder of this section we review the most commonly proposed reasons that point defects and substitutional dopants should attract or repulse each other.

Qualitative discussion of the close-range interactions between point defects and dopants usually follow along two different lines: (1) arguments concerning the Coulombic attraction or repulsion between point defects and ionized dopants, or (2) arguments concerning the "size" of dopant atoms relative to silicon as determined by their covalent bonding radii.

Coulombic attraction between dopants and point defects would occur in reactions such as

$$A^+ + V^- \rightarrow (AV)^\circ \quad \text{or} \quad A^- + I^+ \rightarrow (AI)^\circ . \qquad (6)$$

If Coulombic interactions were the predominant factor in determining the preferred mechanism of diffusion, then all acceptors should diffuse by the same mechanism and likewise all donors should diffuse by the same mechanism. Table I indicates that this isn't the case for the donor dopants of Group V.

Turning to the question of size arguments, we refer to Fig. 3 which displays covalent radii as reported in Pauling's book (Pauling, 1960). Qualitatively, we might expect that atoms with covalent radii greater than Si would pair naturally with vacancies while those atoms with smaller radii would reside more easily near I-type defects. If the size of the atom was the determining factor for diffusion mechanisms, then we would surely expect that the greater the covalent radius, the less likely the atomic species is to prefer an I-type diffusion mechanism. This qualitative statement fits in with the results for Group V elements in Table I, but there is no evidence of this trend for Group III elements.

It can thus be concluded that neither the charge state nor the "size" of a dopant alone determine the mechanism by which it diffuses.

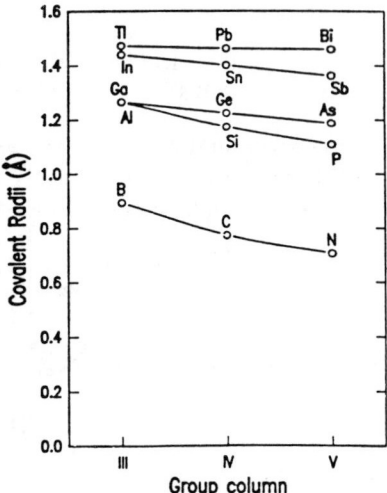

Fig. 3. Covalent radii according to Pauling (1960).

5. VACANCY MECHANISM

It is important to realize that diffusion of a substitutional atom by the vacancy mechanism does *not* occur simply by a dopant atom exchanging lattice sites with a neighboring vacancy; for then the dopant atom and vacancy will continue to occupy alternate sites and no long-range migration of A will take place. After exchange, the vacancy must move some distance away from the dopant so that it does not return to the dopant along the same path. The geometry of the diamond lattice dictates that a vacancy must move to at least a third nearest neighbor site for this condition to be satisfied.

A potential diagram of the vacancy diffusion process is shown in Fig. 4. In this figure, H_V^m is the enthalpy of migration of a vacancy diffusing without interactions with impurities, E_{AV}^b is the binding energy of the AV pair, and ΔE_{AV}^3 is defined as the difference in potential energy between a vacancy far away from the dopant and a vacancy located a third neighbor site away.

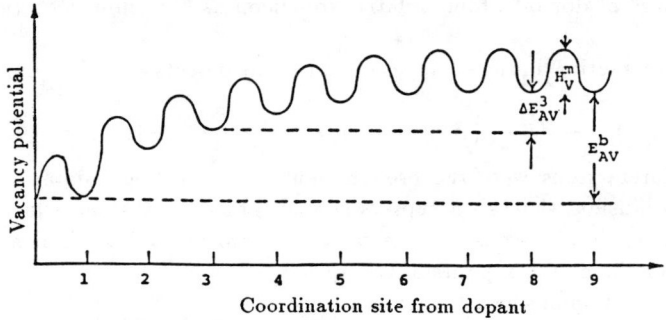

Fig. 4. Vacancy potential near a dopant atom [after Hu (1973)].

The energetics of diffusion by the vacancy mechanism in Si has been analyzed by Hu (1973). The results of his analysis show that the activation energy of diffusion by this process, Q_{AV}, can be viewed as effectively consisting of formation and migration terms,

$$Q_{AV} = \underbrace{H_V^f - E_{AV}^b}_{H_{AV}^f} + \underbrace{H_V^m + E_{AV}^b - \Delta E_{AV}^3}_{H_{AV}^m} . \qquad (6)$$

Since the activation energy of self-diffusion by a vacancy mechanism is given by

$$Q_{SiV} = H_V^f + H_V^m , \qquad (7)$$

it follows that

$$\begin{aligned} Q_{SiV} - Q_{AV} &\equiv \Delta Q_{AV} \\ &= \Delta E_{AV}^3 . \end{aligned} \qquad (8)$$

That is, the activation energy of dopant diffusion by a vacancy mechanism can be an amount $\Delta Q_{AV} = \Delta E_{AV}^3$ lower than the activation energy of self-diffusion by a

Diffusion of Shallow Impurities

vacancy mechanism. The physical interpretation of this result is as follows. While an attractive potential aids the diffusion of A by lowering the formation energy of AV pairs compared to V, it also increases the migration energy of A compared to V because the vacancy must surmount an energy barrier to move away from the dopant. If the $A-V$ interaction potential extends beyond a third nearest neighbor site, the vacancy does not need to surmount the entire barrier to complete a diffusion step. For progressively stronger potentials, dissociation of the AV complex becomes less likely and A and V diffuse as a pair. If the interaction potential does not extend beyond a third nearest neighbor site to the dopant, the activation energy of dopant diffusion should be the same as that for self-diffusion.

It is also of interest to note what differences in activation energy are expected for different dopants. Since many researchers have favored until recently the interpretation of diffusion data in terms of a vacancy mechanism, it is a bit surprising that no one has addressed this question. As an example, we compare P and Sb assuming that both diffuse entirely by a vacancy mechanism. It follows straightforwardly from Eq. 7 that

$$Q_{SbV} - Q_{PV} = \Delta E_{PV}^3 - \Delta E_{SbV}^3 , \qquad (9)$$

and since $Q_{SbV} - Q_{PV} \approx 0.4 \text{eV}$, one would have to conclude that $\Delta E_{PV}^3 > \Delta E_{SbV}^3$; i.e., a vacancy at a third neighbor site and beyond would be more strongly attracted to a P atom than to an Sb atom. But, this is contrary to the usual qualitative arguments that a vacancy would be more stongly attracted to an oversized Sb atom than to an undersized P atom. The results in Table I support the idea that the difference in activation energy between P and Sb arise because these dopants diffuse by different mechanisms (I-assisted and V-assisted, respectively), but the analysis clarifies what features of the dopant-vacancy interaction potential determine activation energies.

6. SUBSTITUTIONAL-INTERSTITIAL(CY) INTERCHANGE MECHANISM

Diffusion by a substitutional-interstitial(cy) interchange mechanism is depicted in Fig. 5. The exact atomistic process leading to an interstitial(cy) component of diffusion (i.e., whether the I-type defects involved are interstitials or interstitialcies) cannot be determined from the diffusion experiments described in this paper. What is important is that the diffusion process takes place without the aid of vacancies.

A single potential diagram cannot be drawn for this mechanism

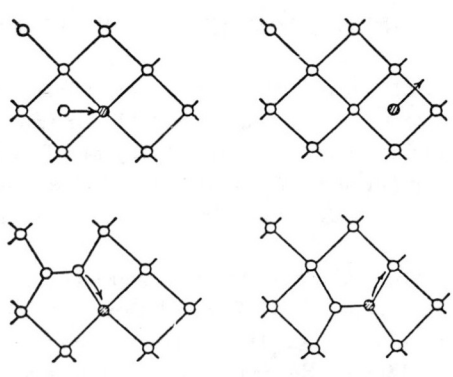

Fig. 5. Interstitial-assisted (top) and interstitialcy-assisted (bottom) dopant diffusion.

as was done for the vacancy mechanism in Fig. 4. An elementary energetics analysis can still be performed, with the basic result that dopant diffusion is not so intimately related to self-diffusion as is the case for the vacancy mechanism. Once the dopant atom is in its diffusing (i.e., nonsubstitutional) state AI, we can view AI as a randomly diffusing particle whose enthalpy of migration H_{AI}^m is determined by its saddle point energy as it diffuses between equivalent lattice sites. The formation enthalpy of an AI defect, H_{AI}^f, can be lower than the formation energy of I, but the term "binding energy" isn't necessarily appropriate. For example, if AI is an interstitial defect, then there is no point defect bound to the dopant atom.

Then in comparing self-diffusion to dopant diffusion we can only write the obvious relationship:
$$\begin{aligned}\Delta Q_{AI} &\equiv Q_{SiI} - Q_{AI} \\ &= (H_I^f - H_{AI}^f) + (H_I^m - H_{AI}^m).\end{aligned} \quad (10)$$

There is no relationship between H_{AI}^f and H_{AI}^m like there is between H_{AV}^f and H_{AV}^m, and no clear relationship between Q_{SiI} and Q_{AI}. There is also no clear expectation of how Q_{AI} should vary among different dopants. Taking In and B as an example and assuming both to diffuse entirely by a substitutional-interstitial(cy) interchange mechanism,
$$Q_{In} - Q_B = (H_{InI}^f - H_{BI}^f) + (H_{InI}^m - H_{BI}^m) \quad (11)$$

But what can we say about this? Using "size" arguments (Sec. 4) we could argue intuitively that compared to an undersized B atom, an oversized In atom (Fig. 3) would require more energy both to form its interstitial-type state and to migrate between lattice sites. In this case activation energies should increase with increasing atomic number; however, there aren't enough good data for diffusivities of group III elements to check this trend. Compared to the vacancy mechanism, not as much can be said about dopant-interstitial(cy) interaction potentials and their relationship to activation energies of diffusion.

7. GERMANIUM AS A PROBE FOR SELF-DIFFUSION

Accurate measurements of Si self-diffusivity have been hindered because the only radioactive isotope readily available for tracer diffusion experiments is ^{31}Si which has a half-life of only 2.6 hours. McVay and DuCharme (1973) first suggested that Ge diffusion might be very similar to self-diffusion since Ge has a covalent bonding radius close to Si (see Fig. 3) and introduces no net charge to the lattice.

Measurements of Ge diffusion over an extended temperature range have been performed by Hettich et al. (1979) and Dorner et al. (1984). While the experiment of Dorner et al. shows a single activation energy of $Q_{Ge} = 5.35$eV over the temperature range $875 - 1300°$C, Hettich et al. reported a non-Arrhenius behavior in Ge diffusivity with a break point at about 1050°C. For temperatures between 1100°C and 1300°C an activation energy of $Q_{Ge} = 5.3$eV is found, while in the temperature range 850 – 1000°C, $Q_{Ge} = 4.1$eV. Recent work by Bouchetout et al. (1986) also indicates a similar break in an Arrhenius behavior for Ge diffusion at about 1050°C, but only one lower

Diffusion of Shallow Impurities

temperature data point (at 1000°C) was generated in this study. There is some evidence that Si self-diffusion shows a similar non-Arrhenius behavior (Demond et al., 1983) changing from about 5 eV for temperatures above 1050°C to about 4 eV for temperatures below 1050°C, but experimental uncertainties are still large. For temperatures above 1050°C, the activation energies and absolute values of diffusivities indicate Ge diffusion to be very similar to Si self-diffusion. If we ignored the work of Dorner et al., present experimental evidence would indicate both Ge and Si self-diffusion to have a lower activation energy below 1050°C. It would be tempting to interpret this change in activation energy as arising from the separate contributions of I and V to self-diffusion. But the work of Dorner et al. is not so easily dismissed; excellent experimental method was employed and the profiling technique using SIMS is as good or better than any of the methods employed in other studies.

8. EVIDENCE FOR BOTH I AND V COMPONENTS OF GE DIFFUSION

A study was recently completed which provides the first experimental evidence on the diffusion mechanisms of Ge in Si (Fahey, Iyer, and Scilla, 1988). Buried marker layers of Ge-doped Si were grown by MBE. These samples were then patterned and exposed to NH_3 (as shown in Fig. 2) at a temperature of 1050°C. Under inert conditions, Ge diffusivities agree very well with the data of Dorner at al. (1984). The effects of I and V injection on Ge diffusion are shown in Fig. 6. Similar results were found for nitridations performed for 3 hours at 1050°C. The fact that Ge diffusion is enhanced by both I and V injection clearly indicates Ge can diffuse by either an I-type or V-type mechanism. Under similar conditions buried marker layers of Ga and Sb, also grown by MBE, show very different behavior. As has been shown in other studies, Sb diffusion was enhanced by V injection and retarded during I injection. Ga displayed the opposite behavior of retarded diffusion during V injection (this is the first experiment to observe Ga under V injection) and enhanced diffusion under I injection.

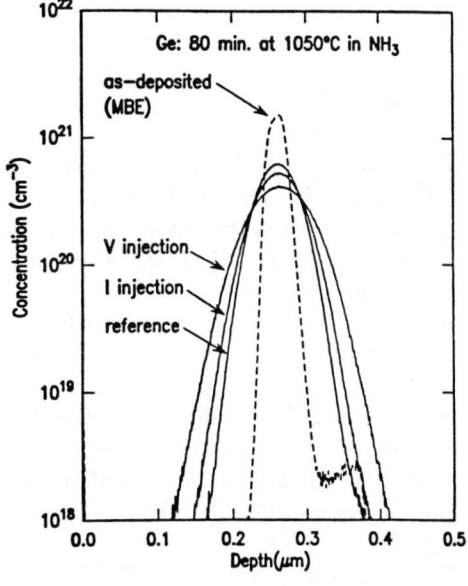

Fig. 6. Effects of excess I or V on Ge diffusion (SIMS profiles).

This experiment clearly shows that at 1050°C: Sb diffusion is dominated by a vacancy mechanism; Ga diffusion is dominated by an I-assisted mechanism; Ge has significant contributions from both I-type and V-type mechanisms to its diffusivity.

9. SUMMARY

Nonequilibrium diffusion studies show clearly that some dopants diffuse preferentially by an interstitial-assisted mechanism while others diffuse primarily by a vacancy mechanism. Experimental results (Table I) show that no single factor, such as charge state of the dopant or the length of its covalent radius, determines which type of mechanism is dominant. Elementary analyses of the substitutional-interstitial(cy) interchange and vacancy diffusion mechanisms give some indication of how the activation energy of dopant diffusion is related to the activation energy of self-diffusion and how activation energies may be determined by the chemical identity of the dopant. But in determining which mechanism, I-type or V-type, is predicted to dominate for a given dopant, these same analyses indicate that there are too many factors involved to expect a simple answer. The possible importance of Ge diffusion as a means to augment our knowledge of self-diffusion was reviewed. Results of the first nonequilibrium study of Ge diffusion in silicon were presented which indicate that Ge has both I and V components of diffusion at 1050°C. This is consistent with results from some previous studies which show a break at 1050°C on an Arrhenius plot of diffusivity if the non-Arrhenius behavior is a result of two different diffusion mechanisms operating simultaneously, but further work at different temperatures is needed to verify this.

REFERENCES

Bouchetout, A. L., N. Tabet, and C. Monty, 1986, *Proc. of the Fourteenth International Conference on Defects in Semiconductors*, edited by H. J. von Bardeleben, (Trans Tech, Switzerland), p. 127.

Demond, F. J., S. Kalbitzer, H. Mannsperger, and H. Damjantschitsch, 1983, *Phys. Lett.*, **93A**, 503.

Dorner, P., W. Gust, P. Predel, and U. Roll, 1984 *Phil. Mag.*, **49**, 557.

Fahey, P., G. Barbuscia, M. Moslehi, and R. W. Dutton, 1985, *Appl. Phys. Lett.*, **46**, 784.

Fahey, P., S. S. Iyer, and G. J. Scilla, 1988, *unpublished*.

Hayafuji, Y., K. Kajiwara, and S. Usui, 1982 *J. Appl. Phys.*, **53**, 8639.

Hettich, G., H. Mehrer, and K. Maier, 1979, in *Defects and Radiation Effects in Semiconductors, 1978*, edited by J. H. Albany (Inst. Phys. Conf. Ser. 46, London), p. 500.

Ho, C. P., S. E. Hansen, and P. M. Fahey, 1984, *Stanford University Technical Report SEL84-001*, Dept. of EE, Stanford University, Stanford, CA.

Hu, S. M., 1973, *Phys. Stat. Sol. (b)*, **60**, 595.

Morehead, F., 1987, *unpublished*.

McVay, G. L., and A. R. DuCharme, 1973, *J. Appl. Phys.*, 44, 1409.

Mizuo, S., T. Kusaka, A. Shintani, M. Nanba, and H. Higuchi, 1983, *J. Appl. Phys.*, **54**, 3860.

Pauling, L., 1960, in *The Nature of the Chemical Bond, Third Edition*, (Cornell University, Ithaca, New York), p. 247.

Diffusion of shallow impurities in silicon

C. S. Nichols, C. G. Van de Walle, and S. T. Pantelides

IBM Research Division,
T. J. Watson Research Center,
Yorktown Heights, New York 10598

Using first-principles total-energy calculations, we investigate the diffusion of dopant impurities in silicon both under conditions of equilibrium and non-equilibrium concentrations of point defects. We find that, under equilibrium conditions, vacancies and interstitials mediate the diffusion of all dopants with comparable activation enthalpies except Sb for which the interstitial component has a high activation enthalpy. The concerted exchange mechanism has a somewhat higher activation enthalpy than the vacancy and interstitial mechanisms in all cases. Under non-equilibrium conditions, e.g. under injection of excess point defects, we derive the relevant expressions for the activation enthalpy for a variety of possible cases. Under oxidation, the calculated values are in excellent agreement with the experimental data.

1. INTRODUCTION

Substitutional dopant impurities in silicon can in principle diffuse by a combination of three mechanisms: vacancy-mediated, self-interstitial-mediated, or by a direct (concerted) exchange. In recent years, it has been observed that a number of surface processing conditions alter the bulk point defect concentration (Fahey et al. 1985). Oxidation, for example, has been shown to inject excess interstitials, while direct nitridation of the surface injects excess vacancies. Although the details of such processes are not completely understood, they afford the possibility of determining the mechanism(s) responsible for impurity diffusion. However, such experiments are usually performed at a single temperature and their conclusions consider only point-defect-mediated mechanisms, are inconsistent with respect to the dominant point defect involved, and are unclear with regards to the atomistic mechanisms. For example, Antoniadis and Moskowitz (1982) observe in oxidation experiments that the phosphorus (P) diffusion coefficient is enhanced with respect to its equilibrium value, although the enhancement is not constant in time. They conclude that P diffuses via a *dual vacancy-interstitialcy mechanism*, but a detailed picture of the mechanism is not provided. Under similar processing conditions, Fahey et al. (1985) find an identical P diffusion enhancement. Under injection of excess vacancies, however, they observe a large retardation of the P diffusion coefficient with respect to its equilibrium value. From these two experiments, they claim that P diffusion is almost *exclusively self-interstitial mediated*. These experiments are also in conflict with a diffusion equation analysis (Mathiot and Pfister 1984) which claims that P is entirely *vacancy-mediated*.

We use the results of first-principles pseudopotential density-functional calculations to investigate vacancy-, interstitial-, and concerted exchange-mediated diffusion pathways of substitutional boron, phosphorus, arsenic and antimony in Si. Activation enthalpies for equilibrium conditions are calculated. In addition, we develop a framework within which activation enthalpies under non-equilibrium conditions

© 1989 IOP Publishing Ltd

can be written down. Using these results, in conjunction with experimental data on diffusion enhancement and retardation, we discriminate between the different mechanisms and we can describe each mechanism in microscopic detail. We find that P, As, and B have large interstitial components, while Sb is vacancy dominated. Theoretical results and experimental data suggest that the CE mechanism has a limited rôle in impurity diffusion.

2. METHODOLOGY

In the calculations we use density-functional theory, the local density approximation (LDA), and norm-conserving pseudopotentials. The supercell method is used to solve the relevant Schrödinger equation (methodology as in Bar-Yam and Joannopoulos 1984). We use a plane-wave basis and include waves up to a kinetic energy of 20 Ry in the expansions for the wave functions and the potentials (plane waves above 10 Ry are included in perturbation theory). For the 32-atom supercells that we used, the distance between defects/impurities in neighboring cells is 9.4 Å. Relaxation of the surrounding Si host network is calculated for every location of the impurity or defect. For high symmetry atomic configurations, a sampling of two special k-points in the irreducible wedge of the Brillouin zone is sufficient, while for lower symmetry configurations, a larger equivalent set is used. For the electron-electron interactions, local-density-functional theory, within the LDA, is utilized. We estimate our total error to be less than 1 eV, depending, of course, upon the atomic configuration and the particular impurity. The majority of the error comes from the LDA uncertainty in the defect- and impurity-related levels. This scheme has a proven reliability in calculating bulk properties of semiconductors, reconstruction of semiconductor surfaces, and general properties of defects.

3. RESULTS AND DISCUSSION

3.1 Equilibrium Diffusion: Results and Discussion

Diffusion processes have been shown to obey the empirical Arrhenius expression

$$D^* = D_o^* \exp(-Q^*/kT) \tag{1}$$

where D^* is the diffusion coefficient, D^*_o is the preexponential which contains a variety of factors, including the entropy of diffusion, Q^* is the activation enthalpy, k is Boltzmann's constant, and T is the temperature. The asterisk denotes equilibrium quantities. For equilibrium diffusion, Q^* is the sum of two terms, the enthalpy of formation (E_f) plus the enthalpy of migration (E_m) for the diffusing species. Specific definitions of the formation enthalpies are given elsewhere (Nichols et al. 1988). Throughout this paper we report results for neutral species only. Charge states usually contribute with only slightly different activation enthalpies (Car et al. 1985).

We consider diffusion pathways mediated by vacancies and by Si self-interstitials, in addition to the concerted exchange (CE) mechanism which involves no point defects. Impurity diffusion mediated by vacancies entails a complex atomistic motion which is described in detail elsewhere (Car et al. 1985). We have not calculated the migration enthalpy for the impurity-vacancy pair because such a calculation requires a supercell size which is beyond present computer capabilities. Instead, we use values from experiment (Hirata et al. 1969). Of the self-interstitial-mediated

pathways which we have investigated, the kick-out mechanism (Car et al. 1985) is dominant. The kick-out involves the coordinated push of a Si self-interstitial on another Si atom along a nearest neighbor bonding direction towards a substitutional impurity. The Si atom is pushed into a neighboring substitutional site, while the impurity is kicked into the channel. The impurity can then migrate along the low-electron-density channel with a relatively small barrier.

A contour plot of the total-energy surface depicting neutral B interstitial diffusion is shown in Fig. 1. The energy values represent the total-energy of the B atom at that particular position in the Si crystal, taking relaxation of the host crystal into account. The details of generating such a surface are described elsewhere (Van de Walle et al. 1988). We highlight several high symmetry sites within the crystal. When placed at the BC site, the two Si neighbors relax outwards by 0.7 Å. Second neighbor relaxation is accounted for as well and it lowers the total energy by 0.3 eV. In the low-electron-density channels, there is very little relaxation of the surrounding Si atoms at the hexagonal (H) site (1% inward relaxation) and at the tetrahedral (T) site (no change).

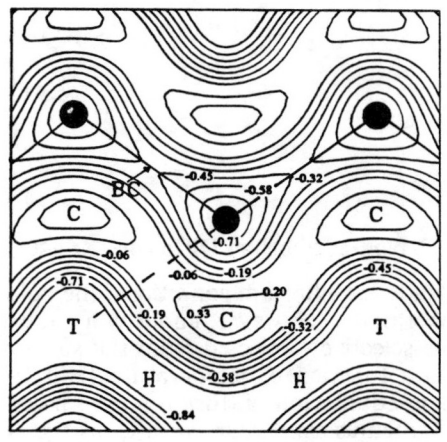

Fig. 1 The total-energy surface for neutral B interstitial diffusion in Si. The Si atoms, denoted by the dark circles, lie in the (110) plane of the diamond structure. The shape of the total-energy surface shows that the BC site is rather high in energy compared to the sites T and H. The lowest energy migration pathway is through the channel where the migration barrier ≈ 0 eV. The kick-out pathway is also evident in the figure with the saddle point $\sim$ two-thirds of the way from the T site to the nearest substitutional site and is denoted by the dashed line.

The CE mechanism was proposed by Pandey (1986) as an alternative to defect-mediated pathways for Si self-diffusion and it was shown to be energetically competitive with such mechanisms. Furthermore, because the saddle point consists of two threefold-coordinated atoms, it was argued that the exchange should be particularly well suited for Group III and Group V impurities. The CE saddle point involves large distortions from the ideal diamond structure atomic configurations, which makes relaxation of the saddle point both difficult but important. We have used our first-principles approach to calculate the saddle point energies without including relaxation. As a first estimate for the relaxation, we use Pandey's calculated relaxation of 0.75 eV for Si self-diffusion. This is probably a lower bound.

The activation enthalpies for vacancy-, interstitial-, and CE-mediated mechanisms for substitutional B, P, As, and Sb diffusion under equilibrium conditions are shown graphically in Fig. 2 (V $\equiv$ vacancy mechanism, I $\equiv$ interstitial mechanism, CE $\equiv$ concerted exchange mechanism). For reference, our calculated values for Si self-diffusion are also given in Fig. 2. For B, P, and As, the differences between V-type

and I-type mechanisms are small (0.4 eV is the maximum). It is not possible, without information regarding the preexponential in Eq. (1) for each mechanism, to distinguish between the two. For Sb, however, the vacancy mechanism activation enthalpy is a full 1.3 eV lower than the interstitial mechanism. The activation enthalpies for the CE mechanism for P, As, and Sb are close to that for defect-mediated pathways. For B, the CE activation enthalpy is considerably larger than that for I-type or V-type mechanisms, but we expect that the small size of B will necessitate further relaxation of the saddle point. Lastly, shown as boxed area on Fig. 2 is a selected range of experimental values. The agreement between our calculated values and the experimental values is quite good, indicating that the proposed microscopic mechanisms provide a realistic description of substitutional impurity diffusion.

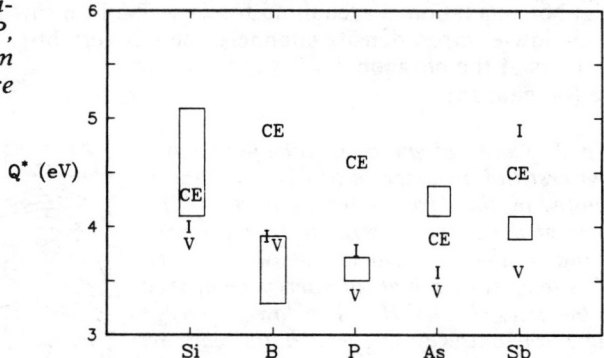

Fig. 2 The calculated activation enthalpies for Si, B, P, As, and Sb under equilibrium conditions. The symbols are explained in the text.

3.2 Non-equilibrium Diffusion: Results and Discussion

Under equilibrium conditions, the point defects responsible for diffusion are created thermally. However, under non-equilibrium diffusion conditions, point defects may be selectively injected from the surface. The limited number of experiments done as a function of temperature (Hill 1981, Bronner and Plummer 1987) show that non-equilibrium diffusion may, similar to equilibrium diffusion, be described by an Arrhenius expression, with Q' as the activation enthalpy. Because the point defects are *supplied* from the surface, the detailed balance of equilibrium is upset and the activation enthalpy for defect-mediated diffusion changes. To define Q' under non-equilibrium conditions requires explicit enumeration of a self-consistent set of assumptions regarding generation and interaction of the point defects and the complexes responsible for diffusion. We briefly outline below the simplest set of such assumptions for diffusion under excess point defect injection. From this general theory, we are able to predict the form for the observed activation enthalpy and to understand diffusion enhancement and retardation on an atomistic scale.

We observe at the outset that point defect injection should have little effect on the CE mechanism. We argue more extensively elsewhere (Nichols et al. 1988) that the magnitude of diffusion retardation observed experimentally places an upper bound on the contribution of the CE mechanism to impurity diffusion. We therefore do not discuss the CE mechanism any further here. Impurities therefore diffuse via the two reactions:

$$I + X_{\text{sub}} \rightarrow X_i \tag{2}$$

$$V + X_{sub} \rightarrow XV \tag{3}$$

where I ≡ self-interstitial, V ≡ vacancy, X_{sub} ≡ any substitutional impurity, the subscript i indicates an interstitial impurity and XV is an impurity-vacancy pair. Diffusion is retarded by any reaction which annihilates point defects or which converts the diffusing species into a substitutional impurity. These reactions are written down explicitly in Nichols *et al.* (1988).

In studying non-equilibrium diffusion, we distinguish two time domains. The transient period is that time during which the diffusion coefficient of point defects and impurity-defect complexes is changing, while the steady-state period means that the diffusion coefficients are constant. We further differentiate between *surface* processes and *bulk* processes. We assume that the external processing condition at the surface only generates one species of point defect, but that the generation of the other species, at another surface or on internal voids, continues as under equilibrium conditions. Bulk processes occur within the buried impurity layer and we assume that there is no impurity concentration gradient in the region on which we focus.

During the transient period, *all* impurities should show enhanced diffusion. This is because the increased concentration of either I or V drives Eq. (2) or (3), respectively, further to the right, building up an excess of diffusing species. For the steady-state time regime, we have considered three special cases. First we assume that impurity diffusion is mediated exclusively by vacancies, that interstitials play no rôle and that vacancies are being injected at the surface. Second we consider vacancy-mediated diffusion under interstitial injection. Note that for these two cases, we can reverse the rôles of vacancies and interstitials and the physics does not change. Space does not permit discussion of the third case here, wherein diffusion is mediated equally by vacancies and interstitials and either point defect is injected. For each case studied, we write down the coupled set of equations for the concentration of each species, assuming different rate constants for the forward and reverse rates for all reactions.

For the case in which *vacancies* exclusively mediate impurity diffusion and we inject *vacancies*, solution of the coupled equations leads to an expression for the concentration of XV pairs. By assuming that local equilibrium between Si vacancies and XV pairs is obtained in the region of the buried layer, we derive the activation enthalpy

$$Q' = E_{inj} - \Delta E + E_m \tag{4}$$

where E_{inj} = the activation enthalpy associated with the injection process (this is zero if the process is athermal), ΔE is the binding energy of the XV pair and E_m is the migration energy of XV. For *interstitial-mediated* diffusion under *interstitial* injection, Q' has an identical form, except that now we identify ΔE as the energy difference between a substitutional X-interstitial Si pair and an X_i. For vacancies, $\Delta E > 0$, whereas for interstitials, ΔE can be either positive or negative.

Equation (4) leads to several predictions. To the best of our knowledge, no values for the activation enthalpy of vacancy injection by direct surface nitridation have been reported. However, Hu (1975) has reported that the activation enthalpy for interstitial injection by oxidation is 2.3 eV. Using this value along with our calcu-

lated values for ΔE and E_m, we find, for example, $Q' = 2.5$ eV for P diffusion under oxidation conditions. This value agrees remarkably well with that found by Hill (2.4 eV). Values of Q' for B and As also agree well with Hill's measured values. Bronner and Plummer (1987) observed P diffusion enhancement for a damaged layer created by Ar implantation as the interstitial source and did not determine E_{inj}. Using their measured Q' of 1.3 eV, we can, however, estimate E_{inj} to be 1.1 eV, a value which can be tested experimentally. Atomistically, enhanced diffusion occurs either through the formation of XV pairs or the kick-out of X_{sub} aided by the injection of either V or I, respectively, and the ineffectiveness of any retardation process. The good agreement between our calculated Q' and Hill's measured values is an indication that B, P, and As have a large interstitial component.

For the case in which *vacancies* mediate diffusion and *interstitials* are injected, after some simplifying assumptions, we find

$$Q' = E_G - (E_{inj} + E_{barrier}) + E_m \tag{5}$$

where E_{inj} and E_m are as above, E_G is the energy required to generate vacancies thermally (not in the bulk) and $E_{barrier}$ is the energy barrier to the annihilation of XV pairs by free interstitials. This expression has a simple and physically transparent interpretation. The first term and the term in parentheses represent the competing costs of generating the point defect responsible for mediating diffusion and of annihilating the diffusing species. The net result from rough estimates for the unknown values of E_G and $E_{barrier}$ is a rather small effective activation enthalpy for retarded diffusion. There exist no experimental data with which we can compare any predictions. Systematic measurement of diffusion coefficients over a range of different temperatures and injection processes is required to evaluate the various terms in Eqs. (4) and (5) and so that the mechanisms of impurity diffusion can be sorted out.

ACKNOWLEDGMENT

This work was supported in part by ONR Contract No. N000014-84-C-0396

4. REFERENCES

Antoniadis D A and Moskowitz I *1982 J. Appl. Phys.* 53 6788
Bar-Yam Y and Joannopoulos J D *1984 Phys. Rev. B* 30 1844
Bronner G B and Plummer J D *1987 J. Appl. Phys.* 61 5286
Car R, Kelly P J, Oshiyama A, and Pantelides S T *1985 Phys. Rev. Lett.* 54 360
Fahey P, Barbuscia G, Moslehi M, Dutton R W *1985 Appl. Phys. Lett.* 46 784
Hill C 1981 in *Semiconductor Silicon 1987,*, edited by H. R. Huff, R. J. Kreigler, and Y. Takeishi (Electrochemical Society, New York, 1981), p. 988
Hirata M, Hirata M, and Saito H *1969 J. Phys. Soc. Japan* 27 405
Hu S M, *1975 Appl. Phys. Lett.* 27 165
Nichols C S, Van de Walle C G, and Pantelides S T *1988 to be submitted to Phys. Rev. B*
Mathiot D and Pfister J C *1984 J. Appl. Phys.* 55 3518
Pandey K C *1986 Phys. Rev. Lett* 57 2287
Van de Walle C G, Bar-Yam Y, and Pantelides S T *1988 Phys. Rev. Lett.* 60 2761

Inst. Phys. Conf. Ser. No 95: Chapter 8
Paper presented at Int. Conf. Shallow Impurities in Semiconductors, Linköping, Sweden, 1988

The effect of heavy doping on complex formation and diffusivity of Sb in Si

A.Nylandsted Larsen, P.Tidemand-Petersson, and P.E.Andersen,
Institute of Physics, University of Aarhus, DK-8000 Aarhus C, Denmark
G.Weyer, EP Division, CERN, CH-1211 Geneva 23, Switzerland

ABSTRACT: The complex formation and diffusivity during rapid thermal annealing of low concentrations of ion-implanted Sb in silicon have been studied as functions of donor concentration by combination of Rutherford backscattering spectroscopy, Hall effect/resistivity measurement combined with layer removal, transmission electron microscopy, and Mössbauer spectroscopy. For extrinsic/intrinsic carrier concentrations $n/n_i \geq 20$ extremely large diffusion coefficients are found exhibiting approximately an $(n/n_i)^4$ dependence. This dependence is correlated to the appearance of a new defect-complex containing Sb as revealed by Mössbauer spectroscopy.

1. INTRODUCTION

Among the group V impurities in Si, Sb has been of particular interest to us because one can do Mössbauer spectroscopy on radioactive Sb in Si (Weyer et al.1980). Thus, together with information on macroscopic parameters such as chemical and carrier profiles, concentration of large defect clusters and precipitates etc. obtained by means of more traditional techniques, one can get information on a microscopic scale of the surroundings of the Sb atoms from Mössbauer spectroscopy and thereby attain to a much more complete description of Sb in silicon (Nylandsted Larsen et al.1986).

Previously we have demonstrated the strength of such a combination of techniques on the elucidation of the behaviour of high concentration of Sb in silicon (Nylandsted Larsen et al.1986). In the present investigation we have focussed on the complex formation and diffusivity of ion implanted Sb in Si as functions of donor concentrations. In particular we have been interested in the possibility of separating the total diffusion coefficient into contributions from different charge states of the vacancy according to $D_{tot} = \Sigma D^r(n_i/n)^r$ where D^r denotes the diffusivity via vacancies in the rth charge state and n_i and n are, respectively, intrinsic and actual carrier concentration at the diffusion temperature.

2. EXPERIMENTAL

The silicon samples were electronic-grade, dislocation free, p-type (boron doped) 150 Ωcm, <100>-oriented, float zone refined monocrystals. The phosphorus background doping was established by implantation of P at an energy of 60 keV to doses between 2×10^{15} and 1.2×10^{16} cm^{-2}. After rapid thermal annealing (RTA) (Nylandsted Larsen and Borisenko 1984) at 1075°C for 20s in an Ar ambient, flat carrier-density profiles extending to a depth of $\sim$2000 Å were obtained with carrier concentrations between 9×10^{19}

© 1989 IOP Publishing Ltd

and 5×10^{20} cm^{-3} (Andersen et al.1988). In one sample the phosphorus predoping was replaced by an arsenic predoping with a carrier concentration of 2.5×10^{20} cm^{-3}. Subsequently, antimony was implanted at an energy of 80 keV to doses of either 2×10^{14} or 5×10^{14} cm^{-2} corresponding to maximum concentrations of 4×10^{19} cm^{-3} and 9×10^{19} cm^{-3}, respectively. All implantations were done in a 7° off-axis direction. Annealing was then done by RTA in an Ar ambient for a time of either 15 or 20 s (including the temperature rise time of ∿5 s) at temperatures between 700 and 1100°C.

The carrier-density profiles were determined by Hall-effect/

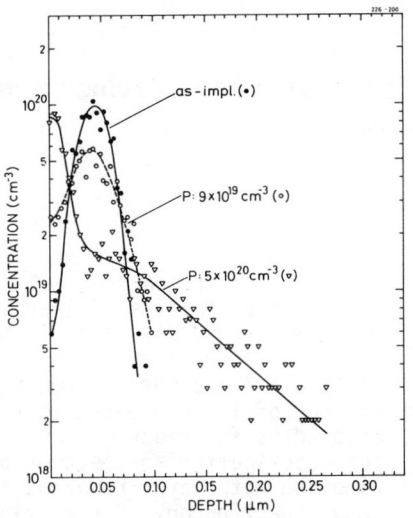

Fig.1. *Chemical profiles of 5×10^{14} cm^{-2} Sb as-implanted and after RTA at 1050°C for 15 s for two different P-background dopings of 9×10^{19} cm^{-3} and 4.5×10^{20} cm^{-3}*

resistivity measurements combined with anodic oxidation and stripping in steps of 100 Å (Galloni and Sardo 1983). The chemical depth profiles were determined by RBS with 2-MeV He$^+$ ions. In some of the samples, in addition to stable Sb a small amount ($\lesssim 1 \times 10^{13}$ cm^{-2}) of radioactive ^{119}Sb was implanted at the ISOLDE facility at CERN, and Mössbauer spectra were measured for the 24-keV γ radiation emitted by the daughter ^{119}Sn by employing resonance-counting technique with CaSnO$_3$ absorbers (Weyer 1976). Transmission electron microscopy was done at a Philips CM12 instrument operating at 120 kV.

3. RESULTS

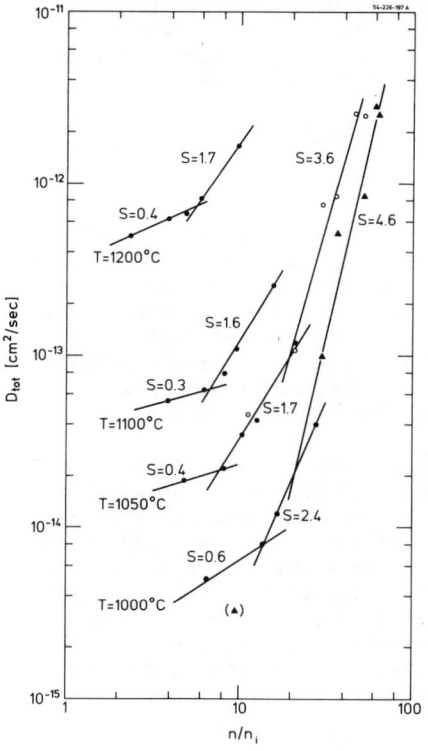

Fig.2. *Total diffusion coefficients as a function of n/n_i for the indicated temperatures. Also given in the figure are slopes extracted from least-squares fits. Points marked with (●) are from Fair et al. (1986).*

Chemical depth profiles of Sb in Si measured as-implanted and after RTA at 1050°C for two different P-background concentrations are shown in Fig.1. A remarkable difference in diffusivity is observed between the sample with low donor concentration and that with high concentration: at the lowest donor concentration, only

a slight broadening is observed as compared to the as-implanted profile, whereas a very large redistribution has occurred at the highest donor concentration resulting in a distribution extending to a depth of ∼0.3 μm and with a significant pile-up at the surface.

Diffusion coefficients extracted from such chemical profiles (Andersen et al. 1988) are displayed in Fig.2 for two different temperatures of 1000 and 1050°C together with diffusion coefficients measured by Fair et al. (1986) for temperatures between 1000 and 1200°C. Three different regions of D_{tot} values versus n/n_i-values can be identified: 1) for small n/n_i values D_{tot} varies approximately as $(n/n_i)^{0.4}$, 2) for n/n_i values of about 10, D_{tot} varies approximately as $(n/n_i)^2$, 3) for n/n_i values larger than 20, D_{tot} varies approximately as $(n/n_i)^4$. The transition from one region to the other is probably not abrupt as drawn in Fig.2, where the straight lines are least squares fits to the points in the different regions, but smooth. From these dependencies, the two first regions have been previously related to diffusion via respectively neutral and doubly charged negative vacancies (Fair et al. 1986 and Nishi et al. 1986). The n/n_i-value of 20 which marks approximately the transition from the $(n/n_i)^2$ to the $(n/n_i)^4$ dependence corresponds to a donor concentration of 1.8×10^{20} at 1050°C.

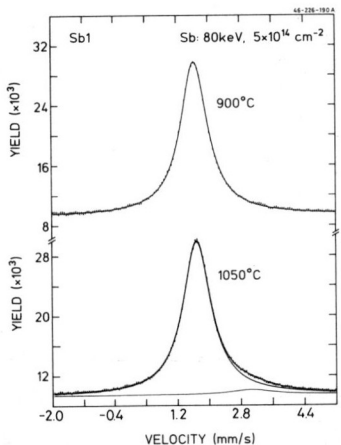

Fig.3. Mössbauer spectra measured at 77K after RTA at the indicated temperatures of a sample without P-background doping. The separation of the spectra into individual lines by the computer analysis is also shown.

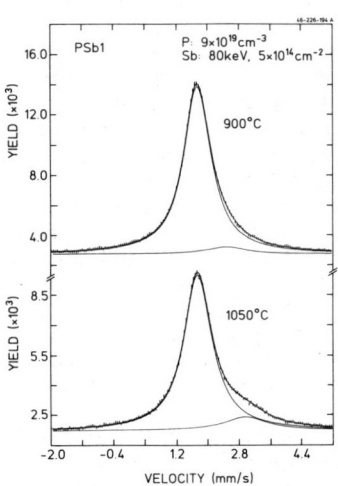

Fig.4. As Fig.3 but with a P-background doping of 9×10^{19} cm^{-3}.

Reference Mössbauer spectra of Sb in silicon after RTA at 900 and 1050°C are shown in Fig.3. They are reference spectra in this context as they stem from a pure Sb implantation without P-predoping. The spectrum after

RTA at 900°C consists of only one line with an isomer shift of 1.78 mm/s (at 77K) and a linewidth of 0.85 mm/s; this line is known to stem from Sb on exact, undisturbed substitutional positions in the Si lattice (Weyer et al. 1980). Thus, for these conditions all the Sb atoms are on substitutional positions. After RTA at 1050°C a small new line with an isomer shift of 3.2 mm/s appears; this line is known to stem from Sb in a thin surface layer as it disappears after a chemical stripping of a 200 Å thick surface layer. From the Mössbauer spectra it is estimated that ~6% of the Sb atoms are contained in these surface positions, the rest is still in substitutional positions. Only minor changes take place when one goes to a sample with a low P-background doping of 9×10^{19} cm³ (Fig.4): there is now a small surface line already after RTA at 900°C which grows after RTA at 1050°C corresponding to a fraction of ~16%. The dominating fraction of Sb, however, is still in substitutional positions. The P-background doping of 9×10^{19} cm^{-3} corresponds to a n/n_i-value in the region where D_{tot} varies as $(n/n_i)^2$. Going to a high P-background doping of 4×10^{20} cm^{-3} results in a dramatic change of the Mössbauer spectrum (Fig.5). In addition to the substitutional line a new line with an isomer shift of 2.53 mm/s has appeared and after RTA at 820°C this line is the dominating line with an intensity equivalent to a Sb-fraction of ~65%. An increase in the RTA temperature to 1080°C results in the appearance of two new lines (labelled

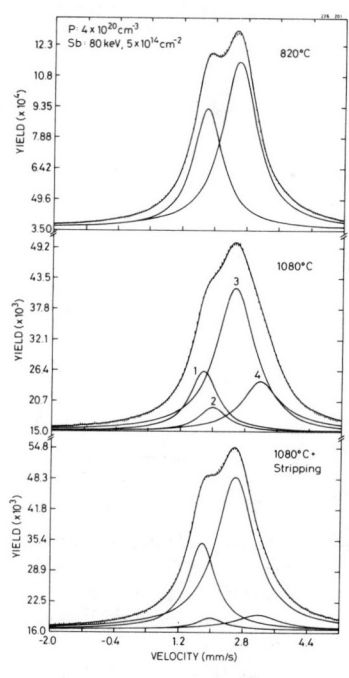

Fig.5. As Fig.3 but with a P-background doping of 4×10^{20} cm^{-3}. The spectrum at the bottom is measured after a chemical stripping of ~200 Å.

line 2 and 4 in Fig.5) of which line 4 is the well-known surface line from Figs.3 and 4. The line 2 is also related to the outermost surface as deduced from the stripping experiment reported in the lower part of Fig.5. These two surface lines will not be discussed any further in this paper. The 2.53 mm/s line (labelled line 3 in Fig.5 and in the following the complex line) is not related to the outermost surface layer which one can deduce from the stripping experiment, however, the intensity ratio between the substitutional line and the complex line is slightly larger after stripping than before, indicating that the distribution of Sb in the positions giving rise to the complex line is slightly more peaked towards the surface than is the distribution of substitutional Sb.

The intensity of the complex line as a function of carrier concentration or n/n_i is shown in Fig.6. The line starts to grow in for a carrier concentration somewhere between 1×10^{20} and 2.5×10^{20} cm³ and has an intensity of about 75% at a carrier concentration of 4.3×10^{20} cm^{-3} corresponding to a Sb fraction of ~80%. By comparison with Fig.2 it appears that the concentration region where the complex line grows in is the same region where D_{tot} varies as $(n/n_i)^4$. One of the points of Fig.6 stems from a sample in which the P-doping was replaced by an As-doping. Also in this case the same complex line is found and as seen from Fig.6 with an intensity comparable to what is found in the P-doped samples.

Diffusion of Shallow Impurities

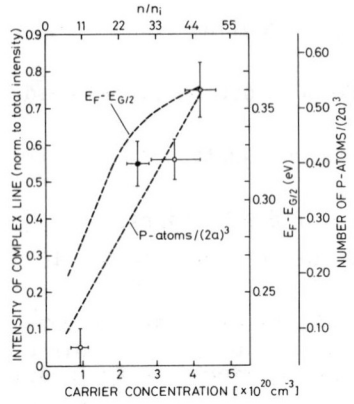

TEM investigations have been performed on the samples to see if the fast diffusion and the appearance of the complex line could be correlated to the existence of dislocations. There is found, however, to be no Sb decoration of dislocation and no correlation between the density of dislocations on one side and the intensity of the defect line and the diffusivity on the other side.

Fig.6. Intensity of the complex line versus carrier concentration of n/n_i after RTA at $1050°C$. The value marked (●) is from a sample with an As-background doping. Also shown in the figure are energy differences between the position of the Fermi level and the middle of the band gap at $1050°C$ from Jain and Van Overstraeten (1973), and the number of P atoms per $(2a)^3$ where a is the Si lattice constant.

4. DISCUSSION and CONCLUSION

The main objective of this investigation has been to study the complex formation and diffusivity of Sb in Si as functions of donor concentration and thereby unveil the effects of different charge states of the vacancy as the Fermi level is raised in the band gap. Three different regions of diffusivity as a function of donor concentration have been identified. Two of these regions have been previously related to diffusion via neutral and doubly charged negative vacancies. In these two regions only substitutional Sb is found except for a small non-substitutional fraction in the outermost surface layer as revealed by Mössbauer spectroscopy. Increasing the donor concentration above $\sim 2 \times 10^{20}$ cm^{-3} at $1050°C$ results in a diffusivity proportional to $(n/n_i)^4$ accompanied by the appearance of a new line in the Mössbauer spectra reflecting the creation of a new defect complex containing Sb. This behaviour has to our knowledge never been observed before. The following conclusions can be made:

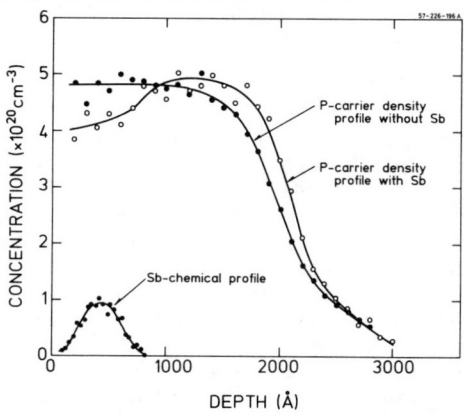

1) The Mössbauer parameters of the complex line (at 77K) is, isomer shift: 2.53(5) mm/s, linewidth: 1.20(8) mm/s and Debye temperature: 189(6)K.

2) The complex line is different from the two non-substitutional lines observed previously in high concentration Sb implanted Si (Nylandsted Larsen et al.1986) which were concluded to stem from Sb in, respectively, a vacancy complex with a broken bond (isomer shift: 2.32(5) mm/s) and in Sb precipitates (isomer shift: 2.74(5) mm/s).

3) The complex is not a result of long range diffusion as it has been observed already after RTA at $700°C$ for a carrier concentration of 5×10^{20} cm^{-3}.

Fig.7. Carrier density profiles without and with Sb implantation after RTA at $850°C$. Also shown is the Sb-chemical profile

4) The complex is a stable defect as it survives the cooling-down from high temperature.
5) The same complex line appears when the P-background doping is replaced by an As-doping.
6) The intensity of the complex line is an increasing function of the diffusion temperature.
7) The complex is not correlated to the existence of dislocations.
8) The $(n/n_i)^4$-dependence has also been observed for Sn (Andersen et al. 1988).
9) From the data it cannot be concluded whether the intensity of the complex line as a function of carrier concentration follows a Fermi-Dirac distribution or just increases linearly with the carrier density (Fig.6).
10) The effect of the defect complex on the carrier concentration is shown in Fig.7, where a drop in carrier concentration is seen to result from the formation of the complex. The number of missing carriers is within uncertainties equal to the number of Sb atoms in these complexes.

From the results of this investigation we are not able to determine the structure and the constituents of this defect complex except that Sb is a constituent. Neither are we able to decide whether this complex is the diffusing species or whether it forms during the cooling-down from the diffusion temperature. However, there are observations which indicate that the complex is formed at the diffusion temperature, e.g., the dependence of the intensity of the defect line on temperature. A plausible structure of the complex is a (neutral) complex of the form Sb-P-vacancy, formed by the interaction of, e.g., a negative E-center and a positively charged Sb^+ ion. It remains to be proven, however, that such a defect can diffuse with such large diffusion coefficients as observed.

ACKNOWLEDGEMENTS

Thanks are due to A.Armigliato, CNR-LAMEL for the TEM measurements. This work has been supported by a NATO Research Grant and by Danish Accelerator Physics Council.

REFERENCES

Andersen P E, Nylandsted Larsen A, Tidemand-Petersson P and Weyer G 1988 Appl.Phys.Lett. in print
Fair R B, Manda M L and Wortman J J 1986 J.Mater.Res.1 705
Galloni R and Sardo A 1983 Rev.Sci.Instrum.54 369
Jain R K and Van Overstraeten R J 1974 IEEE Trans.Elect.Dev.ED21 155
Nishi K, Sakamoto K and Ueda J 1986 J.Appl.Phys.59 4177
Nylandsted Larsen A and Borisenko V E 1984 Appl.Phys.A 33 51
Nylandsted Larsen A,Pedersen F T,Weyer G,Galloni R, Rizzoli R and Armigliato A 1986 J.Appl.Phys.59 1908
Weyer G 1976 Mössbauer Effect Methodology, ed. I.J.Gruverman and C.W. Seidel (New York: Plenum) p 301
Weyer G, Nylandsted Larsen A, Holm N E and Nielsen H L 1980 Phys.Rev.B 21 4939

Excited state spectroscopy of excitonic systems in semiconductors

M.L.W. Thewalt, M. Nissen, D.J.S. Beckett and S. Charbonneau

Department of Physics, Simon Fraser University, Burnaby, B.C., Canada
V5A 1S6

ABSTRACT: In this paper we will review several recent developments in excited state spectroscopy which involve excitonic systems. Examples of the two possible situations will be given: excited state spectroscopy of the bare binding center (the final state of a bound exciton photoluminescence transition), and also excited state spectroscopy of bound excitons themselves. In both cases the emphasis will be on technical advances and new spectroscopic methods.

1. INTRODUCTION

Spectroscopic studies of bound exciton (BE) systems in semiconductors have proliferated remarkably over the past two decades, driven initially by a basic curiosity regarding the diverse phenomena and mechanisms to be found in these systems. More recently, much of the impetus has come from the realization that BE photoluminescence (PL) spectroscopy can be a powerful technique for impurity and defect characterization and quantitative analysis. The establishment of a firm connection between a given PL feature and a known impurity level, characterized for example by donor or acceptor properties, a known thermal activation energy, or series of electronic excited states as determined by far- or mid-infrared absorption, is however quite difficult under many circumstances. This is particularly true in systems where clean, selective doping is impractical, when dealing with ubiquitous impurities or defects in a given system, or when many different BE species are present simultaneously. The only spectroscopic solution to such difficulties is the determination of the binding center excited state spectrum (and thus the ionization energy) directly from the BE PL spectrum. In the first section of this paper we will describe the successful application of this type of technique to a very complicated PL system often observed in GaAs grown by MBE (Künzel and Ploog 1980, Skolnick et al 1981). In the second section we will describe two methods of BE excited state spectroscopy (as opposed to binding center excited state spectroscopy) in connection with two interesting new high-efficiency PL systems recently discovered in S-doped Si (Brown and Hall 1986, Brown et al 1987).

2. DEFECT BE LINES IN MBE GROWN GaAs

Künzel and Ploog (1980) were the first to report on a highly structured series of lines observed in the 1504 to 1511 meV region of the PL spectrum of GaAs grown by MBE, which we will refer to as the KP lines. A representative spectrum of these lines, labelled 1 through 47 after Skolnick et al (1985), is shown in Figure 1. These principal PL lines (by which we mean BE ground-state to binding center ground-state transitions, or alternatively, the most intense PL lines of a given BE) reveal little about the natures of the binding centres, other than that they must be axial defects, since the PL is seen to be polarized (Skolnick et al 1985). This lack of spectroscopic or chemical information as to the nature of the defects responsible for the KP PL lines led to a considerable number of competing models, which cannot be discussed in any detail here.

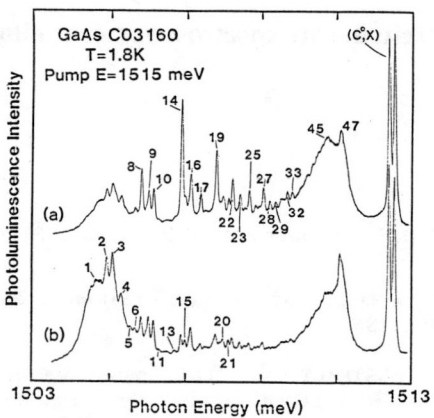

Fig. 1. High resolution PL spectra of the KP system for $\vec{E} \parallel [\bar{1}10]$, (a), and $\vec{E} \parallel [110]$, (b). (C^o,X) is the carbon acceptor BE doublet.

The direct solution to the nature of the KP lines lies in associating a binding center excited state spectrum with each of the principal KP lines. This would reveal whether the KP lines are donor BE (donor-like excited final states), acceptor BE (acceptor-like excited final states), or isoelectronic BE (no electronic excited final states). This information is _in principle_ always contained in the PL spectrum in the form of two-electron (2e) replicas for donors or two-hole (2h) replicas for acceptors (we will henceforth refer only to 2h replicas, since these are what are observed for the KP lines, but in general these methods could be applied equally well to the 2e replicas of donor BE). For an excellent review of BE phenomena up to 1978, see Dean and Herbert (1979).

In a 2h transition, the ground state BE decays into a photon plus an excited acceptor final state, rather than the acceptor ground state which is the final state of the principal BE transition. The 2h replicas thus lie below the principal BE line, the energy separations being equal to the relevant acceptor ground-state to excited-state splitting. The 2h transitions are always much weaker than the principal BE line, since the holes in the BE have a much greater overlap with the acceptor ground-state wave function than with the excited states. For the same reason, the BE 2h replicas are usually dominated by even-parity excited states, and thus can provide useful information which is complementary to that obtained on the odd-parity excited states by the direct infrared absorption spectroscopy of the same acceptors (Lipari et al 1980).

While this simple method of looking for the 2h (or 2e) replica transitions in the non-resonantly excited PL spectrum can work for clean systems which are dominated by a single BE species, it is sure to be impractical for systems such as the KP lines, where each of the >40 principal transitions will have its own series of replicas, all of which will be mixed together on the fairly intense background of the broad donor-acceptor-pair (DAP) PL band which lies below the KP region. Indeed, no resolved 2h transitions can be observed in the ordinary PL spectrum of GaAs samples which have strong KP lines.

An elegant solution to this problem is to resonantly excite a given principal BE transition with a tunable laser, thus creating only a single BE species, and a single 2h replica series. This approach to the KP lines was followed by Skolnick et al (1986), but due to low signal intensities and interference from the DAP band, they were able to reliably identify only a single replica line for each of only 5 of the KP lines. These replica lines were correctly interpreted as 2S 2h replicas, but in the absence of a 2h <u>series</u> for any of the lines the argument was at that time speculative, and based more upon other reasons for thinking that the KP lines were in fact due to acceptor BE (Skolnick et al 1986).

By solving the problems of low signal levels and the interference from the DAP band, we have now been able to observe 2h replica series for 30 of the KP lines, including in many instances excited states up to 5S, thus proving conclusively the acceptor BE nature of the KP lines (Charbonneau et al 1988a, Charbonneau et al 1988b). The sensitivity of our apparatus was greatly improved by using parallel detection obtained with a 2D imaging photon-counting detector (ITT Surface Science Laboratories 'Mepsicron' tube). This detector was coupled to a 3/4 m single spectrometer, with an interference filter for additional rejection of scattered resonant-excitation laser light. The interfering DAP PL band was greatly reduced by a novel time-windowing scheme we developed for the Mepsicron tube, providing simultaneous 2D spatial information and sub-nanosecond timing information for each detected photoevent (McMullan et al 1987). The pulsed excitation was provided by a mode-licked cavity-dumped (4MHz) dye laser using LDS821 and having a linewidth of <0.05 meV. The advantage of the time-windowing in suppressing the long-lived DAP band is evident in Figure 2.

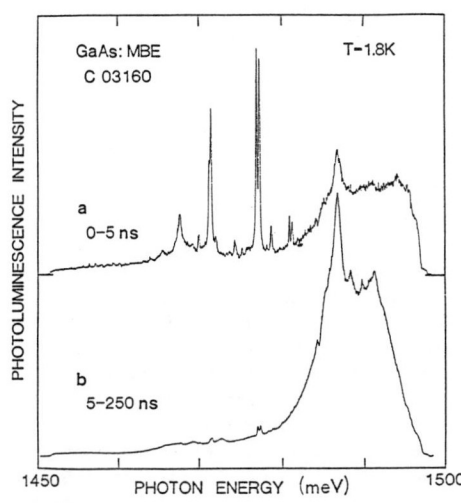

Fig. 2. Two time-windowed selective excitation spectra of the KP two-hole transitions, with excitation at 1508.70 meV. The two-hole transitions are greatly enhanced in the fast time-window (a). The CW spectrum is almost identical to the long time-window spectrum (b).

In Figure 3 some typical KP BE 2h replica spectra are compared to the 2h replicas of the well-known carbon acceptor BE (top spectrum). The spectra have all been shifted by the indicated amounts so as to align the 4S transitions. The striking similarities are clear proof of the acceptor-like nature of the binding centers responsible for the KP lines. The Δ_{LO} transition seen in Figure 3 is associated with the KP transitions (Charbonneau et al 1988a, Charbonneau et al 1988b) but is not a 2h replica, and will not be discussed further here.

The observation of these higher-lying excited states has allowed us to obtain accurate ionization energies for many of the KP acceptors, which are plotted in Figure 4 versus the KP BE localization energies, demonstrating a linear (Haynes' rule) relationship (Haynes 1960). These accurate ionization energies also allow us to prove conclusively the proposal (Skolnick et al 1986, Briones and Collins 1982) that a series of broad PL lines in the 1465 to 1485 meV region, often referred to as the d_n lines, are in fact DAP bands and free-to-bound bands involving the same series of acceptor defects which are responsible for the KP acceptor BE lines. In Figure 5 we demonstrate that the d_n spectra can be very accurately simulated by using the distribution of acceptor ionization energies and populations obtained from our KP 2h spectroscopy results. In concluding this section, it is clear that BE 2h (or 2e) replica spectroscopy can be

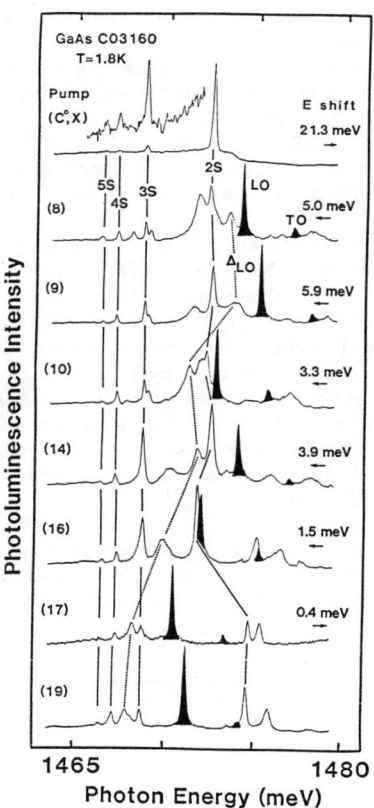

Fig. 3. Resonantly excited two-hole replica spectra of some of the KP lines (numbered on the left) are compared with that of the carbon acceptor BE (top). The 4S and 5S transitions have all been aligned — the energy scale applies to the bottom (line 19) spectrum, while the energy shifts required to align the other spectra are shown on the right. The shaded lines are the LO and TO phonon Raman lines, while Δ-LO is the LO phonon replica of the selectively excited Δ transition.

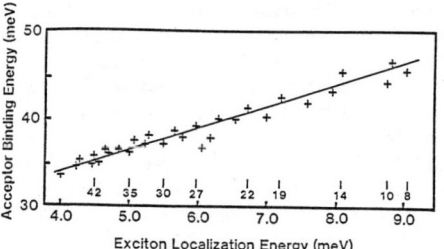

Fig. 4. A comparison of the BE localization energy for the KP lines (numbered at bottom) vs the acceptor binding energies as determined from the 3S two-hole replicas.

Fig. 5. (a) shows a typical PL spectrum for CW above-band-gap pumping, revealing the d_n lines lying below the carbon donor-acceptor (D°,C°) and free-to-bound (e,C°) bands. The fast (e,A°) processes are enhanced in (b) by using pulsed excitation together with a short time window immediately following excitation, while in (c) the long-lived (D°,A°) lines are enhanced (and sharpened) by using a delayed time window. The d_n spectra in (a) and (c) are compared with simulated spectra (involving no adjustable parameters) generated by convolving the observed carbon spectrum with the distribution of defect acceptors as determined from the KP line spectroscopy.

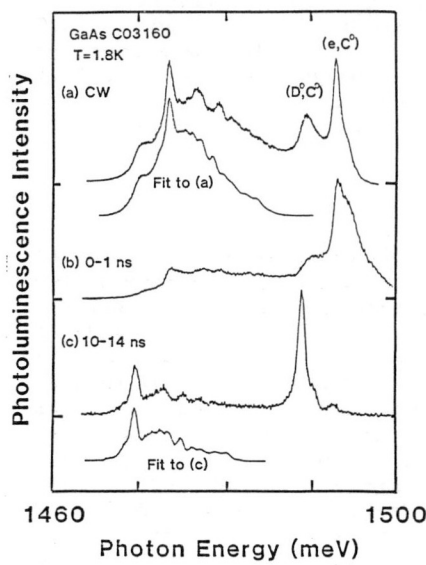

successfully applied to very weak and cluttered systems by combining the advantages of resonant excitation, parallel detection, and transient time-windowing.

3. EXCITED STATES OF SULFUR-RELATED ISOELECTRONIC BE IN Si

The study of the excited states of BE themselves is still an area of more fundamental research, and much has yet to be learned regarding the binding mechanisms and excited state structures of the diverse families of BE in different semiconductors. BE excited state spectroscopy can nevertheless be useful in revealing some aspects of the nature of the binding centers, particularly for isoelectronic BE (IBE), which of course have no binding center excited states or 2e or 2h replicas. One of the most successful models for the structure of IBE states is that of Hopfield, Thomas and Lynch (Hopfield et al 1966), in which the attractive short-range central-cell potential of the binding center strongly localizes a hole (or electron), which in turn binds an electron (hole) in its Coulomb field to form a 'pseudo-donor' ('pseudo-acceptor'). The electronic excited states of the IBE will be those of the Coulomb-bound particle, being donor-like for pseudo-donor IBE, and acceptor-like for pseudo-acceptor IBE. Of course there may be complications due to central-cell effects on the Coulomb-bound particle, or the non-localized nature of the central charge, or due to a possible axial field from the binding center. In an alternate model, the electron and hole states of the IBE are very similar to states at the conduction band and valence band edges, and the IBE localization arises from the local strain field of the binding center (Davis 1984). No detailed theory for the excited states of such an IBE is available, but they would likely resemble those of a free exciton in a strain field of the same symmetry.

Fig. 6. High resolution PL spectra of the S_B system at 8, 20 and 40 K. Selected regions of the 8 and 20 K spectra have been expanded by 10 times for clarity. The labels are as explained in the text.

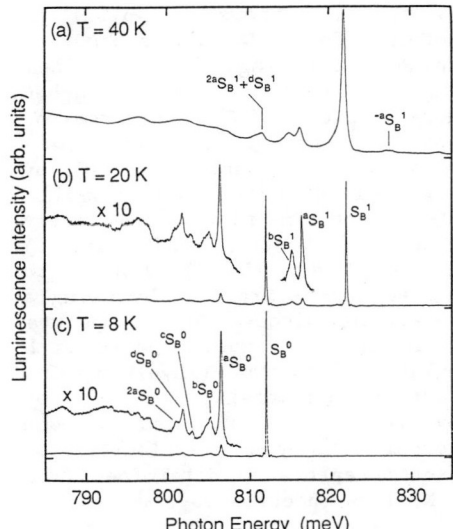

In this paper we will discuss the excited state spectroscopy of two interesting IBE systems recently discovered in Si doped with S (Brown and Hall 1986, Brown et al 1987). These systems are unusual in that one of them maintains its high PL intensity up to quite high temperatures (~100K), raising interesting possibilities of device applications for these or similar centers. Very little spectroscopic work has been done on the PL of these systems, and none on their excited states. In Figure 6 we show the first published high resolution PL spectra of the lower energy system (which we refer to as S_B) versus temperature, while in Figure 7 we show similar spectra for the higher energy S_A system.

The transitions in Figure 6 are labelled $^mS_A^n$, where n identifies the initial IBE state (0 = the IBE ground state) and m identifies any local-mode phonons emitted (positive) or absorbed (negative) during the transition. At least 4 different local mode phonon replicas (a through d) are seen for S_B, with 2a indicating the emission of two a-type phonons. Fewer local-mode phonons are seen for the S_A system in Figure 7, and they are assigned the same labels as the phonons they most closely resemble in the S_B spectrum (namely b and d).

Comparison of Figures 6 and 7 reveals striking similarities between the S_A and S_B systems. Both have ground state transitions (S_A^0 and S_B^0) with very low oscillator strength, as indicated by the total dominance of the higher-lying electronic transitions (S_A^1 and S_B^1) at elevated temperatures. This is a common situation for IBE, due to a splitting of the IBE ground state by e-h j-j coupling which results in a lowest-lying ground state component having a dipole forbidden transition to the crystal ground state (Dean and Herbert 1979). The $S_A^0 - S_A^1$ and $S_B^0 - S_B^1$ splittings are seen to be very similar, as is much of the local-mode phonon structure. The simultaneous appearance of S_A and S_B in all of our samples, which were prepared by diffusion following Brown and Hall (1986), suggests that the binding centers themselves may be almost identical, differing perhaps only in the configurations of the constituents, as is the case for the bistable Tl related IBE in Si (Thewalt et al 1982).

Fig. 7. High resolution PL spectra of the S_A system at 8, 20 and 40 K.

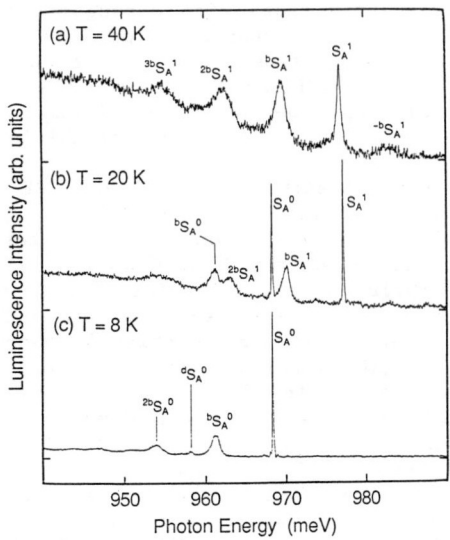

Since the higher-lying S_A^1 and S_B^1 states are revealed in PL by thermally populating them at elevated sample temperatures, one might hope to see the higher-lying excited states in PL by going to even higher temperatures. This approach is almost never successful, due to thermal dissociation of the IBE as well as the transfer of the PL intensity from the sharp no-phonon transitions to the much broader Stokes and anti-Stokes phonon replicas. The IBE excited states could again in principle be observed by sub-band-gap absorption spectroscopy (crystal ground state → IBE excited state), but this is also usually impractical due to the very weak absorption strengths and interference from competing processes.

The 'classical' method of BE excited state spectroscopy is PL excitation spectroscopy (PLE), in which a tunable laser source is scanned through the near-band-gap region where the BE excited states should lie, and the absorption signal is measured by observing PL from the principal BE transition, or one of its phonon replicas (Dean and Herbert 1979). This provides great sensitivity, since the absorption signal now appears on a zero background, as well as selectivity, since only absorption into the excited states of the particular BE whose PL is being monitored will be detected (in the absence of energy transfer processes).

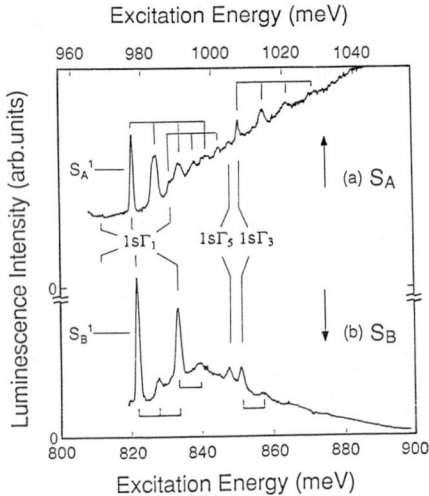

Fig. 8. Photoluminescence excitation spectra of the S_A and S_B systems, shifted so as to align S_A^0 with S_B^0. The multi-element brackets over many of the lines indicate anti-Stokes phonon replicas in the PLE spectrum, where the left-most element in each series is the no-phonon transition.

Nevertheless, PLE spectroscopy does not work very well for many systems, particularly those with strong phonon coupling, whose PLE spectra are dominated by phonon-assisted bands which obscure the weak no-phonon excited state transitions. This is the case for the S_A and S_B systems, as shown in Figure 8. All of the observed structure is attributed to transitions from the ground state manifold, and their phonon replicas. No electronic excited state transitions are evident.

In such situations, a new form of BE excited state spectroscopy which we have developed (Labrie et al 1984, Thewalt et al 1985) can be most useful. In this technique the internal ground-state to excited-state transitions of the BE are detected by photoinduced far-infrared exciton absorption spectroscopy (FIEAS), in close analogy to ordinary donor or acceptor absorption spectroscopy. The FIEAS technique can circumvent the problems of PLE in strongly phonon-coupled systems, since in FIEAS the highly localized particle of the IBE, whose creation in PLE or annihilation in PL results in the strong phonon coupling, does not change its state. Only the weakly (Coulomb) bound particle is promoted from the IBE ground state to an electronic excited state, with no significant phonon effects. Even for systems where PLE produces results, the FIEAS technique is useful in that it provides complementary information. Due to selection rules, PLE usually reveals even-parity excited states, while FIEAS is dominated by odd-parity excited states. FIEAS was first applied to an IBE system in Si (Labrie et al 1984, Thewalt et al 1985), and more recently to the double acceptor BE in Ge (Thewalt et al 1987).

A typical FIEAS spectrum of a Si:S sample giving strong S_A and S_B PL is shown at the top of Figure 9. Bulk excitation was provided by a 1.06 μm Nd:YAG laser, a method which also provides very strong PL. By comparison of the Si:S FIEAS lines to the shifted absorption spectrum of the Si:P donor (Figure 9 bottom), we see that the FIEAS spectrum consists of two series of donor-like excited states separated by 0.93 meV. The transitions are labelled S_A and S_B according to the IBE with which they are associated, as determined by comparing the PL and FIEAS spectra as a function of excitation level and sample temperature. We note that the FIEAS transitions must be due to BE, and not to two new donors, since there is no sign of them in the absence of the 1.06 μm excitation, even though the sample is n-type and absorption due to the shallower P impurities is clearly observed without excitation.

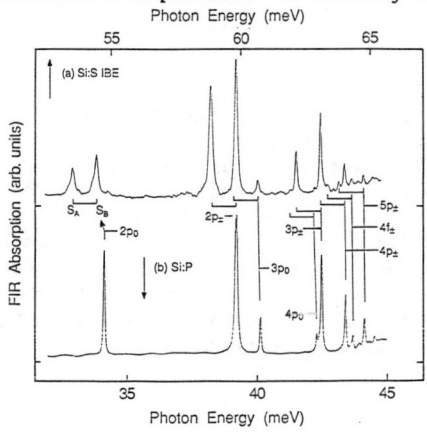

Fig. 9. The photo-induced Si:S bound exciton absorption (top) is compared to the shifted P-donor absorption spectrum (bottom). The top spectrum consists of two donor-like series, split by 0.93 meV, which arise from the S_A and S_B IBE excited states.

By combining the measured IBE transition energies with the calculated (Faulkner 1969) binding energies of the higher donor excited states, one arrives at an electron binding energy of 65.3 meV for the S_A IBE, and 66.2 meV for S_B. The difference could be due to the tighter localization of the effective positive charge of the hole for S_B, although central cell effects cannot be ruled out. One can combine the IBE ground-state to ground-state transition energies (968.24 meV for S_A^0 and 811.96 meV for S_B^0) with these electron binding energies and the Si band gap energy to obtain the binding energies of isolated holes onto these centers, namely 135.7 meV for S_A and 291 meV for S_B.

It is interesting to note that the electron binding energies for both the S_A and S_B IBE are considerably larger than the effective-mass value (Faulkner 1969) or the value obtained for the isocoric P donor. This indicates that the central cell potentials of the two centres are attractive for the electrons as well as the holes. This can cause an increased e-h overlap in the central cell region, and thus leads to unusually large e-h exchange splittings of the IBE ground-state, as is observed here (Monemar et al 1987, Watkins and Thewalt 1986).

The labelling of the S_A and S_B excited states observed in FIEAS is trivial, since the similarity to donor states is so obvious. This leaves the labelling of the split 1S ground-state manifold as observed in PL and PLE. Here we are guided by the fact that these are pseudo-donor IBE, as determined by FIEAS. The S^0 and S^1 PL transitions clearly must be Γ_1 valley-orbit states (Faulkner 1969) of the 1S manifold, split by e-h exchange. The PLE of both systems shown in Fig. 8 reveals another low-lying level, which we interpret as a third exchange-split 1S Γ_1 level. At higher energies the PLE spectra of both systems reveal two weaker transitions whose position relative to the ionization edge identifies them as the Γ_3 and Γ_5 valley-orbit states of the 1S manifold. This concludes the identification of all of the observed S_A and S_B electronic transitions. The spectroscopy of these systems will be described in greater detail elsewhere.

4. CONCLUSIONS

We have shown that new spectroscopic techniques can provide very detailed and useful information regarding the electronic states of both the binding centers associated with BE and of the BE themselves. Unfortunately these results reveal little about the chemical details of the binding centers or their constituents, which remain clouded for both of the systems discussed in this review.

Acknowledgement: This work was supported by the National Sciences and Engineering Research Council of Canada.

REFERENCES

Briones F and Collins D M 1982 *J. Electron. Mater.* **11** 847
Brown T G and Hall D G 1986 *Appl. Phys. Lett.* **49** 245
Brown T G, Bradfield P L and Hall D G 1987 *Appl. Phys. Lett.* **51** 1585
Charbonneau S, McMullan W G, Henry M O and Thewalt M L W 1988a *Mat. Res. Soc. Symp. Proc.* Vol. **104** pp. 549-554
Charbonneau S, McMullan W G and Thewalt M L W 1988b *Phys. Rev. B* **38** 3587

Davies G 1984 *J. Phys. C* **17** 6331
Dean P J and Herbert D C 1979 *Topics in Current Physics* Vol. 14 ed K Cho (Berlin: Springer)
Faulkner R A 1969 *Phys. Rev.* **184** 713
Haynes J R 1960 *Phys. Rev. Lett.* **4** 361
Hopfield J J, thomas D G and Lynch R T 1966 *Phys. Rev. Lett.* **17** 312
Künzel H and Ploog K 1980 *Appl. Phys. Lett.* **37** 416
Labrie D, Timusk T and Thewalt M L W 1984 *Phys. Rev. Lett.* **52** 83
Lipari N O, Thewalt M L W, Andreoni W and Baldereschi A 1980 *J. Phys. Soc. Japan* **49**, Suppl. A, 165
McMullan W G, Charbonneau S and Thewalt M L W 1987 *Rev. Sci. Instrum.* **58** 1626
Monemar B, Lindfeldt U and Chen W M 1987 *Physica* **146B** 255
Skolnick M S, Harris T D, Tu C W, Brennan T M and Sturge M D 1985 *Appl. Phys. Lett.* **46** 427
Skolnick M S, Tu C W and Harris T D 1986 *Phys. Rev. B* **33** 8468
Thewalt M L W, Ziemelis U O, Watkins S P and Parsons R R 1982 *Can. J. Phys.* **60** 1691
Thewalt M L W, Labrie D and Timusk T 1985 *Solid State Commun.* **53** 1049
Thewalt M L W, Labrie D, Booth I J, Clayman B P, Lightowlers E C and Haller E E 1987 *Physica* **146B** 47
Watkins S P and Thewalt M L W 1986 *Phys. Rev. B* **34** 2598

Interband Auger recombination in silicon

D. B. Laks, G.F. Neumark

Div. of Metallurgy and Materials Science, Columbia Univ., New York, NY 10027

A. Hangleiter[1] and S.T. Pantelides

IBM T.J. Watson Research Center, Yorktown Heights, NY 10598

Available theories find pure (electronic) Auger lifetimes in n-Si and p-Si that are significantly larger than experimental values. The accepted explanation is that phonon-assisted recombination dominates. Close examination of these theories, however, reveals that potentially compromising approximations were invoked. Here we report completely new, rigorous calculations for pure Auger recombination in both n-Si and p-Si. For n-Si we find very good agreement with experiment over the entire temperature range; for p-Si theoretical lifetimes are far too large. We conclude that phonon-assisted recombination dominates in p-Si, and pure Auger recombination in n-Si.

INTRODUCTION

Two of the most important electronic properties of a semiconductor are its carrier concentrations and the lifetimes of these carriers. While carrier concentrations in a semiconductor are primarily a function of the dopant concentrations, many recombination mechanisms help determine the carrier lifetimes. Of the recombination mechanisms, those involving deep levels can be eliminated by avoiding impurities that act as recombination centers. Band-to-band recombination processes, which are present even in the perfect crystal, are the ultimate limit to carrier lifetimes. One band-to-band recombination mechanism is the non-radiative Auger effect, in which an electron recombines with a hole and the energy of recombination is transferred to a third carrier (Fig. 1). (Beattie and Landsberg 1958, Landsberg 1987) Understanding the physics of Auger recombination has practical applications; for example, the Auger effect imposes the upper bound to the efficiency of silicon solar cells (Green 1984) and III-V semiconductor lasers. (Takeshima 1985)

Theoretical understanding of Auger recombination, however, has been somewhat limited. In heavily doped silicon, where it is the dominant mechanism, calculations (Hill and Landsberg 1976) indicated that pure Auger recombination is too slow to account for experimental lifetimes. (Dziewior and Schmid 1977) The faster experimental recombination was attributed to phonon-assisted Auger transitions. (Haug 1978, Lochmann and Haug 1980) Calculated phonon-assisted lifetimes (Lochmann and Haug 1980) agree well for p-type silicon but not for n-type. Close examination of available calculations-- for almost any material-- reveals that they entail several

---

[1] Present address: Univ. of Stuttgart, Stuttgart, FR Germany.

© 1989 IOP Publishing Ltd

potentially compromising approximations. For example, prior theories used model energy bands and wave-functions. (Hill and Landsberg 1976, Lochmann and Haug 1980, Takeshima 1983, Takeshima 1984, Bardyszewski and Yevick 1985a, Bardyszewski and Yevick 1985b, Beattie 1985) In the Auger matrix elements, sums over reciprocal lattice vectors (often called "Umklapp terms") were dropped. No estimates were given of the effect of these approximations on the calculated Auger rate.

In this paper we present a complete evaluation of the Auger recombination rate in n-type and p-type silicon. We perform the calculation with accurate energy bands and Bloch functions, and without any approximations of unknown consequence. Fermi-Dirac statistics is used to describe the occupation probability of the majority carriers. We have been careful to include all the terms (including all so-called "Umklapp terms") in the Auger matrix elements. The Auger recombination rate contains an 8 dimensional k-space surface integral; we evaluate this integral numerically over a cubic mesh. We perform separate convergence studies for each of the parameters (such as the number of plane waves used in the Bloch functions and the mesh size in the integral) in the calculation, obtaining convergence to within 1% for each parameter. The computations performed are far more demanding than typical state-of-the-art electronic structure calculations. Results of these calculations are in very good agreement with the experimental rates (Dziewior and Schmid 1977) in highly doped n-Si ($n > 5 \times 10^{18}$ cm^{-3}) from T=77K to T=400K. This agreement between theory and experiment suggests that phonon-assisted recombination is slower than pure Auger recombination in n-Si. In p-Si, the theoretical Auger lifetimes are far larger than experiment, suggesting that phonon-assisted recombination dominates in this case. The band-structure of silicon provides a simple explanation for the difference between the Auger rates in n- and p-type material.

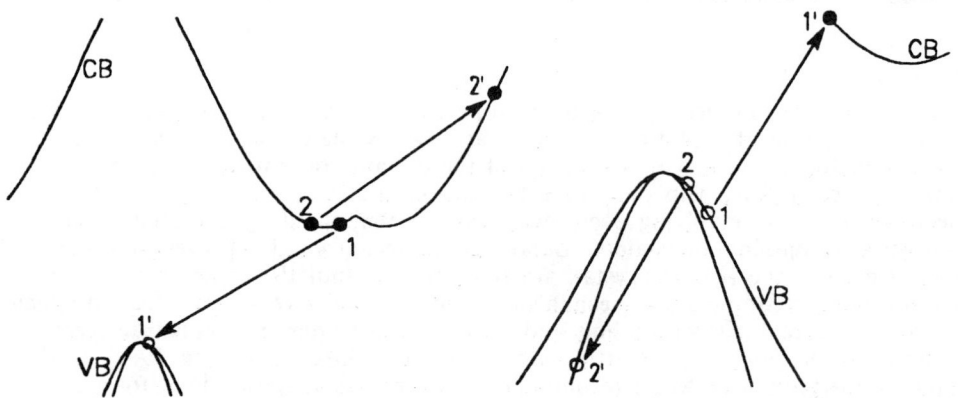

Fig. 1a. e–e–h Auger Recombination   Fig 1b. h–h–e Auger Recombination

THEORY

Auger recombination comes in two types-- electron-electron-hole (e·e·h$^+$) and hole-hole-electron (h$^+$h$^+$e·)-- which dominate in n-Si and p-Si respectively. In e·e·h$^+$ Auger recombination (Fig. 1a) an electron (with wave-vector $\vec{k}_1$ and energy E_1) recombines with a hole ($\vec{k}_{1'}$, $E_{1'}$) while another electron ($\vec{k}_2$, E_2) absorbs the recombination energy and moves to an excited conduction band state ($\vec{k}_{2'}$, $E_{2'}$). In h$^+$h$^+$e· recombination (Fig. 1b) a hole ($\vec{k}_1$, E_1) and an electron ($\vec{k}_2$, E_2) recombine, and a second hole ($\vec{k}_{1'}$, $E_{1'}$) moves deeper into the valence bands ($\vec{k}_{2'}$, $E_{2'}$). The transition is

mediated by the Coulomb repulsion between like carriers. The recombination rate is: (Ridley 1982)

$$R = 2\frac{2\pi}{\hbar}\frac{V^3}{(2\pi)^9}\int_{\vec{k}_1}\int_{\vec{k}_2}\int_{\vec{k}_{1'}}|M|^2 f(E_1)f(E_2)[1-f(E_{1'})][1-f(E_{2'})]\times$$
$$\delta(E_1 + E_2 - E_{1'} - E_{2'})d\vec{k}_1 d\vec{k}_2 d\vec{k}_{1'}$$ (1)

where M is the Auger matrix element, $f(E)$ is the probability that the state with energy E is occupied by an electron and $\delta(E)$ is the energy conserving delta function. M is given by:

$$M = \int\int \phi^*_{\vec{k}_1}(\vec{r}_1)\phi^*_{\vec{k}_2}(\vec{r}_2)v(|\vec{r}_1-\vec{r}_2|)\phi_{\vec{k}_{1'}}(\vec{r}_1)\phi_{\vec{k}_{2'}}(\vec{r}_2)d\vec{r}_1 d\vec{r}_2$$

$$v(r) = \int\frac{d\vec{q}}{(2\pi)^3}\frac{4\pi e^2}{\varepsilon(q)(q^2+\lambda^2)}e^{\vec{q}\cdot\vec{r}}$$

where $\phi_{\vec{k}}(\vec{r})$ is the wave-function of the electron (or hole) with wave-vector $\vec{k}$, $v(r)$ is the screened Coulomb potential, $\varepsilon(q)$ is the dielectric constant of the material, and λ is the electron screening factor. If the Bloch functions are expressed as a Fourier sum over the reciprocal lattice, M becomes:

$$M = \frac{4\pi e^2}{V}\sum_{\vec{G}}\frac{1}{\varepsilon(\vec{k}_1-\vec{k}_{1'}+\vec{G})(|\vec{k}_1-\vec{k}_{1'}+\vec{G}|^2+\lambda^2)}\times$$
$$\sum_{\vec{G}_1}A^*(\vec{k}_1+\vec{G}_1)B(\vec{k}_{1'}+\vec{G}_1-\vec{G})\times$$
$$\sum_{\vec{G}_2}A^*(\vec{k}_2+\vec{G}_2)A(\vec{k}_1+\vec{k}_2-\vec{k}_{1'}+\vec{G}_2+\vec{G})$$ (2)

Here, $\vec{G}_1$, $\vec{G}_2$, and $\vec{G}$ are reciprocal lattice vectors, and the A's and B's are the Fourier components of the electron and hole Bloch functions, respectively. In heavily-doped n-type silicon, where e⁻e⁻h⁺ Auger recombination is dominant, the hole lifetime is $\tau = p/R$. The Auger coefficient C_n, the quantity usually reported in the literature, is defined by $R = C_n n^2 p$. When Boltzmann statistics is used for the electron occupation probabilities and when electron screening is not important, C_n is independent of the carrier concentrations. In p-Si, where h⁺h⁺e⁻ Auger recombination dominates, the relations are completely parallel: $\tau = n/R$ and $R = C_p p^2 n$, where R is now the h⁺h⁺e⁻ recombination rate.

REVIEW OF PRIOR WORK

Many authors have calculated Auger transition rates in a variety of semiconductors. (Hill and Landsberg 1976, Lochmann and Haug 1980, Takeshima 1983, Takeshima 1984, Bardyszewski and Yevick 1985a, Bardyszewski and Yevick 1985b, Beattie 1985) They make a number of drastic approximations, however. The

failure to test the validity of these approximations leads to results with uncontrolled errors. The following major approximations are made by these papers. For the evaluation of the matrix elements (Eq. (2)), the summation over $\vec{G}$ (often called "Umklapp terms"), is omitted and a constant ε, or $\varepsilon = 1$, is used for the dielectric constant. (In fact ε is a function of k, and cannot be factored from the $\vec{G}$ summation.) In Eq. (1), the matrix elements are integrated over at most 6 of the 8 dimensions of the k-space surface integral. Furthermore, $k \cdot p$ perturbation theory is used to determine the energy bands and Bloch functions. Each of these approximations is valid only if all of the k-vectors are very near the band edge. This condition is met in very **narrow-gap** direct semiconductors. Thus in Beattie's calculation (Beattie 1985) of the Auger rate in InSb ($E_g = 0.18$ eV) the approximations may be adequate. For wide-band-gap semiconductors and for indirect semiconductors, such as silicon, $\vec{k}_{2'}$ is very far from the band edge and the approximations fail. The value of C_n for silicon calculated by prior theory (Hill and Landsberg 1976) is twenty times smaller than the experimental value. (Dziewior and Schmid 1977) The higher experimental rates were explained by invoking phonon assistance to lower the energy threshold imposed by the energy and momentum conservation. (Lochmann and Haug 1980) This idea was supported by the temperature dependence of the Auger rate: (Haug 1978, Haug and Schmid 1978) the experimental C_n is virtually constant with temperature, (Dziewior and Schmid 1977, Svantesson and Nilsson 1979) while theory predicts that phonon-assisted recombination should have a weaker temperature dependence than pure Auger recombination. Lochmann and Haug (1980) evaluate the phonon-assisted Auger rate finding C_n one-fourth of the experimental value, and good agreement for C_p.

Another set of papers calculated the matrix elements using more accurate wavefunctions. (Brand and Abram 1984a, Brand and Abram 1984b, Burt et al. 1984) One of these (Erand and Abram 1984b) includes both the q-dependence of ε and the sum over $\vec{G}$ in Eq. (2). Unfortunately, the more accurate matrix elements are evaluated for only a few values of k; no attempt is made to integrate the matrix elements over k (Eq. (1)). Thus, these authors do not calculate the transition rate and their results cannot be compared with experiment.

PRESENT WORK

Our work avoids these problems. The empirical pseudopotential method (with a local potential) (Chelikowsky and Cohen 1974) is used for the band structure and matrix elements. We chose an empirical potential because self-consistent calculations using the local-density approximation give errors in the band-gap and dispersion of the conduction bands. We evaluate all of the terms in the expression for the matrix elements (Eq. (2)): all of the terms in the sum over $\vec{G}$ are retained until the summation converges, the q-dependence of the dielectric constant, ε, is included in the form of Nara and Morita (1966) (these two corrections are very important; each increases the total rate by an order of magnitude), Thomas-Fermi screening is used for λ, and both

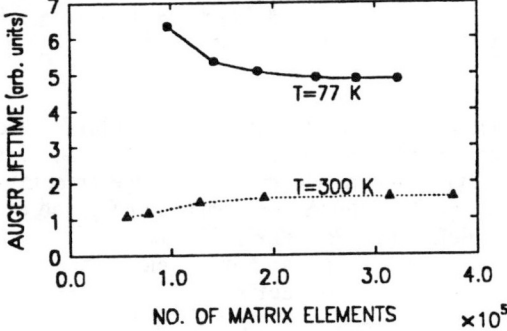

Fig. 2. e–e–h AUGER LIFETIME vs. NO. OF MATRIX ELEMENTS AS MESH SIZE IS CHANGED WITH ALL OTHER PARAMETERS FIXED

the direct and exchange terms of the matrix elements are included. More than 25,000 terms were summed in the evaluation of each matrix element. Fermi-Dirac statistics is used for the occupation probability functions f(E) in Eq. (1).

The k integration (Eq. 1) is evaluated over a 9 dimensional cubic grid. (The grid is the Cartesian product $(\vec{k}_1, \vec{k}_2, \vec{k}_{1'})$ of the k-space grids for the three initial particles; $\vec{k}_{2'}$ is determined by wave-vector conservation.) The energy conservation condition $(E_1 + E_2 = E_{1'} + E_{2'})$ determines an 8 dimensional surface, and the matrix elements and occupation probabilities are evaluated on each cube that intersects this surface. The main advantage of using a cubic grid is that it allows us to evaluate a very large number of matrix elements while determining the wave functions for a much smaller number of points; in a typical calculation we need fewer than 3,000 wave functions to calculate over 500,000 matrix elements. Separate convergence tests are made for each of the parameters that enter the calculation: the number of plane waves used in the pseudopotential, the number of terms included in the summations over the reciprocal lattice for the matrix elements (Eq. (2)), and the mesh size. The calculation was converged to within 1% or less for each of these parameters. Convergence curves of the Auger rate for the mesh size are shown in fig. 2. A converged calculation requires about 500,000 matrix elements in all.

RESULTS AND DISCUSSION

The results of our calculations are shown in Fig. 3. The Auger lifetimes as a function of n (or p) and T are compared to the experimental data of Dziewior and Schmid (1977). (The lifetimes at T = 400 K are almost the same as those for 300 K and are not shown.) Note that, because Fermi-Dirac statistics and Thomas-Fermi screening introduce concentration dependent terms into Eq. (1), the Auger coefficient is itself a function of n (p). This effect is most pronounced in p-Si, where the the lifetimes decrease more rapidly than p^{-2}. The lifetimes for n-Si are in very good agreement with the experimental data over the entire temperature range, suggesting that phonon-assisted Auger recombination is much slower than pure electronic recombination in n-Si. In essentially all recent work on Si the opposite conclusion was reached. The idea that phonons are important was supported by calculations (Hill and Landsberg 1976) that found the pure electronic Auger to be 20 times smaller than experiment, and a subsequent calculation (Lochmann and Haug 1980) that found the phonon-assisted rate to be only four times smaller than experiment. As we

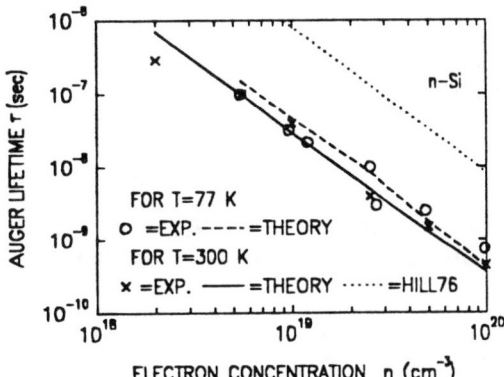

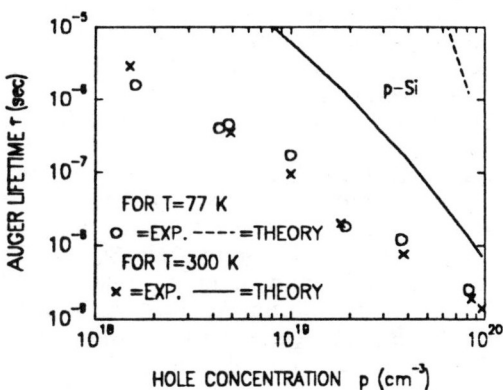

Fig. 3. THEORETICAL AND EXPERIMENTAL AUGER LIFETIMES IN n– AND p–SI

showed earlier, however, at least the electronic part of the above calculations involves unjustified approximations, and is superseded by our results. More specifically, in the phonon-assisted calculation (Lochmann and Haug 1980) an unscreened Coulomb potential was used. Inclusion of appropriate dielectric screening would reduce the resulting rate by an order of magnitude, making it negligible in comparison with the pure electronic rate reported here and with experiment.

Calculation of the phonon-assisted rate at the same level of sophistication as our present work is currently infeasible. Therefore it is worthwhile to give a physical reason why phonon effects should be important in p-Si but not in n-Si. According to previous explanations, phonon-assisted recombination dominates because it relaxes the momentum conservation condition, thereby allowing carriers near the band-edge to participate in recombination. Pure Auger recombination is forbidden by energy and momentum conservation for carriers at the band edge. Our work shows, however, that for $e^-e^-h^+$ processes carriers very near the band edge can indeed participate in pure Auger transitions (Fig. 1a). Under these circumstances, pure Auger recombination-- which is **first order**-- should be faster than phonon-assisted recombination-- which is **second order**. For $h^+h^+e^-$ Auger recombination we find that there is no set of states near the band edges that satisfies both energy and momentum conservation. (See Fig. 1b.) In this case phonon-assisted recombination is favored because it induces transitions among carriers at the heavily populated band edges. It is because of this band-structure effect that the theoretical $e^-e^-h^+$ Auger rate has a much weaker temperature dependence than the $h^+h^+e^-$ rate. Since $e^-e^-h^+$ transitions can occur among particles near the band edge, they are not affected by changing T at a fixed carrier concentration. In $h^+h^+e^-$ transitions increasing T produces more energetic carriers, which are the only ones that can participate in recombination.

We would like to thank E. Yablonovitch and R. Swanson for helpful discussions, and A. Rossi for assistance with the computations. D.B. Laks acknowledges support from an IBM Graduate Fellowship. This work was supported in part by ONR Contract No. N00014-84-C-0396 and a NY State CAT Grant to Columbia U.

REFERENCES
Bardyszewski, W., and Yevick, D. 1985 *J. Appl. Phys.* **57** 4820.
Bardyszewski, W., and Yevick, D. 1985 *J. Appl. Phys.* **58** 2713.
Beattie, A. 1985 *J. Phys. C:* **18** 6501.
Beattie, A., and Landsberg, P. 1958 *Proceedings of the Royal Society A* **249** 16.
Brand, S., and Abram, R. 1984 *J. Phys. C:* **17** L201.
Brand, S., and Abram, R. 1984 *J. Phys. C:* **17** L571.
Burt, M., Brand, S., Smith, C., and Abram, R. 1984 *J. Phys. C:* **17** 6385.
Chelikowsky, J., and Cohen, M. 1974 *Phys. Rev. B* **10** 5095.
Dziewior, J., and Schmid, W. 1977 *Appl. Phys. Lett.* **31** 346.
Green, M. 1984 *IEEE Transactions on Electron Devices* **ED-31** 671.
Haug, A. 1978 *Solid State Comm.* **28** 291.
Haug, A., and Schmid, W. 1978 *Solid State Electron.* **25** 665.
Hill, D., and Landsberg, P. 1976 *Proceedings of the Royal Society A* **347** 547.
Landsberg, P. 1987 *Solid State Electronics* **30** 1107.
Lochmann, W., and Haug, A. 1980 *Solid State Comm.* **35** 553.
Nara, H., and Morita, A. 1966 *J. Phys. Soc. Jap.* **21** 1852.
Ridley, B. 1982 *Quantum Processes in Semiconductors* (Claredon Press) p. 268.
Svantesson, K., and Nilsson, N. 1979 *J. Phys. C:* **12** 5111.
Takeshima, M. 1983 *Jpn. J. Appl. Phys.* **22** 491.
Takeshima, M. 1984 *Phys. Rev. B* **29** 1993.
Takeshima, M. 1985 *J. Appl. Phys.* **58** 3846.

Inst. Phys. Conf. Ser. No 95: Chapter 9
Paper presented at Int. Conf. Shallow Impurities in Semiconductors, Linköping, Sweden, 1988

One- and two-photon acceptor spectra in semiconductors

N.Binggeli[a], A.Baldereschi[a,b] and A.Quattropani[c]

(a) Institut de Physique Appliquée, EPF-Lausanne, Switzerland
(b) Dipartimento di Fisica Teorica & GNSM-CNR Universita di Trieste, Italy
(c) Institut de Physique Théorique, EPF-Lausanne, Switzerland

The intensity of one- and two-photon acceptor spectra are discussed in the framework of the effective mass approximation. The main features are explained by the spherical-model approximation. Inclusion of the valence-band cubic anisotropy is shown to be relevant, especially in the two-photon case. Quantitative values of one- and two-photon oscillator strengths are obtained by taking also into account the coupling to the split-off valence band and central cell-effects. The resulting absorption strengths in Ge and Si are in close agreement with available one-photon experimental spectra.

1. INTRODUCTION

Absorption spectra of donor impurities are in general similar to the hydrogenic spectrum predicted by the simple parabolic band effective mass approximation (EMA). For acceptors, common features are also observed in all semiconductors, but line positions and relative line intensities are completely different from the hydrogenic case.

The energies of acceptor spectra have been explained qualitatively within the spherical model by Baldereschi and Lipari (1973) taking into account light and heavy hole degeneracy at the top of the valence band in a simple one-parameter description. Improved agreement with experiment was later obtained including valence band warping, coupling to the split-off valence band and q-dependent screening (Lipari and Baldereschi 1978).

Most investigations of one-photon acceptor spectra use the position of the line to identify the corresponding transition. However, for final states close in energy the spectral position of the line is not sufficient to determine the transition. Quantitative information on the oscillator strength could help the interpretation of absorption spectra and could also facilitate the determination of the impurity concentration. Unfortunately, experimental and theoretical studies of cross sections are scarce.

Two-photon spectroscopy has been widely used in atomic physic. One of its advantages, which could be exploited for the study of impurities, is the possibility of performing final state as well as intermediate state spectroscopy, and obtain spectral information on states of different parity in the same experiment. To our knowledge, only the two-photon spectrum of donors

© 1989 IOP Publishing Ltd

in GaAs has been measured by Böhm et al. (1981). Theoretical predictions for two-photon spectra have been given for donor impurities by Bassani and Quattropani (1985).

In this work we study the oscillator strength of acceptors in cubic semiconductors in order to explain the measured features of one-photon spectra and to predict two-photon spectra. We present quantitative results for acceptors in Ge and Si, where high resolution one-photon spectra have been measured and accurate host material parameters are available for the calculation.

2. ONE-PHOTON SPECTRA

In the dipole approximation and EMA, the absorption coefficient for acceptor systems at low temperature can be written as:

$$\alpha(\omega) = N_A \frac{2\pi^2 e^2 \gamma_1 \hbar}{ncm} (\mathcal{E}_{eff}/\mathcal{E}_o)^2 \sum_\nu f_{o\nu} \delta(E_\nu - E_o - \hbar\omega) \quad (1)$$

where

$$f_{o\nu} = \frac{2m}{\gamma_1 \hbar^2}(E_\nu - E_o)\frac{1}{g_o}\sum_{i,j}|<\Phi_{\nu j}|z|\Phi_{oi}>|^2 \quad (2)$$

is the oscillator strength of the $o \to \nu$ transition, $\mathcal{E}_{eff}/\mathcal{E}_o$ is the effective/ applied field ratio at the impurity site, n is the refractive index of the crystal, γ_1 the first Luttinger valence-band parameter and N_A the concentration of absorbing centers. The indices i, j label the degeneracy of the initial (ground state with degeneracy g_o) and final acceptor states with energies E_o, E_ν, and envelope functions $\Phi_{oi}, \Phi_{\nu j}$, respectively.

In the spherical approximation, the effective mass hamiltonian depends on a single parameter μ that measures the difference in the curvature of light and heavy hole bands. Its eigenstates have two angular components and can be written as:

$$\Phi_n(L, F, F_Z) = f_L^{nF}(r)|L, J, F, F_Z> + f_{L+2}^{nF}(r)|L+2, J, F, F_Z> \quad (3)$$

where $\mathbf{F} = \mathbf{L} + \mathbf{J}$ is the total momentum, a constant of motion in the spherical approximation, L is the angular momentum of the envelope function and $J = 3/2$ labels the four Bloch functions at the top of the valence band. The ground state $1S_{3/2}$, which is the inital state of the transitions at low temperature, has total momentum $F = 3/2$ and angular components 0 and 2. The radial functions in equation (3) have been determined variationally and the oscillator strength (2) has been computed as a function of μ for transitions from the ground state to the first excited states (Binggeli and Baldereschi 1988). This allowed us to explain why in all acceptor spectra the transition to the first excited state ($2P_{3/2}$, line G) is much weaker than the transition to the doublet slightly higher in energy ($2P_{5/2}$ states, corresponding to the D and C lines), although the final states originate all from hydrogenic 2p states when $\mu = 0$. The difference in oscillator strength between the G and the D,C lines results from the presence of the d-like component in the acceptor ground state wave function, and increases with μ. For all semiconductors $\mu \gtrsim 0.5$, and the oscillator strength of the G line is much lower than that of either of the two $2P_{5/2}$ lines.

The spherical model gives only a qualitative description of the experimental infrared spectra. In order to obtain quantitative results one has to include band warping, coupling to the split-off band and an empirical central cell potential. For the latter we use a short range potential of the form $(A/r)e^{-\alpha r}$ where A is adjusted to fit the experimental ionization energy.

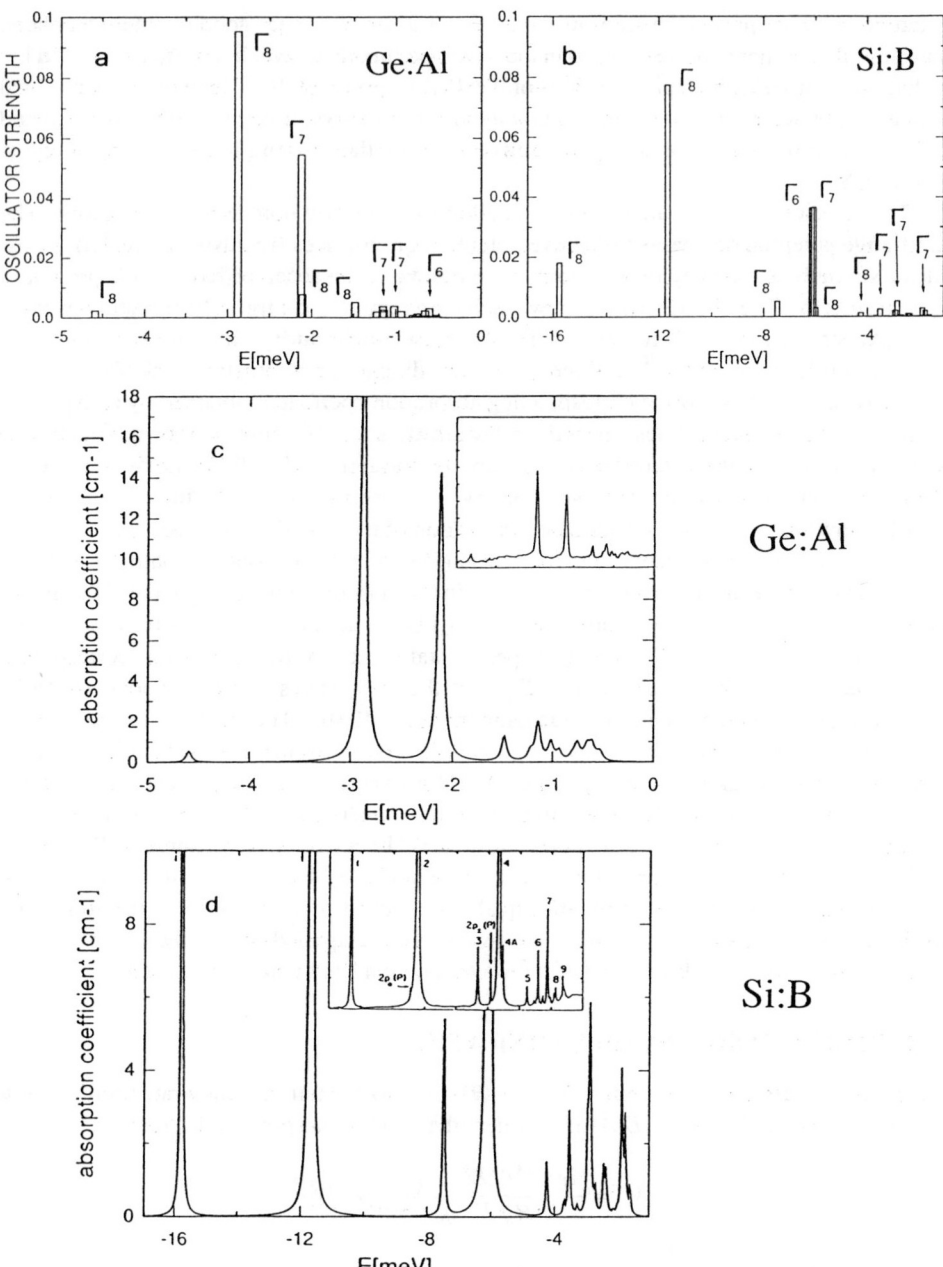

Fig. 1. Oscillator strength and absorption coefficient for Ge:Al ($N_A = 6 \cdot 10^{13} at/cm^3$) and Si:B ($N_A = 8.5 \cdot 10^{14} at/cm^3$), E is the final state binding energy, states bound by less than 0.5 meV in Ge and 1.6 meV in Si are not considered. In the insets (same scale), the experimental absorption coefficients are from Rotsaert et al. (1986) and Jagannath et al. (1981).

Including cubic terms and the split-off band, F and J are no longer good quantum numbers; more angular components are coupled in the envelope functions which have to be labelled according to the irreducible representations of the double group of O_h. Detail on the variational calculation for acceptor states can be found in our first work on line intensities (Binggeli and Baldereschi 1988), where we also give results on the oscillator strength for isocoric acceptors in Ge and Si.

The inclusion of a central cell potential allows us to study non-isocoric acceptors. The short-range potential decreases (attractive potential) or enhances (repulsive potential) the oscillator strength as one may expect from the variation of the acceptor radius. We present in Fig. 1 results for Ge:Al and Si:B both of which have a repulsive central cell correction (weaker for Ge:Al, stronger for Si:B) relative to that of the isocoric impurity. In Figures 1a and 1b we give the oscillator strength and label the peaks according to the symmetry of the final state. In Figures 1c and 1d we show the corresponding absorption coefficient obtained by convoluting the result with a lorentzian lineshape with full width at half maximum $\Gamma = 0.06$ meV and using the acceptor concentrations reported in the work by Rotsaert et al. (1986) and Jagannath et al. (1981). The striking agreement between theory and experiment (see the inset in Figures 1c and 1d) allows us to propose a detailed interpretation of the experimental spectra.

The C line has a complex origin in both Si and Ge, but the components contributing to the line are not the same in the two semiconductors. In Ge most of the intensity of the C line is due to the $2P_{5/2}[\Gamma_7]$ level, the other component ($3P_{3/2}$) being about seven times less intense and too close in energy to be resolved in the experimental spectrum. In Si the C line corresponds, besides the $2P_{5/2}[\Gamma_7]$ final state, to the $2P_{1/2}$ and the $3P_{3/2}$ states. Such components can be seen in a high resolution spectrum by Jagannath et al. (1986). The main peaks close to the continuum in both Ge and Si are mainly due to Γ_7 levels. In Si there is a slight disagreement between the theoretical line splitting in Fig. 1 and the experimental values. This could be due to a small inaccuracy of the Luttinger valence-band parameters employed in our calculations.

The experimental oscillator strength of the D and C lines in Ge:Al measured by Rotsaert et al. (1986) are a factor 0.6 lower than our prediction in Fig. 1. This is likely to be the result of our assumption of an effective field ratio equal to one for the determination of the absorption coefficient, since the local field should be lower than the applied one in the region of high charge density, where the Bloch part of the wave function has higher amplitude.

3. TWO-PHOTON TRANSITION RATE

Under the same assumptions used in Eqs. (1)-(2), the two-photon transition rate for the $o \rightarrow \nu$ transition is $W_o^\nu \delta(\tilde{E}_\nu - \tilde{E}_o - \tilde{E}_1 - \tilde{E}_2)$, where the absorption strength W_o^ν is:

$$W_o^\nu = \frac{\pi e^4 a_o^{*4} \mathcal{E}_1^2 \mathcal{E}_2^2}{2\hbar Ry^{*3}} \frac{1}{g_o} \sum_{i,j} \frac{1}{9} |D_{oi}^{\nu j}|^2 \tag{4}$$

with

$$D_{oi}^{\nu j} = \frac{3}{2}(1 + P_{12}) \sum_{\lambda,k} \frac{<\Phi_{\nu j}|\tilde{z}|\Phi_{\lambda k}><\Phi_{\lambda k}|\tilde{z}|\Phi_{oi}>}{\tilde{E}_\lambda - \tilde{E}_o - \tilde{E}_1} \tag{5}$$

where the tilde indicates quantities in effective atomic units $Ry^* = e^4 m_o/2\hbar^2 \epsilon_o^2 \gamma_1$ and $a_o^* = \hbar^2 \epsilon_o \gamma_1/e^2 m_o$. $\mathcal{E}_1$ and $\mathcal{E}_2$ are the effective fields with photon energy $\tilde{E}_1$ and $\tilde{E}_2$. The operator P_{12} interchanges $\tilde{E}_1$ with $\tilde{E}_2$. Expression (5) refers to two photons polarized in the same direction.

General Properties and Theory

The sum over λ in (5) includes the discrete spectrum and the continuum. The approximate hamiltonian resulting from our representation over a finite basis-set gives discrete states only. The lowest variational states are close to true acceptor bound states, but the same cannot be expected of calculated states at higher energy. We propose to use still all variational eigenstates in (5) to approximate the summation over the continuum. The accuracy of this procedure can be checked from the convergence of the result versus basis size, since the exact result should be obtained with a complete basis set. In our case, we find fast convergence using real exponential or gaussian radial basis functions. We also obtain excellent results in the hydrogenic limit, for which accurate two-photon transition rates have been obtained with other techniques by Bassani et al. (1977) and Tung et al. (1986).

We have computed the two-photon absorption strength W for the $1S_{3/2} \rightarrow 2S_{3/2}$ and $1S_{3/2} \rightarrow 3D_{5/2}[\Gamma_8]$ transitions for the isocoric system Ge:Ga taking into account valence band warping, the split-off band and q-dependent screening. The result is shown in Fig. 2, where W is given as a function of the one-photon energy $x = \tilde{E}_1/(\tilde{E}_1+\tilde{E}_2)$. Resonances occur at allowed intermediate states, i.e. $2P_{3/2}$ for $W_{1S3/2}^{2S3/2}$, followed for $W_{1S3/2}^{3D5/2[\Gamma_8]}$ by $2P_{5/2}[\Gamma_8], 2P_{5/2}[\Gamma_7]$ and $3P_{3/2}$. A strong decrease in intensity occurs in the $1S_{3/2} \rightarrow 3D_{5/2}$ spectrum just before the first peak, and a sharp minimum is present on the high x side of the same peak in the $1S_{3/2} \rightarrow 2S_{3/2}$ spectrum. The corresponding $1s \rightarrow 2s$ and $1s \rightarrow 3d$ spectra in hydrogen have no transparency or sharp minimum. (Bassani et al. 1977, Tung et al. 1986).

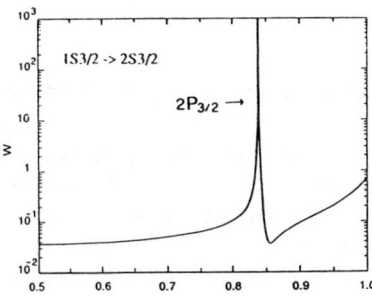
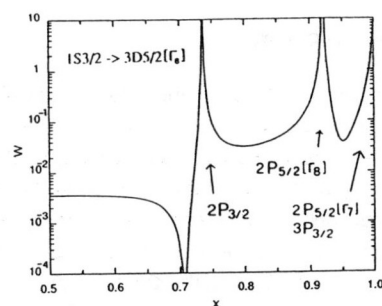

Fig. 2. Two-photon absorption strength W for the $1S_{3/2} \rightarrow 2S_{3/2}$ and $1S_{3/2} \rightarrow 3D_{5/2}[\Gamma_8]$ acceptor transitions. W is given in unit of $\pi e^4 a_o^{*4} \mathcal{E}_1^2 \mathcal{E}_2^2 / 2\hbar Ry^{*3}$ as a function of the one photon energy $x = \tilde{E}_1/(\tilde{E}_1 + \tilde{E}_2)$ where $\tilde{E}_1 + \tilde{E}_2 = 1.87\ Ry^*$ and $2.14\ Ry^*$ for the transition to the $2S_{3/2}$ and $3D_{5/2}[\Gamma_8]$ state, respectively.

Non-hydrogenic features appear already in the spherical model as shown in Fig. 3, where $D_{1S3/2}^{2S3/2}$ and $D_{1S3/2}^{3D5/2[\Gamma_8]}$ have been calculated for the different rows of the Γ_8 ground state. For both spectra the amplitude of D is much lower at x = 0.5 than in the hydrogenic case. For the $1S_{3/2} \rightarrow 3D_{5/2}[\Gamma_8]$ transition, all rows give the same result and a transparency appears between the split $2P_{3/2}$ and $2P_{5/2}$ peaks. For the $1S_{3/2} \rightarrow 2S_{3/2}$ transition, instead, two different D components are associated with the degenerate ground state, and vanish at slightly different energies, producing a sharp minimum in the absorption strength (see Fig. 2). When a small cubic anisotropy is introduced, D for the $1S_{3/2} \rightarrow 3D_{5/2}[\Gamma_8]$ transition splits into two components associated with different rows of Γ_8. The difference between the two is significant only for x higher than the first resonance, where it removes the transparency present in

the spherical model. Near x = 0.5, D is decreased, and zero-crossing occurs almost at the same energy for the two components, which explains the quasi-transparency in Fig. 2. In both cases no qualitative changes are introduced by coupling to the split-off band and q-dependent screening.

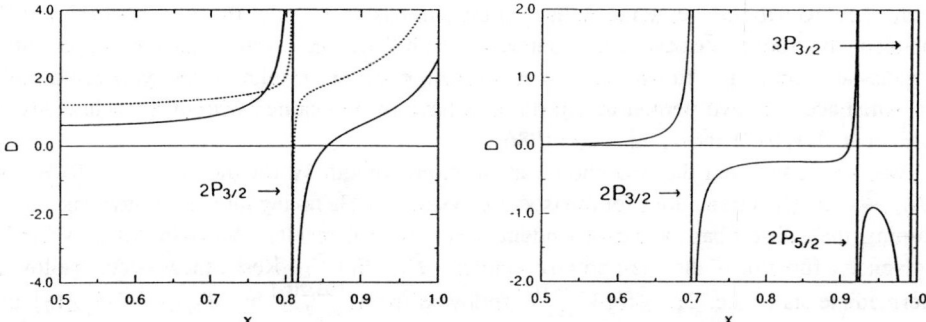

Fig. 3. Two-photon amplitudes D, for $1S_{3/2} \rightarrow 2S_{3/2}$ (left) and $1S_{3/2} \rightarrow 3D_{5/2}[\Gamma_8]$ (right) transitions, evaluated in the spherical model as a function of $x = \tilde{E}_1/(\tilde{E}_1 + \tilde{E}_2)$ where $\tilde{E}_1 + \tilde{E}_2 = 1.58\ Ry^*$ ($1S_{3/2} \rightarrow 2S_{3/2}$) and $1.82\ Ry^*$ ($1S_{3/2} \rightarrow 3D_{5/2}$).

4. CONCLUSIONS

We have examined the one and two-photon absorption spectra of acceptors in semiconductors taking into account the full cubic symmetry of the top of the valence band, the coupling to the split-off valence band and central cell effects. The main features, in particular the relative intensity of the G, D and C lines in the one-photon spectra, can be understood within the spherical model. Other important spectral features, such as transparencies in intermediate state two-photon spectroscopy, are strongly sensitive, instead, to the cubic anisotropy. Our results for Ge and Si describe the experimental one-photon spectra with high accuracy and allow identification of several lines nearly degenerate in energy.

This work was supported by the Swiss National Science Foundation.

REFERENCES

Baldereschi A and Lipari N O 1973 Phys. Rev. B **8** 2697
Bassani F and Quattropani 1985 Solid State Commun. **53** 1077
Bassani F, Forney J J and Quattropani A 1977 Phys. Rev. Lett. **39** 1070
Binggeli N and Baldereschi A 1988 Solid State Commun. **66** 323
Böhm W, Ettlinger E and Prettl W 1981 Phys. Rev, **47** 1198
Jagannath C, Grabowski Z W and Ramdas A K 1981 Phys Rev. B **23** 2082
Lipari N O and Baldereschi A 1978 Solid State Commun. **25** 665
Rotsaert E, Clauws P, Vennik J and van Goethem L 1986 *Shallow Impurity Centers in Semiconductors* (North-Holland Physics Publishing) p. 75-79
Tung J H, Tang A Z, Solomo G J and Chang F T 1986 J. Opt. Soc. Am. B **3** 837

Inst. Phys. Conf. Ser. No 95: Chapter 9
Paper presented at Int. Conf. Shallow Impurities in Semiconductors, Linköping, Sweden, 1988

The ground state of the acceptor molecule A_2 in a cubic semiconductor at large interimpurity distance

Sh.M. Kogan and A.F. Polupanov

Institute of Radioengineering and Electronics of the Academy of Sciences of the USSR, Moscow, 103907, USSR

ABSTRACT: In cubic p-type semiconductors there is a quadrupole-quadrupole interaction between neutral shallow acceptor impurities which are in degenerate Γ_8 ground states. This interaction at large impurity distances $R \gg a$ (a is the effective Bohr radius) falls off as R^{-5} and is therefore greater than other interaction energies. The molecular energy levels that arise from the ground state levels of the acceptor atoms are found.

1. INTRODUCTION

Recently experiments have been made on semiconductors with rather high and almost completely uncompensated impurity concentration. In these materials almost all impurity centres are neutral, the Coulomb fields are small and the interaction between neutral impurities can be studied *per se*.
Let us consider a cubic semiconductor doped only by shallow acceptors. The concentration of impurities N is many times smaller than the critical concentration of the semiconductor-metal transition and so the overlap of the impurities' wave-functions is negligible. But N is assumed to be not very small so the impurities can not be regarded as isolated (intermediate doping).
The peculiarity of shallow acceptors in cubic semiconductors is the four-fold degeneracy of their ground state (representation Γ_8). How do the neutral acceptor impurities interact with each other and what are the low energy states of the impurity system? The solution of this problem is necessary for the understanding of the low-temperature thermodynamic properties of the intermediate-doped p-type semiconductors and for the determination of phonon absorption and scattering spectra in these materials.
Any neutral atoms which are in their ground states and are separated by a large enough distance R interact through the dispersion (van der Waals) interaction (DI) which is proportional to R^{-6}, and through the interaction arising from the wave functions overlap which falls off exponentially. The peculiarity of neutral acceptors in cubic semiconductors is that there is also a quadrupole-quadrupole interaction (QQI) proportional to R^{-5}. It falls off less steeper than the above mentioned interactions. Indeed although the quadrupole moment of the acceptor averaged over all 4 states of the ground level is zero the quadrupole moment in each such state is generally not zero because the wave function includes besides an s-type function (L = 0) also a part with L = 2.

© 1989 IOP Publishing Ltd

If the overlap of the wave functions is negligible the impurity system at zero temperature is in a state that corresponds to the minimum of the sum of QQI and DI energies. The random distribution of acceptors makes the p-type semiconductor a kind of a quadrupole glass in which however the quadrupole moments are not fixed but are determined by the interactions between impurity centres.

The most interesting problem is not the lowering of the ground state energy of the system of interacting impurities but the structure of the energy levels of impurities, i.e. the structure of the impurity band. In contrast to a compensated lightly doped semiconductor in which the width of the impurity band is determined by the fields of the charged impurities the impurity band-width of the uncompensated p-type semiconductor is determined by QQI. We shall find the energy spectrum of a "molecule" of two acceptors that originates from their ground levels when the impurity centres are brought together. The results will be used to evaluate the magnitude of the QQI and DI effects.

2. HAMILTONIAN OF MULTIPOLE INTERACTION

We consider two acceptor impurity centres separated by a distance $R \gg a$ (a is the effective Bohr radius). Let $\mathbf{r}_1$ and $\mathbf{r}_2$ be the radius vectors of the holes measured from the corresponding impurity ions, $x_{1\alpha}$ and $x_{2\alpha}$ are their Cartesian components, X_α - the components of the vector $\mathbf{R}$ between the two centres. The energy of multipole interaction of the impurities is equal to (we restrict ourselves to only two terms):

$$H = H_{dd} + H_{QQ} , \qquad (1)$$

$$H_{dd} = -\frac{e^2}{\varkappa} V_{\alpha\beta} x_{1\alpha} x_{2\beta} , \qquad H_{QQ} = \frac{e^2}{4\varkappa} V_{\alpha\beta\gamma\delta} x_{1\alpha} x_{1\beta} x_{2\gamma} x_{2\delta} , \qquad (2)$$

$$V_{\alpha\beta} = \frac{\partial^2}{\partial X_\alpha \partial X_\beta} \frac{1}{R} , \qquad V_{\alpha\beta\gamma\delta} = \frac{\partial^4}{\partial X_\alpha \partial X_\beta \partial X_\gamma \partial X_\delta} \frac{1}{R} . \qquad (2a)$$

Here H_{dd} is the dipole-dipole, H_{QQ} - the quadrupole-quadrupole interaction, $\varkappa$ is the static dielectric constant.

Let $\Psi_m(r)$ be the wave function of an isolated impurity corresponding to the m-th state of the ground state multiplet. If the overlap of the wave functions of the acceptors has been neglected the molecular states can be taken as linear combinations of wave functions' products:

$$\Psi_{m_1 m_2}(r_1 r_2) \equiv |m_1 m_2\rangle = \Psi_{m_1}(r_1) \Psi_{m_2}(r_2) . \qquad (3)$$

The matrix of an interaction on the basis (3) is a sum of the first-order contribution from H_{QQ} and second-order contribution from H_{dd}:

$$\langle m_1 m_2 | H | m'_1 m'_2 \rangle = \langle m_1 m_2 | H_{QQ} | m'_1 m'_2 \rangle + \langle m_1 m_2 | H_{dis} | m'_1 m'_2 \rangle, \qquad (4)$$

$$\langle m_1 m_2 | H_{dis} | m'_1 m'_2 \rangle = -\sum_{n_1 n_2} \frac{\langle m_1 m_2 | H_{dd} | n_1 n_2 \rangle \langle n_1 n_2 | H_{dd} | m'_1 m'_2 \rangle}{E_{n_1} + E_{n_2} - 2E_0} . \qquad (4a)$$

The indexes n_1 and n_2 correspond to those excited states of the impurities (including the continuous spectrum) to which the dipole transitions from the ground state are allowed, E_n are the corresponding energies. The first

term in (4) is the matrix of QQI, the second is the DI matrix.

3. MATRIX OF QUADRUPOLE-QUADRUPOLE INTERACTION

Using the equation $\Delta(1/R) = 0$ one may represent H_{QQ} in the form:

$$H_{QQ} = \frac{e^2}{4\varkappa} V_{\alpha\beta\gamma\delta} \left[x_{1\alpha} x_{1\beta} - \frac{1}{3} r_1^2 \delta_{\alpha\beta} \right] \left[x_{2\gamma} x_{2\delta} - \frac{1}{3} r_2^2 \delta_{\gamma\delta} \right] . \quad (5)$$

In almost all cubic semiconductors the parameter describing the warping of the valence bands is rather small. Therefore we restrict our calculations to the spherical approximation (SA). In the SA the state of the acceptor is characterized by the parity, the full angular momentum F, and its component F_z (Baldereschi et al. (1973)):

$$\Psi_{FF_z}(r) = R_L(r) |LJFF_z\rangle + R_{L+2} |L+2, JFF_z\rangle .$$

Here J - the pseudospin quantum number ($J=3/2$), R_L and R_{L+2} are the radial functions. In the SA

$$\langle m | x_\alpha x_\beta - \frac{1}{3} r^2 \delta_{\alpha\beta} | m' \rangle = \frac{1}{3} Q (J^{(2)}_{\alpha\beta})_{mm'} , \quad (6)$$

where

$$J^{(2)}_{\alpha\beta} = \frac{1}{2} (J_\alpha J_\beta + J_\beta J_\alpha) - \frac{1}{3} J^2 \delta_{\alpha\beta} ,$$

J_α are the matrices of angular momentum of the order $2F+1$, Q is the quadrupole moment.
In the ground state ($L=0, F=3/2$) (Polupanov (1982))

$$Q = \frac{4}{5} a^2 \int_0^\infty dr \, r^4 R_0(r) R_2(r) . \quad (7)$$

Non-zero quadrupole moment is due to the "mixing" of the states $|\gamma L\rangle$ and $|\gamma, L+2\rangle$ by the "spin-orbital" term in the Hamiltonian of the hole.
Using the equation (6) and the explicit expression for the matrices J_α one can write the matrix H_{QQ} (dimension 16×16) in a rather simple form. Let us define

$$\hat{B} = \begin{pmatrix} 2^{-1/2} B_0 & iB_{-1} & -B_{-2} & 0 \\ iB_1 & -2^{-1/2} B_0 & 0 & -B_{-2} \\ -B_2 & 0 & -2^{-1/2} B_0 & -iB_{-1} \\ 0 & -B_2 & -iB_1 & 2^{-1/2} B_0 \end{pmatrix} \quad (8)$$

Here B_q ($q = 0, \pm 1, \pm 2$) are the components of the irreducible second rank tensor with zero trace composed in a standard manner of the components $\partial/\partial a_\alpha$ of the gradient. Then the matrix of H_{QQ} in (4) is equal to

$$H_{QQ} = \frac{1}{12} E_a \, a \, Q^2 (B \otimes B) R^{-1} . \quad (9)$$

Here $E_a = m_0 e^4 / \hbar^2 \varkappa^2 \gamma_1$, γ_1 being the Luttinger parameter, symbol $\otimes$ denotes

the Kronecker product of matrices.
If the quantization axis is parallel to **R** the only nonzero components of the tensor $V_{\alpha\beta\gamma\delta}$ (which is symmetric in any pair of indexes) are of the type $V_{\alpha\alpha\gamma\gamma}$. The matrix of H_{QQ} becomes very sparse, and it is factorizable and can be diagonalized. We shall however diagonalize the full matrix (4).

4. THE MATRIX OF DDI

In the SA the numerator in (4a) can be represented as a product of two sums in F_z and F_z'. It can be shown that each such sum is equal to

$$\frac{e^2}{\varkappa} \sum_{F_z} \langle m|x_\alpha|nFF_z\rangle \langle nFF_z|x_\gamma|m'\rangle =$$

$$= \frac{1}{2} a^3 E_a \frac{f(nF)}{E_{nF}-E_0} \left\{ \delta_{mm'}\delta_{\alpha\gamma} + \lambda(F)(J^{(2)}_{\alpha\gamma})_{mm'} \right\}. \quad (10)$$

Here $f(nF)$ is the oscillator strength of the dipole optical transition from the ground state to the state nF. In the SA such transitions are allowed only to states with $F = 1/2$, $3/2$, and $5/2$. The corresponding coefficients $\lambda(F)$ are equal to -1, $4/5$, and $-1/5$ respectively. Let us define a matrix B^F with the following elements:

$$B^F_{11} = B^F_{44} = [1 - \frac{1}{2}\lambda(F)](V_{x,x} + V_{y,y}) + [1 + \lambda(F)]V_{z,z} ,$$

$$B^F_{22} = B^F_{33} = [1 + \frac{1}{2}\lambda(F)](V_{x,x} + V_{y,y}) + [1 - \lambda(F)]V_{z,z} ,$$

$$B^F_{12} = -B^F_{34} = \frac{\sqrt{3}}{2}\lambda(F)(V_{y,z} + V_{z,y} + iV_{x,z} + iV_{z,x}), \quad (11)$$

$$B^F_{13} = -B^F_{24} = \frac{\sqrt{3}}{2}\lambda(F)(-V_{x,x} + V_{y,y} + iV_{x,y} + iV_{y,x}),$$

$$B^F_{14} = B^F_{23} = 0 .$$

The quantities $V_{\alpha,\beta}$ introduced here are defined by the product rule: $V_{\alpha,\gamma}V_{\beta,\delta} = V_{\alpha\beta}V_{\gamma\delta}$. From (2), (4a), and (10) it follows that

$$\hat{H}_{dis} = -E_a \sum_{n_1 F_1 n_2 F_2} A(n_1 F_1 n_2 F_2) \hat{B}^{F_1} \otimes \hat{B}^{F_2} , \quad (12)$$

where

$$A = \frac{1}{4} \frac{f(n_1 F_1)}{E_{n_1 F_1} - E_0} \frac{f(n_2 F_2)}{E_{n_2 F_2} + E_0} \frac{1}{E_{n_1 F_1} + E_{n_2 F_2} - 2E_0} . \quad (13)$$

The energies in (13) are measured in units of E_a.
If the quantization axis is oriented along **R** the only non-zero elements $V_{\alpha\beta}$ are the diagonal ones and $V_{xx} = V_{yy} = R^{-3}$, $V_{zz} = -2R^{-3}$. The same elements that are non-zero in the matrix (9) of H_{QQ} are also non-zero in H_{dis} (12). The latter can be expressed using linear combinations of three

dimensionless constans:

$$C_1 = 12 \sum_{n_1 F_1 n_2 F_2} A(n_1 F_1 n_2 F_2), \quad C_2 = 12 \sum_{n_1 F_1 n_2 F_2} \lambda(F_1) A(n_1 F_1 n_2 F_2),$$

$$C_3 = 12 \sum_{n_1 F_1 n_2 F_2} \lambda(F_1) \lambda(F_2) A(n_1 F_1 n_2 F_2).$$
(14)

5. THE ENERGY LEVELS OF THE MOLECULE

Taking into account (9) and (12) one sees that the matrix (4) of the interaction between two acceptors is factorizable and is easily diagonalized. If the interimpurity distance R is large then the interaction energy is much smaller than the spin-orbital interaction in each atom and the "molecule" considered belongs to the case c of Hund's bond. As the molecule consists of identical atoms being in the same state the molecular levels are characterized by the absolute value $|\Omega|$ of the component of full angular momentum parallel to the molecular axis, by the parity relative to the inversion of the coordinates of holes (indexes g and u) and at $\Omega = 0$ also by the sign + or - which correspond to the absence of any change or to the change of the sign of the wave function after reflection in the plane containing the molecular axis.

The full number of levels is $(2F+1)(F+1)=10$: one of each $3u$, $2u$, $2g$, $1g$ and two of each $1u$, $0g^+$, $0u^-$ levels. All levels with $\Omega \neq 0$ are two-fold degenerate and with $\Omega = 0$ are non-degenerate.

Let us define the chracteristic energies:

$$E_{QQ} = \frac{1}{2} E_a \frac{Q^2}{a^4} \left(\frac{a}{R}\right)^5, \quad E_{dis} = \frac{1}{2} E_a \left(\frac{a}{R}\right)^6$$
(15)

The molecular energies are then equal to (in the order of increase of the QQI):

$$E_1(2_u) = -7E_{QQ} - (C_1 - 7C_3)E_{dis},$$

$$E_2(2_g) = -4E_{QQ} - (C_1 - 5C_3)E_{dis},$$

$$E_3(1_u) = -2E_{QQ} - (C_1 - C_3)E_{dis},$$

$$E_4(0_u^-) = 3E_{QQ} - (C_1 + 3C_3)E_{dis} - \left[C_2^2 E_{dis}^2 + (5E_{QQ} - 6C_3 E_{dis})^2\right]^{1/2},$$

$$E_5(0_g^+) = 3E_{QQ} - (C_1 + 3C_3)E_{dis} - \left[C_2^2 E_{dis}^2 + (3E_{QQ} - 2C_3 E_{dis})^2\right]^{1/2}, \quad (16)$$

$$E_6(2_g) = E_{QQ} - (C_1 + C_3)E_{dis},$$

$$E_7(1_u) = 3E_{QQ} - (C_1 - C_2 + 3C_3)E_{dis},$$

$$E_8(3_u) = 3E_{QQ} - (C_1 + C_2 + 3C_3)E_{dis},$$

$$E_9(0_g^+) = 3E_{QQ} - (C_1 + 3C_3)E_{dis} + \left[C_2^2 E_{dis}^2 + (3E_{QQ} - 2C_3 E_{dis})^2\right]^{1/2},$$

$$E_{10}(0_u^-) = 3E_{QQ} - (C_1 + 3C_3)E_{dis} + \left[C_2^2 E_{dis}^2 + (5E_{QQ} - 6C_3 E_{dis})^2\right]^{1/2},$$

As we shall see further at large distances R the DI does not change the order of levels given by (16).
The expresions for the molecular wave functions are rather cumbersome and we'll write explicitly only the wave functions of the two ground states:

$$\psi_1(2_u) = 2^{-1/2}\left\{|\pm\tfrac{3}{2}\rangle_1 \, |\pm\tfrac{1}{2}\rangle_2 + |\pm\tfrac{1}{2}\rangle_1 \, |\pm\tfrac{3}{2}\rangle_2\right\}. \quad (17)$$

Let us evaluate the energies (16) for the case of acceptors in Ge. In this case the SA is appropriate and the quantum-mechanical characteristics (oscillator strengths, polarizabilities etc.) are rather well known (Kogan et al. (1981), Polupanov (1982)). Let us put R equal to the minimum distance Ro at which the overlap of the wave functions can be neglected. According to the well known criterion (Le Roy (1973)) $R_0 = 2[\langle r_A^2\rangle^{1/2} + \langle r_B^2\rangle^{1/2}]$, where $\langle r_A^2\rangle$ is the average distance squared of the hole in atom A measured from the impurity ion. Using the wave functions calculated earlier we find in Ge $R_0 = 4\langle r^2\rangle^{1/2} = 350$ Å. Using also the known value of Q we find that $E_{QQ}(R_0) = 0.8$ μeV, $E_{dis}(R_0) = 3.8$ μeV. Each coefficient C_i can be approximated by a single term of the sum (14) that corresponds to the dominant term in the appropriate polarizability α or α_1 of the impurity atom (α and α_1 have been calculated earlier (Polupanov (1982))). In this approximation

$$C_1 = \tfrac{3}{2}\left(\tfrac{\alpha}{\alpha_0}\right)^2 |E|, \quad C_2 = \tfrac{3}{2}\tfrac{\alpha\alpha_1}{\alpha_0^2}|E|, \quad C_3 = \tfrac{3}{8}\left(\tfrac{\alpha_1}{\alpha_0}\right)^2 |E|. \quad (18)$$

Here $|E|$ is the characteristic energy which is usually taken equal to ionization energy, $\alpha_0 = \pi a^3$. In the case of acceptors in Ge $|E| = 1.25$, $\alpha/\alpha_0 = -0.092$. With these values of parameters the energies (16) of the molecular levels at $R = R_0 = 350$ Å are (in μeV): -6.4, -4.1, -2.51, -2.49, -0.69, -0.14, 1.2, 1.7, 3.8, 5.3. The ground level and the width of the "impurity band" are determined by QQI and are of order 10 μeV. The DI is significant only in the case of levels 5 and 6 and slightly splits the levels 3 and 4, 7 and 8. The difference between two adjacent levels is ~ 1 μeV. We may conclude that in the considered concentration range the effect of QQI on the thermodynamics of p-Ge can be observed only at temperatures lower than 1 K. One expects however that the molecular splitting of impurity levels due to QQI and DI can be observed in the absorption of sound even at higher temperatures.
The relations (6) and (10) and the method of derivation of the matrices of H_{QQ} (9) and H_{dis} (12) can be generalized to the case when the SA is not appropriate and the direction of quantization is determined by cubic axes.

LITERATURE

Baldereschi A and Lipari N O 1973 *Phys. Rev.* **B8** 2697
Kogan Sh M and Polupanov A F 1981 *Sov. Phys. JETP* **53** 201
Le Roy R J 1973 In: *Molecular Spectroscopy. Specialist Periodical Report* ed. Barrow (London: Chem.Soc.) **1** p.113
Polupanov A F 1982 *Physics and Technology of Semiconductors* **16** 27

Influence of electron-phonon coupling on the properties of D^- centers

Janusz Adamowski and Stanisław Bednarek

Zakład Fizyki Ciała Stałego, Akademia Górniczo-Hutnicza,
PL-30059 Kraków, Poland

ABSTRACT: A binding energy and average interparticle distances are calculated for D^- centers in semiconductors taking into account the interaction with LO-phonons. For the strong electron-phonon coupling, the calculated D^- binding energy is of the order of the donor ionisation energy. However, for the intermediate coupling and weak donor potential, the D^- center is either very weakly bound or unstable.

1. INTRODUCTION

The D^- center is a solid state analogue of the negative hydrogen ion H^-. It consists of a positive donor center and two bound electrons. According to the hydrogenlike effective mass approximation, the D^- binding energy is equal to $W(H^-) = 0.0555$ donor rydbergs. The D^- centers with such binding energy were observed only in germanium under high uniaxial stress (Taniguchi and Narita 1977). In Si and Ge crystals without stress, the measured binding energy of D^- is considerably larger than $W(H^-)$, which results from an influence of the multi-valley structure of the conduction bands (Narita 1980). One can hardly find an experimental evidence for D^- in compound semiconductors, in which the coupling with LO-phonons becomes essential. This effect was theoretically considered by Larsen (1981) who found an enhancement of binding of D^- centers due to the electron-phonon coupling, which can take place in the ionic crystals. However, these results do not explain the lack of direct evidence for D^- in semiconducting compounds. The purpose of the present paper is to study the influence of the coupling with the LO-phonons on the properties of D^- centers in the whole range of material parameters. For this aim we calculate both the binding energy and interparticle distances for D^-.

2. METHOD OF CALCULATIONS

We assume that the crystal has a single isotropic conduction band and the potential of an isolated donor impurity is coulombic. We further assume the validity of the effective mass approximation and the Froehlich coupling with the LO-phonons.

© 1989 IOP Publishing Ltd

The Hamiltonian of the D$^-$ center has a form (Larsen 1981)

$$H = -\frac{\hbar^2}{2m_e}(\nabla_1^2+\nabla_2^2) - \frac{e^2}{\varepsilon_o r_1} - \frac{e^2}{\varepsilon_o r_2} + \frac{e^2}{\varepsilon_\infty r_{12}} + H_{ph} + H_{int} , \quad (1)$$

where m_e is the conduction electron band mass, r_1 and r_2 are the electron-donor distances, r_{12} is the electron-electron distance, ε_o and ε_∞ are the static and high-frequency dielectric constants, respectively. The noninteracting LO-phonons are described by

$$H_{ph} = \hbar\omega \sum_k a_k^\dagger a_k \quad (2)$$

and the interaction between the electrons and phonons is given by

$$H_{int} = \hbar\omega \sum_k \left\{ v_k a_k \left[exp(ik \cdot r_1) + exp(ik \cdot r_2) \right] + H.c. \right\} . \quad (3)$$

In (2) and (3) a_k and $a_k^\dagger$ are the annihilation and creation operators of the LO-phonon with the frequency ω and the wave vector k, the amplitude of interaction has the form

$$v_k = -(i/|k|)(4\pi\alpha/\Omega a_p)^{1/2} , \quad (4)$$

where Ω is the crystal volume, a_p is the free polaron radius, and α is the Froehlich electron-phonon coupling constant.

In order to calculate the ground state energy of D$^-$ we appply the method of optimized canonical transformation (Bednarek et al. 1977, Adamowski 1985). In the first step, Hamiltonian (1) is subjected to the transformation

$$U = exp(S-S^\dagger) , \quad (5)$$

where

$$S = \sum_k F_k(r_1,r_2) a_k \quad (6)$$

with the displacement amplitude

$$F_k(r_1,r_2) = v_k \left\{ f_k^{(1)} \left[exp(ik \cdot r_1) + exp(ik \cdot r_2) \right] + f_k^{(2)} \right\} . \quad (7)$$

The phonon amplitudes in (7) are proposed in the form

$$f_k^{(1)} = \frac{\lambda_1}{\varrho_1^2 a_p^2 k^2 + 1} \qquad (7a)$$

and

$$f_k^{(2)} = \frac{\lambda_2}{(\varrho_2^2 a_p^2 k^2 + 1)^2} . \qquad (7b)$$

The variational parameters λ_1, λ_2, ϱ_1, and ϱ_2 in (7a) and (7b) are introduced in order to optimize the choice of U. We obtain — as the result of the first step — the effective Hamiltonian defined as

$$H_{eff} = \langle \chi_o | U^\dagger H U | \chi_o \rangle , \qquad (8)$$

where $|\chi_o\rangle$ is a phonon vacuum state.

In the second step, the ground state energy of the system is calculated from the Schroedinger equation

$$(H_{eff} - E) \Psi = 0 . \qquad (9)$$

Equation (9) is solved with the use of the normalized trial wave function in the form

$$\psi \propto \sum_j d_j (1 + P_{12}) \exp(-a_j r_1 - b_j r_2 - c_j r_{12}) , \qquad (10)$$

which is similar to that proposed by Thakkar and Smith (1977). The sum in (10) contains 16 terms with appropriately chosen variational parameters a_j, b_j, c_j, and d_j. The permutation operator P_{12} interchanges the indices 1 and 2. Trial function (10) has been tested by calculating the ground state energy of the H^- ion and He atom. The results, expressed in the hydrogen rydberg, are -1.055448 for H^- and -5.807435 for He as compared with the exact values -1.055502 and -5.807449, respectively. The ground state energy E of the D^- center is a function of α and R, where $R = R_D/\hbar\omega = m_e e^4/2\hbar^3 \varepsilon_o^2 \omega$ is the ratio of the donor rydberg R_D to the LO-phonon energy $\hbar\omega$. For all physically significant values of α and R, we have calculated E and next the binding energy W of the D^- center, which is defined as

$$W = 2E_{pol} - W_D - E . \qquad (11)$$

The ground state energy E_{pol} of the free polaron is taken from the papers of Larsen (1968) and Adamowski et al. (1980). The ionisation energy W_D of the 1s donor state is calculated in the bound polaron model (Adamowski 1985).

3. RESULTS

The calculated ratio W/W_D is shown in Figure 1 as a function of α and R. The case $\alpha = 0$ corresponds to the H^--like center with the binding energy $W(H^-) = 0.0555\, R_D$. For $R \geq 2$, the ratio W/W_D is a monotonically increasing function of α. For $R < 2$, this ratio firstly decreases and next increases with α taking on a minimum for α from the intermediate coupling region, i.e., $1 < \alpha < 3$. For $R < 0.9$, the minimum value of W is negative, which means that the D^- state is unstable against a dissociation into a neutral donor and a free polaron. E.g., for $R = 0.5$, this instability takes place if $0.6 < \alpha < 3.3$. For the large electron-phonon coupling, i.e., $\alpha > 4$, the D^- center is strongly bound with the binding energy of the order of the donor ionisation energy. This behavior is in agreement with the results of Larsen (1981).

Using trial function (10), we have as well calculated the average interparticle distances in D^-. The results for $R = 2$ are shown in Figure 2. The size of the D^- center is of the order of the donor Bohr radius a_D and decreases with the increasing binding energy. The interparticle separations calculated by us for the H^- ion are $\langle r_1 \rangle / a_D = 2.70$ and $\langle r_{12} \rangle / a_D = 4.40$ as compared with the values 2.71 and 4.41, respectively, which were obtained by Thakkar and Smith (1977).

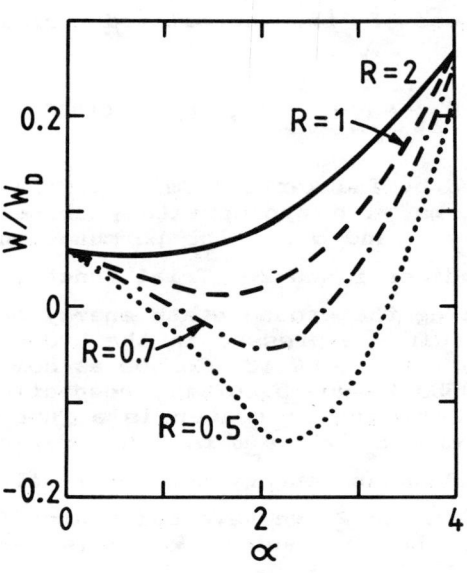

Fig. 1. Calculated ratio W/W_D of the D^- binding energy to the donor ionisation energy as a function of α and R

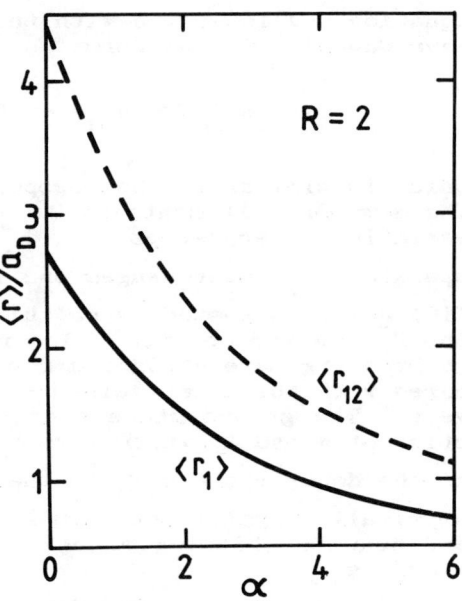

Fig. 2. Calculated average values of the electron-donor $\langle r_1 \rangle$ and electron-electron $\langle r_{12} \rangle$ distances as functions of α for $R=2$

Table 1. Calculated ionisation energy W_D of the donor, binding energy W and expectation values of the electron-donor r_1 and electron-electron r_{12} separations for the D^- center.[1] Energy is expressed in meV and length – in the donor Bohr radii $a_D = \hbar^2 \varepsilon_o / m_e e^2$. α is the electron-phonon coupling constant and R is the ratio of the donor rydberg to the LO-phonon energy.

crystal	α	R	W_D	W	$\langle r_1 \rangle$	$\langle r_{12} \rangle$
CdS	0.53	0.78	31.91	1.17	2.38	3.85
CdTe	0.27	0.67	14.73	0.65	2.55	4.14
ZnSe	0.45	0.92	30.85	1.30	2.40	3.88
AgBr	1.64	1.68	33.36	2.03	1.64	2.62
AgCl	1.90	1.90	62.48	4.92	1.53	2.44
CdF_2	2.53	1.27	96.40	8.34	1.24	1.98

Table 1 shows our results for D^- centers in II-VI semiconductors and ionic crystals. The input parameters in our calculations are listed in the first two columns of Table 1. The very small values of the D^- binding energy in the II-VI semiconductors can be responsible for experimental difficulties in an identification of these centers.

4. DISCUSSION

The optimized canonical transformation method, which was previously applied to the problems of exciton (Bednarek et al. 1977), bound polaron (Adamowski 1985) and bipolaron (Adamowski 1988), appeared to be useful in a description of properties of D^- centers. The calculated ground state energy is lower than that obtained by the method of Larsen (1981) by about 3%, whereas both the methods are variational. It means that our results better estimate the exact energy of the center. The expectation values of the interparticle separations in the D^- state has been calculated for the first time.

The following properties of the D^- center can be infered from our results: For $\alpha \ll 1$, the D^- center is similar to the H^- ion with the binding energy $W \simeq 0.0555\ R_D$. For $R_D > \hbar\omega$, the D^- center is stable for an arbitrary value of the electron-phonon coupling constant. However, for $R_D < \hbar\omega$ and intermediate electron-phonon coupling, the D^- state is very weakly bound or can be even unstable. The donor rydberg R_D is proportional to the strength of the donor potential, which means that this effect occurs when the potential of the donor is relatively weak and the coupling with phonons becomes essential. Since our conclusions are based on the results of the variational calculations, they are not a proof but rather a strong suggestion for the instability.

It is interesting that a similar decrease of the binding energy due to the coupling with phonons was obtained for the exciton-neutral donor complex (Bednarek et al. 1981) if the exciton rydberg is less than the LO-phonon energy. The values of material parameters corresponding to the region of weak binding of the D^- ion are relevant to several compound semiconductors such as TlCl, TlBr, PbS, and PbSe. In these crystals, the D^- ion either has very small binding energy or is unstable.

The present approach is based on the hydrogenlike effective mass approximation and on the assumption that the dominating interaction with phonons is the Froehlich electron-LO-phonon coupling. We do not take into account, e.g., a chemical nature of the donors and details of the band structure. Our assumptions are quite well justified for shallow donor states in semiconducting compounds. Therefore, we believe that our results can be useful in the spectroscopy of D^- centers in these crystals.

ACKNOWLEDGMENTS

This work was supported by the Institute of Physics of the Polish Academy of Sciences under contract CPBP 01.04.

REFERENCES

Adamowski J 1985 Phys. Rev. B32 2588
Adamowski J 1988 Acta Phys. Polon. A73 345
Adamowski J, Gerlach B and Leschke H 1980 *Functional Integration - Theory and Applications* (New York: Plenum Press) pp 291-301
Bednarek S, Adamowski J and Suffczyński M 1977 Solid State Commun. 21 1
Bednarek S, Adamowski J and Suffczyński M 1981 J. Phys. C: Solid State Phys. 14 4405
Larsen D M 1968 Phys. Rev. 172 967
Larsen D M 1981 Phys. Rev. B23 628
Narita S 1980 J. Phys. Soc. Japan 49 173
Taniguchi M and Narita S 1977 J. Phys. Soc. Japan 43 1262
Thakkar A J and Smith Jr V H 1977 Phys. Rev A15 1

Inst. Phys. Conf. Ser. No 95: Chapter 9
Paper presented at Int. Conf. Shallow Impurities in Semiconductors, Linköping, Sweden, 1988

Valence charge distribution around shallow donors in semiconducting compounds

Stanisław Bednarek and Janusz Adamowski

Zakład Fizyki Ciała Stałego, Akademia Górniczo-Hutnicza,
PL-30059 Kraków, Poland

ABSTRACT: A redistribution of valence electrons due to a presence of an impurity in a semiconductor is discussed. It is shown that this redistribution induces a response potential containing a long-range impurity-dependent component, which can explain several questions connected with chemical shifts of shallow donor levels.

1. INTRODUCTION

Energy levels of shallow donors in semiconductors are described by the effective mass approximation in the hydrogenlike model with a rather good precision. This model takes into account the dominating component of the impurity potential, namely, the Coulomb potential which results from a difference of valencies for the interchanged atoms. The real donor additionally introduces into a crystal a short-range potential, which depends on the atomic properties of the impurity core. It leads to deviations from the hydrogenlike spectrum, which are called the chemical shifts of donor levels and depend on the impurity species. In previous papers on this subject, the chemical shift was explained as a result of a direct influence of the impurity potential on the conduction electron wave functions, since the wave function of the donor is their linear combinanation. This approach leads to the so-called central-cell corrections. In this approximation the chemical shift is proportional to a probability of finding of the donor electron in an elementary cell, which contains the donor center, and depends on the difference between the short-range potentials of the impurity and the substituted atom. The theory of central-cell corrections provides a correct description of the chemical shifts for shallow donors in multi-valley semiconductors (Pantelides 1978).

However, in direct-gap semiconductors this method does not yield the correct results. In particular, it does not explain even qualitatively the following effects:
(i)   the ordering of the donor levels does not agree with an increasing atomic number of impurities;
(ii) in II-VI compounds, the slope of the curve $\Delta(E_{1s}-E_{2p})$
      vs donor ionisation energy is considerably smaller than

© 1989 IOP Publishing Ltd

that predicted by the short-range central-cell corrections;

(iii) in GaAs, the chemical shifts predicted by this model are very small due to the large donor radius. Although the observed chemical shifts are small, they are of one order of magnitude larger than those resulting from the model of central-cell corrections.

The above effects suggest that - besides the central-cell corrections - there exists another cause of chemical shifts for the donor levels. In particular, effects (ii) and (iii) suggest an appearance of an additional long-range potential, which depends on the atomic properties of the impurity. Such potential has not been taken into account in the previous papers on shallow donors. The present paper is devoted to a discussion of the impurity-induced change in the valence electron density. We show that this change leads - *via* the Hartree potential - to an occurence of the dielectric response potential, which possesses the desired properties.

2. THEORY

The wave functions of valence electrons in an ideal crystal are the Bloch functions $\psi_{\upsilon q}$, where υ labels the valence subbands and the wave vector q belongs to the first Brillouin zone. The valence charge density $\rho(r)$ is a periodic function with the lattice periodicity. The Fourier transform of $\rho(r)$ is given by

$$\rho(k) = \int d^3 r \, e^{ikr} \rho(r) = \sum_{\upsilon q} \langle \psi_{\upsilon q} | e^{ikr} | \psi_{\upsilon q} \rangle / \langle \psi_{\upsilon q} | \psi_{\upsilon q} \rangle . \qquad (1)$$

Due to its periodicity, $\rho(r)$ has the only non-zero Fourier components for $k = G$, where G is the reciprocal lattice vector. In particular, for small but non-equal zero values of k: $\rho(k) \equiv 0$.

An impurity disturbs the potential of the crystal and changes the distribution of the valence electrons. This change follows from the long-range Coulomb potential and from the short-range potential of the impurity. The first component of the impurity potential results from the difference between valencies of the impurity and the substituted atom, while the second - from the interchange of the atomic cores. This interchange gives rise to either an appearance or disappearance of additional core states, which is manifested in a condition of orthogonality of the valence electron wave functions to the core electron wave functions. This requirement causes a rapid change of a shape of the valence wave functions in the nearest neighbourhood of the impurity core. Although the change of energy levels for valence electrons is small, the corresponding change of the wave functions is substantial and cannot be decribed with the help of the perturbation theory. This problem can however be solved using a concept of pseudopotential.

The general pseudo-Hamiltonian has a form (Austin *et al.* 1962)

$$\mathcal{H} = H + \sum_\lambda |\psi_\lambda\rangle\langle F_\lambda| ,\qquad (2)$$

where H is the Hamiltonian of an unperturbed system, ψ_λ are the core wave functions of all the atoms in the system, and F_λ are arbitrary functions. The eigenfunction ϕ_n of $\mathcal{H}$ is equal to the sum of the eigenfunction ψ_n of H plus a linear combination of the core wave functions, i.e.,

$$\phi_n = \psi_n + \sum_\lambda a_{n\lambda} \psi_\lambda .\qquad (3)$$

The coefficients $a_{n\lambda}$ depend on a choice of F_λ in (2), which leads to a nonuniqueness of pseudowave functions (3). If we know the pseudowave functions, we can calculate the true wave functions ψ_n by projecting ϕ_n on a subspace orthogonal to the core electron states as follows

$$\psi_n = (1-P)\phi_n ,\qquad (4)$$

where the operator P is defined by

$$P = \sum_\lambda |\psi_\lambda\rangle\langle\psi_\lambda| .\qquad (5)$$

The pseudo-Hamiltonian $\tilde{\mathcal{H}}$ of the perturbed system is constructed according to (2) with the help of the perturbed Hamiltonian $\tilde{H}$ and the wave functions $\tilde{\psi}_\lambda$ of the new core. In general case, the perturbed pseudowave function $\tilde{\phi}_n$ is a linear combination of the unperturbed pseudowave functions ϕ_n and core wave functions ψ_λ, which together constitute a complete basis,

$$\tilde{\phi}_n = \sum_m c_{nm} \phi_m + \sum_\lambda c_{n\lambda} \psi_\lambda .\qquad (6)$$

The requirement that the second term in (6) vanishes allows us to introduce a new basis of unperturbed wave functions, which are uniquely determined (Bednarek 1988). These functions are obtained from the eigenfunctions of an arbitrary pseudo-Hamiltonian as follows

$$\bar{\phi}_n = \phi_n - \sum_{\lambda\lambda'} |\psi_\lambda\rangle(V_{\lambda\lambda'})^{-1}\langle\psi_{\lambda'}| V \phi_n ,\qquad (7)$$

where $(V_{\lambda\lambda'})^{-1}$ are the elements of the matrix reciprocal to $\langle\psi_\lambda|V|\psi_{\lambda'}\rangle$ and V is the difference of the Phillips-Kleinman pseudo-Hamiltonians. This pseudo-Hamiltonian has the form

$$\mathcal{H} = H + \sum_\lambda |\psi_\lambda\rangle\langle E_n - \varepsilon_\lambda\rangle\langle\psi_\lambda| \qquad (8)$$

and is a special case of pseudo-Hamiltonian (2). Now, the perturbed pseudowave function can be expressed as a linear combination of only pseudowave functions (7)

$$\tilde{\phi}_n = \sum_m \bar{c}_{nm} \bar{\phi}_m . \qquad (9)$$

The above treatment is valid for an arbitrary system, which is perturbed by either an interchange of atomic cores or their displacement from equilibrium positions. In the present work, we are looking for the perturbed pseudowave functions for the valence electrons in a crystal with an isolated impurity. In order to remove a degeneracy which follows from the crystal symmetry, we diagonalize the perturbed pseudo-Hamiltonian $\tilde{\mathcal{H}}$ in a basis of the Bloch pseudowave functions $\{\bar{\phi}_{v\mathbf{q}}\}$ for the valence electrons. In this way, we obtain a set of new functions

$$\phi_{v\varkappa} = \sum_{v'\mathbf{q}} A^{v'\mathbf{q}}_{v\varkappa} \bar{\phi}_{v'\mathbf{q}} , \qquad (10)$$

where the new label $\varkappa$ takes on the same number of different values as $\mathbf{q}$ in the first Brillouin zone, nevertheless, due to the lack of the translational symmetry in a perturbed crystal, $\varkappa$ cannot be identified with $\mathbf{q}$. In this new basis, we have no intraband matrix elements. The pseudowave function can be obtained from the first order perturbation expression

$$\tilde{\phi}_{v\varkappa} = \phi_{v\varkappa} + \sum_{c\mathbf{q}} \bar{\phi}_{c\mathbf{q}} \langle\bar{\phi}_{c\mathbf{q}}|V|\phi_{v\varkappa}\rangle/(E_{v\varkappa} - E_{c\mathbf{q}}) , \qquad (11)$$

where the index c runs over all the conduction bands. The perturbed valence electron wave functions are constructed similarly as in (4), i.e.,

$$\tilde{\psi}_{v\varkappa} = (1-\tilde{P}) \tilde{\phi}_{v\varkappa} , \qquad (12)$$

where the operator $\tilde{P}$ is given by (5) with the sum running over the core states of the impurity and host atoms. Next, we calculate the valence electron density in the perturbed crystal

$$\tilde{\rho}(\mathbf{r}) = \sum_{v\varkappa} |\tilde{\psi}_{v\varkappa}(\mathbf{r})|^2 , \qquad (13)$$

which has the Fourier components

$$\tilde{\rho}(\mathbf{k}) = \sum_{v\varkappa} \langle\tilde{\phi}_{v\varkappa}|(1-\tilde{P}) e^{i\mathbf{k}\mathbf{r}} (1-\tilde{P})|\tilde{\phi}_{v\varkappa}\rangle/\langle\tilde{\phi}_{v\varkappa}|1-\tilde{P}|\tilde{\phi}_{v\varkappa}\rangle . \qquad (14)$$

The valence electron density depends on the difference $V = \tilde{\mathcal{H}} - \mathcal{H}$ of the pseudo-Hamiltonians, on the difference $\Delta P = \tilde{P} - P$ of the projection operators, and on the quantities, which chara-

cterize the ideal crystal. The dependence on V and ΔP is non-linear and relatively complicated. It takes on a simplified form when the potential of the impurity is weak and the omission of the terms quadratic in V and ΔP is allowed. Then,

$$\tilde{\rho}(k) = \rho_1(k) + \rho_2(k) + \rho_3(k) . \tag{15}$$

The first term consists of the change of valence charge distribution caused by the change of pseudowave functions. It has a form

$$\rho_1(k) = \sum_{\upsilon x} \langle \tilde{\phi}_{\upsilon x} | (1-P) e^{ikr} (1-P) | \tilde{\phi}_{\upsilon x} \rangle / \langle \tilde{\phi}_{\upsilon x} | (1-P) | \tilde{\phi}_{\upsilon x} \rangle \tag{16}$$

and describes the dielectric response to the change of the pseudopotential. The long-range potential connected with this term is a response to the Coulomb potential, which results from the difference of valencies. The second term

$$\rho_2(k) = \sum_{\upsilon q} \langle \phi_{\upsilon q} | \tilde{P} e^{ikr} \tilde{P} - P e^{ikr} P - (\tilde{P}-P) e^{ikr} - e^{ikr} (\tilde{P}-P) | \phi_{\upsilon q} \rangle \tag{17}$$

stems from a reorthogonalisation of valence electron wave functions, which is forced by the change of the core. It contains the change of charge distribution, which induces the short-range as well as long-range potential of the dielectric response. This term does not vanish for small k and, for $k \to 0$, tends to a constant $\rho_2(0) = -Q$, where Q is given by:

$$Q = \sum_{\upsilon q} \langle \phi_{\upsilon q} | \Delta P | \phi_{\upsilon q} \rangle . \tag{18}$$

The third term in (15) has a form

$$\rho_3(k) = \sum_{\upsilon q} \langle \phi_{\upsilon q} | e^{ikr} | \phi_{\upsilon q} \rangle \langle \phi_{\upsilon q} | \Delta P | \phi_{\upsilon q} \rangle . \tag{19}$$

It is a periodic change of the charge distribution, which vanishes for small k (with an exception of $k = 0$). For $k = 0$, this term cancels out the second term in (15), i.e., $\rho_2 + \rho_3 = 0$, which ensures a neutrality of the crystal.

3. DISCUSSION

The purpose of the present paper was to determine a change of the valence electron distribution associated with an isolated impurity in a semiconductor. We have shown that there exists in a crystal the impurity-dependent charge density ρ_2, which is of especial importance in the problem of chemical shifts for shallow donors because of the properties of the potential associated with ρ_2. In the long-range limit, i.e., $r \to \infty$ or $k \to 0$, this potential has a form of the Coulomb potential created by the charge Q, which is the limit of $\rho_2(k)$ for $k \to 0$

and can be treated as a charge effectively localized at an impurity. It results from the requirement of orthogonality of the valence electron wave functions to the wave functions of the impurity core. In an ideal crystal, the wave functions of valence electrons are orthogonal to those of the host-atom core. In a crystal with an impurity, we have to perform a re-orthogonalisation. The charge Q has the same nature as "depletion charge" or "orthogonality hole" in the band structure calculations by the pseudopotential method. The value of Q depends on the impurity and host atom species and on the density of valence electrons around an impurity. Q increases with an increasing atomic number Z of a donor, i.e., an increasing number of core orbitals, since the heavier impurity more effectively repels the valence electrons from the core region. This effect is larger on an anion site than on a cation site due to the larger valence electron density for the anion.

The charge Q takes on negative (positive) values for the impurities, which are lighter (heavier) than the host atom. This charge is an additional source of chemical shifts for shallow impurities. The shallow donor levels should become deeper if the atomic number Z of the impurity increases. The shifts resulting from the central-cell corrections, i.e., from a direct influence of the short-range potential on the conduction band states, exhibit an opposite trend. Both the effects partially compensate, which can lead to a suppression of a correlation between a position of the donor level and the atomic number Z of the impurity. This provides a qualitative explanation of effect (i) listed in Introduction.

Effects (ii) and (iii) can be explained basing on a long-range nature of the potential connected with Q. The central-cell correction yields a shift of only s states of the donor. The long-range correction, which originates in an appearance of the charge Q, provides shifts of both the s and p levels. Therefore, the ratio $\Delta(E_{1s}-E_{2p})/\Delta E_{1s}$ can take on arbitrarily small values. The central-cell corrections are negligibly small if the donor radius is large, which is the case of a semiconductor with a small conduction electron band mass, e.g., GaAs. The effect being discussed in the present paper does not directly depend on the donor size. Therefore, it is a main source of chemical shifts in GaAs, which however small are experimentally accesible.

ACKNOWLEDGMENTS

This work was supported by the Institute of Physics of the Polish Academy of Sciences under contract CPBP 01.04.

REFERENCES

Austin B J, Heine V and Sham L J 1962 Phys. Rev. **127** 276
Bednarek S 1988 Acta Phys. Polon. **A73** 195
Pantelides S T 1978 Rev. Mod. Phys. **50** 797

Monte Carlo simulation as a novel technique in the study of shallow centers

Lino Reggiani and Paolo Lugli *

Dipartimento di Fisica e Centro Interuniversitario di Struttura della Materia dell'Universita' di Modena, Via Campi 213/A, 41100 Modena, Italy

Vladimir Mitin

Institute of Semiconductors, Academy of Sciences of the Ukrainian SSR, Kiev, USSR

ABSTRACT: We present the Monte Carlo technique as a novel numerical method to investigate recombination and generation processes from shallow impurity centers. In particular, we provide a first-principle derivation of the microscopic and macroscopic capture cross-section due to acoustic phonon processes which compares favourably with experimental results.

1. INTRODUCTION

The aim of this paper is to present the Monte Carlo (MC) simulation (Jacoboni and Reggiani 1983) as a novel theoretical technique for the study of shallow centers. One of the basic quantity characterizing shallow impurities is the recombination cross-section. In particular, we are here interested in the acoustic-phonon assisted recombination within the cascade-capture model (Lax 1960, Abakumov et al 1976). In this model, a carrier originally in the conducting band can go through a chain of successive one-phonon transitions along a ladder of excited levels collapsing to the ground state or coming back into the conducting band (see Fig. 1). To exploit the power of this technique, two main approaches will be illustrated.

A first one, which follows a simpler and more traditional way of proceedings, considers both conducting and bound states separately (decoupled treatment) (Reggiani et al 1987). By obtaining an exact solution of the Boltzmann equation in the presence of an electric field F of arbitrary strength the MC analysis of the macroscopic kinetic coefficients (e.g. conductivity, diffusivity, etc.) gives us information on the microscopic capture cross-section itself.

A second one, which we shall call self-consistent, includes a complete treatment of the carriers in the conducting band as well as on the impurity levels (Abakumov and Yassievich 1976). Thus, the simulation enables us to obtain a first principle determination of the microscopic capture cross-section.

2. DECOUPLED TREATMENT

Through the introduction of an energy dependent rate of recombination, the capture process is

treated as an ordinary scattering between an initial k-state in the conducting band and a final k'-state of the electron bound to the impurity.

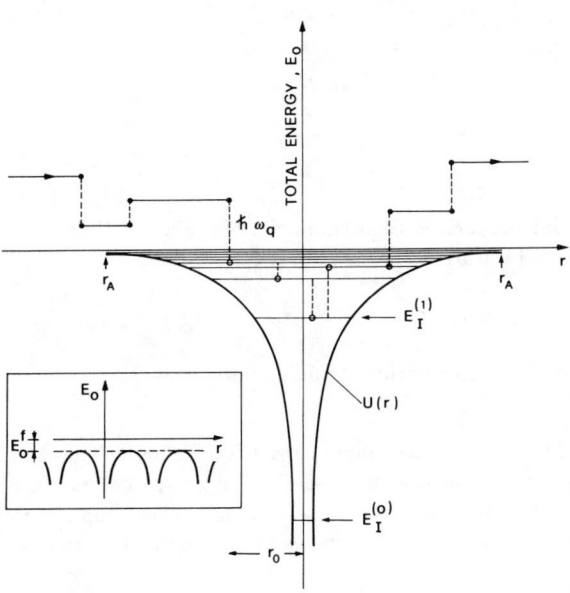

Fig. 1 - Schematic representation of a trajectory of an electron in the (total) energy space at equilibrium. The full dots mark a scattering process via phonons with energy $\hbar\omega_q$ that an electron undergoes in the positive energy region where it moves freely. In the attractive Coulomb field $U(r)$ it is accelerated and at the distance r_0 (measured from the charged center) the electron enters the negative energy region. The open dots mark transitions between the energy levels of the impurity center in the negative energy region where the electron is considered as trapped. All transitions are assumed to occur instantaneously in space and time (transitions in the negative energy region are supposedly to occur at the impurity site). The insert shows the energy level E_0^f associated with the fluctuational Coulomb potential of the impurity.

In the single carrier simulation, when a recombination process occurs, a trapping time is determined exponentially distributed according to the generation rate (during the trapping time the carrier is considered at rest.) Then a generation process follows immediately in which the final energy of the carrier in the conducting band is stochastically assigned according to the probability per unit time of generation from an impurity state to a given energy in the conducting band. In this approach, the impurity levels are not considered in details; rather, only transitions which lead to carrier capturing are accounted for. Therefore, in both capture and generation processes we miss the knowledge to and from which level a transition occurs. The generation recombination probabilities depend basically on the capture cross section, which can be determined from the microscopic interpretation of field dependent transport coefficients. We consider explicitely the case of p-type silicon (Boron doped) at 77 K. Figure 2 shows the different scattering rates used in the MC simulation.

We notice that the recombination is the dominant process at energies below 1 meV. Therefore, a precise knowledge of the distribution function in the low energy tail is needed in order to

obtain reliable results.

Figure 3 shows the field dependent conductivity $\sigma(F)$ for three different doping levels. The systematic decrease of the conductivity $\sigma = en\mu$, (e being the electron charge) is a non-ohmic effect originated from the different field dependence of the carrier concentration, n, and the mobility, μ. Indeed, at increasing electric fields the onset of a hot-electron regime leads to a decrease in the mobility which dominates over a concurrent increase in carrier concentration. The good agreement between theory and experiments (Vaissiere 1986), which also extends to noise-measurements (Reggiani et al 1988a, 1988b), supports the reliability of the present approach. In particular, we have found a recombination cross-section due to acoustic phonons σ_{ac} of 5.5×10^{-13} cm^2 which is within the spread of values quoted in the literature (Abakumov et al 1976).

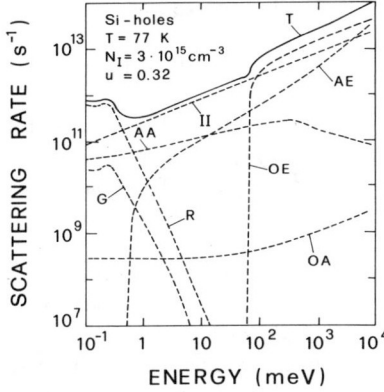

Fig. 2 - Scattering rate as a function of energy for holes in Si at 77 K. Symbols have the following meaning: AA (Acoustic Absorption), AE (Acoustic Emission), OA (Optical Absorption), OE (Optical Emission), II (Ionized Impurity), R (Recombination), G (Generation), T (Total). Notice that for the generation rate the energy refers to the final state of the hole in the valence band.

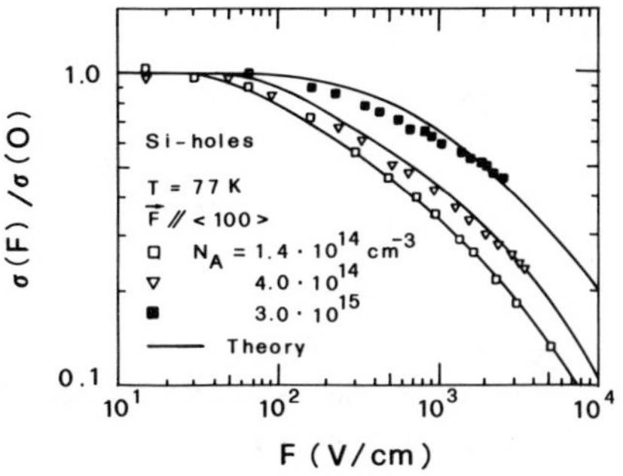

Fig. 3 - Conductivity normalized to its Ohmic value as a function of the electric field for the different acceptor concentration reported. Symbols refer to experiments, the lines to the Monte Carlo simulation.

From the simulation we have obtained the field dependent recombination times which are shown in Fig. 4. Again, the field induced heating of the carriers prevents these from being captured and thus is responsible for the systematic increase of the recombination time. The consequent increase of the fraction of ionized carriers u is shown in Fig. 5. For the case of an acceptor concentration $N_A = 3 \times 10^{15} cm^{-3}$ we have estimated the Poole-Frenkel effect by increasing the equilibrium generation rate through the usal factor $exp(\beta F^{\frac{1}{2}}/KT)$ with $\beta = 3.55 \times 10^{24}$ $J/(V/m)^{\frac{1}{2}}$, K being the Boltzmann constant and T the absolute temperature. The dashed curve in Fig. 5 shows how this effect substantially increases the fraction of ionized carriers above 200 V/cm.

3. SELF-CONSISTENT TREATMENT

In search for a first principle approach to determine the capture cross-section assisted by acoustic phonons, we have devised the following MC procedure at equilibrium (i.e. in the absence of an applied electric field).

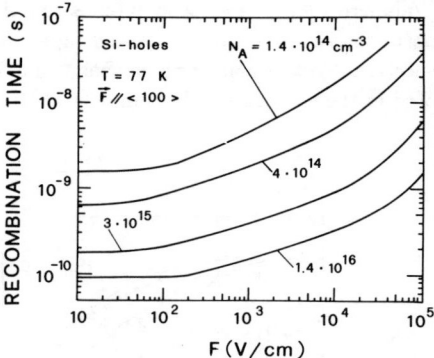

Fig. 4 - Recombination time as a function of the electric field strength for holes in Si at 77 K and for the given acceptor concentration.

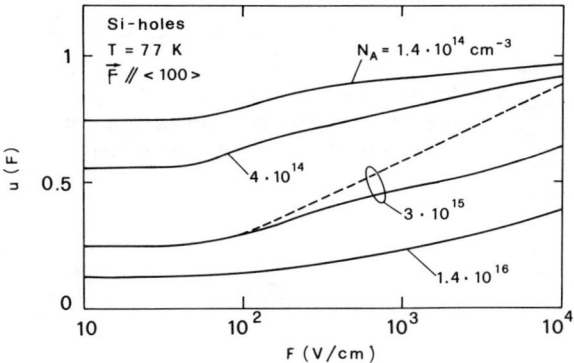

Fig. 5 - Fraction of ionized carriers as a function of the electric field strength for holes in Si at 77 K and for the given acceptor concentration. Dashed curve is calculated by including Poole-Frenkel effect.

By using a total energy scheme, we have accounted for both the kinetic energy of the carrier and the potential energy of the impurities as shown in Fig. 1. Accordingly, the carrier kinetic energy ϵ can be written as:

$$\epsilon = E_0 + E_0^f/x \qquad (1)$$

where E_0 is the carrier total energy (or simply the energy), $E_0^f = \frac{e^2}{4\pi\epsilon_0\epsilon_r}(N_A^-)^{\frac{1}{3}}$ is the fluctuation energy due to the impurity potential, N_A^- being the ionized acceptor concentration, $x = r/r_A$ a dimensionless distance with $r_A = (N_A^-)^{-\frac{1}{3}}$ the mean radius of one impurity. In this scheme, the positive energy region corresponds to free states while the negative energy region to trapped states. The problem is greatly simplified at low temperatures when $KT < E_I^{(1)}$, $E_I^{(1)}$ being the energy of the first excited state, since the differences in energy between the excited states are very small in comparison with both the energy of the ground state $E_I^{(0)}$ and the thermal energy KT. Therefore, the negative energy region may be treated as a continuum spectrum up to, let us say, the first excited level. In this case, the electron distribution function, the density-of-states, and the scattering rates are assumed to depend only on the total energy E_0 (Abakumov and Yassievich 1976).

General Properties and Theory

The general rule to pass from a kinetic energy to a total energy description is to make a space average. Accordingly, for a given function $A(\epsilon)$ the transformation writes:

$$A(E_0) = <A(\epsilon)>_{space} = 3 \int_0^{x_{max}} A(E_0 + E_0^f/x) x^2 dx \qquad (2)$$

where the domain of integration is the volume belonging to one impurity and x_{max} is determined by the constraint $\epsilon \geq 0$, i.e. $x_{max} = 1$ for $E_0 \geq -E_0^f$ and $x_{max} = -\frac{E_0^f}{E_0}$ for $-E_I^{(1)} \leq E_0 \leq -E_0^f$.

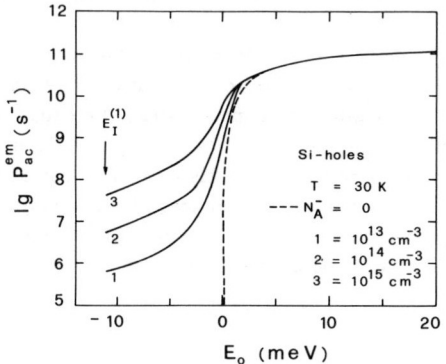 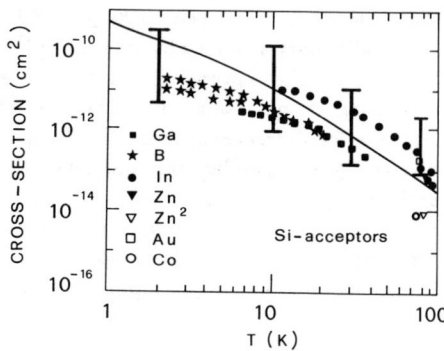

Fig. 6 - Scattering rate for the process of acoustic-phonon emission as a function of the (total) carrier energy for holes in Si at 30 K and for the given acceptor concentration.

Fig. 7 - Holes capture cross-section at equilibrium as a function of temperature. Symbols refer to various acceptors, the line to theoretical calculations of Abakumov et al (1976), the bars to present results where only the values of σ_{1ac} are considered. The width of the bars evidences the dependence with the ionized acceptor concentration.

We have applied the above scheme to the case of p-type silicon at temperatures below 77 K. There, only acoustic scattering enters the problem because of the negligible optical phonon population. Furthermore, being at equilibrium we do not need to follow the momentum of carriers so that only E_0 is important. Therefore, elastic scattering mechanisms, such as with ionized impurities, can be neglected in the simulation.

Figure 6 shows the energy dependence of the scattering rate for acoustic phonon emission. As seen, the presence of the impurities is responsible for a tail of the scattering rate in the negative energy region (the same happens for the case of acoustic phonon absorption). This tail is more pronounced at increasing impurity concentrations. To determine the microscopic capture cross-section we use the following procedure. First we evaluate the recombination rate $1/\tau_r$ from the ratio between the number of transitions which the carrier undergoes in the negative energy region and the time the carrier spends in the valence band. Then, for the band model

assumed we calculate the equilibrium carrier velocity $<v>$ and for the given ionized acceptor concentration obtain finally $\sigma_{ac} = (\tau_r N_A^- <v>)^{-1}$. We remark that in the negative energy region the carrier has two possibilities: (i) to spend some time in the upper excited levels and then come back to the conducting band; (ii) to penetrate far in the negative energy region, that is where $E_0 \leq -KT$. The case (ii) corresponds to the concept of a real capture, and when this occurs we stop the simulation of the given particle and generate a new one in the conducting band in accordance with the actual distribution function in the positive energy region. To analyze the effect of the excited levels on the recombination process we have introduced two capture cross-sections σ_{1ac} and σ_{2ac}. The former accounts for all the transitions to $E_0 \leq 0$ and the latter accounts only for those transitions among the former ones which lead the carrier along the ladder of excited levels to $E_0 \leq -KT$.

Preliminary results obtained for $10^{12} \leq N_A^- \leq 10^{15}\ cm^{-3}$ show the following main features. (i) The capture cross-section depends on the ionized impurity concentration, its value decreasing at increasing concentrations. (ii) For a given impurity concentration the capture cross-section decreases at increasing temperature. (iii) At the higher temperatures $(30 - 77\ K)$ it is $\sigma_{1ac} >> \sigma_{2ac}$ indicating that generation recombination processes mostly occur between the excited levels and the valence band. Figure 7 shows a comparison between present findings and available results in the literature. In view of the uncertainty in the experiments and the still oversimplified model here developed we consider such a comparison satisfactory and in the line to encourage a further development of this approach.

4. CONCLUSIONS

The Monte Carlo simulation has been here presented as a novel numerical method to study the capture cross-section of free carriers at ionized shallow centers. Two main applications have been illustrated leading to a direct comparison of theory with experiments. The satisfactory agreement found gives confidence in predicting a fast development of this technique able to account for further refinements in theory.

This work has been performed on the basis of a Scientific Cooperation Agreement between the Academy of Sciences of the USSR and the National Research Council of Italy. Partial support has been provided by Ministero della Pubblica Istruzione and the Centro di Calcolo of the Modena University.

* Permanent address: Dipartimento di Ingegneria Meccanica, II Universita' di Roma, Italy.

REFERENCES

Abakumov V N and Yassievich I N 1976 Sov. Phys. JETP, **44** 345
Abakumov V N, Perel V I and Yassievich I N 1978 Sov. Phys. JETP, **12** 1
Jacoboni C and Reggiani L 1983 Rev. Mod. Phys. **55** 645
Lax M 1960 Phys. Rev. **119** 1502
Reggiani L, Lugli P and Mitin V 1987 Appl. Phys. Lett. **51** 925
Reggiani L, Lugli P and Mitin V 1988a Phys. Rev. Lett. **60** 736
Reggiani L, Lugli P and Mitin V 1988b presented at the 19th ICPS Conference (Warsaw)
Vaissiere J C 1986 Dissertation University of Montpellier (unpublished)

Electron correlation and disorder effects on the spin susceptibility of doped semiconductors at finite temperature

A Ferreira da Silva

Universidade Federal da Bahia, Instituto de Física, Campus da Federação, 40210 Salvador, Ba., Brazil.

ABSTRACT: We have investigated the spin susceptibility at finite temperature of n-silicon systems. The Gutzwiller variational scheme is used in the calculation. For the sake of comparison, disorder is also introduced in a Hubbard scheme for Si:P. The calculation shows a strongly correlation, temperature and concentration dependence on the susceptibility of n-Si.

1. INTRODUCTION

The investigation of spin susceptibility χ_s, as a function of temperature T and donor concentration N_D, has recently attracted much attention around the metal-nonmetal (MNM) transition critical point N_c in Si:P (Ikehata and Kobayashi 1985, Holcomb 1986, Paalanen et al 1987). One observes that for $10^{18} < N_D < 10^{19}$ cm^{-3}, χ_s shows a strong dependence on T and correlation effect. For $N_D > 10^{19}$ cm^{-3} the T dependence of χ_s slowly disappears and the system undergoes a Pauli susceptibility. For $N_D < 10^{18}$ cm^{-3} the system shows a Curie susceptibility behaviour. The purpose of this work is two fold: i) To calculate χ_s for different n-doped Si systems by means of the Chao and Berggren (1975, 1977) scheme, hereafter denoted CB, which is an extension of the Gutzwiller (1965) method for finite T. As a by-product, it is calculated the values of N_c. ii) To investigate the effects of disorder and correlation on χ_s for Si:P, by a Hubbard-type approach (Ferreira da Silva et al 1981, Kishore et al 1982, Ferreira da Silva and Fabbri 1987).

2. THE APPLICABILITY OF GUTZWILLER'S APPROACH

Gutzwiller (1965) constructed a trial wavefunction for a highly correlated metallic system, by starting with the conventional Bloch state for noninteracting electrons and reducing the amplitude of configurations in which atoms are doubly occupied. Later on, Brinkman and Rice (1970), based in his method derived an expression for χ_s at T=0K. In the wake of the later scheme Berggren (1974) calculated χ_s for Si:P and found a reasonable agreement with the experiment. CB have found

$$\chi_s(T) = \mu_\chi(T)\chi_0(T), \qquad (1)$$

where $\mu_\chi(T)$ is an enhancement factor and $\chi_0(T)$ is the Pauli spin susceptibility. In equation (1) the electron hopping energy t and the intra-atomic Coulomb interection (or correlation) energy U will appear playing an essential role.

© 1989 IOP Publishing Ltd

Using an anisotropic wavefunction to estimate t and adjusting U in order to get the experimental N_c, CB shown clearly the T and N_D dependence on χ_s. Their results are in qualitative agreement with Quirt and Marko (1972, 1973) data. Here, we performed the calculation for χ_s (T=0,1.1 and 77K), in the way of CB, but for the experimental U≈0.96 E_D (E_D being the ionization energy of the system considered) and the many-valley isotropic wavefunction with a variational N_D-dependence (Ferreira da Silva 1986, 1987, 1988).

In Figure 1 we show the results of χ_s for different systems. We found good agreement with the data for T=0,1.1 and 77K in Si:P. From the metallic side, the MNM transition takes place when the number of doubly occupied sites is zero. We found
N_c^a (Si:Sb) = 3.0×10^{18} cm^{-3},
N_c^b (Si:P) = 3.5×10^{18} cm^{-3} and
N_c^d (Si:As) = 6.0×10^{18} cm^{-3}. The experimental values are 3.0, 3.7 and 8.5×10^{18}cm^{-3} respectively (Ferreira da Silva 1988). In the calculation the primary dependence of χ_s turns out to be t and U. For Si:P,As such t is derived as function of the different E_D's present in the system and scaled to E_D of Si:P, with the proper experimental U (Ferreira da Silva 1986, 1987). To this context, χ_s is calculated and gives N_c^c(Si:P,As)= 4.5×10^{18}cm^{-3} (Newman and Holcomb 1983). In the inset of Figure 1 we show the result of χ_s for T=1.1K and the experimental values + (Quirt and Marko 1972, 1973), (Paalanen 1987) and o (Ikehata and Kobayashi 1985).

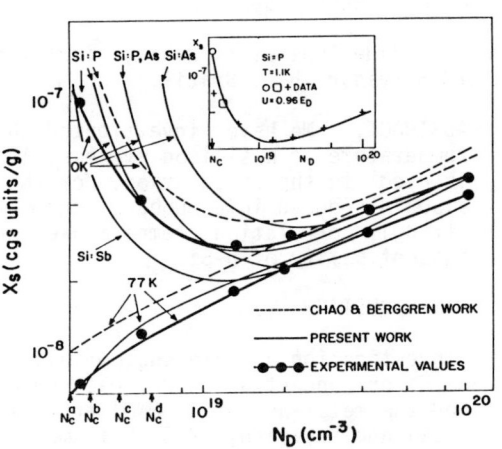

Fig. 1. χ_s of n-systems as a function of N_D. At 77K χ_s is for Si:P.

3. THE INFLUENCE OF DISORDER

The effect of disorder is introduced in the way of a recent theory, described by the Hubbard model for doped semiconductors (Ferreira da Silva et al 1981, Kishore et al 1982, Ferreira da Silva and Fabbri 1987). The χ_s is derived as

$$\chi_s(T) = \lim_{H \to 0} N_D \mu_B \left| m_0(T=0) + m_1 T^2 \right|, \qquad (2)$$

where H is a magnetic field, μ_B is the Bohr magneton and m_0 and m_1 come from the low-temperature expansion of the magnetic moment. The results of χ_s at T=0K, for different U's in Si:P are shown in Figure 2 as curve 1 to 5. It is used the Bohr radius a_0^*=15.2 A. For the sake of comparison we also show the theoretical results of Berggren (1974), for a_0^*=13.5 A, as open circles with error bars and curves a and b for U=1.25 and U≈0.95E_D respectively. The result for the highly correlated electron gas (HCEG)

model (Berggren 1978, Sernelius and Berggren 1981), valid for $N_D > N_C$, is presented as curve 6. The full circles are the experimental data (Quirt and Marko 1972, 1973). It is worthwhile to point out that, the calculation for the disordered system was done neglecting completely the many-valley effects of the host material. We hope that such effects may shift the curves 1-5 to 1'-5', corresponding to the Gutzwiller scheme, i.e., disorder will play a minor role in the calculation.

Considering finite T in equation (2) we find the results of χ_S in Figure 3. The open and full circles are the data of Ue and Maekawa (1971). We note the right behaviour of the calculated χ_S. The discrepancy of these data compared to those of Quirt and Marko still is controvertial (Holcomb 1986).

4. SUMMARY

In this work we have shown, in two ways, that the effects of correlation and temperature play a major role in the enhancement of χ_S for n-Si. We expect that such results may motivate experiments in n-Si systems and in the future a more thorough treatment along the both schemes with a proper many-valley correction.

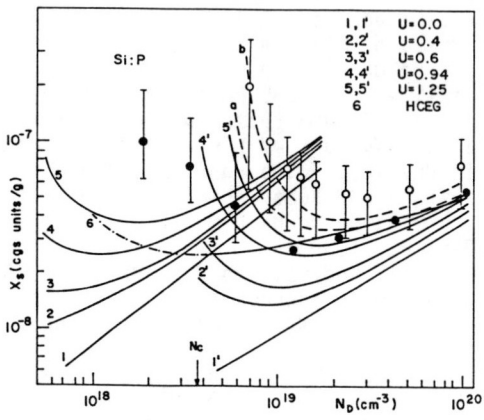

Fig. 2. χ_S of Si:P as a function of N_D for different calculations and various values of U.

ACKNOWLEDGMENTS

The author would like to thank Prof K.-F Berggren for valuable discussions. Financial support is gratefully acknowledged from the Swedish Institute and CAPES (Coordenação de Aperfeiçoamento de Pessoal de Nível Superior - Brazil).

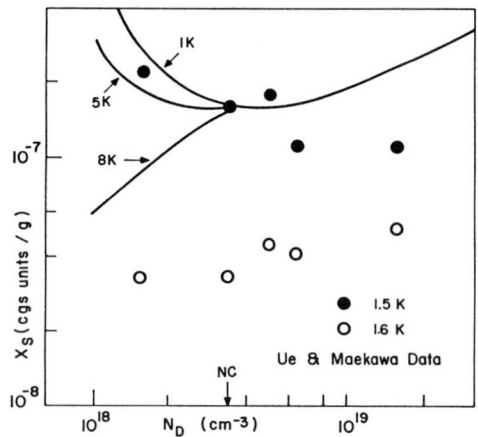

Fig. 3. χ_S of Si:P as a function of N_D. Results from equation (2).

REFERENCES

Berggren K -F 1974 Philos. Mag. 30 1
Berggren K -F 1978 Phys. Rev. B17 2631
Brinkman W F and Rice T M 1970 Phys. Rev. B2 4302
Chao K A and Berggren K -F 1975 Phys. Rev. Lett. 34 880
Chao K A and Berggren K -F 1977 Phys. Rev. B15 1656
Ferreira da Silva A, Kishore R and da Cunha Lima I C 1981 Phys. Rev. B23 4035
Ferreira da Silva A and Fabbri M 1987 (unpublished)
Ferreira da Silva A 1986 Phys. Scr. T14 27
Ferreira da Silva A 1987 Phys. Rev. Lett. 59 1263
Ferreira da Silva A 1988 Phys. Rev. B37 4799
Gutzwiller M C 1965 Phys. Rev. 137 A1726
Holcomb D F 1986 *Localization and Interaction in Desorder Metals and Doped Semiconductors* ed D M Finlayson (Scottish Universities Summer School in Physics, Edinburgh) pp 313-42
Ikehata S and Kobayashi S 1985 Solid State Commun. 56 607
Kishore R, da Cunha Lima I C, Fabbri M and Ferreira da Silva A 1982 Phys. Rev. B26 1038
Newman P F and Holcomb D F 1983 Phys. Rev. Lett. 51 2144
Paalanen M A, Sachdev S and Bhatt R N 1987 *Proc. 18th Int. Conf. on the Physics of Semiconductors* ed O Engstrom World-Scientific, Singapore) pp 1249-52
Quirt J D and Marko J R 1972 Phys. Rev. B5 716
Quirt J D and Marko J R 1973 Phys. Rev. B7 3842
Sernelius B E and Berggren K -F 1981 Philos. Mag. B43 115
Ue H and Maekawa S 1971 Phys. Rev. B3 4232

Dynamical resistivity from shallow-impurity scattering in polar semiconductor films, used as optical coatings

B E Sernelius and M Morling

Department of Physics and Measurement Technology, Linköping University,
S-581 83 Linköping, Sweden.

ABSTRACT: We present numerical results for the dynamical resistivity for Al doped ZnO and Sn doped In_2O_3. The calculation is based on a recent derivation (Sernelius 1987) of the dynamical conductivity in heavily doped, polar semiconductors. Special emphasis is put on the modifications caused by the polar character of these semiconductors.

1. INTRODUCTION

Heavily shallow-impurity doped large-band-gap semiconductors have very interesting optical properties. In some situations they behave as metals, in others as semiconductors, sometimes as neither metals nor semiconductors and in some applications one can benefit from the best from both the metal- and semiconductor behaviors. The work we present here is related to one of such applications, viz. tailoring of optical coatings on highly energy-efficient windows.

Pure large-band-gap semiconductors are transparent for photon energies in the visible region and below. The band gap prevents the carriers from being excited by the impinging photons for photon energies less than the band-gap value. With heavy doping the following happens. The dopant atoms are ionized and the released carriers occupy states at the bottom of the conduction band (at the top of the valence band) in the case of donors (in the case of acceptors). These carriers give metallic properties to the system since no band gap prevents them from having low-energy excitations. A metal can in a simplified way be described as being totally reflecting for energies below the plasmon energy and totally transmitting above. The plasmon energy for the dopant carriers varies with doping concentration but is typically in the range of 0.1-1 eV, while the band-gap for the systems considered here are in the range of 3 -5 eV. Thus, a coating from a heavily doped semiconductor has an optical window ranging basically from the extrinsic plasmon energy up to the band gap value of the semiconductor. The positions of the boundaries of this optical window can be tuned, to give the desired optical properties, by adjustment of the doping concentration.

There are however two other effects from the doping. One is that the optical band gap is changed. The change depends on two competing effects. Many-body effects tend to reduce the gap. This reduction is counteracted by a band-gap widening due to blocking of the lowest states in the conduction band (for n-type doping). This widening is known as the Burstein-Moss effect. In the extreme high doping limit, the Burstein-Moss shift wins over the band-gap narrowing from many-body effects and the gap increases. This moves the upper boundary for the optical window upwards. We have studied this effect earlier for the two

© 1989 IOP Publishing Ltd

systems considered here (Hamberg *et al* 1984, Sernelius *et al* 1988).

The second other effect comes from the presence of the dopant-ion potentials. These potentials are not, like in an ordinary metal, forming a perfect lattice structure. They are randomly distributed. This gives rise to free-carrier absorption which modifies the optical properties near the lower boundary of the optical window. This effect is what we will concentrate on in this work and in particular the modifications due to the polar character of the two semiconductors under consideration.

In Section 2 we very briefly describe the derivation of the expression for the dynamical resistivity in polar semiconductors and give a feeling for the complications from the polar character. The numerical results are presented in Section 3. Finally, in Section 4 we give a brief summary.

2. DYNAMICAL RESISTIVITY IN A POLAR SEMICONDUCTOR

The expression for the dynamic resistivity from impurity scattering in non-polar semiconductors has been derived in several different ways, in the literature, with the following, same result:

$$\rho = -i\frac{m\Omega}{ne^2} + i\frac{2}{3\pi n\Omega}\int_0^\infty dq\, q^2 \left[\frac{1}{\varepsilon_T(q,\Omega)} - \frac{1}{\varepsilon_T(q,0)}\right] \quad (1)$$

where ε_T is the total dielectric function including the background dielectric constant and the polarizability from the dopant carriers. The zero frequency limit of this expression reproduces the Ziman result for the static resistivity. In the derivation of Equation 1 it has been assumed that the dopant ions are randomly distributed, that their concentration is the same as the carrier concentration, n, and that the ion potentials can be approximated by pure Coulomb potentials.

When the expression is generalized to the case of polar semiconductors it is easy to fall into a trap. Intuitively one might think that the thing to do is just to add the polarizability from the phonons to ε_T, the dynamical version in $\varepsilon_T(q,\Omega)$ and the static one in $\varepsilon_T(q,0)$. That the problem is more complicated than that is maybe most easily seen in the derivation with the so-called energy-loss method (see the recent review article by Gerlach (1986)). In this method one makes a transformation into a system in which the electrons are at rest and the ions oscillate with frequency Ω. The electrons respond to the oscillating potentials and the frequency dependent screening from the electrons enters the expressions. The polarizability from the phonons comes from the displacement polarization from the host atoms. If the atoms were at rest in the transformed system there would be no problem, but they are not. This means that there is no trivial way to generalize the expression in Equation 1.

We used (Sernelius 1987) the Kubo formalism and diagrammatic perturbation theory to calculate the high frequency expression for the dynamical conductivity. If this result is interpreted as the high frequency limit of the generalized Drude expression

$$\sigma = \frac{ne^2}{m}\left[\frac{1}{1/\tau(\Omega) - i\Omega}\right] \qquad (2)$$

with a frequency dependent relaxation time, τ, the expression for τ can be identified. Under the assumption that this expression for τ is generally valid one gets an expression for the conductivity that is valid for the full frequency range. The resistivity is then obtained as $1/\sigma$ and is found to be

$$\rho = -i\frac{m\Omega}{ne^2} - i\frac{2}{3\pi n\Omega}\int_0^\infty dq\, q^2 \frac{\left[\alpha(q,\Omega) - \alpha(q,0)\right]\left[\varepsilon_L(\Omega) + \alpha(q,0)\right]}{\varepsilon_T^2(q,0)\,\varepsilon_T(q,\Omega)} \qquad (3)$$

where ε_T now contains the polarizability from the phonons, ε_L is the lattice dielectric function and α the polarizability from the carriers. For frequency independent ε_L and phonon polarizability Equation 3 reduces into the expression for a non-polar semiconductor in Equation 1. Equation 3 is the basic relation for this work. The numerical results are presented in next section. The optical properties can then be obtained in the standard way from the total, transverse dielectric function for the system, $\varepsilon_\perp$, which is related to ρ according to:

$$\varepsilon_\perp(\Omega) = \varepsilon_L(\Omega) + \frac{4\pi i\sigma(\Omega)}{\Omega} = \varepsilon_L(\Omega) + \frac{4\pi i}{\Omega\rho(\Omega)} \qquad (4)$$

3. NUMERICAL RESULTS

In this section we present some numerical results for the dynamical resistivity, from Equation 3, for Al doped ZnO and Sn doped In_2O_3 (ITO). We show only the real part, ρ_1, here since the imaginary part is dominated by the linear contribution from the first term of Equation 3. More information about the numerical calculation and further numerical results, not presented here, can be found elsewhere (Morling 1988).

Figure 1 displays a comparison between the full calculation and two approximate ones, of ρ_1, for Sn doped In_2O_3 at the doping concentration $8.0 \cdot 10^{20}$ cm^{-3}. The result from the full calculation, curve b, is almost constant for energies up to the plasmon energy, which for this doping level is near 1 eV. A structure, associated with the optical phonon energy, is found at the low-energy side of the figure. There is an increase in ρ_1, associated with plasmon excitations, starting at the plasmon energy, followed by a steep and continuing decrease. This decrease means that the free-carrier absorption is only important near the lower boundary of the optical window. The two remaining curves, a and c, are the results when the lattice dielectric function has been approximated by a constant. In curve a its low-frequency limit, $\varepsilon_0 = 9$, has been chosen and in c the high-frequency limit, $\varepsilon_\infty = 4$. As was mentioned in the

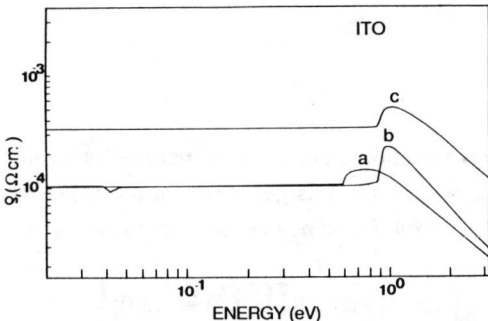

Fig. 1 Real part of the dynamical resistivity of Sn doped In_2O_3 for the doping concentration $8.0 \cdot 10^{20} cm^{-3}$ as a function of energy. Curve b shows the result of the full calculation while curves a and c are results from approximations described in the text.

previous section the results represented by the curves a and c can be obtained from using the simple expression in Equation 1. One finds from the comparison between the curves that curve a is almost identical to curve b for lower frequencies apart from the structure in curve b, absent in curve a, around the phonon frequency. The deviation for higher frequencies comes from that the plasmon frequency is lowered for curve a. Curve c, on the other hand, gives too high values for all frequencies, but has the plasmon frequency in the right position. From this one can conclude that curve a represents a good approximation for the low energy side while curve c places the hump caused by plasmon excitations in the correct energy range. Had one generalized Equation 1 the way described in the beginning of the previous section one would have obtained a result that approached curve a for low frequencies, increased in a range bracketing the phonon energy, and approached curve c for higher frequencies. That behavior is not found experimentally. The experimental results show no drop in ρ_1 when going from energies above to energies below the phonon energy (Hamberg 1984). This supports our derivation.

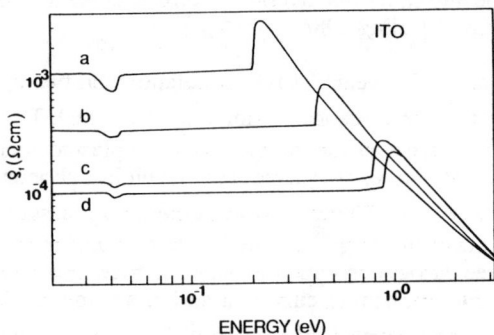

Fig 2. Real part of the dynamical resistivity from the full calculation for Sn doped In_2O_3. Curves a, b, c and d are for the doping densities 0.4, 1.7, 6.2 and $8.0 \cdot 10^{20}$ cm^{-3}, respectively.

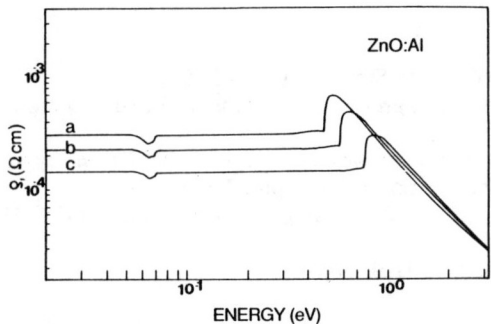

Fig. 3 Same as Figure 2 but now for Al doped ZnO. Curves a, b and c are for the doping densities 1.8, 2.6 and $4.5 \cdot 10^{20}$ cm^{-3}, respectively.

Figure 2 gives the result from the full calculation for ITO for the doping levels 0.4, 1.7, 6.2 and $8.0 \cdot 10^{20}$ cm^{-3}, respectively. The curves all have the same behavior except from that the low-frequency limit decreases with doping level and the position of the plasmon structure moves towards higher energy values.

Figure 3 is similar to Figure 2 and gives the result for ZnO:Al for the doping levels 1.8, 2.6 and $4.5 \cdot 10^{20}$ cm^{-3}, respectively.

In the calculations for In$_2$O$_3$ we have used the values 9.0, 4.0, 46 meV, and 0.35 for ε_0, ε_∞, the longitudinal optical frequency and the electron effective mass, respectively. The corresponding values used for ZnO are 8.0, 3.85, 72 meV and 0.28, respectively.

4. SUMMARY AND CONCLUSION.

We have presented numerical results for the dynamical resistivity and optical properties of optical coatings from heavily doped indium- and zinc oxides. These large band-gap materials are polar and the complications from their polar character, in particular, have been investigated. We found that the results from the expression for the dynamical resistivity, correctly generalized to take the polar character into account, deviated considerably from the corresponding results from a simpler expression, arrived at through an intuitive gerneralization.

5. ACKNOWLEDGMENTS

Research support is acknowledged from the Swedish Natural Science Research Council.

6. REFERENCES

Gerlach E 1986 *J. Phys. C: Solid State Phys.* **19** 4585
Hamberg I, Granqvist C G, Berggren K F, Sernelius B E and Engström L 1984 *Phys. Rev.* B **30** 3240
Hamberg I 1984 *Doctoral thesis at Chalmers University of Technology* (unpublished)
Morling M 1988 unpublished report at Linköping University.
Sernelius B E, Berggren K F, Jin Z C, Hamberg I and Granqvist C G 1988 *Phys. Rev.* B **37** 10244
Sernelius B E 1987 *Phys. Rev.* B **36** 1080

Luminescence enhancement by impurity-impurity interaction in heavily iodine doped AgBr

A.Testa†, W.Czaja‡, A.Quattropani†, P.Schwendimann†
†Institut de Physique Théorique, ‡Institut de Physique Appliquée, Ecole Polytechnique Fédérale de Lausanne, CH-1015 Lausanne, Switzerland

ABSTRACT: We show experimentally that a consequence of impurity-impurity interaction in the system AgBr : I is an enhanced intensity in the LO phonon replicas. This intensity enhancement may be described by a concentration dependent Huang-Rhys factor. This property of the Huang-Rhys factor can be theoretically understood if electron-phonon coupling constants dependent on the electronic state of the impurity are introduced. The choice of the coupling constants is to be justified by model calculations.

The properties of the isolated iodine impurity in AgBr are rather well known, Testa et al. (1987), Testa et al. (1988a). In discussing effects related to impurity - impurity interaction we limit ourselves to the case of weak excitation, i.e. to the case where the interaction between an excited iodine impurity and non excited iodine atoms is dominant. The effect of this interaction on the zero phonon emission has been studied in detail in the paper by Czaja (1983). In Fig. 1, we present experimental data showing a further effect due to the interaction between impurities, an enhancement of the luminescence intensity of the LO phonon replica.
Thus, we have :

$$I_n = I_n(N, N_{ex}) \tag{1}$$

where N is the impurity concentration and $N_{ex} \leq N$ is the concentration of the excited impurities. The dependence of I_n on N_{ex} has been studied elsewhere (Testa et al. 1988b) and it has been shown to be related to the dielectric screening of excited impurities. At a particular value of N, the enhancement of the intensity of LO phonon replicas can be described as a doping dependent Huang-Rhys factor S, as shown in Fig. 2.
We thus have :

$$S(N, N_{ex}) = S(0,0)(1 + \Delta(N, N_{ex})) \tag{2}$$

which reproduces the experimental values as

$$\frac{I_n(N, N_{ex})}{I_0(N, N_{ex})} = \frac{S(0,0)^n}{n!}(1 + n\Delta(N, N_{ex})) \tag{3}$$

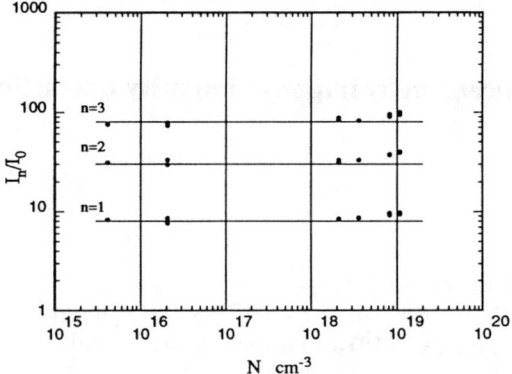

Figure 1: Intensities of the first three phonon replicas normalized to the zero phonon intensity. The horizontal solid lines show the low concentration values corresponding to S=7.8; errors are smaller than the dot size.

with the numerical value from experiment

$$\Delta(N = 10^{19} \text{cm}^{-3}, N_{ex} \sim 0) \simeq 0.1 \ . \tag{4}$$

We stress that this description implies that up to $N = 10^{19}\text{cm}^{-3}$ the spectra have maintained the character of those observed at much lower concentrations. In fact, this has been shown to be the case for luminescence spectra by Czaja (1983) and for absorption spectra by Magne et al. (1981).

The results mentioned above can be represented theoretically as will be shown below. We describe AgBr doped with iodine by the following Hamiltonian

$$\mathbf{H} = \sum_{i,k} E_i^k c_i^{k\dagger} c_i^k + \sum_{q,j,i,k} \hbar\omega_j(\mathbf{q}) \mathbf{a}_{\mathbf{q},j}^\dagger \mathbf{a}_{\mathbf{q},j} c_i^{k\dagger} c_i^k + \\ \sum_{q,j,i,k} h_i^k(\mathbf{q})(\mathbf{a}_{\mathbf{q},j}^\dagger + \mathbf{a}_{\mathbf{q},j}) c_i^{k\dagger} c_i^k \tag{5}$$

c_i^k are fermion operators describing the electron of the atom i in the state k, whereas $a_{q,j}$ are boson operators for phonons of the branch j with momentum q. The phonon energy may in principle depend on the electronic state k resulting in different parabola's in the configuration coordinate model. We will here consider only a state independent phonon energy as it has been shown that the spectrum can be well described under this assumption, Testa et al. (1987). For the excitation energies used in our experiments, only two electronic levels are relevant and we will restrain ourselves to k equal to the ground or first excited state.

Taking into account only LO-phonons in Eq. (5) (j = LO-phonons) we consider the interaction between an excited impurity and a non excited neighbouring impurity, which have only two levels with energies E_{gr} and E_{ex}. We create a new basis

$$\psi_1 = |\downarrow\downarrow>, \quad \psi_2 = |\downarrow\uparrow>, \quad \psi_3 = |\uparrow\downarrow>, \quad \psi_4 = |\uparrow\uparrow> \tag{6}$$

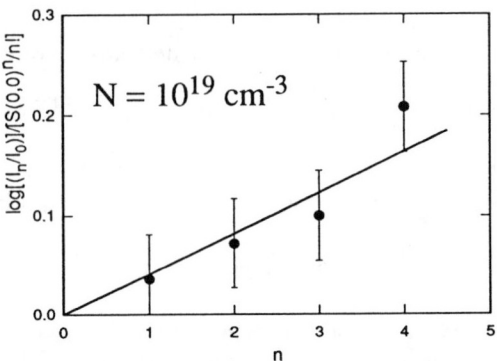

Figure 2: Ratio of the intensities of the nth phonon replica over the zero phonon intensity, the linearity can be explained by a concentration dependent S factor (see text).

where ↑ stands for an impurity in an excited state and ↓ is the symbol for an impurity in the ground state; we then define the projector $P_i = |\psi_i\rangle\langle\psi_i|$. After a suitable unitary transformation which renders the new Hamiltonian diagonal in the new basis, we arrive at

$$\mathbf{H}' = \sum_{i,j}\left[E'_i - \frac{|\alpha_i|^2}{2\hbar\omega_j(\mathbf{q})}\right]P_i + \sum_{\mathbf{q},j,i} 2\hbar\omega_j(\mathbf{q})\mathbf{b}^{i\dagger}_{\mathbf{q},j}\mathbf{b}^{i}_{\mathbf{q},j}P_i \qquad (7)$$

where as usual we have

$$\mathbf{b}^{i}_{\mathbf{q},j} = \mathbf{a}_{\mathbf{q},j} - \alpha_i \qquad (8)$$

we notice that the hamiltonian $\mathbf{H}$ is diagonal in the electronic states ψ_i and degenerate in ψ_2, ψ_3 as $\alpha_2 = \alpha_3$ We now calculate optical transitions between the eigenstates of $\mathbf{H}'$ having as electronic wavefunction ψ_1 and $\frac{1}{\sqrt{2}}\psi_2 + \frac{1}{\sqrt{2}}\psi_3$ and introduce the definitions

$$E'_1 = 2E_{gr}, \quad E'_2 = E_{ex} + E_{gr}, \quad E'_3 = E_{ex} + E_{gr}, \quad E'_4 = 2E_{ex} \qquad (9)$$

with coupling constants independent of the electronic state

$$\alpha_1 = h^1_{gr} + h^2_{gr}, \quad \alpha_2 = h^1_{ex} + h^2_{gr}, \quad \alpha_3 = h^1_{gr} + h^2_{ex}, \quad \alpha_4 = h^1_{ex} + h^2_{ex}. \qquad (10)$$

By using the coupling constant as derived from the Fröhlich Hamiltonian (see Mahan (1981) for example)

$$h^k_i(\mathbf{q}) = F^k(\mathbf{q})e^{i\mathbf{q}\mathbf{r}_i} \qquad (11)$$

it is shown that the intensity I_n of the nth LO-phonon replica has the following functional dependence on n

$$I_n(0) \sim \frac{S(0,0)^n}{n!}[1 + f^n(\mathbf{r}_1 - \mathbf{r}_2)]. \qquad (12)$$

Eq. 12 is at variance with the experimental result.
If we introduce instead of Eq. 10, however, a state k dependent coupling.

$$\beta_1 = h_{gr}^1 + h_{gr}^2, \quad \beta_2 = h_{ex}^1 + \bar{h}_{gr}^2, \quad \beta_3 = \bar{h}_{gr}^1 + h_{ex}^2, \quad \beta_4 = \bar{h}_{ex}^1 + \bar{h}_{ex}^2, \tag{13}$$

in the sense that the presence of an excited neighbour changes the coupling constant in the ground state, we obtain qualitative agreement with the experimental data

$$I_n(0) \sim \frac{S(0,0)^n}{n!}[1 + n\,g(\mathbf{r}_1 - \mathbf{r}_2)]. \tag{14}$$

This generalization (eq. 10) is one possible choice and not necessary the most convenient one. It is suggested by a model (Stoneham and Testa 1988) which gives qualitatively the proper results but not quantitatively. A justification of the particular choice of the coupling Eq. 13 based upon model calculations will be reported elsewhere.

Acknowledgement We acknowledge gratefully many discussions with A.M. Stoneham (Harwell, UK) and A. Khater (Université P. et M. Curie, Paris). This work has been supported in part by the Swiss National Science Foundation.

References

Czaja W 1983 *J.Phys.C: Solid State Phys.* **16** 3197.

Magne G, Czaja W and Hediger H 1981 *Helv.Phys.Acta* **54** 252.

Mahan 1981 G *Many-Particle Physics* Plenum Press, New York.

Stoneham A M and Testa A (1988) in preparation.

Testa A, Czaja W, Quattropani A and Schwendimann P 1987 *J.Phys.C: Solid State Phys.* **20** 1253.

Testa A, Czaja W, Quattropani A and Schwendimann P 1988a *J.Phys.C: Solid State Phys.* **21** 2189.

Testa A, Czaja W, Quattropani A and Schwendimann P 1988b submitted to *J.Phys.C: Solid State Phys.*.

Nitrogen passivation of shallow levels in amorphous silicon

L Martín-Moreno and J A Vergés

Departamento de Física de la Materia Condensada (C-12), Universidad Autónoma de Madrid, 28049 Madrid, Spain

E San-Fabián

Departamento de Química-Física, Universidad de Alicante, 03080 Alicante, Spain

E Louis

Departamento de Física Aplicada, Universidad de Alicante, 03080 Alicante, Spain

ABSTRACT: The electronic structure of isolated N impurities in amorphous Si is studied within the cluster Bethe lattice scheme using a set of tight-binding parameters fitted to the molecular levels of two molecules containing Si-N bonds. While N in a Si tetrahedral environment places an electron in the conduction band, planar threefold-coordinated N produces a doubly-occupied level at $E_v + 0.55$eV. The band originated by this level shrinks the gap for initial stages of a-Si alloying with N.

1. INTRODUCTION

While the role played by standard donor or acceptor impurities like P, As or B in a-Si is more or less well-known, much less is known about impurities like C, O or N that form strong bonds with Si atoms. The purpose of this contribution is the study of the electronic structure of N in the most probable configurations, i.e., planar following the 8-N rule (threefold coordinated) and substitutional (fourfold coordinated).

In order to obtain reliable tight-binding (TB) parameters for the Si-N bond, we have studied two molecules showing the bonding configuration of amorphous silicon nitride. They are $Si(NH_2)_4$ in which Si occupies the center of the molecule surrounded by four NH_2 groups in tetrahedral configuration, and $N(SiH_3)_3$ in which N occupies the center of a triangle formed by Si atoms. In both cases H is used to terminate the molecule preserving in any case the bonding environment of Si (bonds in tetrahedral directions) and N (three bonds in a plane) atoms. Standard unrestricted Hartree-Fock (HF) total energy calculations have been performed for both molecules to optimize dihedral angles whereas bond lenghts have been hold fixed to allow for a proper definition of TB parameters within the two-center approximation (Slater and Koster 1954). Although empty HF levels are also given by the calculation, we do not include them into the fit because typical HF optical gaps are about twice the experimental ones. Due to the relevance of empty levels in the determination of the

semiconductor gap, the first empty *level* of the molecule has been obtained as the difference between the total energy of the molecule in its first electronically excited state (triplet) and the energy of the ground state. In this manner, a correct *molecular optical gap* becomes defined. Ammonia (NH_3) and silane (SiH_4) HF results have been added to the quantum chemistry basedata in order to make well-defined the TB parameters corresponding to N-H and Si-H bonds.

TB parameters for Si, N and H orbital levels and for Si-N, N-H and Si-H interactions have been obtained through a fit to the quantum-chemistry results. Our optimisation algorithm is the following. Firstly, the eigenvalues of NH_3, SiH_4, $Si(NH_2)_4$ and $N(SiH_3)_3$ molecules are calculated by diagonalization of the corresponding TB Hamiltonian matrices constructed for a fixed set of TB parameters. Atomic populations of the molecules are obtained using the standard TB definition. Secondly, the root mean square (r.m.s.) deviation of the TB eigenvalues and charges relative to the HF levels and charges is defined by:

$$\Delta E_{r.m.s.} = \sum_i (E_i^{HF} - E_i^{TB})^2 + \sum_j (q_j^{HF} - q_j^{TB})^2 ,$$

where i runs over all molecular levels of the four molecules included in the fit and j runs over all Si and N atoms. Finally, the subroutine package MINUIT is employed to minimize ΔE_{rms} as a function of the set of nearest-neighbor TB parameters. Sixteen parameters have been optimized with the final result given in Table I.

TABLE I

Slater-Koster tight-binding parameters in eV. $E_{p^*}(N)$ gives the 2p(N) level of nonbonding orbitals, i.e., the orbitals corresponding to *lone pair* electrons both in molecules and amorphous silicon nitride.

SK-parameter	Si-Si	Si-N	Si-H	N-H
$ss\sigma$	-2.075	-4.321	-4.783	-7.203
$sp\sigma$	2.508	3.443	-	-
$ps\sigma$	2.508	5.951	4.466	6.340
$pp\sigma$	2.716	3.385	-	-
$pp\pi$	-0.715	-2.082	-	-
$s^*p\sigma$	2.327	0.699	-	-
$ps^*\sigma$	2.327	-	-	-

Orbital level	Si	N	H
E_s	-12.721	-21.978	-9.083
E_p	-6.806	-9.428	-
E_{p^*}	-	-10.090	-
E_{s^*}	-1.834	-	-

Si-Si parameters of Table I correspond to the values of Vogl et al. (1983). The minimum r.m.s. deviation is 0.31eV for 47 fitted levels, a result that we find quite satisfactory. More details of our fit can be found elsewhere (San-Fabián et al.) together with a thorough comparison of our set of tight-binding parameters with previous sets. Good accord with experiment (Kärcher et al. 1984) and *ab initio*

calculations of crystalline silicon nitride (Ren and Ching 1981) is achieved when the parameters of Table I are used in conjunction with effective medium theories (Louis and Vergés 1986, Gómez-Santos and Vergés 1987 and Martín-Moreno et al. 1987) that allow the calculation of the electronic structure of a-$Si_{1-x}N_x$ for arbitrary alloy compositions.

2. N IMPURITY IN a-SI: PLANAR CONFIGURATION

We have studied the electronic structure of an isolated N atom embedded in pure amorphous silicon choosing the most probable geometries. The first one corresponds to standard impurities following the 8-N rule, i.e., forming as many bonds as it valence number. In this case, N is saturated by three Si Bethe lattices starting with Si atoms bonded with N in a planar trigonal configuration. We work within the cluster Bethe lattice approach (Yndurain et al. 1974) in its simpler version. This should be the most frequent environment of N. The second configuration corresponds to substitutional N. That configuration is much less probable than the first one. Nevertheless, its presence at small N concentration (below 5%) has been detected and characterized in a recent paper by Singh and Budhani(1987) that reanalyses very carefully old results found by other authors. We postpone the study of N in this configuration to the next Section, reporting now our results for isolated threefold N.

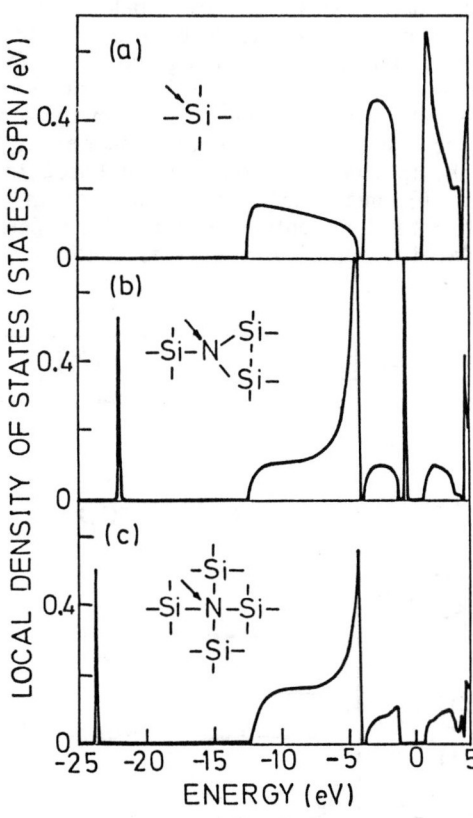

Fig. 1. (a) Density of levels on a Si atom belonging to an ideal Si Bethe lattice. (b) Density of levels on an isolated N impurity saturated by three Si Bethe lattices in a planar trigonal configuration. (c) Same as before but with N saturated by four Si Bethe lattices in a tetrahedral configuration.

The density of levels (DOS) at the N impurity site is given in Figure 1b. Results are shown after an overall shift of the energy scale that locates the silicon dangling bond defect level at $E = 0$. Two localized levels appear near N. The first one is located about -9.5eV below the bottom of the Si valence band. Its character is mainly 2s(N) and its appearance would have been predicted by simple inspection of the molecular levels of both large molecules, $Si(NH_2)_4$ and $N(SiH_3)_3$. Levels formed by 2s(N) orbitals appear about -22.0eV below the Si sp³-hybrid in both cases. The second level is due to a more subtle reason. A planar configuration around N always implies the existence of a non-bonded 2p(N) orbital. The energy of this orbital falls in the middle of a region

of the a-Si spectrum having a large DOS. This region at the top of the Si valence band is mainly formed by 3p(Si) levels. When the N-Si interaction is switched on, a kind of splitting between levels occurs. As a result, a resonance appears below the 3p-part of the valence band and a localized level is produced within the semiconducting gap at $E_v(Si)+0.55eV$. This level is initially occupied by two electrons (T=0K). Therefore, this defect center is deep and diamagnetic.

Both localized levels develop into bands when the N content is increased. Whereas the appearance of a new band below the Si valence band is well stated, the existence of a band **above** the valence band is quite surprising at a first sight. The evolution of this band with N alloying will close the semiconducting gap. If this band would reduce the energy of the optical absorption edge, we were forced to recognize some contradiction with experiments. On the other hand, some experimental facts point to the possibility of existence of such a level/band for minute N contents: (i) The tails in the optical absorption coefficient seem to increase when N is added to the sample, and, (ii) some increases in conductivity and/or photoconductivity have been reported for small N contents. More detailed experimental studies of the electronic levels at isolated N impurities are necessary to solve this apparent paradox.

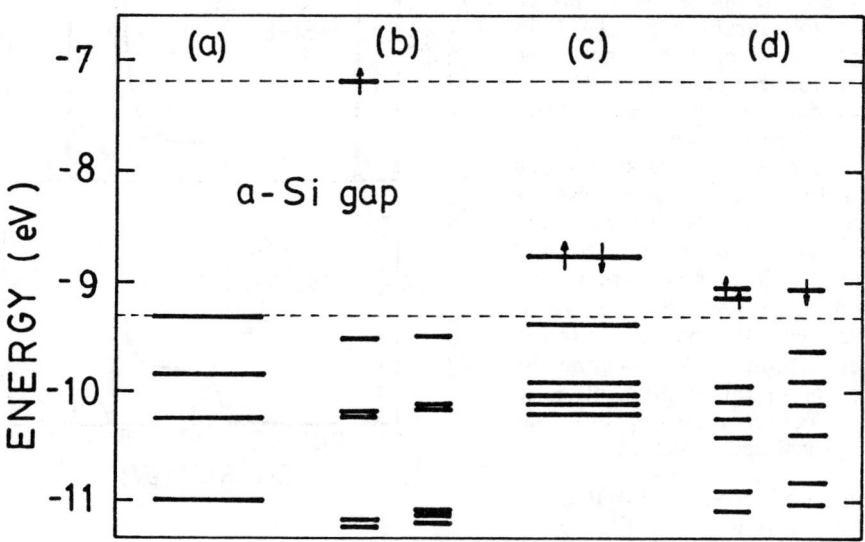

Fig. 2. Molecular levels of (a) $Si_{17}H_{36}$, (b) $NSi_{16}H_{36}$, (c) $NSi_{12}H_{27}$ and (d) $NBSi_{11}H_{27}$ clusters in their optimised geometry as obtained by the semiempirical MNDO method. Long lines are associated to double spin occupation whereas short lines are associated to single spin occupation.

Due to the relevance of the existence of a N level in the gap of a-Si, we have done further calculations following a *quantum chemistry* point of view. The molecular levels of large clusters of Si containing a N impurity have been obtained using the MNDO semiempirical method (Dewar and Thiel 1977 and Stewart 1982). Figure 2 gives the upper molecular levels. $Si_{17}H_{36}$ defines the top of the valence band whereas $NSi_{16}H_{36}$ (isolated substitutional N) allows the approximate knowledge of the bottom of the conduction band. As before threefold coordinated N ($NSi_{12}H_{27}$ cluster) produces a doubly-occupied level within the gap (0.57eV above the valence band for

this calculation). Complex many-body relaxations take place when one of the three Si atoms bonded to N is substituted by B (NBSi$_{11}$H$_{27}$ cluster). Nevertheless, the final result can be simply interpreted saying that the N level becomes shallower due to the loss of electronic charge whereas the hole produced by B in the valence band enters into the gap due to the Coulomb interaction. In any case, no shallow levels are available for thermal excitation around the strong N-B bond.

3. N IMPURITY IN a-SI: TETRAHEDRAL CONFIGURATION

As said above, substitutional N can be introduced into a-Si up to concentrations about 5%. The electronic structure of substitutional N is completely different from the electronic structure of threefold N. Our calculation follows again a cluster Bethe lattice approach. Now, N is linked through four N-Si tetrahedral bonds to four Bethe lattices modelling a-Si. As before, TB parameters of Table I are used. The resulting DOS is given in Figure 1c. The following features can be stressed: i) A new level appears below the valence band at -23.45eV; as before, it is mainly of 2s(N) character. ii) A huge resonance appears in the sp-part of the silicon band mainly due to 2p(N)-3p(Si) interactions. iii) Four of the five N electrons are accomodated in the valence band whereas the fifth has to occupy a conduction band level. iv) No levels appear within the spectral a-Si gap.

As general comment, the DOS around substitutional N (Figure 1c) differs little from the DOS around threefold N (Figure 1b) except for energies within the spectral gap. The existence of four bonds -no bonds of the continuous random network need be broken to accomodate N- prevents the appearance of defect levels in the gap. Appart from the existence or not of a level within the gap, both DOS are very similar. As a consequence, the fifth N electron is forced into the conduction band giving rise to doping. This situation is more clearly appreciated in Figure 2b. In fact, substitutional N should give rise to a shallow donor level due to the long tail part of the impurity Coulomb potential; TB methods are nonetheless inadequate for the description of localised levels that do not originate from the central cell potential and, consequently, this shallow level does not appear in our calculation.

4. CONCLUSIONS

Let us discuss briefly the main consequences of the existence of the new level associated with threefold coordinated N. More detailed discussion of some points can be found elsewhere (Martín-Moreno et al. 1987 and San-Fabián et al.).

Although, our quantum-chemistry procedure leads to a set of tight-binding parameters for the Si-N bond that allow a good overall description of the electronic structure of silicon nitride alloys, some difficulties still remain. Presently, the main point of disagreement between theory and experiment is the defect level that appears above the Si valence band for isolated threefold-coordinated N impurities and that give rise to an impurity band that diminishes the spectral gap for N/Si relations up to 0.25. While the appearance of this fully occupied level/band is theoretically well established, clear experimental support is still missing. The influence of small N contents on the optical and electrical properties of amorphous silicon is presently a subject of strong controversy (Singh and Budhani 1987). Although, it is widely accepted that N improves transport properties of a-Si, the mechanism -substitutional position, reduction of the activation energy or other possibilities- is not yet known. Therefore, the existence of a N band that reduces the spectral gap (or the distance between mobility edges in an alternative language) cannot be discarded on the basis of transport measurements.

On the other hand, optical measurements are normally interpreted according with Tauc's formula (Tauc et al. 1966) that leads to an *optical gap* that summarizes in some way the optical behaviour of the sample. This procedure is not justified for silicon nitride samples. Figure 3 shows that an eventual N impurity band leads to the presence of two different and well defined absorption regions: one for photon energies allowing excitations from the N impurity band to the conduction band and the second one for photon energies giving rise to normal excitations from the valence band to the conduction band. As the inset of Figure 3 and several comments scattered through the experimental bibliography asserting that Tauc's law does not work for some silicon nitride specimens suggest, experimental measurements could be analysed alternatively as having two different edges. Only after such a kind of analysis is carried out, could the existence of the predicted new band inside the a-Si gap be confirmed or discarded.

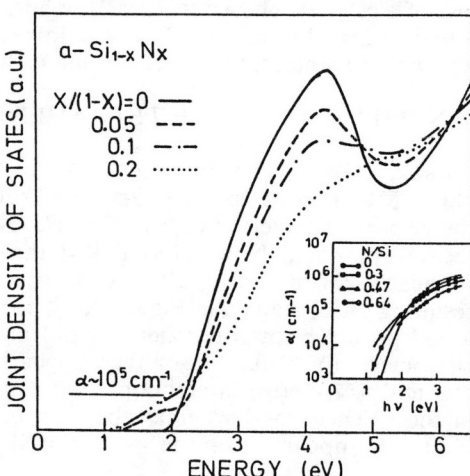

Fig. 3. Evolution of the joint density of levels of a-$Si_{1-x}N_x$ for small N contents (x/(1-x) ≤ 0.2). The approximate threshold 10^5 cm^{-1} for the optical absorption coefficient has been signalised by an arrow. The inset shows experimental data by Herak et al. (1984).

Summarising, we are proposing two lines of study : i) defect levels of **isolated** N impurities, and, ii) reanalysis of the optical and transport properties of samples with a small N content in terms of the predicted existence of a N impurity band within the gap.

REFERENCES

Dewar M J S and Thiel W 1977 *J. Am. Chem. Soc.* **99** 4899
Gómez-Santos G and Vergés J A 1987 *J. Phys. C* **20** 5501
Herak T V, McLead R D, Kao K C, Card H C, Watanabe H, Katoh K, Yasui M and Shibata Y 1984 *J. Non-Cryst. Solids* **69** 39
Kärcher R, Ley L and Johnson R L 1984 *Phys. Rev. B* **30** 1896
Louis E and Vergés J A 1986 *Solid State Commun.* **60** 157
Martín-Moreno L M, Martínez E, Vergés J A and Yndurain F 1987 *Phys. Rev. B* **35** 9683
Ren S Y and Ching W Y 1981 *Phys. Rev. B* **23** 5454
San-Fabián E, Louis E, Martín-Moreno L and Vergés J A *Phys. Rev. B (in press)*
Singh J and Budhani R C 1987 *Solid State Commun.* **64** 34
Slater J C and Koster G F 1954 *Phys. Rev.* **94** 1498
Stewart J 1982 *MOPAC Quantum Chemistry Program Exchange* Program No. 455
Tauc J, Grigorovici R and Vancu A 1966 *Phys. Stat. Sol.* **15** 627
Vogl P, Hjalmarson H P and Dow J D 1983 *J. Phys. Chem. Solids* **44** 365
Yndurain F, Joannopoulos J D, Cohen M L and Falicov L M 1974 *Solid State Commun.* **15** 617

Photoluminescence and tail states in amorphous semiconductors

T M Searle and W A Jackson

Department of Physics, The University of Sheffield, Sheffield UK

ABSTRACT: The relationship between the conduction band tail density of states and the temperature dependence of photoluminescence is discussed in terms of carrier ionisation and thermalisation. The results are compared with photoconductivity and drift mobility data on a-Si:H.

1. INTRODUCTION

The localised states pulled out of the conduction and valence bands by topological or chemical disorder in amorphous semiconductors dominate their transport properties and play an important role in their optical behaviour. Photoemission experiments (Winer et al, 1987) in a-Si:H suggest that the valence band tail at least is exponential over 3 decades with an inverse logarithmic slope of ~45meV. Convolution with the conduction band tail leads (Cody et al, 1981) to an Urbach like optical absorption edge with characteristic slope U~60meV. The shape of the conduction band tail is still a matter for dispute. Whilst recent photoemission results (Winer et al, 1988) on doped samples suggest exponential tails with a slope of ~30meV, the electron drift mobility seems to require a drop in the density of states around 140meV below the mobility edge (Marshall et al, 1986), who also argued that the mobility data are fitted better by a density of states which increases linearly rather than exponentially above this threshold.

Since the photoluminescence (PL) excited states are also tail states, from which carriers can be thermally ionised, the temperature dependence of the PL efficiency $\eta(T)$ can also in principle lead to information on the conduction band tail density of states. The holes are generally supposed to be more deeply trapped in the broader valence band tails. However, the first attempts to do this lead (Collins and Paul, 1982) to puzzling results, in particular requiring the excited state density to have a maximum below the mobility edge, rather than increasing monotonically towards it. The discussion which follows is an attempt to clarify the situation by examining the results of a more careful averaging procedure, by the use of more recent models of the PL process and by comparison with the drift mobility data.

2. EXPERIMENTAL DATA

Early discussions were limited to measurements of $\eta(T)$ in a-Si:H films, in which the low temperature efficiency can be changed by doping but show a limited range of gaps and Urbach slopes. A wider range of data can be obtained from alloys in which both gap and slope may vary widely depending on composition. Figure 1 shows such a set from our work on the a-Si:N:H

© 1989 IOP Publishing Ltd

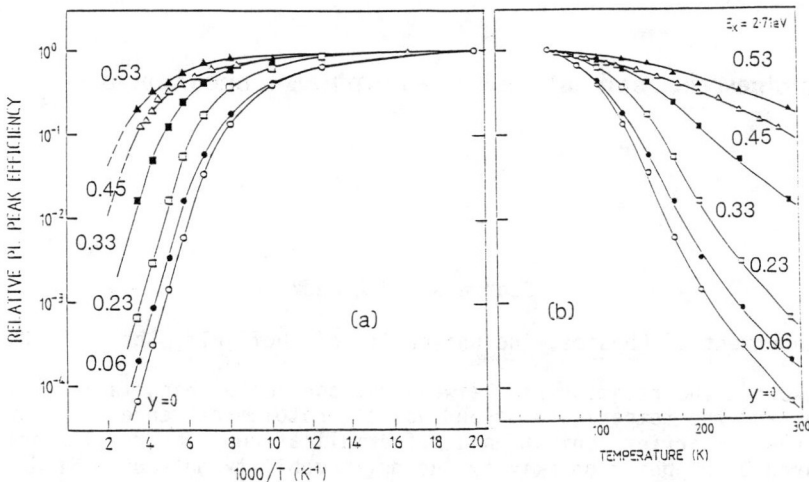

FIGURES 1a & b. Log PL efficiency vs 1/T (a) and T (b) for a series of a-Si$_{1-y}$N$_y$ films. The efficiencies have been normalised at T=0.

system. Experimental details were given in Austin et al (1985). We show the data in two ways here, one an activated plot (log (η(T) vs 1/T) and the other with a linear abscissa (log η(T) vs T). The reason for the inclusion of both will become clear below. These films cover the composition range 0<y<0.53, where y is the nitrogen fraction. We do not intend to attempt a detailed analysis of these data here, but include them as they provide a good guide to the typical features of η(T). The activated plots are compatible at high T with an activation energy of ~140 meV, independent of y, but the broad shoulder moves to higher T with increasing nitrogen. The linear plots show that at high T the data can also be fitted by η(T)$\simeq$ exp(-T/T$_0$). Indeed, Collins et al (1980) showed that the results from several laboratories on y=0 material could be plotted in this way to yield straight lines over some 3-5 decades.

3. ANALYSIS OF THE DATA

All attempts to understand these features start with the supposition that the electron may be thermally ionised from its excited state in the tail into states above the mobility edge. In the simplest model it is then captured into a non-radiative centre. This results immediately in an expression for the efficiency of a single centre.

$$\eta_{sc}(T) = 1/(1 + B \cdot \exp(-E/k.T)) \quad \ldots \ldots (1),$$

where B is the ratio of the ionisation pre-exponential to the radiative rate, and E the depth of the radiative excited state below the mobility edge. In a slightly more refined picture which allows for recapture into either the radiative or nonradiative centre with total (density dependent) cross sections s_r and s_n, one gets

$$\eta'(T) = 1/(1 + S \cdot (1 + B \cdot \exp(-E/k.T))) \quad \ldots \ldots (2),$$

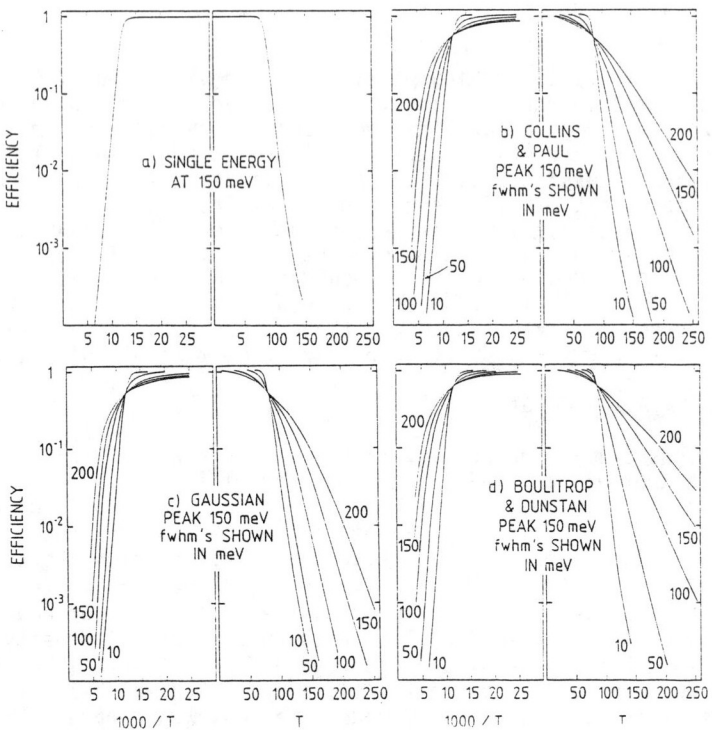

FIGURES 2. Calculated log efficiencies versus 1/T and T. B= 1.10^9.

with $S=s_n/s_r$. We now have the proper low temperature branching ratio. The questions are how to extend these results to a situation where there is a distribution of excited states, and whether information about this distribution can be obtained from the experimental $\eta(T)$. Until now it has been assumed that the correct approach was to start from equation (1) and average it with a density of states distribution G(E), giving

$$\eta_{av}(T) = \int_0^\infty \eta_{sc}(T).G(E).dE \quad .....(3).$$

Collins and Paul (1982) pointed out that for equation 3 to describe the T independent low temperature region, G(E) has to be a decreasing function of energy towards the mobility edge. They used an approximation to the integral which enabled them to find such a function which also approximately generated the high temperature $\exp(-T/T_0)$ form. We evaluate their expression on figure 2b, where it is compared with the single activation energy efficiency (equation 1) of figure 2a. In the plots of figure 2 we have throughout held the energy of the most probable activation energy constant at 150 meV, and examined the effects of broadening the distribution. We have plotted the calculated efficiencies against both T and 1/T, as in figure 1. The comparison of figures 2a and 2b shows clearly how the sharp shoulder of the single activation energy is broadened and shifted to higher temperatures by the spread of energies of figure 2b. Despite a range of activation energies with fwhm exceeding the most probable depth, the 1/T plots yield a well defined activation energy close

to the latter.

It turns out that there is nothing particularly special about the Collins and Paul distribution. Accurate log ($\eta(T)$) vs T behaviour depends on an exponential G(E) at large E, and the shoulder breadth on the fwhm. Other distributions approximate this behaviour: figure 2c shows an evaluation of equation 3 using a Gaussian distribution of activation energies, again with a peak 150meV below the mobility edge and increasing fwhm. At first sight the two sets of curves are very similar, though a detailed examination shows that the Gaussian distribution produces less linear T plots. As before the 1/T plots show a broadened shoulder at higher T, but yield the peak energy in G(E). Similar effects are seen with other distributions: it is necessary to go to very extended functions, such as a Lorentzian to move significantly the observed activation energy below that of the peak.

4. THE INTERPRETATION OF G(E)

G(E) in equations 1-3 is the density of PL excited states. One obvious question is whether all the tail states are potentially such states, i.e. is the tail state distribution g(E) = G(E)? In a model due to Higashi et al (1979) (see also the review by Street (1981)) this is the case. They assume that all states are occupied below the level at which the thermal ionisation rate equals that of radiative recombination. As Street (1981) points out, this is equivalent to replacing the denominator of equation 3 with a step function at E_{HK}=k.T.ln(B). With an exponential density of states and with g(E)=G(E) below E_{HK} one regains the η(T)=exp(T/T_0) form, but **not** the low temperature constant efficiency region which still requires a peaked density of states. In this model the peak would have to be a real feature of the total density of states which photoemission and other techniques do not show. Street argued in favour of this interpretation that it could also explain the rapid red shift of the PL band in a-Si:H. However, our work shows that his mechanism fails in the nitrogen alloys (Austin et al 1985). If the model of Higashi et al is not to apply, then there must be rapid thermalisation down the tails to levels below E_{HK}, which sets an upper limit to the occupation of g(E). We think therefore that it is a balance between radiative recombination rate and thermalisation rate, rather than ionisation rate which determines the levels that electrons reach as their excited states for PL.

The lowest level that might be reached during thermalisation depends not only on the form of g(E), but also on the spatial density fluctuations at E. As electrons trickle downwards, they become isolated in regions of below average density. Eventually it becomes easier for them to recombine than to thermalise out of this region of volume V. Dunstan and Boulitrop (1984) have discussed a related model, finding that

$$G(E)=g(E).\exp(-V.\int_{\infty}^{E}g(E)dE)\ldots\ldots(4)$$

They attempted unsuccessfully to explain the PL spectral shape with this distribution (it underestimates the width by a factor ~2), but did not discuss its application to the question of η(T). In our view, it may provide an approximate guide to G(E). It is a peaked asymmetric distribution which with an exponential g(E) of logarithmic inverse slope U has width 2.45.U and peaks at -U.C, C being a V dependent constant which Dunstan and Boulitrop estimated to be 5.6. With U~ 30meV for the conduction band tail (Winer et al 1988) equation (4) predicts a peak energy (170meV) close to that seen in activated plots like figure 1a (140meV).

Figure 2d shows the evaluation of equation 3 with the distribution of equation 4. Since $G(E) \to g(E)$ deep in the tail and the latter is exponential, log ($\eta(T)$) is linear in T in Figure 2d as in 2b. The importance of equation 4 is not that it provides a definitive distribution that explains the data of figure 1, for there are detailed problems in understanding how on alloying the activation energy might remain constant (figure 1a) whilst the width, as measured via T_0 of figure 1b, increases. The point here is that even simple statistical models predict peaked forms of G(E) with monotonically increasing g(E).

5. THE ENSEMBLE AVERAGE

The discussion above, and in particular the emphasis on the temperature independent region of η (T) assumes the averaging of equation 3 is correct. It is not difficult to calculate the ensemble average of equation 2 on the assumption of spatially random tail states. We find

$$<\eta(T)> = \eta_{av}(T)/(1+S \cdot \eta_{av}(T)) \ldots \ldots (5).$$

Equation 5 allows for the fact that, whilst ionisation of the shallower states is rapid, recapture into them may also be fast if g(E) is monotonic and they are the most numerous centres. Since $\eta_{av}(T) \to 1$ at low T, $<\eta(T)>$ will also $\to 1$ if S>>1. That is, the PL efficiency may become T-independent if it is low. The experimental evidence argues against such an interpretation, since a-Si:H:N the shoulder in efficiency plots (figure 1a) moves to higher temperatures though the absolute value of $\eta(0)$ falls with increasing nitrogen (note that in figure 1 all the low temperature efficiencies have been normalised to unity). Even in undoped films, S is probably only ~2 since $\eta(0) \sim 0.3$. We therefore do not think that the inclusion of retrapping can provide an escape from Collins and Paul's conclusion that G(E) is a peaked function.

6. CONNECTIONS WITH PHOTOCONDUCTIVITY (PC) AND MOBILITY

The PL data suggest that electrons thermalise rapidly into states which at least locally define the bottom of the tails. Above T~120K electrons normally leave this group of states, some 140meV below the mobility band edge by thermal ionisation to it. At lower temperatures, the electron either recombines radiatively or tunnels (hops) through the localised levels. This last process, which may normally transport the electron to other radiative centres of the group can be seen even at 15K (Searle et al 1987) via changes in the PL lifetime distribution. The implication is that the PC lifetimes are determined by the PL rates, and we have found very similar broad distributions in both kinds of measurement. There is a non-radiative component, but the radiative decays show lifetimes from ~100μs to more than 20ms. We also observed strong intensity related changes in these distributions, which we associate with tail state filling produced by the long lifetimes. As the tails fill the electrons move in less localised states and the recent observation by Cloude et al (1986) of an intensity driven increase in mobility is not surprising. It cannot occur at higher temperatures where ionisation competes with saturation.

At the higher temperatures where the photocurrent is predominately carried in the extended states, the situation is reminiscent of crystalline systems eg the F centre in alkali halides, apart from the broadened excited state. The competition between PL and PC requires that the sum of their efficiencies is roughly constant as observed by Jahn et al (1987): their

relative magnitude may be altered by either varying T or by the application of an external electric field. We stress this result is not dependent on any "geminate" pairing as assumed by these authors: it is a direct result of the competition between the radiative and ionisation processes.

The relationship between PL and transport can also be seen in drift mobility measurements. Between about 300-150K transits in a-Si:H are essentially dispersionless, and show an activated mobility with activation energy 130-140meV (Spear et al 1983). They interpretted their data as trap limited transport in the extended states: we suggest that the traps are the PL excited states, now thermally emptied much faster than the PL lifetime. Oddly, the association of these two energies seems not to have been made before. At lower temperatures the transits become increasingly dispersive: we note that the straight exponential region of the nitrogen free film of figure 1a ends at ~150K, and we enter the shoulder region where shallower levels and retrapping become important.

7. CONCLUSIONS

We have shown that PL, like mobility data provides information on tail state distributions. PL yields G(E), the density of occupied levels after thermalisation. The peak energy of G(E) is readily found from η(T) at high T, but the details lie in the inversion of the shoulder region. The inversion strictly requires the use of equation 5, but Collins and Paul's equation 3 may suffice. The range of radiative lifetimes complicates the inversion, through S.

We also suggest, for the first time, that the centres around the peak of G(E), the tail bottom states, control the mobility, PC and PL at T<250K. The transport path and the PL efficiency is determined by the rate of ionisation out of these levels some 140meV below the mobility edge. Above 120K, these levels act as traps from which ionisation occurs readily and PL rarely. At lower T ionisation is unlikely and carriers hop amongst tail bottom states where PL limits their lifetime.

8. REFERENCES

I G Austin, W A Jackson, T M Searle, P K Bhat & R A Gibson, Phil Mag B52, 271 (1985)
C Cloude, W E Spear, P G LeComber & A C Hourd, Phil Mag B54, L113 (1986)
G Cody, T Tiedje, B Abeles, B Brooks & Y Goldstein, Phys Rev Let 47, 1480 (1981)
R W Collins & W Paul, Phys Rev B25, 5257 (1982)
R W Collins, M A Paesler & W Paul, Solid State Comms 34, 833 (1980)
D J Dunstan & F Boulitrop, Phys Rev B30, 5945 (1984)
G S Higashi & M Kastner, J Phys C12, L812 (1979)
K Jahn, R Carius & W Fuhs, J Non Cryst Sol 97 & 98, 575 (1987)
T M Searle, M Hopkinson, M Edmeades, S Kalem, I Austin & R Gibson, "Disordered Semiconductors", Ed M Kastner et al, Plenum (1987), p 357
J M Marshall, R A Street & M J Thompson, Phil Mag B51, 60 (1986)
R A Street, Adv in Phys 30, 593 (1981)
W E Spear, J Non Crys Sol 59 & 60, 1 (1983)
K Winer & L Ley, Phys Rev B36, 6072 (1987)
K Winer, I Hirabayashi & L Ley, Phys Rev Lett 60, 2697 (1988)

Author Index

Abdullin Kh A, *477*
Achtziger N, *149*
Adamowski J, *533, 539*
Ambros S, *289*
Ambrosy A, *131*
Ammerlaan C A J, *191*
Andersen P E, *499*
Anderson J R, *371*
Assali L V C, *453*

Baldereschi A, *521*
Balk P, *289*
Bar-Yam Y, *405*
Baranovskii S D, *271*
Barbier E, *63*
Bassani F, *51, 107*
Baurichter A, *471*
Beall R B, *315*
Bech Nielsen B, *101*
Beckett D J S, *505*
Bednarek S, *533, 539*
Beinikhes I L, *161*
Bekman H H P Th, *191*
Bergman K, *447*
Bhatt R N, *69*
Binggeli N, *521*
Boukerche M, *351*
Briddon P, *459*
Briggs A, *143*
Buczko R, *51, 107*
Burghoorn J, *277*

Chamberlain J M, *243, 295*
Chambers F A, *21*
Charbonneau S, *505*
Chikhrai E V, *477*
Choi J B, *341*
Chroboczek J A, *143*
Clauß W, *167*
Clauws P, *185*
Claybourn M, *211*
Corbett J W, *221*
Czaja W, *561*

Dannefaer S, *283*
Davies G, *125*
Davies M, *295*
Deak P, *221*
Deicher M, *155*
Denteneer P J H, *405*

Deubler S, *149, 471*
Devane G P, *21*
Dmochowski J E, *325*
Dmowski L, *315*
do Carmo M C, *125*
Dohlus P, *149*
Donckers M C J M, *255*
Drew H D, *341*

Eaves L, *315*

Fahey P, *483*
Faurie J P, *351*
Ferreira da Silva A, *551*
Fisher P, *95*
Forkel D, *149, 471*
Foster T J, *315*
Foxon C T, *277*
Fraizzoli S, *51*

Gel'mont B L, *271*
Gienger M, *173*
Gillmann G, *63*
Glaser E, *233*
Goldman V J, *341*
Golubev V G, *271*
Gomes V M S, *453*
Gossard A C, *27*
Grattepain C, *63*
Gregorkiewicz T, *191*
Grimes R T, *295*
Grimmeiss H G, *79, 101*
Groß P, *173*
Grossmann G, *79*
Grübel G, *155*

Haller E E, *179, 425*
Halliday D P, *295*
Ham F S, *89*
Hangleiter A, *515*
Harris J J, *315*
Hart R M, *89*
Haug R, *119*
Hauser M, *11*
Hayden S R, *465*
Hayes T, *447*
Heiblum M, *315*
Heinecke H, *289*
Henini M, *295*
Henry A, *227*

Herring C, *415*
Hill G, *295*
Hiraki A, *377*
Holtz P O, *27*
Huber A M, *63*
Huebener R P, *167*
Hughes O H, *295*

Ichimura M, *301*
Ivanov-Omskii V I, *271*

Jackson W A, *571*
Jantsch W, *325*
Jaworowski A E, *465*
Johnson N M, *415*
Jones R, *459*

Kamimura H, *259, 265*
Kamp M, *289*
Kanehisa M A, *383*
Katayama-Yoshida H, *395*
Keilmann F, *179*
Keller R, *155*
Kennedy T A, *233*
Kerr D, *283*
Klaassen T O, *277*
Kleverman M, *79, 101*
Klitzing K v, *11*
Koch F, *1*
Kochereshko V P, *39*
Kogan Sh M, *161, 527*
Kop'ev P S, *39, 57*
Köpf A, *131*
Kurz H, *289*

Laks D B, *515*
Langer J M, *325*
Lappe J J, *249*
Laßmann K, *131, 173*
Leite J R, *453*
Lewis R A, *95*
Liu X, *33*
Lopata J, *447*
Louis E, *565*
Lugli P, *545*

Maan J C, *265*
Mani R G, *371*
Marfaing Y, *389*
Martín-Moreno L, *565*
Mascher P, *283*
Maude D K, *315*
Mayer K M, *167*
McCombe B D, *21, 33*
McLean N A, *95*

Meilwes N, *201*
Merz J L, *27*
Meyer B K, *249*
Michel J, *201*
Mitin V, *545*
Molnar B, *233*
Monemar B, *227*
Morikawa H, *301*
Morling M, *555*
Mukashev B N, *477*
Murray R, *315*

Najda S P, *265*
Nakata H, *113*
Nathan M I, *315*
Natori A, *259, 265*
Neumark G F, *515*
Newman R C, *211, 315*
Nichols C S, *493*
Niklas J R, *201*
Nissen M, *505*
Nylandsted Larsen A, *499*

Olajos J, *79, 101*
Oshiyama A, *335*
Osutin A V, *271*
Otsuka E, *113*

Paalanen M A, *69*
Pajot B, *437*
Pantelides S T, *405, 493, 515*
Parisi J, *167*
Peale R E, *89*
Pearton S J, *447*
Peinke J, *167*
Petrou A, *33*
Ploog K, *11*
Polupanov A F, *527*
Portal J C, *315*
Prigge H, *155*

Quattropani A, *521, 561*

Ralston J, *33*
Ramdas A K, *361*
Rau U, *167*
Recknagel E, *155*
Reeder A A, *21*
Reggiani L, *545*
Rikken G L J A, *243*
Robison J H, *465*
Röhricht B, *167*
Rotsaert E, *185*

Said M, *383*

San-Fabián E, *565*
Sasaki T, *395*
Schmidt J, *255*
Schnegg A, *155*
Schulz N, *155*
Schwendimann P, *561*
Searle T M, *571*
Sernelius B E, *137, 555*
Shayegan M, *341*
Sievers A J, *89*
Sigg H, *11*
Sigmund E, *119*
Skudlik H, *155*
Snyder L C, *221*
Spaeth J-M, *201, 249*
Stanaway M B, *295*
Stavola M, *447*
Steeds J W, *45*
Stradling R A, *265*
Sundaram M, *27*
Svob L, *389*

Taguchi T, *377*
Takeda A, *301*
Takeda M, *301*
Tamendarov M F, *477*
Tashenov T B, *477*
Testa A, *561*
Theiler T, *179*
Theis T N, *307*
Thewalt M L W, *505*
Tidemand-Petersson P, *499*
Tokmoldin S Z, *477*

Toombs G A, *243*
Tu C W, *45*

Uddin A, *113*
Uhrmacher M, *471*
Uraltsev I N, *39, 57*
Ustinov V M, *57*

Van de Walle C G, *405, 493*
Van Goethem L, *185*
van Klarenbosch A, *277*
Vasil'ev A M, *57*
Vennik J, *185*
Vergés J A, *565*

Wada T, *301*
Wang J, *45*
Weman H, *227*
Wenckebach W Th, *277*
Weyer G, *499*
Weyers M, *289*
Wichert Th, *155*
Wicks G, *33*
Witthuhn W, *149, 471*
Wolf H, *149, 471*
Wolter K, *289*
Wu R, *221*
Wyder P, *243*

Yakovlev D R, *39*
Yamada Y, *377*

Zrenner A, *1*